PACIFIC SYMPOSIUM ON BIOCOMPUTING 2024

PACIFIC SYMPOSIUM ON BIOCOMPUTING 2024

Kohala Coast, Hawaii, USA,
3 – 7 January 2024

Edited by

Russ B. Altman
Stanford University, USA

Lawrence Hunter
University of Colorado Health Sciences Center, USA

Marylyn D. Ritchie
University of Pennsylvania, USA

Tiffany Murray
Stanford University, USA

Teri E. Klein
Stanford University, USA

World Scientific

NEW JERSEY • LONDON • SINGAPORE • BEIJING • SHANGHAI • HONG KONG • TAIPEI • CHENNAI • TOKYO

Published by

World Scientific Publishing Co. Pte. Ltd.
5 Toh Tuck Link, Singapore 596224
USA office: 27 Warren Street, Suite 401-402, Hackensack, NJ 07601
UK office: 57 Shelton Street, Covent Garden, London WC2H 9HE

Online ISSN: 2335-6936
Print ISSN: 2335-6928

Library of Congress Control Number: 2023949005

British Library Cataloguing-in-Publication Data
A catalogue record for this book is available from the British Library.

BIOCOMPUTING 2024
Proceedings of the Pacific Symposium

ISBN 978-981-12-8642-1 (ebook)
ISBN 978-981-12-8641-4 (print)

PRECISION MEDICINE: INNOVATIVE METHODS FOR ADVANCED UNDERSTANDING OF MOLECULAR UNDERPINNINGS OF DISEASE

WORKSHOPS

ERRATUM

In Loving Memory of

Lucas Benjamin Arsenault

February 16, 2021 – August 2, 2023

PACIFIC SYMPOSIUM ON BIOCOMPUTING 2024

2024 marks the 29th Pacific Symposium on Biocomputing (PSB). We gather once again on the Big Island to share the latest progress and challenges in biocomputing. 2023 was a year of remergence for Artificial Intelligence (AI). Large language models (LLMs) and particularly ChatGPT brought AI into the public consciousness in a way not previously seen. At the same time, advances in AI have fueled great progress in the analysis of structural and functional molecular data, omics data sets, electronic health records, biobanks, and many other areas of biocomputation. LLMs themselves have clear applications in both clinical medicine and in basic research, and we are experiencing an explosion of creative uses of these powerful (but still imperfect) tools. In addition to LLMs for human language, there are powerful LLMs built on protein and DNA sequence which show remarkable utility in representing these molecules and detecting signals and correlations between sequence and structure/function. These areas provide a rich background for PSB 2024. Of course, not all biocomputing is AI and there is still need for important efforts in traditional algorithms, informatics, data science, statistics and (importantly) in the understanding of the social setting in which our tools are used. We are more aware than ever that the choice of problems to address, the representativeness of the data that we use, and the ways we evaluate the success of our computational artifacts should all be considered with intention and sensitivity to considerations of justice, autonomy, beneficence, and non-maleficence.

In addition to being published by World Scientific and indexed in PubMed, the proceedings from all PSB meetings are available online at http://psb.stanford.edu/psb-online/. Since 1996, all PSB papers are indexed in PubMed. These papers are routinely cited in archival journal articles and routinely represent important early contributions in new subfields—many times before there is an established literature in more traditional journals; for this reason, many papers have garnered hundreds of citations.

The social media handle for PSB is @PacSymBiocomp and the hashtag for PSB 2024 is #PSB24.

The efforts of a dedicated group of session organizers have produced an outstanding program. The sessions of PSB 2024 and their hard-working organizers are as follows:

Artificial Intelligence in Clinical Medicine: Generative and Interactive Systems at the Human-Machine Interface
Organizers: Sajjad Fouladvand, Emma Pierson, Ivana Jankovic, David Ouyang, Jonathan H. Chen, Roxana Daneshjou

Digital Health Technology Data in Biocomputing: Research Efforts and Considerations for Expanding Access
Organizers: Jessilyn Dunn, Michelle Holko, Chris Lunt

Drug-Repurposing and Discovery in the Era of "Big" Real-World Data: How the Incorporation of Observational Data, Genetics, and Other -omic Technologies Can Move Us Forward
Organizers: Megan M. Shuey, Jacklyn N. Hellwege, Nikhil Khankari, Marijana Vujkovic, Todd L. Edwards

Overcoming Health Disparities in Precision Medicine
Organizers: Francisco M. De La Vega, Kathleen C. Barnes, Keolu Fox, Alexander Ioannidis, Eimear Kenny, Rasika A. Mathias, Bogdan Pasaniuc

Precision Medicine: Innovative Methods for Advanced Understanding of Molecular Underpinnings of Disease
Organizers: Yana Bromberg, Hannah Carter, Steven E. Brenner

We are also pleased to present five workshops in which investigators with a common interest come together to exchange results and new ideas in a format that is more informal than the peer-reviewed sessions. For this year, the workshops and their organizers are:

Large Language Models (LLMs) and ChatGPT for Biomedicine
Organizers: Zhiyong Lu, Steven E. Brenner, Cecilia Arighi

Practical Approaches to Enhancing Fairness, Social Responsibility and the Inclusion of Diverse Viewpoints in Biomedicine
Organizers: Daphne O. Martschenko, Nicole Martinez-Martin, Meghan Halley

Risk prediction: Methods, Challenges, and Opportunities
Organizers: Rui Duan, Lifang He, Ruowang Li, Jason H. Moore

Statistical Analysis of Single-Cell Protein Data
Organizer: Brooke Fridley

Tools for Assembling the Cell: Towards the Era of Cell Structural Bioinformatics
Organizers: Emma Lundberg, Trey Ideker, Andrej Šali

The PSB 2024 keynote speakers are Scott Penberthy (Science keynote) and Andrea Roth (Ethical, Legal and Social Implications keynote).

Tiffany Murray has managed the peer review process and assembly of the proceedings since 2001 and plays a key role in many aspects of the meeting. We are grateful for the support of the National Institutes of Health[1], ISCB, and Cleveland Institute for Computational Biology. The Research Parasite Awards benefit from support from GigaScience, Jeff Stibel, Mr. and Mrs. Stephen Canon, and Drs. Casey and Anna Greene. The Research Symbiont Awards benefit from support from the Wellcome Trust and the DragonMaster Foundation.

We are particularly grateful to the PSB staff Al Conde, Paul Murray, Ryan Whaley, Mark Woon, BJ Morrison McKay, Cynthia Paulazzo, Jackson Miller, Heather Miller, and Nicholas Murray for their assistance. We also acknowledge the many busy researchers who reviewed the submitted manuscripts on a very tight schedule. The partial list following this preface does not include many who wished to remain anonymous, and of course we apologize to any who may have been left out by mistake.

We look forward to a great meeting and to seeing you on the Big Island. Aloha!

Pacific Symposium on Biocomputing Co-Chairs,
October 9, 2023

Russ B. Altman
Departments of Bioengineering, Genetics, Medicine & Biomedical Data Science, Stanford University

Lawrence Hunter
Department of Pharmacology, University of Colorado Health Sciences Center

Marylyn D. Ritchie
Department of Genetics and Institute for Biomedical Informatics, University of Pennsylvania

Teri E. Klein
Departments of Biomedical Data Science & Medicine, Stanford University

Thanks to the reviewers...

Finally, we wish to thank the scores of reviewers. PSB aims for every paper in this volume to be reviewed by three independent referees. Since there is a large volume of submitted papers, paper reviews require a great deal of work from many people. We are grateful to all of you listed below and to anyone whose name we may have accidentally omitted or who wished to remain anonymous.

Tiffany Amariutta
Marcus Bagedley
Nuno Bandeira
Chinmayi Bankar
Alfonso Barajas
Kathleen Barnes
Joshua Barrios
Kyle Beauchamp
Michael Beer
Jamie Bennett
Erik Bergstrom
Shreyas Bhave
Alexander Bick
Nathaniel Bloodworth
Yana Bromberg
Dan Brown
Andrew Carroll
Hannah Carter
Jin Chen
Yile Chen
Min Chiu
Young-Rae Cho
Min Choi
Andrew Cirincione
Wyatt Clark
Joseph Colonel
Hejie Cui
Matteo D'Antonio
Roxana Daneshjou
Jishnu Das
Francisco De La Vega
Clara De Paolis Kaluza
Roozbeh Dehghannasiri
Yi Ding
Grant Duffy
Jessilyn Dunn
Todd Edwards
Jesse Engreitz
Jaclyn Eissman
Alex Flynn
Sunyang Fu
Lana Garmire
Chris Gignoux
Federico Gomez
Wilfredo Gonzalez-Rivera
Benjamin Greenbaum
Sasha Gutfraind
Prashnna Gyawali
Melissa Gymrek
Bryan He
Dominik Heider
Jacklynn Hellwege
Rachel Hoffing
Michelle Holko
Weston Hughes
Alexander Ioannidis
Atishay Jain
Elizabeth Jasper
Linda Kachuri
Rajas Kale
Vipina Keloth
Nikhil Khankari
Nicholas Kiefer
Adam Klie
Mikhail Kolmogorov
Michael Lamkin
Erica Landis
Leslie Lange
Binglan Li
Yan Chak Li
Yun Li
Liz Litkowsky
Alejandro Lozano
Chris Lunt
Clarence Mah
Simon Mallal

Hiral Master
Rasika Matthias
Brian Mautz
Rachel Mester
Tyne Miller
Sean Mooney
Yves Moreau
Genevieve Mortensen
Fateme Nateghi
Ashwin Nayak
Khaliq Newaz
Madelena Ng
Ziad Obermeyer
Samaneh Omranian
David Ouyang
Kivilcim Ozturk
Bogdan Paisanuc
Guarav Pandey
Alice Popejoy
Yannick Pouliot
Sing Prem
Amy Price
Amy Price
Predrag Radivojac
Protiva Rahman
Daniele Raimondi
Prabakaran Ramakrishnan
Rashika Ramola
Ojas Ramwala
Brooke Rhead
Manuel Rivas
Genevieve Roberts
Cassianne Robison-Cohen
Yuki Sahashi
Nidhi Sahni
Tom Savage
Amand Florian Schmidt
Ruhollah Shemirani
Megan Shuey
Sarah Stallings
James Talwar
Artem Trotsyuk
Eli Van Allen
Olivia Veatch
Nora Verplaetse
Karin Verspoor
Ha Vu
Marijana Vujkovic
Milos Vukadinovic
Wei Wang
Gary Weissman
John Witte
Gary Wu
Su Xian
Chao Yan
Alaa Yousef
Zhi Yu
Neal Yuan
Simone Zaccaria
Daniel Zeiberg
Pengfei Zhang
Shilin Zhao
Tianming Zhou
Xiaomin Zhu
Maryam Zolnoori

[1]Funding for this conference was made possible (in part) by R13LM006766 from the National Library of Medicine. The views expressed in written conference materials or publications and by speakers and moderators do not necessarily reflect the official policies of the Department of Health and Human Services; nor does mention by trade names, commercial practices, or organizations imply endorsement by the U.S. Government.

Artificial Intelligence in Clinical Medicine: Generative and Interactive Systems at the Human-Machine Interface

Sajjad Fouladvand
Real-World Evidence and Advanced Analytics, Johnson and Johnson
Brisbane, CA, USA
Email: Sfouladv@its.jnj.com

Emma Pierson
Jacobs Technion-Cornell Institute, Cornell Tech
New York City, NY, USA
Email: ep432@cornell.edu

Ivana Jankovic
Oregon Health and Sciences University
Portland, OR, USA
Email: jankovii@ohsu.edu

David Ouyang
Department of Cardiology, Cedars-Sinai Medical Center
Los Angeles, CA, USA
Email: david.ouyang@cshs.org

Jonathan H. Chen
Department of Medicine and Center for Biomedical Informatics Research, Stanford University
Stanford, CA, USA
Email: jonc101@stanford.edu

Roxana Daneshjou
Biomedical Data Science, Stanford University
Stanford, CA, USA
Email: roxanad@stanford.edu

Artificial Intelligence (AI) models are substantially enhancing the capability to analyze complex and multi-dimensional datasets. Generative AI and deep learning models have demonstrated significant advancements in extracting knowledge from unstructured text, imaging as well as structured and tabular data. This recent breakthrough in AI has inspired research in medicine, leading to the development of numerous tools for creating clinical decision support systems, monitoring tools, image interpretation, and triaging capabilities. Nevertheless, comprehensive research is imperative to evaluate the potential impact and implications of AI systems in healthcare.

At the 2024 Pacific Symposium on Biocomputing (PSB) session entitled "Artificial Intelligence in Clinical Medicine: Generative and Interactive Systems at the Human-Machine Interface", we spotlight research that develops and applies AI algorithms to solve real-world problems in healthcare.

Keywords: Artificial Intelligence, clinical medicine, decision support systems.

1. Introduction

Recent progress in AI has led to the development of advanced large language models (LLMs), image, genomic and tabular data analysis tools (Huang et al., 2023; Movva et al., 2023, 2023; Omiye et al., 2023; OpenAI, 2023; Singhal et al., 2023; Tate et al., 2023; Wehbe et al., 2023). Leveraging these AI models for real-world biomedical data analysis is critical for enhancing diagnostic accuracy, predicting patient outcomes, and personalizing treatment plans, ultimately contributing to improved patient care and health outcomes. However, systematic evaluation of the potentials and limitations of AI algorithms within the medical domain is crucial to ensure the efficacy, safety, and reliability of AI-driven healthcare solutions and interventions (Wornow et al., 2023).

Here, we highlight the accepted submissions for the Artificial Intelligence in Clinical Medicine: Generative and Interactive Systems at the Human-Machine Interface session at the Pacific Symposium on Biocomputing (PSB) 2024. A goal of this session is to showcase research that has identified a clinical need that can be addressed by AI methods. Accepted submissions include use cases of using generative and classical AI models for analyzing different clinical data modalities and for a variety of applications such as answering medical questions, medical image analysis, clinical note analysis, cognitive monitoring, digital twins, and other decision support systems.

2. Artificial Intelligence in Clinical Medicine

2.1. Medical text and clinical notes analysis

There have recently been numerous successful applications of LLMs in ingesting medical text and clinical notes to extract vital information and insights for enhanced patient care. Lozano et al. (2024) proposed Clinfo.ai: an open-source retrieval-augmented LLM system for answering medical questions using scientific literature. The authors evaluated Clinfo's performance (along with the performance of other question-answering systems, which the proposed method improves on) on a benchmark the authors made publicly available. Systems like this highlight the potential of large language models to help clinicians stay abreast of the enormous (and growing) medical literature. Jiang et al. (2024) proposed VetLLM, a large language model for predicting diagnosis from veterinary notes. They evaluated whether LLMs can be used to extract diagnoses from unstructured veterinary notes. This approach can more easily facilitate broad veterinary research given that previous work often relies on customized, specialized models for each diagnosis. The paper revealed that, even without fine-tuning, open-source LLMs like Alpaca-7B show promising performance in diagnosis extraction tasks; performance is further improved when the model is fine-tuned on datasets of veterinary notes.

Pakhomov et al. (2024) proposed a conversational agent for early detection of neurotoxic effects of medications through automated intensive observation. This paper presents an AI system for monitoring cognitive symptoms of neurotoxicity which can occur in response to some immunotherapies. The system, a conversational agent, conducts a cognitive assessment over the phone including both spontaneous speech and neurocognitive tests. The authors present the results of a pilot study. Such systems have the potential to allow for intensive monitoring of patients while reducing the burden on them and medical staff (since automated monitoring can be conducted while the patient remains at home).

2.2. Medical image analysis

Another compelling avenue where AI has shown promising results is in the realm of medical image processing. This domain has witnessed a remarkable transformation, with AI algorithms now capable of efficiently analyzing a wide range of medical images, including ultrasound, X-rays, MRI scans, and CT scans, to yield faster and more accurate diagnoses. Duffy et al. (2024) used convolutional neural networks (CNNs) to evaluate the performance of AI models on 2D and 3D cardiac ultrasound datasets. Generally recorded as 2D video data, newer ultrasound transducers allow the collection of 3D data that can be post processed into standard 2D view videos. Using previously published CNNs for echocardiography (Ouyang et al., 2020), Duffy et al. showed that biases in 2D data (foreshortening and off axis views) can be simulated from the 3D data and have important impacts on model output.

Li et al. (2024) proposed BrainSTEAM, a practical pipeline for connectome-based fMRI analysis towards subject classification. This work addressed the overfitting problem in Graph Neural Networks (GNNs) used for analyzing structured network data. BrainSTEAM uses a spatio-temporal module that includes an EdgeConv GNN model, an autoencoder, and a strategy to dynamically segment time series signals, construct correlation networks, capture regions of interest (ROIs) connectivity structures, denoise data, and enhance model training. BrainSTEAM was evaluated on two real-world neuroimaging datasets, ABIDE for autism prediction and HCP for gender prediction, showing superior performance compared to existing models. This framework is potentially applicable to other studies for connectome-based fMRI analysis, promising enhanced reliability for clinical applications. Finally, recognizing the variation in human-quantified phenotypes, Vukadinovic et al. (2024) show that different ways of assessing left ventricular ejection fraction, including variation within the range of clinician-to-clinician variability, can cause significant impact on downstream analyses, including genome wide association studies, where less precise measurements have a substantial impact on signal for genetic loci. Compared with sample size variation, 1% less precision in measurements resulted in the equivalent loss of power as a 10% decrease in cohort sample size.

2.3. Neurobiology and cognitive function

Prantzalos et al. (2024) presented MaTiLDA, which serves as an integrated machine learning and topological data analysis platform for brain network dynamics. Brain activity is recorded via electroencephalograms (EEGs); however, analyzing large volumes of recordings can be difficult. They introduced and publicly shared MaTiLDA to enable the use of machine learning with topological data analysis on EEG data. They then showed how their platform could be used to analyze EEG data from neurological disorders such as epilepsy.

Yang et al. (2024) showed that DNNs on brain MRI images can be used to detect and distinguish between normal subjects and subjects with cognitive impairments like Alzhiemer's disease. Javedani Sadaei et al. (2024) proposed Zoish: a novel feature selection approach leveraging Shapley additive values for machine learning applications in healthcare. They present a feature selection python package leveraging Shapley additive values to simplify feature selection for a variety of healthcare prediction tasks. As an illustrative example, Zoish was applied to a predictive model on Parkinson's progression as measured by the Montreal Cognitive Assessment (MOCA) and showed not only greater predictive performance overall but also improved interpretability compared to another feature selection method. As AI models attempt to move away from the "black box", tools such as Zoish can help clinicians better understand how the models produce predictions

2.4. Human-machine interface

Moore et. al. (2024) proposed SynTwin: a graph-based approach for predicting clinical outcomes using digital twins derived from synthetic patients. SynTwin introduces a novel methodology for generating and utilizing digital twins for clinical outcome prediction in precision medicine. The approach begins by estimating the distance between subjects based on their features, and then uses these distances to construct a network. Communities of subjects are defined, and a population of synthetic patients is generated. Digital twins, selected from this synthetic patient population, are used to enhance the prediction of clinical endpoints. When applied to a population-based cancer registry, the SynTwin approach significantly improved the prediction of mortality compared to using real data alone, demonstrating the potential of this method in advancing precision medicine efforts. Patel et. al. (2024) proposed optimizing computer-aided diagnosis with cost-aware deep learning models. They propose a deep learning computer-aided diagnosis system to address the common situation in healthcare in which a false negative is more serious than a false positive. Whereas traditional computer-aided diagnosis systems penalize both types of misclassification equally, the cost-aware neural net model described here shows how using cost as a hyperparameter can boost sensitivity while largely maintaining overall accuracy.

3. Conclusion

Submissions accepted at the Artificial Intelligence in Clinical Medicine: Generative and Interactive Systems at the Human-Machine Interface session underscore the expanding role of AI in clinical medicine. The array of studies, spanning from advancements in AI-driven medical text and clinical notes analysis to breakthroughs in medical image processing, neurobiology, and human-machine interfaces, highlights the potential of generative and classical AI to improve healthcare. The consistent theme across all submissions is the emphasis on practical, real-world applications, showing AI's capability to enhance diagnostic accuracy, monitor cognitive symptoms, analyze diverse data types, and augment clinical decision-making processes. Despite these advancements, the need for identifying clinical problems and ongoing evaluation and assessment of AI technologies in healthcare to ensure their safety, efficacy, and reliability remains paramount. The works presented herein contribute significantly to this ongoing dialogue, showcasing both the possibilities and the remaining challenges in integrating AI into the healthcare landscape.

References

Duffy, G., Christensen, K., & Ouyang, D. (2024). Leveraging 3D Echocardiograms to Evaluate AI Model Performance in Predicting Cardiac Function on Out-of-Distribution Data. *Pacific Symposium on Biocomputing (PSB).*

Huang, Z., Bianchi, F., Yuksekgonul, M., Montine, T. J., & Zou, J. (2023). A visual–language foundation model for pathology image analysis using medical Twitter. *Nature Medicine, 29*(9), Article 9. https://doi.org/10.1038/s41591-023-02504-3

Javedani Sadaei, H., Loguercio, S., & Shafiei Neyestanak, M. (2024). Zoish: A Novel Feature Selection Approach Leveraging Shapley Additive Values for Machine Learning Applications in Healthcare. *Pacific Symposium on Biocomputing (PSB).*

Jiang, Y., Irvin, J. A., Ng, A. N., & Zou, J. (2024). VetLLM: Large Language Model for Predicting Diagnosis from Veterinary Notes. *Pacific Symposium on Biocomputing (PSB).*

Li, A., Yang, Y., Cui, H., & Yang, C. (2024). BrainSTEAM: A Practical Pipeline for Connectome-based fMRI Analysis towards Subject Classification. *Pacific Symposium on Biocomputing (PSB).*

Lozano, A., Fleming, S. L., Chiang, C.-C., & Shah, N. (2024). Clinfo.ai: An Open-Source Retrieval-Augmented Large Language Model System for Answering Medical Questions using Scientific Literature. *Pacific Symposium on Biocomputing (PSB).*

Moore, J. H., Li, X., Chang, J. H., Tatonetti, N. P., Theodorescu, D., Chen, Y., Asselbergs, F. W., Venkatesan, M., & Wang, Z. P. (2024). SynTwin: A graph-based approach for predicting clinical outcomes using digital twins derived from synthetic patients. *Pacific Symposium on Biocomputing (PSB).*

Movva, R., Balachandar, S., Peng, K., Agostini, G., Garg, N., & Pierson, E. (2023). *Large language models shape and are shaped by society: A survey of arXiv publication patterns* (arXiv:2307.10700). arXiv. https://doi.org/10.48550/arXiv.2307.10700

Omiye, J. A., Gui, H., Rezaei, S. J., Zou, J., & Daneshjou, R. (2023). *Large language models in medicine: The potentials and pitfalls* (arXiv:2309.00087). arXiv. https://doi.org/10.48550/arXiv.2309.00087

OpenAI. (2023). *GPT-4 Technical Report* (arXiv:2303.08774). arXiv. https://doi.org/10.48550/arXiv.2303.08774

Ouyang, D., He, B., Ghorbani, A., Yuan, N., Ebinger, J., Langlotz, C. P., Heidenreich, P. A., Harrington, R. A., Liang, D. H., Ashley, E. A., & Zou, J. Y. (2020). Video-based AI for beat-to-beat assessment of cardiac function. *Nature, 580*(7802), Article 7802. https://doi.org/10.1038/s41586-020-2145-8

Pakhomov, S., Solinsky, J., Michalowski, M., & Bachanova, V. (2024). A Conversational Agent for Early Detection of Neurotoxic Effects of Medications through Automated Intensive Observation. *Pacific Symposium on Biocomputing (PSB).*

Patel, C., Wang, Y., Ramaraj, T., Tchoua, R., Furst, J., Raicu, D. (2024). Optimizing Computer-Aided Diagnosis with Cost-Aware Deep Learning Models. *Pacific Symposium on Biocomputing (PSB).*

Prantzalos, K., Upadhyaya, D., Shafiabadi, N., Fernandez-BacaVaca, G., Gurski, N., Yoshimoto, K., Sivagnanam, S., Majumdar, A., Sahoo, S. (2024). MaTiLDA: An Integrated Machine Learning and Topological Data Analysis Platform for Brain Network Dynamic. *Pacific Symposium on Biocomputing (PSB).*

Singhal, K., Azizi, S., Tu, T., Mahdavi, S. S., Wei, J., Chung, H. W., Scales, N., Tanwani, A., Cole-Lewis, H., Pfohl, S., Payne, P., Seneviratne, M., Gamble, P., Kelly, C., Babiker, A., Schärli, N., Chowdhery, A., Mansfield, P., Demner-Fushman, D., … Natarajan, V. (2023). Large language models encode clinical knowledge. *Nature, 620*(7972), Article 7972. https://doi.org/10.1038/s41586-023-06291-2

Tate, S., Fouladvand, S., Chen, J. H., & Chen, C.-Y. A. (2023). The ChatGPT therapist will see you now: Navigating generative artificial intelligence's potential in addiction medicine research and patient care. *Addiction.* https://doi.org/10.1111/add.16341

Vukadinovic, M., Renjith, G., Yuan, V., Kwan, A., Cheng, S. C., Li, D., Clarke, S. L., & Ouyang, D. (2024). Impact of Measurement Noise on Genetic Association Studies of Cardiac Function. *Pacific Symposium on Biocomputing (PSB).*

Wehbe, R. M., Katsaggleos, A. K., Hammond, K. J., Hong, H., Ahmad, F. S., Ouyang, D., Shah, S. J., McCarthy, P. M., & Thomas, J. D. (2023). Deep Learning for Cardiovascular Imaging: A Review. *JAMA Cardiology*. https://doi.org/10.1001/jamacardio.2023.3142

Wornow, M., Xu, Y., Thapa, R., Patel, B., Steinberg, E., Fleming, S., Pfeffer, M. A., Fries, J., & Shah, N. H. (2023). The shaky foundations of large language models and foundation models for electronic health records. *Npj Digital Medicine*, *6*(1), Article 1. https://doi.org/10.1038/s41746-023-00879-8

Yang, Y., Sathe, A., Schilling, K., Shashikumar, N., Moore, E., Dumitrescu, L., Pechman, K. R., Landman, B. A., Gifford, K. A., Hohman, T. J., Jefferson, A. L., & Archer, D. B. (2024). A deep neural network estimation of brain age is sensitive to cognitive impairment and decline. *Pacific Symposium on Biocomputing (PSB)*.

Clinfo.ai: An Open-Source Retrieval-Augmented Large Language Model System for Answering Medical Questions using Scientific Literature

Alejandro Lozano[1,*,†], Scott L. Fleming[1,*], Chia-Chun Chiang[2,3], Nigam Shah[4,5,6]
[1]*Department of Biomedical Data Science, Stanford University, Stanford, CA, USA*
[2]*Department of Neurology, Mayo Clinic, Rochester, MN*
[3]*Human Centered Artificial-Intelligence Institute, Stanford University, Stanford, CA, USA*
[4]*Department of Medicine, Stanford School of Medicine, Stanford, CA, USA*
[5]*Clinical Excellence Research Center, Stanford School of Medicine, Stanford, CA, USA*
[6]*Technology and Digital Solutions, Stanford Health Care, Palo Alto, California, USA*
[*]*Equal Contribution*
[†]*E-mail: lozanoe@stanford.edu*

The quickly-expanding nature of published medical literature makes it challenging for clinicians and researchers to keep up with and summarize recent, relevant findings in a timely manner. While several closed-source summarization tools based on large language models (LLMs) now exist, rigorous and systematic evaluations of their outputs are lacking. Furthermore, there is a paucity of high-quality datasets and appropriate benchmark tasks with which to evaluate these tools. We address these issues with four contributions: we release Clinfo.ai, an open-source WebApp that answers clinical questions based on dynamically retrieved scientific literature; we specify an information retrieval and abstractive summarization task to evaluate the performance of such retrieval-augmented LLM systems; we release a dataset of 200 questions and corresponding answers derived from published systematic reviews, which we name PubMed Retrieval and Synthesis (PubMedRS-200); and report benchmark results for Clinfo.ai and other publicly available OpenQA systems on PubMedRS-200.

Keywords: Large Language Models, Abstractive Summarization, Artificial Intelligence, Clinical Medicine, Generative AI, Interactive Systems, ChatGPT

1. Introduction

The aggregation and distribution of medical knowledge, facilitated by platforms such as PubMed or Cochrane, enables healthcare professionals and medical researchers to stay abreast of the latest scientific discoveries and make informed decisions based on up-to-date scientific evidence.[1] However, the staggering influx of more than 1 million papers each year into PubMed alone (equivalent to two papers per minute as of 2016)[2] highlights the daunting task of keeping up with scientific findings.[3] This is especially true for practicing clinicians, who face the challenge of keeping track of the most updated research findings in all areas related to their patient care duties.[4]

Existing technologies fail to adequately satisfy the information needs of health care professionals and researchers. In daily practice, clinicians have on average one care-related question for every other patient seen[5] and they refer to sources like PubMed or UpToDate to obtain summarized information answering these questions.[6] Questions that cannot be answered within 2 to 3 minutes are often abandoned, potentially negatively impacting patient care and outcomes.[5,7] While systematic review (SR) articles can provide quick answers to clinical questions, many questions are not answerable through existing reviews. On the other hand, manually synthesizing findings from multiple primary sources without the help of a published review article can be extraordinarily time consuming. Review articles take on average 67.3 weeks to complete,[8] and those written reviews may not even include the most updated research published in the literature. Question-answering tools that leverage frequently updated external electronic resources would enable researchers and clinicians to obtain up-to-date information in a more efficient way that benefits scientific discovery and quality of patient care.[9–13]

In previous decades, applications that integrated clinical systems with on-line information to answer users' information needs (e.g., "infobuttons")[14] were typically driven by semantic networks. Other works such as CHiQA proposed a combination of knowledge-based, machine learning, and deep learning approaches to develop a question-answering system using patient-oriented resources to answer consumer health questions.[15]

The new capabilities of agents powered by large language models (LLM) has accelerated the development of automated literature summarization tools. Most of these solutions tend to be privately developed, closed-source solutions based on retrieval-augmented[16] (RetA) LLMs[17] (e.g. Scite,[18] Elicit,[19] GlacierMD,[20] Consensus,[21] OpenEvidence,[22] Statpearls semantic search[23]). However, the paucity of publicly available technical reports describing these systems and the lack of appropriate guidelines, regulations, and evaluations to ensure their safe and responsible usage is an urgent concern.[24]

This Natural Language Generation (NLG) problem has been exacerbated by a lack of (1) representative datasets and associated tasks, and (2) automated metrics for evaluating RetA LLMs on said tasks.

Fortunately, developments in the LLM evaluation space have shown that a number of automated metrics correlate moderately with human preference, even in domain-specific scenarios (including medicine).[25–27]

Building on these advancements, we provide four contributions:

(1) Clinfo.ai [a], the first publicly available, open-source, end-to-end retrieval-augmented LLM-based system for querying and synthesizing the clinical literature. The system is hosted as a publicly available WebApp at https://www.clinfo.ai/.
(2) An open information retrieval and abstractive summarization task specification designed to evaluate an algorithm's ability to both retrieve relevant information and adequately synthesize it. In the task setup, both the information retrieval and abstractive summarization sub-tasks are compared to gold standard (human generated but pragmatically retrieved)

[a]https://github.com/som-shahlab/Clinfo.AI

references and answers. Furthermore, our task is defined to truly resemble RetA deployment conditions (enabling the evaluation of already deployed but potentially closed-source systems).

(3) PubMed Retrieval and Synthesis (PubMedRS-200), a publicly available dataset of 200 questions structured in Open QA format, paired with answers derived from systematic reviews and corresponding references.

(4) Benchmark results for Clinfo.ai and other publicly available OpenQA systems on PubMedRS-200).

2. Related Work

LLMs in healthcare The remarkable performance of LLMs in the general domain has brought about a revolution in the field of natural language processing,[28] showcasing exceptional capabilities in tasks like summarization, question-answering, and NLG.[29] Given their wide utility, researchers are now actively exploring applications of LLMs in healthcare.[30–33] Several LLMs have achieved human-level performance on numerous medical professional licensing exams such as the United States Medical Licensing Exam (USMLE).[34] Other works have demonstrated promise in various healthcare-inspired tasks, such as automated clinical note generation and reasoning about public health topics.[30–33] However, NLG tasks and publicly available benchmarks that directly address true medical needs are still underrepresented in the literature. Such tasks and benchmarks are especially important for estimating the capabilities and risks of LLMs in the clinical domain.

LLMs have several documented disadvantages and risks. First, updating LLMs with new knowledge and information is challenging and inefficient.[35] Second, the training objective of LLMs to predict the most probable next token can cause these models to generate inaccurate information (hallucination), requiring costly and imperfect post-hoc model adjustments like reinforcement learning with human feedback (RLHF).[36] More importantly, most popular consumer-facing LLMs (e.g., OpenAI's GPT-4,[29] Meta's Llama 2,[37] Anthropic's Claude 2[38]) do not provide references pointing to their source of information, even when the model's output is factual. This can engender distrust with users in many scientific domains, including healthcare. Prior work has proposed ReTA LLMs[16] to solve the information provenance issue and have shown promising results. These ReTA LLMs do not require post-hoc model editing in order to incorporate new knowledge.

Retrieval Augmentation Question Answering LLMs in Medicine Hiesinger et al.[39] introduced Almanac, a novel LLM integrated with a vector database and calculator, designed to answer 130 clinical questions generated by a panel of five board-certified clinicians and resident physicians. The results showed that Almanac surpassed a standard LLM (GPT-4) in factuality, safety, and correctness, indicating that retrieval systems lead to more accurate and reliable responses to clinical inquiries. Soong et al.[40] evaluated GPT-3.5 and GPT-4 models against a custom RetA LLM using a set of 19 questions. The evaluation, based solely on human judgments, revealed that both GPT-3.5 and GPT-4 exhibited more hallucinations in all 19 responses compared to the RetA model. While these works on RetA LLM systems represent significant progress, they suffer from at least two shortcomings: (1) they typically require human evaluation, making systematic benchmarking of new systems challenging and unscaleable; (2) they often focus solely on evaluating an LLM's output, disregarding the relevance of the information retrieved to generate an answer. Deciding which "relevant" sources should be summarized can be just as challenging as generating the actual summary. Hence there is a need for a benchmark that enables integrated evaluation of both a system's ability to select relevant documents as well as its ability to summarize these documents.

3. Materials and Methods

3.1. *Dataset Generation*

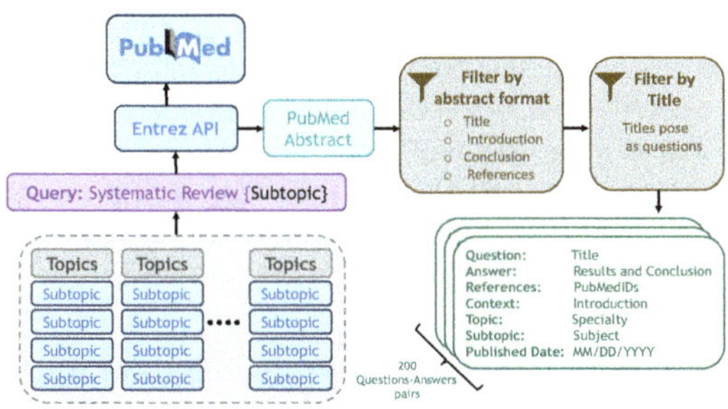

Fig. 1: Schematic Representation of the Protocol for Retrieving Abstracts from PubMed and Generating Title-Based Questions

PubMed is a free resource supporting search and retrieval of biomedical literature. As prior work has demonstrated, a large quantity of research papers available in this index are phrased as questions, and it is possible to structure them in a question-answer format.[41,42] Extending this idea, we created an open information retrieval and abstractive summarization dataset, using SR as a proxy for inquiries of medical interest. The rationale is that SRs are structured reviews written by human experts which summarize the pertinent literature related to a question of interest in an evidence-based manner.[43] In writing a SR, experienced authors (1) screen the published literature in a systematic way and include studies in a standardized manner; (2) critically evaluate methodology and reported outcomes of the included studies; and (3) carefully extract data, summarize original research findings, and in some instances, conduct additional statistical analysis of extracted results from studies including randomized controlled trials, observational cohort studies, case series and other qualitative studies on a specific topic. Furthermore, SRs are extensively used to provide evidence for various purposes, including policy-making, clinical practice guidelines, health technology assessment, and decision making in healthcare.[44] As SRs unify and present a comprehensive overview of a given subject by human experts, we chose to leverage published SRs as gold standards when building our database.

To populate such a dataset, we employed E-utilities, a public API to the NCBI Entrez system[45] , to access PubMed and construct question-answer pairs with their respective references. Figure 1 illustrates our process in detail. First, we established a comprehensive selection of medical specialties and subspecialties. Second, we formulated a query to retrieve Systematic Reviews relevant to each medical specialty/subspecialty. Upon constructing the specialty-specific queries and retrieving associated abstracts, we retrieved all papers structured in a format that can be easily converted to questions-answer pairs (as noted by Jin et al 2019[41]) namely Title, Introduction, Conclusion, and References. Third, we applied another filtering process, narrowing down to solely those publications whose titles included an explicit question (i.e., publications whose titles including question marks). The questions from these titles were extracted.

Finally, two human evaluators (AL and SF) manually reviewed the retrieved questions and extracted an answer to each question using minimally modified text from the results and conclusions section of the corresponding SR abstract. Concretely, in order to generate each answer, the human reviewers removed from the Results and Conclusions section of the abstract any text describing the structure or design of the systematic review (e.g., "We used PubMed to retrieve 100 papers"), leaving only text that directly addressed the question extracted from the SR's title. In the process, abstracts that were lacking substantive results and abstracts that merely described research proposals (e.g. descriptions of future work) were entirely removed.

3.2. *Clinfo.ai: An LLM Chain for Information Retrieval and Synthesis*

Our proposed RetA LLM system, Clinfo.ai, consists of a collection of four LLMs working conjointly (an LLM chain[46]) coupled to a Search Index (either PubMed or Semantic Scholar) as depicted in Figure 2. Previous works have observed that very large language models (e.g., 100B parameters or more) exhibit zero-shot reasoning capabilities, where task-specification prompts can be used to guide the LLM output without further fine-tuning.[47,48] We leverage the zero-shot reasoning capabilities of two LLMs, specifically OpenAI's GPT-3.5 and GPT-4 models, to complete each step in the LLM chain depicted in Figure 2. All prompts used in each step of the chain are available in the supplemental material [b]. We use LangChain's API to send prompts and receive outputs from GPT-3.5 and GPT-4. While different models could technically be used through this entry point, our experiments are limited to OpenAI's GPT-3.5 and GPT-4 models (snapshots gpt-3.5-turbo-0613 , gpt-4-0613 respectively). For both models, we employ a temperature of 0.5 and a max token generator limit of 1024.

3.2.1. *Query Generator*

In our Clinfo.ai system, the input is the question submitted by the user. Once a question is submitted, the primary task of the query generator (labeled "Question2Query" in Figure 2) is to construct a PubMed (or Semantic Scholar) query that efficiently retrieves a substantial number of relevant articles pertaining to the posed question. This is achieved by instructing

[b]https://github.com/som-shahlab/Clinfo.AI/tree/main/SupplementalMaterial

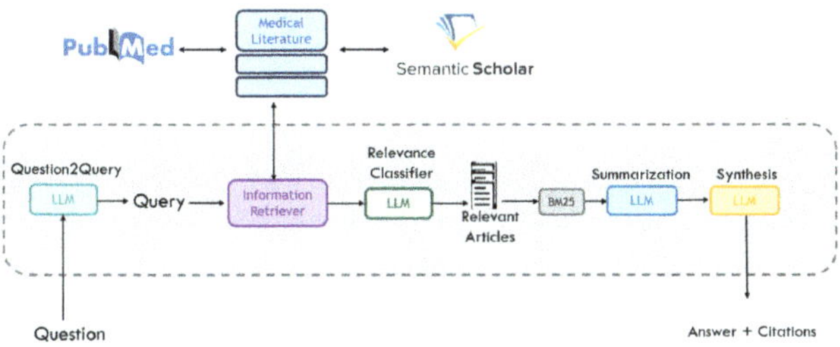

Fig. 2: Clinfo.ai: A RetA LLM system for retrieving and summarizing scientific articles

the model to incorporate the most crucial and relevant keywords that accurately represent the query's context and requirements.

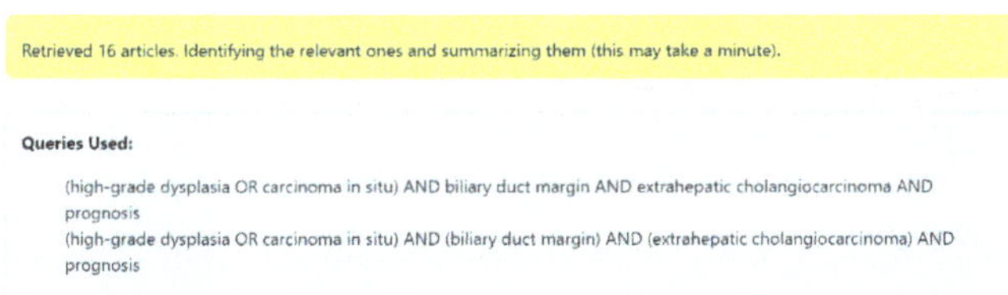

Fig. 3: Query Generated by Clinfo.ai for question: "Does high-grade dysplasia/carcinoma in situ of the biliary duct margin affect the prognosis of extrahepatic cholangiocarcinoma?"

3.2.2. *Information Retriever*

In a similar fashion to the Dataset Generation process, we utilize the Entrez API to fetch abstracts from PubMed using the output generated by the Query Generator. By leveraging the Entrez API, we are able to programmatically access and retrieve the relevant abstracts that match the constructed PubMed queries. Because LLM output is stochastic and different queries may capture different aspects of the literature, we take the union of all papers returned by three LLM-generated queries (each with the same prompt but different seeds).

3.2.3. *Relevance Classifier*

Since the query generator emphasizes recall over precision (i.e., it retrieves as many potentially relevant articles as possible), it is crucial to classify the relevancy of the retrieved articles. To achieve this, we adopt an LLM-enabled binary classification approach, wherein each article is categorized as either relevant or not relevant to the posed question using GPT-3.5. Once the relevant articles are identified, we make use of the full abstract metadata of each article to construct their citations in the IEEE format. If more than 35 relevant articles are deemed relevant, the user can decide to re-rank and filter them using BM25.[49]

3.2.4. *Summarization*

The penultimate step in Clinfo.ai uses an LLM to summarize each relevant abstract within the context of the user-submitted question.

3.2.5. *Synthesis*

In the final step of Clinfo.ai, the relevant article summaries are organized as an ordered list, with each number in the list corresponding to a citation. This structured list of article summaries is then fed to a LLM with the task of constructing a concise and informative summary. The LLM is also instructed to utilize only the provided article summaries and no other additional information, relying on the structured list of citations to reference and accurately attribute each finding.

3.3. *www.clinfo.ai: A Clinfo.ai User Interface via Web Application*

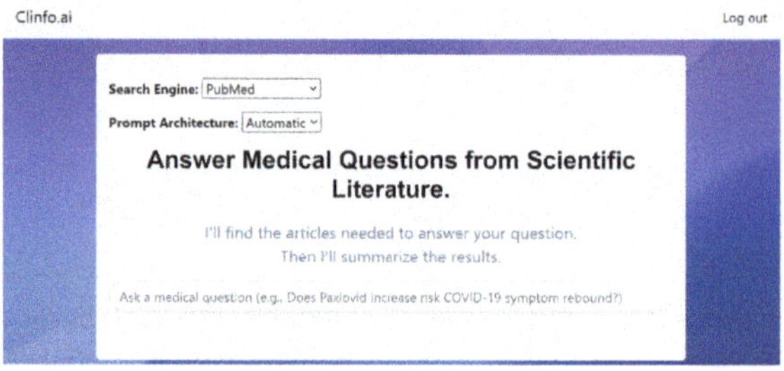

Fig. 4: Clinfo.ai user interface

To facilitate interaction with our system, we developed a web application that allows users to submit their own questions and/or customize the prompts. The latter enables users to tailor the system according to their individual preferences and needs, as illustrated in Figure 4.The entire process provides real-time access, displaying the queries generated during the search (as shown in Figure 3), the number of retrieved articles, a concise summary of each important article, and a final "Literature Summary" (or "Synthesis", to distinguish it from the individual article summaries) accompanied by an abbreviated answer to the question ("TL;DR"). Additionally, the references are presented as hyperlinks, enabling users to verify both the validity of the reference and the information captured from it. It is possible that even after summarizing an article's abstract, Clinfo.ai may not include that article in final Literature Summary or "TL;DR". Nevertheless, we ensure that all relevant articles are presented to the user so that they can access and explore them as needed. An example of a final Literature Review constructed with Clinfo.ai is shown in Figure 5.

3.4. *Task Description and Evaluation*

The task is defined in a three step manner:

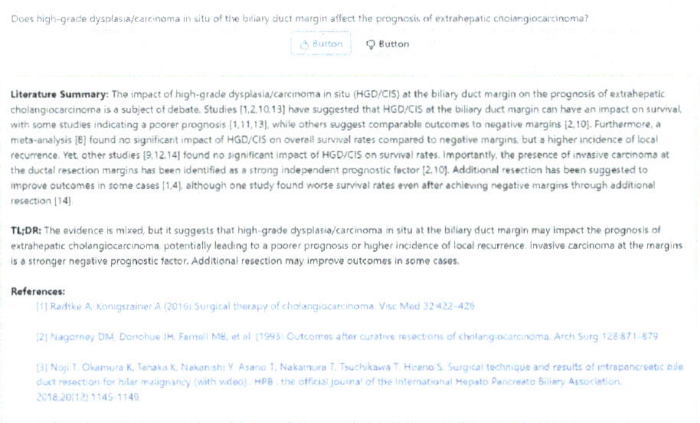

Fig. 5: "Literature Summary" (Synthesis) and "TL;DR" constructed with Clinfo.ai for the question, "Does high-grade dysplasia/carcinoma in situ of the biliary duct margin affect the prognosis of extrahepatic cholangiocarcinoma?" (not all references are included in figure)

(1) Given a question, generate a query to retrieve a set of articles;
(2) Given the provided articles, determine their relevancy to the question;
(3) Given relevant articles, summarize the findings.

Step (2) is evaluated based on precision and recall. Considering the set of all documents D, $RET(D,k)$ denotes the set of k retrieved documents deemed relevant and $REL(D,q)$ the set of all documents referenced by a SR. We define precision and recall in this context as follows:

$$\text{precision} = \frac{|RET(D,k) \cap REL(D,q)|}{|RET(D,k)|} \tag{1}$$

$$\text{recall} = \frac{|RET(D,k) \cap REL(D,q)|}{|REL(D,q)|} \tag{2}$$

Step (3) is conducted using both source-free (SF) and source-augmented (SA) automated metrics. Source-free metrics compare a model's output to a gold standard reference summary, without including any information from the articles used to generated the gold standard summary. For our evaluation purposes, the gold standard is the human-curated answer (derived from conclusions and/or results of each SR). On the other hand, SA metrics additionally consider relevant context to evaluate the quality of model-generated outputs. For our experiments, context is constructed by concatenating a SR's introduction, results, and conclusion sections. The SA metrics we employed (and the LMs they use) include UniEval[26] (T5 -large), COMET (XLM-RoBERTa),[50] and CTC Summary Consistency (BERT).[51]

UniEval is a multi-dimensional evaluator designed for summarization tasks and takes into account four key dimensions (and their corresponding overall average):

- **Coherence:** Assesses whether the summary forms a cohesive and rational body of text;

- **Consistency:** Evaluates the factual alignment between the information presented in the summary and the content of the source document;
- **Fluency:** Assesses the readability and linguistic fluency of a summary;
- **Relevance:** Measures whether the summary contains only the important information from the source document.

COMET is an evaluation metric developed to assess the quality of Machine Translation (MT) systems. Despite being trained on multilingual MT outputs, it performs remarkably well in monolingual settings, when predicting summarization output quality.[52] CTC is an evaluation framework, based on information alignment between input, output, and context, for compression (e.g summary), transduction (e.g translation), and creation (e.g. conversation).

Finally we perform an evaluation using SF metrics, including BERTScore,[53] ROUGE-L,[54] METEOR,[55] chrF[56] , GoogleBLEU, CTC Summary (without providing context) , and CharacTer.[57] The majority of these metrics have shown moderate correlation with human preference and are widely reported in NLG tasks.[25,26]

The multi-dimensional evaluation based on source-augmented metrics makes the assumption that an LLM+RetA model is able to (1) retrieve abstracts of works that were deemed relevant by an author of a SR and (2) synthesize them in a similar fashion. We acknowledge that if this assumption is not met, the evaluation would heavily penalize the output. Conversely, if the system retrieves an article that was not considered by a SR but bears a similar semantic meaning to an article present in the references of a SR, the evaluation would not penalize the generated text. For our proposed method, both behaviors are desired.

4. Baselines and Experiments

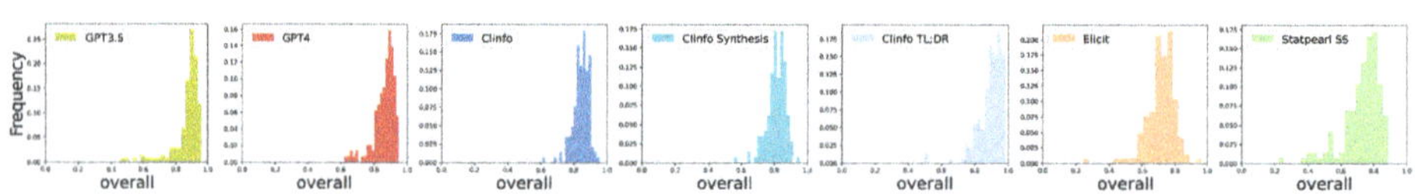

Fig. 6: UniEval Overall Score of 146 questions (unconstrained by published date) from PubMedRS-200 distribution across Unrestricted Search (GPT3.5 and GPT4 zero-shot performance is added)

Using our proposed task, we evaluated the performance of GPT-4 and GPT-3.5 without retrieval augmentation, Clinfo.ai (our GPT-enabled RetA LLM system), and two deployed tools: Elicit (an AI research assistant based on LLMs, designed for facilitating literature review generation, accessed on 07-02-2023), and Statpearls Semantic Search (a free search tool for medical knowledge, accessed on 07-25-2023). While other automated literature summarization systems are available, at the time of this study the vast majority require a subscription to answer multiple questions. Additionally, a subset of these systems refused to provide an answer to a significant number of the PubMedRS-200 questions as posed, making evaluation for these systems fraught and difficult to interpret. We exclude these systems from our analysis.

Table 1: Performance on 146 questions from PubMedRS-200 using source-augmented (SA) metrics: UniEval (T5-large), COMET (XLM-RoBERTa), CTC summary (BERT)

	Unified Multi-Dimensional Evaluator (UniEval)						CTC (SA)	
Model	Coherence ↑	Consistency ↑	Fluency ↑	Relevance ↑	Overall ↑	COMET ↑	Consistency ↑	Avg. Length
LLM								
GPT-3.5	0.908 (0.149)	0.694 (0.144)	0.947 (0.059)	0.939 (0.101)	0.872 (0.082)	0.676 (0.075)	0.865 (0.017)	104.834 (47.778)
GPT-4	0.915 (0.099)	0.655 (0.145)	0.942 (0.051)	0.929 (0.078)	0.86 (0.062)	0.677 (0.075)	0.866 (0.017)	84.214 (39.772)
LLM + RetA								
Restricted Search								
Synthesis & TL;DR	**0.949 (0.065)**	0.466 (0.105)	0.903 (0.104)	**0.964 (0.053)**	0.82 (0.055)	0.704 (0.055)	0.84 (0.014)	205.579(46.181)
Synthesis	0.925 (0.066)	0.394 (0.11)	0.893 (0.119)	0.939 (0.101)	0.788 (0.059)	0.693 (0.057)	0.842 (0.015)	165.814 (40.749)
TL;DR	0.866 (0.143)	0.787 (0.161)	0.954 (0.018)	0.826 (0.159)	0.858 (0.098)	0.665 (0.078)	0.874 (0.018)	38.766 (11.682)
Source Dropped								
Synthesis & TL;DR	0.942 (0.092)	0.465 (0.104)	0.918 (0.085)	0.962 (0.059)	0.822 (0.055)	0.706 (0.056)	0.843 (0.014)	204.248 (38.394)
Synthesis	0.925 (0.066)	0.398 (0.112)	0.912 (0.096)	0.943 (0.055)	0.795 (0.055)	0.695 (0.059)	0.845 (0.016)	164.938 (33.221)
TL;DR	0.829 (0.202)	0.763 (0.197)	0.953 (0.029)	0.796 (0.194)	0.835(0.13)	0.672 (0.078)	0.876 (0.017)	38.31 (10.726)
Unrestricted Search								
Our Models								
Synthesis & TL;DR	0.945 (0.064)	0.539 (0.127)	0.912 (0.096)	0.962 (0.059)	0.84 (0.052)	**0.721 (0.055)**	0.852 (0.017)	214.338 (44.173)
Synthesis	0.916 (0.092)	0.48 (0.142)	0.904 (0.098)	0.935 (0.069)	0.809 (0.06)	0.712 (0.057)	0.855 (0.019)	173.379 (38.492)
TL;DR	0.896 (0.123)	**0.81 (0.159)**	**0.955 (0.012)**	0.857 (0.135)	**0.88 (0.081)**	0.681 (0.072)	**0.88 (0.016)**	39.959 (11.754)
Deployed Models								
Elicit[19]	0.854 (0.136)	0.352 (0.147)	0.743 (0.151)	0.902 (0.117)	0.713 (0.085)	0.7 (0.066)	0.866 (0.017)	130.566 (22.946)
Statpearls SS[23]	0.753 (0.225)	0.383 (0.129)	0.93 (0.053)	0.845 (0.159)	0.728 (0.112)	0.633 (0.075)	0.841 (0.016)	118.172 (26.603)

Lastly, since our framework generates two outputs — "TL;DR" and "Literature Summary" (also referred to as "Synthesis") — we conducted evaluations of three forms of Clinfo.ai's output: (1) the synthesis of the articles retrieved and deemed relevant ("Synthesis"); (2) the abbreviated summary distilling the proposed "Synthesis" into one or two sentences ("TL;DR"); (3) the combined "Synthesis" and "TL;DR".

We recognize that the usage of scientific literature to extract question-answer pairs comes with the possibility that an answer deemed correct at the time of acquisition may be incorrect as new discoveries are published. To ensure that a system is not rewarded for simply copy-pasting the text of a retrieved source SR nor penalized when new relevant articles are published, we consider three evaluation regimes:

(1) **Restricted Search (RS)**: The retrieval process is constrained to include publications up to one day before the publication date. While this approach may not guarantee the retrieval of all publications considered important by the authors of each source systematic review, it effectively narrows down the search space to the subset of publications that could have been retrieved and deemed relevant during the review's preparation.

(2) **Source Dropped (SD)**: The retrieval process can retrieve articles published both before and after the source systematic review. However, if the source SR is retrieved, it is removed from the set of relevant articles and not used in the subsequent steps of the summarization process.

(3) **Unrestricted Search (US)** No restriction is applied; the source SR may (but need not)

Table 2: Performance on 146 questions from PubMedRS-200 using source-free (SF) metrics

Model	BERTScore ↑	ROUGE-L ↑	METEOR ↑	chrF ↑	GoogleBLEU ↑	CTC (SF) ↑	CharacTer↓	Avg. Length
LLM								
GPT-3.5	0.781 (0.037)	0.165 (0.053)	0.181 (0.073)	30.2 (10.5)	0.077 (0.036)	0.575 (0.065)	0.912 (0.102)	104.834 (47.778)
GPT-4	0.78 (0.037)	0.157 (0.049)	0.192 (0.07)	31.6 (9.06)	0.074 (0.031)	0.571 (0.064)	0.89 (0.099)	84.214 (39.772)
LLM + RetA								
Restricted Search								
Synthesis & TL;DR	0.77 (0.028)	0.135 (0.043)	0.121 (0.055)	21.5 (9.98)	0.058 (0.03)	0.527 (0.059)	0.993 (0.029)	205.579(46.181)
Synthesis	0.773 (0.028)	0.141 (0.044)	0.133 (0.059)	24.3 (10.4)	0.063 (0.032)	0.533 (0.06)	0.976 (0.056)	165.814 (40.749)
TL;DR	0.784 (0.041)	0.145 (0.068)	0.221 (0.089)	32.7 (7.67)	0.061 (0.043)	0.594 (0.068)	0.833 (0.086)	38.766 (11.682)
Source Dropped								
Synthesis & TL;DR	0.773 (0.028)	0.136 (0.037)	0.119 (0.054)	21.4 (9.69)	0.057 (0.028)	0.53 (0.06)	0.989 (0.036)	204.248 (38.394)
Synthesis	0.775 (0.026)	0.143 (0.038)	0.132 (0.057)	24.1 (9.91)	0.061 (0.043)	0.536 (0.06)	0.976 (0.056)	164.938 (33.221)
TL;DR	0.787 (0.041)	0.148 (0.064)	0.218 (0.078)	33 (6.98)	0.06 (0.039)	0.6 (0.066)	0.83 (0.092)	38.31 (10.726)
Unrestricted Search								
Our Models								
Synthesis & TL;DR	0.786 (0.029)	0.167 (0.06)	0.145 (0.073)	23.5 (11.2)	0.079 (0.046)	0.546 (0.067)	0.989 (0.036)	214.338 (44.173)
Synthesis	0.789 (0.03)	0.178 (0.067)	0.164 (0.084)	26.7 (12)	0.088 (0.051)	0.555 (0.07)	0.975 (0.065)	173.379 (38.492)
TL;DR	0.793 (0.038)	0.169 (0.076)	**0.252 (0.092)**	**35.5 (7.95)**	0.076 (0.049)	**0.61 (0.067)**	**0.825 (0.094)**	39.959 (11.754)
Deployed Models								
Elicit[19]	**0.807 (0.04)**	**0.218 (0.095)**	0.206 (0.093)	31.6 (12.5)	**0.127 (0.085)**	0.596 (0.07)	0.938 (0.096)	130.566 (22.946)
Statpearls SS[23]	0.77 (0.028)	0.136 (0.037)	0.149 (0.057)	26.5 (9.8)	0.062 (0.026)	0.536 (0.06)	0.939 (0.09)	118.172 (26.603)

Table 3: Clinfo.ai Precision and Recall on PubMedRS-200

Evaluation Regime	Precision ↑	Recall ↑	Source Included
Restricted Search	0.224 (0.239)	0.057 (0.061)	0.0 (0.0)
Source Dropped	0.186 (0.22)	0.064 (0.064)	0.0 (0.0)
Unrestricted Search	0.162 (0.175)	0.052 (0.064)	0.965 (0.185)

be included in the set of relevant articles retrieved by the system. Because we could not control the set of articles retrieved and summarized by closed-source tools like Elicit and Statpearls SS, they effectively fall within this evaluation regime.

Finally, to ensure that conformity with the SD regime would not prevent direct comparison with the other evaluation regimes, we removed questions from all other training regimes for which Clinfo.ai could only retrieve the source article (resulting in zero articles remaining after exclusion under the SD regime). This yielded 145 SRs (80 after October 2021 and 65 before).

5. Experimental Results and Analysis

Is RetA associated with significant improvements in automated metric evaluation?

As reported in previous studies,[34,39,58] both GPT-3.5 and GPT-4 without RetA demonstrated strong zero-shot performance using both source-augmented (Table 1) and source-free (Table 2) metrics. Notably, there was no substantial performance drop observed when these models were presented with questions based on source SRs published after September 2021 (Comparing Table 1 and Table S1 in the Supplement). While more studies are necessary, we postulate that this can be attributed to the models' exposure to prior published works during training. Since SRs are built upon existing literature ranging across multiple years, it is plausible that the models have been trained on relevant information that aids them in providing accurate responses to questions based on newer research. However, comparing all LLM against LLM + RetA models, the inclusion of RetA leads to a slight improvement in the overall performance of the models when evaluated with SF and SA automated metrics, irrespective of the publication date of the source SR. Previous works based on human evaluation have observed a similar trend, corroborating our automated evaluation framework.

How does Clinfo.ai perform compared to other systems?

As depicted in Table 1, Clinfo.ai exhibited better performance in overall UniEval compared to other RetA systems, irrespective of the chosen output strategy (Synthesis, TL;DR, or a concatenation of the two). This improvement in performance remained consistent regardless of the average length of the output, with Clinfo.ai achieving better results for both approximately 3x shorter (TL;DR) and around 2x longer outputs (Synthesis). Furthermore, this performance persisted across all different evaluation regimes, even when the source SR was dropped. This improvement amounted to at least 6.2% and at most 14.9% in UniEval Overall performance. These results suggest two significant points: (1) Our system is not merely copying and pasting information from an SR review. Instead, it demonstrates a genuine ability to process and present the information effectively, resulting in enhanced performance compared to other available tools; and (2) even in the absence of a source SR, Clinfo.ai can still provide conclusions that are better aligned with a source SR's conclusion (compared to tools that might include the source SR).

TL;DR or Synthesis?

Clinfo.ai TL;DR demonstrates significantly better performance compared to Synthesis and Synthesis & TL;DR, even though they all utilize the same relevant retrieved articles. It is worth noting that while Synthesis provides evidence to answer the question based on the retrieved articles, this evidence may not align with the original evidence reported by a Systematic Review (SR). However, the increased performance of TL;DR could be attributed to the LLM's capability to correctly identify the most salient points of the relevant articles and effectively summarize them. On the other hand, using only source-free (SF) metrics (Table 2), Elicit performs better under BERTScore, ROUGE-L and GoogleBLEU, while Clinfo.ai TL;DR performs better under METEOR, chrF, CTC (SF), and CharacTer.

These results highlight a potential limitation of automated evaluation . For instance, SF metrics tend to reward short responses, which may not necessarily be accurate or comprehensive. On the other hand, several SA metrics can assign the best score to considerably larger generations (UniEval's Coherence and Relevance, and COMET), acknowledging their quality and relevance. This discrepancy in evaluation metrics raises concerns about the fair assessment of model performance and emphasizes the need for a comprehensive evaluation approach.

Comparing different evaluation regimes, the best performance was observed under the Unrestricted Search evaluation regime, possibly due to the fact that the source SR was retrieved on 96.5% of the questions. As expected given the restricted set of retrievable documents, Clinfo.ai's precision was highest under the Restricted Search regime (Table 3).

6. Conclusion

The rapidly expanding medical literature and the capabilities of LLMs to process and summarize vast amounts of information have led to the development of several tools that utilize LLMs to generate on-demand summaries of published scientific literature. However, the lack of high-quality datasets and appropriate benchmarking tasks has hindered rigorous evaluations of these tools. To address this gap, we have introduced Clinfo.ai, an open-source end-to-end LLM-chain workflow designed to query, evaluate, and synthesize medical literature into concise summaries for answering questions on demand. Additionally, we introduce a unique dataset, PubMedRS-200, which consists of questions and answers extracted from systematic reviews, enabling automatic evaluation of LLM performance in Retrieval Augmentation Question Answering. Our tools and benchmarking dataset are publicly available to ensure reproducibility and to facilitate further research in harnessing LLMs for Retrieval Augmentation Question Answering tasks.

7. Limitations

In this study, we employed automated metrics that have demonstrated moderate-to-high correlation with human preferences, but we did not explicitly solicit human preferences to evaluate the RetA LLM systems considered. Future work should consider including human evaluation to ensure alignment of automated metrics and human preferences. Lastly, it is worth noting that prior studies have reported that LLMs demonstrate the ability to generate accurate Boolean operators and syntax, effectively adhering to PubMed query formats. However, our observations revealed that these models also generated hallucinated MeSH terms, which could potentially lead to the exclusion of relevant studies. To overcome this limitation, future research efforts should prioritize improving the query generation process, ensuring that generated MeSH terms are reliable and relevant for better precision and recall in medical literature search tasks.

8. Acknowledgments

AL is funded by Arc Institute. SF is supported by a Stanford Graduate Fellowship. This effort was supported in part by the Mark and Debra Leslie endowment for AI in Healthcare. We thank Will Haberkorn for his aid with Figure S1.

References

1. K. I. Bougioukas, E. C. Bouras, K. I. Avgerinos, T. Dardavessis and A.-B. Haidich, How to keep up to date with medical information using web-based resources: A systematised review and narrative synthesis, *Health Information & Libraries Journal* **37**, 254 (2020).
2. E. Landhuis, Scientific literature: Information overload, *Nature* **535**, 457 (2016).
3. R. Van De Schoot, J. De Bruin, R. Schram, P. Zahedi, J. De Boer, F. Weijdema, B. Kramer, M. Huijts, M. Hoogerwerf, G. Ferdinands *et al.*, An open source machine learning framework for efficient and transparent systematic reviews, *Nature machine intelligence* **3**, 125 (2021).

4. J. E. Andrews, K. A. Pearce, C. Ireson and M. M. Love, Information-seeking behaviors of practitioners in a primary care practice-based research network (pbrn), *Journal of the Medical Library Association* **93**, p. 206 (2005).
5. G. Del Fiol, T. E. Workman and P. N. Gorman, Clinical questions raised by clinicians at the point of care: a systematic review, *JAMA internal medicine* **174**, 710 (2014).
6. A. Daei, M. R. Soleymani, H. Ashrafi-Rizi, A. Zargham-Boroujeni and R. Kelishadi, Clinical information seeking behavior of physicians: A systematic review, *International journal of medical informatics* **139**, p. 104144 (2020).
7. J. W. Ely, J. A. Osheroff, M. L. Chambliss, M. H. Ebell and M. E. Rosenbaum, Answering physicians' clinical questions: obstacles and potential solutions, *Journal of the American Medical Informatics Association* **12**, 217 (2005).
8. R. Borah, A. W. Brown, P. L. Capers and K. A. Kaiser, Analysis of the time and workers needed to conduct systematic reviews of medical interventions using data from the prospero registry, *BMJ open* **7**, p. e012545 (2017).
9. D. A. Cook, M. T. Teixeira, B. S. Heale, J. J. Cimino and G. Del Fiol, Context-sensitive decision support (infobuttons) in electronic health records: a systematic review, *Journal of the American Medical Informatics Association* **24**, 460 (2017).
10. D. Lobach, G. D. Sanders, T. J. Bright, A. Wong, R. Dhurjati, E. Bristow, L. Bastian, R. Coeytaux, G. Samsa, V. Hasselblad *et al.*, Enabling health care decisionmaking through clinical decision support and knowledge management., *Evidence report/technology assessment* , 1 (2012).
11. P. A. Bonis, G. T. Pickens, D. M. Rind and D. A. Foster, Association of a clinical knowledge support system with improved patient safety, reduced complications and shorter length of stay among medicare beneficiaries in acute care hospitals in the united states, *International journal of medical informatics* **77**, 745 (2008).
12. T. Isaac, J. Zheng and A. Jha, Use of uptodate and outcomes in us hospitals, *Journal of hospital medicine* **7**, 85 (2012).
13. D. A. Reed, C. P. West, E. S. Holmboe, A. J. Halvorsen, R. S. Lipner, C. Jacobs and F. S. McDonald, Relationship of electronic medical knowledge resource use and practice characteristics with internal medicine maintenance of certification examination scores, *Journal of general internal medicine* **27**, 917 (2012).
14. J. J. Cimino, G. Elhanan and Q. Zeng, Supporting infobuttons with terminological knowledge., in *Proceedings of the AMIA annual fall symposium*, 1997.
15. D. Demner-Fushman, Y. Mrabet and A. Ben Abacha, Consumer health information and question answering: helping consumers find answers to their health-related information needs, *Journal of the American Medical Informatics Association* **27**, 194 (2020).
16. P. Lewis, E. Perez, A. Piktus, F. Petroni, V. Karpukhin, N. Goyal, H. Küttler, M. Lewis, W.-t. Yih, T. Rocktäschel *et al.*, Retrieval-augmented generation for knowledge-intensive nlp tasks, *Advances in Neural Information Processing Systems* **33**, 9459 (2020).
17. Q. Jin, R. Leaman and Z. Lu, Pubmed and beyond: Recent advances and best practices in biomedical literature search, *arXiv preprint arXiv:2307.09683* (2023).
18. J. M. Nicholson, M. Mordaunt, P. Lopez, A. Uppala, D. Rosati, N. P. Rodrigues, P. Grabitz and S. C. Rife, Scite: A smart citation index that displays the context of citations and classifies their intent using deep learning, *Quantitative Science Studies* **2**, 882 (2021).
19. Ought, Elicit: The ai research assistant (2023).
20. GlacierMD, Glaciermd - a modern physician reference (2023).
21. Consensus, Consensus (2023).
22. OpenEvidence, Openevidence: Making medical knowledge more useful, open, accessible, and understandable (2023).
23. Hippocratic AI, statpearls semantic search (2023).

24. M. Sallam, Chatgpt utility in healthcare education, research, and practice: Systematic review on the promising perspectives and valid concerns, *Healthcare* **11** (2023).
25. I. Ni'mah, M. Fang, V. Menkovski and M. Pechenizkiy, Nlg evaluation metrics beyond correlation analysis: An empirical metric preference checklist, *arXiv preprint arXiv:2305.08566* (2023).
26. M. Zhong, Y. Liu, D. Yin, Y. Mao, Y. Jiao, P. Liu, C. Zhu, H. Ji and J. Han, Towards a unified multi-dimensional evaluator for text generation, *arXiv preprint arXiv:2210.07197* (2022).
27. S. L. Fleming, A. Lozano, W. J. Haberkorn, J. A. Jindal, E. P. Reis, R. Thapa, L. Blankemeier, J. Z. Genkins, E. Steinberg, A. Nayak *et al.*, Medalign: A clinician-generated dataset for instruction following with electronic medical records, *arXiv preprint arXiv:2308.14089* (2023).
28. M. Bakker, M. Chadwick, H. Sheahan, M. Tessler, L. Campbell-Gillingham, J. Balaguer, N. McAleese, A. Glaese, J. Aslanides, M. Botvinick *et al.*, Fine-tuning language models to find agreement among humans with diverse preferences, *Advances in Neural Information Processing Systems* **35**, 38176 (2022).
29. S. Bubeck, V. Chandrasekaran, R. Eldan, J. Gehrke, E. Horvitz, E. Kamar, P. Lee, Y. T. Lee, Y. Li, S. Lundberg *et al.*, Sparks of artificial general intelligence: Early experiments with gpt-4, *arXiv preprint arXiv:2303.12712* (2023).
30. M. Sallam, Chatgpt utility in healthcare education, research, and practice: systematic review on the promising perspectives and valid concerns, in *Healthcare*, (6)2023.
31. G. Eysenbach *et al.*, The role of chatgpt, generative language models, and artificial intelligence in medical education: a conversation with chatgpt and a call for papers, *JMIR Medical Education* **9**, p. e46885 (2023).
32. M. Cascella, J. Montomoli, V. Bellini and E. Bignami, Evaluating the feasibility of chatgpt in healthcare: an analysis of multiple clinical and research scenarios, *Journal of Medical Systems* **47**, p. 33 (2023).
33. S. L. Fleming, K. Morse, A. M. Kumar, C.-C. Chiang, B. Patel, E. P. Brunskill and N. Shah, Assessing the potential of usmle-like exam questions generated by gpt-4, *medRxiv* , 2023 (2023).
34. T. H. Kung, M. Cheatham, A. Medenilla, C. Sillos, L. De Leon, C. Elepaño, M. Madriaga, R. Aggabao, G. Diaz-Candido, J. Maningo *et al.*, Performance of chatgpt on usmle: Potential for ai-assisted medical education using large language models, *PLoS digital health* **2**, p. e0000198 (2023).
35. E. Mitchell, C. Lin, A. Bosselut, C. D. Manning and C. Finn, Memory-based model editing at scale, in *International Conference on Machine Learning*, 2022.
36. W. X. Zhao, K. Zhou, J. Li, T. Tang, X. Wang, Y. Hou, Y. Min, B. Zhang, J. Zhang, Z. Dong *et al.*, A survey of large language models, *arXiv preprint arXiv:2303.18223* (2023).
37. H. Touvron, T. Lavril, G. Izacard, X. Martinet, M.-A. Lachaux, T. Lacroix, B. Rozière, N. Goyal, E. Hambro, F. Azhar *et al.*, Llama: Open and efficient foundation language models, *arXiv preprint arXiv:2302.13971* (2023).
38. Anthropic, Claude 2 (2023).
39. W. Hiesinger, C. Zakka, A. Chaurasia, R. Shad, A. Dalal, J. Kim, M. Moor, K. Alexander, E. Ashley, J. Boyd *et al.*, Almanac: Retrieval-augmented language models for clinical medicine (2023).
40. D. Soong, S. Sridhar, H. Si, J.-S. Wagner, A. C. C. Sá, C. Y. Yu, K. Karagoz, M. Guan, H. Hamadeh and B. W. Higgs, Improving accuracy of gpt-3/4 results on biomedical data using a retrieval-augmented language model, *arXiv preprint arXiv:2305.17116* (2023).
41. Q. Jin, B. Dhingra, Z. Liu, W. W. Cohen and X. Lu, Pubmedqa: A dataset for biomedical research question answering, *arXiv preprint arXiv:1909.06146* (2019).
42. H. Scells and G. Zuccon, Generating better queries for systematic reviews, in *The 41st international ACM SIGIR conference on research & development in information retrieval*, 2018.
43. M. Pourreza and F. Ensan, Towards semantic-driven boolean query formalization for biomedical

systematic literature reviews, *International Journal of Medical Informatics* , p. 104928 (2022).
44. K. Khan, R. Kunz, J. Kleijnen and G. Antes, *Systematic reviews to support evidence-based medicine* (Crc press, 2011).
45. E. W. Sayers, E. E. Bolton, J. R. Brister, K. Canese, J. Chan, D. C. Comeau, C. M. Farrell, M. Feldgarden, A. M. Fine, K. Funk *et al.*, Database resources of the national center for biotechnology information in 2023, *Nucleic acids research* **51**, D29 (2023).
46. T. Wu, M. Terry and C. J. Cai, Ai chains: Transparent and controllable human-ai interaction by chaining large language model prompts, in *Proceedings of the 2022 CHI conference on human factors in computing systems*, 2022.
47. T. Brown, B. Mann, N. Ryder, M. Subbiah, J. D. Kaplan, P. Dhariwal, A. Neelakantan, P. Shyam, G. Sastry, A. Askell, S. Agarwal, A. Herbert-Voss, G. Krueger, T. Henighan, R. Child, A. Ramesh, D. Ziegler, J. Wu, C. Winter, C. Hesse, M. Chen, E. Sigler, M. Litwin, S. Gray, B. Chess, J. Clark, C. Berner, S. McCandlish, A. Radford, I. Sutskever and D. Amodei, Language models are few-shot learners, in *Advances in Neural Information Processing Systems*, eds. H. Larochelle, M. Ranzato, R. Hadsell, M. Balcan and H. Lin (Curran Associates, Inc., 2020).
48. T. Kojima, S. S. Gu, M. Reid, Y. Matsuo and Y. Iwasawa, Large language models are zero-shot reasoners, in *Advances in Neural Information Processing Systems*, eds. S. Koyejo, S. Mohamed, A. Agarwal, D. Belgrave, K. Cho and A. Oh (Curran Associates, Inc., 2022).
49. A. Trotman, A. Puurula and B. Burgess, Improvements to bm25 and language models examined, in *Proceedings of the 19th Australasian Document Computing Symposium*, 2014.
50. R. Rei, C. Stewart, A. C. Farinha and A. Lavie, Comet: A neural framework for mt evaluation, *arXiv preprint arXiv:2009.09025* (2020).
51. M. Deng, B. Tan, Z. Liu, E. P. Xing and Z. Hu, Compression, transduction, and creation: A unified framework for evaluating natural language generation, *arXiv preprint arXiv:2109.06379* (2021).
52. K. Mateusz and P. Pecina, From comet to comes–can summary evaluation benefit from translation evaluation?, in *Proceedings of the 3rd Workshop on Evaluation and Comparison of NLP Systems*, 2022.
53. T. Zhang, V. Kishore, F. Wu, K. Q. Weinberger and Y. Artzi, Bertscore: Evaluating text generation with bert, *arXiv preprint arXiv:1904.09675* (2019).
54. C.-Y. Lin, Rouge: A package for automatic evaluation of summaries, in *Text summarization branches out*, 2004.
55. S. Banerjee and A. Lavie, Meteor: An automatic metric for mt evaluation with improved correlation with human judgments, in *Proceedings of the acl workshop on intrinsic and extrinsic evaluation measures for machine translation and/or summarization*, 2005.
56. M. Popović, chrf: character n-gram f-score for automatic mt evaluation, in *Proceedings of the tenth workshop on statistical machine translation*, 2015.
57. W. Wang, J.-T. Peter, H. Rosendahl and H. Ney, CharacTer: Translation edit rate on character level, in *Proceedings of the First Conference on Machine Translation: Volume 2, Shared Task Papers*, (Association for Computational Linguistics, Berlin, Germany, August 2016).
58. H. Nori, N. King, S. M. McKinney, D. Carignan and E. Horvitz, Capabilities of gpt-4 on medical challenge problems, *arXiv preprint arXiv:2303.13375* (2023).

A Conversational Agent for Early Detection of Neurotoxic Effects of Medications through Automated Intensive Observation

Serguei Pakhomov[†], Jacob Solinsky, Martin Michalowski, Veronika Bachanova

University of Minnesota,
Minneapolis, MN 55108, USA
[†]*E-mail: pakh0002@umn.edu*

We present a fully automated AI-based system for intensive monitoring of cognitive symptoms of neurotoxicity that frequently appear as a result of immunotherapy of hematologic malignancies. Early manifestations of these symptoms are evident in the patient's speech in the form of mild aphasia and confusion and can be detected and effectively treated prior to onset of more serious and potentially life-threatening impairment. We have developed the Automated Neural Nursing Assistant (ANNA) system designed to conduct a brief cognitive assessment several times per day over the telephone for 5-14 days following infusion of the immunotherapy medication. ANNA uses a conversational agent based on a large language model to elicit spontaneous speech in a semi-structured dialogue, followed by a series of brief language-based neurocognitive tests. In this paper we share ANNA's design and implementation, results of a pilot functional evaluation study, and discuss technical and logistic challenges facing the introduction of this type of technology in clinical practice. A large-scale clinical evaluation of ANNA will be conducted in an observational study of patients undergoing immunotherapy at the University of Minnesota Masonic Cancer Center starting in the Fall 2023.

Keywords: Large language models, artificial intelligence, speech, language, immunotherapy, Immune effector cell-associated neurotoxicity syndrome

1. Introduction

Immune effector cell-associated neurotoxicity syndrome (ICANS) represents a unique complication of immune effector therapy particularly in patients treated with chimeric antigen receptor T-cell therapy (CAR-T) cells for hematologic malignancies. ICANS incidence varies from 40-60% depending on specific CAR-T product and grading using a 4-point scale, with 1 being the mild manifestation and 4 the most severe. ICANS usually presents 3-5 days after CAR-T infusion and about 20% of events present at grade 3 or higher. The clinical centers administering approved CAR-T therapies have to comply with the Risk Evaluation and Mitigation Strategeis (REMS) mandated by the Food and Drug Administration (FDA). These include monitoring and prompt treatment of ICANS symptoms. The purpose of ICANS monitoring and detection after the CAR-T infusion and prompt treatment is to halt ICANS progression and minimize the risk of brain edema/herniation, the most feared sequelae of ICANS resulting in severe cognitive impairment, coma, ICU stay, intubation, and, in rare cases, death.[1–3]

Clinical manifestations of ICANS typically begin with word-finding difficulty, headaches, confusion, dysphasia, aphasia, impaired fine motor skills resulting in agraphia, and somnolence[4] and, if untreated, can progress to the more severe sequalae. Expressive aphasia has been found to be the most specific symptom of ICANS. It starts as impaired ability to name objects, paraphasia errors, hesitant speech, and verbal perseveration, which can then proceed to global aphasia (inability to speak or respond to commands) with increasing ICANS severity.[4] In fact, initial expressive aphasia is highly prevalent (86%) in patients that then go on to develop severe neurotoxicity.[5] Low-grade ICANS is managed predominantly by supportive care or low dose dexamethasone, whereas severe ICANS is usually treated with high doses of corticosteroids and anakinra which can partially block the cascade of inflammation leading to pathology.[6] Recently emerging clinical evidence suggests that early intervention with a short course of corticosteroids such as dexamethasone in patients with low-grade ICANS can resolve these symptoms completely and thereby prevent progression to more severe ICANS.[7] However, administration of corticosteroids as prophylaxis of ICANS in all patients undergoing CAR-T therapy is not desirable as corticosteroids may have a negative impact on the effectiveness of CAR-T therapy itself, have short and long-term side-effects, increase risk of infections and therefore lower dose and short course is desirable.[8]

The existing methods for detecting neurotoxicity of immunotherapy (as described in the National Comprehensive Cancer Network (NCCN) guidelines) consists of administering brief cognitive assessment tools such as the Immune Effector Cell-Associated Encephalopathy (ICE) Assessment Tool or the CAR-T Cell Toxicity Tool (CARTOX-10). Both are loosely based on the Mini-Mental State Examination (MMSE) originally developed for the diagnosis of dementia and include several brief cognitive instruments. The CARTOX-10 consists of the following 4 categories: Orientation: orientation to year, month, city, hospital, president of country of residence (5 pts); Naming: ability to name 3 objects (e.g., point to clock, pen, button) (3 pts); Writing: ability to write a standard sentence (e.g., "Our national bird is the bald eagle") (1 pt); Attention: ability to count backwards from 100 by 10 (1 pt). ICE adds one more category to the CARTOX-10 instrument: Following commands: ability to follow simple commands (e.g., "Close your eyes and stick out your tongue") (1 pt).

These tools are widely used for screening for ICANS, are brief and easy to use at bedside, and are highly specific for ICANS but lack scientifically rigorous evaluation. These tools inherited low sensitivity from the MMSE on which they were based, as evidence from practice suggests that patients in the early stages of ICANS may pass the ICE assessment (especially if they are able to memorize it due to its frequent administration) while displaying some of the more subtle ICANS symptoms.[9] Another major drawback of the existing screening tools is that while these paper-and-pencil tests are not particularly difficult to administer and score, their administration requires a qualified healthcare provider and is time consuming. Since post CAR-T therapy follow-up requires intensive daily monitoring usually for up to 14 days, that introduces a significant burden on clinical personnel and healthcare resources. This routine practice limits the frequency and depth to which patients can be feasibly monitored with ICE and, consequently, may lead to missing the onset of early symptoms in between assessments. Using technology to help in administering and facilitating more frequent follow-up of patients would be a significant advance in assuring safer CAR-T therapies by enabling earlier detection of more subtle symptoms. There is also an increasing trend to administer CAR-T therapy in the outpatient setting. At-home monitoring of ICANS symptoms is highly desirable as it offers the potentially more timely intervention while providing patients with more comfort and convenience.

Early detection of ICANS would allow using lower doses of corticosteroids but would also require intensive monitoring of cognitive function (e.g., 3 times per day vs. the typical once per day frequency). Infrequent monitoring for ICANS (once a day or less) is likely to miss the early onset of subtle symptoms, as demonstrated by a study of 133 patients undergoing CAR-T therapy.[10] Fifty-one of these patents developed ICANS and 27 of the 51 patients (53%) presented already with Grade >= 2 ICANS as the initial diagnosis. According to the ASTCT Consensus Grading guidelines, Grade 2 ICANS is diagnosed when the patient scores in the range 3-6 (out of 10 possible points). In practical terms, to score 3-6 on the ICE test, the patient would have to be significantly impaired (i.e., unable to tell what year, month it is, which city or hospital they are in, who the president is, and/or name three basic objects). The fact that over half of the patients with ICANS are initially diagnosed with Grade 2 or higher, combined with the fact that ICANS can develop in a matter of hours, indicates high likelihood that milder symptoms were present earlier but were missed either due to poor sensitivity of ICE, its relatively infrequent administration, or both.

Limitations of the standard-of-care approaches to ICANS detection combined with the availability of highly effective therapy to prevent its further progression[7] create the urgent need for a validated, low provider burden, and well-tolerated by patients solution for early identification of neurotoxicity. Deploying such a solution will potentially result in preventing an estimated 40-70% of cancer patients who are at risk of ICANS from severe and potentially debilitating symptoms. An effective solution will also reduce the total dose and duration of steroids, mitigate the steroid effect on CAR-T function and response, and can potentially improve CAR-T outcomes, enable easier access to CAR-T for older people, and facilitate outpatient administration and management after CAR-T therapy.

In this paper, we provide a description of the design and implementation of an Automated Neural Nursing Assistant (ANNA) system designed to address the limitations of the standard-of-care approaches by automating the administration and analysis of speech-based neurocognitive tests [a]. We also discuss the challenges specific to this particular clinical use case of intensive monitoring for cognitive changes associated with neurotoxic effects of immunotherapy, as well as other emerging areas where such intensive monitoring may be needed. We also report on the results of a small preliminary functional evaluation study designed to evaluate user experience with the system and collect feedback to determine areas for improvement prior to conducting a clinical study scheduled to begin in the Fall of 2023.

[a] A live demo version of ANNA has been presented at the 2023 Interspeech symposium and is currently available at +1 (612)-682-6292. Note: the phone number may change over time - to obtain the current number for the demo, please contact the authors

2. System Description

ANNA consists of a multi-platform app (iOS, Android, telephony) that administers neurocognitive tests, collects voice responses, and securely uploads them to a web service that stores the audio and automatically scores the tests. The implementation described in this paper operates via the telephone interface. To make the conversation as natural as possible, the system is implemented to work in full-duplex audio mode in which both the patient and the system can speak at the same time without the need for the patient to signal the end of utterances by pressing a button.

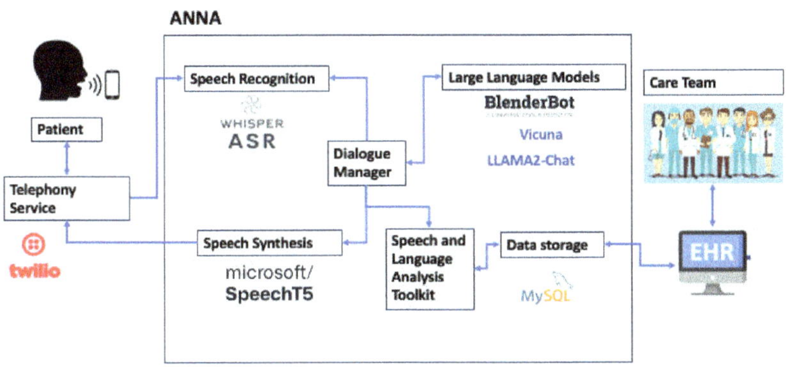

Fig. 1: Illustration ANNA system architecture and data flow.

ANNA's architecture illustrated in Figure 1 consists of two independent components, a dialogue manager and a cognitive assessment test battery. As the conversation manager goes through the script of the phone call, it transcribes the patient's speech, responds with synthesized speech, plays audio, and listens for pauses and cue words from the patients speech to allow ANNA to take turns in conversation in a natural manner. For speech transcription it employs OpenAI's state-of-the-art Whisper transcriber, which we have found to produce acceptable transcriptions of audio recordings from even the lowest end consumer phones. For speech generation we use the the pre-trained SpeechT5 model.[11] Twilio [b] currently provides telephony services to ANNA, however, the Dialogue Manager can easily be reconfigured to accept input from and produce input for other audio recording and playback devices, allowing us to reuse it in our group's other voice application projects. The Dialogue Manager currently consists of a set of rules that are designed to walk the patient through the process of

[b]https://www.twilio.com/

participating in the cognitive assessment. The conversation manager can either read directly from a script, which is how it conducts the word-list cognitive assessments or it can prompt a Large Language Model (LLM) using the patient's last utterance to generate a response. For this purpose we currently use the blenderbot-400M-distill[12] model and insert the responses to the input received from the patient at each conversation turn. Blenderbot is pre-configured to understand dialog and uses no prompt, responding directly to the patient's utterance. We also continue to experiment with other pre-trained LLMs (limited to those that can be used in a local HIPAA-compliant environment) including the Vicuna Chat[13] and, most recently, Llama2 Chat[14] models.

We developed ANNA as an easily deployed set of Docker images which can be deployed within an on-site server when provided with a phone number, web address, and a GPU with at least 24Gb of VRAM (e.g., NVIDIA RTX 3090 Ti). The current demo implementation is running on a server with two NVIDIA RTX 4090 cards. We have also constructed an alternative implementation of ANNA which does not use Docker images for any components which require access to a GPU, as the fully containerized application can have difficulty accessing the GPU in some environments.

2.1. *Spontaneous Speech and Language Elicitation*

We programmed ANNA to make a phone call to the patient's phone (smartphone or landline) and administer the following tasks: a brief conversation with a conversational agent based on a LLM that asks the patient to describe how they are feeling and conducts a brief conversation on one of a set of pre-defined topics such as favorite pastime, books, movies, etc. Topics are currently randomly drawn from a pre-defined list without replacement to alleviate practice effects.

2.2. *Cognitive Testing: Word List Recall*

The conversation is followed by a series of brief cognitive tasks including a word list learning task in which the patient is presented with a list of 6 words and is asked to recall as many of these words as the patient can immediately after the presentation (immediate recall) and a few minutes later (delayed recall). The word list recall task is vulnerable to practice effects in serial testing.[15] Practice effects can mask subtle cognitive changes due to early stages of ICANS; therefore, we developed a mechanism for generating multiple alternative lists of 6 words to minimize the effects of repeated test administration. To ensure that the lists of words are roughly equivalent across multiple presentations, we developed an approach for automatically generating lists of words that are equivalent in their lexical properties of frequency, concreteness, and imageability using the MRC Psycholinguistic database.[16]

2.3. *Cognitive Testing: Verbal Fluency*

Two verbal fluency tests are administered between the immediate and delayed recall tasks. The verbal fluency tests consist of a category fluency test in which the patient is asked to name as many animals as they can think of in 30 seconds, followed by a letter fluency test asking to name as many words beginning with the letter "F" as they can think of also in 30 seconds. The verbal fluency task also suffers from practice effects; however, prior work of other researchers and our own preliminary data show that in these generative tasks the practice effects are small and plateau after several presentations in individuals with cognitive impairment.[16] The rationale for selecting verbal fluency and list learning tasks rests on the evidence that they are particularly sensitive to a broad-spectrum of cognitive impairment effects caused by a wide variety of acute and chronic conditions including effects of medications,[17] are quick to administer, and lend themselves well to automation.

We selected the abbreviated versions of the list learning and verbal fluency tests to make them less burdensome for patients undergoing cancer treatment. The abbreviated version have been shown to have similar psychometric properties to their full counterparts (10 words for the list learning and 60 seconds for the verbal fluency tests).[18,19]

2.4. *Speech and Language Analysis*

The speech collected with ANNA is first subjected to automatic speech recognition to produce a verbatim transcript of everything the patient said during the interaction with the system. The current implementation of ANNA relies on the pre-trained Whisper neural transformer model (large-v1).[20] The transcribed speech is analyzed to extract the following language characteristics: syntactic complexity and language model perplexity. Syntactic complexity is measured using technology we previously developed to characterize language changes in patients with dementia.[21] Measures of syntactic complexity include the mean number of clauses, various measures of the depth of syntactic trees obtained from a syntactic constituency parser, and the mean syntactic dependency length obtained from a dependency parser. Perplexity is a measure of how many different equally most probable words can follow any given word based on probabilities obtained from a probabilistic or a neural language model. High mean perplexity computed over an utterance that did not participate in training the model indicates a poor fit between a language model and the text of the utterance. This measure has been shown to be useful for distinguishing between speech of individuals with probable Alzheimer's Disease and healthy controls.[22] Both the syntactic complexity and the language model perplexity have been included in an attempt to capture early signs of confusion and changes in language patterns that have been noted in patients starting to develop ICANS. In addition to the language characteristics described above, we also extract the following paralinguistic speech characteristics: initial pause duration prior to onset of speech, mean pause duration, perseveration and hesitation density. These characteristics have also been noted (anecdotally) by oncology treatment teams as being observed in early stages of ICANS.

2.5. *Motivation for Selecting Analytical Measures*

Rigorous prior work by other investigators and by our group demonstrated that the category verbal fluency task (animal naming) discriminates between individuals with Alzheimer's disease dementia and healthy controls with sensitivity of 0.88 and specificity of 0.96.[23] The list learning task has also been previously shown to have excellent psychometric properties for detecting mild (mean MMSE score = 22.1) memory impairment (sensitivity and specificity of

90%) and high test-retest reliability.[24] In our own prior work, we have demonstrated that using deep neural modeling of the spontaneous speech patterns produced during a picture description task is able to discriminate between Alzheimer's disease dementia and controls with 87% accuracy.[22] Our team has been engaged in validation studies in which we have demonstrated that neurocognitive tests of verbal fluency (semantic and phonemic) are highly sensitive to the neurotoxic effects of psychoactive medications such as lorazepam and topiramate,[25] as well as the effects of chronic traumatic encephalopathy,[26] effects of nicotine withdrawal[27] and neurodegeneration due to Alzheimer's disease.[28] While ICANS is distinctly different from the slowly progressing symptoms in dementia and chronic traumatic encephalopathy, it is similar to the rapid (on the order of hours) changes in cognition observed as a result of acute effects of psychoactive medications and nicotine withdrawal.

2.6. *Pilot Evaluation Study*

We conducted a small functional evaluation study of ANNA to elicit initial feedback from healthy individuals that could inform any further changes in system design and help us debug the system. We asked 10 Amazon Mechanical Turk workers to place an anonymous call to ANNA, interact with the system, and respond to a brief survey shown in Figure 3.

- **ANNA**: Hello, this call will be recorded for research use. Because I am a robot, I can be slow sometimes. Please don't hang up if I take too long to respond or say things that don't make much sense. I am going to ask you a few questions. Please listen carefully and answer them in as much detail as you can. Are you ready?
- **CALLER**: Yes.
- **ANNA**: Okay. First, I would like to ask you how you are feeling. For example, have there been any changes in how you feel in the last few hours?
- **CALLER**: No, I think I feel great. I have been enjoying the weather and I think overall, day's been going great and no changes in how I am feeling.
- **ANNA**: That's great! I'm glad you're having a good day. What's the weather like where you are?
- **CALLER**: Right now it is 97 degrees Fahrenheit. It's sunny. We're supposed to get triple digit 100 degree weather probably for the next seven days. But I mean, besides that, I'm in a cool environment and trying to limit my time outside.
- **ANNA**: Thank you for sharing. Now, I would like to ask you to tell me as much as you can about your most memorable trip. Feel free to take your time in answering.
- **CALLER**: Okay, well I would say most memorable trip is, this is a long time ago, this is back in 1999, in December of 1999, took a trip to Miami, Florida.
- **ANNA**: That sounds like a lot of fun. I've never been to Florida. What did you do there?
- **CALLER**: Well, went to Disney World, went to the beach, went to Miami, went to Orlando.
- **ANNA**: Thank you. Now, I am going to ask you to remember a list of six words. You will hear six words being spoken to you. Please repeat each word aloud as it is being spoken to you. Later, I will ask you to recall all six words. Are you ready?
- **CALLER**: Yeah, that's what we – it's pretty much what we did. Yes.
- **ANNA**: Here is the list: turtle, spray, lion, yard, queen, land. Now repeat as many of these words as you remember and say next when you are done
- **CALLER**: Turtle. Prey. Lion. Yard. Plane. Land. Turtle. Prey. Land. Farm. Trying to think of the other ones.
- **ANNA**: Thank you. Now, I will give you a letter of the alphabet. I am going to ask you to name words that begin with that letter, as fast as you can. For example, if I give you the letter S, as in sam, you can say soft, smile, and so on. Do not use the same word with a different ending such as smiling, or smiles. Are you ready?
- **CALLER**: Yes.
- **ANNA**: Okay. Your letter is the letter F, as in foxtrot. Please name all the words that you can think of that begin with the letter F. You have thirty seconds. Please begin.
- **CALLER**: Field, final, fantasy, fortune, fox, frisbee,
- **ANNA**: Please stop. Now, I will give you a category. I am going to ask you to name as fast as you can all the things that belong to that category. For example, if I give you the category of articles of clothing, you can say shirt, or jacket, or pants, and so on. Are you ready?
- **CALLER**: Yes.
- **ANNA**: Okay. Your category is animals. Begin naming as many animals as you can think of. You have thirty seconds.
- **CALLER**: Go ahead.
- **ANNA**: Please stop. We are almost done. Just one last thing. A few minutes ago I read a list of six words to you. Please try to recall as many of these words as you can and say them aloud as you remember them. You have thirty seconds. Please begin.
- **CALLER**: Tiger. Field prey, turtle.
- **ANNA**: Thank you. This concludes our session. Until next time. Goodbye.

Fig. 2: Example of an interaction between an Amazon Mechanical Turk worker and ANNA. This example shows the actual unedited transcription of the caller's voice with the Whisper transcriber.

3. Results

An example interaction between a functional evaluation study participants and ANNA is shown in Figure 2. This example shows a verbatim transcript of the interaction which illustrates the performance of all ANNA components including the automatic speech recognition and large language models. The quantitative results of the functional evaluation are summarized in Table 1.

The Duration column in Table 1 reflects the amount of time it took the evaluator to interact with ANNA and complete the evaluation survey. The mean duration for the 10 evaluators was 11 minutes. All evaluators had a 100% approval rating on the Amazon Mechanical Turk system (i.e., they were approved for payment for all human intelligence tasks that they performed in the past). All evaluators were able to get to the end of the interaction with ANNA successfully (Completed column in Table 1). The mean audibility rating was 4.1 (SD: 0.74), the sensibility of ANNA's responses to evaluators was rated as 3.7 (SD: 0.95), and the latency of system responses was rated as 2.0 (0.92).

Table 1: Results of pilot functional evaluation.

Evaluator	**Duration** (sec.)	**Completed** Y/N	**Audibility** (1-5)	**Sensibility** (1-5)	**Latency** (1-5)	**Comments**
1	660	Y	5	3	1	–
2	669	Y	4	3	4	Good exprience
3	660	Y	5	3	1	–
4	557	Y	3	5	2	Make it a little bit faster in response.
5	784	Y	4	5	2	–
6	808	Y	4	3	3	It could be a bit more human-like, now it sounds too machine like.
7	782	Y	3	5	2	Just improving response time would be a huge upgrade.
8	744	Y	5	3	1	–
9	748	Y	4	3	2	GOOD
10	588	Y	4	4	2	It should be able to restate the instructions instead of just waiting on an affirmative to being.
Mean (SD)	700 (86.18)	Y	4.1 (0.74)	3.7 (0.95)	2.0 (0.92)	–

4. Discussion

The broad clinical need that ANNA is designed to address arises from the limitations of the healthcare system in which intensive monitoring of patients' cognitive function (multiple times per day) by a human healthcare professional is not feasible and is cost-prohibitive. Monitoring cognitive function is unlike monitoring of physiologic function in that the former requires symbol-mediated interaction, which is typically achieved through the use of language. Many years of research in language technology and artificial intelligence yielded a number of conversational agents designed for use in healthcare applications.[29] However, recent developments in speech and language technology and, in particular, the introduction of large language models such as ChatGPT and Whisper can potentially move these efforts to a new level by making these systems simpler an more accurate in recognizing the incoming speech and producing more natural and flexible responses.

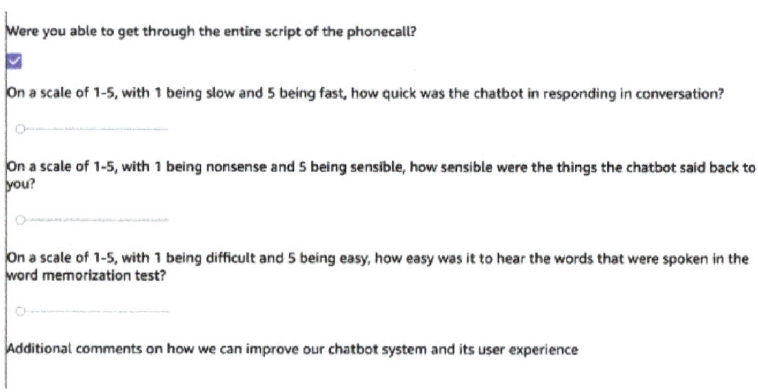

Fig. 3: Evaluation survey administered to Amazon Mechanical Turk workers.

The proposed automated cognitive assessment methods address the limitations of the existing manual methods by using a series of brief, validated and easy to administer neurocognitive tests that use speech as the input modality to detect expressive aphasia deficits - the most specific symptom of ICANS. The key innovative aspect of our approach is the use of AI technologies such as LLMs and automatic speech recognition based on deep learning to convert spoken responses to text that can subsequently be used to compute traditional scores as well as novel speech and language-based measures to further improve the sensitivity and specificity of these instruments. The use of AI, large language models and automatic speech recognition and synthesis, as well as scoring algorithms tailored to the neurocognitive tests at hand, is what sets us apart from other commercial and academic computerized neurocognitive testing approaches. To the best of our knowledge none of the current computerized approaches to neurocognitive assessment use conversational AI technology to elicit speech from patients and to analyze the resulting speech for cognitive impairment due to immunotherapy with machine learning. Another innovation is that we use extensively validated and recognized neurocognitive tests in a novel, accessible, and fully automated way that can also enable at-home and/or remote monitoring for ICANS, which could improve the accessibility of immunotherapy in rural and other settings away from medical centers.

In addition to the immunotherapy used to treat certain types of cancer, other newly emerging therapies that leverage the immune system have been recently approved by the US Food and Drug Administration for treatment of Alzheimer's disease. The most recent approval was granted in July 2023 to lecanemab, an immunotherapy agent that was demonstrated to remove Alzheimer's disease biomarkers from the brain and significantly (albeit moderately) slow down the disease progression as compared to other treatments. However, serious side effects including brain edema in 12.6% of the participants in the active arm of the clinical trial of this medication were observed.[30] Therefore, similarly to ICANS, early detection and clinical management of these side-effects in the treatment of Alzheimer's disease may potentially benefit from intensive cognitive monitoring. Another potential clinical application area for systems like ANNA is in automating the monitoring for post-operative delirium. Proactive monitoring for post-operative delirium and early intervention has been shown to shorten length of hospital stay and improve surgical outcomes.[31]

Pending successful demonstration of ANNA's feasibility and validity for early detection of ICANS, as we move outside of the realm of research and into wide adoption of ANNA in clinical practice, ANNA is ready to be integrated into a wide variety of clinical settings as a laboratory service using already existing technology and informatics standards including the Health Level 7 (HL7 v2) and FHIR protocols to interface with EHR via the standard lab test results route. One of the challenging issues that we expect to face has to do with handling of critical values. Critical values or failure to do the ANNA assessment will need to be communicated to the care team verbally by phone also using a standard protocol (ISO 15189) for communicating critical lab values. To this end, ANNA would need to have an interface (voice or graphical) to enable the care team to configure the system for each individual patient. The configuration will need to include telephone numbers for the patient and the care team as well as some of the patients' preferences (e.g., topics for the conversational part of ANNA's assessments, do-not-call times, system voice and personality preferences).

4.1. *Limitations*

ANNA currently has several technical limitations. Due to the use of multiple large neural models, the response latency can vary between less than a second for short turns (e.g., confirmations) to 3-4 seconds for longer turns in which ANNA has to convert longer input utterances to text and then also generate a response text and synthesize it into a spoken utterance. The evaluators in the pilot study clearly noted this as something that should be improved. We found that neural text-to-speech generation is the biggest contributor to response latency; however, other modules can be optimized as well. We plan to reduce the response latency in the production version of ANNA by a) switching to a faster version of the Whisper model [c] which has been benchmarked to be about 5 times faster than the OpenAI version, and b) distributing the LLM and TTS models across multiple GPU cards.

Another potential limitation that ANNA inherits from the pre-trained large language models is the potential for going off-topic (a.k.a. "hallucinating") during the initial conversational part of the assessment. To minimize this potential risk, we limit the amount of text produced by the models in response to the user input to 1-2 utterances. In the near future, we also plan to implement a set of guardrails to prevent ANNA from responding in inappropriate or offensive manner [d]. This limitation was not noted in the pilot study as the sensibility of ANNA's responses was rated as fairly high (mean 3.7 out of 5) and none of the 10 evaluators commented on any specific nonsensical responses.

The racial, cultural, gender and ethnic biases learned by large language models from training data is a major concern with applications of AI in medicine in general[32] and is a potential concern in our application as well. Given the nature of the interactions between patients and ANNA and the focus on eliciting as much speech from patients as possible over as few conversational turns as possible, we do not anticipate any such biases to have a chance to manifest themselves in any discernible fashion to the patients. Nonetheless, since inherent bias in language models is a known issue and we plan to examine the data collected with ANNA for any signs of bias or unfairness and experiment with current de-biasing methods.

ANNA's use case also has a distinct strength with respect to one of the biggest known limitations of large language models - variable trustworthiness of the information they generate. The lack of confidence in the information provided by these models is currently one of the major barriers to their adoption for clinical applications as primary sources of clinical knowledge.[33] ANNA's clinical use case, however, does not rely on large language models for knowledge. We rely on these models only to support a chatbot application used to elicit speech from patients for subsequent analysis and not to inform either patients or clinicians. As such, ANNA currently represents one of the safest and most immediate ways of using large language models in a clinical context.

5. Next Steps

We have developed and submitted an observational study protocol to the University of Minnesota Institutional Review Board. In this prospective clinical study University of Minnesota Masonic Cancer Center patients undergoing CAR-T therapy for hematological cancers will be monitored for ICANS with ANNA concurrently with the standard of care ICE testing. The following primary endpoints will be evaluated: a) acceptability of the frequency of ANNA administration; b) quality and quantity of audio collected from patients; and c) naturalness and ease of interaction with automated ANNA assessments. As we test the central feasibility hypothesis, we will also seek to understand the reasons why ANNA administration may not have occurred (examples: unable, refused, too tired, ill, forgot, technical reason, app not

[c]https://github.com/guillaumekln/faster-whisper
[d]https://github.com/NVIDIA/NeMo-Guardrails

working, battery out, other). We will also evaluate ANNA's usability characteristics that are not central to its feasibility but may affect t he f easibility i ndirectly s uch a s n aturalness of interactions with patients, convenience, and patients' perceptions of ease of use.

Prior to conducting the clinical study, we plan to address the system latency limitation pointed out by the pilot study evaluators as well as experiment with the more recently released chat models such as Llama2 to improve the sensibility of the initial conversations with the patient.

We also plan to enhance the language analysis of the conversations collected with ANNA by adding language coherence measures using a recently developed Time-series Augmented Representations for Detection of Incoherent Speech (TARDIS) method that relies on a time-series analysis of coherence features computed using semantic relatedness between words in a given piece of discourse [e]. The TARDIS approach has been used successfully to characterise disordered speech in patients with schizophrenia[34] and may prove to be useful for detecting possible thought disturbances caused by early ICANS.

One of our current concerns with using ANNA for intensive monitoring of cognitive changes in cancer patients is that even the abbreviated version of the cognitive tests we have currently implemented may present a burden for the patients who are likely to experience significant distress and fatigue as a result of therapy. Our ultimate goal in the forthcomeing clinical study is to determine if we can reliably ascertain the onset of ICANS based entirely on the analysis of the brief conversation between ANNA and the patient. If we can successfully do so, then we would likely be able to dispense with the more formal word list learning and verbal fluency tests, which would make intensive monitoring much less burdensome for patients.

References

1. M. S. Topp, T. van Meerten, R. Houot, M. Minnema, N. Milpied, P. J. Lugtenburg, C. Thieblemont, M. Wermke, K. Song, I. Avivi, J. Kuruvilla, U. Dührsen, R. Chu, L. Zheng, V. Plaks, A. Kerber and M. J. Kersten, Earlier Steroid Use with Axicabtagene Ciloleucel (Axi-Cel) in Patients with Relapsed/Refractory Large B Cell Lymphoma (R/R LBCL), *Biology of Blood and Marrow Transplantation* **26**, p. S101 (March 2020).
2. S. S. Neelapu, F. L. Locke, N. L. Bartlett, L. J. Lekakis, D. B. Miklos, C. A. Jacobson, I. Braunschweig, O. O. Oluwole, T. Siddiqi, Y. Lin, J. M. Timmerman, P. J. Stiff, J. W. Friedberg, I. W. Flinn, A. Goy, B. T. Hill, M. R. Smith, A. Deol, U. Farooq, P. McSweeney, J. Munoz, I. Avivi, J. E. Castro, J. R. Westin, J. C. Chavez, A. Ghobadi, K. V. Komanduri, R. Levy, E. D. Jacobsen, T. E. Witzig, P. Reagan, A. Bot, J. Rossi, L. Navale, Y. Jiang, J. Aycock, M. Elias, D. Chang, J. Wiezorek and W. Y. Go, Axicabtagene Ciloleucel CAR T-Cell Therapy in Refractory Large B-Cell Lymphoma, *New England Journal of Medicine* **377**, 2531 (December 2017).
3. J. Gust, R. Ponce, W. C. Liles, G. A. Garden and C. J. Turtle, Cytokines in CAR T Cell-Associated Neurotoxicity, *Frontiers in Immunology* **11**, p. 577027 (2020).
4. D. W. Lee, B. D. Santomasso, F. L. Locke, A. Ghobadi, C. J. Turtle, J. N. Brudno, M. V. Maus, J. H. Park, E. Mead, S. Pavletic, W. Y. Go, L. Eldjerou, R. A. Gardner, N. Frey, K. J. Curran, K. Peggs, M. Pasquini, J. F. DiPersio, M. R. van den Brink, K. V. Komanduri, S. A. Grupp

[e]TARDIS is available open source at https://github.com/LinguisticAnomalies/Coherence

and S. S. Neelapu, ASTCT Consensus Grading for Cytokine Release Syndrome and Neurologic Toxicity Associated with Immune Effector Cells, *Biology of Blood and Marrow Transplantation* **25**, 625 (April 2019).

5. B. D. Santomasso, J. H. Park, D. Salloum, I. Riviere, J. Flynn, E. Mead, E. Halton, X. Wang, B. Senechal, T. Purdon, J. R. Cross, H. Liu, B. Vachha, X. Chen, L. M. DeAngelis, D. Li, Y. Bernal, M. Gonen, H.-G. Wendel, M. Sadelain and R. J. Brentjens, Clinical and Biological Correlates of Neurotoxicity Associated with CAR T-cell Therapy in Patients with B-cell Acute Lymphoblastic Leukemia, *Cancer Discovery* **8**, 958 (August 2018).
6. M. V. Maus, S. Alexander, M. R. Bishop, J. N. Brudno, C. Callahan, M. L. Davila, C. Diamonte, J. Dietrich, J. C. Fitzgerald, M. J. Frigault, T. J. Fry, J. L. Holter-Chakrabarty, K. V. Komanduri, D. W. Lee, F. L. Locke, S. L. Maude, P. L. McCarthy, E. Mead, S. S. Neelapu, T. G. Neilan, B. D. Santomasso, E. J. Shpall, D. T. Teachey, C. J. Turtle, T. Whitehead and S. A. Grupp, Society for Immunotherapy of Cancer (SITC) clinical practice guideline on immune effector cell-related adverse events, *Journal for ImmunoTherapy of Cancer* **8**, p. e001511 (December 2020).
7. N. Möhn, V. Bonda, L. Grote-Levi, V. Panagiota, T. Fröhlich, C. Schultze-Florey, M. P. Wattjes, G. Beutel, M. Eder, S. David, S. Körner, G. Höglinger, M. Stangel, A. Ganser, C. Koenecke and T. Skripuletz, Neurological management and work-up of neurotoxicity associated with CAR T cell therapy, *Neurological Research and Practice* **4**, p. 1 (December 2022).
8. P. Strati, S. Ahmed, F. Furqan, L. E. Fayad, H. J. Lee, S. P. Iyer, R. Nair, L. J. Nastoupil, S. Parmar, M. A. Rodriguez, F. Samaniego, R. E. Steiner, M. Wang, C. C. Pinnix, S. B. Horowitz, L. Feng, R. Sun, C. M. Claussen, M. C. Hawkins, N. A. Johnson, P. Singh, H. Mistry, S. Johncy, S. Adkins, P. Kebriaei, E. J. Shpall, M. R. Green, C. R. Flowers, J. Westin and S. S. Neelapu, Prognostic impact of corticosteroids on efficacy of chimeric antigen receptor T-cell therapy in large B-cell lymphoma, *Blood* **137**, 3272 (June 2021).
9. M. M. Herr, G. L. Chen, M. Ross, H. Jacobson, R. McKenzie, L. Markel, S. R. Balderman, C. M. Ho, T. Hahn and P. L. McCarthy, Identification of Neurotoxicity after Chimeric Antigen Receptor (CAR) T Cell Infusion without Deterioration in the Immune Effector Cell-Associated Encephalopathy (ICE) Score, *Biology of Blood and Marrow Transplantation* **26**, e271 (November 2020).
10. J. Gust, K. A. Hay, L.-A. Hanafi, D. Li, D. Myerson, L. F. Gonzalez-Cuyar, C. Yeung, W. C. Liles, M. Wurfel, J. A. Lopez, J. Chen, D. Chung, S. Harju-Baker, T. Özpolat, K. R. Fink, S. R. Riddell, D. G. Maloney and C. J. Turtle, Endothelial Activation and Blood-Brain Barrier Disruption in Neurotoxicity after Adoptive Immunotherapy with CD19 CAR-T Cells, *Cancer Discovery* **7**, 1404 (December 2017).
11. J. Ao, R. Wang, L. Zhou, C. Wang, S. Ren, Y. Wu, S. Liu, T. Ko, Q. Li, Y. Zhang, Z. Wei, Y. Qian, J. Li and F. Wei, Speecht5: Unified-modal encoder-decoder pre-training for spoken language processing (2022).
12. K. Shuster, J. Xu, M. Komeili, D. Ju, E. M. Smith, S. Roller, M. Ung, M. Chen, K. Arora, J. Lane, M. Behrooz, W. Ngan, S. Poff, N. Goyal, A. Szlam, Y.-L. Boureau, M. Kambadur and J. Weston, Blenderbot 3: a deployed conversational agent that continually learns to responsibly engage (2022).
13. L. Zheng, W.-L. Chiang, Y. Sheng, S. Zhuang, Z. Wu, Y. Zhuang, Z. Lin, Z. Li, D. Li, E. P. Xing, H. Zhang, J. E. Gonzalez and I. Stoica, Judging llm-as-a-judge with mt-bench and chatbot arena (2023).
14. H. Touvron, L. Martin, K. Stone, P. Albert, A. Almahairi, Y. Babaei, N. Bashlykov, S. Batra, P. Bhargava, S. Bhosale, D. Bikel, L. Blecher, C. C. Ferrer, M. Chen, G. Cucurull, D. Esiobu, J. Fernandes, J. Fu, W. Fu, B. Fuller, C. Gao, V. Goswami, N. Goyal, A. Hartshorn, S. Hosseini, R. Hou, H. Inan, M. Kardas, V. Kerkez, M. Khabsa, I. Kloumann, A. Korenev, P. S. Koura, M.-A. Lachaux, T. Lavril, J. Lee, D. Liskovich, Y. Lu, Y. Mao, X. Martinet, T. Mihaylov, P. Mishra,

I. Molybog, Y. Nie, A. Poulton, J. Reizenstein, R. Rungta, K. Saladi, A. Schelten, R. Silva, E. M. Smith, R. Subramanian, X. E. Tan, B. Tang, R. Taylor, A. Williams, J. X. Kuan, P. Xu, Z. Yan, I. Zarov, Y. Zhang, A. Fan, M. Kambadur, S. Narang, A. Rodriguez, R. Stojnic, S. Edunov and T. Scialom, Llama 2: Open foundation and fine-tuned chat models (2023).

15. B. E. Gavett, A. S. Gurnani, J. L. Saurman, K. R. Chapman, E. G. Steinberg, B. M. Martin, C. E. Chaisson, J. Mez, Y. Tripodis and R. A. Stern, Practice effects on story memory and list learning tests in the neuropsychological assessment of older adults, *PLoS ONE* **11** (2016).
16. M. Wilson and I. Division, Mrc psycholinguistic database: Machine usable dictionary, version 2.00., *Behav Res Methods* **20** (06 1997).
17. M. D. Lezak and M. D. Lezak (eds.), *Neuropsychological assessment*, 4th ed edn. (Oxford University Press, Oxford ; New York, 2004).
18. J. D. Herrera-García, I. Rego-García, V. Guillén-Martínez, M. Carrasco-García, C. Valderrama-Martín, R. Vílchez-Carrillo, S. López-Alcalde and C. Carnero-Pardo, Discriminative validity of an abbreviated Semantic Verbal Fluency Test, *Dementia & Neuropsychologia* **13**, 203 (June 2019).
19. E. S. Gromisch, V. Zemon, R. H. Benedict, N. D. Chiaravalloti, J. DeLuca, M. A. Picone, S. Kim and F. W. Foley, Using a highly abbreviated California Verbal Learning Test-II to detect verbal memory deficits, *Multiple Sclerosis Journal* **19**, 498 (April 2013).
20. A. Radford, J. W. Kim, T. Xu, G. Brockman, C. McLeavey and I. Sutskever, Robust speech recognition via large-scale weak supervision (2022).
21. S. Pakhomov, D. Chacon, M. Wicklund and J. Gundel, Computerized assessment of syntactic complexity in alzheimer's disease: a case study of iris murdoch's writing, *Behavior research methods* **43**, p. 136—144 (March 2011).
22. T. Cohen and S. Pakhomov, A tale of two perplexities: Sensitivity of neural language models to lexical retrieval deficits in dementia of the alzheimer's type, in *Proceedings of the 58th Annual Meeting of the Association for Computational Linguistics, ACL 2020, Online, July 5-10, 2020*, eds. D. Jurafsky, J. Chai, N. Schluter and J. R. Tetreault (Association for Computational Linguistics, 2020).
23. S. D. Canning, L. Leach, D. Stuss, L. Ngo and S. E. Black, Diagnostic utility of abbreviated fluency measures in Alzheimer disease and vascular dementia, *Neurology* **62**, 556 (February 2004).
24. T. A. C. Thompson, P. H. Wilson, P. J. Snyder, R. H. Pietrzak, D. Darby, P. Maruff and H. Buschke, Sensitivity and Test-Retest Reliability of the International Shopping List Test in Assessing Verbal Learning and Memory in Mild Alzheimer's Disease, *Archives of Clinical Neuropsychology* **26**, 412 (August 2011).
25. S. Marino, S. Pakhomov, S. Han, K. Anderson, M. Ding, L. Eberly, D. Loring, C. Hawkins-Taylor, J. Rarick, I. Leppik, J. Cibula and A. Birnbaum, The effect of topiramate plasma concentration on linguistic behavior, verbal recall and working memory, *Epilepsy & Behavior* **24**, 365 (July 2012).
26. S. V. Pakhomov, S. E. Marino, S. Banks and C. Bernick, Using automatic speech recognition to assess spoken responses to cognitive tests of semantic verbal fluency, *Speech Communication* **75**, 14 (December 2015).
27. S. V. S. Pakhomov, W. Teeple, A. M. Mills and M. Kotlyar, Use of an automated mobile application to assess effects of nicotine withdrawal on verbal fluency: A pilot study., *Experimental and Clinical Psychopharmacology* **24**, 341 (October 2016).
28. S. V. S. Pakhomov, L. E. Eberly and D. S. Knopman, Recurrent perseverations on semantic verbal fluency tasks as an early marker of cognitive impairment, *Journal of Clinical and Experimental Neuropsychology* **40**, 832 (September 2018).
29. L. Laranjo, A. G. Dunn, H. L. Tong, A. B. Kocaballi, J. Chen, R. Bashir, D. Surian, B. Gallego, F. Magrabi, A. Y. S. Lau and E. Coiera, Conversational agents in healthcare: a systematic review,

Journal of the American Medical Informatics Association **25**, 1248 (07 2018).
30. C. H. van Dyck, C. J. Swanson, P. Aisen, R. J. Bateman, C. Chen, M. Gee, M. Kanekiyo, D. Li, L. Reyderman, S. Cohen, L. Froelich, S. Katayama, M. Sabbagh, B. Vellas, D. Watson, S. Dhadda, M. Irizarry, L. D. Kramer and T. Iwatsubo, Lecanemab in early alzheimer's disease, *New England Journal of Medicine* **388**, 9 (2023), PMID: 36449413.
31. B. Naughton, S. Saltzman, F. Ramadan, N. Chadha, R. Priore and J. Mylotte, A multifactorial intervention to reduce prevalence of delirium and shorten hospital length of stay, *Journal of the American Geriatrics Society* **53**, 18 (02 2005).
32. P. Schramowski, C. Turan-Schwiewager, N. Andersen, C. Rothkopf and K. Kersting, Large pre-trained language models contain human-like biases of what is right and wrong to do, *Nature Machine Intelligence* **4**, 258 (03 2022).
33. K. Singhal, S. Azizi, T. Tu, S. S. Mahdavi, J. Wei, H. W. Chung, N. Scales, A. Tanwani, H. Cole-Lewis, S. Pfohl, P. Payne, M. Seneviratne, P. Gamble, C. Kelly, A. Babiker, N. Schärli, A. Chowdhery, P. Mansfield, D. Demner-Fushman, B. Agüera Y Arcas, D. Webster, G. S. Corrado, Y. Matias, K. Chou, J. Gottweis, N. Tomasev, Y. Liu, A. Rajkomar, J. Barral, C. Semturs, A. Karthikesalingam and V. Natarajan, Large language models encode clinical knowledge, *Nature* (July 2023).
34. W. Xu, W. Wang, J. Portanova, A. Chander, A. Campbell, S. Pakhomov, D. Ben-Zeev and T. Cohen, Fully automated detection of formal thought disorder with time-series augmented representations for detection of incoherent speech (TARDIS), *J. Biomed. Informatics* **126**, p. 103998 (2022).

Leveraging 3D Echocardiograms to Evaluate AI Model Performance in Predicting Cardiac Function on Out-of-Distribution Data*

Grant Duffy, Kai Christensen and David Ouyang
Department of Cardiology, Smidt Heart Institute, Cedars-Sinai Medical Center,
127 S San Vicente Blvd A3600
Los Angeles, CA 90048
Email: David.Ouyang@cshs.org

Advancements in medical imaging and artificial intelligence (AI) have revolutionized the field of cardiac diagnostics, providing accurate and efficient tools for assessing cardiac function. AI diagnostics claims to improve upon the human-to-human variation that is known to be significant[1–3]. However, when put in practice, for cardiac ultrasound, AI models are being run on images acquired by human sonographers whose quality and consistency may vary. With more variation than other medical imaging modalities[4], variation in image acquisition may lead to out-of-distribution (OOD) data and unpredictable performance of the AI tools. Recent advances in ultrasound technology has allowed the acquisition of both 3D as well as 2D data, however 3D has more limited temporal and spatial resolution and is still not routinely acquired[5]. Because the training datasets used when developing AI algorithms are mostly developed using 2D images, it is difficult to determine the impact of human variation on the performance of AI tools in the real world. The objective of this project is to leverage 3D echos to simulate realistic human variation of image acquisition and better understand the OOD performance of a previously validated AI model[2]. In doing so, we develop tools for interpreting 3D echo data and quantifiably recreating common variation in image acquisition between sonographers. We also developed a technique for finding good standard 2D views in 3D echo volumes. We found the performance of the AI model we evaluated to be as expected when the view is good, but variations in acquisition position degraded AI model performance. Performance on far from ideal views was poor, but still better than random, suggesting that there is some information being used that permeates the whole volume, not just a quality view. Additionally, we found that variations in foreshortening didn't result in the same errors that a human would make.

Keywords: 3D Echo; AI; Machine Learning; Echocardiology.

---

*This work is supported by NIH R00 HL157421-01

1. Introduction

Echocardiography, or cardiac ultrasound, is the most prevalent imaging modality[6]. Cardiac ultrasound is able to provide an accurate, noninvasive views of the heart in real time with limited equipment and with high temporal resolution[7]. In traditional transthoracic echocardiology, a sonographer will acquire 2D images and videos of the heart in standard orientations or views. Two standard views are the apical four chamber (A4C) and apical two chamber (A2C) views which are both views taken along the major axis of the heart from its apex. These views are crucial for assessing cardiac function, diagnosing heart failure and cardiac hypertrophy[1,6,8–15]. These two views are in theory only separated by a probe rotation of roughly 60 degrees, however this depends on sonographer judgement for the view quality and probe placement.

Recent advances in ultrasound technology have increased the temporal and spatial resolution of images acquired. Wide field of view allows for 3D images to be acquired with the same probes and hardware, however at lower resolution[7,8]. In addition to the standard TTE views, sometimes additional 3D images are acquired to better characterize complex cardiac structures and provide holistic evaluates of cardiac form and function. Focused images of the heart valves as well as the left ventricle can be used to accurately assess metrics that might be challenging to measure in 2D images.

One example of acquisition error in 2D images is foreshortening, where inappropriate or suboptimal images of the left ventricle can cause overestimation of the cardiac function[16,17]. Apical views depend on being placed near the apex of the left ventricle, which should not contract in, however off-axis foreshortened views will show contraction of the left ventricle that exaggerate the left ventricular function. The result of this error is the underestimate of LV volume at systole and ultimately an overestimate of ejection fraction[17]. Although foreshortening is known to be a common source of measurement error, it is difficult to know how prevalent it is because it is difficult to quantify foreshortening in 2D images. There have been attempts to automatically detect foreshortening using machine learning or other algorithms[16,18,19]. These algorithms need to be run in real-time on the ultrasound machine or trained on other modalities limiting their practicality.

Although adding 3D acquisitions to a study may add value in these cases, it also takes additional time and training. The result is that 3D echo images are much less prevalent. In the Cedars Sinai Medical Center (CSMC), apical 3D echo images are outnumbered by other video acquisitions roughly 11,000 to 1 making 3D echo datasets of reasonable size rare.

There is a large, and quickly growing, body of research dedicated to AI in medicine and specifically cardiology. Several models aim to automate echo measurements or diagnosis[1–3,20]. These models show promise in revolutionizing how echocardiology is performed. Because of the large disparity in prevalence of 2D vs 3D echos and the often-proprietary data format of 3D images, AI models in this field are almost exclusively trained and evaluated on 2D TTE images. 2D datasets curated in this way contain only images acquired by human sonographers in specific views and do not span the full distribution of possible echo images.

It is known that machine learning models can perform unpredictably on out of distribution data[21]. Training methods including data augmentations that translate, rotate and resize images attempt to broaden the coverage of the datasets and mitigate these risks. But these augmentations can only simulate the transformation of an image constrained to the 2D plane. Real ultrasound acquisitions can include rotations and translation in 3D. One of the main goals of AI in medicine is the mitigation of human error. For models that do not perform well with 3D view transformations, the performance of the model could be strongly dependent on the sonographer's acquisition quality.

In this research, we propose methods for evaluating AI model performance on off-axis views by introducing realistic 3D spatial transformations to the acquisition plane in 3D volumes. Although 3D echos remain relatively rare in the CSMC system, searching over 16 years, we curated a dataset of 1,528 apical 3D images. Through reverse engineering, we were able to decode the Phillips 3D DICOM data format these images are stored in. We developed functions for slicing 3D data into 2D images and simulating realistic transformations that could be introduced by sonographer motion. We use a deep learning image view classifier, trained specifically for this task, to find the ideal view to compare performance vs. distance from ideal view.

To test these methods, we chose to evaluate the EchoNet-Dynamic model[1] for measurement of left ventricular ejection fraction (LVEF) as the downstream tasks. LVEF is the ratio of the diastolic LV volume to the systolic LV volume as a percentage of volume ejected. It is an important measurement for assessing cardiac function and heart failure[11,22,23]. Typically, LVEF measurements are made by tracing the LV for systolic and diastolic frames in an A4C view video. EchoNet-Dynamic is a ResNet derived regression model that was trained on 144,184 videos from SHC. These images are primarily of the apical-4-chamber and apical-2-chamber views. It has been well validated on external datasets and even a randomized clinical trial[2]. We evaluate the performance of this model on synthetically produced 2D images with simulated probe rotation, translation, and foreshortening to draw conclusions about the robustness of this model in the real world and dependence on view quality.

2. Methods

To realize the impact of this research, several challenges were to be overcome. One of the largest challenges is simply working with 3D echo. To be able to make use of the 3D echo data, we first needed to pull the DICOM images from the hospital dataset, reverse engineer the proprietary data format, and develop tools for interpreting and slicing 3D volumes. The next crucial step was to align the 3D volumes along standard views so that they could be analyzed together. This was done using a view classifier that we trained just for this project. Finally, we evaluate the performance of the EchoNet EF prediction model.

2.1. *Working With 3D Echo*

The 3D echo dataset used in this research is a subset of all of the echos in CSMC's database between 2012 and 2022, nearly 15 million images. Of these images, 1,349 of these are 3D acquisitions taken in the apical position. The apical 3D echos were used because of their ability to generate A4C and A2C 2D views with relatively benign artifacts from the slicing process. All 3D echos were captured on Philips EPIQ CVx ultrasounds. A breakdown of the relative size relevant factors for the CSMC 2D and 3D datasets can be found in Table 1.

Table 1. Breakdown of CSMC image types from 2006-2022

Dataset	N Studies	N Images	Mean EF	Frame Rate	Acquisition Duration
All Echos	369,306	14,922,383	69.10%	26.42 fps	2.75 seconds
3D Apical Echos	1528	1,349	56.26%	18.32 fps	2.81 seconds

Like standard 2D echos, 3D echos are stored in DICOM format. Unlike 2D echo, the data stored in the "pixel data" tag in the DICOMs is only a snapshot of the volume that the sonographer chose to capture and not the full 3D echo data. The full data is stored in a proprietary compressed format under other tags that we were able to reverse engineer. The decompressed data consists of voxel data and physical bounds for the captured volume. Unlike voxel data captured in MRI and 3D formats, this voxel data is not rectilinear - instead it is defined by a spherical coordinate system, as shown in figure Fig. 1. This coordinate system is parameterized by one linear dimension (ρ), and two rotational dimensions (φ and θ). For each of these axes, the physical bounds given in the DICOM define a section of a sphere containing the scanned region that called the frustum. For convenience, we will also be using a 3D cartesian coordinate system with the origin at the probe on the surface of the skin and the x axis pointing parallel to the probe into the body.

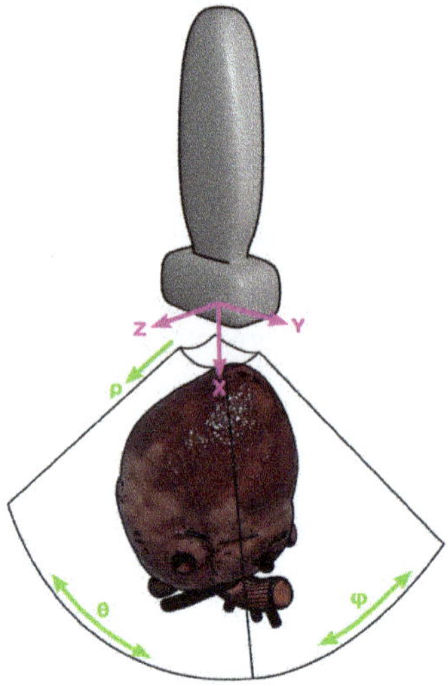

Fig. 1. Diagram showing the 3D world and spherical coordinate systems.

To generate 2D slices of 3D videos, we must first define points on a plane corresponding to the 2D view that we wish to sample. Although there are many degrees of freedom and ways to slice a 3D volume, we decided to constrain our slices to just 4 degrees of freedom to ensure relatively realistic looking slices and clinical relevance. We first define a square region on the x-y plane centered at the center of the volume and whose width is the max width of the volume to ensure that any slice will be centered and reasonably zoomed. We rotate this plane around the x-axis and then translate it forward or backward through the volume. A translation of 1 corresponds to all the way forward through the volume and -1 corresponds to all the way backward through the volume. Translations of roughly -0.5 to 0.5 result in reasonable slices. Two additional degrees of freedom were added to simulate foreshortening. A horizontal axis is defined on the plane and the slice is rotated forward or backward. We found that an axis location of 30% from the top of the plane to the bottom is reasonable for simulating foreshortening in our dataset.

Once we have defined the plane that we wish to slice, we then define a grid of points on that plane resulting in an array with a shape of (n, m, 3) where the last dimension contains the XYZ location of each point. We then transform these points into spherical coordinates using the following equations resulting in an array with the same shape but whose last dimension contains ρ, φ, θ.

$$\rho = \sqrt{x^2 + y^2 + z^2}$$
$$\varphi = tan^{-1}\left(\frac{z}{x}\right)$$
$$\theta = tan^{-1}\left(\frac{y}{\sqrt{x^2+z^2}}\right)$$

Eq. 1

Because the spherical coordinates are aligned with the voxel data, we can obtain the voxel indices for each point on the plane by simply renormalizing them using the volume bounds.

$$i = \frac{\rho - \rho_{min}}{\rho_{max} - \rho_{min}}$$
$$j = \frac{\varphi - \varphi_{min}}{\varphi_{max} - \varphi_{min}}$$
$$k = \frac{\theta - \theta_{min}}{\theta_{max} - \theta_{min}}$$

Eq. 2

To generate a 2D image all we need to do is round each index to the nearest integer and lookup its value in the voxel data. Any indices out of bounds of the volume result in an intensity of 0. Although this sampling method works, the relatively low-resolution voxel data results in voxel artifacts due to the relatively low resolution of 3D data. To mitigate this problem, we implemented trilinear interpolation between voxels which results in much smoother images as shown in Fig. 2.

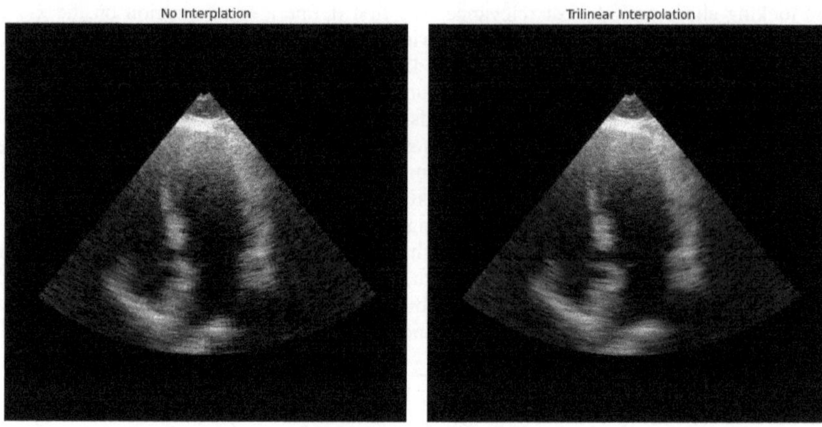

Fig. 2. The impact using trilinear interpolation when generating 2D slices from 3D echo.

2.2. *View Classifier*

With the slicing algorithm that we developed, we are able to accurately simulate the motion of a human moving a probe around a heart, but to characterize a particular view as being a quantifiable rotation and translation away from an optimal view, we need to first define the optimal view. To do this we trained a 2D image view classifier on a standard 2D echo dataset of known standard views. This dataset contains 30,045 echo videos labeled as A4C, A2C, PLAX, Subcostal, or Other views from Stanford Healthcare (SHC). The breakdown of label frequencies can be found in Table 2. During training, random frames are selected from videos in the dataset. Because when running inference on the 3D dataset this model would encounter images unlike anything in the training dataset, we attempted to increase the coverage of the training dataset by adding random mirroring augmentation and additional labels for mirrored A4C, A2C, PLAX and Subcostal. We used a ResNet18[24] image classifier architecture and cross-entropy loss to train the view classifier. The view classifier achieved an AUC of 0.997 for both A4C and A2C views on the SHC test set.

Table 2. Distribution of labels in the view classifier training dataset.

View	N Total	N Train	N Val	N Test
A4C	5,036	4,054	499	483
A2C	3,224	2,577	318	329
PLAX	4,059	3,239	403	417
Subcostal	2,726	2,166	283	276
Other	15,000	12,000	1,500	1,500

2.3. *EF Inference*

The EchoNet model we evaluated has been shown to be accurate on several datasets and even in a randomized clinical trial situation, but it is not known how sensitive it is to small changes in view quality due to poor probe placement and foreshortening.

We addressed this problem by running inference on slices of 3D volumes while varying the rotation, translation and foreshortening from the ideal view. For each 3D volume, we ran both EF and view inference on every combination of translations -0.5 to 0.5 and rotation 0 to 360 degrees. After the best A4C slice, we introduced foreshortening to this view, -40 to 40 degrees, and ran EF inference again. With these results, we were able to draw conclusions about the performance of the EF model as a function of rotation, translation, and foreshortening from the ideal A4C view.

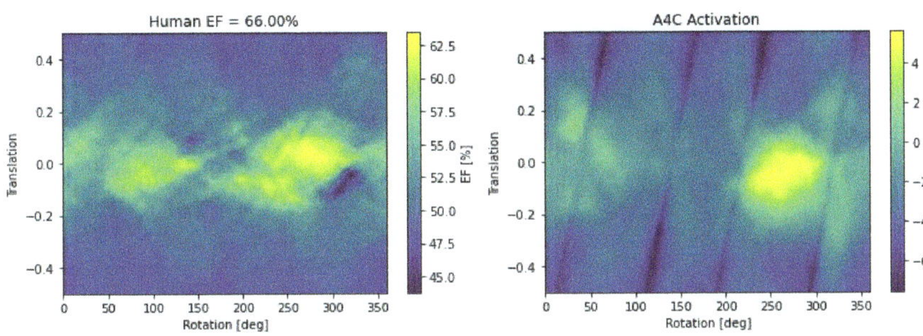

Fig. 3. Phase diagram showing A4C EF prediction and view activation for every combination of rotation and translation. The human measured EF for this patient is 66%.

3. Results

We constrained the slice degrees of freedom to rotation and translation and generated view and EF predictions for every combination of rotation and translation. These predictions were then plotted as a 2D image that summarizes how the model predictions change as the slices are rotated and

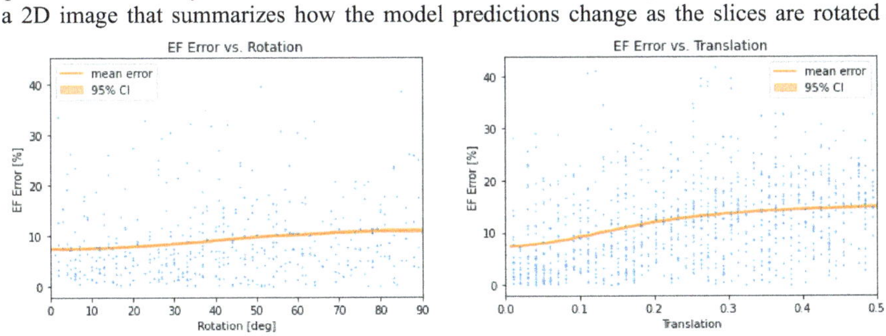

Fig. 4. MAE performance across dataset as view is rotated and translated.

translated shown in Fig. 3. In these plots we can see that regions of high activation for A4C correspond to regions of more accurate EF predictions. The point of maximum A4C activation on this plot for each example is considered to be the optimal view for subsiquent analysis.

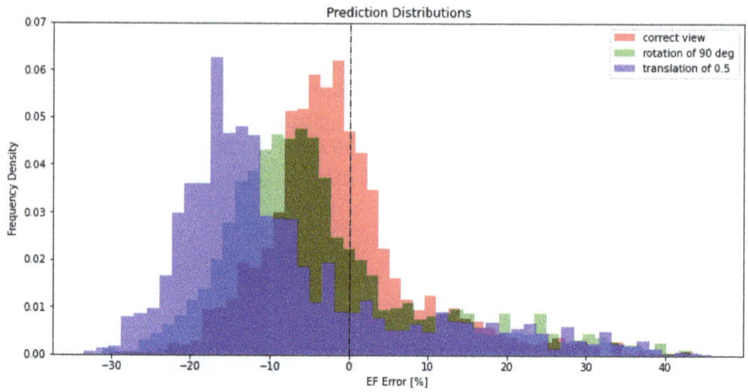

Fig. 5. EF Error distribution for best view, 90 degrees of rotation, and a translation of 0.5.

When EF inference was run on optimal view slices, the mean absolute error (MAE) was 7.3 (7.0-7.7%). Although this is worse than the claimed performance of this model (6.3% comparing model to human or 2.8% comparing model to final value in clinical trial)[2], it is consistent with interobserver variability and the variability between 2D and 3D echo[4,25]. As shown in Fig. 4, when we introduce either rotation or translation to the slice, the error increases. The MAE for rotation increases to 10.9% (10.6-11.1%) while the MAE for translation increases to 14.7% (14.6-14.9%) suggesting that there is more information being used near the center of the volume than near the edges as represented by slices with larger translation error.

One characteristic we noticed was a relatively high frequency of low error, regardless of view quality, especially for patients with near normal EF. This led us to hypothesize that when faced with a poor view, the model makes a guess near the mean of the dataset. We investigate this hypothesis by looking at the prediction trends in various situations. In Fig. 5, we compare the EF prediction distributions of 90-degree rotations and translations of 0.5 to the ideal view slices. We can see that when the view is near ideal, the distribution is relatively tight, and centered around zero. For both introduced rotation and translation, we see that the distributions are shifted to the left, corresponding to underestimates of EF on poorly oriented views. This underestimate cannot be explained by a difference in mean LVEF for the EchoNet training set compared to our 3D dataset. Both datasets have mean LVEF values of roughly 55%[1].

We also analyzed the subset of patients with human measured EF of greater than 70% and the subset of EF less than 30%. In these subsets, the increase in MAE due to introduced rotation and translation is much greater as shown in Fig. 6. This is because for patients with extremely abnormal EF, the model is not able to achieve high accuracy predictions when the view is poor by predicting a value near the mean. For these patients, this effect is stronger than the tendency of the model to underpredict. Therefore, for patients with an LVEF < 30%, the model tends to overpredict EF when the view is poor. An interesting consequence of these two effects is that for low EF patients, there is a threshold where increasing translation decreases error because low EF predictions are nearer to the human measurements for these patients. Fig. 7 illustrates how EF and A4C predictions vary with rotation and translation for a patient with a high human measured EF.

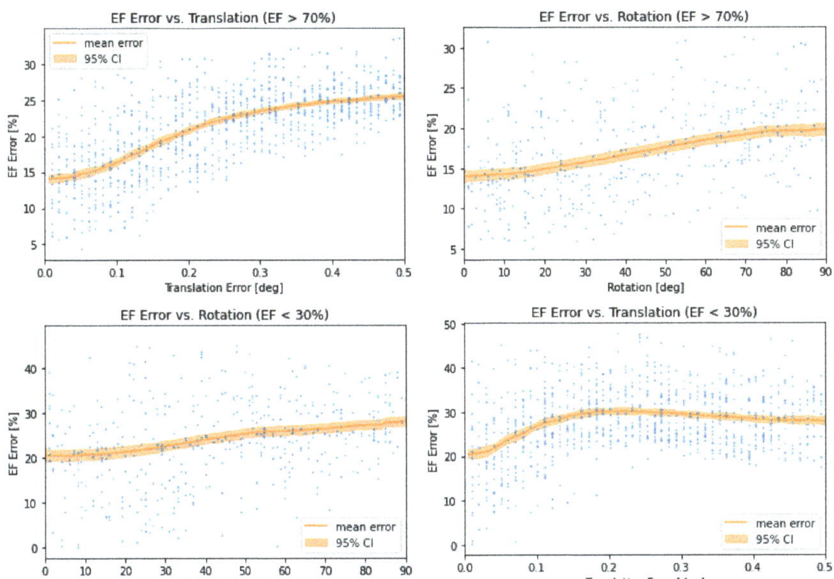

Fig. 6. EF model performance for the >70% and <30% EF subsets.

When looking at foreshortening specifically, we might expect the model to overpredict EF if it calculates EF in the same way as human sonographers, but we see a similar trend as with rotation and translation. This suggests that when predicting EF, the AI model is not segmenting the LV and calculating LV volume to determine EF the way a sonographer would. Fig. 8 shows the results for varying foreshortening from ideal views.

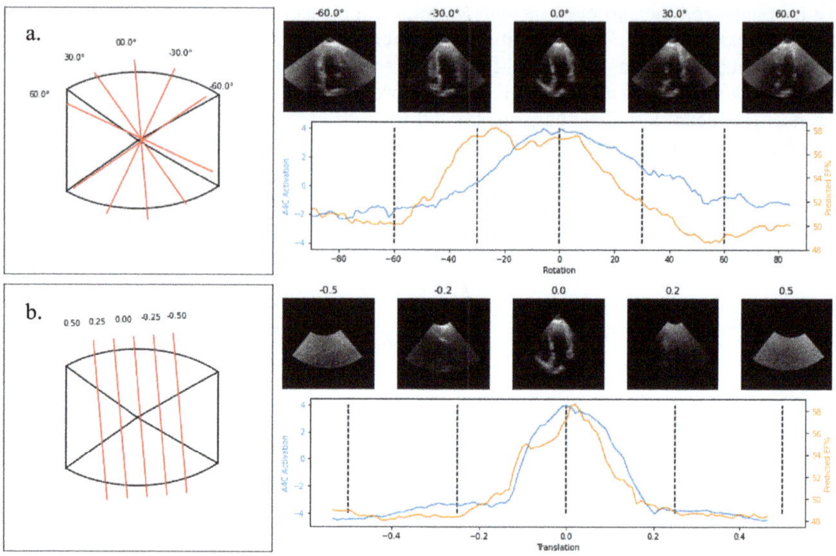

Fig. 7. Example slices and predictions for a range of (a.) rotations and (b.) translations for a selected example volume.

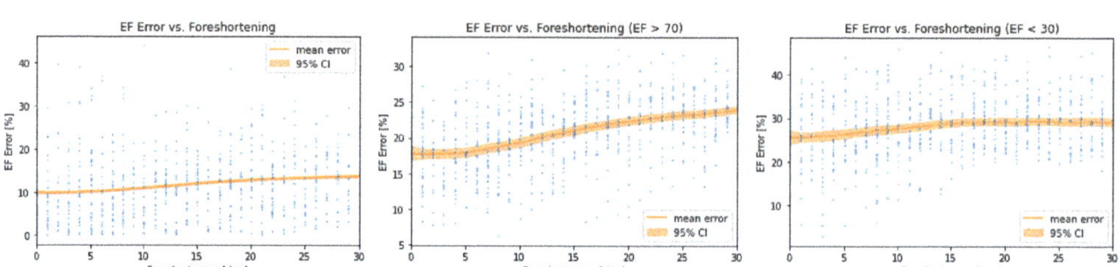

Fig. 8. Performance figures for slices with introduced foreshortening.

4. Discussion

This work demonstrates how 3D echos can be used to evaluate the performance of AI models on realistically out-of-distribution data that these models would likely encounter in real world applications. Understanding distribution shifts and model performance in real world applications may be necessary to understand how AI truly performs in clinical practice, a major barrier in AI research adoption in medicine[26,27]. We presented the methods used for interpreting and utilizing 3D echo data and evaluated the performance of an established AI model predicting LVEF with these methods.

We found that the EF model we evaluated performed well when the ideal slice is viewed, but error increases as we introduce rotation, translation, or foreshortening. The overall behavior of the model when subjected to OOD data is to guess a value, usually a little below the mean of the training dataset. This overall result of this is a tendency to underestimate EF when the view is poor. The model tends to overpredict EF for patients with very low EF and underestimate EF for patients with very high EF. These trends extent to foreshortening where humans would overestimate EF. Although it makes intuitive sense for the model to guess somewhere near the mean of the dataset when faced with OOD data, the mechanism causing underestimates for OOD data would require further investigation to explain. We hypothesize that the model is gauging the overall amount of motion in the heart to predict EF and for poor views there is a lack of apparent motion, thus the videos look more similar to ones of patients with low EF.

The performance of the EF model even on ideal view slices from 3D echo has lower performance than on 2D videos in prior work. There are several factors that may contribute to this error. First, 3D echo has fundamentally lower spatial and temporal resolution. While the frame rate of standard 2D echos is usually around 30-50 frames per second, 3D echos are much slower, in the range of 13-24 frames per second, with higher framerates associated with lower spatial resolution. Second, the 3D dataset might be comprised of a different distribution of patients than the general population due to selection bias for patients needing additional 3D echos. This is likely, given the average EF of the 3D dataset is 13% lower than the overall CSMC population. Finally, the view classifier we use to find the "ideal" slice is not perfect. It is trained on a dataset of human acquired images that aren't always perfect. Our classifier also only has 4 standard views when in reality there are many more views and several different views may have been grouped together under "A4C". Like the EchoNet model, the view classifier was only trained on 2D images and performance on OOD 3D slices might not be reliable. This would result in the ideal slice for predicting EF not being found.

There is significant opportunity for future research in this field with the use of 3D echo data. An improved view classifier would allow more accurate identification of ideal view orientation. For models trained on clinical 2D datasets, like the EchoNet-Dynamic dataset, it is difficult to quantify the amount of foreshortening and perturbances present. Future work could use 3D echo data to train a model that is able to predict the amount of foreshortening or perturbance in a 2D slice. This would allow us to retrospectively evaluate the view quality and distribution of datasets models are trained on. Additionally, with better tools to simulate and evaluate 3D distribution shifts, there is an opportunity to develop new data augmentations and normalization techniques addressing the spatial nature of echocardiology. Ultimately, as 3D echo data becomes more prevalent, future models could use these techniques to train on 2D slices of 3D data in addition to standard 2D views. These proposed methods would further our understanding and improve the robustness of AI models in echocardiology.

When black box AI models are deployed in healthcare, clinicians may have no sense of whether a model is performing within its operating domain and could lead to either overreliance or mistrust of the AI. In this study, we show how relatively subtle changes to the input data can significantly impact model performance. This has significant impact as with more AI models getting integrated into healthcare systems, it is important to consider how the deployment environment can be different from the environment they were trained and validated in. We show how identifying, simulating, and evaluating these hypothetical distribution shifts can lead to a better understanding of our AI systems and their performance in the real world.

References

1. Ouyang, D. *et al.* Video-based AI for beat-to-beat assessment of cardiac function. *Nature* **580**, 252–256 (2020).
2. He, B. *et al.* Blinded, randomized trial of sonographer versus AI cardiac function assessment. *Nature* **616**, 520–524 (2023).
3. Johnson, K. W. *et al.* Artificial Intelligence in Cardiology. *J. Am. Coll. Cardiol.* **71**, 2668–2679 (2018).
4. Farsalinos, K. E. *et al.* Head-to-Head Comparison of Global Longitudinal Strain Measurements among Nine Different Vendors: The EACVI/ASE Inter-Vendor Comparison Study. *J. Am. Soc. Echocardiogr.* **28**, 1171–1181, e2 (2015).
5. Hung, J. *et al.* 3D echocardiography: a review of the current status and future directions. *J. Am. Soc. Echocardiogr.* **20**, 213–233 (2007).
6. Papolos, A., Narula, J., Bavishi, C., Chaudhry, F. A. & Sengupta, P. P. U.S. Hospital Use of Echocardiography: Insights From the Nationwide Inpatient Sample. *J. Am. Coll. Cardiol.* **67**, 502–511 (2016).
7. Feigenbaum, H. Evolution of echocardiography. *Circulation* **93**, 1321–1327 (1996).
8. Ziaeian, B. & Fonarow, G. C. Epidemiology and aetiology of heart failure. *Nat. Rev. Cardiol.* **13**, 368–378 (2016).
9. WRITING COMMITTEE MEMBERS *et al.* 2013 ACCF/AHA guideline for the management of heart failure: a report of the American College of Cardiology Foundation/American Heart Association Task Force on practice guidelines. *Circulation* **128**, e240-327 (2013).
10. Heidenreich, P. A. *et al.* Forecasting the future of cardiovascular disease in the United States: a policy statement from the American Heart Association. *Circulation* **123**, 933–944 (2011).
11. Koh, A. S. *et al.* A comprehensive population-based characterization of heart failure with mid-range ejection fraction. *Eur. J. Heart Fail.* **19**, 1624–1634 (2017).
12. Shah, K. S. *et al.* Heart Failure With Preserved, Borderline, and Reduced Ejection Fraction: 5-Year Outcomes. *J. Am. Coll. Cardiol.* **70**, 2476–2486 (2017).
13. Foppa, M., Duncan, B. B. & Rohde, L. E. P. Echocardiography-based left ventricular mass estimation. How should we define hypertrophy? *Cardiovasc. Ultrasound* **3**, 17 (2005).
14. Angeli, F. *et al.* Day-to-day variability of electrocardiographic diagnosis of left ventricular hypertrophy in hypertensive patients. Influence of electrode placement. *J. Cardiovasc. Med.* **7**, 812–816 (2006).
15. Ghorbani, A. *et al.* Deep learning interpretation of echocardiograms. *NPJ Digit Med* **3**, 10 (2020).
16. Poon, J., Leung, J. T. & Leung, D. Y. 3D Echo in Routine Clinical Practice - State of the Art in 2019. *Heart Lung Circ.* **28**, 1400–1410 (2019).
17. Ünlü, S. *et al.* Impact of apical foreshortening on deformation measurements: a report from the EACVI-ASE Strain Standardization Task Force. *Eur. Heart J. Cardiovasc. Imaging* **21**, 337–343 (2020).
18. Kim, W.-J. C. *et al.* Automated Detection of Apical Foreshortening in Echocardiography Using Statistical Shape Modelling. *Ultrasound Med. Biol.* **49**, 1996–2005 (2023).
19. Labs, R. B., Zolgharni, M. & Loo, J. P. Echocardiographic image quality assessment using deep neural networks. *arXiv [eess.IV]* (2022).
20. Poplin, R. *et al.* Prediction of cardiovascular risk factors from retinal fundus photographs via deep learning. *Nat Biomed Eng* **2**, 158–164 (2018).

21. Sehwag, V. *et al.* Analyzing the Robustness of Open-World Machine Learning. in *Proceedings of the 12th ACM Workshop on Artificial Intelligence and Security* 105–116 (Association for Computing Machinery, 2019).
22. Chioncel, O. *et al.* Epidemiology and one-year outcomes in patients with chronic heart failure and preserved, mid-range and reduced ejection fraction: an analysis of the ESC Heart Failure Long-Term Registry. *Eur. J. Heart Fail.* **19**, 1574–1585 (2017).
23. Malm, S., Frigstad, S., Sagberg, E., Larsson, H. & Skjaerpe, T. Accurate and reproducible measurement of left ventricular volume and ejection fraction by contrast echocardiography: a comparison with magnetic resonance imaging. *J. Am. Coll. Cardiol.* **44**, 1030–1035 (2004).
24. He, K., Zhang, X., Ren, S. & Sun, J. Deep Residual Learning for Image Recognition. *arXiv [cs.CV]* (2015).
25. Yuan Neal *et al.* Systematic Quantification of Sources of Variation in Ejection Fraction Calculation Using Deep Learning. *JACC Cardiovasc. Imaging* **0**,.
26. Gerke, S., Babic, B., Evgeniou, T. & Cohen, I. G. The need for a system view to regulate artificial intelligence/machine learning-based software as medical device. *NPJ Digit Med* **3**, 53 (2020).
27. Wu, E. *et al.* How medical AI devices are evaluated: limitations and recommendations from an analysis of FDA approvals. *Nat. Med.* **27**, 582–584 (2021).

BrainSTEAM: A Practical Pipeline for Connectome-based fMRI Analysis towards Subject Classification

Alexis Li
Hamilton High School,
E-mail: li.alexis1111@gmail.com

Yi Yang
Duke University,
E-mail: owen.yang@duke.edu

Hejie Cui
Department of Computer Science, Emory University,
E-mail: hejie.cui@emory.edu

Carl Yang
Department of Computer Science, Emory University,
E-mail: j.carlyang@emory.edu

Functional brain networks represent dynamic and complex interactions among anatomical regions of interest (ROIs), providing crucial clinical insights for neural pattern discovery and disorder diagnosis. In recent years, graph neural networks (GNNs) have proven immense success and effectiveness in analyzing structured network data. However, due to the high complexity of data acquisition, resulting in limited training resources of neuroimaging data, GNNs, like all deep learning models, suffer from overfitting. Moreover, their capability to capture useful neural patterns for downstream prediction is also adversely affected. To address such challenge, this study proposes BrainSTEAM, an integrated framework featuring a spatio-temporal module that consists of an EdgeConv GNN model, an autoencoder network, and a Mixup strategy. In particular, the spatio-temporal module aims to dynamically segment the time series signals of the ROI features for each subject into chunked sequences. We leverage each sequence to construct correlation networks, thereby increasing the training data. Additionally, we employ the EdgeConv GNN to capture ROI connectivity structures, an autoencoder for data denoising, and mixup for enhancing model training through linear data augmentation. We evaluate our framework on two real-world neuroimaging datasets, ABIDE for Autism prediction and HCP for gender prediction. Extensive experiments demonstrate the superiority and robustness of BrainSTEAM when compared to a variety of existing models, showcasing the strong potential of our proposed mechanisms in generalizing to other studies for connectome-based fMRI analysis.

Keywords: Brain Connectome Analysis; Neuroimaging Studies; Synthetic Data Generation

1. Introduction

Functional brain networks illustrate the dynamic connectivity patterns between anatomical regions of interest (ROIs) for different cognitive states and different responses to disease or injury.[1] The study of functional brain networks provides insights into the underlying mechanisms of human consciousness, developmental processes, and the neural bases of various neurological and psychiatric disorders such as autism, ADHD, depression, and schizophrenia.[2] However, existing computational tools often extract a single static graph structure based on correlations among full BOLD signals, which ignores the dynamic changes of functional connectivity.[3–5]

Compared with other deep learning paradigms such as Convolutional Neural Networks (CNNs),[6] and Recurrent Neural Networks (RNNs),[7] Graph Neural Networks (GNNs)[8,9] provide unique benefits in functional brain network analysis due to its capability in modeling connectivity structures.[10–17] However, most GNN-based frameworks resort to static correlation networks as data instances, and they are prone to unstable performances due to large data noises in the BOLD signals and overfitting due to limited data labels of clinical outcomes. This is especially true for the ABIDE dataset as the images come from 17 international sites with differing imaging protocol, as well as heterogeneity within the dataset.[18]

To address the challenges above, this study proposes BrainSTEAM, an integrated pipeline that features a spatio-temporal module, for brain connectome analysis on dynamic fMRI networks. Specifically, we propose a temporal chunking approach to dynamically segment the BOLD signals of each subject into partitioned sequences based on a tunable sliding window to capture the local connectivity structures at different scales, which are further modeled by EdgeConv. An autoencoder is devised to discover the important connectivity patterns during ROI pooling through learnable dropout, where the objective is to reconstruct the full connectivity patterns only based on the important ones. Mixup is applied to further stabilize and enhance training of the whole framework through linear data augmentation to prevent the model from memorizing certain data points.

Extensive experiments conducted in this study demonstrate that our BrainSTEAM model outperforms state-of-the-art models on both mental disorder prediction and gender classification, indicating its effectiveness in modeling functional brain networks and also highlighting its flexibility and versatility. It is also promising to apply BrainSTEAM to the analysis of functional brain networks for other clinical applications, as well as other dynamic graphs extracted from time-series data. For clinical applications in particular, this model would help to address the limitations of MRI data collection as there are limited scans due to the expensive nature of MRIs and constant exposure for the patient. Decreasing information loss can make the model more robust, providing a more reliable aid for those in a clinical setting.

2. Related Work

2.1. *Data Augmentation*

Mixup utilizes the principles of vicinal risk minimization across different classes, constructing new data as a combination of existing data points.[19] Graph Mixup techniques often involve creating synthetic graphs samples connected subgraphs or reorder the original graph structure.

Previous works such as G-Mixup use probability matrices to predict if an edge exists between two nodes, and Graph Transplant samples the top nodes in a graph and then appends a partial K-hope subgraph to predict edges.[20,21] However, sampling subgraphs and appending them to the original graph becomes problematic when considering the fixed nature of brain ROIs.

Additionally, previous studies have used temporal based augmentation techniques to improve model generalization. STDAC proposed a module using random discontinuous sampling period with a tensor fusion method to combine it with the spatial model.[22] Multi-Head GAGNN modeled both spatio and patterns of functional brain networks simultaneously to fully utilize their characteristics.[23] These methods are still often limited by small sizes, thus there lies potential in combining a spatio-temporal data augmentation technique with mixup.

2.2. *Graph Pooling*

Previous Graph Pooling methods use hierarchical graph clustering methods, following the principle of local neighborhoods with nodes.[24] This has extended to deterministic clustering algorithms and attention based mechanisms to increase the quality of assigning the clusters.[25,26] Other methods include node drop pooling to decrease the time and space required for the process by simply selecting a subset of nodes to construct the coarsened graph. Traditional pooling methods include selecting the top-k nodes, using self-attention networks, and a gated structured aware approach.[27–29] Yet, these methods are also limited by small sample sizes and are prone to focusing on local structures rather than the graph as a whole.

3. The Proposed Model

3.1. *Capturing Dynamic Connectivity via Temporal Chunking and EdgeConv Analysis*

We define a directed graph as $G = \{V, E\}$ for each brain network subject, where V is the set of nodes with a time series and $\boldsymbol{E}$ represents the weighted connectivity. Temporal Chunking is defined by looking at a smaller window of time in the subject's time series data at any one point in time rather than aggregating it as a whole. The window sizes vary from 128 to 50 to 64 and the starting points of the window are randomly generated for each epoch. For each generated window, the partial correlation matrices are extracted to form the adjacency matrix. This dramatically increases the variety of the data the model has to work with, allowing for more robust model at the end of training. It helps to combat the issue of overfititng that previous models have cited as limitations. The model is relevant for clinical use as it can better adapt to the small sample sizes that are commonly seen in MRI datasets and can better adapt to new patients as well. Its novelty lies in its integration of BrainGMixup which ensures maximal data variation by accounting for both spatial and temporal based data augmentation.

Edge features are defined as $\boldsymbol{e}_{ij} = h_\theta(\boldsymbol{x}_i, \boldsymbol{x}_j)$ with $h_\theta = \mathbb{R}^F \times \mathbb{R}^F \to \mathbb{R}^{F'}$ as the MLP for the model with a nonlinear function parameterized by a set of learnable parameters. $\boldsymbol{x}_i$ represents the embedding of node i and $\boldsymbol{x}_j$ represents the embeddings of all the neighbors of node i, including the node itself. In this case, a sum aggregation operation is performed over all the edge features to get the final embedding for the node' and its neighbor's edges represented by

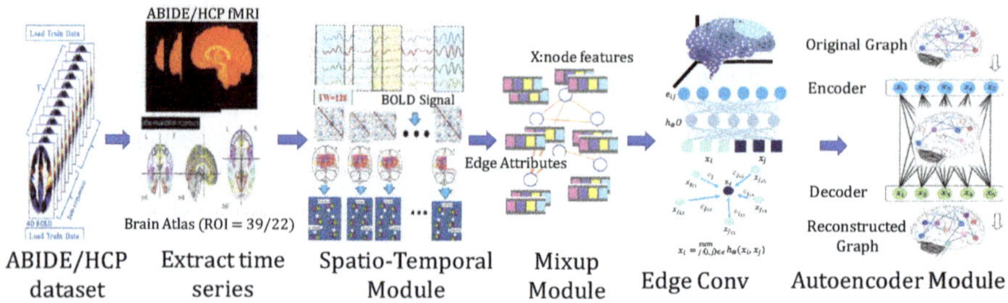

Fig. 1: Overview of the proposed BrainSTEAM architecture.

the following equation:

$$\boldsymbol{x}_i' : \underset{j_i(i,j)\in E}{sum} h_\theta(\boldsymbol{x}_i || \boldsymbol{x}_j - \boldsymbol{x}_i) \tag{1}$$

EdgeConv[30] allows for the extraction of neighborhood-level features within the overall topological structure of the network. Different aggregation methods can be used across the embeddings of the node and the neighbors. By determining the pairwise distance matrices for the characteristics and selecting the k nearest neighbors for each point, the graph is also dynamically updated, where k is a hyper-parameter that can be varied to obtain desirable results.

3.2. *Discovering Important Connectivity via Autoencoder-based Pooling*

Graph pooling is a key component to compress the predictions of multiple nodes into a graph-level classification. To discover important connectivity, we adapted Graph Autoencoder[31] technique where the node dropping is performed to measure the importance of the node for reconstructing the topological structure without labels. The new graph generated by the pooled graph can be defined as:

$$G' = \mathrm{POOL}(G), \tag{2}$$

where the pooling method SAGPool[28] acts as the encoder of the autoencoder. The SAGPool first generates scores for all the nodes from convolution and performs pooling by only taking the top k scoring nodes, with the pooling ratio determined by a hyperparameter k. Those nodes are then used to compose a new coarsened graph by learning the attribute and adjacency matrices:

$$\begin{aligned} \boldsymbol{Z}^{(l+1)} &= \boldsymbol{Z}^{(l)}_{idx^{(l)}} \odot \boldsymbol{S}^{(l)}_{idx^{(l)}} \in \mathbb{R}^{n^{(l+1)} \times 1}, \\ \boldsymbol{A}^{(l+1)} &= \boldsymbol{A}^{(l)}_{(idx^{(l)}, idx^{(l)})} \in \{0,1\}^{n^{(l+1)} \times n^{(l+1)}}, \end{aligned} \tag{3}$$

where idx serves at the indexing operator for the top-k significant scoring nodes, $\boldsymbol{Z}^{(l)}_{idx^{(l)}}$ is the row wise indexed embedding matrix, and $\odot$ is the broadcast elementwise product. $\boldsymbol{A}$ is displayed as the row-wise and column-wise adjacency matrix. $\boldsymbol{Z}^{l+1}$ and $\boldsymbol{A}^{(l)}_{(idx^{(l)}, idx^{(l)})}$ are respectively the new attribute and adjacency matrices. $\boldsymbol{S}^{(l)}_{idx^{(l)}}$ represents the score matrix of the top k selected nodes at layer l. The score matrix was calculated by inputting the adjacency matrix and node embedding matrix at the layer l into a Graph Convolution Network (GCN).

The decoder reconstructs the embeddings of the dropped nodes, which includes the creation of an empty attribute matrix with the pooled node embeddings to reconstruct a new embedding matrix, with zero padding operations performed. To measure the validity of this reconstructed matrix, the Euclidean distance is calculated between the reconstructed attribute matrix and the original input matrix. This becomes a loss function L_f. The Euclidean distance is shown below:

$$L_f^{(l)} = \left\| \boldsymbol{X} - \psi_a(\hat{x}^{(l)}) \right\|_F^2, \tag{4}$$

$$L_d^{(l)} = \left\| \boldsymbol{D}^{(l)} - \psi_d(Z^{(l)}) \right\|_F^2, \tag{5}$$

where $L_f^{(l)}$ represents the loss of the node attributes for the lth layer, $\|\|\|_F$ is the Frobenius norm, and $\boldsymbol{X}$ represents the node feature matrix. An additional L_d is adopted to regularize the distance between the true degree values and the reconstructed ones. This determines how close the pooling mechanism reconstruction came to the original graph of the subject. $\psi_a(\hat{x}^{(l)})$ represents the the reconstructed node attribute matrix. This method of pooling is preferrable to the typical mean, max, or summation pooling as it identifies the most structurally important nodes and reduces the number of noisy nodes allowing for more focused analysis.

3.3. *Enhancing Model Training and Stability via Mixup*

It is difficult for GNNs to properly analyze the underlying signals in functional brain images with the overfitting and memorization of noise in specific training data.[32] Vicinal risk minimization[33] rather than empirical risk minimization[19] techniques have been applied to improve generalization capability. Vicinal risk minimization referring to creating virtual examples of training data based on their neighborhood of data.

This paper proposes BrainGMixup[34] which utilizes 2D feature vectors from the node and edge features rather than the 1D feature vectors for other forms of data such as CNN networks. This requires interpolation between the rows/ROIs of the graph rather than between individual feature columns. This differs from traditional Graph Mixup approaches as it involves interpolation instead of concatenation of smaller sub graphs. Mixup intends to take two subjects and combine their feature and edge index information to create a new node for the model to train on. The fixed nature of ROIs in the data allows for the mixup to be applied across rows for the node feature matrix and edge index matrix,

$$\begin{aligned} \tilde{\boldsymbol{X}} &= \lambda \boldsymbol{X}_i + (1-\lambda)\boldsymbol{X}_j, \ \ where\ i,j = 1,...,N, i \neq j, \\ \tilde{\boldsymbol{E}} &= \lambda \boldsymbol{E}_i + (1-\lambda)\boldsymbol{E}_j, \ \ where\ i,j = 1,...,N, i \neq j, \end{aligned} \tag{6}$$

$$\tilde{y} = \lambda y_i + (1-\lambda) y_j. \tag{7}$$

N represents the number of ROIs defined in the node feature matrix, $\boldsymbol{E}$ is the edge index matrix, and y is the corresponding label. $\tilde{\boldsymbol{X}}$, $\tilde{y}$, and $\tilde{\boldsymbol{E}}$ are the mixup-augmented samples of

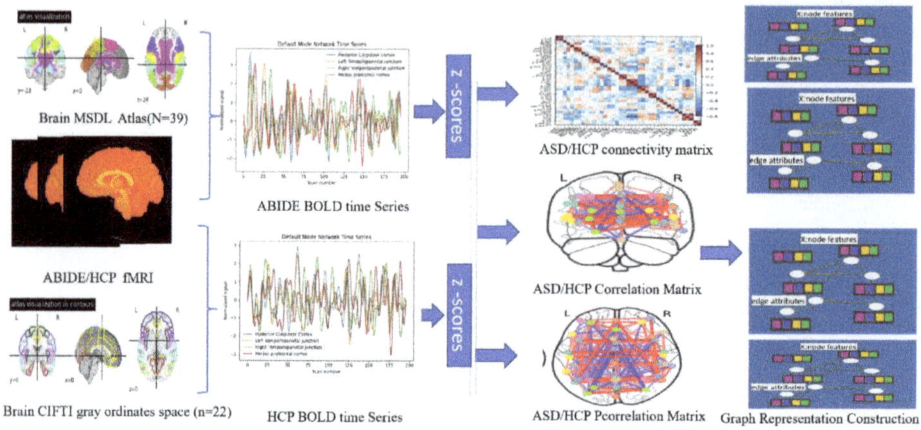

Fig. 2: The process of brain network construction.

the corresponding matrices. It involves the interpolation of the previous graph samples to cover for in-between brain network variations

$$l = \lambda \cdot c(p, y_a) + (1 - \lambda) \cdot c(p, y_b). \tag{8}$$

The mixup loss criterion L_m utilizes the Vicinity distribution[33] to find the chance that a particular feature target is in the near area of that graph to generate L_m. The differentianting lambda values ensures that even if data is taken at a similar timepoint, the resulting graph will not be the same. The hyperparameter a is used to determine the degree of interpolation between the different ROI regions and edge connectives. This serves as an efficient and effective way of accounting for the heterogeneous, scarce, and noisy nature of brain networks. The total overall loss is $L_{all} = \alpha * L_f + \beta * L_d + L_m$. With, *alpha* and *beta* serving as hyperparameters to determine the weight of the feature loss and degree loss. Mixup serves as an additional sample size increase alongside temporal chunking to provide the model with more training modules.

4. Experiments

Datasets. We evaluate our framework using two publicly available real-world neuroimaging datasets, the Autism Brain Imaging Data Exchange (ABIDE)[35]on the ASD prediction task and the Human Connectome Project (HCP)[36] on the gender classification task. The CPAC[37] preprocessed ABIDE dataset is a collection of 4D resting-state functional MRI scans from a total of 1,112 individuals with 539 Autism Spectrum Disorder (ASD) and 573 typical health controls. The preprocessed HCP dataset, on the other hand, is a large-scale dataset that includes resting-state fMRI scans from 1095 subjects with the gender split being 595 females and 500 males with about 1200 frames in each scan.

Brain network construction. The brain Blood Oxygenation Level Dependent (BOLD) signal time series is then extracted from those fMRI subset data with MSDL brain Probabilistic atlas which defines soft parcellations of the brain to 39 ROI on ABIDE and CIFTI(Connectivity Informatics Technology Initiative) (ROI=22) on HCP to produce ABIDE time series matrix (196 $\times$ 39) and HCP time series matrix (1200 $\times$ 22). These were determined from previous experimentation and papers on the appropriate number depending on the condition.[38]

Then, the brain connectivity matrices among ROI are calculated from time series data with partial correlation and correlation matrix, followed by z-scores normalization . Non-zero adjacency matrices mean a pair of ROI nodes share an edge, and the values of adjacency matrices indicate edge weights between nodes. The sparse partial correlation matrix can help to avoid the over-smoothing issue commonly seen in GNN applications. Node features are initialized with the corresponding rows in the edge weight matrix.

The temporal windows/chunks of each subject is constructed using a graph representation object as seen in Figure 2's third step. Each graph representation object then goes through mixup to create a new graph that is a interpolation of two different subjects via the BrainMixup module. This data is then fed into the EdgeConv model to train, and pooling is conducted by the AutoEncoder.

The EdgeConv model contains three block which each block containing a dyanmic EdgeConv layer, a batch normalization layer, and a relu activation layer. Each block also includes the feature decoder and degree decoder layers as a part of the AutoEncoder module. The loss is calculated as seen in the methods section with different weights applied to the loss of the model and Autoencoder loss in regards to the reconstructed feature and degrees in comparison to their ground truth values.

Baselines. We compare our proposed BrainSTEAM with baseline model MAGE,[39] SVM-MTFS,[40] MISO-DNN,[41] e-STAGIN,[41] MAGIN,[42] IMAGIN[42] on the ABIDE dataset, and with ST-GCN,[38] LTSM,[43] GCN,[8] GC-LSTM,[43] STAGIN-SERO[44] and DECENNT[45] on the HCP dataset.

Experimental settings. This study performs training and testing in 5-fold cross-validation, and dynamically construct graph data object for each sub-sequences of different window sizes with fixed optimum W as 128. The learning rate is set as 10^{-4}, epochs as 10000 for ABIDE and 30000 for HCP. All reported results are averaged of five runs of five-fold cross-validation. Additional details regarding the experiment settings can be found in the supplementary materials.

Prediction performance. The overall prediction results presented in Table 2 show that BrainSTEAM outperformed the baseline model MAGE by 9.38%, IMAGIN by 8.25% on the ABIDE dataset, and achieves 7.71% improvements over ST-GCN and 3.21% improvements over STAGIN-SERO on the HCP dataset. The results demonstrate the superiority of BrainSTEAM in neuropsychiatric disorder prediction and gender classification compared to other state-of-the-art models.

	ABIDE					HCP			
	Accuracy	AUC	Precision	Recall		Accuracy	AUC	Precision	Recall
MAGE	75.86	83.14	71.53	79.24	ST-GCN	83.7	-	-	-
SVM+MTFS	$76.7_{\pm 2.7}$	$81_{\pm 0.31}$	$72.5_{\pm 3.2}$	$76.7_{\pm 2.7}$	LTSM	81.7	-	-	-
MISO-DNN	$77.73_{\pm 4.26}$	-	$76.73_{\pm 4.11}$	$77.16_{\pm 3.72}$	GCN	83.98	-	84.59	87.78
e-STAGIN(Sch	$75.81_{\pm 1.70}$	$81.12_{\pm 0.30}$	$78.03_{\pm 2.34}$	$79.06_{\pm 0.89}$	GC-LSTM	81.50	-	-	-
MAGIN	$78.12_{\pm 1.91}$	$85.72_{\pm 0.2}$	$78.37_{\pm 2.11}$	$79.55_{\pm 1.62}$	STAGIN-SERO	$88.20_{\pm 1.33}$	$92.96_{\pm 1.87}$	-	-
IMAGIN	$79.25_{\pm 2.33}$	$86.44_{\pm 0.24}$	$81.03_{\pm 3.47}$	$79.06_{\pm 0.89}$	DECENNT	86.00	93.6	87.2	88.6
BrainSTEAM	$\mathbf{87.5}_{\pm 0.99}$	$89.23_{\pm 0.88}$	$82.24_{\pm 2.48}$	$96.11_{\pm 2.47}$	BrainSTEAM	$\mathbf{91.41}_{\pm 0.02}$	$93.67_{\pm 0.01}$	$100_{\pm 0.00}$	$78.78_{\pm 0.04}$

Table 1: Overall performance (%) comparison on two datasets. Results with - were not provided in the original work.

	ABIDE					HCP			
	Accuracy	AUC	Precision	Recall		Accuracy	AUC	Precision	Recall
BrainSTEAM	$\mathbf{87.5}_{\pm 0.99}$	$89.23_{\pm 0.88}$	$82.24_{\pm 2.48}$	$96.11_{\pm 2.47}$	BrainSTEAM	$\mathbf{91.41}_{\pm 0.02}$	$93.67_{\pm 0.01}$	$100_{\pm 0.00}$	$78.78_{\pm 0.04}$
BrainEAM	$62.86_{\pm 0.87}$	$62.36_{\pm 0.78}$	$67.23_{\pm 0.09}$	$63.95_{\pm 1.70}$	BrainEAM	$77.20_{\pm 1.35}$	$80.15_{\pm 2.27}$	$87.43_{\pm 4.49}$	$66.59_{\pm 5.82}$
BrainEM	$63.66_{\pm 1.45}$	$62.50_{\pm 1.66}$	$68.09_{\pm 2.14}$	$71.24_{\pm 1.69}$	BrainEM	$74.42_{\pm 0.01}$	$74.48_{\pm 0.01}$	$77.11_{\pm 0.01}$	$73.79_{\pm 0.01}$
BrainE	$59.43_{\pm 1.48}$	$59.24_{\pm 1.64}$	$60.22_{\pm 0.61}$	$63.98_{\pm 1.34}$	BrainE	$67.85_{\pm 0.01}$	$68.01_{\pm 0.01}$	$68.97_{\pm 0.01}$	$70.46_{\pm 0.02}$

Table 2: The ablation study with different model variants: BrainSTEAM is the full version with all components, BrainEAM removes the temporal chunking, BrainEM removes both the temporal chunking and autoencoder, and BrainE is only equipped with Edgeconv.

We further investigate the influence of each proposed component by removing each at a time. The results are shown in Table 2. Results show the temporal module contributes to the greatest increase in model prediction accuracy performance, improving about 23.84% on ABIDE, and about 14.21% on HCP. The autoencoder module provides more stability to the network as seen by the decrease in the standard deviation.

Key hyperparameter studies are shown in Fig. 3. (a) shows performance is about 1.87% higher when k=10 than k=15; (b) shows performance is about 3.12% higher when window=128 than window=50; (c) shows performance is 7.03% higher when loss alpha and loss beta is set to 0.3 vs 0.1; (d) and (e) show performance increase dramatically when epoch increases from 1k, 5k to 10k/30k with BrainSTEAM, on the contrary, performance stays flat for BrainEAM when epoch increase accordingly both on ABIDE and HCP.

5. Interpretation Analysis

As summarized above, the proposed BrainSTEAM is shown to significantly outperform baseline models. We claim the fundamental reason is that the other baselines only obtain one graph from the subject full range of time series thus only resulting in 1112 graphs for ABIDE and 1095 graphs for HCP. With our proposed time series temporal chunk combined with the mixup, an exponential increase in the number of new graphs can be generated. Specifically, the model is trained on the same 1000 subjects but the generation of time series chunks with 5-fold cross-validation for 30,000 epochs leads to 150,000 different graphs. Hyperparameter tuning with epochs reveals that the BrainEAM model hits a training accuracy of 99% in 200 epochs,

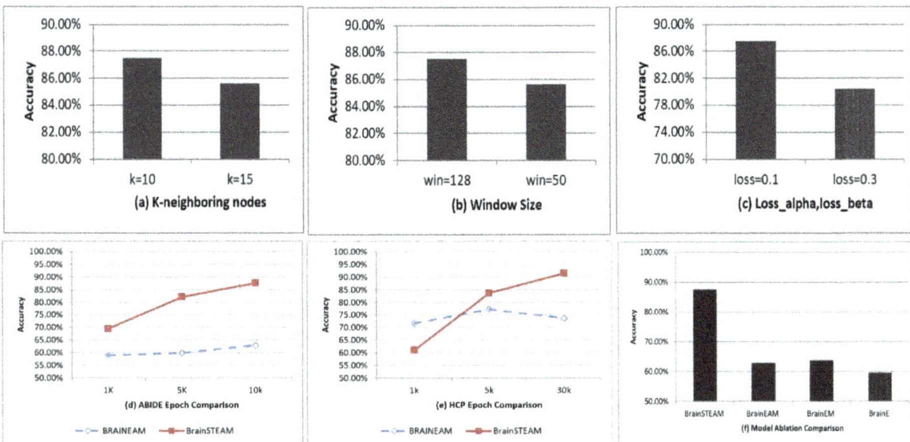

Fig. 3: The hyperparameter study for BrainSTEAM on the ABIDE dataset.

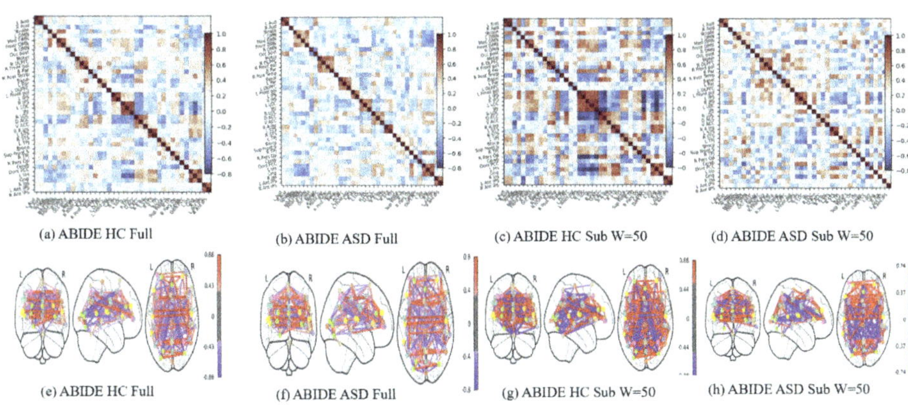

Fig. 4: The visualization of brain connectome, where the subfigure (a) & (e) represent the connectome of ABIDE Health Control (HC) with the full sequence of time series; (b) & (f) represent that of Autism; (c) & (g) represent HC with time series subsequence of window size 50; and (d) & (h) represent Autism with time series subsequence of window size 50.

but the validation accuracy stays in the low 60s indicating a typical sign of overfitting. When applying temporal chunking, the training and validation accuracy scale more evenly with an 18.13% increase of validation accuracy as the number of epochs varies from 1000 to 10000. The temporal chunking results as visualized in Fig. 4 demonstrates that graphs generated at different time windows have significantly different levels of connectivity between ROIs. This stood true for both health control and patients diagnosed with Autism. The level of interactions for the health control is far more pronounced, as noted with the increase in deep red boxes, than for the Autism patient. This fine-grained interaction difference is not expressed within the graph generated from an average of the entire time series. This demonstrates that the proposed temporal chunking method is able to better capture time specific interactions in the brain and will generate more robust generalization patterns.

The model outperforms ST-GCN, demonstrating that only the temporal module might not be comprehensive enough to cover all the issues that create overfitting and accuracy deficits. A combination of retaining connectivity information and performing self-supervised node dropping is needed to create the most robust version of the model.

6. Conclusion

This study proposes a dynamic functional brain network analysis framework BrainSTEAM, which integrates the temporal sliding window module with EdgeConv, Autoencoder and Mixup for the first time. Extensive experiments on two real-world neuroimaging datasets exhibit significant performance improvement over the state-of-the-art. This study also shows the contribution of each component to the system, demonstrating the temporal chunking approach as the major contributor to performance improvement, which allows for the representation of functional brain connectivity within smaller time windows to capture unique fine-grained ROI interactions. In the meantime, the study also shows EdgeConv helps in capturing the connectivity structures of the brain networks, autoencoder helps in reducing data noise and identifying the most relevant connectivity patterns, and mixup helps in enhancing the model training through linear interpolation. For future work, we look to improve BrainSTEAM with explainability, such as identifying meaningful biomarkers linked to neuropsychiatric disorders and mental development, understanding which neural systems contribute most to the prediction of a specific disease, applying the model to other datasets and tasks, and exploring its potential applications in clinical settings.

References

1. E. Bullmore and O. Sporns, The economy of brain network organization, *Nat. Rev. Neurosci.* (2012).
2. S. L. Bressler and V. Menon, Large-scale brain networks in cognition: emerging methods and principles, *Trends Cogn. Sci.* (2010).
3. S. M. Smith, K. L. Miller, G. Salimi-Khorshidi, M. Webster, C. F. Beckmann, T. E. Nichols, J. D. Ramsey and M. W. Woolrich, Network modelling methods for FMRI, *NeuroImage* **54** (2011).
4. S. L. Simpson, F. D. Bowman and P. J. Laurienti, Analyzing complex functional brain networks: fusing statistics and network science to understand the brain, *Stat. Surv.* **7**, p. 1 (2013).
5. X. Kan, Y. Kong, T. Yu and Y. Guo, Bracenet: Graph-embedded neural network for brain network analysis, in *IEEE Big Data*, 2022.
6. J. Kawahara, C. Brown, S. Miller, B. Booth, V. Chau, R. Grunau, J. Zwicker and G. Hamarneh, Brainnetcnn: Convolutional neural networks for brain networks; towards predicting neurodevelopment, *NeuroImage* **146**, 1038 (2017).
7. Y. Cui, S. Zhao, H. Wang, L. Xie, Y. Chen, J. Han, L. Guo, F. Zhou and T. Liu, Identifying

brain networks at multiple time scales via deep recurrent neural network, *IEEE J Biomed Health Inform* **23**, 2515 (2018).
8. T. N. Kipf and M. Welling, Semi-supervised classification with graph convolutional networks, *ICLR* (2016).
9. P. Veličković, G. Cucurull, A. Casanova, A. Romero, P. Liò and Y. Bengios, Graph attention networks, *ICLR* (2018).
10. X. Li, Y. Zhou, N. Dvornek, M. Zhang, S. Gao, J. Zhuang, D. Scheinost, L. Staib, P. Ventola and J. Duncan, Braingnn: Interpretable brain graph neural network for fmri analysis, *Med. Image Anal.* **74** (2021).
11. H. Cui, W. Dai, Y. Zhu, X. Li, L. He and C. Yang, Interpretable graph neural networks for connectome-based brain disorder analysis, in *MICCAI*, 2022.
12. X. Kan, H. Cui, J. Lukemire, Y. Guo and C. Yang, Fbnetgen: Task-aware gnn-based fmri analysis via functional brain network generation, in *MIDL*, 2022.
13. X. Kan, W. Dai, H. Cui, Z. Zhang, Y. Guo and C. Yang, Brain network transformer, in *NeurIPS*, 2022.
14. H. Cui, W. Dai, Y. Zhu, X. Kan, A. A. C. Gu, J. Lukemire, L. Zhan, L. He, Y. Guo and C. Yang, A benchmark for brain network analysis with graph neural networks, *IEEE TMI* (2022).
15. Y. Zhu, H. Cui, L. He, L. Sun and C. Yang, Joint embedding of structural and functional brain networks with graph neural networks for mental illness diagnosis, in *EMBC*, 2022.
16. Y. Yang, Y. Zhu, H. Cui, X. Kan, L. He, Y. Guo and C. Yang, Data-efficient brain connectome analysis via multi-task meta-learning, in *KDD*, 2022.
17. Y. Yang, H. Cui and C. Yang, Ptgb: Pre-train graph neural networks for brain network analysis, in *CHIL*, 2023.
18. A. Abraham, M. P. Milham, A. D. Martino, R. C. Craddock, D. Samaras, B. Thirion and G. Varoquaux, Deriving reproducible biomarkers from multi-site resting-state data: An autism-based example, *NeuroImage* **147**, 736 (2017).
19. H. Zhang, M. Cisse, Y. N. Dauphin and D. Lopez-Paz, mixup: Beyond empirical risk minimization, *ICLR* (2018).
20. X. Han, Z. Jiang, N. L. Liu and X. Hu, G-mixup: Graph data augmentation for graph classification, *ICML* (2022).
21. J. Park, H. Shim and E. Yang, Graph transplant: node saliency-guided graph mixup with local structure preservation, *Proceedings of the First MiniCon Conference* (2022).
22. Q. Liu, Y. Zhang, L. Guo and Z. Wang, Spatial-temporal data-augmentation-based functional brain network analysis for brain disorders identification, *Frontiers in Neuroscience* (2023).
23. J. Yan, Y. Chen, Z. Xiao, S. Zhang, M. Jiang, T. Wang, T. Zhang, J. Lv, B. Becker, R. Zhang, D. Zhu, J. Han, D. Yao, K. M. Kendrick, T. Liu and X. Jiang, Modeling spatio-temporal patterns of holistic functional brain networks via multi-head guided attention graph neural networks (multi-head gagnns), *Medical Image Analysis* **80** (2022).
24. Z. Zhang, J. Bu, M. Ester, J. Zhang, Z. Li, C. Yao, H. Dai, Z. Yu and C. Wang, Hierarchical multi-view graph pooling with structure learning," in ieee transactions on knowledge and data engineering, *IEEE TKDE* **35**, 545 (2023).
25. A. Duval and F. Malliaros, Higher-order clustering and pooling for graph neural networks, in *CIKM*, 2022.
26. Z. Peng, H. Liu, Y. Jia and J. Hou, Attention-driven graph clustering network, *ACM International Conference on Multimedia* , 935 (2021).
27. H. Gao and S. Ji, Graph u-nets, *IEEE TPAMI* **44**, 4948 (2022).
28. J. K. Junhyun Lee, Inyeop Lee, Self-attention graph pooling, *ICML* (2019).
29. H. Yu, J. Yuan, H. Cheng, M. Cao and C. Wang, Graph u-nets, in *IJCNN*, 2021.
30. Y. Wang, Y. Sun, Z. Liu, S. Sarma, M. Bronstein and J. Solomon, Dynamic graph cnn for

learning on point clouds, *ACM Trans. on Graphics* **38**, 1 (2018).
31. C. Liu, Y. Zhan, X. Ma, D. Tao, B. Du and W. Hu, Masked graph auto-encoder constrained graph pooling, *ECML-PKDD* (2022).
32. K. Zhou, Y. Dong, W. S. Lee, B. Hooi, H. Xu and J. Feng, Effective training strategies for deep graph neural networks, *arXiv* (2020).
33. O. Chapelle, J. Weston, L. Bottou and V. Vapnik, Vicinal risk minimization, *NeurIPS* (2000).
34. A. Li, Brainmixup: Data augmentation for gnn-based functional brain network analysis, *IEEE Big Data* (2022).
35. A. D. Martino, C.-G. Yan, Q. Li, E. Denio, F. X. Castellanos, K. Alaerts, J. S. Anderson, M. Assaf, S. Y. Bookheimer, M. Dapretto, B. Deen, S. Delmonte, I. Dinstein, B. Ertl-Wagner, D. A. Fair, L. Gallagher, D. P. Kennedy, C. L. Keown, C. Keysers, J. E. Lainhart, C. Lord, B. Luna, V. Menon, N. J. Minshew, C. S. Monk, S. Mueller, R.-A. Müller, M. B. Nebel, J. T. Nigg, K. O'Hearn, K. A. Pelphrey, S. J. Peltier, J. D. Rudie, S. Sunaert, M. Thioux, J. M. Tyszka, L. Q. Uddin, J. S. Verhoeven, N. Wenderoth, J. L. Wiggins, S. H. Mostofsky and M. P. Milham, The autism brain imaging data exchange: towards a large-scale evaluation of the intrinsic brain architecture in autism., *Mol Psychiatry* (2022).
36. D. C. V. Essen, S. M. Smith, D. M. Barch, T. E. Behrens, E. Yacoub and K. Ugurbil, The wu-minn human connectome project: An overview., *NeuroImage* (2013).
37. C. Craddock, S. Sikka, B. Cheung, R. Khanuja, S. S. Ghosh, C. Yan, Q. Li, D. Lurie, J. Vogelstein, R. Burns *et al.*, Towards automated analysis of connectomes: The configurable pipeline for the analysis of connectomes (c-pac), *INCF Congress of Neuroinformatics.* (2014).
38. S. Gadgil, Q. Zhao, A. Pfefferbaum, E. V. Sullivan, E. Adeli and K. M. Pohl, Spatio-temporal graph convolution for resting-state fmri analysis, *National Library of Medicine* (2020).
39. U. Pervaiz, D. Vidaurre, C. Gohil, S. Smith and M. Woolrich, Multi-dynamic modelling reveals strongly time-varying resting fmri correlations, *Medical Image Analysis* **77** (2022).
40. A. S. Karampasi, A. D. Savva, V. C. Korfiatis, I. Kakkos and G. K. Matsopoulos, Informative biomarkers for autism spectrum disorder diagnosis in functional magnetic resonance imaging data on the default mode networkl networks, *Appl. Sci.* (2021).
41. T. M. Epalle, Y. Song, Z. Liu and Others, Multiatlas classification of autism spectrum disorder with hinge loss trained deep architectures: Abide i results, *Applied Soft Computing* **107**, p. 107375 (2011).
42. A. S. James Orme-Rogers, Spatio-temporal attention in multi-granular brain chronnectomes for detection of autism spectrum disorder, *arXiv* (2022).
43. P. Cao, G. Wen, L. Li, X. Liu, J. Yang and O. Zaiane, Temporal graph representation learning for autism spectrum disorder brain networks, *BIBM* (2022).
44. J.-J. K. Byung-Hoon Kim, Jong Chul Ye, Learning dynamic graph representation of brain connectome with spatio-temporal attention, *AAAI* (2021).
45. U. Mahmood, Z. Fu, V. Calhoun and S. Plis, Deep dynamic effective connectivity estimation from multivariate time series, *IJCNN* (2022).

MaTiLDA: An Integrated Machine Learning and Topological Data Analysis Platform for Brain Network Dynamics

Katrina Prantzalos, Dipak Upadhyaya
Department of Population and Quantitative Health Sciences, Case Western Reserve University, Cleveland, OH 44106, USA
Email: Katrina.prantzalos@case.edu; Dipak.upadhyaya@case.edu

Nassim Shafiabadi, Guadalupe Fernandez-BacaVaca
Department of Neurology, University Hospitals Cleveland Medical Center, Cleveland, OH 44106, USA
Email: Nassim.Shafiabadi@uhhospitals.org; Guadalupe.Fernandez-BacaVaca@uhhospitals.org

Nick Gurski
Department of Mathematics, Case Western Reserve University, Cleveland, OH 44106, USA
Email: Nick.gurski@case.edu

Kenneth Yoshimoto, Subhashini Sivagnanam, Amitava Majumdar
San Diego Supercomputer Center, University of California, San Diego, CA, USA
Email: kenneth@sdsc.edu; sivagnan@sdsc.edu; majumdar@sdsc.edu

Satya S. Sahoo
Department of Population and Quantitative Health Sciences, Case Western Reserve University, Cleveland, OH 44106, USA
Email: Satya.sahoo@case.edu

Topological data analysis (TDA) combined with machine learning (ML) algorithms is a powerful approach for investigating complex brain interaction patterns in neurological disorders such as epilepsy. However, the use of ML algorithms and TDA for analysis of aberrant brain interactions requires substantial domain knowledge in computing as well as pure mathematics. To lower the threshold for clinical and computational neuroscience researchers to effectively use ML algorithms together with TDA to study neurological disorders, we introduce an integrated web platform called MaTiLDA. MaTiLDA is the first tool that enables users to intuitively use TDA methods together with ML models to characterize interaction patterns derived from neurophysiological signal data such as electroencephalogram (EEG) recorded during routine clinical practice. MaTiLDA features support for TDA methods, such as persistent homology, that enable classification of signal data using ML models to provide insights into complex brain interaction patterns in neurological disorders. We demonstrate the practical use of MaTiLDA by analyzing high-resolution intracranial EEG from refractory epilepsy patients to characterize the distinct phases of seizure propagation to different brain regions. The MaTiLDA platform is available at: https://bmhinformatics.case.edu/nicworkflow/MaTiLDA

Keywords: Epilepsy Seizure Network; Topological Data Analysis (TDA); Machine Learning; Neurological Disorders

1. Introduction

The increasing availability of multimodal brain activity recordings highlights an emergent demand for accurate and reliable analytical methods to characterize brain interaction dynamics to meet clinical research goals and to improve patient care[1]. The analysis of brain recordings provide insights into the dynamics of interaction patterns involving specialized brain regions that may be responsible for higher-order brain functions[2]. Understanding disruptions in brain interaction patterns is crucial to characterizing neurological disorders, revealing pathophysiological mechanisms, and defining biomarkers for clinical diagnoses[1–3]. These research goals are particularly important in epilepsy, which is a complex neurological disorder affecting over 50 million individuals worldwide[4]. Epilepsy is characterized by recurrent seizures stemming from abnormal electrical discharges that spread throughout the brain[4]. Similar to other disease domains, there has been a rapid increase in the use of machine learning (ML) algorithms to study brain interaction dynamics in epilepsy patients[5,6]. ML algorithms such as support vector machines (SVM) have used features extracted from neurophysiological signal data, such as electroencephalogram (EEG), to lateralize seizure onset zone for subsequent surgical intervention [5,6].

Graph-based models of networks are commonly applied to characterize interaction patterns in the brain; however, recent studies have used rigorous algebraic topology methods to analyze brain recordings to address several limitations of graph-based models[5,7–10]. Topological data analysis (TDA) is a quantitative framework that can be used to characterize higher-dimensional interaction patterns by using robust, scale-invariant methods, such as persistent homology[11]. Specifically, quantitative measures generated from persistent homology values, such as persistence landscapes, persistence images, and persistent entropy, have highlighted the promise of applying TDA methods to analyze EEG data with respect to seizure (ictal) activity[5,7,9,10] and to distinguish seizure onset from preictal activity[5,7]. Moreover, TDA methods have been integrated with ML algorithms for several applications[12], including characterizing brain interaction dynamics[5].

The development and use of an integrated ML and TDA tool to characterize brain interaction dynamics is a resource-intensive endeavor that demands expertise in domains such as mathematics, neurology, and computing. Therefore, there is a high entry barrier for the wider neuroscience community to use TDA methods and ML algorithms together for research studies[13,14]. To address this critical barrier, we introduce MaTiLDA as the first integrated web platform for TDA methods and ML algorithms to analyze neurophysiological recordings. We demonstrate the practical utility of MaTiLDA by characterizing brain interaction dynamics in refractory epilepsy patients using high resolution intracranial EEG (iEEG) recordings.

2 Background

2.1 The Neuro-Integrative Connectivity platform

Over the past decade we have developed an integrated neuroinformatics workflow tool called the Neuro-Integrative Connectivity (NIC) platform to automate the multi-step methods used to characterize brain interaction dynamics using signal data[15–17]. The NIC platform is a modular tool that supports addition of new modules in a flexible manner as support for new functionalities, including ML, are added. One module transforms neurophysiological signal recording stored in European Data Format (EDF) into a JSON- based human-readable format with semantic annotations using an epilepsy domain ontology that is more suitable for storage and analysis[15]. A second module computes signal coupling measures using both frequency and amplitude features of the signal data[16]. A third module computes a variety of graph model-based metrics[17]. A fourth module supports persistent homology functions using open source libraries such as GUDHI[18]. We refer to our previous work for additional details of the NIC tool[15–17]. MaTiLDA is an extension of the NIC tool to enable users to use TDA with ML algorithms for integrated analysis of signal data.

2.2 Topological data analysis of EEG

Brain functions are often characterized by interaction between multiple brain regions[2]; therefore, TDA is well-suited to characterizing these interaction patterns with high dimensionality, which cannot be easily represented using graph models[14]. Persistent homology is a TDA method that has been successfully used to identify brain states by analyzing multi-dimensional interactions across brain regions[5,7,9,14]. Specifically, studies applying persistent homology to neurophysiological signal data have shown the promise of TDA in characterizing aberrant brain interaction dynamics in neurological disorders[5,7,9,14]. In this section, we briefly describe the terminology associated with TDA methods to facilitate understanding of the subsequent sections of the paper.

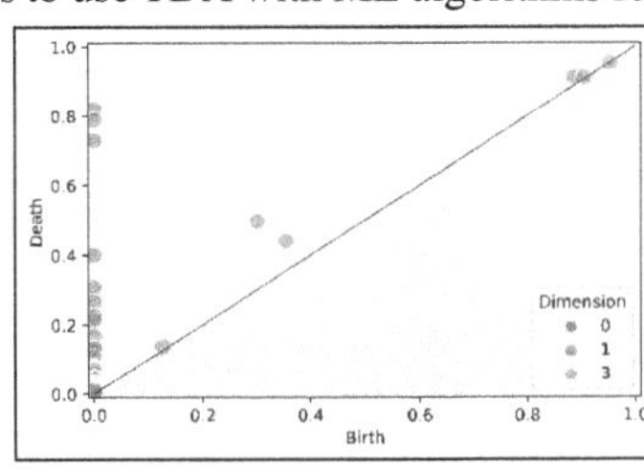

Figure 1: A persistence diagram from our analysis (section 2.6). A persistence diagram is a visualization of the results from persistent homology, where each point represents one homology class.

Persistent homology is a TDA method used to quantify the presence of topological structures, called homology classes, across various thresholds, or filtration values[14,19,20]. A homology class is a boundary composed of simplices, defined as the convex hull of a set of p+1 vertices[20]. A simplex has dimension p, and is referred to as a p-simplex, if it has a cardinality of p+1[13]. Persistent homology tracks the filtration at which each homology class is created (birth), the filtration at which it is terminated (death), and dimension of each homology class. These values can be visualized with a persistence diagram (Figure 1), a plot representing birth along the x axis and death along the y axis[11,13,19]. The lifespan, (death minus birth) of homology classes, as displayed in the persistence diagram, can be analyzed across various periods of neurophysiological signal recording to identify changes in topological structures and gain insights into the topology of brain networks[11,13,14]. We refer interested readers to Edelsbrunner and Harer[11] for further descriptions of persistent homology.

3. Methods

The computation and analysis of topological features from neurophysiological signal data entails multiple stages of processing, which include extraction of signal data, computation of signal coupling measures, TDA of signal coupling, data cleaning, and comparative analysis of topological features (Figure 2). Scientific workflow systems like the NIC platform have been used to automate these multi-step processes[17]. In this paper, we describe MaTiLDA as an extension of the NIC platform to implement integrated support for TDA and ML algorithms for brain interaction studies.

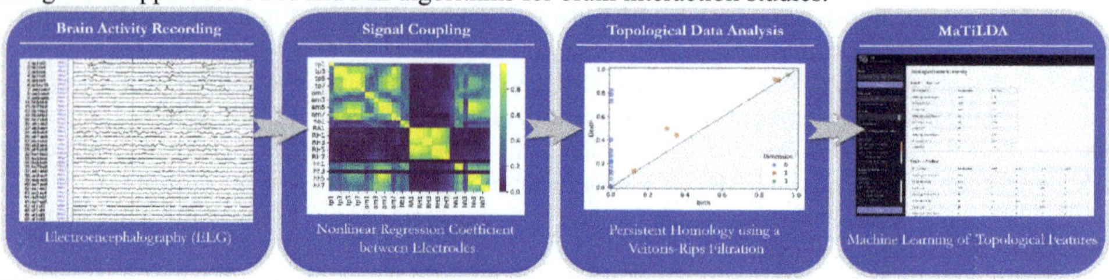

Figure 2: *Our framework for computing and comparing topological features from neurophysiological recordings. EEG from intracranial electrodes is used to extract signal data during epileptic seizures. Signal coupling is calculated using the nonlinear regression coefficient developed by Pijn et al.*[21]*. Persistent homology is applied to the signal coupling values using a Vietoris-Rips filtration as implemented in GUDHI*[18]*. MaTiLDA then allows users to select specialized data structures such as persistence landscapes or persistence images to use as input for user-selected machine learning classification such as SVM.*

3.1 MaTiLDA architecture and development

The MaTiLDA platform was built using the Django web application framework, which uses the Python programming language and features several libraries and modules that support a variety of data processing and analysis tasks including libraries for ML and TDA. MaTiLDA adopts the Model View Template (MVT) approach, with user inputs managed by an object relational data component (Model), the user interface handled by the View component, and user interaction mediated by the Template component.

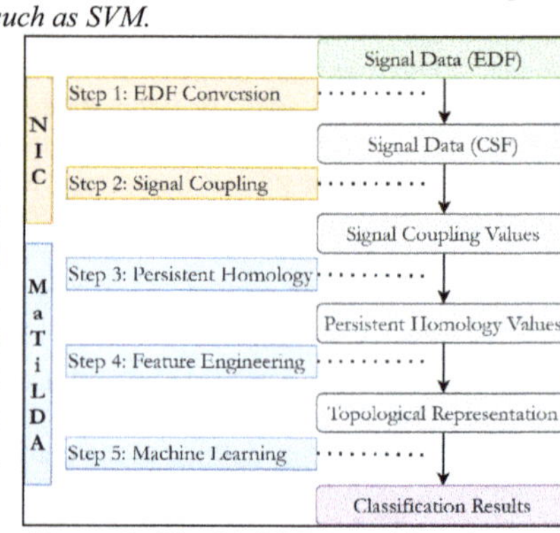

Figure 3: The MaTiLDA workflow leverages the NIC workflow to compute signal coupling. MaTiLDA applies persistent homology to the coupling values and allow users to select representations of the resulting persistent homology values for input into machine learning classifications of their choice.

3.2 A framework for classifying brain states

MaTiLDA leverages modules from the NIC tool and maintains a modular analysis process (Figure 3). Before analysis with MaTiLDA, neurophysiological recordings such as those from EEG are processed with the NIC tool to convert from EDF to CSF and to compute signal coupling measures that can be used as input into MaTiLDA for a desired ML classification task. Users are required to provide a set of folders each containing a set of coupling measure values (Figure 4). Users can subsequently apply MaTiLDA's persistent homology function, using a Vietoris-Rips filtration, to each input using the GUDHI[18] library. The persistent homology values are transformed into a specialized data structure as requested; these data structures are used as input values for ML models selected by the user. A ML model is trained using an 80% data partition. Labels are predicted for the remaining 20% data partition as a test set. The test set accuracy score is reported alongside the precision, recall, and the area under the receiver operating characteristic (ROC) curve. Accuracy scores are calculated as the number of correctly identified predictions out of total predictions[22]. Precision is calculated as the number of true positive predictions divided by the number of positive predictions[22,23]. Recall, or true positive rate, is calculated as the number of true positive predictions divided by the number of positive samples[22,23]. The ROC curve is a plot of the true positive rate along the y-axis against the false positive rate along the x-axis for varying vales of a threshold used to classify samples[23].

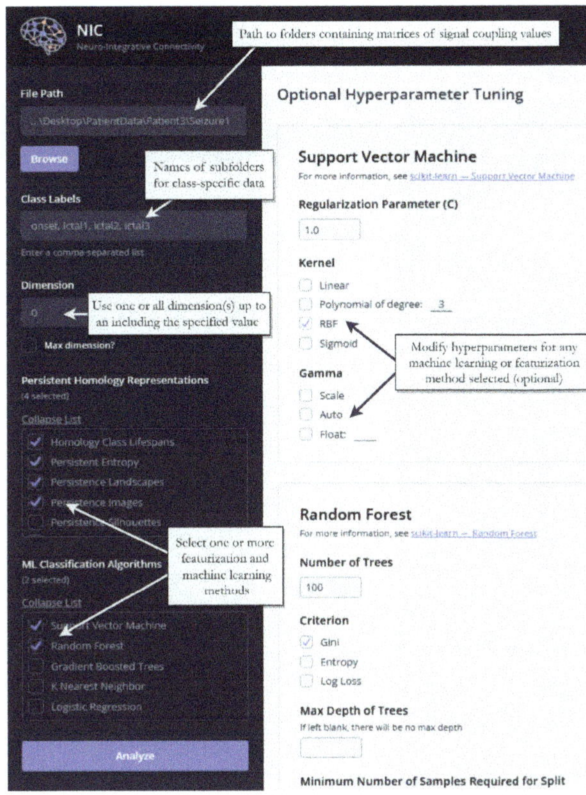

Figure 4: MaTiLDA supports various representations of persistent homology values in specialized data structures and ML algorithms with optional hyperparameter inputs. Users provide a folder including subfolders of outputs from the NIC correlator module, a list of all class labels (subfolder names), and a dimension for analysis. Users may select multiple data structures and multiple machine learning classification algorithms for their analysis using the checkboxes. For any selected representation or machine learning algorithm, a set of hyperparameters will appear in the left of the screen. The user may refine these parameters or use the preselected defaults. MaTiLDA will run each combination of representation-algorithm pairs selected for analysis. In the example provided above, the results from 8 analyses will be given.

The area under the ROC curve (AUC) measures the average classification accuracy across all thresholds[23]. A separate ML model is run for each combination of selected data structures and ML algorithms. By default, all ML models are implemented using default model parameters from Scikit-learn and GUDHI; however, users have the option to modify these parameters.

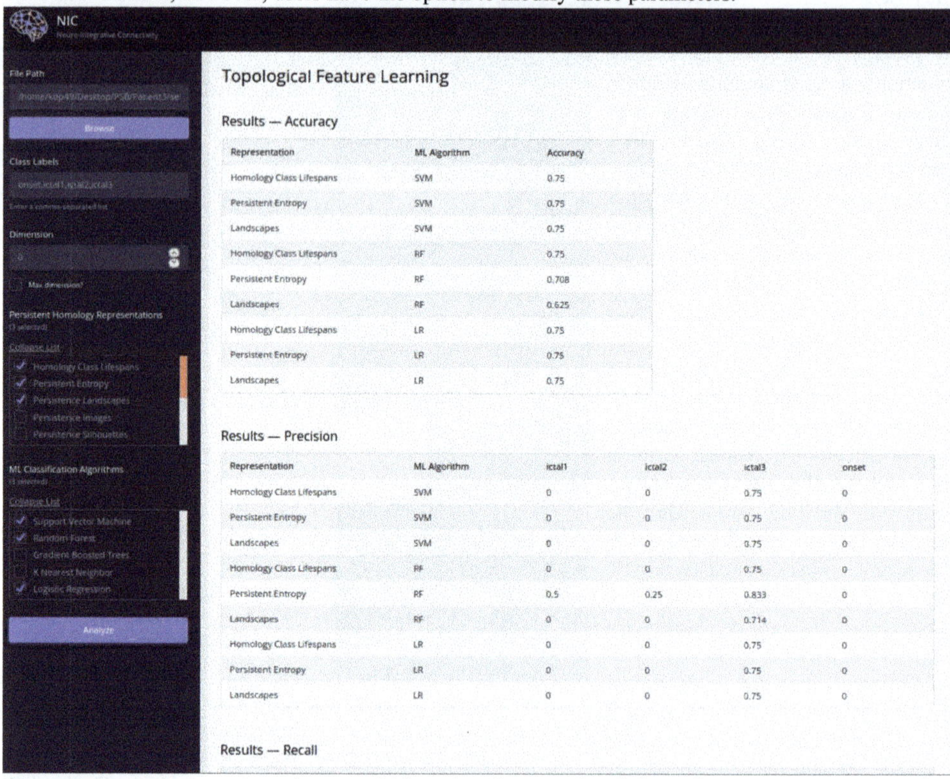

Figure 5: *Results for one seizure from a multiclass classification of ictal phases for patient one using homology class lifespans, persistent entropy, persistence landscapes, or persistence images as input to SVM, random forest, and logistic regression models.*

3.3 MaTiLDA user interface

The MaTiLDA user interface (Figure 4) consists of an intuitive data entry module and a minimal results table (Figure 5). MaTiLDA requires users to specify a directory containing several subdirectories, each of which should contain signal coupling values derived from neurophysiological signal data. MaTiLDA internally manages all data preprocessing, expecting signal coupling values to be in the format produced by the NIC tools. A list of labels must be specified by the user; these labels will be matched to the subdirectory names to select and label signal coupling data from the main directory provided. Users must select a dimension for analysis; they may limit analysis to homology classes of that dimension, or they may analyze homology classes of dimension 0 through that dimension. Users may select several specialized data structures as representations for persistent homology values as well as several ML algorithms from a set of available options and may refine parameters for each selection using simple radio buttons and numeric input fields. Results are generated for all representation-algorithm pairs selected. The results table displays the representation chosen, the ML algorithm used, the model's accuracy in testing data, the true positive rate, the false negative rate, and the AUC.

3.4 Topological feature representation for machine learning

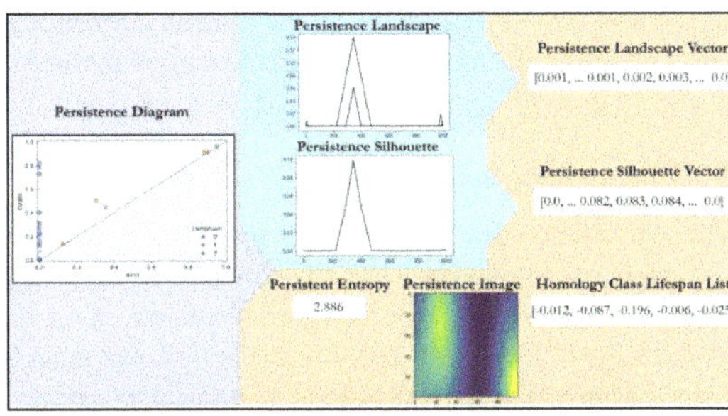

Figure 6: MaTiLDA offers several options for representing persistent homology values as vectors in Euclidean space, including persistence landscapes, persistence silhouettes, persistence images, persistent entropy, and homology class lifespans. Homology class lifespans create a list of values from the lifespans of all homology classes in a persistence diagram. Persistence landscapes and silhouettes transform persistence diagrams and apply a tent function before sampling uniformly across the transformed axis to create a list of values. Persistence images convert a persistence diagram into a two-dimensional image where each pixel represents a rectangular area of the diagram, and the intensity of the image represents the frequency of occurrence of homology classes. Persistent entropy is the Shannon

A key challenge for applying persistent homology lies in the difficulty of statistical interpretation of results[13]. Persistent homology values lack geometric properties that would allow for the definition of basic statistical concepts such as mean or median[13]. While persistence diagrams are an intuitive visualization method for representing the attributes of topological structures, the visual component of persistence diagrams makes it challenging to use statistical methods to quantitatively analyze them[12,13,19]. Additionally, persistence diagrams are not vectors in a Hilbert or Banach space and thus a unique mean cannot be established to define statistical measures[12,13]. Moreover, persistent homology values, and the persistence diagrams representing them, do not maintain a consistent number of homology classes, which creates a challenge for conducting balanced comparisons[12]. Consequently, a range of quantitative methods have been devised to facilitate the integration of persistence diagrams and persistent homology values into ML classifications. These methods for feature engineering can be used to represent persistent homology values as specialized data structures that can be used as input to ML models[12,19]. We provide the necessary background for the five quantitative methods for persistent homology value representation that have been implemented in the initial version of MaTiLDA: homology class lifespans, persistence landscapes, persistence silhouettes, persistence images, and persistent entropy (Figure 6). In this work, we show how MaTiLDA can be used to intuitively conduct analyses by using these quantitative methods to represent persistent homology values derived from coupling measures computed from neurophysiological recordings and using the resulting features as input into ML algorithms.

3.4.1 Homology class lifespan

We calculate the lifespan for each homology class resulting from persistent homology and store the values in a list. Lifespan lists are ordered based on the lifespan values such that the first value in the lifespan list is the longest lifespan within that list. The lifespan list has a length equivalent to the sum of the Betti numbers (the number of homology classes) from all dimensions included in analysis. We create the input features for ML algorithms using tensor data structures that are padded with zero values to account for varying length of the tensors corresponding to different homology class lifespan values. Our methods are similar to the work described in the study by Bendich et al.[24]; however, unlike Bendich et al., we do not limit the number of lifespan values included in a list.

3.4.2 Persistence landscapes & silhouettes

The persistence landscape is a sequence of piecewise-linear functions, $\lambda_1, \lambda_2, \dots : \mathbb{R} \rightarrow \mathbb{R}$, that map persistent homology values to a vector space, where λ_k refers to the kth persistence landscape function[25]. The persistence landscape can be calculated using Eq 1, where t denotes the filtration value, *kmax* denotes the kth largest element in the set of persistent homology values, *I*, and each homology class in *I* has a birth b_i and a death d_i[25].

$$\lambda(k,t) = kmax\{\max(0, \min(birth_i + t, death_i - t)\}_{i \in I} \tag{1}$$

The persistence landscape is plotted with the filtration along the x axis and the persistence landscape value $\lambda(k,t)$ along the y axis (Figure 6). A vector is created by uniformly sampling points along the x-axis and calculating the maximum of the persistence landscape functions at that point[19]. A persistence silhouette is a variation of the persistence landscape in which a vector is created by taking the weighted average of the functions, rather than the maximum[19,26]. The advantages of persistence landscapes and silhouettes are that they are invertible, parameter-free, nonlinear, and have desirable properties for statistical modeling including a unique mean[19,25].

3.4.3 Persistence images

To create a persistence image, a Gaussian function is applied to each homology class resulting from persistent homology[27]. The weighted sum of Gaussian functions are discretized to define a grid, and a matrix of pixel values is created by taking the integral of this grid on each grid box[27]. Consequently, each pixel value in the persistence image represents a rectangular area of the persistence diagram, and the intensity of the image represents the frequency of occurrence of homology classes[19,27]. Persistence images require a distribution, a resolution, and a weighting function to calculate[19]. The advantages of persistence images are that they are stable, interpretable, and computationally efficient representations in $\mathbb{R}^n$[19,27].

3.4.4 Persistent entropy

Persistent entropy is a single value representing the Shannon entropy of a probability distribution obtained from persistent homology[28]. The persistent entropy of a set of persistent homology values can be calculated using Eq (2), where l_i is the lifespan of a topological structure[28].

$$\sum -\frac{l_i}{\sum -l_i}\log\left(\frac{l_i}{\sum -l_i}\right) \quad (2)$$

3.5 Machine Learning of Topological Features

In the MaTiLDA pipeline (Figure 4), persistent homology is applied to signal coupling values derived from neurophysiological signal recordings. Based on user specification (section 2.3), feature engineering is applied to the resulting persistent homology values to create specialized data structures (section 3.4) to be used as input features for ML models. Five common algorithms for ML classification were selected to be implemented in the initial version of MaTiLDA: support vector machines, random forest, gradient boosted trees, K-nearest neighbor, and logistic regression. In this section, we provide a brief introduction to each of these algorithms.

3.5.1 Support vector machine

Support vector machine (SVM) is a supervised learning algorithm that aims to find the best-separating function, called a kernel, to classify data into different categories[22]. While kernels do not naturally distinguish between more than two classes, SVM can be extended to multi-class classification problems using approaches such as the one-vs-one and one-versus-rest approaches[22]. For MaTiLDA, multi-class classifications using SVM are handled using the one-versus-rest approach. In the one-versus-rest approach, for a classification of K classes, SVM will fit K kernels where each kernel will compare one of the K classes to the remaining K-1 classes[22].

3.5.2 Random forest and gradient boosted trees

Random forest (RF) is a form of decision tree bagging (generating several training sets by sampling from the original training set with replacement) that focuses on making the ensemble of decision trees more diverse[29]. As in bagging, an ensemble of trees is built based on bootstrapped training samples[22]. However, rather than varying the training sets, a random sampling of attributes is selected at each split point in the tree; of this sample, the attribute with the highest information gain is selected as the split[29]. A majority vote from the tree-specific predictions is used to classify each example[29].

Gradient boosted trees (GBT), like random forest, is a powerful learning algorithm that can learn complex, non-linear relationships[29]. GBT is a boosting algorithm using gradient descent[29]. While bagging builds trees on bootstrapped data independently of other trees, boosting uses a modified version of the original dataset to sequentially grow trees such that each tree is grown using information from previously grown trees[22].

3.5.3 K-nearest neighbor

K-nearest neighbor (KNN) is a non-parametric, supervised learning classifier that facilitates classification for observations by leveraging their proximity to the K nearest datapoints, or neighbors, in the training data[22,29]. The classification decision is made through a majority voting scheme among the K nearest neighbors[29]. KNN has a high computational cost due to performing distance calculations for each observation[29].

3.5.4 Logistic regression

Logistic regression (LR) models the probability that an observation belongs to a particular class[22]. By employing a logistic function, a linear combination of predictors is mapped to the range [0, 1], allowing LR to estimate the probability of class membership using maximum likelihood estimation[22].

3.6 Validation of MaTiLDA

Epilepsy is the second most common neurological disorder[4] and presents a unique opportunity for the application of TDA to study aberrant brain interaction dynamics. Epilepsy is characterized by recurrent seizures stemming from abnormal electrical discharges that spread throughout the brain and disrupt normal functioning[4,30]. Most significant changes to brain interactions during seizures occur during the spread of aberrant activity to new brain regions (referred to as ictal phases such as ictal 1 phase, ictal 2 phase, etc.) and the termination of a seizure[30]. One approach to understanding these changes in brain interaction dynamics is the classification of these ictal phases. To validate the use of the MaTiLDA interface for characterizing aberrant brain interaction dynamics using TDA and ML, we apply the MaTiLDA pipeline to analyze neurophysiological signal data from a cohort of four refractory epilepsy patients undergoing pre-surgical evaluation in the epilepsy monitoring unit (EMU) at University Hospitals Cleveland Medical Center's level 4 epilepsy facility that regularly performs epilepsy surgery. All patients were between the ages of 25 and 50 and had refractory epilepsy; 75% of the patients were women. ***Table 1*** shows the characteristics of these patients. Using MaTiLDA, we applied TDA and ML to analyze iEEG recordings from two seizures from each of these patients to classify ictal phases including seizure onset and propagation to different brain regions.

3.6.1 Study Data

We selected iEEG recordings from two seizures each from four refractory epilepsy patients undergoing pre-surgical evaluation. Intracranial electrodes are implanted based on a presurgical protocol described in work by Wu et al.[31]. Retrospective visual analyses of EEG recordings were conducted using a Nihon-Kohden Neurofax system (Nihon Kohden America, Foothill Ranch, CA, U.S.A.) with AC amplifiers, a high sampling rate of 2,000 Hz, and an acquisition rate spanning 0.016-300 Hz[31,32]. The EEG was filtered at 600 Hz with a 0.03s time constant and sensitivity ranging from 30-100 μV based on optimal seizure visibility for each implant[31,32]. A 60 Hz notch filter was applied to all EEG recordings[31]. Clinicians defined seizure onset as the earliest distinctive occurrence of rhythmic sinusoidal activity or repetitive spikes; the region of activity was noted as the seizure onset zone[31]. Ictal phases were defined as the subsequent spread of seizure activity to new brain regions. EEG sequences were broken down into one second epochs and features were computed for each epoch.

Table 1: Characteristics of two seizures from four randomly selected refractory epilepsy patients.

Patient	Age Range	Sex	Epileptogenic Zone	Medication	Seizure Duration (s)	Active Electrodes	Ictal Phases	Seizure Semiology
1	25-30	F	Left Hemisphere	Trileptal, Keppra	48	IM1, IM8-9, SM1-3, IL6-8, ML1-8, SP2-5, IP1-3, MP1-3, HH1-10	2	Aura → mouth and hand automatisms → mild combativeness & amnesia
					43	IM1, IM8-9, SM1-3, IL6-8, ML1-8, SP2-5, IP1-3, MP1-3, HH1-10	2	Aura
2	45-50	M	Bitemporal	Lamotrigine, Phenytoin, Valproic Acid	90	TP1-8, AM1-8, HB1-2, RA1-2 RH1-8, HH1-8	2	Aura → postictal aphasia
					120	TP1-8, AM1-4, HB1-2	2	Aura → postictal aphasia
3	20-25	F	Left Mesial Temporal	Trileptal, Vimpat	120	HH1-3, HB1-3, AM1-3, MI1-12, PI1-12, IA1-12, IM1-12, SA1-12, MA1-12	4	Abdominal aura.
					120	HH1-3, HB1-3, AM1-3, MI1-12, PI1-12, IA1-12, IM1-12, SA1-12, MA1-12	4	Abdominal & gustatory aura
4	30-35	F	Right Mesial Temporal	Keppra, Lacosamide	60	HH2-3, EM8-9, HH1-12, HB1-12, TT1-12, OF1-12	4	After stimulating AM3 with 50Hz, 4.6mA, 3s, patient felt "oozy"
					60	AM1-2, EM9-10, HH1-12, HB1-12, TT1-12, OF1-12	4	After stimulating AM4 with 5Hz, 7mA, 3 seconds, patient felt funny

3.6.2 Study Design

All seizure data was preprocessed using the NIC tools. For each seizure, we used MaTiLDA to apply persistent homology to signal coupling values from one-second epochs of iEEG data and to create data structures representing the resulting persistent homology values that were used as input into ML models to classify epochs as belonging to an ictal phase. Of the eight seizures selected, four seizures were analyzed in binary classification tasks to classify seizure onset from ictal 1 phase, and the remaining four seizures were analyzed in multiclass classification tasks to

Table 2: The sample size of each class is equal to the duration of the associated ictal phase.

Patient	Seizure	Duration of Ictal Phase			
		Onset	Ictal 1	Ictal 2	Ictal 3
1	1	15	33	-	-
	2	15	28	-	-
2	1	10	80	-	-
	2	5	115	-	-
3	1	10	15	5	90
	2	10	15	5	90
4	1	10	15	5	30
	2	10	15	5	30

classify ictal phases (seizure onset, ictal 1 phase, ictal 2 phase, ictal 3 phase, and ictal 4 phase). Each seizure was analyzed separately. The number of one-second epochs in each ictal phase of each seizure, equivalent to the sample sizes of each class label in each seizure-specific analysis, is provided in Table 2. Default parameters were used for all representations of persistent homology values and for all ML algorithms in the analysis of each of the eight seizures to show the baseline capabilities of MaTiLDA.

4. Results

To validate the use of the MaTiLDA interface, we aimed to classify ictal phases within a seizure for eight seizures from four refractory epilepsy patients, as described in section 2.6. For brevity, we present only the results from the analysis of persistent homology values in dimension 0.

Binary classifications were used to compare seizure onset and ictal 1 phase for the four seizures from patient one and patient two, as these seizures were limited to these two ictal phases. Due to space constraints, we review only the results for RF, SVM, and LR models using either the lifespan or persistence landscape methods. ROC curves can be seen for each of these models for all four seizures in Figure 7. Model performance varied across all seizures, and no ML algorithm or representation of persistent homology values outperformed others to consistently distinguish seizure onset and ictal 1 phase (Figure 8). This may be due to imbalanced class sizes (Table 2). For example, the 20% test partition of patient two's second seizure contained only one epoch from seizure onset, and only four epochs from seizure onset were included in the 80% training partition. For all combinations of ML algorithms and representations of persistent homology values, this one epoch was misclassified as belonging to ictal 1 phase, resulting in precision and recall values of 0 and an AUC of 0.50 but an accuracy

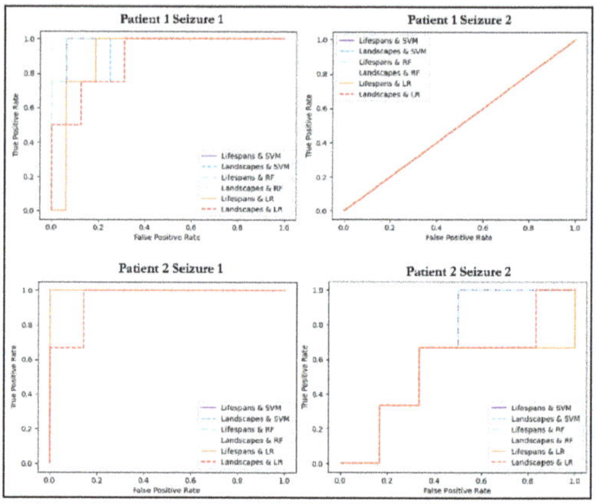

Figure 7: *ROC curves for each seizure from the binary classifications for seizures from patients one and two using lifespans or persistence landscapes in SVM, RF, or LR.*

Patient	Seizure	Algorithm	Featurization	Accuracy	Precision	Recall	AUC
1	1	RF	Lifespan	1	1	1	1
			Landscape	0.9	1	0.67	0.83
		SVM	Lifespan	0.9	1	0.67	0.83
			Landscape	0.8	0.67	0.67	0.76
		LR	Lifespan	0.8	1	0.33	0.67
			Landscape	0.8	1	0.33	0.67
	2	RF	Lifespan	0.67	0.5	0.33	0.58
			Landscape	0.67	0.5	0.33	0.58
		SVM	Lifespan	0.67	0.5	0.33	0.58
			Landscape	0.67	0.5	0.33	0.58
		LR	Lifespan	0.56	0	0	0.42
			Landscape	0.56	0.5	0.33	0.58
2	1	RF	Lifespan	0.95	0.8	1	0.97
			Landscape	1	1	1	1
		SVM	Lifespan	0.8	0	0	0.5
			Landscape	0.9	1	0.5	0.75
		LR	Lifespan	0.8	0	0	0.5
			Landscape	0.85	0.67	0.5	0.72
	2	RF	Lifespan	0.96	0	0	0.5
			Landscape	0.96	0	0	0.5
		SVM	Lifespan	0.96	0	0	0.5
			Landscape	0.96	0	0	0.5
		LR	Lifespan	0.96	0	0	0.5
			Landscape	0.96	0	0	0.5

Figure 8: *MaTiLDA's model performance for RF, SVM, and LR using lifespans or persistence landscapes for the four seizures from patients one and two.*

of 0.96. Increasing the number of samples from seizure onset may improve the ML models (as seen for patient two's first seizure). MaTiLDA's implementation of data augmentation, however, is still under development.

Multiclass classifications were used to classify seizure phases for each of the remaining four seizures from patients three and four which included multiple ictal phases (seizure onset, ictal 1 phase, ictal 2 phase, ictal 3 phase, and ictal 4 phase). Due to space constraints, we limit our results to the RF models using the lifespans and persistence landscapes (Figure 9). No algorithm or representation of persistent homology values consistently outperformed others to classify ictal phases, and there was high variation in model performance within and across seizures (Figure 9).

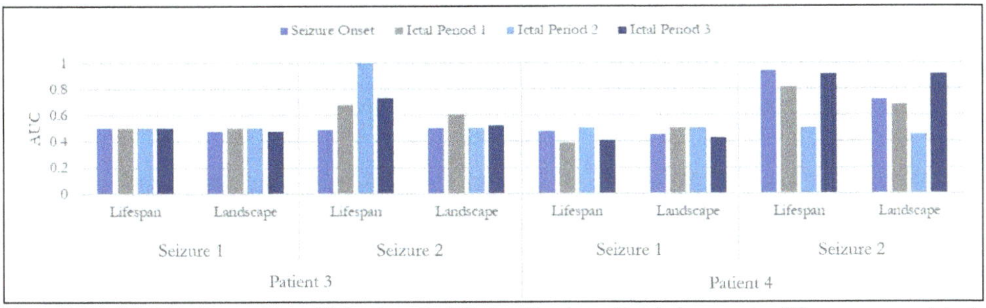

***Figure 9**: MaTiLDA's One-vs-Rest AUC values for RF classification of ictal phases using lifespans or persistent landscapes for each of the four seizures from patients three and four show high variation in model performance within and across seizures.*

5. Discussion & Conclusion

The results of this evaluation demonstrate that MaTiLDA is an effective tool for analyzing complex topological features, enabling the detection of changes in brain interactions during seizures. We have developed a novel pipeline that can classify brain states, such as the ictal phases of several seizures in this study, using various common TDA methods and ML algorithms. The MaTiLDA platform provides a robust, accessible, and reliable framework for applying TDA and ML algorithms to datasets from neurophysiological recordings to characterize brain interaction dynamics in neurological disorders. MaTiLDA enables the wider neuroscience research community, who have limited experience in both TDA and ML algorithm implementation to use ML and TDA algorithms to analyze the increasingly large volumes of brain activity recordings and characterize brain interaction dynamics. We believe that the MaTiLDA tool can be used in future research to investigate complex brain interaction patterns in neurological disorders such as epilepsy, and allow clinicians and researchers to characterize neurological disorders, understand pathophysiological mechanisms, and identify biomarkers for clinical diagnoses.

References

1. Bassett, D. S. & Sporns, O. Network neuroscience. Nat Neurosci 20, 353–364 (2017).

2. Menon, V. Large-scale brain networks and psychopathology: a unifying triple network model. Trends in Cognitive Sciences 15, 483–506 (2011).

3. Bullmore, E. T. & Bassett, D. S. Brain Graphs: Graphical Models of the Human Brain Connectome. Annual Review of Clinical Psychology 7, 113–140 (2011).

4. World Health Organization. Epilepsy. World Health Organization Epilepsy Fact Sheet https://www.who.int/news-room/fact-sheets/detail/epilepsy (2023).

5. Caputi, L., Pidnebesna, A. & Hlinka, J. Promises and pitfalls of topological data analysis for brain connectivity analysis. NeuroImage 238, 118245 (2021).

6. Grinenko, O. et al. A fingerprint of the epileptogenic zone in human epilepsies. Brain 141, 117–131 (2018).

7. Merelli, E., Piangerelli, M., Rucco, M. & Toller, D. A topological approach for multivariate time series characterization: the epileptic brain. in Proceedings of the 9th EAI International Conference on Bio-inspired Information and Communications Technologies (formerly BIONETICS) (ACM, 2016). doi:10.4108/eai.3-12-2015.2262525.

8. Zhang, J. et al. Characterizing Brain Network Dynamics using Persistent Homology in Patients with Refractory Epilepsy. AMIA Annu Symp Proc 2021, 1244–1253 (2022).

9. Wang, Y., Ombao, H. & Chung, M. K. Topological Data Analysis of Single-Trial Electroencephalographic Signals. Ann Appl Stat 12, 1506–1534 (2018).

10. Piangerelli, M., Rucco, M., Tesei, L. & Merelli, E. Topolnogical classifier for detecting the emergence of epileptic seizures. BMC Res Notes 11, 392 (2018).

11. Edelsbrunner, H. & Harer, J. Computational Topology: An Introduction. (American Mathematical Society, 2009).

12. Pun, C. S., Lee, S. X. & Xia, K. Persistent-homology-based machine learning: a survey and a comparative study. Artif Intell Rev 55, 5169–5213 (2022).

13. Otter, N., Porter, M. A., Tillmann, U., Grindrod, P. & Harrington, H. A. A roadmap for the computation of persistent homology. EPJ Data Sci. 6, 1–38 (2017).

14. Sizemore, A. E., Phillips-Cremins, J. E., Ghrist, R. & Bassett, D. S. The importance of the whole: Topological data analysis for the network neuroscientist. Netw Neurosci 3, 656–673 (2019).

15. Jayapandian, C. et al. A scalable neuroinformatics data flow for electrophysiological signals using MapReduce. Frontiers in Neuroinformatics 9, (2015).

16. Gershon, A. et al. Computing Functional Brain Connectivity in Neurological Disorders: Efficient Processing and Retrieval of Electrophysiological Signal Data. AMIA Jt Summits Transl Sci Proc 2019, 107–116 (2019).

17. Sahoo, S. S. et al. NeuroIntegrative Connectivity (NIC) Informatics Tool for Brain Functional Connectivity Network Analysis in Cohort Studies. AMIA Annu Symp Proc 2020, 1090–1099 (2021).

18. Maria, C., Boissonnat, J.-D., Glisse, M. & Yvinec, M. GUDHI library. GUDHI library https://project.inria.fr/gudhi/software/ (2014).

19. Barnes, D., Polanco, L. & Perea, J. A. A Comparative Study of Machine Learning Methods for Persistence Diagrams. Front Artif Intell 4, 681174 (2021).

20. Giusti, C., Ghrist, R. & Bassett, D. S. Two's company, three (or more) is a simplex: Algebraic-topological tools for understanding higher-order structure in neural data. J Comput Neurosci 41, 1–14 (2016).

21. Pijn, J. P. & Lopes da Silva, F. Propagation of Electrical Activity: Nonlinear Associations and Time Delays between EEG Signals. in Basic Mechanisms of the EEG (eds. Zschocke, S. & Speckmann, E.-J.) 41–61 (Birkhäuser Boston, 1993). doi:10.1007/978-1-4612-0341-4_4.

22. James, G., Witten, D., Hastie, T., Tibshirani, R. & Taylor, J. An Introduction to Statistical Learning: with Applications in Python. (Springer International Publishing, 2023). doi:10.1007/978-3-031-38747-0.

23. Zou, K. H., O'Malley, A. J. & Mauri, L. Receiver-Operating Characteristic Analysis for Evaluating Diagnostic Tests and Predictive Models. Circulation 115, 654–657 (2007).

24. Bendich, P., Marron, J. S., Miller, E., Pieloch, A. & Skwerer, S. Persistent Homology Analysis of Brain Artery Trees. Ann Appl Stat 10, 198–218 (2016).

25. Bubenik, P. The Persistence Landscape and Some of Its Properties. 15, 97–117 (2020).

26. Chazal, F., Fasy, B. T., Lecci, F., Rinaldo, A. & Wasserman, L. Stochastic convergence of persistence landscapes and silhouettes. Journal of Computational Geometry 6, 140–161 (2015).

27. Adams, H. et al. Persistence Images: A Stable Vector Representation of Persistent Homology. Journal of Machine Learning Research 18, 1–35 (2017).

28. Rucco, M., Castiglione, F., Merelli, E. & Pettini, M. Characterisation of the idiotypic immune network through persistent entropy. in Springer Proceedings in Complexity (Springer, Cham, 2015). doi:https://doi.org/10.1007/978-3-319-29228-1_11.

29. Mitchell, T. M. Machine Learning. (McGraw-Hill, 1997).

30. Bartolomei, F. et al. Defining epileptogenic networks: Contribution of SEEG and signal analysis. Epilepsia 58, 1131–1147 (2017).

31. Wu, S. et al. Role of ictal baseline shifts and ictal high-frequency oscillations in stereo-electroencephalography analysis of mesial temporal lobe seizures. Epilepsia 55, 690–698 (2014).

32. Diehl, B. & Lüders, H. O. Temporal Lobe Epilepsy: When Are Invasive Recordings Needed? Epilepsia 41, S61–S74 (2000).

Zoish: A Novel Feature Selection Approach Leveraging Shapley Additive Values for Machine Learning Applications in Healthcare

Hossein Javedani Sadaei, Salvatore Loguercio, Mahdi Shafiei Neyestanak, and Ali Torkamani †

Scripps Research Translational Institute, and
Department of Integrative Structural and Computational Biology
Scripps Research, La Jolla, CA 92037, USA
E-mail: hjavedani@scripps.edu
E-mail: loguerci@scripps.edu
E-mail: mshafiei@scripps.edu
† *E-mail: atorkama@scripps.edu*
www.scripps.edu

Daria Prilutsky

Takeda Development Center Americas, Inc., Cambridge, MA, 02139, USA
E-mail: daria.prilutsky@takeda.com
www.takeda.com

In the intricate landscape of healthcare analytics, effective feature selection is a prerequisite for generating robust predictive models, especially given the common challenges of sample sizes and potential biases. Zoish uniquely addresses these issues by employing Shapley additive values—an idea rooted in cooperative game theory—to enable both transparent and automated feature selection. Unlike existing tools, Zoish is versatile, designed to seamlessly integrate with an array of machine learning libraries including scikit-learn, XGBoost, CatBoost, and imbalanced-learn.

The distinct advantage of Zoish lies in its dual algorithmic approach for calculating Shapley values, allowing it to efficiently manage both large and small datasets. This adaptability renders it exceptionally suitable for a wide spectrum of healthcare-related tasks. The tool also places a strong emphasis on interpretability, providing comprehensive visualizations for analyzed features. Its customizable settings offer users fine-grained control over feature selection, thus optimizing for specific predictive objectives.

This manuscript elucidates the mathematical framework underpinning Zoish and how it uniquely combines local and global feature selection into a single, streamlined process. To validate Zoish's efficiency and adaptability, we present case studies in breast cancer prediction and Montreal Cognitive Assessment (MoCA) prediction in Parkinson's disease, along with evaluations on 300 synthetic datasets. These applications underscore Zoish's unparalleled performance in diverse healthcare contexts and against its counterparts.

Keywords: Feature Selectors, Zoish, SHapley Additive exPlanations

1. Introduction

Healthcare datasets, despite being typically sparse and heterogeneous, are a treasure trove of rich information. However, their high-dimensionality, often paired with smaller sizes, presents obstacles to building predictive models, with overfitting and extensive training time being common concerns.[1–3] Feature selection becomes an essential strategy in this context, aimed at pruning redundant or less important features. This helps to minimize information loss, enhance model interpretability, and curtail computational demands.

Although traditional feature selection methods, grounded in statistical concepts like correlation analysis or chi-square tests, are widely used, they tend to fall short in offering detailed insights into feature importance. This shortcoming, along with the manual effort required, can lead to a time-intensive cycle of feature selection and performance evaluation, thus requiring expert intervention.[4,5]

Our proposed feature selection tool, Zoish, aims to overcome these limitations by utilizing the mathematical framework of additive Shapley values. Originating from cooperative game theory, Shapley values offer detailed understanding of feature importance,[6] thereby enhancing both local (instance-level) and global interpretability. Moreover, the integration of Zoish with our scalable hyperparameter optimization package, Lohrasb,[7] facilitates building models with optimal feature sets, all the while maintaining an industry-ready, user-friendly design.

The structure of the paper is as follows. The first section sheds light on the core concepts of additive Shapley values and delves into the mathematical principles vital to Zoish. Subsequent to this, a section introducing a user guide for Zoish is presented. Lastly, we demonstrate the adaptability of Zoish through a variety of use-cases, experiments on large synthetic datasets, and a closing discussion.

2. Theoretical Foundations of Zoish

Our exploration into Zoish begins with the foundation of its theoretical structure, built upon Shapley additive values. First proposed in the field of cooperative game theory, Shapley additive values have proven to be a potent tool for understanding the contribution of each feature to a prediction made by a machine learning model. Simply put, the Shapley value of a feature represents the average marginal contribution of that feature, factored across all possible combinations of features.

2.1. *Shapley Additive Values and Feature Selection: A Game Theoretical Approach*

In the realm of cooperative game theory, the Shapley value denotes each player's payoff based on their marginal contribution across all possible coalitions. In machine learning, 'players' correspond to the features and 'game' to the prediction task.[6]

An additive cooperative game assumes that the value of any coalition equals the sum of its members' independent values. This idea leads us to Shapley additive values, where the Shapley value of a feature equals its average marginal contribution across all feature subsets.

Mathematically, the Shapley value for a feature i in an additive game is:

$$\phi_i(v) = \sum_{S \subseteq N \setminus \{i\}} \frac{|S|!(|N| - |S| - 1)!}{|N|!}(v(S \cup \{i\}) - v(S)) = v(\{i\})$$

In this formula, $|N|$ denotes the total number of features, S a subset of features excluding feature i, and $|S|$ the number of features in subset S. The terms $|S|!(|N| - |S| - 1)!$ and $|N|!$ calculate the number of possible permutations of features. The expression $v(S \cup \{i\}) - v(S)$ computes the marginal contribution of feature i when added to subset S.

In feature selection, the Shapley value presents a way to distribute the model's prediction among the features, based on their marginal contribution.[8,9] Shapley Additive exPlanations (SHAP) values, maintaining additivity, provide a unified measure of feature importance, attributing the difference between the model's actual and expected output to each influencing feature.[10] High Shapley or SHAP values indicate significant feature importance, while values near zero suggest negligible predictive power.[11] This correlation aids in reducing data dimensionality and enhances model interpretability, marking a significant stride in feature selection.

2.2. *Properties of Shapley Additive Values*

The Shapley additive values satisfy a number of properties that make them particularly useful for interpreting machine learning models:

- **Efficiency**: The sum of the Shapley values of all features is equal to the difference between the prediction for an instance and the average prediction over all instances.
- **Symmetry**: If two features contribute equally to all possible combinations of features, they have the same Shapley value.
- **Additivity**: Given two games (or in our context, two models), the Shapley value of the combined game is the sum of the Shapley values of the individual games.
- **Nullity (Dummy)**: If a feature does not improve the prediction for any combination of features, its Shapley value is zero.

2.2.1. *Proof of Nullity (Dummy)*

The Nullity (Dummy) property states that if a feature does not change the prediction model, i.e., its contribution is always zero, then its Shapley value is also zero. Let f be the prediction model and d be such a dummy feature.

According to the definition of Shapley values, the Shapley value of a feature is the average of its marginal contributions across all possible subsets of features. Therefore, the Shapley value $s_f(d)$ for the dummy feature d is:

$$s_f(d) = \frac{1}{M} \sum_{S \subseteq N \setminus \{d\}} |S|!(|N| - |S| - 1)!(f(S \cup \{d\}) - f(S))$$

In the above formula, M is the total number of possible subsets of N that can be formed when the dummy feature d is excluded. The term $|S|!(|N| - |S| - 1)!$ is the number of permutations of N in which the dummy feature d and the features in subset S appear together, and

$f(S \cup \{d\})$ and $f(S)$ are the values of the game f when the dummy feature d is added to subset S and when it is not, respectively.

Since d is a dummy feature, adding it to any subset S does not change the value of the game f. Thus, we have $f(S \cup \{d\}) = f(S)$ for all $S \subseteq N \setminus \{d\}$, which simplifies the above formula to:

$$s_f(d) = \frac{1}{M} \sum_{S \subseteq N \setminus \{d\}} |S|!(|N| - |S| - 1)!(0)$$

$$s_f(d) = 0$$

This confirms that the Shapley value of a feature that does not contribute to the prediction model is indeed zero, thus proving the Nullity (Dummy) property.[9]

2.3. *Leveraging Shapley Additive Values for Efficient Feature Selection*

Shapley additive values have become increasingly prominent in feature selection for machine learning due to their robustness, efficiency, and power. Verdinelli et al. examined the explainability of machine learning models, focusing on methods such as LOCO and Shapley Values for assessing feature importance. Although their research indicated that Shapley Values do not eliminate feature correlation, they proposed new, statistically sound axioms for measuring feature importance.[12] In a separate study, Karczmarz et al. compared Shapley and Banzhaf values in the context of explaining tree ensemble models. They found Banzhaf values to be more intuitive, efficient, and numerically robust, and introduced faster algorithms for both methods to improve computational efficiency.[13] The SHAP (SHapley Additive exPlanation) library serves as a prime example of effectively leveraging the additivity and efficiency inherent in Shapley values.[14] A fundamental advantage of Shapley values lies in their additivity, which enables fast and efficient computation, especially in the context of tree-based models. The SHAP and FastTreeShap libraries employ Tree SHAP, a highly efficient and accurate algorithm designed for tree ensembles.[14,15] Given the innate additivity of ensemble tree models, which amalgamate multiple decision trees, this characteristic ensures swift and precise computation of Shapley values.[6] The efficacy of Shapley values is further underscored by their intrinsic efficiency. This is manifested in the fact that the sum of the Shapley values for all features equals the difference between the prediction for a specific instance and the average prediction across all instances. This aspect permits a meaningful distribution of the "credit" for a prediction across features, hence illuminating their relative importance. The Zoish package,[16] designed to optimize feature selection, taps into the beneficial properties of Shapley values. It employs the Nullity (or Dummy) property to eliminate features with Shapley values close to or exactly zero, indicating their minimal predictive relevance. To assist users in setting the cut-off level, two methods are offered: one involves setting an internal parameter called *threshold*, while the other entails defining the number of desired features to retain in the model. By removing these non-influential features, the package enables dimensionality reduction of the model without sacrificing prediction quality. The symmetry property of Shapley values is also exploited by Zoish. This property mandates that if two features contribute equally to all possible subsets of other features, they must have identical Shapley values. By identifying and discarding these redundant features, Zoish facilitates the construction of models that are simpler and more interpretable, with no compromise on predictive power. By utilizing the SHAP and FastTreeShap libraries to incorporate Shapley additive values, Zoish implicitly benefits from its advantages, including the mathematical robustness and beneficial properties of Shapley values. Therefore, these libraries and the Zoish package present themselves as potent instruments for feature selection, spanning a wide range of machine learning tasks.

3. Feature Selection Approaches

Zoish is a versatile package designed to enhance the evaluation of feature importance and improve the overall performance of machine learning models.[16] While Zoish can function effectively as a standalone tool for feature selection, it is engineered to be highly extensible and can seamlessly integrate with hyperparameter optimization packages to further refine its capabilities. One such potent integration is with the Lohrasb package,[7] which provides advanced tuning methods to optimize the feature selection process. However, it's worth noting that users are not confined to using Lohrasb; Zoish's flexible architecture allows for easy integration with other hyperparameter optimization tools as well.

3.1. *Optimization and Flexibility in Zoish*

Zoish's integration with Lohrasb serves a dual purpose: it not only optimizes the tree-based estimator used for feature selection but also offers a choice of hyperparameter tuning methods, including Optuna, GridSearchCV,[17] RandomizedSearchCV,[18] OptunaSearchCV, tune-sklearn,[19] and Ray's Tune.[20] This optimization is crucial for enhancing Zoish's feature selection capabilities, as represented in Fig 1. However, the use of Lohrasb is optional, giving users the freedom to employ other tree-based estimators or hyperparameter tuning engines. Even without hyperparameter optimization, Zoish maintains its core functionality, allowing for a balance between efficiency and interpretability. The importance of hyperparameter optimization for feature selection is further elaborated in Section 6. Therefore, while Lohrasb's role is significant for optimal performance, users have the flexibility to choose an approach that best suits their specific needs.

3.2. *Workflow explanation*

Within a machine learning pipeline, Zoish functions as a feature selection component. The pipeline commences by cleaning and splitting the original dataset into training and validation subsets. A tree-based estimator, which is compatible with Zoish, is trained on the training subset. If hyperparameter tuning is applied, tools such as Lohrasb optimize the estimator against a specific metric, as shown in Fig 1.

Once the estimator is optimized, it becomes an input to Zoish along with a set of parameters, such as cross-validation settings, Shapley value calculation algorithms, and feature importance thresholds. Zoish computes Shapley values via either the SHAP library for smaller datasets, due to its exhaustive computational approach, or FastTreeShap for larger datasets, owing to its computational efficiency.

Based on the calculated Shapley values, Zoish automatically selects the highest-ranking features. The training set is then narrowed down to these selected features. Subsequently, these refined training and validation sets are channeled to the next steps in the pipeline, which usually involve fitting another predictive model.

To ensure robustness in feature selection, Zoish employs multiple rounds of cross-validation on the same training set, regulated by a parameter named *n_iter*.

Fig. 1. Zoish workflow

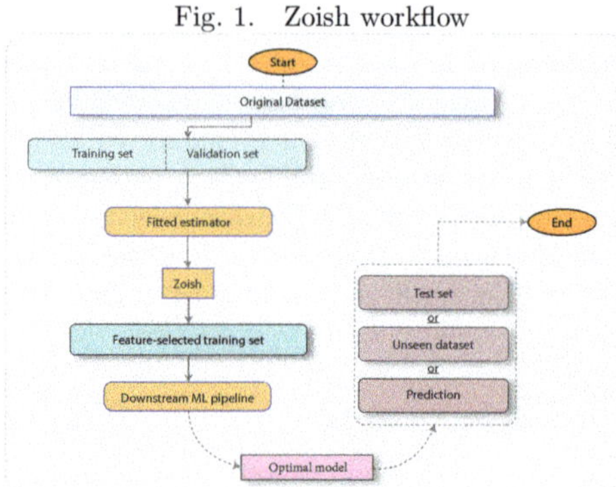

Documentation and code examples elucidating these operational details can be found in the Zoish repository.

4. Source Code, installation and usage example

The public repository of Zoish is available on GitHub alongside examples for end users is `https://github.com/TorkamaniLab/zoish`. Zoish package is available on PyPI and can be installed with pip:

```
pip install zoish
```

A straightforward example demonstrates how Zoish can be effectively combined with hyperparameter optimizers. Both this example and the comprehensive documentation in the repository highlight the package's flexibility and adaptability across various scenarios.

```
import xgboost as xgb
from sklearn.model_selection import KFold, GridSearchCV
from zoish.feature_selectors.shap_selectors import ShapFeatureSelector
from sklearn.pipeline import Pipeline
from sklearn.linear_model import LogisticRegression
```

```
X_train, X_test, y_train, y_test = ...  # Your dataset here

grid = GridSearchCV(xgb.XGBClassifier(), {'n_estimators': [100, 150], 'max_depth':\
[6, 10], 'gamma': [0.5, 1.0]}, cv=5, n_jobs=-1, scoring='accuracy').\
fit(X_train, y_train)

shap_selector = ShapFeatureSelector(grid.best_estimator_, \
num_features=15, cv=KFold(10), n_iter=5, direction="maximum", \
scoring="accuracy", algorithm='auto', use_faster_algorithm=True)

pipeline = Pipeline(steps=[("s", shap_selector), \
("m", LogisticRegression())]).fit(X_train, y_train)
```

5. Use cases and applications

5.1. *Use case 1: Application to UCI breast cancer dataset - comparison with related XAI work*

To demonstrate the value of the Zoish feature selector in a real use case from the biomedical domain, we applied it to the openly available breast cancer dataset from the UCI Archive,[21] and compared the results with a recent study evaluating different feature importance measures for the same dataset.[22]

The UCI dataset includes benign and malignant samples from 569 patients, 212 with cancer and 157 with fibrocystic breast masses. Each sample includes thirty features - ten real valued features for each cell nucleus (*radius, texture, perimeter, area, smoothness, compactness, concavity, concave points, symmetry, fractal dimension*) each reported as Mean, Standard Error (SE) and Worst.[23] As the classes in this dataset are almost linearly separable, classification per se is not a difficult task; however, the most important features generally differ depending on the technique used.[24] To further investigate this aspect, Saarela et al.[22] compared different feature importance measures using both linear (logistic regression) and non-linear (random forest) models and local interpretable model-agnostic explanations for the same dataset. In Fig. 2 we show the top 20 important features for the UCI Breast Cancer dataset computed with Zoish by training a XGboost classifier over ten folds of cross validation. The AUC for the trained classifier was 0.96, similar to the mean AUC reported in[22] (0.99+-002). Overall, the most important features in Zoish/XGboost agree well with the set of nine statistically significant features for both RF and LR reported in[22] – where, for each method, significance was computed through a procedure based on permutation tests – i.e. by shuffling class labels in the training data over hundreds of runs. Seven out of nine features deemed statistically significant in[22] were found in Fig. 2 (*Mean concave points, Worst concave points, Worst Area, Worst Radius, Worst Perimeter, Mean Concavity, Mean Area*), with five of the most significant features near the top of the list (Table 1). Only one feature was labeled as not significant by both RF and LR (*Worst Compactness*) and such feature has consistently a zero Shap value in Zoish/XGboost (not shown).

Fig. 2. Shap summary plot of the top 20 important features for the UCI Breast Cancer dataset - computed with Zoish by training a XGboost classifier.

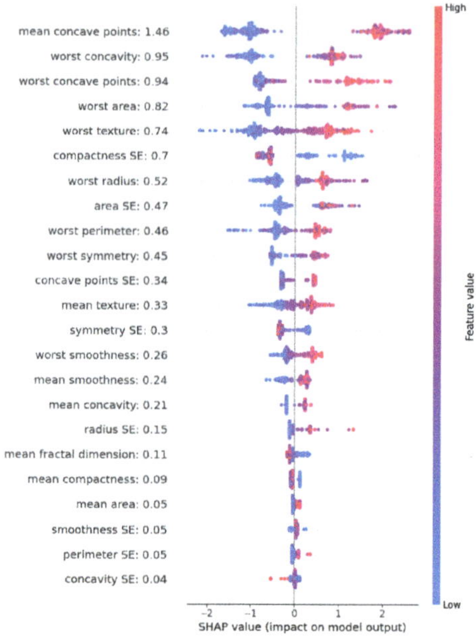

It is worth noting that certain features which rank very high in Zoish/XGboost (*Worst concavity, Worst texture, Compactness SE*) appeared not to be significant in RF classification, hence not reported in the set of nine common, statistically significant features for breast cancer classification in the UCI dataset. Why did Zoish / XGboost select them up then? A very interesting hint comes from the analysis of local importance measures for a specific set of observations in the UCI dataset. Again in,[22] LIME (local interpretable model-agnostic explanations,[25]) was used to estimate local importances for the four most interesting observations, (i.e. correctly classified as benign with highest probability, correctly classified as malignant with highest probability, misclassified as benign with highest probability, and misclassified as malignant with highest probability). Strikingly, the features ranking high in Zoish/XGboost but absent in RF were also important features in LIME, especially for the observations misclassified as benign (false negatives), which are critical for medical purposes (Table 1). The final recommendation in[22] was to combine several explanation techniques in order to provide more reliable and trustworthy results, but this advice can often be impractical. Conversely, The Zoish/XGboost feature selector appears to select relevant features both at the global and local level, adding more detailed explanations of feature importance (i.e., not just the magnitude but also the direction of change), while being fast and straightforward to use.

Table 1. Comparison of top features in the UCI Breast Cancer dataset

Zoish features	Significant features in LR and RF	LIME - Correctly classified benign (RF)	LIME - Misclassified benign (RF)
concave points 1	concave points 3	area 3	perimeter 3
concavity 3	area 3	perimeter 3	area 3
concave points 3	concave points 1	radius 3	radius 3
area 3	area 1	concave points 3	texture 3
texture 3	perimeter 3	texture 3	concavity 3
compactness 2	radius 3	concave points 1	area 2
radius 3	concavity 1	concavity 3	smoothness 3
area 2	perimeter 1	area 2	area 1
perimeter 3	radius 1	texture 1	concave points 1
symmetry 3	area 2		
concave points 2	concavity 3		
texture 1	texture 3		
symmetry 2	texture 1		
smoothness 3	compactness 3		
smoothness 1	radius 2		
concavity 1	perimeter 2		
radius 2	compactness 1		
fractal dimension 1	smoothness 3		
compactness 1	symmetry 3		
area 1	fractal dimension 3		

This Table is about comparison of top features in the UCI Breast Cancer dataset computed by Zoish/XGboost vs. Random Forest / Logistic Regression / LIME. For all features, 1=Mean, 2=Standard Error, 3=Worst. Top 20 features in Column 2 are ranked based on permutation p-value for RF. In red are features found in LIME but not considered significant in RF/LR.

5.2. *Use case 2: Predict short-term PD progression status using the Montreal Cognitive Assessment (MoCA)*

Our model, Zoish, was put to another practical test where we aimed to predict short-term PD progression status using the Montreal Cognitive Assessment (MoCA) total scores for patients in baseline. MoCA was developed as a tool to screen patients who present with mild cognitive complaints and usually perform in the normal range on the MMSE (Mini-Mental State Examination).[26] For this prediction task, we utilized the AMP-PD dataset, which is a comprehensive collection of data from various sources, including clinical information, genetic data, imaging data, and other biomarkers from individuals with Parkinson's disease. The dataset consists of eight cohorts, making it a large and harmonized resource. Access to the data was obtained under the AMP-PD Data Use Agreement, and the information was retrieved from the website: `https://amp-pd.org/`. Our prediction model incorporated several essential features from the datasets, such as "family history," genetic information (PRS), "medical history," "smoking and alcohol history," and demographic information of the participants from the eight cohorts. After fitting the model, we evaluated its performance using the coefficient of determination, commonly known as R-squared, and achieved an R-squared value of 23.6 percent on the test dataset. Additionally, we calculated the Mean Squared Error (MSE) of the model to be 2.49 for the test dataset. Furthermore, the Mean Absolute Error (MAE) was found to be 13.04. The MAE represents the average absolute difference between the predicted values and the actual total score values in the test dataset. List of selected features by Zoish can be seen in Fig. 3

In order to draw a comparison with another prevalent feature selector, specifically, **SelectFromModel** from the sklearn library, we applied it to the same dataset under identical

Fig. 3. List of selected features and their importance by Zoish

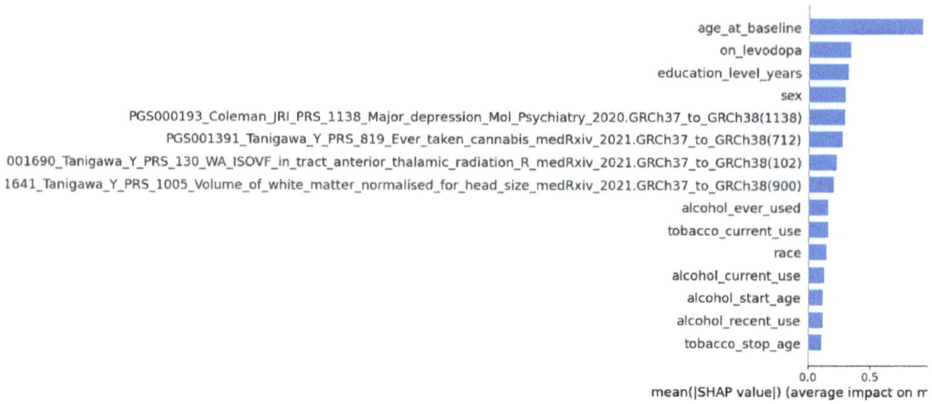

conditions. This application yielded a Mean Absolute Error (MAE) of 15.91, a Mean Squared Error (MSE) of 2.69, and an R-squared value of 0.12. The features selected by this approach are depicted in Fig. 4.

Fig. 4. List of selected features and their importance by SelectFromModel

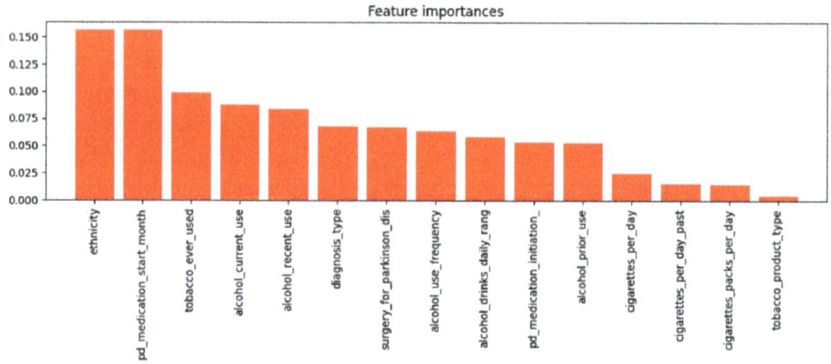

As observed, Zoish not only outperforms its counterparts in terms of prediction accuracy, but it also excels in the selection of meaningful features. Notably, the Polygenic Risk Scores (PRSs) selected by Zoish have demonstrated substantial relevance to the Montreal Cognitive Assessment (MoCA). A prime example among these is *PGS001641*, which is renowned for its strong correlation with the volume of white matter, normalised for head size. This particular PRS underscores the genetic predisposition towards the volume of white matter, a crucial neuroimaging measurement that relates directly to cognitive functions evaluated in the MoCA test. Therefore, the selection of this PRS by Zoish validates its capability in discerning features with profound implications for cognitive assessment.

6. Evaluations and Performance Analysis

To offer a comprehensive evaluation of Zoish, we performed an array of tests ranging from comparative analyses to hyperparameter optimization and scalability assessments.

Comparative Analysis: We initiated our evaluation with a rigorous comparison involving 300 synthetic datasets tailored to mirror the complexities of healthcare data. These datasets span regression, binary classification, and multi-label classification tasks. Zoish was compared against six established feature selection techniques from Scikit-learn under identical conditions.Our findings suggested that Zoish surpassed other selectors in 77% of regression problems, while in multi-label classification and binary classification tasks, Zoish outperformed in 53% and 57% of the cases, respectively (refer to Table 2).

Table 2. Performance comparison of Zoish with other feature selectors

Selector	Regression	Binary Classification	Multi-label Classification
Zoish	**77%**	**57%**	**53%**
VarianceThreshold	2%	3%	6%
SelectKBest	2%	3%	6%
SelectPercentile	2%	2%	2%
RFE	5%	10%	6%
RFECV	7%	10%	11%
SelectFromModel	5%	15%	16%

Hyperparameter Optimization: While Zoish itself is powerful, coupling it with a hyperparameter optimization tool like Lohrasb significantly improves performance. We performed 100 runs comparing Zoish's efficacy with and without Lohrasb, and found marked improvements when paired with Lohrasb (see Fig. 5).

Scalability: Our most recent update introduces a faster algorithm for Shapley value computation, making Zoish efficient on large datasets. In our trials, Zoish selected 500 features from a dataset with 10,000 samples in under 2 minutes on a machine with a 2.3 GHz Quad-Core Intel Core i7 processor and 32 GB RAM.

Fig. 5. The importance of Hyperparameter Optimization for Better Feature Selection

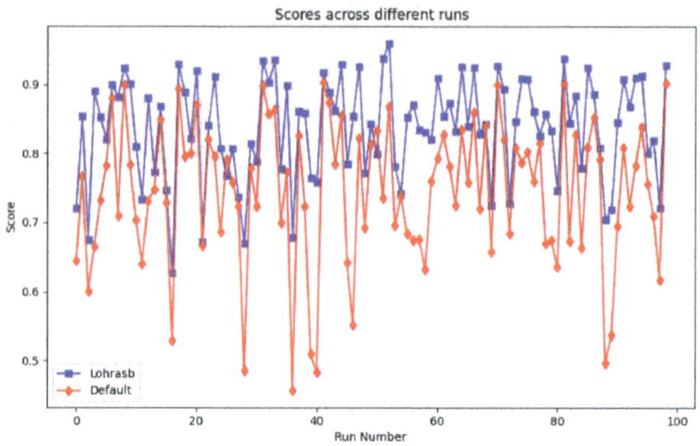

All the code for our tests is available in the public repository, allowing for independent verification and further exploration of Zoish's capabilities.

7. Discussion and Limitations

This paper introduces Zoish, a feature selection tool built on cooperative game theory principles.[16] Zoish has gained traction in the community, as evidenced by a significant number of downloads from pip-trends (`https://piptrends.com/package/zoish`). The tool specializes in optimizing predictive models, particularly in the healthcare sector, and leverages Shapley additive values for a comprehensive view of feature importance at both local and global scales.[10] Through its Nullity property, Zoish effectively minimizes model complexity by omitting features with negligible Shapley values, thereby retaining model performance.[3] The tool is further enriched by integration with the Lohrasb package, which aids in achieving optimal estimators and hyperparameter settings.[7]

While Zoish's capabilities are robust, some limitations are noteworthy. Firstly, its computational efficiency may be compromised when dealing with exceptionally large datasets. Secondly, the Shapley values employed assume feature independence and local linearity—assumptions that may not be fully met in complex applications like healthcare. These limitations are partially mitigated by Zoish's tree-based modeling approach, which is robust to feature correlation and can capture non-linear relationships.[6]

The flexibility and interpretability of Zoish make it a promising tool for future applications in other high-dimensional data fields, including finance and e-commerce. Additional functionalities could be incorporated to broaden its applicability further.

Future work will focus on extending Zoish's utility to various high-dimensional domains and incorporating more algorithms and tools for an even more robust feature selection process. We have conducted extensive tests on synthetic datasets mimicking real-world complexities in healthcare, which are detailed in Section 6. These tests demonstrate Zoish's reliability and adaptability, even under challenging conditions.

8. Author contributions statement

Author contributions include; conceptualization (HJS, AT), methodology and software (HJS), validation and data curation (HJS, SL, MSN, AT, and DP), writing (HJS, SL, MSN, and AT), and funding (AT).

9. Acknowledgments

This work is supported by UL1RR025774 and R01HG010881. Also We would like to express our gratitude to Dua, D. and Graff, C.[21] for generously providing us with the Breast Cancer Data used in this study. Without their contribution, this research would not have been possible.

References

1. I. M. Johnstone and D. M. Titterington, Statistical challenges of high-dimensional data (2009).
2. M. Muja and D. G. Lowe, Scalable nearest neighbor algorithms for high dimensional data, *IEEE transactions on pattern analysis and machine intelligence* **36**, 2227 (2014).
3. H. J. Sadaei, A. Cordova-Palomera, J. Lee, J. Padmanabhan, S.-F. Chen, N. E. Wineinger, R. Dias, D. Prilutsky, S. Szalma and A. Torkamani, Genetically-informed prediction of short-term parkinson's disease progression, *npj Parkinson's Disease* **8**, p. 143 (2022).
4. J. Tang, S. Alelyani and H. Liu, Feature selection for classification: A review, *Data classification: Algorithms and applications* , p. 37 (2014).
5. M. A. Hall, Correlation-based feature selection for machine learning (1999).
6. D. Fryer, I. Strümke and H. Nguyen, Shapley values for feature selection: The good, the bad, and the axioms, *Ieee Access* **9**, 144352 (2021).
7. H. Javedani Sadaei and A. Torkamani, Lohrasb: A scalable estimator optimization tool compatible with scikit-learn APIs (April 2023).
8. T. P. Michalak, K. V. Aadithya, P. L. Szczepański, N. R. Jennings and R. Narayanam, Efficient computation of the shapley value for game-theoretic network centrality, *Journal of Artificial Intelligence Research* **46**, 607 (2013).
9. L. S. Shapley, A value for n-person games, *Contributions to the Theory of Games* **2**, 307 (1953).
10. S. M. Lundberg and S.-I. Lee, A unified approach to interpreting model predictions, *Advances in neural information processing systems* **30** (2017).
11. C. Molnar, *Interpretable machine learning* (Lulu.com, 2020).
12. I. Verdinelli and L. Wasserman, Feature importance: A closer look at shapley values and loco, *arXiv preprint arXiv:2303.05981* (2023).
13. A. Karczmarz, A. Mukherjee, P. Sankowski and P. Wygocki, Improved feature importance computations for tree models: Shapley vs. banzhaf, *arXiv preprint arXiv:2108.04126* (2021).
14. S. Lundberg, S.-I. Lee *et al.*, SHAP (SHapley Additive exPlanations) `https://github.com/slundberg/shap`, (2023), Accessed: 2023-09-25.
15. J. Yang, Fast treeshap: Accelerating shap value computation for trees, *arXiv preprint arXiv:2109.09847* (2021).
16. H. Javedani Sadaei and A. Torkamani, Zoish: Automated feature selectoion tools (July 2023), If you use this software, please cite it as below.
17. S. Varma and R. Simon, Parameter estimation using grid search with cross-validation, *XGBoost: A Scalable Tree Boosting System* **12**, p. 48 (2012).
18. J. Bergstra and Y. Bengio, Random search for hyper-parameter optimization, *Journal of Machine Learning Research* **13**, 281 (2012).
19. R. Liaw, E. Liang, R. Nishihara, P. Moritz, J. E. Gonzalez and I. Stoica, Tune: A research platform for distributed model selection and training in pytorch, *arXiv preprint arXiv:1807.05118* (2018).
20. R. Liaw, E. Liang, R. Nishihara, P. Moritz, J. E. Gonzalez and I. Stoica, Tune: A research platform for distributed model selection and training in pytorch, *arXiv preprint arXiv:2002.02613* (2020).
21. D. Dua and C. Graff, UCI machine learning repository (2017).
22. S. J. Mirka Saarela, Comparison of feature importance measures as explanations for classification models, *SN Applied Sciences* **3** (2021).
23. O. M. WH Wolberg, WN Street, Machine learning techniques to diagnose breast cancer from image-processed nuclear features of fine needle aspirates, *Cancer Letters* **77**, p. 163–171 (1994).
24. A. S. E Aličković, Breast cancer diagnosis using ga feature selection and rotation forest, *Neural Computation Applications* **28**, p. 753–763 (2017).

25. M. T. Ribeiro, S. Singh and C. Guestrin, "why should i trust you?": Explaining the predictions of any classifier (2016).
26. Z. S. Nasreddine, N. A. Phillips, V. Bédirian, S. Charbonneau, V. Whitehead, I. Collin, J. L. Cummings and H. Chertkow, The montreal cognitive assessment, moca: a brief screening tool for mild cognitive impairment, *Journal of the American Geriatrics Society* **53**, 695 (2005).

SynTwin: A graph-based approach for predicting clinical outcomes using digital twins derived from synthetic patients

Jason H. Moore[1,2], Xi Li[1], Jui-Hsuan Chang[1], Nicholas P. Tatonetti[1,2], Dan Theodorescu[2], Yong Chen[4], Folkert W. Asselbergs[3], Mythreye Venkatesan[1], Zhiping Paul Wang[1]

[1]Department of Computational Biomedicine, Cedars-Sinai Medical Center, West Hollywood, CA
[2]Cedars-Sinai Cancer, Cedars-Sinai Medical Center, Los Angeles, CA
[3]Department of Cardiology, Amsterdam University Medical Center, Amsterdam, The Netherlands
[4]Department of Biostatistics, Epidemiology, and Informatics, University of Pennsylvania, Philadelphia, PA
Email: jason.moore@csmc.edu

The concept of a digital twin came from the engineering, industrial, and manufacturing domains to create virtual objects or machines that could inform the design and development of real objects. This idea is appealing for precision medicine where digital twins of patients could help inform healthcare decisions. We have developed a methodology for generating and using digital twins for clinical outcome prediction. We introduce a new approach that combines *synthetic* data and network science to create digital *twins* (i.e. SynTwin) for precision medicine. First, our approach starts by estimating the distance between all subjects based on their available features. Second, the distances are used to construct a network with subjects as nodes and edges defining distance less than the percolation threshold. Third, communities or cliques of subjects are defined. Fourth, a large population of synthetic patients are generated using a synthetic data generation algorithm that models the correlation structure of the data to generate new patients. Fifth, digital twins are selected from the synthetic patient population that are within a given distance defining a subject community in the network. Finally, we compare and contrast community-based prediction of clinical endpoints using real subjects, digital twins, or both within and outside of the community. Key to this approach are the digital twins defined using patient similarity that represent hypothetical unobserved patients with patterns similar to nearby real patients as defined by network distance and community structure. We apply our SynTwin approach to predicting mortality in a population-based cancer registry (n=87,674) from the Surveillance, Epidemiology, and End Results (SEER) program from the National Cancer Institute (USA). Our results demonstrate that nearest network neighbor prediction of mortality in this study is significantly improved with digital twins (AUROC=0.864, 95% CI=0.857-0.872) over just using real data alone (AUROC=0.791, 95% CI=0.781-0.800). These results suggest a network-based digital twin strategy using synthetic patients may add value to precision medicine efforts.

Keywords: Digital twins; Precision medicine; Artificial intelligence; Synthetic data.

1. Introduction to Digital Twins

The concept of a digital twin came from the engineering, industrial, and manufacturing domains and refers to the creation of virtual objects or machines that can inform the design and development of real objects (Grieves & Vickers 2017). The promise of this approach in manufacturing is to reduce costs, improve efficiency, reduce waste, and minimize variability among products (Attaran et al. 2023). This is accomplished by enumerating and evaluating design parameters of the digital twin of a physical product with some measurable outcome that can then be applied to manufacturing. Use cases in industry include product design, process design and optimization, supply chain management, preventive system maintenance, farm management, weather modeling, soil management, facility and operations design, construction, etc. (Attaran & Celik 2023). Consider the use case of monitoring weed pressure and crop growth (Verdouw et al. 2021). Data on crops, weeds, weather, and soil conditions are collected from crop sensors. These data are used to build a digital twin of the crops where parameters for a weeding machine can be enumerated and evaluated. Optimized parameters from the digital twin can then be put into practice for weed management with benefits including crop weight, size, and yield.

The successful use of digital twins in industry has opened the door for their use in medicine and healthcare where they represent virtual or simulated patients that could be used to inform health outcomes or treatment decision for real patients (Acosta et al. 2022). This idea of using digital twins in precision medicine has been explored for asthma management (Drummond et al. 2023), the treatment of immune-mediated diseases (Benson 2023), and dementia care (Wickramasinghe et al. 2022), for example. Despite the interest in this area, the development of computational methods and open-source software for creating and using digital twins has been slow to emerge. This is likely due to the industry focus on creating twins of mechanical objects using principles of physics and engineering that do not exist with enough detail to create simulated patients with molecular, cellular, physiological, and anatomical realness and appropriate environmental and societal context. Some of these challenges have been previously discussed (Benson 2023).

The goal of the present study was to create a computational methodology for generating digital twins based on synthetic patients rather than biophysics. The generation of synthetic data is becoming a mature field (Gonzales et al. 2023) and lends itself well to the digital twin strategy. The working hypothesis is that the correlation structure of clinical variables among patients can inform the creation of digital twins that represent unobserved individuals. In other words, patient relationships might be able to serve as a surrogate for biophysical realizations. The advantage of this surrogate approach is that it can be implemented and evaluated today while we wait for better and more complete biophysical models that could take decades to develop and validate.

We introduce here a new approach that combines *synthetic* data and network science to create digital *twins* (i.e. SynTwin) for precision medicine. Our approach starts by estimating the distance between all subjects based on their available features. We explore here several different distance metrics. Second, the distances are used to construct a network with subjects as nodes and edges defining distance less than the percolation threshold. Third, communities or cliques of subjects are identified using a Multilevel community detection algorithm. Fourth, a large population of synthetic patients or subjects are generated. Several synthetic data generators were evaluated. Fifth, digital twins are selected from the synthetic patient population that are within a given distance defining a subject community in the network. By design, the digital twins represent unobserved hypothetical patients with similar clinical profiles as their real patient counterparts. Finally, we compare and contrast community-based prediction of clinical endpoints using real subjects, digital twins, or both. This is compared to predictive performance using real patients outside the community as a baseline. We apply our synthetic digital twin (SynTwin) approach to predicting mortality in a population-based cancer registry (n=87,674) from the Surveillance, Epidemiology, and End Results (SEER) program from the National Cancer Institute (USA). Bootstrapping is used to assess the standard error of all performance metrics and to estimate 95% confidence intervals for hypothesis testing. Our results demonstrate that nearest network neighbor prediction of mortality in the SEER breast cancer data is significantly improved with digital twins. These results support a growing number of studies highlighting the benefit of synthetic data in other applications.

2. Methods

We describe here the data used and the detailed methods for the SynTwin approach.

2.1. *Cancer Registry Data*

We chose a population-based cancer registry from the Surveillance, Epidemiology, and End Results (SEER) program from the National Cancer Institute (USA) for this study due its large sample size and ease of access by simple registration with an email address to allow for reproducibility. We utilized SEER Stat Version 8.4.1 for data retrieval.

To extract patient data specifically for breast cancer, we applied the following filters:

Database name: Incidence - SEER Research Data, 17 Registries, Nov 2021 Sub (2000-2019) - Linked To County Attributes - Time Dependent (1990-2019) Income/Rurality, 1969-2020 Counties.

Additional filter criteria included:

Site recode ICD-O-3/WHO 2008 = 'Breast' AND Year of diagnosis = '2010', '2011', '2012', '2013', '2014', '2015' AND {Vital status recode (study cutoff used) = 'Alive' OR {Vital status recode (study cutoff used) = 'Dead' AND SEER cause-specific death classification = 'Dead {attributable to this cancer dx}'}}

We chose to exclude data from the years 2015-2019 due to the significant imbalance observed within that period. Specifically, the data exhibited a notable disparity between the number of surviving patients and the number of deceased cases. More than 80% of the patients within that timeframe were still alive, rendering the dataset heavily skewed. Our criteria yielded 324,117 patient records. Removing redundant entries resulted in 231,930 records, consisting of 188,093 Alive cases and 43,837 Dead cases. Subsequently, we conducted a stratified sampling based on vital status to create a balanced dataset for prediction purposes. We retained all Dead cases (n=43,837) and randomly undersampled the same number of Alive cases (n=43,837). This process yielded a total of 87,674 records for our final dataset. We partitioned this sample into a training dataset (n=57,674) and a validation dataset of approximately 1/3 of the sample (n=30,000) to assess internal validity of the results. The training data was used to generate the digital twins while the validation dataset was held out for making predictions using the real patient data and their network and communities. The data processing steps are outlined by the flowchart in Figure 1.

Features included age, year of diagnosis, sex, race, ICDO3, tumor grade, laterality, primary site, survival in months, tumor sequence, diagnostic confirmation, ICCC site, combined summary stage, and vital status (Alive or Dead). The last feature was used as the clinical outcome of class variable for prediction.

2.2. *Bootstrapping*

A central goal of this study was to compare and contrast different methods for estimating patient distances, different methods for generating synthetic data, and different approaches to using digital twins to predict outcome. In order to generate a sampling distribution of all objective functions we carried out 1000-fold bootstrapping by sampling 90% of patients in the holdout or validation data with replacement in each community of size 10 or greater a total of 1000 times. Each performance measure was estimated using all 1000 replications to derive its empirical distribution. This allowed 95% confidence intervals to be estimated for all performance metrics. These were used for uncertainty quantification, statistical comparisons, and hypothesis testing.

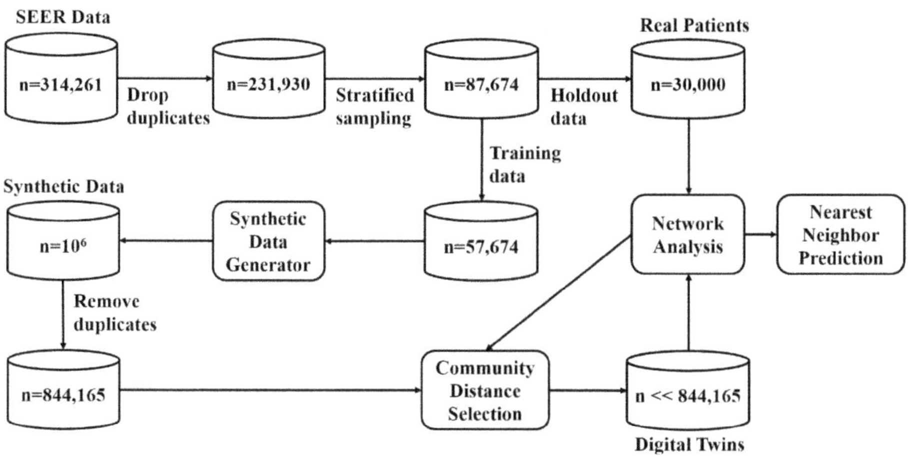

Fig. 1. Flowchart for data processing and analysis.

2.3. *The SynTwin Algorithm for Network-Based Generation of Digital Twins from Synthetic Data*

We describe here our six-step algorithm for generating digital twins and using them for predicting mortality in the SEER data. This involves computing patient distances based on clinical features, constructing a network using distances based on the percolation threshold, identifying patient communities, generating synthetic patients, selection of digital twins, a nearest neighbor (i.e. within community) prediction of mortality.

2.3.1. *Distance Measures*

The first step is to estimate the distances between patients. We evaluated four different distance metrics. These include Euclidian, Manhattan, Cosine (Lee et al. 2015), and Gower (Gower 1971). Each has different strengths and weaknesses. For example, Gower is appealing because it is scale-invariant and works well with both discrete and continuous data. Further, as shown in the results, this distance measure yielded the best results.

2.3.2. *Network Construction*

The second step is to build a network with patients as nodes and edges with weights based on the estimated distances in the first step. To prevent an uninformative fully connected network, we used a percolation threshold equal to the first upward inflection point of the convex part of the sigmoid relationship between edge weight (X axis) and network size (Y axis) as an objective approach to filtering edges.

2.3.3. *Community Detection*

The third step is to detect communities of patients (i.e. cliques or modules) in the network. There are many different community detection algorithms for large networks. We selected the Multilevel algorithm (Blondel et al. 2008) for this study. This algorithm uses a heuristic for modularity optimization and is designed specifically for large networks. The Multilevel algorithm was shown to outperform other community detection algorithms available at the time and with better time complexity (Blondel et al. 2008). Further, a more recent study compared this algorithm with seven others on several graph benchmarks and showed that the Multilevel algorithm was best for both accuracy and time complexity (Yang et al. 2016). In our study, we varied the resolution parameter settings to maximize the number of communities with at least 10 subjects. This yielded between 11,000 and 19,000 communities across the four different distance metrics we investigated.

2.3.4. *Synthetic Data Generation*

The fourth step is to generate synthetic patients to be used as the population to select digital twins from. We evaluated three synthetic data generation algorithms. The first, categorical latent Gaussian process (CLGP), uses continuous latent variables to represent categorical variables that can then be modeled using a Gaussian process (Gal et al. 2015). Here, synthetic data can be generated by sampling from the posterior distribution of the latent variables. The second, mixture of product of multinomials (MPoM), uses a probabilistic model to generate synthetic data with similar statistical properties to the original data (Dunson & Xing 2009). The third, multi-categorical extension of a medical generative adversarial network (MC-MedGAN), uses two adversarial neural networks to generate synthetic data (Choi et al. 2017). Here, The first network learns to generate realistic synthetic data, and the second one attempts to distinguish between real and synthetic data generated by the first network. Autoencoders are used to transform the multivariate categorical data to continuous values, which are then used by the GAN to generate synthetic data.

All three of these methods were recently evaluated and compared (Goncalves et al. 2020). We used the following performance metrics highlighted in this study to evaluate each approach: pairwise correlation difference (PCD), log-cluster (LC), support coverage (SC), and cross-classification (CrCl). The PCD metric is computed as the Frobenius norm difference between Pearson correlation matrices of real and synthetic datasets. It measures how well a method captures the correlation between variables. The LC metric assesses the similarity in latent structure between real and synthetic datasets using k-means clustering. The SC metric quantifies the extent to which the variables support in real data is captured in synthetic data. It is calculated as the ratio of the cardinalities of number of levels (support) for each variable in real and synthetic data. CrCl assesses how accurately a synthetic dataset replicates the statistical dependence found in real data using a classifier.

We used the best hyperparameters reported for "small-set" in the study (Goncalves et al. 2020) to set up our synthetic data generation algorithms considering the smaller number of variables in our dataset. For CLGP we used 100 inducing points and 5-dimensional latent space. For MPoM we set the number of clusters (k) to 30, concentration parameter (α) to 10, Gibbs sampling steps to 10,000, and burn-in steps to 1,000. For MC-MedGAN we used a learning rate 1e-3 and batch size 100 samples. We applied L-2 regularization on the weights of the neural network with λ=1e-3 and set temperature parameter for Gumbel-Softmax trick to τ=0.666. The autoencoder part was built with a code size 64, two encoder layers (hidden size – 256 and 128), and two decoder layers (hidden size – 256 and 128). The GAN part consisted of one generator step with two generator layers (hidden size – 64 and 64) and two discriminator steps each with two discriminator layers (hidden size – 256 and 128). The autoencoder and the GAN were trained for 100 and 500 epochs, respectively.

2.3.5. *Selection of Digital Twins*

The fifth step is to select digital twins from a population of synthetic patients. For a synthetic twin to be a digital twin it must be within some distance of one or more real patients such that the clinical features can represent realistic unobserved measures and outcomes. For each community we selected those synthetic patients whose distances places them within that community. We refer to these virtual patients as digital twins of the real patients in the community. Only those digital twins in a community are used for prediction of mortality.

2.3.6. *Prediction of Mortality*

The final step is to use features from real patients and/or digital twins to predict mortality (Alive or Dead) using a majority vote using the study design described in the next section (2.4). This prediction strategy resembles k-nearest neighbor classification. We estimated six different classification performance measures for predicting mortality across 1000 bootstrapped samples of the holdout data sampled with replacement from each community with at least 10 patients. These included accuracy, balanced accuracy area under the receiver operating characteristic curve (AUROC), precision, recall, and F1. The mean of each performance metric across the 1000 bootstrapped datasets was reported along with the bootstrapped 95% confidence interval (CI).

2.4. *Study Design and Analysis.*

A central goal of this study is to evaluate whether digital twins add any value to predicting mortality beyond that provided by data from the real patients. To answer this question, we developed the following study design (Figure 2). Here, we evaluated prediction of mortality in target patients (black circle) using real patients (A), digital twins (B), real patients and digital twins (C), the closest digital twins equal to the number of real patients in the community (D), real patients and closest digital twins (E), and real patients outside the community (F) as a control for the value of considering communities. Here, each subject in a community alternate as the target patient in a leave-one-out style analysis.

A total of one million synthetic patients were generated from the training data using the best synthetic data generation algorithm (MPoM). We predicted target patient mortality using the nearest neighbor majority vote classification method in the holdout or validation dataset. We estimated 95% confidence intervals for each of the classification performance metrics and statistically compared distance metrics, synthetic data generation algorithms, and study designs.

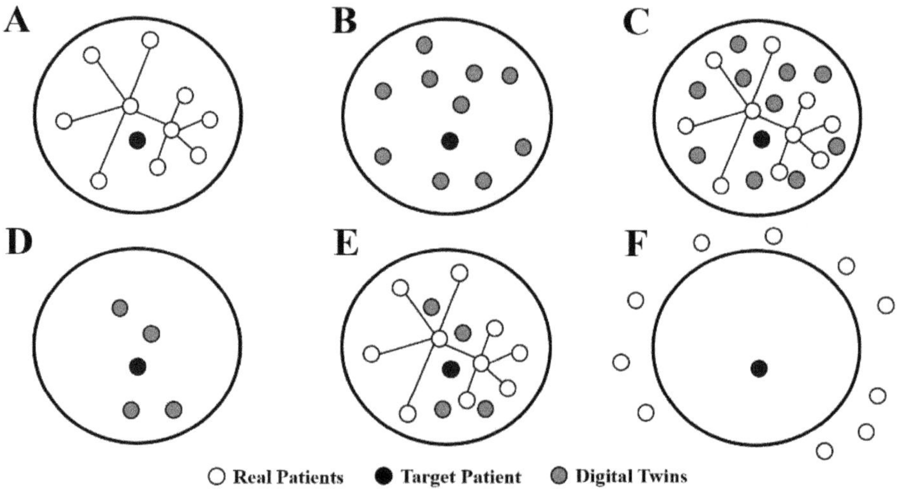

Fig. 2. Study design for comparing outcome prediction using real patients and/or digital twins. The large circles represent a community within the patient network. Prediction of the target patient is carried out using real patients (A), digital twins (B), real patients and digital twins (C), the closest digital twins (D), real patients and closest digital twins (E), and real patients outside the community (F).

3. Results

Table 1 summarizes the performance metrics for the three synthetic data generators considered. Across all metrics, mixture of product of multinomials (MPoM) performed significantly better than the other two methods with nonoverlapping 95% confidence intervals. Consider for example that MPoM had a cross-classification (CrCl) of 0.982 indicating a very high degree of correlation between the same features in the real dataset and in the synthetic dataset. This was significantly higher than the CrCl for MC-MedGAN (0.759) and CLGP (0.645) with nonoverlapping confidence intervals when compared to MPoM. This was true for the other metrics. The only exception was the categorical latent Gaussian process (CLGP) for coverage which was comparable to MPoM. These results mirror a previous evaluation of these algorithms using the SEER data where MPoM outperformed the MC-MedGAN adversarial neural network approach (Goncalves et al. 2020). Therefore, we selected MPoM as our synthetic data generator and used it for the remainder of the study.

Table 1. Comparison of synthetic data algorithms (columns) for four performance metrics (rows). Bolded metric values are significantly better than the others.

	CLGP		MC-MedGAN		MPoM	
Metric	Mean	95% CI	Mean	95% CI	Mean	95% CI
CrCl	0.645	0.533, 0.757	0.759	0.634, 0.884	**0.982**	**0.941, 1.022**
LC	-1.545	-1.594, -1.496	-2.941	-4.211, -1.672	**-5.191**	**-6.146, -4.236**
SC	**1.000**	**0.999, 1.001**	0.830	0.625, 1.036	**0.989**	**0.978, 1.000**
PCD	1.723	1.453, 1.992	2.772	1.007, 4.538	**1.012**	**0.720, 1.305**

Table 2. Comparison of study design performance as measured by AUROC for each distance measure. Bolded metric values are significantly better than the others.

	Cosine		Euclidean		Gower		Manhattan	
Design*	mean	95% CI	mean	95% CI	mean	95% CI	mean	95% CI
A	0.800	0.792, 0.808	0.807	0.800, 0.814	0.791	0.781, 0.800	0.800	0.792, 0.807
B	0.793	0.785, 0.801	0.799	0.792, 0.806	0.784	0.774, 0.794	0.792	0.784, 0.800
C	0.793	0.785, 0.801	0.798	0.791, 0.805	0.783	0.773, 0.793	0.791	0.783, 0.798
D	0.840	0.833, 0.847	0.848	0.842, 0.854	**0.864**	**0.857, 0.872**	0.852	0.845, 0.858
E	0.840	0.833, 0.847	0.845	0.839, 0.852	**0.852**	**0.844, 0.860**	0.846	0.839, 0.852
F	0.510	0.500, 0.521	0.512	0.503, 0.522	0.494	0.482, 0.507	0.485	0.475, 0.495

*Real patients (A), digital twins (B), real patients and digital twins (C), closest digital twins (D), real patients and closest digital twins (E), and real patients outside the community (F).

Table 2 summarizes the AUROC for predicting mortality in the holdout or validation data for each of the four distance metrics and each of the six study designs (A-F, see Figure 1). Study designs D and E had significantly higher AUROCs than the others but were not significantly better than each other given overlapping confidence intervals. Unique to study designs D and E are the presence of digital twins selected to be close to the target patient being predicted. The performance of D and E was significantly higher for the Gower distance than Cosine, Euclidian, or Manhattan. Therefore, we are reporting the mean AUROCs for Gower distance. These patterns of significance were similar for accuracy, balanced accuracy, and the other performance metrics (tables not shown). For example, the Gower accuracies for D and E were 0.788 (95% CI=0.780-0.797) and 0.781 (95% CI=0.772-0.790), respectively. The Gower accuracy for just the real patients (A) in the community was 0.719 (95% CI=0.710-0728). The mean balanced accuracies were very similar for D (0.789), E (0.783), and A (0.721) suggesting that there were no biased accuracies due to imbalanced data. Thus, the accuracies associated with including close digital twins within communities was significantly higher than that for just real patients within communities.

Interestingly, the performance of A (real patients only), B (digital twins only), and C (real patients and digital twins) were not significantly different from one another across the different distance metrics including Gower. It is important to note that F (real patients outside the community) had an AUROC of approximately 0.50 as might be expected by chance given these patients have a distance that exceeds the percolation threshold and places them outside the community. Thus, the distance from the target patient being considered for prediction plays an important role in predictive accuracy and is highly relevant for precision medicine where context is a key consideration. An example network of real patients for three communities is shown in Figure 3 along with the corresponding digital patients.

4. Discussion

We have developed a new digital twin approach to improve the prediction of clinical endpoints. Our approach combines network science to model patient similarity and *synthetic* data generation to generate digital *twins* (SynTwin). Key to SynTwin is using patient similarity to synthesize nearby digital twins that represent hypothetical unobserved patients with clinical data correlations that are consistent with real patients. This distance-based approach is different than the digital twin approaches from industry that rely on well-known physical principles that govern a complex system (Attaran & Celik 2023; Attaran et al. 2023; Grieves & Vickers 2017). Biophysical properties governing health are not well known and are often only available for certain cellular or physiological processes. Indeed, simulating a single cell is quite challenging for a number of reasons including the lack of physics-based models (Thornburg et al. 2022). It is our working hypothesis that distance-based digital twins will be useful for informing patient outcomes above and beyond that provided by the observed clinical data. Indeed, our results suggest that generating and selecting digital twins close to the target patient whose outcome is being predicted significantly improves predictive performance above and beyond the real patients in the community. Choosing real patients outside the community for predicting target patients inside a community was not better than flipping a coin.

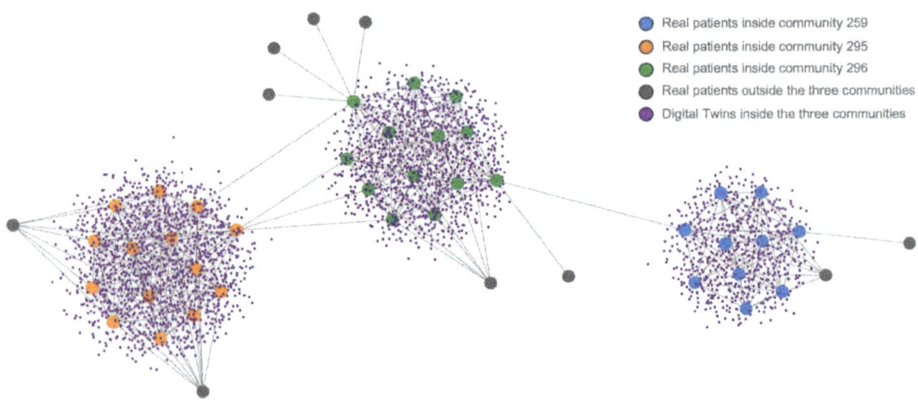

Fig. 3. Section of the network showing three communities of real patients (orange, green, and blue circles). Also shown are the digital patients (small purple circles) and real patients outside the communities (grey circles).

The generation and use of synthetic data for biomedical research is in and of itself not new. A recent review highlighted more than 70 published papers representing at least seven different use cases for synthetic data (Gonzales et al. 2023). Most of the use cases involve generating a synthetic dataset that can be used to avoid the privacy and security concerns of real data. For example, a synthetic dataset could be distributed to students to use for learning objectives without fear of identifying real patients. Other use cases involve using synthetic data to benchmark algorithms, evaluate information technology software, and public release of data. A very specific use case is to allow investigators to test a hypothesis without the need for Institutional Review Board (IRB) approval and the time it takes to retrieve data from an electronic health record which is a process that can take months depending on the complexity of the data and the wait time for available qualified personnel. Any interesting patterns found in the easily available synthetic data could then justify the time and expense of retrieving real data to confirm the finding before publication as has been suggested (Foraker et al. 2018). This approach was recently evaluated by comparing statistical and machine learning results obtained from real patient data and a synthetic derivative generated using a commercially available platform (Foraker et al. 2020). Similar results were seen when using a large integrated data resource (Foraker et al. 2021). In each case, the authors were able to draw the same conclusions from the analytical results using both real and synthetic datasets.

The use of synthetic data to generate digital twins was not mentioned in the review by Gonzales et al. (2023). However, using synthetic data to improve the sample size of a real dataset for improving predictive accuracy was specifically discussed. A study evaluating the addition of synthetic data to a real dataset showed that variance improved and five machine learning algorithms had improved prediction of heart disease (Aljaaf et al. 2016). The idea that synthetic data can improve machine learning performance has been observed in the image analysis domain. For example, a synthetic image generation approach using general adversarial networks (GANs) has been shown to improve image segmentation when the number of training examples is small (Thambawita et al. 2022). This approach may have clinical applications. For example, a recent study showed that synthetic colonoscopy images with polyps can improve the sensitivity of a deep learning neural network to detect polyps in real images (Adjei et al. 2022). This may be true in ophthalmology as well (You et al. 2022). Our observation that synthetic data may improve the performance of predictive accuracy is consistent with these studies. More studies are needed to validate this phenomenon.

Most synthetic data generation studies have focused on generating and using an entire synthetic dataset and checking to make sure the patterns detected by a machine learning algorithm are similar (Gonzales et al. 2023). Our SynTwin digital twin approach is different in the sense that we are using patient similarity and network community structure to select synthetic patients (i.e. twins) that can inform clinical outcome prediction. This is a more targeted approach that is much more consistent with the goals of precision medicine where treatment decisions and clinical outcomes are assessed in patient subgroups with similar characteristics. As such, this represents a fundamental shift in how synthetic data are used and may be more informative for clinical decision support.

Despite progress in this area, there are some possible limitations and challenges for moving forward. First, we applied our method to a dataset with a large sample size and a small number of features. On one hand, this was an ideal dataset to evaluate a new approach. Further, this dataset is publicly available, has been carefully curated, and has been well studied for understanding cancer risk and outcomes. However, the question remains of how the SynTwin approach will scale to hundreds or thousands of features or how it will behave when the synthetic data are generated from a dataset with small sample size. Further, the validation data was derived from the same cohort. Second, SynTwin is highly dependent on the community structure of the network. Not every patient is part of a community and prediction of outcomes in those patients may need to be performed using standard machine learning (ML) methods. Thus, a hybrid SynTwin-ML approach may need to be developed to make sure those patients are considered fully. Thirdly, it is of great interest to develop formal statistical inferential procedures to quantify the uncertainty of the subsequent analyses including estimation and prediction. Intuitively the generated digital twins should be weighted differently from the real patients in the precision of the downstream analyses. Finally, our implementation of SynTwin relied on bootstrapping to assign confidence intervals to performance metrics. This adds 1000-fold more computation time which might be prohibitive for larger datasets with more features. Future studies will need to balance the need for statistical inference with computing resources that are available. This study benefited from access to a 2000-core high-performance computing system to carry out all computations.

Precision medicine relies heavily on artificial intelligence and machine learning methods to develop models for predicting disease risk and patient outcomes in a manner that takes into account the uniqueness of the patient in question and other patients with similar profiles (Rajpurkar et al. 2022). The SynTwin digital twin strategy we presented here takes a step toward the use of synthetic data to augment the prediction of clinical outcomes by generating hypothetical unobserved patients to be used alongside real patients. The use of digital twins in medicine and biomedical research is in its infancy. We have a lot to learn from industrial uses of this approach and will need to develop new algorithms and software that consider the unique aspects of patients and their data. We agree with others who have speculated that digital twins will have a big impact on research and patient care but that new biophysical, computational, and statistical methods are needed (Acosta et al. 2022; Armeni et al. 2022; Attaran & Celik 2023; Attaran et al. 2023; Kamel Boulos & Zhang 2021).

5. Acknowledgments

This work was supported by NIH grants LM010098 and AG066833.

References

Acosta JN, Falcone GJ, Rajpurkar P, Topol EJ. 2022. Multimodal biomedical AI. *Nat Med.* 28(9):1773–84

Adjei PE, Lonseko ZM, Du W, Zhang H, Rao N. 2022. Examining the effect of synthetic data augmentation in polyp detection and segmentation. *Int J CARS.* 17(7):1289–1302

Aljaaf AJ, Al-Jumeily D, Hussain AJ, Fergus P, Al-Jumaily M, Hamdan H. 2016. Partially Synthesised Dataset to Improve Prediction Accuracy. *Intelligent Computing Theories and Application*, pp. 855–66. Cham: Springer International Publishing

Armeni P, Polat I, De Rossi LM, Diaferia L, Meregalli S, Gatti A. 2022. Digital Twins in Healthcare: Is It the Beginning of a New Era of Evidence-Based Medicine? A Critical Review. *J Pers Med.* 12(8):1255

Attaran M, Attaran S, Celik BG. 2023. The impact of digital twins on the evolution of intelligent manufacturing and Industry 4.0. *Adv Comput Intell.* 3(3):11

Attaran M, Celik BG. 2023. Digital Twin: Benefits, use cases, challenges, and opportunities. *Decision Analytics Journal.* 6:100165

Benson M. 2023. Digital Twins for Predictive, Preventive Personalized, and Participatory Treatment of Immune-Mediated Diseases. *Arterioscler Thromb Vasc Biol.* 43(3):410–16

Blondel VD, Guillaume J-L, Lambiotte R, Lefebvre E. 2008. Fast unfolding of communities in large networks. *J. Stat. Mech.* 2008(10):P10008

Choi E, Biswal S, Malin B, Duke J, Stewart WF, Sun J. 2017. Generating Multi-label Discrete Patient Records using Generative Adversarial Networks. *Proceedings of the 2nd Machine Learning for Healthcare Conference*, pp. 286–305. PMLR

Drummond D, Roukema J, Pijnenburg M. 2023. Home monitoring in asthma: towards digital twins. *Curr Opin Pulm Med.* 29(4):270–76

Dunson DB, Xing C. 2009. Nonparametric Bayes Modeling of Multivariate Categorical Data. *Journal of the American Statistical Association.* 104(487):1042–51

Foraker R, Guo A, Thomas J, Zamstein N, Payne PR, et al. 2021. The National COVID Cohort Collaborative: Analyses of Original and Computationally Derived Electronic Health Record Data. *J Med Internet Res.* 23(10):e30697

Foraker R, Mann DL, Payne PRO. 2018. Are Synthetic Data Derivatives the Future of Translational Medicine? *JACC: Basic to Translational Science.* 3(5):716–18

Foraker RE, Yu SC, Gupta A, Michelson AP, Pineda Soto JA, et al. 2020. Spot the difference: comparing results of analyses from real patient data and synthetic derivatives. *JAMIA Open.* 3(4):557–66

Gal Y, Chen Y, Ghahramani Z. 2015. Latent Gaussian Processes for Distribution Estimation of Multivariate Categorical Data. *Proceedings of the 32nd International Conference on Machine Learning*, pp. 645–54. PMLR

Goncalves A, Ray P, Soper B, Stevens J, Coyle L, Sales AP. 2020. Generation and evaluation of synthetic patient data. *BMC Medical Research Methodology*. 20(1):108

Gonzales A, Guruswamy G, Smith SR. 2023. Synthetic data in health care: A narrative review. *PLOS Digital Health*. 2(1):e0000082

Gower JC. 1971. A General Coefficient of Similarity and Some of Its Properties. *Biometrics*. 27(4):857–71

Grieves M, Vickers J. 2017. Digital Twin: Mitigating Unpredictable, Undesirable Emergent Behavior in Complex Systems. In *Transdisciplinary Perspectives on Complex Systems: New Findings and Approaches*, eds. F-J Kahlen, S Flumerfelt, A Alves, pp. 85–113. Cham: Springer International Publishing

Kamel Boulos MN, Zhang P. 2021. Digital Twins: From Personalised Medicine to Precision Public Health. *Journal of Personalized Medicine*. 11(8):745

Lee J, Maslove DM, Dubin JA. 2015. Personalized mortality prediction driven by electronic medical data and a patient similarity metric. *PLoS One*. 10(5):e0127428

Rajpurkar P, Chen E, Banerjee O, Topol EJ. 2022. AI in health and medicine. *Nat Med*. 28(1):31–38

Thambawita V, Salehi P, Sheshkal SA, Hicks SA, Hammer HL, et al. 2022. SinGAN-Seg: Synthetic training data generation for medical image segmentation. *PLOS ONE*. 17(5):e0267976

Thornburg ZR, Bianchi DM, Brier TA, Gilbert BR, Earnest TM, et al. 2022. Fundamental behaviors emerge from simulations of a living minimal cell. *Cell*. 185(2):345-360.e28

Verdouw C, Tekinerdogan B, Beulens A, Wolfert S. 2021. Digital twins in smart farming. *Agricultural Systems*. 189:103046

Wickramasinghe N, Ulapane N, Andargoli A, Ossai C, Shuakat N, et al. 2022. Digital twins to enable better precision and personalized dementia care. *JAMIA Open*. 5(3):ooac072

Yang Z, Algesheimer R, Tessone CJ. 2016. A Comparative Analysis of Community Detection Algorithms on Artificial Networks. *Sci Rep*. 6(1):30750

You A, Kim JK, Ryu IH, Yoo TK. 2022. Application of generative adversarial networks (GAN) for ophthalmology image domains: a survey. *Eye Vis (Lond)*. 9(1):6

Optimizing Computer-Aided Diagnosis with Cost-Aware Deep Learning Models

Charmi Patel[1†], Yiyang Wang[2], Thiruvarangan Ramaraj[1], Roselyne Tchoua[1], Jacob Furst[1], Daniela Raicu[1]

[1] *DePaul University, Chicago, IL, 60604, U.S.A*

[2] *Milwaukee School of Engineering, Milwaukee, WI, 53202, U.S.A*

[†] *E-mail: cpatel54@depaul.edu*

Classical machine learning and deep learning models for Computer-Aided Diagnosis (CAD) commonly focus on overall classification performance, treating misclassification errors (false negatives and false positives) equally during training. This uniform treatment overlooks the distinct costs associated with each type of error, leading to suboptimal decision-making, particularly in the medical domain where it is important to improve the prediction sensitivity without significantly compromising overall accuracy. This study introduces a novel deep learning-based CAD system that incorporates a cost-sensitive parameter into the activation function. By applying our methodologies to two medical imaging datasets, our proposed study shows statistically significant increases of 3.84% and 5.4% in sensitivity while maintaining overall accuracy for Lung Image Database Consortium (LIDC) and Breast Cancer Histological Database (BreakHis), respectively. Our findings underscore the significance of integrating cost-sensitive parameters into future CAD systems to optimize performance and ultimately reduce costs and improve patient outcomes.

Keywords: Misclassification errors; Cost-sensitive activation function; Convolutional neural network

1. Introduction

Machine learning (ML) models have been developed to identify patterns in data across various domains, including computer-aided diagnosis,[1] public health,[2] and defect detection.[3] Generally, these ML models are optimized based on overall prediction accuracy across all classes and data points, assuming that misclassification errors are equal.[4] However, this assumption can be perilous in classification problems where misclassifying a positive instance carries a higher cost than misclassifying a negative instance. Particularly in the medical domain, a false negative error will likely have much greater consequences than a false positive.

To address this challenge, we propose a novel cost-aware deep learning-based CAD system that incorporates different cost values into the activation function to boost the model's sensitivity. By fine-tuning the cost values associated with false positive and false negative instances, we can significantly increase true positives. Our contributions are twofold: 1) a CAD training framework designed to enhance sensitivity while maintaining overall accuracy, and 2) a proof-of-concept demonstrating the value of incorporating cost values as hyperparameters in future CAD systems.

2. Related Work

2.1. *Cost-Sensitive Learning in Non-Medical Domains*

In recent years, cost-sensitive learning has gained popularity as a valuable tool in non-medical domains to tackle class imbalance[4–10] and to address its associated costs of misclassification.

Prior research by Zhu and Wan[11] proposed a cost-sensitive learning method for semi-supervised hit-and-run analysis to handle the data imbalance issue which significantly improved model's performance even with a small proportion of labeled historical data. Khoshgotaar et al.[12] introduced cost-sensitive learning into Software Defect Prediction (SDP) and used a boosting method to build software quality models. Le et al.[13] implemented a hybrid approach by combining oversampling techniques and cost-sensitive learning, which significantly improved bankruptcy prediction performance. Devi et al.[14] proposed a cost-sensitive weighted random forest algorithm for effective credit card fraud detection. The model assigns more weight to minority instances during training, resulting in improved performance compared to existing random forest techniques. Xiao et al.[15] integrated a group handling neural network-based cost-sensitive semi-supervised selective ensemble model for credit-scoring problems.

Other prior work focused on improving the overall prediction performance by modifying the loss function to consider different cost values for various misclassifications. Li et al.[16] proposed a pixel-based adaptive weighted cross-entropy loss function to facilitate road crack detection. Wang et al.[17] also introduced a novel cost-sensitive loss function for semantic segmentation of remote sensing images. More recently, Li et al.[18] constructed a new cost-sensitive loss function that incorporates the cost difference caused by misclassification between different classes proving its ability to enhance the model's effectiveness.

2.2. *Cost-Sensitive Learning Applied to Medical Diagnosis*

Research studies on the ML application to medical diagnosis typically employ traditional ML algorithms and advanced algorithms via ensemble learning,[19] evolutionary algorithms,[20] sparse autoencoders (SAE).[21] However, few research works have conducted cost-sensitive learning in medical diagnosis. Recently, Manop[5] developed a cost-sensitive XGBoost model for breast cancer detection and evaluated it on four breast cancer datasets with uneven class distribution, achieving accuracy ranging from 95.99% to 96.43%. Ali et al.[6] developed a method that combines cost-sensitive learning and ensemble learning techniques to predict breast cancer. The ensemble learning methods include GentleBoost, Bagging, and Adaptive Boosting, resulting in a 3.91% improvement. Zieba et al.[7] proposed the combination of ensemble learning and cost-sensitive Support Vector Machine (SVM) to address the lung cancer patients' post-operation life expectancy. They observed that patients not covered by the minority rules have a 97% chance of surviving the considered survival period. Ali et al.[22] applied cost-sensitive ensemble methods in the classification of chronic kidney disease (CKD) which incorporates feature ranking capabilities instead of enhancing predictive accuracy. Cost-sensitive deep neural networks (CSDNN) have also been developed by Wang et al.[23] to predict hospital readmission, achieving a significant improvement in accuracy of 6% and 4% for 1-year and 30-day readmission prediction, respectively. Mienye and Sun[10] developed robust cost-sensitive classifiers for predicting medical diagnosis by modifying the objective functions of algorithms such as logistic regression, decision trees, extreme gradient boosting, and random forest. They tested these classifiers on four medical datasets and demonstrated that cost-sensitive methods yield improvements ranging from 1% to 4% compared to the standard algorithms.

These works[5–7,10,23] suggest that incorporating cost-sensitive learning with specific cost values during misclassifications into ML models can improve overall classification performance. However, these studies focus solely on improving overall classification results without determining the impact of different cost values on sensitivity and specificity metrics. Our study expands the integration of the misclassification costs into the learning algorithm by optimizing the diagnostic interpretation sensitivity without sacrificing the overall diagnostic performance.

3. Methodology

3.1. *Convolutional Neural Networks*

A classical CNN model (Figure 1) comprises convolutional layers and fully connected layers. The convolutional layers are designed to extract features, while the fully connected layers are responsible for classification. For our study, we focus on the binary classification problem distinguishing between malignant and benign cases. To achieve this, we employed a single neuron in the output layer with the sigmoid activation function. The output of the sigmoid activation function determines the predicted label.

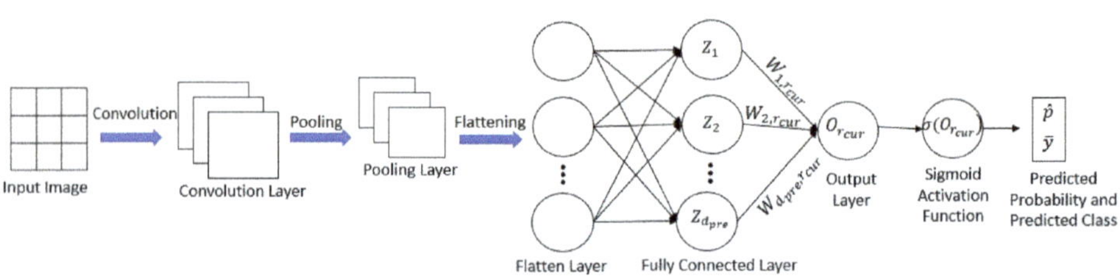

Fig. 1: CNN architecture overview.

Let $O^{(t)}_{r_{\text{cur}}}$ be the output of the current layer's neuron with index r_{cur} at epoch t for any image X_i:

$$O^{(t)}_{r_{\text{cur,i}}} = w^{(t)}_{0,r_{\text{cur,i}}} + \sum_{r_{\text{pre}}=1}^{d_{\text{pre}}} g(z^{(t)}_{r_{\text{pre}}}) w^{(t)}_{r_{\text{pre}} r_{\text{cur,i}}} \tag{1}$$

where $w^{(t)}_{r_{\text{pre}} r_{\text{cur}}}$ represents the weights between the neuron indexed as r_{pre} in the previous layer and the neuron indexed as r_{cur} in the current layer at epoch t. d_{pre} represents the number of neurons in the previous layer, and g represents the activation function applied to each output value of a neuron $z^{(t)}_{r_{\text{pre}}}$ from the previous layer r_{pre}.

Given an image X_i, where i represents the image index, we denote y_i as its actual label and $\hat{p}_i$ as its prediction probability:

$$\hat{p}_i^{(t)} = \sigma(O_{r_{cur,i}}^{(t)}) = \frac{1}{1 + e^{-O_{r_{cur,i}}^{(t)}}} \tag{2}$$

where σ represents the sigmoid activation function applied to the output layer neuron. The predicted $\bar{y}_i$ is calculated as follows:

$$\bar{y}_i = \begin{cases} 1, & \text{if } \sigma(O_{r_{cur,i}}^{(t)}) \geq 0.5 \\ 0, & \text{if } \sigma(O_{r_{cur,i}}^{(t)}) < 0.5 \end{cases} \tag{3}$$

Using the actual label y_i and its predicted probability $\hat{p}_i$, the loss for instance X_i at epoch t is calculated using Binary Cross Entropy (BCE) loss:

$$Loss_{BCE_i}^{(t)} = -[y_i \log_2(\hat{p}_i) + (1 - y_i)\log_2(1 - \hat{p}_i)] \tag{4}$$

During the backpropagation, the model updates its weights to minimize the overall loss $L_{ERM}^{(t)}$ (Equation 5) at epoch t, which is the average loss across all N training instances.

$$L_{ERM}^{(t)} = \frac{1}{N}\sum_{i=1}^{N} Loss_{BCE_i}^{(t)}(\hat{p}_i) \tag{5}$$

where ERM denotes Empirical Risk Minimization.

3.2. *Cost-aware CNN model*

To direct the performance of the CNN model towards the true positives (malignant cases denoted by 1), we propose a cost-sensitive activation function $\sigma(O_{r_{cur}}^{(t)}, c(y, \bar{y})^{(t)})$ that penalizes more false negatives than the false positives by assigning higher costs to outcomes that are misclassifications of true positives and lower costs for misclassifying true negatives (benign cases denoted by 0). Grounded in the work by Li et al.,[18] we define an inverse relationship between the cost of false negatives $c(1, 0)$ and the cost of false positives $c(0, 1)$ and restrict the values of $c(1, 0)$ to be greater than 1:

$$c(0, 1) = \frac{1}{c(1, 0)} \tag{6}$$

The activation function remains the same for correctly classified cases, and therefore, the costs for true positives $c(1, 1)$ are equal to the costs for true negatives $c(0, 0)$ and are equal to 1.

By integrating the cost values into the activation function, Equation (2) is transformed into Equation (7), denoting that the predicted probabilities $\hat{p}^{(t)}$ at epoch t are now influenced by the costs associated with each type of output:

$$\hat{\hat{p}}^{(t)} = \sigma(O_{r_{cur}}^{(t)}, c(y, \bar{y})^{(t)}) = \frac{1}{1 + e^{-O_{r_{cur}}^{(t)} \cdot c(y, \bar{y})^{(t)}}} \tag{7}$$

Implicitly, the new $L_{ERM\_Cost}^{(t)}$ loss function at epoch t, which is based on the predicted probabilities, will be dependent on the cost values:

$$L_{ERM\_Cost}^{(t)} = \frac{1}{N}\sum_{i=1}^{N} Loss_{BCE_i}^{(t)}(\hat{\hat{p}}_i) \tag{8}$$

3.3. *Cost Analysis w.r.t. Sensitivity and Specificity*

Maximizing both sensitivity and specificity simultaneously is not possible as they are inversely related.[24,25] However, introducing cost-values in the activation function of the outcome layer allows optimizing the performance of each without a decline in the overall accuracy. Here we illustrate two examples that show the impact of the costs on the performance of a CNN model.

3.3.1. *Handling False Negative Cases*

In a false negative case, where the actual label $y = 1$ and the predicted label $\bar{y} = 0$, to achieve $\bar{y} = 0$, $O_{r_{\text{cur}}}$ must be a negative value. Using Equations (2) and (4), and if $c(1,0) = 1$, which represents no cost value being used, we obtain the loss function:

$$Loss_{BCE_i^{(t)}} = -\left[\log_2\left(\frac{1}{1+e^{-O^{(t)}_{r_{\text{cur},\text{i}}}}}\right)\right] \tag{9}$$

If we introduce $c(1,0) > 0$ in the activation function, which is a cost value associated with the false negative situation, we obtain the modified loss function:

$$Loss_{BCE\_Cost_i^{(t)}} = -\left[\log_2\left(\frac{1}{1+e^{-O^{(t)}_{r_{\text{cur},\text{i}}}\cdot c(y,\bar{y})^{(t)}}}\right)\right] \tag{10}$$

$$Loss_{BCE\_Cost_i^{(t)}} > Loss_{BCE_i^{(t)}} \tag{11}$$

After introducing a cost value of $c(1,0)$, the loss value obtained from Equation (10) demonstrates an increase, as shown in Equation (11). This cost value impacts the 'false negative' case during the training process, leading to a decrease in false negative cases and an increase in sensitivity.

3.3.2. *Handling False Positive Cases*

In a false positive case, where $y = 0$ and $\bar{y} = 1$, to achieve $\bar{y} = 1$, $O_{r_{\text{cur}}}$ must be a positive value. If $c(0,1) = 1$, which means no cost values being used, we obtain the loss function:

$$Loss_{BCE_i^{(t)}} = -\left[1-\log_2\left(\frac{1}{1+e^{-O^{(t)}_{r_{\text{cur},\text{i}}}}}\right)\right] \tag{12}$$

If we introduce $c(0,1) = \frac{1}{c(1,0)}$ with $c(1,0) > 1$, which represents a cost value associated with the false positive situation, we obtain the modified loss function:

$$Loss_{BCE\_Cost_i^{(t)}} = -\left[1-\log_2\left(\frac{1}{1+e^{-O^{(t)}_{r_{\text{cur},\text{i}}}\cdot c(y,\bar{y})^{(t)}}}\right)\right] \tag{13}$$

$$Loss_{BCE\_Cost_i^{(t)}} < Loss_{BCE_i^{(t)}} \tag{14}$$

After introducing a cost value of $c(0,1)$, the loss value obtained from Equation (13) demonstrates a decrease, as shown in Equation (14). This cost value influences the training process, leading to a decrease in false positive cases and a reduction in specificity.

By incorporating cost-sensitive learning for both "false negative" and "false positive" cases, the model can effectively update its weights to improve performance based on the specific misclassifications encountered during training.

4. Applications

We apply the cost-sensitive activation function approach to deep learning CAD models for lung cancer and breast cancer.

For the lung cancer application, we use the NIH/NCI LIDC[26] data and for the breast cancer application, we use the BreakHis.[27] For both applications, the data was split into training, validation and testing sets using stratified random sampling, with proportions of 70%, 10%, and 20%, respectively. To ensure more robust results, we repeated the process of data splitting and model development for 30 times. The classification performance on the testing set was reported with the mean value across all 30 trials and a 95% confidence interval.

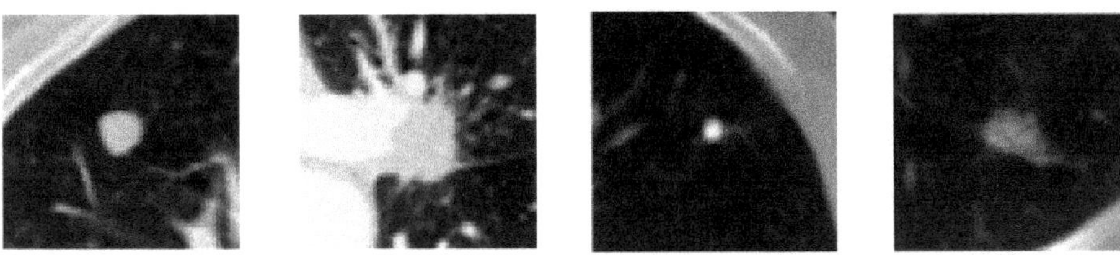

Fig. 2: Multiple visual appearances of cropped lung nodules: The two nodules on the left exhibit malignant features, characterized by spiculated contours and larger size, while the two on the right are benign, displaying smaller, smoother nodules indicative of non-malignancy.

4.1. *LIDC Dataset*

The LIDC[26] dataset contains 2,680 distinct nodules found in Computed Tomography (CT) scans from 1,010 patients. In this study, we implemented the following data preprocessing steps: First, we cropped nodules into images of size 71 x 71 (Figure 2), which is the size of the largest nodule in the dataset. Third, we assigned malignancy classification labels, where nodules with malignancy ratings of 1 (highly unlikely) and 2 (moderately unlikely) were labeled as 'Benign', while nodules with malignancy ratings of 4 (moderately suspicious) and 5 (highly suspicious) were labeled as 'Malignant'. After data pre-processing, which includes normalization and removal of indeterminate nodules with malignancy rating 3, we were left with 1,605 nodules. This dataset is imbalanced, comprising 699 malignant and 906 benign ones.

4.2. *BreakHis Dataset*

BreakHis[27] comprises of 7,909 histological images of breast tumor tissue collected from 82 patients using varying magnification levels (40X, 100X, 200X, and 400X). It contains imbalanced data with 2,480 benign and 5,429 malignant images. The images were 3-channel RGB, 8-bit depth each channel and dimension of 700 x 460 pixels. In this study, we normalized images using min-max normalization. A few sample images are shown in Figure 3.

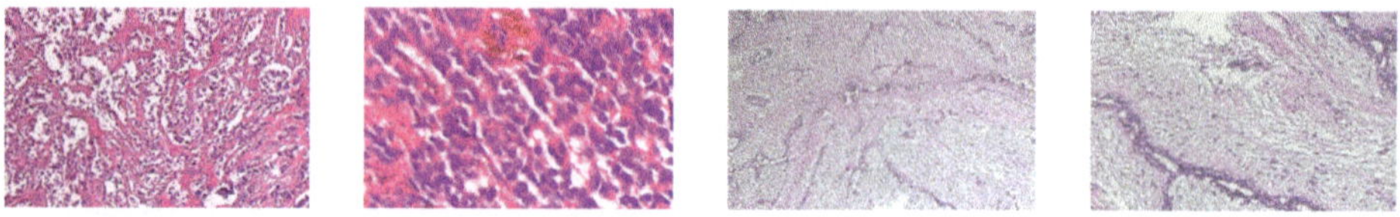

Fig. 3: Breast cancer histological images from the BreakHis Dataset. The left two images depict malignancy, while the right two show benign samples, illustrating critical distinctions for diagnosis.

4.3. *Design and Architecture of the Deep CNN Model*

The transfer learning method overcomes the limitation of having a small amount of training data by initially pre-training a deep learning model on a publicly available large dataset. As part of our study's cost-sensitive algorithm classification model, we fine-tune[28] a pre-trained ResNet18 convolutional neural network from ImageNet[29] on our own dataset.

For this study, we followed Nibali et al.'s recommendation[30] and utilized the ResNet18 CNN architecture.[31] In the intermediate layers of the architecture, the Rectified Linear Unit (ReLU) activation function is used. Given the objective of addressing a binary classification problem (malignant vs. benign), we implemented a single neuron with the sigmoid activation function as the output layer. Consequently, during training, if the output of the sigmoid activation function is greater than or equal to 0.5, we classify the instance as Malignant (positive class); otherwise, we classify it as Benign (negative class).

4.4. *Experimental Results*

The accuracy, sensitivity, and specificity assessment of the two datasets across 30 trials are tabulated in Table 1 through Table 4, showing the performance when the classifier is trained with different misclassification costs on the LIDC and BreakHis datasets. The first column and second column indicate the cost value given for false positives and false negatives, while the various metrics are listed from the third to the last columns. The numbers in bold indicate a significant difference compared to the results obtained without any cost values, whereas non-bold numbers indicate no significant difference.

From the experimental results, the cost-sensitive classifier indeed supports the analysis in Section 3.3. Table 1 reveals that we can enhance sensitivity while maintaining the same level of accuracy. Notably, in both datasets, sensitivity increases with a decrease in specificity for higher values of $c(1,0)$, which is associated with false negatives, and lower values of $c(0,1)$, which is associated with false positives. For the LIDC dataset, the highest sensitivity was achieved when $c(0,1) = 0.33$ and $c(1,0) = 3$, with accuracy and specificity remaining unaffected. We observed a significant improvement of 3.84% compared to the baseline model, where $c(0,1) = c(1,0)$. Table 2 presents the classification performance of the BreakHis dataset, showing a 5.4% significant improvement when $c(0,1) = 0.067$ and $c(1,0) = 15$. Additionally, we observe a trend of significantly increasing sensitivity with decreasing specificity while keeping the overall accuracy within the range of 86.55% to 88.10%. In Tables 1 to 4, numbers within

Table 1: The classification performance on the LIDC testing data was assessed through 30 trials with increasing $c(1,0)$ cost values.

$c(0,1)$	$c(1,0)$	Accuracy	Sensitivity	Specificity
1	1	85.26% (84.63%, 85.89%)	76.22% (74.62%, 77.83%)	91.25% (90.33%, 92.17%)
0.5	2	85.25% (84.39%, 86.11%)	76.89% (75.17%, 78.61%)	90.79% (89.91%, 91.67%)
0.33	3	85.80% (85.19%, 86.41%)	**80.06% (78.81%, 81.30%)**	89.61% (88.54%, 90.69%)
0.2	5	84.93% (84.37%, 85.48%)	78.22% (76.76%, 79.69%)	**89.37% (88.39%, 90.35%)**
0.1	10	84.81% (84.06%, 85.55%)	**78.97% (77.33%, 80.62%)**	**88.67% (87.31%, 90.04%)**
0.067	15	84.56% (83.90%, 85.23%)	**78.58% (76.83%, 80.33%)**	**88.53% (87.23%, 89.83%)**

Table 2: The classification performance on the BreakHis testing data was assessed through 30 trials with increasing $c(1,0)$ cost values.

$c(0,1)$	$c(1,0)$	Accuracy	Sensitivity	Specificity
1	1	88.10% (88.03%, 88.16%)	90.38% (90.29%, 90.47%)	83.10% (82.98%, 83.21%)
0.5	2	**88.38% (88.32%, 88.45%)**	**91.84% (91.75%, 91.93%)**	**80.81% (80.63%, 80.99%)**
0.33	0.3	**88.39% (88.33%, 88.46%)**	**93.79% (93.69%, 93.88%)**	**76.58% (76.40%, 76.76%)**
0.2	5	**87.82% (87.74%, 87.91%)**	**95.05% (94.92%, 95.18%)**	**72.00% (71.76%, 72.23%)**
0.1	10	**87.15% (87.04%, 87.26%)**	**95.71% (95.58%, 95.84%)**	**68.42% (68.00%, 68.83%)**
0.067	15	**86.55% (86.51%, 86.79%)**	**95.78% (95.64%, 95.92%)**	**66.66% (66.09%, 67.23%)**

parentheses represent 95% confidence intervals, with bold numbers indicating significant distinctions compared to the baseline model (c(0,1) = c(1,0)).

An increase in sensitivity is observed with decreasing c(0,1) values, which shows significant improvement compared to the baseline models. Table 3 illustrates the results, focusing on increasing c(0,1) values, which will lead to a decrease in sensitivity significantly and an increase

Table 3: The classification performance on the LIDC testing data was assessed through 30 trials with increasing $c(0,1)$ cost values.

$c(0,1)$	$c(1,0)$	Accuracy	Sensitivity	Specificity
1	1	85.26% (84.63%, 85.89%)	76.22% (74.62%, 77.83%)	91.25% (90.33%, 92.17%)
2	0.5	84.83% (84.13%, 85.52%)	75.06% (73.24%, 76.88%)	91.31% (90.24%, 92.38%)
3	0.33	84.39% (83.70%, 85.07%)	**73.25% (71.44%, 75.06%)**	91.77% (90.66%, 92.88%)
5	0.2	83.00% (82.53%, 84.08%)	**69.56% (67.24%, 71.88%)**	92.41% (91.17%, 93.65%)
10	0.1	81.93% (81.03%, 82.82%)	**64.11% (61.89%, 66.33%)**	**93.74% (92.70%, 94.78%)**
15	0.067	80.48% (79.38%, 81.58%)	**60.69% (57.28%, 64.11%)**	**93.59% (92.36%, 94.83%)**

Table 4: The classification performance on the BreakHis testing data was assessed through 30 trials with increasing $c(0,1)$ cost values.

$c(1,0)$	$c(0,1)$	Accuracy	Sensitivity	Specificity
1	1	88.10% (88.03%, 88.16%)	90.38% (90.29%, 90.47%)	83.10% (82.98%, 83.21%)
2	0.5	**87.43% (87.35%, 87.50%)**	**88.75% (88.64%, 88.87%)**	**84.52% (84.33%, 84.71%)**
3	0.33	**84.70% (80.93%, 88.46%)**	**84.43% (78.48%, 90.39%)**	**85.27% (84.22%, 86.32%)**
5	0.2	**81.34% (76.26%, 86.41%)**	**79.02% (70.99%, 87.05%)**	**86.41% (85.02%, 87.80%)**
10	0.1	**70.62% (63.16%, 78.10%)**	**61.39% (49.71%, 73.08%)**	**90.85% (89.06%, 92.65%)**
15	0.067	**69.41% (62.17%, 76.66%)**	**59.33% (48.02%, 70.63%)**	**91.50% (89.76%, 93.20%)**

in specificity. We observe that with c(0,1) = 15 and c(1,0) = 0.067, the highest specificity is achieved, albeit with a trade-off in sensitivity. When we examine the impact of decreasing c(0,1) values as illustrated in Table 4, we observe a reverse relationship, with decreasing specificity and increasing sensitivity. This leads to a notable 8.4% improvement in specificity; however, it comes at the expense of a reduction in sensitivity.

For both the LIDC and BreakHis datasets, we observe an increasing trend of sensitivity with increasing c(1,0) values, and an increasing trend of specificity with increasing c(0,1) values. Notably, in both datasets, sensitivity increases with the decrease in specificity for higher c(1,0) values and lower c(0,1) values.

Our results indicate that by tuning the cost values, we can achieve higher sensitivity or specificity. Significantly, the overall accuracy remains consistent in the majority of cases.

5. Conclusion and Future Work

In this study, we proposed the incorporation of a cost-sensitive values into the activation function for deep learning-based CAD systems. Specifically, it addresses one of the most common problems in CAD, which is improving true positives measured using sensitivity, by adjusting the cost values without significantly impacting accuracy. The effectiveness and robustness of the model are demonstrated through theoretical analysis and experiments on different datasets. Compared with previous work on LIDC and BreakHis, this is the first study that utilizes cost values in activation functions to enhance sensitivity. Our findings strongly suggest that incorporating cost values as hyperparameters in future CAD systems holds promising benefits, with statistically significant increases of 3.84% and 5.4% in sensitivity, while maintaining overall accuracy, for LIDC and BreakHis Data.

While our study has yielded valuable insights, certain constraints, limited to the datasets used in this research, may impact the generalizability of our findings. Furthermore, our study predominantly relied on a single activation function and the use of binary cross entropy loss.

Future investigations can involve the inclusion of datasets from Medical Imaging and Data Resource Center[32] (MIDRC), The Cancer Imaging Archive[33] (TCIA), and other medical-related databases, which can provide a broader perspective on model performance. To improve model performance and address imbalanced datasets, we can explore various activation functions and loss functions, such as focal loss.[34] In addition to this, we will also explore different thresholds for classifying instances as malignant or benign based on the sigmoid activation function output. Currently, we use a default threshold of 0.5. We will also investigate the integration of the proposed cost values with the group Distributionally Robust Optimization[35] (gDRO) algorithm. This approach aims to enhance the worst group performance while preserving the overall CAD system's effectiveness. Additionally, we plan to delve into the impact of cost-sensitive activation functions on multi-class classification. These endeavors collectively aim to improve the accuracy and effectiveness of the cost-sensitive learning approach in medical diagnosis, ultimately benefiting diagnostic decision-making and patient outcomes.

References

1. H. Li, N. Zeng, P. Wu and K. Clawson, Cov-net: A computer-aided diagnosis method for recognizing covid-19 from chest x-ray images via machine vision, *Expert Systems with Applications* **207**, p. 118029 (2022).
2. P. Wu, H. Li, N. Zeng and F. Li, Fmd-yolo: An efficient face mask detection method for covid-19 prevention and control in public, *Image and Vision Computing* **117**, p. 104341 (2022).
3. N. Zeng, P. Wu, Z. Wang, H. Li, W. Liu and X. Liu, A small-sized object detection oriented

multi-scale feature fusion approach with application to defect detection, *IEEE Transactions on Instrumentation and Measurement* **71**, 1 (2022).
4. C. X. Ling and V. S. Sheng, Cost-sensitive learning and the class imbalance problem, *Encyclopedia of Machine Learning* **2011**, 231 (2008).
5. M. Phankokkruad, Cost-sensitive extreme gradient boosting for imbalanced classification of breast cancer diagnosis, in *2020 10th IEEE International Conference on Control System, Computing and Engineering (ICCSCE)*, 2020.
6. S. Ali, A. Majid, S. G. Javed and M. Sattar, Can-csc-gbe: Developing cost-sensitive classifier with gentleboost ensemble for breast cancer classification using protein amino acids and imbalanced data, *Computers in Biology and Medicine* **73**, 38 (2016).
7. M. Zieba, J. M. Tomczak, M. Lubicz and J. Światek, Boosted svm for extracting rules from imbalanced data in application to prediction of the post-operative life expectancy in the lung cancer patients, *Applied Soft Computing* **14**, 99 (2014).
8. N. Thai-Nghe, Z. Gantner and L. Schmidt-Thieme, Cost-sensitive learning methods for imbalanced data, in *The 2010 International Joint Conference on Neural Networks (IJCNN)*, 2010.
9. S. Wang, W. Liu, J. Wu, L. Cao, Q. Meng and P. J. Kennedy, Training deep neural networks on imbalanced data sets, in *2016 International Joint Conference on Neural Networks (IJCNN)*, 2016.
10. I. D. Mienye and Y. Sun, Performance analysis of cost-sensitive learning methods with application to imbalanced medical data, *Informatics in Medicine Unlocked* **25**, p. 100690 (2021).
11. S. Zhu and J. Wan, Cost-sensitive learning for semi-supervised hit-and-run analysis, *Accident Analysis & Prevention* **158**, p. 106199 (2021).
12. T. M. Khoshgoftaar, E. Geleyn, L. Nguyen and L. Bullard, Cost-sensitive boosting in software quality modeling, in *7th IEEE International Symposium on High Assurance Systems Engineering, 2002. Proceedings*, 2002.
13. T. Le, M. T. Vo, B. Vo, M. Y. Lee and S. W. Baik, A hybrid approach using oversampling technique and cost-sensitive learning for bankruptcy prediction, *Complexity* **2019**.
14. D. Devi, S. K. Biswas and B. Purkayastha, A cost-sensitive weighted random forest technique for credit card fraud detection, in *2019 10th International Conference on Computing, Communication and Networking Technologies (ICCCNT)*, 2019.
15. J. Xiao, X. Zhou, Y. Zhong, L. Xie, X. Gu and D. Liu, Cost-sensitive semi-supervised selective ensemble model for customer credit scoring, *Knowledge-Based Systems* **189**, p. 105118 (2020).
16. K. Li, B. Wang, Y. Tian and Z. Qi, Fast and accurate road crack detection based on adaptive cost-sensitive loss function, *IEEE Transactions on Cybernetics* (2021).
17. E. Wang, Y. Jiang, Y. Li, J. Yang, M. Ren and Q. Zhang, Mfcsnet: Multi-scale deep features fusion and cost-sensitive loss function based segmentation network for remote sensing images, *Applied Sciences* **9**, p. 4043 (2019).
18. D. Li, X. Ma, Y. Ren and S.-W. Teng, Rectified softmax loss with all-sided cost sensitivity for age estimation, *IEEE Access* **8**, 32551 (2020).
19. I. D. Mienye, Y. Sun and Z. Wang, An improved ensemble learning approach for the prediction of heart disease risk, *Informatics in Medicine Unlocked* **20**, p. 100402 (2020).
20. H. Yaghoobi, E. Babaei, B. M. Hussen and A. Emami, Ebst: an evolutionary multi-objective optimization based tool for discovering potential biomarkers in ovarian cancer, *IEEE/ACM Transactions on Computational Biology and Bioinformatics* **18**, 2384 (2020).
21. I. D. Mienye, Y. Sun and Z. Wang, Improved sparse autoencoder based artificial neural network approach for prediction of heart disease, *Informatics in Medicine Unlocked* **18**, p. 100307 (2020).
22. S. I. Ali, H. S. M. Bilal, M. Hussain, J. Hussain, F. A. Satti, M. Hussain, G. H. Park, T. Chung and S. Lee, Ensemble feature ranking for cost-based non-overlapping groups: A case study of chronic kidney disease diagnosis in developing countries, *IEEE Access* **8**, 215623 (2020).

23. H. Wang, Z. Cui, Y. Chen, M. Avidan, A. B. Abdallah and A. Kronzer, Predicting hospital readmission via cost-sensitive deep learning, *IEEE/ACM Transactions on Computational Biology and Bioinformatics* **15**, 1968 (2018).
24. R. Parikh, A. Mathai, S. Parikh, G. C. Sekhar and R. Thomas, Understanding and using sensitivity, specificity and predictive values, *Indian Journal of Ophthalmology* **56**, p. 45 (2008).
25. D. M. Naeger, M. P. Kohi, E. M. Webb, A. Phelps, K. G. Ordovas and T. B. Newman, Correctly using sensitivity, specificity, and predictive values in clinical practice: how to avoid three common pitfalls., *AJR. American Journal of Roentgenology* **200**, W566 (2013).
26. S. G. Armato III, G. McLennan, L. Bidaut, M. F. McNitt-Gray, C. R. Meyer, A. P. Reeves, B. Zhao, D. R. Aberle, C. I. Henschke, E. A. Hoffman *et al.*, The lung image database consortium (lidc) and image database resource initiative (idri): a completed reference database of lung nodules on ct scans, *Medical Physics* **38**, 915 (2011).
27. F. A. Spanhol, L. S. Oliveira, C. Petitjean and L. Heutte, A dataset for breast cancer histopathological image classification, *IEEE Transactions on Biomedical Engineering* **63**, 1455 (2015).
28. W. Rawat and Z. Wang, Deep convolutional neural networks for image classification: A comprehensive review, *Neural Computation* **29**, 2352 (2017).
29. O. Russakovsky, J. Deng, H. Su, J. Krause, S. Satheesh, S. Ma, Z. Huang, A. Karpathy, A. Khosla, M. Bernstein *et al.*, Imagenet large scale visual recognition challenge, *International Journal of Computer Vision* **115**, 211 (2015).
30. A. Nibali, Z. He and D. Wollersheim, Pulmonary nodule classification with deep residual networks, *International Journal of Computer Assisted Radiology and Surgery* **12**, 1799 (2017).
31. K. He, X. Zhang, S. Ren and J. Sun, Deep residual learning for image recognition, in *Proceedings of the IEEE Conference on Computer Vision and Pattern Recognition*, 2016.
32. N. Baughan, H. M. Whitney, K. Drukker, B. Sahiner, T. Hu, M. McNitt-Gray, K. Myers, M. L. Giger *et al.*, Sequestration of imaging studies in midrc: a multi-institutional data commons, in *Medical Imaging 2022: Image Perception, Observer Performance, and Technology Assessment*, 2022.
33. K. Clark, B. Vendt, K. Smith, J. Freymann, J. Kirby, P. Koppel, S. Moore, S. Phillips, D. Maffitt, M. Pringle *et al.*, The cancer imaging archive (tcia): maintaining and operating a public information repository, *Journal of digital imaging* **26**, 1045 (2013).
34. M. Mulyanto, M. Faisal, S. W. Prakosa and J.-S. Leu, Effectiveness of focal loss for minority classification in network intrusion detection systems, *Symmetry* **13**, p. 4 (2020).
35. S. Sagawa, P. W. Koh, T. B. Hashimoto and P. Liang, Distributionally robust neural networks for group shifts: On the importance of regularization for worst-case generalization, *arXiv preprint arXiv:1911.08731* (2019).

VetLLM: Large Language Model for Predicting Diagnosis from Veterinary Notes

Yixing Jiang, Jeremy A. Irvin, Andrew Y. Ng and James Zou†
Stanford University, Stanford, CA, United States †*E-mail: jamesz@stanford.edu*

Lack of diagnosis coding is a barrier to leveraging veterinary notes for medical and public health research. Previous work is limited to develop specialized rule-based or customized supervised learning models to predict diagnosis coding, which is tedious and not easily transferable. In this work, we show that open-source large language models (LLMs) pretrained on general corpus can achieve reasonable performance in a zero-shot setting. Alpaca-7B can achieve a zero-shot F1 of 0.538 on CSU test data and 0.389 on PP test data, two standard benchmarks for coding from veterinary notes. Furthermore, with appropriate fine-tuning, the performance of LLMs can be substantially boosted, exceeding those of strong state-of-the-art supervised models. VetLLM, which is fine-tuned on Alpaca-7B using just 5000 veterinary notes, can achieve a F1 of 0.747 on CSU test data and 0.637 on PP test data. It is of note that our fine-tuning is data-efficient: using 200 notes can outperform supervised models trained with more than 100,000 notes. The findings demonstrate the great potential of leveraging LLMs for language processing tasks in medicine, and we advocate this new paradigm for processing clinical text.

Keywords: Diagnosis Extraction, Veterinary Notes, Veterinary Medicine, Large Language Models, LLM, Foundation Models.

1. Introduction

Most veterinary records are in free-text forms without structured diagnostic codes, making it difficult to use for medical research, public health monitoring or quality-improvement programs.[1] For example, the eligibility criteria for many clinical trails include diagnosis history. It is challenging to accurately identify certain cohorts which meet specific diagnostic criteria for translational research without structured diagnosis codes for each individual animal. A small number of large veterinary centers hire dedicated coding staff to manually apply disease codes to clinical records, which is labor-intensive, while most veterinary clinics do not code the notes.[1] One potential solution that previous works have explored is to develop systems which automatically code veterinary notes. However, these approaches have been limited to specialized rule-based or machine learning-based models, which are tedious to design and do not easily generalize well to new formats of reports.

Large language models (LLMs) have the potential to serve as an effective method for veterinary information extraction. There is rising interest in studying large language models, commonly referred to as types of "foundation models" (models which can be adapted to many different tasks). LLMs have a large number of parameters and are typically pre-trained on a large text corpus. They have shown promising performances on many NLP tasks, even in zero-shot and few-shot settings.[2,3] However, there is no study on how well those LLMs perform on analyzing veterinary notes. Besides, veterinary notes have shifted styles compared with general text available on Internet. For example, the vocabulary used is different and many acronyms are included. Given the pre-training corpus for most LLMs was sourced from Internet, veterinary notes are good examples to evaluate the performance of LLMs on atypical text.

Our contributions can be summarised as follows:

(1) We develop VetLLM to extract diagnostic information from veterinary notes and investigate its performance on dataset portions from two veterinary practices. Specifically we assess performance of models trained without any fine-tuning and with fine-tuning.
(2) We empirically show that LLMs can achieve promising performance on the task of diagnosis extraction from veterinary notes. Base LLMs without finetuning can achieve reasonable performances under zero-shot settings. For example, Alpaca-7B can achieve an zero-shot F1 of 0.538 when evaluated on CSU test data.
(3) Fine-tuned VetLLM perform better by a large margin compared with strong state-of-the-art supervised models. When evaluating on external test data, VetLLM outperforms the VetTag model by 21% and 8% in F1 score and exact match score respectively.
(4) We find finetuning LLM for the diagnosis extraction task is data-efficient. More specifically, using 200 notes can outperform supervised models trained with more than 100,000 notes in terms of F1 score.
(5) We detail a new paradigm for processing medical text in section 5.3, and the findings show the superiority of this new paradigm which leverages LLMs. Code will be available at `https://github.com/stanfordmlgroup/VetLLM`.

2. Related Work

Many previous studies have studied the automatic information extraction from clinical notes, including MetaMap,[4] statistical modeling,[5] text CNN,[6] and long-short-term memory network (LSTM).[7] More specifically, DeepTag[1] and VetTag[8] are some previous work on this dataset. DeepTag extended a bidirectional LSTM architecture with a hierarchical loss function and achieved better performances.[1] VetTag further leveraged transformers architecture[9] and conducted large-scale pre-training on veterinary text, leading to the current state-of-the-art performances on this dataset.[8]

There has been some recent studies showing many of those LLM are "generalist" in the sense that they can perform reasonably well on a large variety of tasks across domains.[2,3] In the medical domain, LLM have shown promising performances for many tasks, including information extraction,[10] medical Q&A,[3,11,12] generating USMLE-style questions[13] and radiology reports.[14] There are also some commentaries on the potential and regulation of LLM for medical use cases.[15–18]

3. Methods

The task is to extract diagnosis from veterinary notes which are in the free-text form. The extraction task can be formulated as a multi-class multi-label classification problem. Specifically, for each disease, the model should output whether there is positive mention in the veterinary clinical note.

The development pipeline included data cleaning, model selection, prompt design, resolver design, model finetuning, and system evaluation.

3.1. *Data*

The DeepTag[1] dataset was used for the project. It contains over 100K expert labeled veterinary notes from the Colorado State University (CSU) and a private practice clinic (PP). Both CSU portion and PP portion used here were previously used for VetTag, and it's noteworthy that VetTag was also pre-trained on another much larger dataset. In this project, we selected nine most prevalent diseases for analysis due to computational constraints. These nine diseases covered at least one diagnosis in around 90% cases in both CSU and PP portion, and they covered around 60% to 70% of all top-level disease labels. We removed incomplete reports which are shorter than 200 characters after manual review to ensure data quality.

The CSU portion contains 112,557 veterinary notes from the Colorado State University College of Veterinary Medicine and Biomedical Sciences. Each note was labeled with a set of SNOMED-CT codes by veterinarians at Colorado State. Colorado State is a tertiary referral center with an active and nationally recognized cancer center. We kept the same train/val/test as VetTag for fair comparison.

The PP portion contains 586 discharge summaries curated from a commercial veterinary practice located in Northern California, and six notes were removed due to incompleteness. Two veterinary experts applied SNOMED-CT codes to these records. Records with coding discrepancies were reviewed by both coders to reach a consensus on each record. This dataset is drastically different from the CSU dataset. PP notes are written often in an informal style, evidenced by their shorter length and usage of abbreviations. The PP data also has a different diagnosis distribution compared to a specialized academic cancer center CSU. It is of note that all notes in the PP portion are used for testing serving as an external validation dataset.

Table 1 shows the details of the dataset. Here is one example of veterinary note from PP portion together with the labels: *cried at home 8 body condition not drinking excess not urinating more frequently appetite is normal energy level is good skin is normal heart auscultates normal abnormal findings pain over l-s pain pulling hips back x rays ventral dorsal hips cauda equina looked perfect* **Expert annotated diseases:** 'Hypersensitivity condition', 'Propensity to adverse reactions'

Given both of these datasets are private and the data usage agreement prohibits data sharing with third parties, only models hosted locally can be used to analyze the data.

3.2. *Models*

Alpaca-7B and VetLLM Alpaca-7B[19] was used as the base LLM model as it has been instruction fine-tuned and is publicly available. Furthermore, a subset of CSU training split

Table 1. Descriptive statistics of the DeepTag dataset

	CSU	PP
# of notes	112,557	580
Size of test split	5483	580
Avg # of words	368	253

was used to further fine-tune Alpaca-7B using low-rank adaption,[20] leading to VetLLM. The details of fine-tuning was discussed in Section 3.4. The temperature for both Alpaca-7B and VetLLM was set to zero to allow reproducibility.

VetTag The supervised baseline model was the one developed in the VetTag paper, achieving state-of-the-art performances on the dataset.[8] It was pre-trained on a large corpus of unlabelled veterinary notes (917,665 notes) using casual language modeling, and then fine-tuned using the training split (101,301 notes) of CSU portion. The prediction logits from VetTag were obtained from the VetTag team, and the logits corresponding to the nine diseases were extracted to calculate the metrics.

KeywordMatch Another baseline model was to use keyword matching. The synonyms of the diseases were retrieved using WordNet, and fuzzy matching with the partial ratio metric was used. The model would return positive if the partial ratio between the veterinary note and the disease names was above 80%.

In short, four models would be compared: Alpaca-7B (LLM baseline), VetLLM (fine-tuned LLM), VetTag (supervised baseline) and KeywordMatch.

3.3. *Prompts and Resolvers*

The guiding principle for prompt design is to follow the format of the instruction tuning set and to be clear and specific. We just tried a small number of prompts on ten notes from the CSU validation split, and the main metric was whether the output was easily resolvable. Figure 1 shows the prompt used along with one example input and output from the VetLLM. The prompt queried the LLM with one disease each time rather than querying the LLM to list down all diagnosis. This design choice greatly simplified the resolver design. The query was conducted on two A4000 GPUs.

After getting the response from LLM, a resolver was utilized to convert the text response into a structured prediction. The resolver used in this study was simple, and it first converted the decoded text response into lower case and stripped any trailing space on the left. A positive prediction was rendered if the resultant string started with "yes", and a negative prediction was rendered if it started with "no". Otherwise, the case was rendered as un-resolvable.

3.4. *Finetuning*

A subset of 5,000 notes were randomly sampled from the CSU training split, and this subset was used for fine-tuning Alpaca-7B using low-rank adaption (LoRA)[20] and four A4000 GPUs. We chose LoRA as it generally provides superior performances and induces no extra inference overhead. The fine-tuning samples were generated using the same prompt template described

Prompt Template

Below is an instruction that describes a task, paired with an input that provides further context. Write a response that appropriately completes the request.

Instruction:
Answer the following yes/no question based on the veterinary note delimited by triple backticks:

Input:
```{text}```

Does this animal have {disease}?

Response:

Example Input to VetLLM

Below is an instruction that describes a task, paired with an input that provides further context. Write a response that appropriately completes the request.

Instruction:
Answer the following yes/no question based on the veterinary note delimited by triple backticks:

Input:
```tt w=12.9 lbs itching constantly flea allergy dermatitis 7 body condition not drinking excess not urinating more frequently appetite is normal energy level is good skin is normal heart auscultate s normal abnormal findings lots of hair loss fleas found other cat treatment , prescription dex dro ps disc flea allergy dermatitis and not spraying yard roos lmor 8 / 28```

Does this animal have Hypersensitivity condition?

Response:

Example Output from VetLLM

yes</s>

Fig. 1. Prompt Template with Example Input and Output from VetLLM

in the previous section. The hyper-parameters were included in Appendix A. A subset of 200 notes were randomly sampled from the CSU validation split to form the validation set for fine-tuning. An early stopping callback with a patience of five was added.

To study the data efficiency of fine-tuning, the 5,000 notes subset was further sampled into 2000, 1000, 500 and 200 notes sequentially. Consequently, each subset of a smaller size is strictly a subset of the one of a larger size. And these subsets were each used as the fine-tuning set. In short, five fine-tuned Alpaca-7B models were trained.

3.5. *Evaluation Metrics*

As a multi-class multi-label classification problem, there were metrics for both the overall prediction and each individual class. More specifically, each model was evaluated based on exact match (EM, the fraction of notes where the algorithm's predicted diagnoses exactly

Table 2. Quantitative evaluation on classification

Model	CSU				PP			
	Exact Match	Precision	Recall	F1	Exact Match	Precision	Recall	F1
VetLLM	53.5% ± 0.7%	0.726	0.774	0.747 ± 0.004	38.0% ± 2.1%	0.661	0.630	0.637 ± 0.015
Alpaca-7B (zero shot)	34.0% ± 0.6%	0.604	0.527	0.538 ± 0.005	22.0% ± 1.7%	0.485	0.375	0.389 ± 0.017
VetTag (supervised)	49.3% ±0.7%	0.798	0.492	0.592 ±0.006	30.1% ±1.9%	0.680	0.344	0.422 ±0.018
KeywordMatch	29.1%	0.442	0.002	0.003	24.9%	0.050	0.006	0.010

match the expert diagnoses), precision (the fraction of notes with positive predictions that match the expert diagnoses), recall (the fraction of notes where the expert diagnoses are successfully retrieved), and F1 (the harmonic mean of precision and recall). The last three metrics was macro-averaged across classes to get the overall metrics. The standard deviations of those metrics were calculated using bootstrapping with 1,000 re-samples.

4. Results

4.1. *Overall Evaluation on Classification*

Table 2 shows the quantitative evaluation results averaged across classes. When evaluating on the CSU portion, Alpaca-7B performs reasonably in a zero-shot manner, with only 6% gap in F1 compared with the supervised baseline. With VetLLM which was fine-tuned using 5,000 notes, the performances greatly improve, leading to a 21% boost in F1 and 19% boost in exact match score.

4.2. *Stratified Evaluation on Classification*

Figure 2 and 3 show the F1 metrics of three models evaluated on each class. They show the VetLLM model, fine-tuned from Alpaca-7B, outperforms the supervised VetTag model in each single class on both in-distribution data (CSU portion) and out-of-distribution data (PP portion). They also show significant improvements in performances in most classes after fine-tuning, and there is no degradation in any class after fine-tuning.

4.3. *Data-efficiency of Fine-tuning*

Figure 4 shows how the performance improves as the number of fine-tuning samples increase. It shows only using fewer than 200 notes can exceed performances of the supervised model, demonstrating the data-efficiency of fine-tuning LLM. It is of note the X-axis represents the number of veterinary notes used, so the size of fine-tuning set is nine times that the number of notes.

5. Discussion

The results show the promise of VetLLM for diagnosis extraction task from veterinary notes, which is inspiring. More broadly, they demonstrate the great potential of leveraging LLMs for processing medical text which is detailed in Section 5.3.

Although the performances are promising, they are sensitive to prompt design such as problem formulation, order of information and presence of extra information. In some cases,

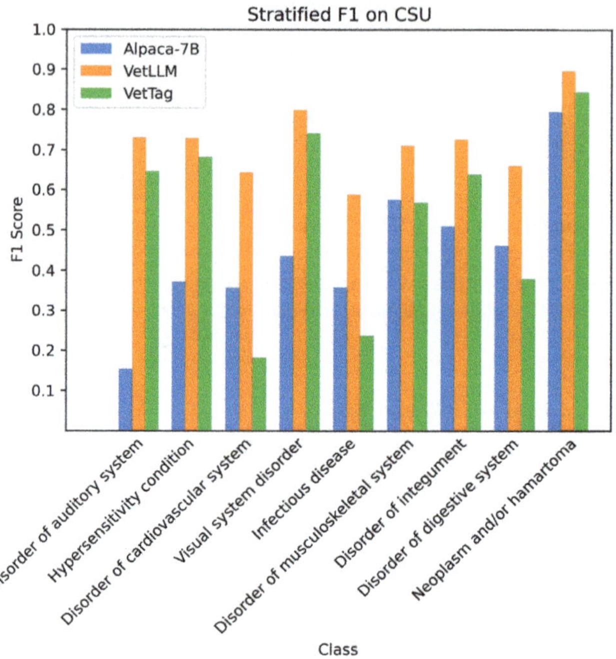

Fig. 2. Stratified F1 on CSU test data. Alpaca-7B is the base LLM model, and VetLLM is Alpaca-7B fine-tuned with 5,000 notes. VetTag is the state-of-the-art supervised model.

adding trailing spaces at the beginning of each line also affects the performances. It seems there is still no well-established systematic way of assessing LLM's sensitivity towards prompt designs, but the development of LLM is likely to benefit from ongoing research on AI alignment. Therefore, more comprehensive evaluation must be conducted or some post-hoc quality control measures must be taken if this system is to be deployed.

Also, the evaluation in this paper is limited to datasets from two centers in the United States. Veterinary notes from other veterinary medicine centers are likely to have different distributions which might affect performances.

5.1. *Error Analysis*

To gauge the knowledge embedded in LLM, the Alpaca-7B model was prompted to explain the top fourteen diseases. The responses from Alpaca-7B were included in Appendix B and manually reviewed in terms of relevance and factuality. The results indicate Alpaca can provide highly relevant and factually correct descriptions of diseases, hinting that the pre-training corpus might contain high-quality medical text describing various diseases.

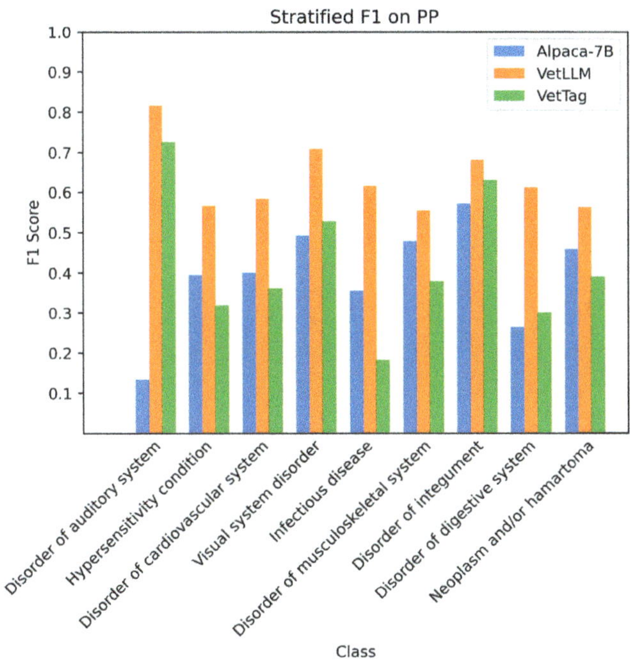

Fig. 3. Stratified F1 on PP test data. Alpaca-7B is the base LLM model, and VetLLM is Alpaca-7B fine-tuned with 5,000 notes. VetTag is the state-of-the-art supervised model.

Furthermore, the correlation between note length and performances were analyzed using the two portions, and the results are shown in Figure 5 and 6. The note lengths were binned into five quantiles. Based on the results, the exact match score is negatively correlated with the note length, while the trends for F1 score seem inconsistent.

5.2. *Computational Costs*

In the era of large models, computational costs and environmental impact of model training and inference have become more concerning. All estimates in this section are in the settings of four NVIDIA RTX A4000 GPUs launched in April 2021, and each A4000 has 16GB GPU memory. VetLLM was fine-tuned from Alpaca-7B using 5,000 notes, and the fine-tuning took around 48 hours with a micro batch size of one.

One limitation of VetLLM is it requires multiple pass for multi-class classification, while traditional supervised models can generate multi-class predictions with a single pass by using multiple neurons in the last layer. A single inference pass for VetLLM takes around 0.3 seconds, and the model loading before first use takes around 15 seconds. It means VetLLM is likely to

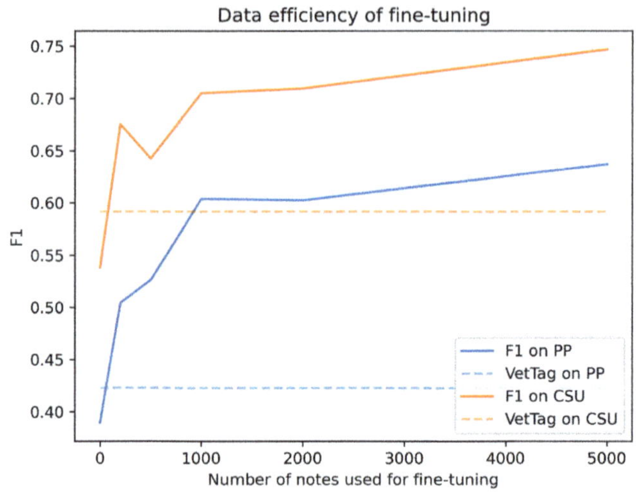

Fig. 4. Data efficiency plot. The number of notes here refer to the number of notes in the CSU training split used for fine-tuning Alpaca-7B. PP portion was only used for test. It is of note that VetTag used over 100,000 notes for fine-tuning.

have slower inference speed compared with traditional supervised models. Given the significant boost in performances and the application does not have strong real-time requirement, we think the increased inference time is reasonable. One mitigation, which we leave as future work, is to utilize a multi-label approach such as asking multiple questions in a single turn or asking the model to select all diseases present in the veterinary note.

5.3. *New Paradigm for Processing Medical Text*

In this paper, we demonstrate the potential of a new paradigm for processing medical text: starting with pre-trained large language models (LLM), then designing a prompt and resolver. The resolver interprets the output from the LLM and transforms the raw output into structured answers. After designing the prompt and resolver, the next step is to conduct a quick evaluation in a zero-shot or few-shot setting. If the performance is satisfactory, it is a good idea to proceed with more comprehensive evaluation and iteratively improve the prompt and resolver. If the performance is poor, it might be worth curating a small fine-tuning dataset and utilizing data-efficient techniques like LoRA to fine-tune the LLM. Evaluation and iterative refinement can be conducted after the fine-tuning.

Thanks to the great contributions from various communities, most of these steps have been implemented in various library packages or are available as API calls, thereby speeding up the entire pipeline. The traditional pipeline for processing medical text involves curating a large training dataset and training a specialized model via supervised learning. This pipeline

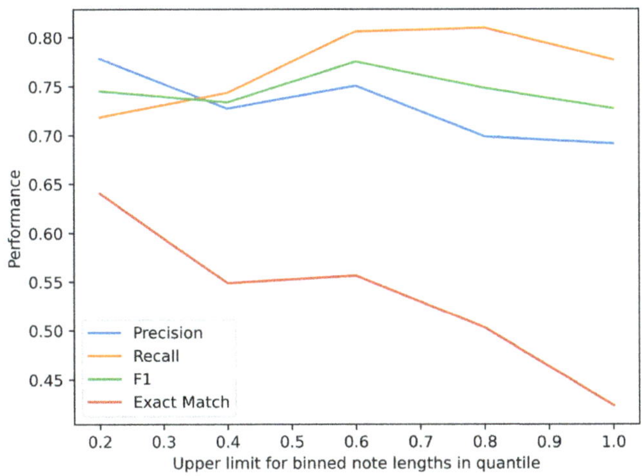

Fig. 5. Performances of VetLLM in terms of note length on CSU test data. The X-axis refers to the upper quantile of note length for each binned group. Exact match is a baseline evaluation metric, and F1 is used more frequently in practice.

tends to require significant resources, with the process often spanning several months or even years. This new paradigm might lower the barriers to building some interesting applications, with many potentially developed within weeks.

Beyond the great performance and fast iteration, another advantage of this new paradigm is the ability to easily expand classification categories. For example, the prompt can be modified to extract diagnosis of other diseases. In contrast, traditional supervised models might require extensive fine-tuning to include new classes.

6. Conclusion

In this study, large language models were used for diagnosis extraction task from veterinary notes. With fine-tuning only on a small number of notes, VetLLM outperform strong supervised models significantly. Given the time constraints, simple prompts and resolvers were used in the study. Richer prompt strategies can be explored, and robustness towards prompt variations should be examined.

In a broader sense, this project has shown the potential of LLMs to work on clinical data and be efficiently fine-tuned to achieve strong performances on downstream tasks. Although this study is limited to veterinary notes, we believe the new paradigm detailed in section 5.3 is generally applicable. Therefore, it is interesting to evaluate the performances of base LLMs and fine-tuned LLMs in other medical applications including human clinical notes. Furthermore, it might be interesting to conduct similar assessment using more advanced models or more domain-specific ones.

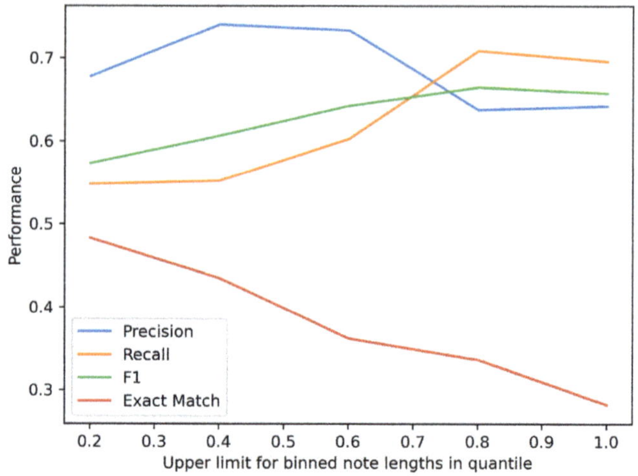

Fig. 6. Performances of VetLLM in terms of note length on PP test data. The X-axis refers to the upper quantile of note length for each binned group. Exact match is a baseline evaluation metric, and F1 is used more frequently in practice.

Acknowledgments

Yixing Jiang is supported by National Science Scholarship (PhD). We want to thank Anh (Tom) Nguyen and Tulika Jha for helping with some experiments. Besides, we want to thank Yuhui Zhang, Allen Nie, Federico Bianchi and Bryan He for helping with dataset and model access.

Appendix A. Hyper-parameters used for fine-tuning

Table A1 shows the hyper-parameters used for LoRA fine-tuning.

Table A1. Hyper-parameters used for fine-tuning

Name	Value
batch_size	32
max_epochs	3
cutoff_len	2048
train_on_inputs	False
lora_r	8
lora_alpha	16
lora_dropout	0.05
lora_target_modules	["q_proj", "v_proj"]

Appendix B. Alpaca-7B' responses to disease definition questions

- Disorder of auditory system : Disorder of the auditory system is a broad term that refers to any disorder of the ear or hearing system that results in a hearing impairment or loss. This can include conditions such as conductive hearing loss, sensorineural hearing loss, mixed hearing loss, and other disorders that affect the ability to hear.
- Hypersensitivity condition : Hypersensitivity is an abnormal increase in the body's reactivity to a substance or stimulus, such as an allergy or an autoimmune disorder. Hypersensitivity reactions can range from mild to severe and can affect the skin, respiratory tract, gastrointestinal tract, and other organs. Symptoms may include rash, hives, eczema, congestion, coughing, wheezing, and other respiratory symptoms, as well as digestive issues.
- Disorder of cardiovascular system : Disorder of the cardiovascular system is any abnormality of the heart, blood vessels, or blood that affects how the body circulates blood. This can include conditions such as high blood pressure, heart disease, stroke, and blood clots.
- Visual system disorder : Visual system disorder is a condition that affects the way the eyes send and process visual information. It can cause problems with vision, such as blurred vision, double vision, or difficulty with eye coordination and tracking. It can also lead to headaches, eyestrain, and other vision-related problems.
- Infectious disease : Infectious disease is a broad term used to describe any disease that is caused by the presence, growth, and/or activity of a pathogen, such as a virus, bacteria, parasite, or fungus. These pathogens can cause illnesses ranging from mild to severe, and can even be fatal. Infectious diseases are spread through contact with infected individuals, contact with contaminated surfaces, or contact with contaminated food or water.
- Disorder of musculoskeletal system : Disorder of the musculoskeletal system is a general term that describes any injury or disease that affects the muscles, bones, joints, and associated structures. Examples of disorders of the musculoskeletal system include fractures, sprains, strains, tendinitis, bursitis, and muscle pain.
- Disorder of integument : Disorder of the integument is a general term used to describe a group of skin disorders that affect the structure or function of the skin. These disorders can range from mild forms such as dry skin or eczema, to more severe forms such as psoriasis or dermatitis.
- Disorder of digestive system : Disorder of the digestive system is any problem with the digestive tract that impairs its functioning and can cause symptoms such as abdominal pain, nausea, vomiting, diarrhea, constipation, and other complications. Common disorders of the digestive system include gastroenteritis, irritable bowel syndrome, inflammatory bowel disease, and cancer.
- Neoplasm and/or hamartoma : Neoplasm is a general term for a tumor or growth that is caused by abnormal cellular proliferation, while hamartoma is a benign tumor composed of normal tissue.

References

1. A. Nie, A. Zehnder, R. L. Page, Y. Zhang, A. L. Pineda, M. A. Rivas, C. D. Bustamante and J. Zou, Deeptag: inferring diagnoses from veterinary clinical notes, *NPJ digital medicine* **1**, p. 60 (2018).
2. T. Brown, B. Mann, N. Ryder, M. Subbiah, J. D. Kaplan, P. Dhariwal, A. Neelakantan, P. Shyam, G. Sastry, A. Askell *et al.*, Language models are few-shot learners, *Advances in neural information processing systems* **33**, 1877 (2020).
3. H. Nori, N. King, S. M. McKinney, D. Carignan and E. Horvitz, Capabilities of gpt-4 on medical challenge problems, *arXiv preprint arXiv:2303.13375* (2023).
4. A. R. Aronson and F.-M. Lang, An overview of metamap: historical perspective and recent advances, *Journal of the American Medical Informatics Association* **17**, 229 (2010).
5. M. Subotin and A. R. Davis, A method for modeling co-occurrence propensity of clinical codes with application to icd-10-pcs auto-coding, *Journal of the American Medical Informatics Association* **23**, 866 (2016).
6. Y. Zhang and B. Wallace, A sensitivity analysis of (and practitioners' guide to) convolutional neural networks for sentence classification, *arXiv preprint arXiv:1510.03820* (2015).
7. Z. C. Lipton, D. C. Kale, C. Elkan and R. Wetzel, Learning to diagnose with lstm recurrent neural networks, *arXiv preprint arXiv:1511.03677* (2015).
8. Y. Zhang, A. Nie, A. Zehnder, R. L. Page and J. Zou, Vettag: improving automated veterinary diagnosis coding via large-scale language modeling, *NPJ digital medicine* **2**, p. 35 (2019).
9. A. Vaswani, N. Shazeer, N. Parmar, J. Uszkoreit, L. Jones, A. N. Gomez, Ł. Kaiser and I. Polosukhin, Attention is all you need, *Advances in neural information processing systems* **30** (2017).
10. M. Agrawal, S. Hegselmann, H. Lang, Y. Kim and D. Sontag, Large language models are few-shot clinical information extractors, in *Proceedings of the 2022 Conference on Empirical Methods in Natural Language Processing*, 2022.
11. K. Singhal, S. Azizi, T. Tu, S. S. Mahdavi, J. Wei, H. W. Chung, N. Scales, A. Tanwani, H. Cole-Lewis, S. Pfohl *et al.*, Large language models encode clinical knowledge, *arXiv preprint arXiv:2212.13138* (2022).
12. T. H. Kung, M. Cheatham, A. Medenilla, C. Sillos, L. De Leon, C. Elepaño, M. Madriaga, R. Aggabao, G. Diaz-Candido, J. Maningo *et al.*, Performance of chatgpt on usmle: Potential for ai-assisted medical education using large language models, *PLoS digital health* **2**, p. e0000198 (2023).
13. S. L. Fleming, K. Morse, A. M. Kumar, C.-C. Chiang, B. Patel, E. P. Brunskill and N. Shah, Assessing the potential of usmle-like exam questions generated by gpt-4, *medRxiv* , 2023 (2023).
14. D. Van Veen, C. Van Uden, M. Attias, A. Pareek, C. Bluethgen, M. Polacin, W. Chiu, J.-B. Delbrouck, J. M. Z. Chaves, C. P. Langlotz *et al.*, Radadapt: Radiology report summarization via lightweight domain adaptation of large language models, *arXiv preprint arXiv:2305.01146* (2023).
15. R. Li, A. Kumar and J. H. Chen, How chatbots and large language model artificial intelligence systems will reshape modern medicine: Fountain of creativity or pandora's box?, *JAMA Internal Medicine* (2023).
16. A. J. Thirunavukarasu, D. S. J. Ting, K. Elangovan, L. Gutierrez, T. F. Tan and D. S. W. Ting, Large language models in medicine, *Nature Medicine* , 1 (2023).
17. B. Meskó and E. J. Topol, The imperative for regulatory oversight of large language models (or generative ai) in healthcare, *NPJ Digital Medicine* **6**, p. 120 (2023).
18. S. Gilbert, H. Harvey, T. Melvin, E. Vollebregt and P. Wicks, Large language model ai chatbots require approval as medical devices, *Nature Medicine* , 1 (2023).
19. R. Taori, I. Gulrajani, T. Zhang, Y. Dubois, X. Li, C. Guestrin, P. Liang and T. B.

Hashimoto, Stanford alpaca: An instruction-following llama model `https://github.com/tatsu-lab/stanford_alpaca`, (2023).
20. E. J. Hu, Y. Shen, P. Wallis, Z. Allen-Zhu, Y. Li, S. Wang, L. Wang and W. Chen, Lora: Low-rank adaptation of large language models, *arXiv preprint arXiv:2106.09685* (2021).

Impact of Measurement Noise on Genetic Association Studies of Cardiac Function*

Milos Vukadinovic[1,2,4], Gauri Renjith[1], Victoria Yuan[1,2], Alan Kwan[1,4], Susan C. Cheng[1], Debiao Li[4], Shoa L. Clarke[5,†], David Ouyang[1,6,†]

1. *Department of Cardiology, Smidt Heart Institute, Cedars-Sinai Medical Center, Los Angeles, CA*
2. *Department of Bioengineering, University of California Los Angeles, Los Angeles, CA*
3. *Department of Medicine, University of California Los Angeles, Los Angeles, CA*
4. *Biomedical Imaging Research Institute, Cedars-Sinai Medical Center, Los Angeles, CA*
5. *Department of Medicine, Division of Cardiovascular Medicine, Stanford University, Stanford, CA*
6. *Division of Artificial Intelligence in Medicine, Cedars-Sinai Medical Center, Los Angeles, CA*

†. Co-senior author

Email: shoa@stanford.edu, David.ouyang@cshs.org

1 Abstract

Recent research has effectively used quantitative traits from imaging to boost the capabilities of genome-wide association studies (GWAS), providing further understanding of disease biology and various traits. However, it's important to note that phenotyping inherently carries measurement error and noise that could influence subsequent genetic analyses. The study focused on left ventricular ejection fraction (LVEF), a vital yet potentially inaccurate quantitative measurement, to investigate how imprecision in phenotype measurement affects genetic studies. Several methods of acquiring LVEF, along with simulating measurement noise, were assessed for their effects on ensuing genetic analyses. The results showed that by introducing just 7.9% of measurement noise, all genetic associations in an LVEF GWAS with almost forty thousand individuals could be eliminated. Moreover, a 1% increase in mean absolute error (MAE) in LVEF had an effect equivalent to a 10% reduction in the sample size of the cohort on the power of GWAS. Therefore, enhancing the accuracy of phenotyping is crucial to maximize the effectiveness of genome-wide association studies.

Keywords: Precision phenotyping; Genome-Wide association study; Left ventricular ejection fraction; Cardiac magnetic resonance imaging; UK Biobank

---

* This work is partially supported by the National Institutes of Health NIH K99 HL157421.

2 Introduction

Cardiovascular disease is the leading cause of death in the world, and significant work has been undertaken to understand the mechanisms of disease and develop preventive measures. By studying the human genome, insights have been obtained to understand pathways and mechanisms of function and disease risk, and in recent studies, researchers have moved beyond binary labels of disease diagnosis to quantitative phenotypes to obtain greater power in assessing the relationship between genotype and phenotype[1–4]. From quantitative laboratory biomarkers elucidating the relationship between hypercholesterolemia and coronary artery disease[5] to imaging characteristics in population cohorts[4] revealing the genetic determinants of cardiovascular development [6,7], quantitative assessments of health provide additional signal compared to conventional binary labels of disease.

Despite its relative frequency, critical public health importance, and often penetrant inheritance, heart failure has relatively few known genetic risk factors. Early classic genetic studies were not able to identify many genetic associations with measurements determined by echocardiography[8]. Recent studies with larger cohorts and measurements from cardiac MRI have been able to find additional loci of relevance and reaffirm previously suspected variants[2], suggesting both larger sample sizes, as well as improvements in phenotyping precision, can improve our understanding of the human disease.

While quantitative traits often have more power than binary labels of disease, the issue of measurement error in quantitative traits is a known problem[9]. For example, left ventricular ejection fraction (LVEF) as measured by echocardiography can have measurement variation up to 7 - 10%[10,11], impacting downstream analyses. We use LVEF, the most prevalent metric of cardiac function, as an example of an important but noisy measurement to explore the impact of measurement variability on downstream genetic association studies. We compare various methods to obtain the same phenotypic measurement as well as introduce simulated noise in the phenotype measurement to evaluate the relative impact of measurement noise and sample size on downstream genetic studies.

Table 1. Cohort baseline characteristics

Characteristic	Mean or n
N	39624
Age at MRI	54.9 ± 7.47
Male	18933 (47.8%)
Self-identified White British	33726 (85.1%)
Body mass index (kg/m^2)	26.5 ± 4.19
Hypertension	2487 (6.3%)
Pulse rate	67.9 ± 10.9
LV ejection fraction (%)	55.4 (6.78)
LV end diastolic volume (mL)	141
LV end systolic volume (mL)	64.1

3 Methods

3.1 Cohort

The UK Biobank is a population-based cohort that links genetic and phenotypic data for approximately 500,000 adult participants from the United Kingdom [12,13]. We focused on 39,624 participants who had InlineVF measured LVEF [14], cardiac MRI, and genetic data available. Before running Genome-Wide Association Studies this cohort was passed through additional quality check filters **(Figure A1)**.

3.2 Multiple Approaches to Measure LVEF

Multiple methods of calculating LVEF from the same underlying imaging data were used to assess the impact of phenotyping precision on downstream analyses. First, the UKB provides automated LVEF measurements derived from MRI using Inline VF software[15], however, this is presented without manual quality control. To compare alternative automated approaches, we also derived LVEF from MRIs using the deep learning segmentation approach suggested by Bai et al [6]. From the short-axis view videos, segmentation was performed, we calculated the LV volume for each frame with Simpson's method and used the following LVEF formula:

$$\frac{ED\ Volume - ES\ Volume}{ED\ Volume} \times 100 \qquad (1)$$

To simulate reader variability, additional experiments were performed introducing Gaussian noise with a mean of 0 and a standard deviation (sd) ranging from [1,10]. We generated multiple phenotypic measurements from the same underlying imaging data, gradually incrementing Gaussian noise, and performed GWAS on each to investigate how measurement error/imprecision affects genetic associations.

Additionally, we further compared results with two final approaches to assess LVEF. When visually assessing LVEF, clinicians often round the value to the nearest 5%, thus we generated a set of phenotype labels by rounding LVEF values to the nearest multiple of 5. For the final comparison, we generated binary LVEF labels by categorizing values as normal or abnormal, with normal values ranging from 52-72 for males and 54-74 for females.

3.3 Genome-wide association study

We used the UKB imputed genotype calls in BGEN v1.2 format. Samples were genotyped using the UK BiLEVE or UK Biobank Axiom arrays. Imputation was performed using the Haplotype Reference Consortium panel and the UK10K+1000 Genomes panel[12]. We used the QC files provided by UKB to create a GWAS cohort consisting of subjects who did not withdraw, were of inferred European ancestry, and were unrelated. Subjects with a genotype call rate < 0.98 were also removed. We considered variants with a minor allele frequency (MAF) ≥ 0.01, and we required genotyped variants to have a call rate ≥ 0.95 and imputed variants to have an INFO score ≥ 0.3. Variants with a Hardy-Weinberg equilibrium P value $< 1\text{x}10^{-20}$ were excluded. After variant filtering, we were left with 9774199 filtered variants. GWAS was done on a Spark 3.1.1

cluster, using the library Hail 0.2 with Python version 3.6. The GWAS was adjusted for age at MRI and sex. We used the conventional P value of 5×10^{-8} as the threshold for defining genome-wide significance.

3.4 Assessing Association Power's Relationship with Cohort Size

Apart from noise in phenotype measurements, we also evaluate the effect of cohort decrease on GWAS results. We generated 6 different phenotype files where, starting from the original LVEF cohort (39,624), we keep 90% (35,661), 80% (31,699), 70% (27,736), 60% (23,774), 50% (19812), and 40% (15850) of the samples. Cohort decrease was performed before GWAS QC, and for each step the selection of samples to be excluded was random. Inspecting the effect of cohort decrease helps us define the relationship between the number of LVEF samples and GWAS power.

3.5 SNP-based accuracy

We use an accuracy metric to determine the amount of overlap in significant SNPs between the baseline GWAS results and noise-modified GWAS results. First, we remove all non-significant SNPs by excluding SNPs with a p-value less than 5×10^{-8}, which is the Bonferroni corrected p-value threshold. Then, we consider significant SNPs found in both the base results and noise-modified results as true positives (TP), the SNPs found only in the noise-modified results as false positives (FP), and the SNPs not found in the noise-modified results but found in the base results as false negatives (FN). We then calculate

$$SNP_{accuracy} = \frac{TP}{TP+FP+FN} \tag{2}$$

3.6 GWAS Sensitivity

Sensitivity determines the amount of overlap in significant loci between the baseline GWAS results and noise-modified GWAS results. Specifically, given that $peaks_{base}$ is the number of significant loci in base GWAS, and $peaks_{correct}$ is the number of significant loci that persisted in noise GWAS then

$$Sensitivity = \frac{peaks_{correct}}{peaks_{base}} \tag{3}$$

The number of loci and their position can be determined by manual inspection, but we also developed an automatic method. Our automatic method applies a hierarchical clustering algorithm on SNPs above the significance threshold line to determine the number and the position of loci from both GWAS, which we then use to compute $peaks_{base}$ and $peaks_{correct}$.

3.7 Heritability

Heritability is a measure of the level of influence genetic variation has on a given trait's phenotypic variation. To estimate SNP heritability based on GWAS summary statistic we use command line tool LDSC[16]. LDSC performs LD score regression between GWAS test statistic χ_j^2 and per SNP LD scored which allows for the estimation of h_g^2

4 Results

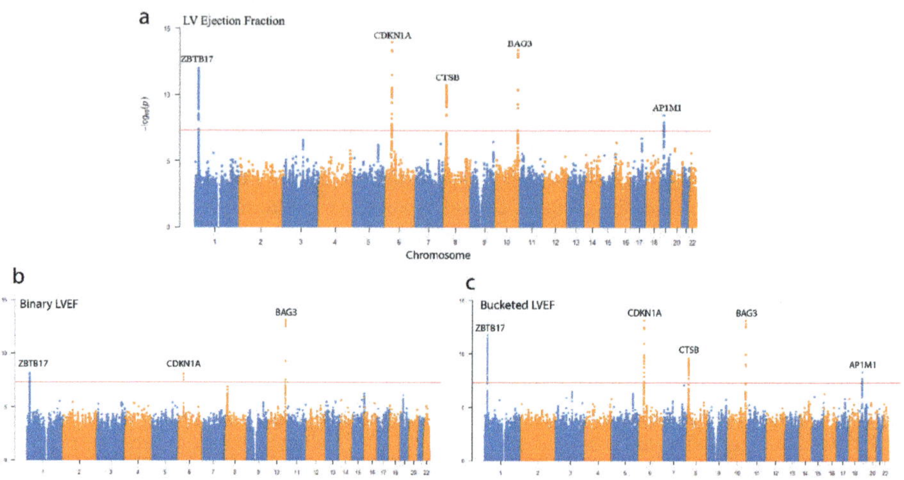

Figure 1 Manhattan plots for genome-wide association studies on UK Biobank reported left ventricular ejection fraction **a,** GWAS on continuous LVEF measurements **b,** GWAS on Normal/Abnormal LVEF where the range for normal is 52-72 in male and 54-74 female population **c,** GWAS on LVEF bucketed to the nearest multiple of 5

4.1 Quantitative phenotypes improve power of association studies

The study cohort for all analyses consisted of 39,624 adult unrelated subjects of European ancestry (**Table 1**). As a baseline, we first conducted a GWAS of the LVEF phenotype released with the UKBB cardiac MRI data. We identified 5 loci at genome-wide significance on chromosomes 1, 6, 8, 10, and 19 near genes *ZBTB17*, *CDKN1A*, *CTSB*, *BAG3*, and *AP1M1* (**Figure 1**). In comparison, for an LVEF phenotype binarized to simply abnormal or normal, multiple previously detected loci lost genome-wide significance (including loci for *CTSB* and *AP1M1*). Similarly, recognizing the inherent variation present in measuring LVEF, we additionally compared the results if the LVEF was bucketed to 5% bins and showed such imprecision decreased statistical power in all SNPs in the association study compared to the continuous LVEF baseline phenotype.

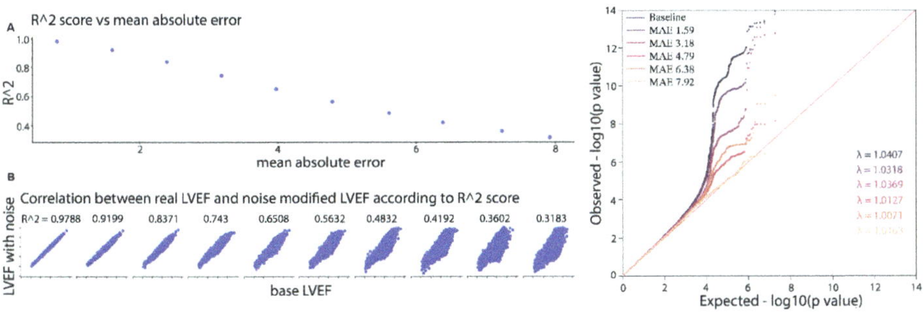

Figure 2 Impact of noise in LVEF on GWAS **a,** Visualizing r2 score, mean absolute error, and the distribution of noise-modified-LVEF with respect to the baseline LVEF. **b**, Q-Q plots of P values from GWAS summary statistics for different levels of noise

4.2 Phenotype noise degrades power of association studies

To investigate the effect of measurement imprecision on GWAS power, we performed a series of association studies while introducing noise in the range of known clinician variation **(Figure 2)**. Simulated variation to the LVEF measurement naturally increases in mean absolute error. Noise with a gaussian standard deviation of 5 results in a mean absolute error of 3.97% and R2 of 0.65 (**Table A1**), and results in the loss of genome-wide significance for the *AP1M1* loci on chromosome 19. As we increase phenotypic noise in the range of clinical variation, heritability and power gradually declines and the noise equivalent to 7.92% MAE results in a complete loss of genomic-wide significance (**Table 2**). Given echocardiography is known to have a clinician-to-clinician variation of the same or greater MAE[10], such measurement imprecision could contribute to the limited hits in historial echocardiography-derived GWAS[8].

Table 2. Metrics of genetic signal for each increase in SD

Noise SD	SNP Accuracy	Loci Sensitivity	Heritability
0%	1.0	1.0	0.1114 (00357)
1%	0.9377	1.0	0.1055 (0.0332)
2%	0.8547	1.0	0.0878 (0.0352)
3%	0.3675	1.0	0.1003 (0.0265)
4%	0.2537	0.8	0.089 (0.0256)
5%	0.3921	0.8	0.1208 (0.0355)
6%	0.0228	0.4	0.0179 (0.0271)
7%	0.0307	0.4	0.0482 (0.0247)
8%	0.0145	0.4	0.022 (0.0349)
9%	0.0020	0.2	0.0355 (0.0204)
10%	0	0	0.0477 (0.0214)

4.3 Comparison of Impact of Phenotype Noise vs Cohort Size

Given the summary statistics from 16 different GWAS, we modeled the relationship between noise and GWAS power (**Figure 3, Figure A4)**. There is a linear relationship between the increase in MAE and the decrease in GWAS power. We calculated that an increase of 1% in MAE causes the loci sensitivity to decrease by 13% (p=5.5e-6) and the SNP accuracy by 14% (p=6.6e-5). Experiments with other methods of introducing noise in assessing LVEF similarly show a decrease in genetic association with more imprecise measurements **(Figure A3, Table A2)**. A similar effect occurs with reductions in cohort size, as a 1% decrease in cohort size results in a 1.3% decrease in loci sensitivity (p=0.01) and a 1.9% decrease in SNP-based accuracy (p=0.0007). We found that a 1% MAE increase has the same effect on loci sensitivity as a 10.3% cohort decrease and the same effect as a 7.2% cohort decrease on SNP accuracy.

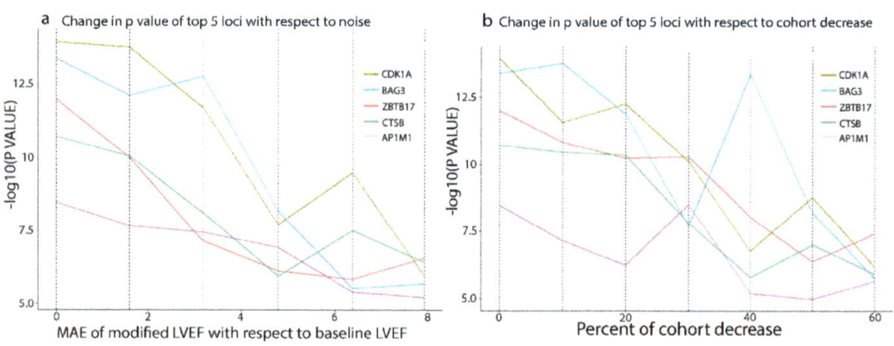

Figure 3 a, Slope chart shows the change in the P value of the top 5 loci with respect to mean absolute error; **b,** Slope chart shows the change in a P value of top 5 loci with respect to the cohort decrease; each locus is named after the closest gene

4.4 Improving phenotyping augments downstream genetic analyses

Cardiac MRI provides clinicians and researchers with a plethora of high-resolution imaging, with even the abbreviated 20-min UK Biobank cardiac MRI protocol resulting in 9 sequences with over 30,000 images per study[17]. With so many images and patients, the released UKBB measurements were generated using a fully automated workflow (with Siemens inLineVF) without quality inspection and bias correction. When compared with manual clinician evaluation, the automated measurements of LVEF result in a mean absolute error (MAE) of 3.4%, R2 of 0.348, and ICC of 0.521 for LVEF[15]. Imprecision in the inline LVEF can be partially addressed by linear adjustment[18] and doing so slightly increases genetic signal, within the difference in identified loci with MAE of 1% **(Figure A2)**. To evaluate the role of imprecision, we applied a deep learning-based method of obtaining LVEF and analyzed downstream results. Using a previously published deep learning segmentation model[6], we independently derived LV segmentation-based calculated LVEF and found a MAE of 6.1%, R2 of 0.335, and ICC of 0.431

for LVEF compared to the automated measurements from UKBB, and MAE of 5.3%, R2 of 0.60, and ICC of 0.518 compared to the linearly adjusted LVEF (**Figure 5**). However, with these deep learning segmentation derived LVEF measurements, the same cohort identified more loci of interest with significant experimental data backing its relevance. In particular, loci on chromosomes 2, 5, and 8 near genes TTN, DNAJC18, and ZNF572 were not previously identified using the released UKBB LVEF measurements but able to be picked up with our quality-controlled measurements. While we could not directly compare the segmentation-derived LVEF measurements to clinical labels due to the absence of manual labels, the stronger genetic signal and higher association with linearly adjusted LVEF suggest that deep learning derived LVEF is less noisy.

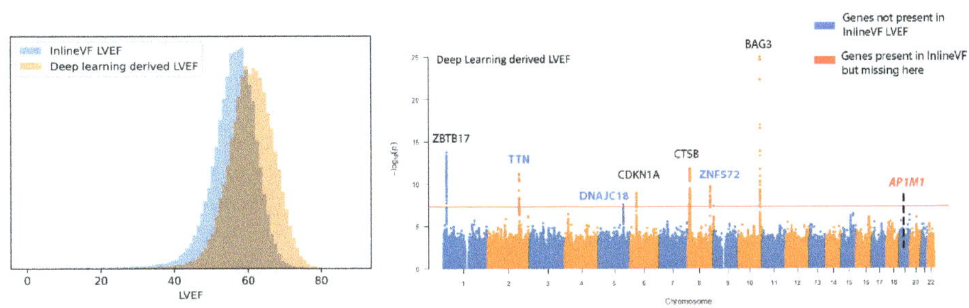

Figure 5 Differences in distribution and GWAS summary statistics between two methods of obtaining LVEF from MRI **a,** Histograms of InlineVF derived LVEF and Deep Learning derived LVEF **b,** Manhattan plot from GWAS performed on Deep Learning derived LVEF; genes colored in blue don't appear in InlineVF LVEF GWAS (Figure 1a); genes colored in red appear in InlineVF LVEF GWAS but not in deep learning derived LVEF GWAS

5 Discussion

In this study, we assessed the impact of measurement noise on genetic associations with LVEF and found substantially impaired power in downstream GWAS analysis with even slight increases in measurement imprecision. Even slight phenotyping variation can significantly impact downstream genetic associations, often to a greater extent than changes in cohort size. As measurement variation is present in many clinical measurements, efforts to improve the precision of measurements can potentially be a cost-effective way to maximize the yield of genetic association studies.

Cardiac function as measured by LVEF is an important clinical measurement that defines disease and identifies patients who are eligible for life-prolonging therapeutics as implantable devices. In echocardiography, human test-retest evaluation of LVEF can range between 7-10%, with slight changes in annotation as well as timing that can significantly impact calculations[10,19]. Few variability studies have been undertaken in cardiac MRI, although similar degrees of manual measurement variability have been found[20]. Prior studies have suggested that polygenic risk scores of LVESV have more power than polygenic risk scores of LVEF[2], consistent with our analyses that more precise measurements correspond to stronger genetic associations. Our analysis suggests that a substantial and primary gain in signal comes from the improvement of noisy measurements that can affect the power and accuracy of downstream analyses.

Noise in measurements can appear in both semi-automated and fully automated workflows[11], and by improving the precision of measuring LVEF, we also improve the accuracy and robustness of downstream GWAS results. The relatively large improvement in yield of genetic association with more precise phenotyping was substantial in comparison to the marginal benefit of increasing the cohort size. As more genetic analyses are undertaken with automated measurements or assessments[4,7,21,22], an additional evaluation must be taken to assess the variability and quality of the phenotyping. Such insights ideally will be confirmed with orthogonal measurements of similar phenotypes. Some of the first genetic association studies were performed on quantitative traits like height, but it should be recognized that many imaging-based phenotypes do not have the same precision and accuracy as the assessment of height on a population.

In summary, genetic association studies on imaging phenotypes allow researchers to discover many associations that help understand the underlying biology of the disease and structure[23]. For LVEF, even advanced imaging has variability in measurements that can substantially impact downstream association studies. The impact of such variability is even more profound than significant changes in cohort size, suggesting improvement in imaging precision and precise phenotyping in general has significant additional value in improving the power of genetic association studies.

Our study offers key insights into measurement noise's effect on genetic associations with LVEF. However, a few considerations remain. The impact of measurement noise could vary for different quantitative phenotypes, and thus future studies should investigate its influence on various phenotypes for a broader understanding. Secondly, our GWAS methodology could be further enhanced by using a linear mixed model method[24], shown to produce more significant associations. Lastly, while our deep learning LVEF method showed a high GWAS signal, we could not compare it to manual clinical labels due to their unavailability.

6 Appendix

Table A1. Mapping between Gaussian Noise SD and MAE

SD	MAE	R2
0	0	1
1	0.797489	0.9788
2	1.594416	0.9199
3	2.386753	0.8371
4	3.183924	0.743
5	3.974958	0.6508
6	4.793956	0.5632
7	5.604129	0.4832
8	6.380848	0.4192
9	7.228321	0.3602
10	7.920860	0.3183

Table A2. Metrics of genetic signal for each decrease in cohort size

Cohort decrease	SNP Accuracy	GWAS Sensitivity	Heritability
0%	1.0	1.0	0.1114 (00357)
10%	0.8744	0.8	0.1071 (0.0397)
20%	0.8713	0.8	0.0867 (0.037)
30%	0.3436	1.0	0.082 (0.0332)
40%	0.1392	0.4	0.0497 (0.0216)
50%	0.0477	0.4	0.039 (0.0287)
60%	0.0019	0.2	0.0384 (0.0288)

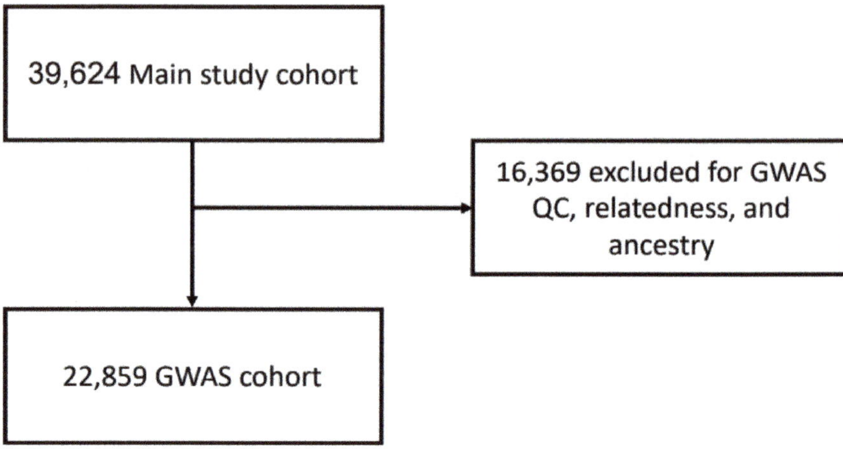

Figure A1. Cohort diagram

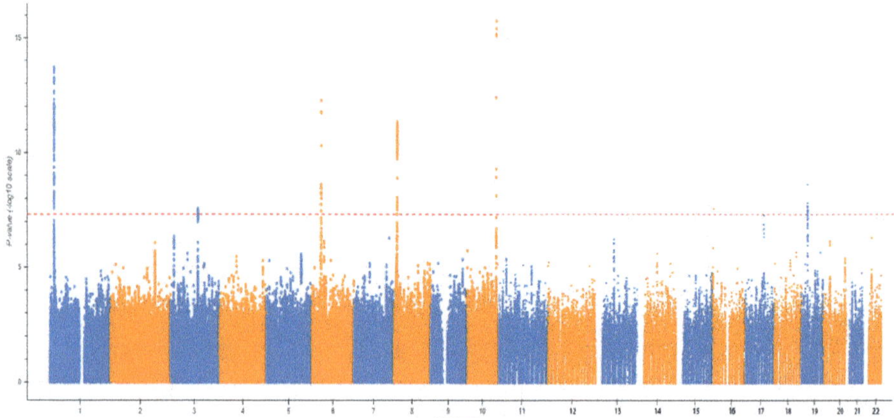

Figure A2. Manhattan plot for genome-wide association study on corrected left-ventricular ejection fraction.

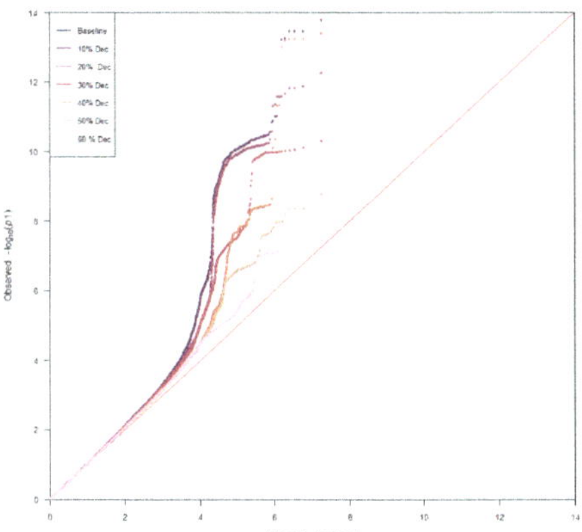

Figure A3. Q-Q plots of P values from GWAS summary statistics for different percentages of cohort decrease

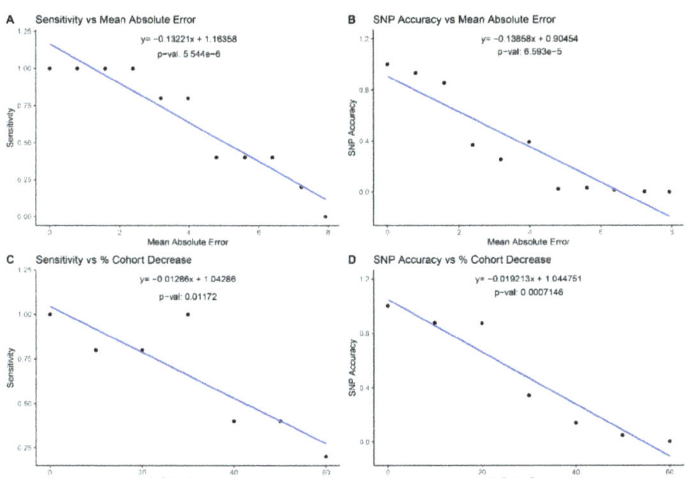

Figure A4 Impact of cohort decrease and noise generation on GWAS power. **a**, Regression analysis on the impact of measurement error quantified by a mean absolute error on sensitivity. **b**, Regression analysis on the impact of the mean absolute error on SNP accuracy. **c**, Regression analysis of the impact of cohort size decline on sensitivity. **d**, Regression analysis of the impact of cohort size decline on SNP accuracy

References

1. Pirruccello, J. P. *et al.* Genetic analysis of right heart structure and function in 40,000 people. *bioRxiv* (2021) doi:10.1101/2021.02.05.429046.
2. Pirruccello, J. P. *et al.* Analysis of cardiac magnetic resonance imaging in 36,000 individuals yields genetic insights into dilated cardiomyopathy. *Nat. Commun.* **11**, 2254 (2020).
3. Agrawal, S. *et al.* Inherited basis of visceral, abdominal subcutaneous and gluteofemoral fat depots. *Nat. Commun.* **13**, 3771 (2022).
4. Haas, M. E. *et al.* Machine learning enables new insights into genetic contributions to liver fat accumulation. *Cell Genom* **1**, (2021).
5. Liu, D. J. *et al.* Exome-wide association study of plasma lipids in >300,000 individuals. *Nat. Genet.* **49**, 1758–1766 (2017).
6. Bai, W. *et al.* Automated cardiovascular magnetic resonance image analysis with fully convolutional networks. *J. Cardiovasc. Magn. Reson.* **20**, 65 (2018).
7. Meyer, H. V. *et al.* Genetic and functional insights into the fractal structure of the heart. *Nature* **584**, 589–594 (2020).
8. Vasan, R. S. *et al.* Genetic variants associated with cardiac structure and function: a meta-analysis and replication of genome-wide association data. *JAMA* **302**, 168–178 (2009).
9. Carroll, R. J. *et al.* Nonparametric Prediction in Measurement Error Models [with Commants]. *J. Am. Stat. Assoc.* **104**, 993–1014 (2009).
10. Farsalinos, K. E. *et al.* Head-to-Head Comparison of Global Longitudinal Strain Measurements among Nine Different Vendors: The EACVI/ASE Inter-Vendor Comparison Study. *J. Am. Soc. Echocardiogr.* **28**, 1171–1181, e2 (2015).
11. O'Dell, W. G. Accuracy of Left Ventricular Cavity Volume and Ejection Fraction for Conventional Estimation Methods and 3D Surface Fitting. *J. Am. Heart Assoc.* **8**, e009124 (2019).
12. Littlejohns, T. J. *et al.* The UK Biobank imaging enhancement of 100,000 participants: rationale, data collection, management and future directions. *Nat. Commun.* **11**, 2624 (2020).
13. Bycroft, C. *et al.* The UK Biobank resource with deep phenotyping and genomic data. *Nature* **562**, 203–209 (2018).
14. Petersen, S. E. *et al.* UK Biobank's cardiovascular magnetic resonance protocol. *J. Cardiovasc. Magn. Reson.* **18**, 8 (2016).
15. Suinesiaputra, A. *et al.* Fully-automated left ventricular mass and volume MRI analysis in the UK Biobank population cohort: evaluation of initial results. *Int. J. Cardiovasc. Imaging* **34**, 281–291 (2018).
16. Bulik-Sullivan, B. *et al.* An atlas of genetic correlations across human diseases and traits. *Nat. Genet.* **47**, 1236–1241 (2015).
17. Petersen, S. E. *et al.* UK Biobank's cardiovascular magnetic resonance protocol. *J. Cardiovasc. Magn. Reson.* **18**, 1–7 (2016).
18. Sanghvi, M. M. *et al.* Automatic left ventricular analysis with Inline VF performs well compared to manual analysis: results from Barts Cardiovascular Registry. *J. Cardiovasc. Magn. Reson.* **18**, 1–2 (2016).

19. Yuan, N. *et al.* Systematic Quantification of Sources of Variation in Ejection Fraction Calculation Using Deep Learning. *JACC Cardiovasc. Imaging* **14**, 2260–2262 (2021).
20. Augusto, J. B. *et al.* Diagnosis and risk stratification in hypertrophic cardiomyopathy using machine learning wall thickness measurement: a comparison with human test-retest performance. *Lancet Digit Health* **3**, e20–e28 (2021).
21. Zekavat, S. M. *et al.* Deep Learning of the Retina Enables Phenome- and Genome-Wide Analyses of the Microvasculature. *Circulation* **145**, 134–150 (2022).
22. Kosaraju, A., Goyal, A., Grigorova, Y. & Makaryus, A. N. Left Ventricular Ejection Fraction. in *StatPearls* (StatPearls Publishing, 2022).
23. Uffelmann, E. *et al.* Genome-wide association studies. *Nature Reviews Methods Primers* **1**, 1–21 (2021).
24. Loh, P.-R., Kichaev, G., Gazal, S., Schoech, A. P. & Price, A. L. Mixed-model association for biobank-scale datasets. *Nat. Genet.* **50**, 906–908 (2018).

A deep neural network estimation of brain age is sensitive to cognitive impairment and decline

Yisu Yang[1], Aditi Sathe[1], Kurt Schilling[3,7], Niranjana Shashikumar[1], Elizabeth Moore[1], Logan Dumitrescu[1,2], Kimberly R. Pechman[1], Bennett A. Landman[1,3,4,5,6], Katherine A. Gifford[1], Timothy J. Hohman[1,2], Angela L. Jefferson[1,7], Derek B. Archer[1,2,†]

[1]*Vanderbilt Memory and Alzheimer's Center, Vanderbilt University School of Medicine, Nashville, TN, USA, 37212*
[2]*Vanderbilt Genetics Institute, Vanderbilt University Medical Center, Nashville, TN, USA, 37212*
[3]*Vanderbilt University Institute of Imaging Science, Vanderbilt University Medical Center, Nashville, TN, USA, 37212*
[4]*Department of Biomedical Engineering, Vanderbilt University, Nashville, TN, USA, 37212*
[5]*Department of Electrical and Computer Engineering, Vanderbilt University, Nashville, TN, USA, 37212*
[6]*Department of Radiology & Radiological Sciences, Vanderbilt University Medical Center, Nashville, TN, USA, 37212*
[7]*Department of Medicine, Vanderbilt University Medical Center, Nashville, TN, USA, 37212*
[†]*Email: derek.archer@vumc.org*

The greatest known risk factor for Alzheimer's disease (AD) is age. While both normal aging and AD pathology involve structural changes in the brain, their trajectories of atrophy are not the same. Recent developments in artificial intelligence have encouraged studies to leverage neuroimaging-derived measures and deep learning approaches to predict brain age, which has shown promise as a sensitive biomarker in diagnosing and monitoring AD. However, prior efforts primarily involved structural magnetic resonance imaging and conventional diffusion MRI (dMRI) metrics without accounting for partial volume effects. To address this issue, we post-processed our dMRI scans with an advanced free-water (FW) correction technique to compute distinct FW-corrected fractional anisotropy (FA_{FWcorr}) and FW maps that allow for the separation of tissue from fluid in a scan. We built 3 densely connected neural networks from FW-corrected dMRI, T1-weighted MRI, and combined FW+T1 features, respectively, to predict brain age. We then investigated the relationship of actual age and predicted brain ages with cognition. We found that all models accurately predicted actual age in cognitively unimpaired (CU) controls (FW: $r=0.66$, $p=1.62\times10^{-32}$; T1: $r=0.61$, $p=1.45\times10^{-26}$, FW+T1: $r=0.77$, $p=6.48\times10^{-50}$) and distinguished between CU and mild cognitive impairment participants (FW: $p=0.006$; T1: $p=0.048$; FW+T1: $p=0.003$), with FW+T1-derived age showing best performance. Additionally, all predicted brain age models were significantly associated with cross-sectional cognition (memory, FW: $\beta=-1.094$, $p=6.32\times10^{-7}$; T1: $\beta=-1.331$, $p=6.52\times10^{-7}$; FW+T1: $\beta=-1.476$, $p=2.53\times10^{-10}$; executive function, FW: $\beta=-1.276$, $p=1.46\times10^{-9}$; T1: $\beta=-1.337$, $p=2.52\times10^{-7}$; FW+T1: $\beta=-1.850$, $p=3.85\times10^{-17}$) and longitudinal cognition (memory, FW: $\beta=-0.091$, $p=4.62\times10^{-11}$; T1: $\beta=-0.097$, $p=1.40\times10^{-8}$; FW+T1: $\beta=-0.101$, $p=1.35\times10^{-11}$; executive function, FW: $\beta=-0.125$, $p=1.20\times10^{-10}$; T1: $\beta=-0.163$, $p=4.25\times10^{-12}$; FW+T1: $\beta=-0.158$, $p=1.65\times10^{-14}$). Our findings provide evidence that both T1-weighted MRI and dMRI measures improve brain age prediction and support predicted brain age as a sensitive biomarker of cognition and cognitive decline.

Keywords: Alzheimer's disease, free-water correction, deep neural network, cognition

1. Introduction

Alzheimer's disease (AD) is a progressive neurodegenerative disorder whose greatest known risk factor is advancing age. Both normal aging and AD are accompanied by structural changes in the brain, but they follow distinct trajectories. Specifically, healthy aging typically exhibits global reductions in gray matter volume[1,2] characterized by volume loss in frontal and temporal lobes[3,4] and enlargement of ventricles[3,5], whereas AD-related brain atrophy typically starts in the hippocampus and gradually spreads to the entire brain[6,7]. Additionally, studies have shown that AD brains undergo deterioration more rapidly than healthy brains[8]. Given these differences, there arose recent efforts of using neuroimaging-derived measures of gray matter volume from T1-weighted magnetic resonance imaging (MRI) and white matter microstructure from diffusion MRI (dMRI) to predict an individual's "brain age" via machine learning approaches[9–12], which can differ from their chronological age and predict cognitive decline[13–15]. These models were trained on cognitive unimpaired individuals to learn common patterns in healthy aging, which then allowed them to detect aging-related abnormalities such as those associated with AD. A larger difference between brain age and chronological age indicates that the individual is on an accelerated trajectory compared with normal aging and is typically seen in individuals with cognitive impairment (e.g., mild cognitive impairment [MCI], AD)[16–18], suggesting the potential of brain age as a sensitive biomarker along the AD continuum. Moreover, the development of the free-water (FW) correction post-processing technique[19] has enabled the partition of a conventional fractional anisotropy (FA) map into a FW-corrected FA map (FA_{FWcorr}) and a FW map; the FA_{FWcorr} and FW metrics individually describe tissue and fluid, thereby enhancing the biological specificity of dMRI scans. Recently, our group has demonstrated that abnormal FW-corrected dMRI metrics are associated with higher rates of longitudinal cognitive decline and diagnosis along the AD clinical continuum[20,21]. These findings suggest that incorporating FW-corrected metrics into models of predicted brain age may provide more sensitive associations with cognitive impairment and decline.

The present study leveraged neuroimaging data from a longitudinal cohort of aging to build three densely connected neural networks using FW-corrected dMRI, T1-weighted MRI, and combined FW+T1 features to predict participant brain age. To evaluate model performance, we examined the relationship between predicted brain age and chronological age. We then investigated the association between predicted brain age and two domains of cognition (memory and executive function performance at baseline and over time). We hypothesized that FW-, T1-, and FW+T1-derived models would all accurately predict participant brain age, with the FW+T1-derived model showing the best performance as it incorporates both gray and white matter regions. We also hypothesized that all predicted brain age models would predict baseline and longitudinal memory and executive function performance, with FW+T1-derived brain age showing the strongest associations.

2. Methods

2.1. Participants

All data leveraged in the present study were obtained from the Vanderbilt Memory and Aging Project (VMAP)[22], a longitudinal observational study that was launched in 2012 and recruited individuals 60 years and older who speak English, have adequate auditory and visual capacity for testing, and have a stable study partner. Participants underwent comprehensive neuropsychological assessment and were categorized into cognitively unimpaired (CU) or MCI status; MCI participants were age-, sex-, and race-matched with CU controls. Cognitive (memory, executive function) measures were obtained from all participants and neuroimaging (T1-weighted MRI, dMRI) measures were obtained from a subset of participants. Only participants who had all necessary cognitive and neuroimaging data were included in the present study (n=295). All protocols for VMAP were approved by the IRB at Vanderbilt University Medical Center and all participants gave voluntary informed consent prior to enrollment. Data from the VMAP cohort can be freely accessed following approval (vmacdata.org). **Table 1** summarizes demographic and clinical information for the present cohort.

Table 1. Vanderbilt Memory and Aging Project Cohort Information

Measure	Diagnosis at Baseline		p-value
	CU	MCI	
Cohort Characteristics			
Number of participants	168	127	-
Total number of visits	568	372	-
Longitudinal follow-up (years)	3.10 (1.44)	2.96 (1.43)	0.230
Demographics and Health Characteristics			
Age at baseline (years)	73.10 (7.16)	73.67 (7.41)	0.504
Sex (% female)	42.26	42.52	1.000
Education (years)	16.39 (2.48)	15.09 (2.75)	**<0.001**
Race (% non-Hispanic White)	86.90	86.61	1.000
APOE-ε4 (% positive)	29.76	44.09	**0.016**

Mean (standard error) are provided unless otherwise indicated. Abbreviations: CU, cognitively unimpaired; MCI, mild cognitive impairment; *APOE*, apolipoprotein. Boldface signifies $p<0.05$ unless otherwise indicated.

2.2. Neuroimaging data acquisition and preprocessing

T1-weighted MRI images (repetition time: 8.9 ms, echo time: 4.6 ms, resolution: 1 mm isotropic) were obtained from each participant on 3T Philips Achieva using an 8-channel SENSE reception coil and underwent multi-atlas segmentation to calculate the volumes of 132 regions of interest (ROI)[23]. All measures were normalized by total intracranial volume, calculated as the volumetric sum of all 132 segmented ROIs. dMRI images (resolution: 2 mm isotropic, b-values: 0, 1000 s/ mm^2, number of directions: 32) were obtained from each participant using the previously described scanner and preprocessed using *PreQual*[24]. FW and FW-corrected metrics were calculated in MATLAB from the preprocessed images, as previously described[19]. The FW and FA_{FWcorr} maps were transformed by a non-linear warp using the ANTs package to create a standardized space representation. Finally, publicly available tractography templates (https://github.com/VUMC-VMAC/Tractography_Templates) were applied to the FW and FA_{FWcorr} maps to quantify white matter microstructure within 48 tracts.

T1-weighted MRI and FW-corrected dMRI metrics (FA_{FWcorr}, FW) were harmonized separately using *Longitudinal Combat*[25] in R (version 4.1.2), controlling for age at baseline, education, sex, race/ethnicity, *APOE*-ε4 positivity, *APOE*-ε2 positivity, and the interaction of age at baseline with time interval from baseline. We also included the random effects of intercept and time interval from baseline for each participant and a batch variable that accounted for all combinations of image acquisition. The batch variable was *scanner x software x coil* for T1 metrics and *site x scanner x protocol* for FW-corrected metrics.

2.3. Neuropsychological metrics calculation

Participants completed comprehensive neuropsychological testing administered by experienced technicians which assessed multiple cognitive domains, including memory and executive function. Psychometrically sound memory and executive function composite scores were calculated from item-level data. Longitudinal cognitive measures (memory slope, executive function slope) for each participant were obtained by calculating the random effect coefficient using a linear mixed-effects model where the fixed effect was time interval from baseline and the outcome was composite score.

2.4. Brain age prediction model architecture

In the present study, we used a densely connected neural network to predict participants' brain age based on neuroimaging regions (i.e., features) and created three separate models using FW, T1, and combined FW+T1 features. **Figure 1** shows an overview of model workflow. Each model consists of four layers: an input layer whose dimensions correspond to the number of features (FW: 96 features, T1: 132 features, FW+T1: 228 features), two densely connected layers with rectified linear unit (ReLU) activation whose number of nodes equals half and a quarter of the number of features, respectively, and an output layer with a single node and linear activation for brain age prediction.

All models were trained on baseline neuroimaging data from the VMAP cohort by subsetting all imaging sessions to the first visit of CU participants. We minimized the loss function as characterized by mean absolute error (MAE) while monitoring the mean squared error (MSE) and root mean squared error (RMSE). We conducted ten-fold cross-validation where 90% of the data were used for training and 10% of the data were reserved for testing in each fold, repeating this process ten times until the entire dataset had been tested only once. Within the training data for each fold, 80% were used to train the model and 20% were used to validate model performance. During each fold, training was stopped when the loss function on the validation dataset had not improved for 15 epochs and only the best model was saved. For each set of features (FW, T1, FW+T1), saved models were compared across folds and the one which yielded the lowest validation loss was selected as the final model. All models were developed in Python (version 3.9.13) using the Keras library (version 2.9.0) with Tensorflow backend (version 2.9.1). We used the three final models to generate FW, T1, and FW+T1 predicted brain ages for all participants (CU, MCI) at all timepoints (baseline, longitudinal follow-ups).

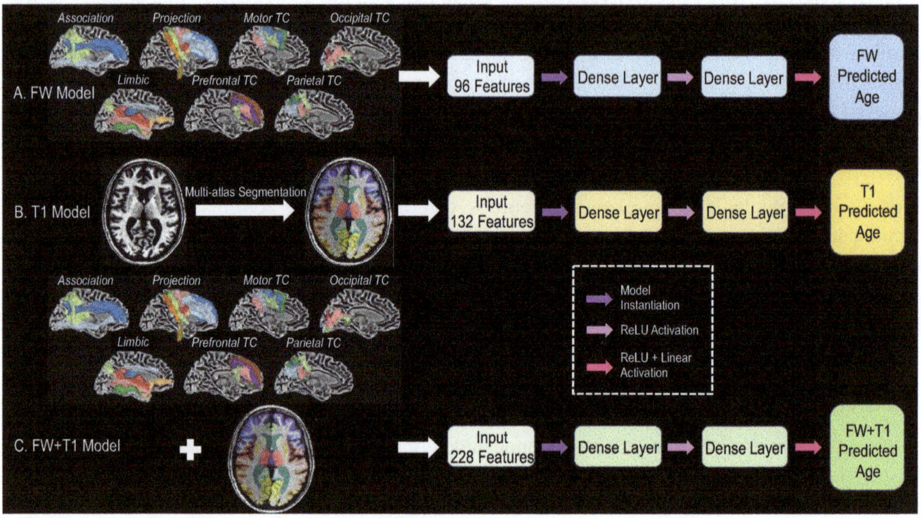

Figure 1. Model workflow for brain age prediction. We created three separate, densely connected neural networks to predict brain age, including FW-derived (**A**), T1-derived (**B**), and FW+T1-derived (**C**) models.

For each model, we computed SHAP (SHapley Additive exPlanation) values for all relevant neuroimaging features to quantify their contribution to age prediction. **Figure 2** shows the top 10 most important features for each model based on mean SHAP value.

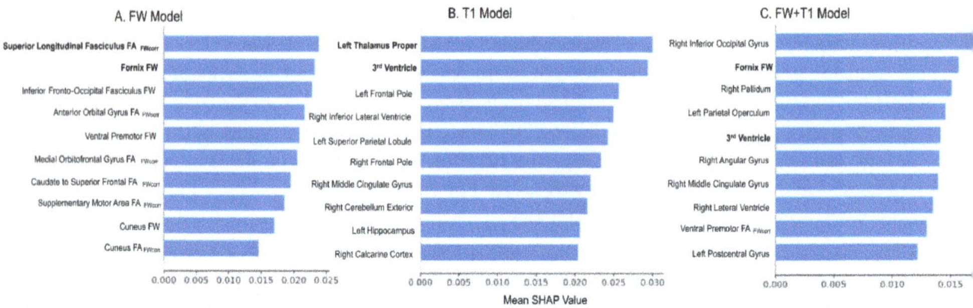

Figure 2. Top 10 most important features for FW-derived (**A**), T1-derived (**B**), and FW+T1-derived (**C**) models. Boldface signifies top features involved in aging and AD, including superior longitudinal fasciculus (SLF) FA_{FWcorr}, fornix FW for the FW-derived model and left thalamus proper, 3rd ventricle for the T1-derived model.

2.5. Statistical analyses

All statistical analyses were conducted in Python (version 3.9.13) and R (version 4.1.2). We first performed simple linear regression between actual age and each predicted age to assess model performance as well as independent groups t-tests to compare the mean actual age and mean predicted brain ages of CU and MCI participants. We also conducted logistic regression analyses, using actual and each predicted brain age as direct predictor of diagnostic category, then evaluated model performance using area under the receiver operator characteristic curve (ROC-AUC) and DeLong's test. Next, we conducted a series of linear models and competitive model analyses to assess actual and predicted brain age association with cognition. All models covaried for diagnosis, race/ethnicity, sex, education, and *APOE*-ε4 positivity. Significance was set *a priori* at α=0.05. For baseline cognition, actual age and predicted brain ages (FW, T1, FW+T1) were included in a general linear model individually to determine their main effects on baseline memory and executive function. We then introduced age-by-diagnosis interaction terms to the linear models to investigate the potential modifying effect of age on baseline memory and executive function scores. Finally, we conducted post-hoc competitive model analysis to determine the unique variance in baseline memory and executive function contributed by FW, T1, and FW+T1 predicted brain age, beyond that contributed by covariates and actual age. The described analyses were repeated for longitudinal cognition (longitudinal memory slope, longitudinal executive function slope).

3. Results

Participant characteristics of the VMAP cohort are shown in **Table 1**. There were no significant differences in longitudinal follow-up interval, age at baseline, sex, or race between diagnostic groups (CU, MC). The CU group had more years of education and lower *APOE*-ε4 positivity than the MCI group.

3.1. Combined model using free-water (FW) and T1 features showed best performance

Figure 3 shows the agreement between predicted brain age measures (FW, T1, FW+T1) and actual age; model performance was characterized using average mean absolute error (MAE_{avg}) and average mean squared error ($RMSE_{avg}$) across folds and Pearson's correlation through ten-fold cross

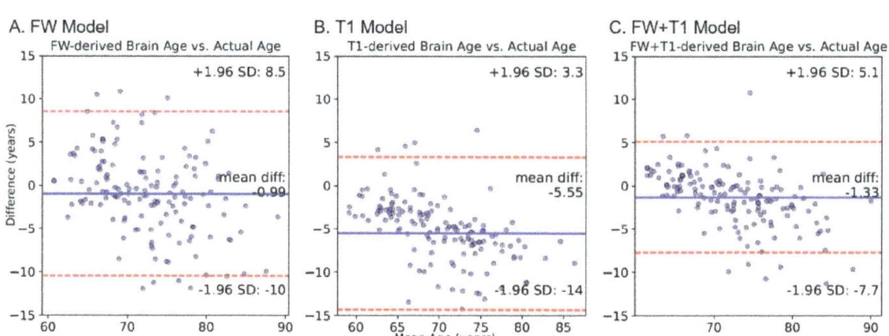

Figure 3. Bland Altman plots for FW-derived age (**A**), T1-derived age (**B**), and FW+T1-derived age (**C**). All models accurately predict age. FW+T1-derived age is most significantly associated with actual age, in comparison with FW-derived age and T1-derived age.

validation. While all predicted brain ages significantly predicted actual age (FW: MAE_{avg}= 0.115, $RMSE_{avg}$=0.129, r=0.66, p=1.62x10^{-32}; T1: MAE_{avg}=0.106, $RMSE_{avg}$=0.114, r=0.61, p=1.45x10^{-26}), the combined FW+T1 model yielded the best performance with highest r as well as lowest MAE_{avg} and $RMSE_{avg}$ (MAE_{avg}=0.072, $RMSE_{avg}$=0.087, r=0.77, p=6.48x10^{-50}).

We then compared means of actual age and predicted brain ages between CU and MCI participants. While there was no difference in actual age between CU and MCI groups (age_{CU}=73.07±7.24, age_{MCI}=72.83±6.92, p=0.792), all predicted brain ages for the MCI group were significantly higher than those for the CU group (FW: age_{CU}=72.08±5.55, age_{MCI}=74.18±6.16, p=0.006; T1: age_{CU}=67.52±4.96, age_{MCI}=68.82±5.27, p=0.048), with the combined FW+T1 model showing the largest difference (age_{CU}=71.74±5.58, age_{MCI}=73.93±5.67, p=0.003).

Figure 4 shows the Receiver Operating Characteristic curves for actual and predicted brain ages in predicting diagnostic category (CU, MCI). Pairwise comparisons revealed that ROC-AUC values for all predicted brain ages were significantly greater than that of actual age (FW-actual: p=0.003; T1-actual: p=0.030; FW+T1-actual: p=0.004); however, no differences were found between the predicted brain ages (all p>0.05).

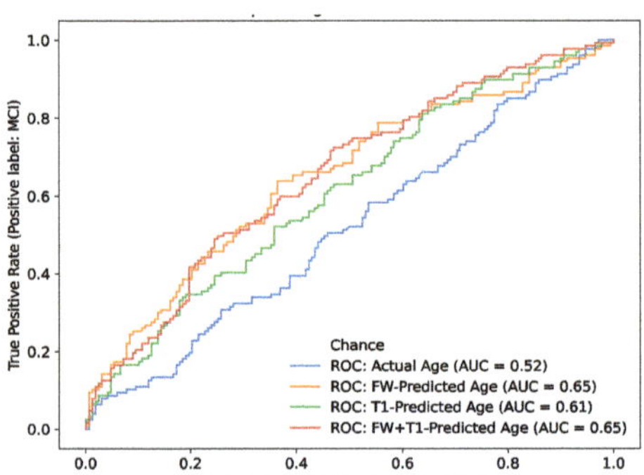

Figure 4. Receiver Operating Characteristic curves for actual, FW-predicted, T1-predicted, and FW+T1-predicted age in predicting diagnostic category. All predicted ages performed significantly better than actual age, but no difference in performance was found between predicted ages.

3.2. Predicted brain age association with baseline cognition

Actual age and predicted brain age (FW-derived, T1-derived, FW+T1-derived) associations with cross-sectional cognition (memory, executive function) are shown in **Figure 5**. While all models significantly predicted memory score at baseline (Actual: R_{adj}^2=0.497, p=1.23x10^{-34}; FW: R_{adj}^2=0.481, p=4.14x10^{-33}; T1: R_{adj}^2=0.481, p=4.26x10^{-33}), the combined FW+T1 model showed the most robust performance (R_{adj}^2=0.513, p=2.31x10^{-36}). Similarly, all models significantly predicted executive function score at baseline (Actual: R_{adj}^2=0.472, p=3.22x10^{-32}; FW: R_{adj}^2=0.445, p=1.24x10^{-29}; T1: R_{adj}^2=0.422, p=1.69x10^{-27}) and the combined FW+T1 model was the most robust (R_{adj}^2=0.519, p=5.81x10^{-37}). When examining main effect associations of each respective age variable, we saw that actual age and all predicted brain ages each had a significant main effect on baseline memory score (**Figure 5A**; Actual: β=-1.162, p=1.58x10^{-8}; FW: β=-1.094, p=6.32x10^{-7}; T1: β=-1.331, p=6.52x10^{-7}), with the combined FW+T1 predicted brain age showing the strongest relationship (β=-1.476, p=2.53x10^{-10}). Likewise, we saw significant age effects for actual and all predicted ages on baseline executive function score (**Figure 5B**; Actual: β=-1.371, p=2.98x10^{-12}; FW: β=-1.276, p=1.46x10^{-9}; T1: β=-1.337, p=2.52x10^{-7}), with the combined FW+T1 predicted brain age showing the strongest relationship (β=-1.850, p=3.85x10^{-17}). We found no significant interactions between actual or predicted brain ages and diagnostic status on baseline memory or executive function.

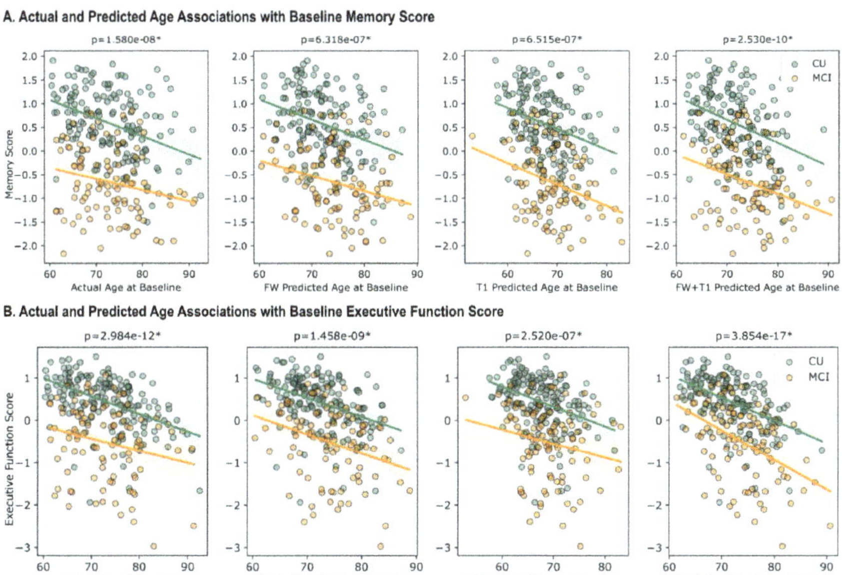

Figure 5. Actual and predicted age associations with baseline cognition. Actual and all derived ages are significantly associated with baseline memory (**A**) and executive function performance (**B**); FW+T1-derived age shows highest associations. Datapoint colors: green=CU; orange=MCI.

Table 2 summarizes results of the competitive model analysis on cross-sectional cognition. We found that covariates alone explained approximately 43% of the variance in baseline memory score (R_{adj}^2=42.60%) and the addition of actual age led to an increase in overall model performance (ΔR^2_{adj}=6.92%). We then iteratively added predicted brain ages to this model to determine whether FW, T1, or FW+T1 predicted brain age contributed to any unique variance beyond covariates and actual age. While FW and T1 predicted brain ages were not found to be a significant contributor to baseline memory score, we observed that the combined FW+T1 predicted brain age significantly added to the model and led to increased R_{adj}^2 (FW+T1: ΔR^2_{adj}=1.74%). Similarly, covariates alone explained approximately 36% of the variance in baseline executive function score (R_{adj}^2=35.70%) and the addition of actual age led to a drastic increase in model performance (ΔR^2_{adj}=11.58%). When iteratively adding each predicted brain age to the model to determine its unique contribution beyond covariates and actual age, we observed that both FW and FW+T1 predicted brain age explained additional variance in baseline executive function score, with FW predicted brain age leading to a small increase in R_{adj}^2 (ΔR^2_{adj}=0.63%) and FW+T1 predicted brain age leading to a large increase in R_{adj}^2 (ΔR^2_{adj}=4.59%). However, T1 predicted brain age did not provide a significant increase to the model.

Table 2. Comparison of Actual and Predicted Age Main Effects on Baseline Cognition

	Baseline Memory Score					**Baseline Executive Function Score**				
	β	SE	t	*p*	ΔR^2_{adj}	β	SE	t	*p*	ΔR^2_{adj}
Covariates +										
Actual age	-1.162	0.199	-5.852	**<0.001**	6.915	-1.371	0.186	-7.355	**<0.001**	11.576
Covariates + actual age +										
Predicted age										
FW	-0.474	0.287	-1.652	0.100	0.358	-0.531	0.269	-1.978	**0.049**	0.630
T1	-0.643	0.327	-1.964	0.051	0.590	-0.376	0.308	-1.219	0.224	0.106
FW+T1	-1.132	0.365	-3.105	**0.002**	1.743	-1.628	0.333	-4.892	**<0.001**	4.585

3.3. Predicted brain age association with longitudinal cognition

Actual age and predicted brain age associations with longitudinal cognition are shown in **Figure 6**. While all models significantly predicted longitudinal memory slope (Actual: R_{adj}^2=0.427, p=5.08x10^{-28}; FW: R_{adj}^2=0.439, p=4.90x10^{-29}; T1: R_{adj}^2=0.412, p=1.16x10^{-26}), the combined FW+T1 model showed the most robust performance (R_{adj}^2=0.444, p=1.50x10^{-29}). Similarly, all models significantly predicted longitudinal executive function slope (Actual: R_{adj}^2=0.424, p=9.20x10^{-28}; FW: R_{adj}^2=0.404, p=5.89x10^{-26}; T1: R_{adj}^2=0.420, p=2.38x10^{-27}) and the combined FW+T1 model was the most robust (R_{adj}^2=0.446, p=1.13x10^{-29}). When examining the age effect, we saw that actual age and all predicted brain ages each had a significant main effect on longitudinal memory slope (**Figure 6A**; Actual: β=-0.082, p=5.30x10^{-10}; FW: β=-0.091, p=4.62x10^{-11}; T1: β=-0.097, p=1.40x10^{-8}), with the combined FW+T1 model showing the strongest relationship (β=-0.101, p=1.35x10^{-11}). Likewise, we saw significant main effects for actual and all predicted brain ages on longitudinal executive function slope (**Figure 6B**; Actual: β=-0.128, p=1.58x10^{-12}; FW: β=-0.125, p=1.20x10^{-10}; T1: β=-0.163, p=4.25x10^{-12}), with the combined FW+T1 model showing the strongest relationship (β=-0.158, p=1.65x10^{-14}). We found no significant interactions between actual age or predicted brain ages and diagnostic status on longitudinal memory or executive function.

Table 3 summarizes results of the competitive model analysis on longitudinal cognition. We found that covariates alone explained approximately 33% of the variance in longitudinal memory slope (R_{adj}^2=33.10%) and the addition of actual age led to an increase in overall model performance (ΔR^2_{adj}=9.68%). We then added predicted brain ages to this model one at a time to determine whether FW, T1, or FW+T1 predicted brain age contributed to any unique variance beyond covariates and actual age. We observed that all predicted brain ages were significant contributors to longitudinal memory slope and led to increases in R_{adj}^2 (FW: ΔR^2_{adj}=2.36%; T1: ΔR^2_{adj}=1.17%;

FW+T1: ΔR^2_{adj}=2.01%). Similarly, covariates alone explained approximately 30% of the variance in longitudinal executive function slope (R_{adj}^2=29.50%) and the addition of actual age led to a drastic increase in model performance (ΔR^2_{adj}=12.99%). When iteratively adding each predicted brain age to the model to determine its unique contribution beyond covariates and actual age, we observed

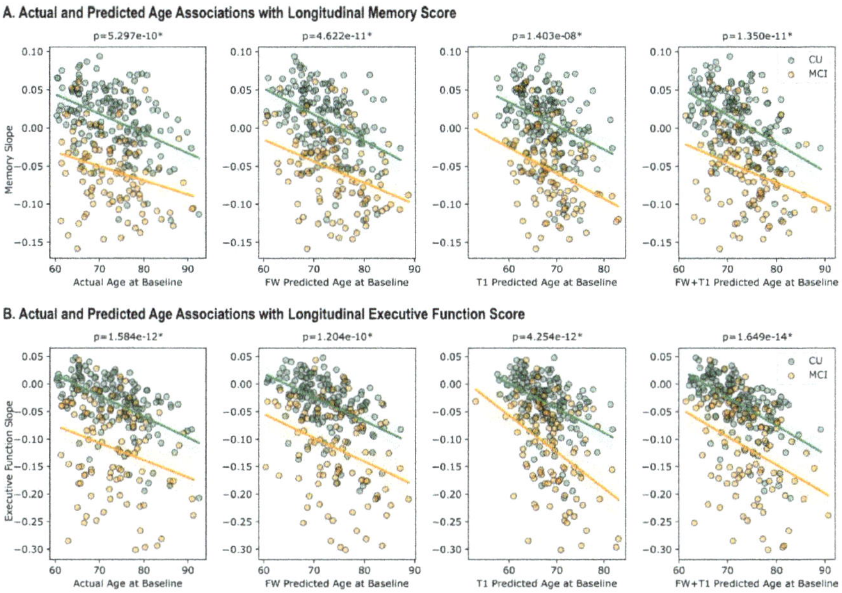

Figure 6. Actual and predicted age associations with longitudinal cognition. Actual and all derived ages are significantly associated with longitudinal memory (**A**) and executive function performance (**B**); FW+T1-derived age shows highest associations.

Table 3. Comparison of Actual and Predicted Age Main Effects on Longitudinal Cognition

	Longitudinal Memory Slope					**Longitudinal Executive Function Slope**				
	β	SE	t	*p*	ΔR^2_{adj}	β	SE	t	*p*	ΔR^2_{adj}
Covariates + Actual Age	-0.082	0.013	-6.474	<0.001	9.681	-0.128	0.017	-7.458	<0.001	12.989
Covariates + Actual Age + Predicted Age										
FW	-0.060	0.018	-3.372	0.001	2.362	-0.060	0.025	-2.441	0.015	1.160
T1	-0.057	0.016	-3.558	<0.001	1.174	-0.097	0.028	-3.482	0.001	2.539
FW+T1	-0.073	0.023	-3.128	0.002	2.014	-0.112	0.031	-3.564	<0.001	2.665

that all predicted brain ages explained additional variance in longitudinal executive function slope and led to increases in R_{adj}^2 (FW: ΔR^2_{adj}=1.16%, T1: ΔR^2_{adj}=2.54%; FW+T1: ΔR^2_{adj}=2.67%).

4. Discussion

The present study created 3 densely connected neural network models to predict brain age using FW, T1, and combined FW+T1 neuroimaging features, respectively. We evaluated model performance by comparing actual age with FW, T1, and FW+T1 predicted brain age then investigated the relationships between different age variables with cross-sectional and longitudinal cognitive performance (memory, executive function). Specifically, we examined age effects on baseline and longitudinal memory and executive function performance and conducted post hoc competitive model analyses to determine the unique contribution provided by each predicted brain age to variance in cognitive function. We report 3 main findings. First, we found that predicted brain ages from all 3 deep learning models using different sets of neuroimaging features (FW, T1, FW+T1) were highly associated with actual age; top neuroimaging features shown in model SHAP plots (**Figure 2**) were also biologically relevant to aging and cognitive decline, such as superior longitudinal fasciculus (SLF) FA_{FWcorr} and fornix FW in the FW model and thalamus and 3rd ventricle in the T1 model. Second, we found that all predicted brain ages differentiated CU from MCI participants and significantly predicted both cross-sectional and longitudinal cognitive performance. Finally, we found that, among all 3 models, FW+T1 predicted brain age was the strongest predictor of cross-sectional and longitudinal cognitive performance and contributed the largest unique variance in these outcome variables.

4.1. Densely connected neural network robustly predicts age using neuroimaging features

We found that predicted brain ages generated by a densely connected neural network using 3 distinct sets of neuroimaging features (FW-corrected dMRI, T1-weighted MRI, combined FW+T1) all showed high correlation with actual age in baseline CU participants, which confirms findings from previous literature that have accurately predicted chronological age of healthy adults using neuroimaging-derived measures with machine learning approaches including deep learning[11,12,17,26–29]. Importantly, the top-contributing neuroimaging features identified for each model (**Figure 2**) provide biological interpretability as they include brain regions that have been associated with both normal aging and AD neuropathology. For instance, previous evidence has shown that thalamic volume, the most important feature identified in the T1 model, decreases with advancing age[30] independently from total brain volume loss and correlates with cognitive speed and verbal memory performance[31,32]. Similarly, the identification of 3rd ventricle volume as the second most important feature in the T1 model is consistent with prior literature which demonstrated that ventricular expansion is associated with normal aging and expands at an accelerated rate in individuals with cognitive impairment (MCI, AD)[33,34] or AD-related pathology[35]. Among top features identified for the FW is the SLF, which is a white matter tract projecting from the occipital, parietal, and temporal lobes to the frontal cortex and is involved in language, attention, and memory[36]. Specifically, conventional FA within the SLF has been shown to undergo stable decline between ages 30-65 and accelerated decline after age 65[37]. Likewise, integrity of the fornix – a limbic white matter tract projecting from the hippocampus[38]– has been shown to decline with normal aging[39] and to predict episodic memory[40] and executive function performance[41] in both healthy older adults and individuals with neurological disorders. Most existing literature on brain age prediction using machine learning techniques has leveraged T1-weighted MRI measures or conventional dMRI metrics. One significant advance in the present study is that we developed models using both T1-weighted and FW-corrected diffusion MRI data, and our results suggest that multi-modal MRI models may more accurately quantify brain age.

4.2. Predicted age is a more sensitive measure than actual age and predicative of cognition

We found that FW, T1, and FW+T1 predicted brain ages all differentiated CU from MCI patients by providing a significantly higher brain age for MCI patients even though the two groups did not differ in actual age, suggesting that predicted brain age may be a sensitive biomarker to AD clinical staging. This is consistent with previous research which computed predicted age difference (i.e., predicted brain age subtracted by chronological age) from T1-weighted MRI scans and found significantly larger predicted age difference in amnestic MCI participants compared with healthy controls[16]. Moreover, individuals with a higher predicted brain age at baseline were more likely to convert from MCI to AD[42] or develop dementia later in life[18]. Studies generating predicted age difference from structural MRI scans of healthy controls have also found correlations with performance on traditional screening tools for AD (e.g., Mini-Mental State Examination, Clinical Dementia Ratio), anatomical measurements such as cortical thickness and hippocampal volume[43], AD neuropathology such as β-amyloid positivity[16,26], and AD risk factors such as *APOE*-ε4 carrier status[16,26].

We also found that all predicted brain ages were robustly associated with cross-sectional and longitudinal cognitive function including baseline memory and executive function scores and longitudinal memory and executive function slopes. This agrees with prior literature that has found predicted age difference to be associated with memory and executive function impairment[16] as well as early signs of cognitive decline[14]. However, the relationship between predicted age and both baseline and longitudinal cognitive function needs further clarification as one previous study found negative associations with psychomotor speed at baseline but no significant association with delayed recall performance or general cognitive status at baseline[44]. The present study supports predicted brain age as a sensitive biomarker along the AD continuum as it distinguishes between CU and MCI participants and is associated with memory and executive function performance at baseline and longitudinally.

4.3. Application of neural networks in clinical medicine

Deep learning algorithms, particularly neural networks, offer remarkable clinical utility by enabling researchers to harness complex patterns from large-scale data and consolidate this information into easy-to-use platforms. Prior neuroimaging studies have used deep learning methods to predict brain age[9,45–47]; however, the present study is the first to combine T1-weighted and FW-corrected diffusion MRI data, shedding light on the potential of using multi-modal MRI to accurately predict brain age and use it as an endophenotype for cognitive impairment and decline, especially in the context of aging and AD. Importantly, our neural networks add weight to the idea that both gray and white matter features are important to consider in aging and AD.

4.4. Strengths and limitations

The present study has several strengths, including a well-characterized longitudinal cohort with multi-modal MRI data with paired cognitive data. Regarding our neuroimaging analysis, one major strength is that we incorporated T1-weighted data in conjunction with FW-corrected diffusion MRI data, and this data was used as input into densely connected neural networks. Importantly, our data driven approach found that several aging related features (e.g., fornix integrity) were some of the highest contributing factors in our models. One limitation of this study is that it used a well-educated, mostly non-Hispanic white population, thus limiting our networks' versatility. Future studies should incorporate more diverse populations to ensure that the neural networks are more generalizable. Moreover, although we have a large population with extensive longitudinal follow-up, one major limitation is that we only used data from a single cohort. Future studies leveraging multiple cohorts would drastically enhance our ability to predict brain age and likely improve its utility as an endophenotype for cognitive impairment and decline.

4.5. Conclusions

This study provided evidence that deep neural networks can be used to predict brain age, and that this predicted age is a strong predictor of cross-sectional cognitive impairment and future cognitive decline. Our findings provide evidence that using both T1-weighted and FW-corrected diffusion MRI data improves our ability to predict brain age; thus, future studies should consider both gray and white matter features when building deep learning models in aging and AD.

5. Acknowledgements

This study was supported by several funding sources, including K01-EB032898 (KGS), K01-AG073584 (DBA), U24-AG074855 (TJH), 75N95D22P00141 (TJH), R01-AG059716 (TJH), UL1-TR000445 and UL1-TR002243 (Vanderbilt Clinical Translational Science Award), S10-OD023680 (Vanderbilt's High-Performance Computer Cluster for Biomedical Research). The research was supported in part by the Intramural Research Program of the National Institutes of Health, National Institute on Aging. Study data were obtained from the Vanderbilt Memory and Aging Project (VMAP). VMAP data were collected by Vanderbilt Memory and Alzheimer's Center Investigators at Vanderbilt University Medical Center. This work was supported by NIA grants R01-AG034962 (PI: ALJ), R01-AG056534 (PI: ALJ), R01-AG062826 (PI: KAG), and Alzheimer's Association IIRG-08-88733 (PI: ALJ).

References

1. Courchesne, E. *et al.* Normal Brain Development and Aging: Quantitative Analysis at in Vivo MR Imaging in Healthy Volunteers. *Radiology* **216**, 672–682 (2000).
2. Good, C. D. *et al.* A Voxel-Based Morphometric Study of Ageing in 465 Normal Adult Human Brains. *NeuroImage* **14**, 21–36 (2001).
3. Coffey, C. E. *et al.* Quantitative cerebral anatomy of the aging human brain: a cross-sectional study using magnetic resonance imaging. *Neurology* **42**, 527–536 (1992).
4. Sowell, E. R. *et al.* Mapping cortical change across the human life span. *Nat Neurosci* **6**, 309–315 (2003).
5. Resnick, S. M. *et al.* One-year age changes in MRI brain volumes in older adults. *Cerebral Cortex* **10**, 464–472 (2000).

6. Jack, C. R. *et al.* NIA-AA Research Framework: Toward a biological definition of Alzheimer's disease. *Alzheimer's & dementia : the journal of the Alzheimer's Association* **14**, 535–562 (2018).
7. Coupé, P., Manjón, J. V., Lanuza, E. & Catheline, G. Lifespan Changes of the Human Brain In Alzheimer's Disease. *Sci Rep* **9**, 3998 (2019).
8. Fotenos, A. F., Snyder, A. Z., Girton, L. E., Morris, J. C. & Buckner, R. L. Normative estimates of cross-sectional and longitudinal brain volume decline in aging and AD. *Neurology* **64**, 1032–1039 (2005).
9. Bermudez, C. *et al.* Anatomical context improves deep learning on the brain age estimation task. *Magn Reson Imaging* **62**, 70–77 (2019).
10. Chen, C.-L. *et al.* Generalization of diffusion magnetic resonance imaging–based brain age prediction model through transfer learning. *NeuroImage* **217**, 116831 (2020).
11. Jiang, H. *et al.* Predicting Brain Age of Healthy Adults Based on Structural MRI Parcellation Using Convolutional Neural Networks. *Frontiers in Neurology* **10**, (2020).
12. Lombardi, A. *et al.* Brain Age Prediction With Morphological Features Using Deep Neural Networks: Results From Predictive Analytic Competition 2019. *Frontiers in Psychiatry* **11**, (2021).
13. Boyle, R. *et al.* Brain-predicted age difference score is related to specific cognitive functions: A multi-site replication analysis. *Brain Imaging Behav* **15**, 327–345 (2021).
14. Elliott, M. L. *et al.* Brain-age in midlife is associated with accelerated biological aging and cognitive decline in a longitudinal birth cohort. *Mol Psychiatry* **26**, 3829–3838 (2021).
15. Karim, H. T. *et al.* Independent replication of advanced brain age in mild cognitive impairment and dementia: detection of future cognitive dysfunction. *Mol Psychiatry* **27**, 5235–5243 (2022).
16. Huang, W. *et al.* Accelerated Brain Aging in Amnestic Mild Cognitive Impairment: Relationships with Individual Cognitive Decline, Risk Factors for Alzheimer Disease, and Clinical Progression. *Radiology: Artificial Intelligence* **3**, e200171 (2021).
17. Lee, J. *et al.* Deep learning-based brain age prediction in normal aging and dementia. *Nat Aging* **2**, 412–424 (2022).
18. Biondo, F. *et al.* Brain-age is associated with progression to dementia in memory clinic patients. *NeuroImage: Clinical* **36**, 103175 (2022).
19. Pasternak, O., Sochen, N., Gur, Y., Intrator, N. & Assaf, Y. Free water elimination and mapping from diffusion MRI. *Magn Reson Med* **62**, 717–30 (2009).
20. Yang, Y. *et al.* White matter microstructural metrics are sensitively associated with clinical staging in Alzheimer's disease. *Alzheimer's & Dementia: Diagnosis, Assessment & Disease Monitoring* **15**, e12425 (2023).
21. Archer, D. B. *et al.* Free-water metrics in medial temporal lobe white matter tract projections relate to longitudinal cognitive decline. *Neurobiol Aging* **94**, 15–23 (2020).
22. Jefferson, A. L. *et al.* The Vanderbilt Memory & Aging Project: Study Design and Baseline Cohort Overview. *Journal of Alzheimer's Disease* 1–20 (2016).
23. Huo, Y. *et al.* Consistent cortical reconstruction and multi-atlas brain segmentation. *Neuroimage* **138**, 197–210 (2016).
24. Cai, L. Y. *et al.* PreQual: An automated pipeline for integrated preprocessing and quality assurance of diffusion weighted MRI images. *Magnetic Resonance in Medicine* **86**, 456–470 (2021).
25. Beer, J. C. *et al.* Longitudinal ComBat: A method for harmonizing longitudinal multi-scanner imaging data. *NeuroImage* **220**, 117129 (2020).
26. Cumplido-Mayoral, I. *et al.* Biological brain age prediction using machine learning on structural neuroimaging data: Multi-cohort validation against biomarkers of Alzheimer's disease and neurodegeneration stratified by sex. *eLife* **12**, e81067 (2023).
27. Jonsson, B. A. *et al.* Brain age prediction using deep learning uncovers associated sequence variants. *Nat Commun* **10**, 5409 (2019).
28. Cole, J. H. *et al.* Predicting brain age with deep learning from raw imaging data results in a reliable and heritable biomarker. *NeuroImage* **163**, 115–124 (2017).

29. Yin, C. *et al.* Anatomically interpretable deep learning of brain age captures domain-specific cognitive impairment. *Proceedings of the National Academy of Sciences* **120**, e2214634120 (2023).
30. Hughes, E. J. *et al.* Regional changes in thalamic shape and volume with increasing age. *NeuroImage* **63**, 1134–1142 (2012).
31. Van Der Werf, Y. D. *et al.* Thalamic volume predicts performance on tests of cognitive speed and decreases in healthy aging: A magnetic resonance imaging-based volumetric analysis. *Cognitive Brain Research* **11**, 377–385 (2001).
32. Philp, D. J., Korgaonkar, M. S. & Grieve, S. M. Thalamic volume and thalamo-cortical white matter tracts correlate with motor and verbal memory performance. *NeuroImage* **91**, 77–83 (2014).
33. Carmichael, O. T. *et al.* Cerebral Ventricular Changes Associated With Transitions Between Normal Cognitive Function, Mild Cognitive Impairment, and Dementia. *Alzheimer Dis Assoc Disord* **21**, 14–24 (2007).
34. Todd, K. L. *et al.* Ventricular and Periventricular Anomalies in the Aging and Cognitively Impaired Brain. *Frontiers in Aging Neuroscience* **9**, (2018).
35. Silbert, L. C. *et al.* Changes in premorbid brain volume predict Alzheimer's disease pathology. *Neurology* **61**, 487–492 (2003).
36. Kamali, A., Flanders, A. E., Brody, J., Hunter, J. V. & Hasan, K. M. Tracing superior longitudinal fasciculus connectivity in the human brain using high resolution diffusion tensor tractography. *Brain Struct Funct* **219**, 269–281 (2014).
37. Westlye, L. T. *et al.* Life-span changes of the human brain white matter: diffusion tensor imaging (DTI) and volumetry. *Cerebral cortex* **20**, 2055–2068 (2010).
38. Saunders, R. C. & Aggleton, J. P. Origin and topography of fibers contributing to the fornix in macaque monkeys. *Hippocampus* **17**, 396–411 (2007).
39. Burzynska, A. Z. *et al.* White Matter Integrity Declined Over 6-Months, but Dance Intervention Improved Integrity of the Fornix of Older Adults. *Frontiers in Aging Neuroscience* **9**, (2017).
40. Douet, V. & Chang, L. Fornix as an imaging marker for episodic memory deficits in healthy aging and in various neurological disorders. *Frontiers in Aging Neuroscience* **6**, (2015).
41. Srisaikaew, P. *et al.* Fornix Integrity Is Differently Associated With Cognition in Healthy Aging and Non-amnestic Mild Cognitive Impairment: A Pilot Diffusion Tensor Imaging Study in Thai Older Adults. *Frontiers in Aging Neuroscience* **12**, (2020).
42. Gaser, C. *et al.* BrainAGE in Mild Cognitive Impaired Patients: Predicting the Conversion to Alzheimer's Disease. *PLOS ONE* **8**, e67346 (2013).
43. Beheshti, I., Maikusa, N. & Matsuda, H. The association between "Brain-Age Score" (BAS) and traditional neuropsychological screening tools in Alzheimer's disease. *Brain and Behavior* **8**, e01020 (2018).
44. Wrigglesworth, J. *et al.* Brain-predicted age difference is associated with cognitive processing in later-life. *Neurobiology of Aging* **109**, 195–203 (2022).
45. Peng, H., Gong, W., Beckmann, C. F., Vedaldi, A. & Smith, S. M. Accurate brain age prediction with lightweight deep neural networks. *Medical Image Analysis* **68**, 101871 (2021).
46. Mouches, P., Wilms, M., Rajashekar, D., Langner, S. & Forkert, N. D. Multimodal biological brain age prediction using magnetic resonance imaging and angiography with the identification of predictive regions. *Hum Brain Mapp* **43**, 2554–2566 (2022).
47. Liu, X. *et al.* Brain age estimation using multi-feature-based networks. *Computers in Biology and Medicine* **143**, 105285 (2022).

Digital health technology data in biocomputing: Research efforts and considerations for expanding access (PSB2024)

Michelle Holko
International Computer Science Institute at Berkeley
Bethesda, MD 20817, USA
Email: michelle.holko@gmail.com

Chris Lunt
National Institutes of Health
Bethesda, MD 20892, USA
Email: chris.lunt@nih.gov

Jessilyn Dunn
Biomedical Engineering, Duke University
Durham, NC 27708, USA
Email: jessilyn.dunn@duke.edu

Data from digital health technologies (DHT), including wearable sensors like Apple Watch, Whoop, Oura Ring, and Fitbit, are increasingly being used in biomedical research. Research and development of DHT-related devices, platforms, and applications is happening rapidly and with significant private-sector involvement with new biotech companies and large tech companies (e.g. Google, Apple, Amazon, Uber) investing heavily in technologies to improve human health. Many academic institutions are building capabilities related to DHT research, often in cross-sector collaboration with technology companies and other organizations with the goal of generating clinically meaningful evidence to improve patient care, to identify users at an earlier stage of disease presentation, and to support health preservation and disease prevention. Large research consortia, cross-sector partnerships, and individual research labs are all represented in the current corpus of published studies. Some of the large research studies, like NIH's *All of Us* Research Program, make data sets from wearable sensors available to the research community, while the vast majority of data from wearable sensors and other DHTs are held by private sector organizations and are not readily available to the research community. As data are unlocked from the private sector and made available to the academic research community, there is an opportunity to develop innovative analytics and methods through expanded access. This is the second year for this Session which solicited research results leveraging digital health technologies, including wearable sensor data, describing novel analytical methods, and issues related to diversity, equity, inclusion (DEI) of the research, data, and the community of researchers working in this area. We particularly encouraged submissions describing opportunities for expanding and democratizing academic research using data from wearable sensors and related digital health technologies.

Keywords: digital health technologies; wearables; sensors; waveform data; time-series data; algorithms.

1. Background

Wearable devices and other digital health technologies (DHTs), such as smartwatches, fitness trackers, and smart rings, are becoming increasingly popular for tracking and monitoring a wide range of health and fitness metrics [1]. Figure 1 below reflects the growth of scientific publications with the word "wearable" with 5,713 papers published in 2022. A similar pattern is seen when searching for "digital health technology." These devices can collect data on everything from heart rate and sleep patterns to activity levels and blood oxygen levels. In recent years, there has been a growing interest in using wearable devices and DHTs for health research. Wearable devices offer a number of advantages over traditional research methods, such as questionnaires and surveys. For example, wearable devices can collect data continuously and over long periods of time, providing researchers with a more complete picture of an individual's health and well-being. Additionally, wearable devices can be used to collect data in real-world settings, rather than in laboratory environments, which can provide more insights into how people behave in their everyday lives.

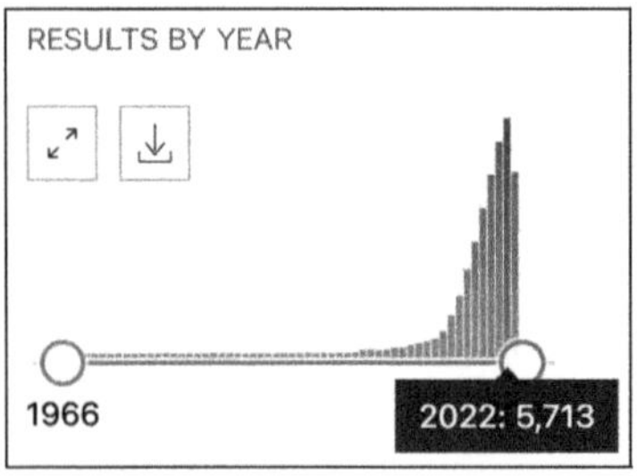

Fig. 1. Number of publications with "wearable" in PubMed from 1966-2023, highlighting exponential growth of this subject and 5,713 papers published in 2022.

Some specific examples of disease areas with active DHT and wearables research include:

- **Cardiovascular disease:** Apple watch devices have been used to study the relationship between heart rate and physical activity levels, and to develop algorithms to predict the risk of heart disease [2].
- **Respiratory disease:** Fitbit devices have been used to study the effects of different types of air pollution on lung function, and to develop algorithms to detect early signs of asthma attacks [3].

- **Metabolic disease:** Wearable devices tracking blood glucose and activity have been used to study gestational diabetes, and to develop algorithms to predict the risk of developing diabetes during pregnancy [4].
- **Mental health:** Wearable devices have been used to study the relationship between physical activity levels and mood, and to develop algorithms to predict symptom trajectory for bipolar disorder [5].

In addition to these specific examples, wearable devices and DHTs are also being used to study a wide range of other health conditions, such as cancer, infectious diseases, and chronic pain. But there are gaps in who has access to data and devices, who is performing the research and algorithm development, and therefore who the new technologies are poised to help improve health outcomes. Reviews of the current landscape of DHT research studies in the National Center for Biotechnology Information (NCBI)'s Clinical Trials database (clinicaltrials.gov), and of studies published by the top-20 funded private sector DHT companies, highlight several patterns and limitations:

1. **Small sample size:** Aside from a few large studies, most of the published clinical trials utilizing DHT have been relatively small, and are largely under-powered. "Nearly half the studies - 829, or 46.5% - had less than 100 enrollees. Only 8% had more than 1,000 [6]."
2. **Narrow Health Focus:** The majority of published DHT studies focus on cardiometabolic health and mental health/wellness, while relatively little published research examines critical healthcare burden diseases like stroke, chronic obstructive pulmonary disease (COPD), and diabetes [7].
3. **Narrow Population Focus:** Of studies published by the top 20 funded DHT private-sector companies, the majority (72%) include only healthy volunteers, rather than high-risk populations with comorbid conditions [8]. The breadth and diversity of the study population(s), including socioeconomic, healthcare status, and racial diversity, may be the most critical component of building AI-based DHT algorithms. This diversity is lacking in current published research, likely leading to biased results [9]. The "bring your own device" model has been used by many research studies, but this design may result in biased selection of participants, and therefore biased results [10].
4. **Limited Outcome Assessments:** Only 15% of published DHT studies measured clinical effectiveness, and only in relation to the patient outcomes and did not evaluate healthcare cost or access to care [11]. As healthcare cost and access are two of the most pressing needs in healthcare, it is important to expand research to examine these outcomes.
5. **Insufficient Reporting and Data Publishing:** Importantly, not only is reporting in clinicaltrials.gov not required for observational DHT trials, there is also no public database

for DHT data and algorithms. This complicates the ability to understand the full range of DHT "real world evidence" (RWE)-based research, and undermines research reproducibility and validation. The lack of a consensus DHT database also means that DHT data curation, feature (e.g., digital biomarker) discovery, and algorithm development is limited to those who have data, which is largely the private sector DHT companies. One attempt to develop standardized pipelines and data repositories for digital health data, the Digital Health Data Repository as part of the Digital Biomarker Discovery Pipeline [12], developed by co-organizer Jessilyn Dunn's lab, is still not fully funded.

6. **Bridging the Regulatory Gap and Moving to Clinical Implementation:** Despite tremendous progress in DHT research and development, there is still a lot of work to be done along the research → regulatory → clinical implementation continuum. The *All of Us* Research Program is uniquely situated within NIH to interact with FDA colleagues and assist in developing regulatory standards for this new and uncharted territory. The FDA also has a Center for Digital Health Excellence, and there is a Digital Health Consortium, housed within the Office of the National Coordinator, for senior leaders within the federal government to convene across the digital health continuum. The Digital Medicine Society is a professional organization that has been working across sectors with the community to support innovation and standardization, in part via the Digital Health Measurement Collaborative Community (DATAcc) [13] and the Digital Health Playbook [14]. For clinical implementation, HumanFirst has built the Atlas precision measures platform, a cloud-based platform with endpoints and measures being researched using DHTs across the industry to help pharma and clinicians decide on which devices and how they can be used in clinical research and healthcare [15].

The above limitations don't begin to address potential bias in algorithm development due to a limited pool of researchers interacting with these data. The purpose of this Session is to provide a forum for current research, address issues related to Diversity, Equity and Inclusion (DEI) in terms of the types of research and the researchers engaged, and ultimately to energize non-commercial research in the area. Our motivating question is how can this community work together to create more equitable research in the digital health tech space to benefit the research community and resulting impact?

2. Relevance to biocomputing

Computational biology approaches and algorithm development are critical enablers to the use of wearable devices and DHTs for biomedical research and health. Computational biologists are developing new methods for extracting meaningful insights from the large and complex datasets collected by these devices; algorithm developers are developing new algorithms to improve the accuracy and reliability of wearable devices and DHTs. The continuous or near-continuous data streams from DHTs are ripe for machine learning and artificial intelligence (ML/AI) research. Algorithms developed for detecting anomalies and other biomedically-related phenomena in wearable sensor data are increasingly being incorporated into research and moving into clinical practice and other health adjacent applications.

Despite the many advantages of using wearable devices and DHTs for health research, there are also a number of challenges that need to be addressed. One challenge is that the data collected by these devices can be noisy and complex with significant levels of missing data, making it difficult to extract meaningful insights. Another challenge is that the algorithms used to analyze this data need to be carefully validated to ensure that they are accurate and reliable. There are also many different devices, and the community doesn't yet have robust standards to compare between and among signals and data from different devices.

In this session, we bridge these gaps across sectors and domains to identify opportunities for researchers in the PSB community to contribute to the growing biomedical research leveraging wearables and DHT to understand and improve health. In prior years of PSB, there has been good representation of a variety of data types, including genomics, imaging and clinical data sets; there has been limited coverage of wearable sensors and digital health technologies research. Last year, PSB2023, we hosted the first year of this Session [16]. We wanted to continue to support this conversation and topic area as this field continues to grow and obstacles to academic research continue to need to be overcome. Many of the other conferences where DHT computational researchers are more focused on the clinical aspects and clinical trials, and not as much on the computational biology or biomedical research aspects of DHT data analysis and algorithm development.

The goal of this information sharing and discussion opportunity for participants and the community is to expand awareness and access to these data and tools, to enrich computational biology research, and bridge DEI gaps. The session includes a range of voices from academia, government, and private sector. It's important to represent private sector voices in this discussion since much of the research is currently happening in tech companies developing digital health devices. Creating a forum for dialogue across sectors is important for bridging gaps in awareness and understanding, and encouraging more researchers to participate in developing computational methods and analysis of data from digital health tech. The papers and discussion will focus on key challenges facing the field, and participants are encouraged to contribute ideas to potential solutions and initiate lasting collaborations with researchers and communities in this area. The session will also provide an opportunity to discuss as a community what is needed to truly enable cross-sector and expanded research for digital health technologies.

3. Session overview

The organizers will introduce the session, providing a background of the topic area, goals, and motivation for holding the session. There will then be a series of brief talks from the authors of the papers that were accepted for inclusion in the proceedings, a keynote by Vik Kheterpal from Care Evolution, ending with a panel discussion to include voices from academia, industry, and government including Q&A with attendees. The accepted papers/talks include causal data analysis of observational wearable device data, analysis of wearable silicone wristbands for chemical exposure, and digital biomarkers for detecting mild cognitive impairment. The talks are original research for publication, are widely varied, and the titles are listed below:

- Expanding the access of wearable silicone wristbands in community-engaged research through best practices in data analysis and integration
- Subject Harmonization of Digital Biomarkers: Improved Detection of Mild Cognitive Impairment from Language Markers
- Scalar-Function Causal Discovery for Generating Causal Hypotheses with Observational Wearable Device Data
- FedBrain: Federated Training of Graph Neural Networks for Connectome-based Brain Imaging Analysis (poster presentation only)

Following the original research talks, the keynote will be offered by Vik Kheterpal, the CEO and founder of Care Evolution. Vik is a nationally recognized expert in the area of healthcare informatics who has been focused on healthcare data exchange and interoperability for the past 11 years. He brings to the conversation the perspective of a serial entrepreneur working across IT, healthcare, and research sectors, and a go-to expert on real world data, healthcare IT, product design and usability, business, and leadership. After the keynote, attendees will be offered an opportunity to recommend DHT data collections and analysis methods that will help advance precision medicine research. This information will be shared with groups, such as the *All of Us* Research Program, that are collecting research data for the sake of advancing precision medicine.

The session will conclude with a panel discussion and audience Q&A. The panelists will feature speakers from industry, academia, and government; the session organizers will be joined by the keynote speaker and paper authors for a moderated discussion and Q&A from the participants. Session attendees are encouraged to participate in an interactive discussion on the current research, current challenges, and opportunities to expand access and use of DHT/wearables data in research.

References

1. https://www.pewresearch.org/fact-tank/2020/01/09/about-one-in-five-americans-use-a-smart-watch-or-fitness-tracker/
2. AI detection of cardiac dysfunction from consumer watch ECG recordings. Nat Med 28, 2478–2479 (2022). https://doi.org/10.1038/s41591-022-02079-5
3. Tsang KCH, Pinnock H, Wilson AM, et alPredicting asthma attacks using connected mobile devices and machine learning: the AAMOS-00 observational study protocolBMJ Open 2022;12:e064166. doi: 10.1136/bmjopen-2022-064166
4. H. Y. Lu et al., "Digital Health and Machine Learning Technologies for Blood Glucose Monitoring and Management of Gestational Diabetes," in IEEE Reviews in Biomedical Engineering, doi: 10.1109/RBME.2023.3242261.
5. Bennett, C.C., Ross, M.K., Baek, E. et al. Smartphone accelerometer data as a proxy for clinical data in modeling of bipolar disorder symptom trajectory. npj Digit. Med. 5, 181 (2022). https://doi.org/10.1038/s41746-022-00741-3
6. https://jamanetwork.com/journals/jamainternalmedicine/article-abstract/2725079
7. https://www.healthaffairs.org/doi/full/10.1377/hlthaff.2018.05081
8. https://bmcmedicine.biomedcentral.com/articles/10.1186/s12916-019-1377-7
9. https://preprints.jmir.org/preprint/29510/accepted
10. https://www.healthaffairs.org/doi/full/10.1377/hlthaff.2018.05081
11. https://github.com/DigitalBiomarkerDiscoveryPipeline/Digital_Health_Data_Repository
12. https://datacc.dimesociety.org/
13. https://playbook.dimesociety.org/
14. https://pubmed.ncbi.nlm.nih.gov/33948242/
15. https://www.pnas.org/doi/full/10.1073/pnas.2119097118
16. Session Introduction: Digital health technology data in biocomputing: Research efforts and considerations for expanding access. M Holko, C Lunt, J Dunn - Pacific Symposium on Biocomputing, 2023

Expanding the access of wearable silicone wristbands in community-engaged research through best practices in data analysis and integration

Lisa M. Bramer[1], Holly M. Dixon[2], David J. Degnan[1], Diana Rohlman[3], Julie B. Herbstman[4], Kim A. Anderson[2], Katrina M. Waters[1,2]

[1]Biological Sciences Division, Pacific Northwest National Laboratory, 902 Battelle Blvd Richland, WA 99354, United States
Email: lisa.bramer@pnnl.gov, david.degnan@pnnl.gov, katrina.waters@pnnl.gov

[2]Environmental and Molecular Toxicology, Oregon State University, 1007 Agriculture & Life Sciences Building, Corvallis, OR 97331, United States
Email: holly.dixon@oregonstate.edu, kim.anderson@oregonstate.edu

[3]College of Health, Oregon State University, 103 SW Memorial Place, Corvallis, OR 97331, United States
Email: diana.rohlman@oregonstate.edu

[4]Department of Environmental Health Sciences, Mailman School of Public Health, Columbia University, 722 West 168th Street, New York City, NY 10032, United States
Email: jh2678@cumc.colmbia.edu

Wearable silicone wristbands are a rapidly growing exposure assessment technology that offer researchers the ability to study previously inaccessible cohorts and have the potential to provide a more comprehensive picture of chemical exposure within diverse communities. However, there are no established best practices for analyzing the data within a study or across multiple studies, thereby limiting impact and access of these data for larger meta-analyses. We utilize data from three studies, from over 600 wristbands worn by participants in New York City and Eugene, Oregon, to present a first-of-its-kind manuscript detailing wristband data properties. We further discuss and provide concrete examples of key areas and considerations in common statistical modeling methods where best practices must be established to enable meta-analyses and integration of data from multiple studies. Finally, we detail important and challenging aspects of machine learning, meta-analysis, and data integration that researchers will face in order to extend beyond the limited scope of individual studies focused on specific populations.

Keywords: Silicone Wristbands, Wearables, Exposome, Environmental Health, Exposure Science, Public Health.

1. Introduction

Silicone wearables as passive sampling devices have emerged as a powerful and versatile personalized exposure assessment tool, allowing researchers to characterize chemical exposures for a wide variety of organic chemicals and study the impact of exposures on human health [1]. The use of silicone wearables in research, especially wristbands, has grown substantially since the first publication in 2014 [2]. Thousands of participants from several countries on six continents have worn wristbands [3] and there have been over 60 peer-reviewed papers published to date [1]. Since wristbands are easy-to-wear, do not require in-person consultation or training, and can be transported at ambient temperature in the mail back to the laboratory for analysis, they are a convenient choice for researchers and study participants alike even in challenging scenarios like disasters or pandemics [4-6].

However, despite the growing use of wristbands in research, the majority of individual wristband studies are limited due to small sample size and narrow population focus. In addition, no established best practices for analyzing wristband data across multiple studies exist, thereby limiting impact and access of these data for larger meta-analyses. Dixon et al. is the only study that has taken wristbands from multiple studies and reported trends in chemical exposure patterns across the globe [7]. In this paper, authors took wristband extracts from 14 different communities on three continents and re-ran those extracts on the same analytical method for the presence and absence of 1530 chemicals. Authors identified common chemical mixtures between geographically diverse participants. Dixon et al. also reported that wristbands worn in Texas post-Hurricane Harvey had the highest mean number of chemical detections compared with the other study locations, illustrating that comparing wristband studies from a diverse set of communities and geographical areas can highlight populations with unique chemical exposure profiles and therefore unique health risk profiles.

Re-running wristband extracts from different studies on the same analytical method as done in Dixon et al. [7] is not a sustainable strategy for using wristband data to better understand broad exposure patterns and trends. We need new data analysis strategies to combine wristband data from multiple studies or use meta-analysis procedures, which would increase data access and interoperability. The growing number of individual wristband studies can be leveraged by combining data across studies to uncover patterns about personal chemical exposure, which can lead to new human health discoveries and can be used to direct research, interventions, and policy resources towards communities with higher exposure burdens or unique exposure patterns. In this manuscript, we present key considerations for analyzing wristband data and combining data collected from multiple studies. We use datasets from three studies to highlight challenges associated with data structure and missingness and the consequences of varying analysis techniques and choices between studies, which are often overlooked or not addressed in individual studies.

2. Methods

2.1. *Study design and data collection*

Our paper illustrates data analysis and integration challenges using chemical exposure data from 616 wristbands worn by participants in two study cohorts, one in New York City and one in Eugene,

Oregon. The New York (NY) wristband data was collected as part of an ongoing longitudinal birth cohort study at the Columbia Center for Children's Environmental Health. Individuals pregnant with a singleton who were 18 years and older wore a wristband for 48 hours in their third trimester of pregnancy [8]. There are two sets of wristband data from the NY cohort that are included in this report, one set includes 22 wristbands from a pilot study collected between 2013 and 2015 [8] (referred to as "NY Pilot") and the second set includes 168 wristbands from a larger study between 2015 and 2019 (referred to as "NY"). We also include data from 426 wristbands worn by study participants in Oregon in 2017 and 2018 (referred to as "OR"). Study participants were asked to wear wristbands for seven consecutive days in two seasons (summer and winter), wearing a new wristband each day of the study. Study participants had to be 18 years or older, be diagnosed with mild to moderate asthma, be a current non-smoker, and live near Eugene, Oregon.

All participants provided informed written consent in accordance with the Columbia University Institutional Review Board (IRB; #AAAK6753) for the NY cohort and in accordance with the Oregon State University IRB (#8058) for the OR cohort.

We prepared, cleaned, and extracted all the wristbands as previously described [5]. We also created and analyzed several quality control samples throughout the wristband preparation, transport, and laboratory processing steps, which is described in Dixon et al. [5]. We analyzed the New York wristband extracts for 61 organic chemicals with an Agilent 7890B gas chromatograph (GC) paired with a 7000C triple-quadrupole mass spectrometer (MS/MS) [5]. We analyzed the OR wristband extracts for 94 organic chemicals using an Agilent 7890A GS interfaced with an Agilent 5975B MS. Further analytical details can be found in Anderson et al. [9].

2.2. *Data Processing*

We converted chemical concentrations to moles per gram wristband and applied a log transformation ($\log_2$ pmol/g wristband). We set the concentration value for a given chemical equal to NA if there was matrix interference (Section 3.1) [5]. We conducted analyses using the statistical software R, version 4.1.2 [10]. For each dataset, we filtered out chemicals which were not detected in any wristbands; this resulted in 53, 44, and 69 chemicals in the NY Pilot, NY, and OR datasets, respectively. We masked chemical names in the results as part of our de-identification process.

3. Data Properties

3.1. *Types of missing and censored data*

Missing values in data can arise for a variety of reasons and are handled differently depending on the type of missing data. Missing data types are commonly grouped into three categories: missing completely at random (MCAR), missing at random (MAR), and not missing at random (MNAR). When data are MCAR, the probability of an observation being missing is unrelated to any other observed or unobserved factors. Missing values that can be completely explained by another observed variable or variables are MAR. When data are MNAR, the probability of an observation going missing is related to an unobserved variable or variables. There are two primary types of missing data in wristband studies: observations that are below the limit of detection (LOD) and observations that are impacted by matrix interference (MI). Data values below LOD arise from a combination of all three missing data categories. The absence of a quantifiable peak for a sample and chemical of interest (MAR) will result in a missing annotation and results in a majority of the below LOD missing values. Much less frequently, a human error in data processing, such as deletion of a peak in quantification software (MCAR), or a participant's failure to comply with study protocols, such as removing a wristband for part of a day resulting in low levels of measured chemicals not representative of true exposure may cause data to be annotated as missing and below LOD. Alternatively, MI occurs when a deuterated surrogate peak, used for quantification, is masked; this can arise when a wristband sample contains compounds from personal care products or sweat (MNAR).

3.2. *Handling missing data*

A majority of statistical and machine learning methods require complete observations (i.e. no missing values). Therefore, to leverage these techniques effectively, researchers must decide how to handle missing data, especially when using chemical concentrations from wristbands in a multivariate manner. One solution is to filter any samples or chemicals that contain missing values. We calculated the percentage of chemicals with complete observations across all samples for each of the three studies. Then, for each study we iteratively removed the sample with the most missing values and recalculated the percentage of chemicals with complete data. We summarized the percentage of chemicals with complete data at varying numbers of wristbands. The percentage of chemicals completely observed across all wristbands drops to 50% with only 4, 2, and 3 wristbands for NY Pilot, NY, and OR, respectively. A total of 13, 16, and 21 wristbands with the fewest missing values result in 25% of chemicals with complete observations for NY Pilot, NY, and OR, respectively. When study sizes grow to more than 165 wristbands for NY and OR, only one chemical is observed across all wristbands.

Further, in targeted analytical methods, there is high confidence in the LODs and information about what chemicals are below LOD in wristband extracts contains meaningful data about what people are not being exposed to or are exposed to in very small amounts [9]. Thus, the large number of missing values in wristband data means data removal approaches will significantly diminish the size and information in the data and may introduce bias if missing observations are MNAR.

An alternative approach is imputation of missing observations. The most common imputation approach taken in wristband studies is the replacement of the below LOD missing values with a constant value, such as half the LOD (e.g. [8, 11-13]). Unlike many other mass spectrometry-based measurement fields (e.g. proteomics), a vast majority of below LOD values are due to the true absence of a quantifiable peak, thus half the LOD values are reasonably close to the true unobservable values. However, imputation using a constant value likely does not reflect the true values if they could be measured and can significantly affect the covariance structure of the data, resulting in differences in common downstream analyses, such as principal component analyses (e.g. [11, 14]). As an example, we ran principal component analysis via projection pursuit (PPCA) [15] on the NY Pilot data without imputing missing values (Fig. 1A) and ran k-means clustering [16] based on the first two principal component scores, setting k=4 based on the optimal number of clusters as determined by evaluating the silhouette score [17]. Additionally, we ran PPCA on the NY Pilot data where missing values below the LOD were imputed with half the LOD (Fig. 1B) with samples colored by the clusters assigned based on PPCA results without imputation. The percentage of variability explained by each component is considerably different for the two analyses, and although some samples clustered similarly, several samples formed much different clusters when PPCA was run with imputed values. For example, samples 15, 16, and 22 cluster at the top left in Fig. 1B and the same behavior is not observed in Fig. 1A. Further, we examined the loadings of each chemical on the first principal component (PC1) as seen in Fig. 1C. Large differences in loadings were observed with many chemicals having very little influence on scores in the non-imputed PPCA but having large positive loadings in the half LOD PPCA.

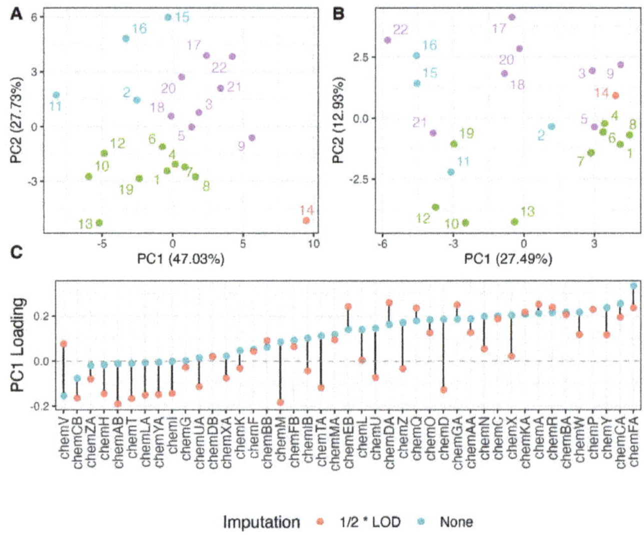

Fig. 1. PC1 and PC2 loadings from PPCA for the NY Pilot data for (A) no imputation of missing values and (B) imputation of missing values with half the LOD. Samples are colored by cluster as determined by k-means clustering based on the PC1 scores from PPCA results without imputation. (C) PC1 loadings for each chemical when missing values were not imputed (green dots) and when missing values were imputed with half the LOD (red dots).

Alternatively, missing values can be imputed using more complex algorithms. These methods provide the benefit of introducing variability in imputed values, unlike imputation of half the LOD. A few wristband studies have utilized these approaches to date (e.g. [18]) for all types of missing values. We applied two such example imputation methods to the OR dataset. We imputed missing values using two different methods, random forest imputation [19] and multiple imputation by chained equations (MICE) using the predictive mean metric [20], for chemicals with no more than 40% missing observations. Fig. 2 shows a comparison of imputed values generated by the two methods for observations that were missing due to MI and being below LOD. While some imputed values are very close to one another for the two methods, there are many values that differ by orders of magnitude. Further, the imputation methods were unable to differentiate between the missing value mechanisms, as below LOD and MI observations overlap, nor were they able to impute values below the LOD in nearly all cases. In general, empirically-driven imputation methods are insufficient for imputation of observations below LOD as imputed values are often much larger than the LOD (Fig. 2), which is not consistent with the results of the chemical analysis. Even methods aimed at imputation of left-censored data (e.g. [21]) rely on minimum observed values in the dataset and are still orders of magnitude larger than known LODs, as these algorithms have been designed for different application areas. On the other hand, these imputation methods use the structure of the data to fill in missing observations, and can be useful for resolving missing observations due to MI. The choice of imputation method and underlying assumptions should be carefully considered, as they can lead to significantly different interpretation of a dataset in downstream analyses. Further, no guidance exists nor have any thorough reviews been conducted to determine the threshold of detection rate for a chemical to be included in an analysis and imputed. Researchers have used a wide range of thresholds ranging from 20% [22] to 75% [23] of observations detected for a chemical to be included in downstream analyses. At a minimum, researchers should conduct sensitivity analyses to evaluate the effect of chosen threshold on their results.

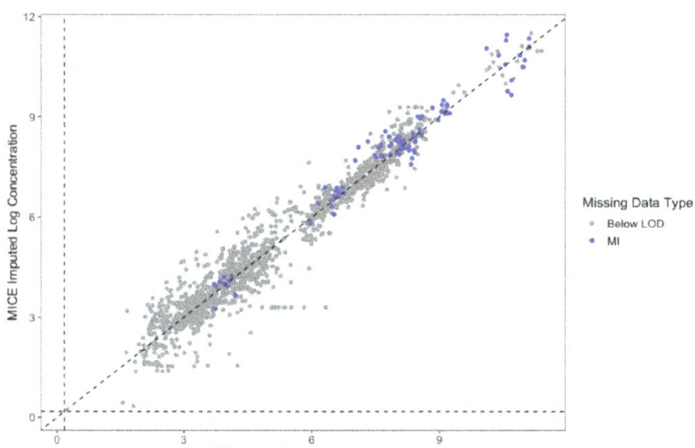

Fig. 2. Comparison of MICE and RF imputation methods for two types of missing chemical data, data that is below the LOD (gray dots) and MI (blue dots), in the OR study. The dotted black vertical and horizontal lines represent the median LOD across chemicals.

The final potential solution is to develop or use novel analysis techniques which are tolerant to missing or that can partition sources of variability, so they are not skewed by large numbers of constant values near LOD. For example, in the proteomics field, a statistical analysis technique which combined quantitative and qualitative (presence/absence) models was developed to accommodate and utilize missing observation information [24]. Further research and development of new data analysis models and methods for wristband data is needed to specifically address the challenges described here for wearable wristband data.

3.3. *Distributional properties of concentrations*

Understanding the underlying distribution of chemical concentrations measured by wristbands is of fundamental importance to select appropriate statistical models and analyses to conduct. Fig. 3 shows the distribution of three chemicals from the OR study, with half the LOD filled in for observations below LOD. Within and across chemical observations from wristbands vary by orders of magnitude resulting in distributions heavily skewed to the right (Fig. 3A). Log transformation is a common technique to stabilize variances and transform skewed data distributions to approximately normal distributions and is commonly used in wristband study analyses (e.g. [13, 25]). On a log-transformed scale, chemical concentrations above LOD can be reasonably approximated by a normal distribution (Fig. 3B-D). However, the full distribution of log concentrations is bimodal even for small numbers of observations below LOD, and the distance between observations above and below LOD depends on the chemical.

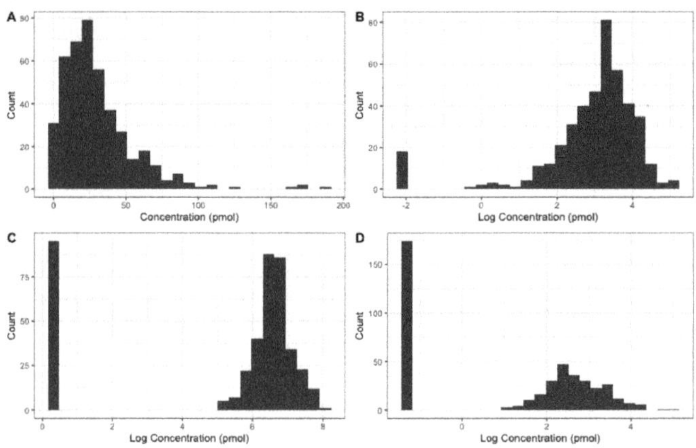

Fig. 3. Distribution of three chemicals from the OR study, with half the LOD filled in for observations below LOD (A and B) chemH, (C) chemPB, and (D) chemCB, with chemH visualized on both the (A) linear scale and (B) log scale.

4. Data Analysis Methods

4.1. *Statistical methods and summaries*

After conducting a survey of existing literature detailing wristband studies, with a broad range of research applications and hypotheses, and their data analysis methods, several prevalent statistical methods and models emerge, including correlation, linear regression, basic hypothesis testing between groups, and logistic regression. However, despite a small number of statistical models and methods being used in the field, nuances and details of how data preprocessing and model fitting are carried out vary widely. This will create significant problems as the number of wristband studies continues to grow and the research community attempts to combine results from multiple studies through techniques such as meta-analysis [26]. Meta-analysis can be a powerful tool for assessing the consistency and generalizability of results across multiple studies and study populations. However, study combination approaches require consistent statistical data processing and testing procedures.

Perhaps the most common summary statistic used in wristband studies is the calculation of correlations between chemical concentrations measured by wristbands, between a chemical's concentration profile from wristbands and the profile of another exposure assessment methodology (e.g. urine biomarkers) or to health outcomes. A majority of researchers utilize a non-parametric Spearman correlation, recognizing an assumption of both variables following a normal distribution, as required by metrics such as Pearson correlation, was not appropriate. However, some researchers have calculated Spearman correlation by imputing below LOD observations with half the LOD, or a similar small constant value (e.g. [8, 23, 27, 28]), while other researchers have chosen to calculate correlation using only observations above LOD (e.g. [29]). In the case of linear regression, chemical concentrations are used as the dependent variable or independent variable(s) depending on the research question. Some researchers choose to log-transform concentrations, impute half LOD values for below LOD observations, and limit chemicals used in analyses based on percentage of observations above LOD in an effort to avoid violation of the normality assumption of errors (e.g. [12, 23]). Other researchers restrict linear regression models to chemical concentrations above LOD [30]. Further, the choice of the minimum percentage of detections required per chemical for inclusion in analyses varies widely across studies. When testing for differences in concentrations between groups of interest, some researchers leverage non-parametric tests such as the Wilcoxon Test (e.g. [29]) and others conduct parametric statistics such as a t-test on log-transformed concentrations (e.g. [31]).

Although differences in analyses mentioned above appear small or trivial in nature, the implications on results and conclusions drawn across studies using techniques such as meta-analysis are potentially large. For example, when testing differences in chemical concentrations between groups of interest, a t-test is testing for differences in the mean concentration levels, while the Wilcoxon Test conducts a test for differences in the distribution of values and is typically sensitive to differences in the median, depending on the sample size, but often not the mean unless the sample size is very large [32]. As an illustrative example of the downstream effect of differences in data treatment and modeling, we used the OR dataset. We considered two pairs of chemicals chemUB & chemT and chemH & chemT, for which nearly all wristbands had pairs of observations above LOD (94.8% and 93.7% of 426 wristbands, respectively). We first calculated the Spearman correlation between each pair of chemicals' concentrations for all complete pairs of observations chemUB & chemT and chemH & chemT as 0.157 and 0.959, respectively. Then for each pair of chemicals, we synthetically introduced missing below LOD values into the data. At each iteration an additional observation was set to below LOD for 1 to 420 (0.2% to 99% of the data) wristbands and the Spearman correlation was calculated 1) with half the LOD imputed for missing values and 2) ignoring missing below LOD observations. Fig. 4 shows the correlation values for the two pairs of chemicals. The treatment of observations below LOD causes the Spearman correlation to differ considerably between the two methods even with small percentages of missing values. As the percentage of detections decreases, correlation calculated using imputation with half LOD inflates and effectively becomes a metric of correspondence between detections and non-detections between the two chemicals rather than measuring the strength of quantitative association, even above thresholds of filtering seen in literature (e.g. 70% marked by the vertical dashed line in Fig. 4). When ignoring missing values from the Spearman correlation calculation, values are centered around the true correlation value but have high variability as the percentage of below LOD observations gets larger. Fig. 4 clearly shows that these two methods for computing correlation are representative of different properties of the data when not all observations are above LOD.

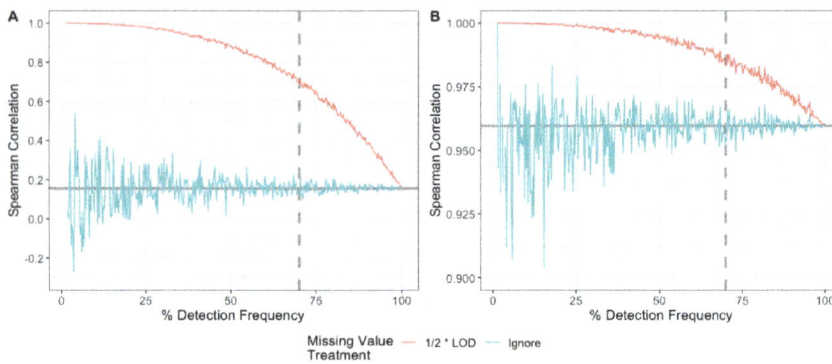

Fig. 4. Spearman correlation for (A) chemUB & chemT and (B) chemH & chemT at varying levels of simulated missing values. Vertical lines indicate a common threshold for filtering at 70% detection, and horizontal lines represent the correlation with complete data.

4.2. *Utilizing machine learning*

Machine learning (ML) offers a promising avenue for advanced multivariate analyses, such as discovery of associations between multiple chemicals and a particular health outcome or prediction of a chemical exposure level based on behavior and environmental factors. Despite the promise of ML models, they are most powerful in cases where large sample sizes are available. As the technology grows in popularity and laboratories become more established in the methodology, many studies such as the OR and NY datasets presented here and other studies (e.g. [33]) are reaching sample sizes where ML is a viable option. One limitation of ML is that a majority of methods require no missing values, requiring researchers to again consider and establish best practices when dealing with missing observations due to being below LOD, MI, etc., for wristband data.

If a researcher uses chemical detection status (i.e. detected or below LOD) as the response variable in their model, it is important to note that ML classification model performance can suffer when outcome category frequencies are highly imbalanced [34]. In this case, ML models learn characteristics of the majority class only when prediction accuracy is being optimized. The use of alternative metrics and techniques such as down sampling and upweighting [35] may help alleviate this issue if large enough sample sizes are available. When using a quantitative outcome, a large majority of ML methods, particularly those used on smaller sample sizes (e.g. <1,000), assume that the response variable approximately follows a normal distribution. Therefore, if a researcher uses chemical concentrations as the response variable and imputes below LOD observations, methods such as discriminant analysis, naive Bayes, and support vector machines will be inappropriate. Tree-based methods such as regression trees [36] and random forest regression [37] do not make distributional assumptions and provide more promise for use with wristband data. However, even for these models, the ratio of detects and non-detects, the distance between LOD and detected values, and the optimization or loss function used must be considered carefully. For example, if the mean-squared error is used the model can effectively become a classification model between detections and non-detections with no ability to differentiate large differences in concentrations, because they are still much smaller than the large distance between LOD and observed concentrations. If chemical concentrations are used as predictor variables, some ML also assume normally distributed explanatory variables (e.g. discriminant analysis). Random forest regression models utilize resampling with replacement to grow multiple regression trees. The importance of sample size and even distributional properties, such as number of below LOD observations, becomes important as the resampling method may have a difficult time representing the underlying distribution well, particularly for bimodal distributions. Fig. 5 shows the original distribution of concentrations with half LOD imputed values for chemZ from the NY Pilot dataset. Blue densities show 25 resampled distributions drawn by the random forest model. Some random draws represent the original distribution well, while others do not sample any below LOD observations at all, because of the small sample size. Additional research into nuances of ML methods and establishment of best practices is of fundamental importance as study sizes grow and combination of study datasets becomes possible.

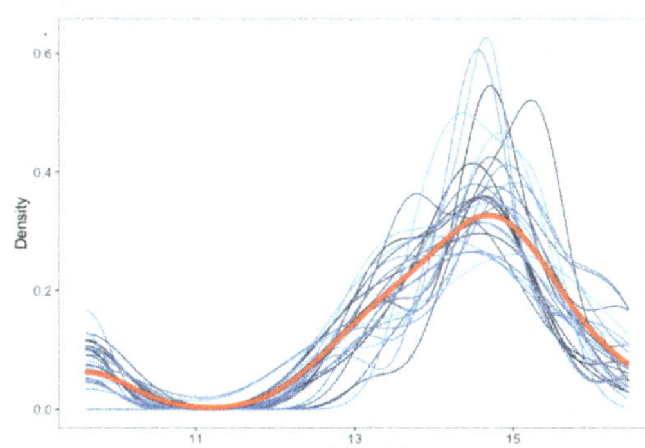

Fig. 5. Density of $\log_2$ chemZ concentrations from NY Pilot data (red) and densities from 25 random forest resample draws (blue).

5. Utilizing Data from Multiple Studies

Combining data from multiple wristband studies is crucial to allow researchers to uncover patterns of personal chemical exposure that correlate with potential health impacts across diverse communities. A shared understanding of fundamental wristband data properties and a set of statistically sound strategies to analyze the data is needed in order to pave the way for combining data from multiple studies. However, there are also several other factors that hinder combining wristband data across studies [1, 38, 39]; some are due to research gaps and some are due to differences in how researchers report their data. For example, the way that authors report chemical concentrations varies. Some authors report chemical mass per entire wristband (e.g. ng/wristband) and others report chemical mass per unit mass of the wristband (e.g. ng/g wristband or pmol/g wristband) [1]. When these concentration reporting differences are present and wristband masses are not reported alongside the data, then this inhibits the comparison of chemical concentrations across studies [39]. In addition, more research is needed to understand the variables that influence the rate and amount of chemicals entering into wristbands [1], which could lead to strategies to normalize wristband data across studies where wristbands were worn for different lengths of time and in different environmental conditions. As described in Samon et al., chemical uptake into silicone wristbands is not consistent over time and is dependent on each chemical's physical-chemical properties and environmental conditions during the study [1]. Silicone post-deployment cleaning and extraction methods also differ between laboratories [3, 39] and it is unknown how these differences may affect quantified concentrations across studies in different laboratories. To reduce additional sources of variability in the data, researchers need to agree on best practices for communicating study protocols to participants to address potential misconceptions up front. For example, participants may think that if they wear the wristband longer than requested, they are helping the research goal and providing more data, when instead they are introducing more variability and complicating interpretation of study results.

While further research into sources of variability due to differences in study designs is needed, how chemical concentrations might be normalized across studies remains an open research question. In the meantime, researchers can consider alternate ways to utilize data from multiple studies. For example, the comparison of study locations can be made by looking for differences in detection frequency, if other important factors are reasonably controlled or equitable between populations. However, different analytical methods for chemical identification and quantification have been developed and used, even for studies coming out of the same research laboratory. In these cases, the chemicals targeted differ. For example, the NY dataset was measured for 61 chemicals, and the OR dataset was measured for 94 chemicals. Of these chemicals, a total of 45 were measured in both studies. The joining of these datasets would result in 110 chemicals in total, and more missing values would be introduced into the joined dataset. However, unlike previous missing data, these missing values would be MAR and downstream statistics would need to account for the additional mechanism for missing values. When comparing the exposure of individuals between study populations, the detection frequency across all chemicals analyzed within each study relative to a measure of central tendency summary across other participants in the same study would give a sense of total exposure level for an individual. Further, many studies have discretized chemical concentration values into categories such as tertiles (e.g. [5, 28]). This concept could be used within a given study to derive chemical tertile profiles for each wristband. Then categorical data analysis strategies such as multiple correspondence analysis [40] could be used to perform clustering of wristband samples across studies to find samples with common patterns. Alternatively, the actual percentile within a study could be recorded and used to visualize and begin to understand exposure patterns across studies. However, the treatment of missing values, either due to a study-based MAR source or a MNAR below LOD source needs to be carefully considered and treated differently. For example, Fig. 6 shows empirical cumulative density curves for chemCB in both the NY and OR datasets with tertile thresholds denoted by gray dashed lines. For this chemical, the proportion of wristbands with values below LOD in the OR study is greater than 0.33. It would be nonsensical to assign some of the wristbands with non-detections to the lower tertile and others to the middle tertile. If all wristbands below LOD were assigned to the lower tertile for OR, a researcher would need to consider if tertiles composed of different proportions of samples are still comparable, and what proportion of LOD observations comparisons are no longer meaningful.

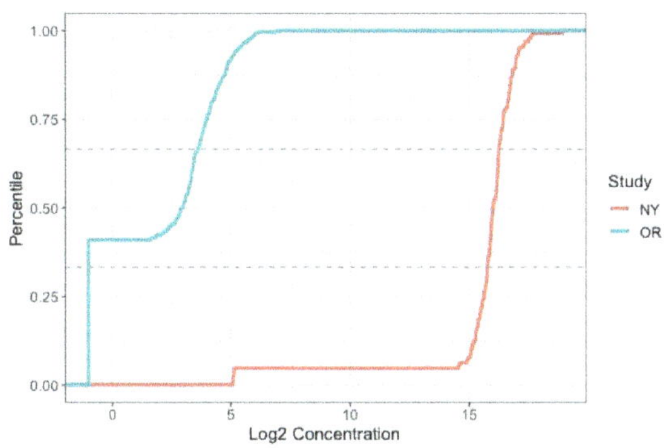

Fig. 6. Empirical cumulative density plots of chemCB with tertile thresholds denoted by gray lines.

6. Recommendations and Conclusions

The use of silicone wristbands in research studies has rapidly grown over the past few years, especially in community-engaged research [1, 6]. Currently, researchers are using a wide range of data processing and statistical approaches to analyze data from silicone wristbands with a focus on within-study interpretation, but these approaches jeopardize the ability to use the datasets for larger meta-analyses. The need for better guidance and established best practices is evident in the examples shown here, where minor differences in data handling and modeling can lead to vastly different conclusions and interpretation. Some key takeaways and guidance from example analyses presented here are as follows:

- The imputation of half the LOD, or other small constant values, can greatly affect the covariance structure of wristband data (Fig. 1). Even when scaling features, imputation of below LOD values is not recommended as analyses utilizing the covariance structure will be determined by detection rates rather than the intended quantitative information.
- Many dimension reduction techniques for exploratory data analysis have implementations that do not require imputation of missing values, such as projection pursuit PCA. These algorithms are preferable to imputation of below LOD values for wristband data.

- Wristband data chemical concentrations where below LOD values have been imputed with half the LOD do not follow a normal distribution, even when log-transformed, and are bimodal. Methods such as linear regression are not appropriate for this data, and alternative methods such as Gaussian mixture models [41] should be considered.
- Missing value treatment strategies must account for the different missing data mechanisms in wristbands studies, rather than blindly implementing one imputation method on multiple types of missing values (Fig. 2). For example, the use of data-driven imputation algorithms, such as MICE, on below LOD observations can result in imputed values considerably larger than LOD.
- Non-parametric hypothesis tests are not equivalent to parametric hypothesis tests at sample sizes typical of wristband studies. Although methods such as the Wilcoxon Test can be used to skirt the assumption of normally distributed data with half LOD imputed data, even at small percentages of missingness, the tests are effectively determined by detection rates rather than concentration values (Fig. 4).

When presented with data from a wristband study or multiple wristband studies, researchers should first prioritize evaluating the units reported, amount of missing data the effect on the structure of specific chemicals analyzed. If data from multiple studies are present, concentration units should be standardized. Further, only chemicals commonly detected between studies should be considered in downstream analyses. Sensitivity analyses should then be conducted to determine which chemicals should be analyzed using quantitative concentrations or using detected/not-detected information based on the detection frequency of each compound. If data is to be used quantitatively, concentrations should be log-transformed. Tree-based ML methods, which can handle detection information and concentrations simultaneously through LOD imputation, may be considered with large enough sample sizes. Sensitivity analyses looking at the resampled distribution of chemicals should be examined before considering these ML methods. Finally, if the overall exposure profile is of interest, researchers should evaluate if techniques, such as PPCA, provide an interpretable reduced dimension option to represent wristband data. Often in wristband studies, chemicals will all have loadings in the same direction on one of the principal components leading to one potential metric of overall exposure.

This is the first paper to summarize data properties, current data analysis approaches and their issues, and important areas where best practices are needed for wristband data. We demonstrate there is a need for standardized and thorough wristband data analysis methods from the research community, which will create more opportunities to combine wristband data from multiple studies or use meta-analysis procedures, leading to increased data access and interoperability. In addition, more research is needed to understand other factors that hinder the combination of data from individual wristband studies (e.g. how to normalize for differences in wristband wear time and environmental conditions). Overall, a combination of these efforts will enable research to move beyond the narrow population focus of individual studies, leading to new discoveries about personal chemical exposure and potential impacts to human health.

Acknowledgements

We thank the study participants for their willingness to engage with our research team. Research reported in this publication was supported by the National Institute of Environmental Health Sciences (NIEHS) under award numbers R21/R33ES024718, P30ES030287, and P42ES016465. The content is solely the responsibility of the authors and does not necessarily represent the official views of the NIEHS. PNNL is a multi-program laboratory operated by Battelle for the U.S. Department of Energy under contract DEAC05-76RL01830.

Data and Code Availability

- De-identified wristband data is available for download at https://data.pnnl.gov/group/nodes/dataset/33672.
- R code to reproduce all analyses and plots is on GitHub at https://github.com/PNNL-Superfund-Research-Center/PSB_Wristband_Analyses/.

Declaration of Competing Interest

Kim A. Anderson and Diana Rohlman, authors of this research, disclose a financial interest in MyExposome, Inc., which is marketing products related to the research being reported. The terms of this arrangement have been reviewed and approved by Oregon State University in accordance with its policy on research conflicts of interest. The authors have no other relevant financial or non-financial interests to disclose.

References

1.Samon SM, Hammel SC, Stapleton HM, Anderson KA. Silicone wristbands as personal passive sampling devices: current knowledge, recommendations for use, and future directions. Environ Int. 2022: 107339.

2.O'Connell SG, Kincl LD, Anderson KA. Silicone wristbands as personal passive samplers. Environ Sci Technol. 2014; 48(6): 3327-35.

3.Dixon HM, Poutasse CM, Anderson KA. Silicone wristbands and wearables to assess chemical exposures. In: Phillips K, Yamamoto D, Racz L, editors. Total exposure health: An introduction. First ed: CRC Press; 2020. p. 139-60.

4.Anderson KA, Points III GL, Donald CE, Dixon HM, Scott RP, Wilson G, et al. Preparation and performance features of wristband samplers and considerations for chemical exposure assessment. J Expo Sci Environ Epidemiol. 2017; 27: 551.

5.Dixon HM, Bramer LM, Scott RP, Calero L, Holmes D, Gibson EA, et al. Evaluating predictive relationships between wristbands and urine for assessment of personal PAH exposure. Environ Int. 2022; 163: 107226.

6.Rohlman D, Samon S, Allan S, Barton M, Dixon H, Ghetu C, et al. Designing equitable, transparent community-engaged disaster research. Citizen science: theory and practice. 2022; 7(1).

7.Dixon HM, Armstrong G, Barton M, Bergmann AJ, Bondy M, Halbleib ML, et al. Discovery of common chemical exposures across three continents using silicone wristbands. Royal Society Open Science. 2019; 6(2): 181836.

8.Dixon HM, Scott RP, Holmes D, Calero L, Kincl LD, Waters KM, et al. Silicone wristbands compared with traditional polycyclic aromatic hydrocarbon exposure assessment methods. Anal Bioanal Chem. 2018; 410(13): 3059-71.

9.Anderson KA, Szelewski MJ, Wilson G, Quimby BD, Hoffman PD. Modified ion source triple quadrupole mass spectrometer gas chromatograph for polycyclic aromatic hydrocarbon analyses. J Chromatogr A. 2015; 1419: 89-98.

10.R Development Core Team. R: A language and environment for statistical computing. Vienna, Austria: R Foundation for Statistical Computing; 2023.

11.Donald CE, Scott RP, Blaustein KL, Halbleib ML, Sarr M, Jepson PC, et al. Silicone wristbands detect individuals' pesticide exposures in West Africa. Royal Society Open Science. 2016; 3(8): 160433.

12.Hammel SC, Hoffman K, Webster TF, Anderson KA, Stapleton HM. Measuring personal exposure to organophosphate flame retardants using silicone wristbands and hand wipes. Environ Sci Technol. 2016; 50(8): 4483-91.

13.Kassotis CD, Herkert NJ, Hammel SC, Hoffman K, Xia Q, Kullman SW, et al. Thyroid receptor antagonism of chemicals extracted from personal silicone wristbands within a papillary thyroid cancer pilot study. Environ Sci Technol. 2020; 54(23): 15296-312.

14.Paulik LB, Hobbie KA, Rohlman D, Smith BW, Scott RP, Kincl L, et al. Environmental and individual PAH exposures near rural natural gas extraction. Environ Pollut. 2018; 241: 397-405.

15.Webb-Robertson B-JM, Matzke MM, Metz TO, McDermott JE, Walker H, Rodland KD, et al. Sequential projection pursuit principal component analysis–dealing with missing data associated with new-omics technologies. BioTechniques. 2013; 54(3): 165-8.

16.Hartigan JA, Wong MA. Algorithm AS 136: A k-means clustering algorithm. Journal of the royal statistical society series c (applied statistics). 1979; 28(1): 100-8.

17.Rousseeuw PJ. Silhouettes: a graphical aid to the interpretation and validation of cluster analysis. Journal of computational and applied mathematics. 1987; 20: 53-65.

18.Doherty BT, McRitchie SL, Pathmasiri WW, Stewart DA, Kirchner D, Anderson KA, et al. Chemical exposures assessed via silicone wristbands and endogenous plasma metabolomics during pregnancy. J Expo Sci Environ Epidemiol. 2022; 32(2): 259-67.

19.Stekhoven DJ, Bühlmann P. MissForest—non-parametric missing value imputation for mixed-type data. Bioinformatics. 2012; 28(1): 112-8.

20. Van Buuren S, Groothuis-Oudshoorn K. mice: Multivariate imputation by chained equations in R. Journal of statistical software. 2011; 45: 1-67.

21. Lazar C, Burger T. imputeLCMD: a collection of methods for left-censored missing data imputation. R package, version. 2022; 2.1.

22. Poutasse CM, Haddock CK, Poston WSC, Jahnke SA, Tidwell LG, Bonner EM, et al. Firefighter exposures to potential endocrine disrupting chemicals measured by military-style silicone dog tags. Environ Int. 2022; 158: 106914.

23. Wang S, Romanak KA, Hendryx M, Salamova A, Venier M. Association between thyroid function and exposures to brominated and organophosphate flame retardants in rural central appalachia. Environ Sci Technol. 2019; 54(1): 325-34.

24. Webb-Robertson B-JM, McCue LA, Waters KM, Matzke MM, Jacobs JM, Metz TO, et al. Combined statistical analyses of peptide intensities and peptide occurrences improves identification of significant peptides from MS-based proteomics data. J Proteome Res. 2010; 9(11): 5748-56.

25. Reddam A, Tait G, Herkert N, Hammel SC, Stapleton HM, Volz DC. Longer commutes are associated with increased human exposure to tris (1, 3-dichloro-2-propyl) phosphate. Environ Int. 2020; 136: 105499.

26. Rothman KJ, Greenland S, Lash TL. Modern epidemiology: Wolters Kluwer Health/Lippincott Williams & Wilkins Philadelphia; 2008.

27. Levasseur JL, Hoffman K, Herkert NJ, Cooper E, Hay D, Stapleton HM. Characterizing firefighter's exposure to over 130 SVOCs using silicone wristbands: A pilot study comparing on-duty and off-duty exposures. Sci Total Environ. 2022; 834: 155237.

28. Hammel SC, Phillips A, Hoffman K, Stapleton HM. Evaluating the use of silicone wristbands to measure personal exposure to brominated flame retardants. Environ Sci Technol. 2018.

29. Romano ME, Gallagher L, Doherty BT, Yeum D, Lee S, Takazawa M, et al. Inter-method reliability of silicone exposome wristbands and urinary biomarker assays in a pregnancy cohort. Environ Res. 2022; 214: 113981.

30. Bergmann AJ, North PE, Vasquez L, Bello H, Ruiz MdCG, Anderson KA. Multi-class chemical exposure in rural Peru using silicone wristbands. J Expo Sci Environ Epidemiol. 2017; 27(6): 560-8.

31. Quintana PJ, Hoh E, Dodder NG, Matt GE, Zakarian JM, Anderson KA, et al. Nicotine levels in silicone wristband samplers worn by children exposed to secondhand smoke and electronic cigarette vapor are highly correlated with child's urinary cotinine. J Expo Sci Environ Epidemiol. 2019; 29(6): 733-41.

32. Divine G, Norton HJ, Hunt R, Dienemann J. A review of analysis and sample size calculation considerations for Wilcoxon tests. Anesth Analg. 2013; 117(3): 699-710.

33.Samon S, Rohlman D, Tidwell L, Hoffman P, Oluyomi A, Walker C, et al. Determinants of exposure to endocrine disruptors following hurricane Harvey. Environ Res. 2023; 217: 114867.

34.Japkowicz N, Stephen S. The class imbalance problem: A systematic study. Intelligent data analysis. 2002; 6(5): 429-49.

35.Susan S, Kumar A. The balancing trick: Optimized sampling of imbalanced datasets—A brief survey of the recent State of the Art. Engineering Reports. 2021; 3(4): e12298.

36.Breiman L. Classification and regression trees. First ed: Routledge; 1984.

37.Breiman L. Random forests. Machine learning. 2001; 45: 5-32.

38.O'Connell SG, Anderson KA, Epstein MI. Determining chemical air equivalency using silicone personal monitors. J Expo Sci Environ Epidemiol. 2022; 32(2): 268-79.

39.Wacławik M, Rodzaj W, Wielgomas B. Silicone wristbands in exposure assessment: analytical considerations and comparison with other approaches. Int J Env Res Public Health. 2022; 19(4): 1935.

40.Le Roux B, Rouanet H. Geometric data analysis: from correspondence analysis to structured data analysis: Springer Science & Business Media; 2004.

41.Reynolds DA. Gaussian mixture models. Encyclopedia of biometrics. 2009; 741(659-663).

Subject Harmonization of Digital Biomarkers: Improved Detection of Mild Cognitive Impairment from Language Markers

Bao Hoang[1,‡], Yijiang Pang[1,‡], Hiroko H. Dodge[2], Jiayu Zhou[1,†]
[1] *Department of Computer Science and Engineering, Michigan State University, East Lansing, MI 48824, USA*
[2] *Department of Neurology, Massachusetts General Hospital, Harvard Medical School, Boston, MA02129, USA*
[†] *Corresponding E-mail: jiayuz@msu.edu*
[‡] *Equal contribution*

Mild cognitive impairment (MCI) represents the early stage of dementia including Alzheimer's disease (AD) and is a crucial stage for therapeutic interventions and treatment. Early detection of MCI offers opportunities for early intervention and significantly benefits cohort enrichment for clinical trials. Imaging and *in vivo* markers in plasma and cerebrospinal fluid biomarkers have high detection performance, yet their prohibitive costs and intrusiveness demand more affordable and accessible alternatives. The recent advances in digital biomarkers, especially language markers, have shown great potential, where variables informative to MCI are derived from linguistic and/or speech and later used for predictive modeling. A major challenge in modeling language markers comes from the variability of how each person speaks. As the cohort size for language studies is usually small due to extensive data collection efforts, the variability among persons makes language markers hard to generalize to unseen subjects. In this paper, we propose a novel subject harmonization tool to address the issue of distributional differences in language markers across subjects, thus enhancing the generalization performance of machine learning models. Our empirical results show that machine learning models built on our harmonized features have improved prediction performance on unseen data. The source code and experiment scripts are available at https://github.com/illidanlab/subject_harmonization.

Keywords: Mild Cognitive Impairment; Harmonization Algorithm

1. Introduction

Alzheimer's disease (AD) is a major type of dementia and ranks as the seventh-leading cause of death in the United States in 2020.[1] Mild Cognitive Impairment (MCI) is the prodromal stage of dementia, including AD, characterized by minor problems with memory, language, or judgment. Early detection of MCI is critical for early intervention and cohort enrichment. *In vivo* biomarkers such as Aβ-amyloid identified by cerebrospinal fluid Aβ42 or PET amyloid imaging are sensitive to the early or pre-clinical stage. Yet, it is not easily accessible nor affordable for massive screening of general older adults, especially those with limited healthcare access.

Recently developed digital biomarkers have offered an affordable and non-intrusive alternative. Especially language markers,[2–4] linguistic and speech variables derived from conversations, both structured[5] or semi-structured,[4] have shown a significant correlation with the cognitive capability of the subjects and are recently used for MCI detection.[6] Digital biomarkers are generally derived and utilized in a data-driven fashion. For example, language markers are derived from carefully designed cohorts[4,7] to build predictive models that take language features as input and clinical variables as output.

One significant challenge of digital biomarkers is the limited cohort sample size, where specially designed collection protocols and devices must be deployed for data collection. For example, in the studies of language markers, the I-CONECT study[4] collected semi-structured conversation data from 74 subjects in a five-year clinical trial, and the ADReSS data from DementiaBank has spontaneous speech of 158 subjects.[7] As the small sample size greatly limits the machine learning models that can be used for analysis, a standard to enrich the sample size is constructing multiple data points from the same subject and associated with the same clinical label of the subject as the prediction target. In sensor studies, for example, by using a fixed time window, multiple time series are derived from the same subject as data points.[8,9] Another example is in language marker studies, where linguistic and speech markers are derived from one conversation, and thus multiple conversations from the same subject are treated as different data points.[2–4]

Even though these treatments greatly increased the sample size for predictive modeling, they have violated the basic assumption of most analytic approaches, that data points should be independent and identically distributed (i.i.d.). The non-*i.i.d.* is complicated by another challenge of digital biomarkers, which usually have high individual variability compared to other biomarkers, leading to unstable prediction performance and poor generalization performance to unseen subjects.[10] Again use language markers as an example: the way people speak can be drastically different, and such differences are much more outstanding than subtle differences characterizing cognitive capabilities. The intuitive idea is to harmonize the distributional bias from subjects, similar to the harmonization that removes confounding factors from demographic data or eliminates batch effects. However, *subject harmonization* has drastically distinguished itself from eliminating typical confounding variables: the subjects in the testing/inference stage are not accessible during the training, and the embedding of subject information is implicit and may be non-linearly correlated with multiple dimensions in the original feature representations. Therefore, the existing harmonization approach cannot be used to quantify and remove the subject effects.

In this paper, we propose a novel framework for subject harmonization. The proposed approach uses an auxiliary classification task on the subjects to learn a deep harmonization network, which eliminates both linear and non-linear effects in differentiating subjects. Our empirical results show that the language markers harmonized by the proposed approach can improve MCI detection performance.

2. Related Works

Detection of MCI. There are many approaches developed for detecting MCI using a combination of clinical information,[11] brain imaging,[12–15] and genetics.[16,17] For example, machine learning models built on brain imaging such as MRI and FDG PET have been shown effective for capturing structural and metabolism information of the brain and are strongly associated with the development of AD.[14,18] Yet these biomarkers are often expensive and instructive, making them hard to screen general older adults. More recently, digital biomarkers[2–4] have offered a promising affordable, and non-intrusive alternative for broader adoption. The development of language markers is still in its early stage. Digital markers derived from the behavior are highly variable and different language markers derived from limited data often yield unstable detection models and are hard to generalize to unseen populations.

Data Harmonization. A fundamental challenge of data analysis is the harmonization of confounding variables, i.e., eliminating the effects from confounding variables.[19,20] With explicit confounding variables, common harmonization approaches eliminate confounding variables' influence on the original input features or output.[21,22] Recent deep learning models require the harmonization of non-linear effects, leading to the development of end-to-end frameworks that cooperate with the task prediction loss and a penalty loss that usually minimizes dependence between confounders and prediction outcomes.[23–26] Meanwhile, fair machine learning schemes exploit distributional robust optimization to control implicit demographic confounding effects (bias).[27–29] From another aspect, the underlying variables can be considered as some strong signal in the original features but is irrelevant to our prediction goal, then feature engineering helps reduce the effects.[30] Most existing harmonization approaches need confounding variables to be accessible during the training and secure the generalization to unseen groups. However, in digital biomarker studies where subjects are treated as a confounding variable, the challenging arises when testing subjects are not seen during the training and demands a generalizable harmonization on subjects.

3. Methods

3.1. *Data*

We use semi-structured conversational data from a clinical trial I-CONECT (Clinicaltrials.gov: NCT02871921). The data is available upon request at https://www.i-conect.org/. This clinical trial aims to investigate the potential benefits of regular video chat conversations on the cognitive functions and psychological well-being of individuals aged 75 and older. The dataset has 6771 conversation sessions from 74 participants, with 36 participants being cognitively normal (NL) and 38 diagnosed with mild cognitive impairment (MCI). Each conversational session is about 30 minutes in length. Table 1 shows the participants' demographic information.

Table 1. Demographics of Participants

Variable	All (n = 74)	NL (n = 36)	MCI (n = 38)
Age	80.7 ± 4.6	79.7 ± 3.9	81.7 ± 5.0
Gender (%women)	71.6	77.8	65.8
Years of education	15.2 ± 2.5	15.4 ± 2.5	15.1 ± 2.5
Number of Conversations	91.5 ± 37.2	92.4 ± 35.8	90.7 ± 38.4

3.2. *Language Markers*

We derived a total of 99 feature variables for each conversation as language markers, including four types: Linguistic Inquiry and Word Count (LIWC), Syntactic Complexity, Lexical Diversity, and Response Length.

Linguistic Inquiry and Word Count (LIWC): For the LIWC feature variables, we use the 2007 English version of Linguistic Inquiry and Word Count.[31] This tool categorizes English words into 64 different "LIWC categories". These categories cover a wide range of linguistic, psychological, and topical aspects, enabling us to gain insights into various social, cognitive, and affective processes. To obtain the LIWC features, we follow:[3] We first generate a 64-dimensional LIWC feature vector for every word in each conversation, with each dimension corresponding to a specific LIWC category (1 = word belongs to the category, 0 = word does not belong); we then sum over the feature vectors of all words in the conversation, resulting in a single 64-dimensional feature vector representing the linguistic feature of that conversation.

Syntactic complexity represents the range and intricacy of grammatical structures employed in language production.[32] We used the L2 Syntactic Complexity Analyzer[33] to extract the syntactic complexity feature. This tool is specifically designed to automate the analysis of syntactic complexity in English language texts produced by advanced learners of English. We extract a 23-dimensional vector from each conversation representing the syntactic complexity of conversation, with each dimension corresponding to a specific English syntactic complexity measure from the tool.

Lexical Diversity is the range of different words within a given text, wherein a wider range indicates greater diversity.[34] Given a text input, lexical diversity has been measured using the type-token ratio (TTR),[35] obtained by dividing the total number of unique words by the overall word count. To adopt this in our study, we extract the TTR from participants' conversational responses, as well as its variations, such as the moving average type-token ratio (MATTR)[36] and the mean segmental type-token ratio (MSTTR). We also use additional lexical diversity measures, including the Hypergeometric distribution D (HD-D) and the measure of textual Lexical Diversity (MTLD).[37] In total, we derive a 10-dimensional vector representing conversations' lexical diversity, with each dimension corresponding to one of the aforementioned lexical diversity measures and its respective variation.

Response length: Our analysis suggests that NL individuals tend to provide lengthier responses to questions posed by interviewer than MCI individuals, showing great potential for distinguishing between MCI and NL individuals. We extract the mean and variance of participants' response lengths within each conversation.

3.3. *Generalized Least Squares*

Generalized least squares is a widely used harmonization approach to remove linear effects given confounding variables, such as age, gender, and education.[21,38] For each conversation's extracted language marker features x_i, we assume that these features are linearly biased by three confounding variables age, sex, and education of the subject, denoted

$c_i = [\text{age}, \text{sex}, \text{education}]$, such that:

$$x_i = w \cdot c_i^T + x_i^{\text{harmonized}}$$

where w is weight matrix and $x_i^{\text{harmonized}}$ is our goal harmonized language markers. The objective function for generalized least square method is given by:

$$\min_w \sum_{i=1}^{n} (wc_i^T - x_i)^2$$

After obtaining weight matrix w by solving the above objective function, the harmonized language markers is derived by:

$$x_i^{\text{harmonized}} = x_i - w \cdot c_i^T$$

3.4. *Subject Harmonization for Non-linear Predictive Modeling*

Unlike other types of *in-vivo* biomarkers, digital markers show great individual variability. In language markers, for instance, how one speaks a language can differ greatly, even if they are all native speakers. The differences can be visualized by checking the distributions of language features. Our empirical results in Sec. 4.1 show that the feature variables have clear clustering structures w.r.t. subjects. As such, successful analysis and predictive modeling need careful harmonization to eliminate individual variability. Generalized least squares's harmonization mechanism eliminates the linear subspace that is predictive of these confounding variables and uses the orthogonal complement subspace as the harmonized features. Though all linear effects are removed through the harmonization approach, the approach does not remove any non-linear effects from data. For example, if the multiplication of two confounding variables (e.g., age and gender) has effects on the data, such effects will not be removed and will be picked up by non-linear models such as random forest and deep learning models. Another challenge comes from the generalization of harmonization, where digital biomarkers demand a unique harmonization procedure that can be generalized to unseen subjects.

To address the above challenge, we propose a deep harmonization network to facilitate analytics with digital biomarkers. In the context of the prediction of MCI from language markers, we are given a set of conversations collected from a set of different subjects and we would like to build a predictive model for MCI using these conversations. We follow the last section to extract features for *each conversation* and form a feature vector for each conversation. The setting of predictive modeling is to classify each conversation/feature vector into a label (MCI or not), which will be later aggregated into a prediction of the subject. The feature vectors of one subject will be either used in training or testing but not both. The goal of harmonization is to remove the confounding factor of subjects in the feature vectors. The proposed approach has two stages: in the first stage, we construct an auxiliary task to learn the deep harmonization network; in the second, the learned harmonization network is used to transform the data points, and the harmonized data is then used for building a downstream classifier of MCI.

The design of a deep harmonization network is based on two intuitions: 1) a good harmonization should remove all linear and non-linear effects from subjects, and therefore the harmonized features should not be able to differentiate subjects under deep models; 2) the harmonized features should be as close to the original feature as possible (otherwise, the harmonization admits a trivial solution where all features are wiped and set to the same value). Following these intuitions, the proposed approach seeks to minimize the subject differentiation between data points obtained from different subjects and minimizes the differences between harmonized and original language markers. Generally, for M pairs of extracted language features and corresponding subject labels $(\mathbf{x}_i, \mathbf{y}_i^s)$, we denote $f_{\text{FH}}(\cdot) : \mathbf{x} \rightarrow \bar{\mathbf{x}}$ as the feature harmonization network parameterized with θ_{FH}, $f_{\text{s}}(\cdot) : \bar{\mathbf{x}} \rightarrow \mathbf{s}$ as the auxiliary subject classifier parameterized with θ_s. The composite function $f_s \circ f_{\text{FH}}$ denotes a classifier f_s using harmonized features f_{FH}. The objective for learning feature harmonization is given by:

$$\min_{\theta_{FH}, \theta_s} \frac{1}{M} \sum\nolimits_{i=1}^{M} -\ell_{ent}(f_s \circ f_{\text{FH}}(\mathbf{x}_i), \mathbf{y}_i^s) + \ell_{mse}(f_{\text{FH}}(\mathbf{x}_i), \mathbf{x}_i), \tag{1}$$

where $\ell_{ent}(\cdot)$ is the cross-entropy loss and minimizing $-\ell_{ent}(\cdot)$ encourages the harmonized features cannot be differentiated by subject identities, and $\ell_{mse}(\cdot)$ is the mean square error which encourages the similarity between the original features and the harmonized features. Note that we do not restrict the type of classifier to be used in f_s, but a non-linear model is preferred due to the design of deep harmonization. In our study, we use a 3-layer MLP for the harmonization network.

3.5. *MCI Detection using Harmonized Features*

After the harmonization process, we use the harmonized features with confounding effects removed for the downstream task of MCI detection. The MCI detection can be modeled by two classification tasks: a) *conversation classification* that identifies whether a given conversation is from an MCI subject or an NL subject using language markers extracted from the conversation, and b) *subject classification*, which collectively uses the results from the conversation classification on conversations from one subject and predict if a subject is an MCI subject or an NL subject. We model conversation classification as a standard machine learning task that seeks a classifier that takes language markers as an input and outputs a binary prediction. Formally, we have M pairs of extracted features and corresponding cognitive status label $(\mathbf{x}_i, \mathbf{y}_i^c)$. We denote $f_{\text{t}}(\cdot) : \bar{\mathbf{x}} \rightarrow \mathbf{t}$ as the MCI classifier parameterized with θ_t. In our study, we use two classifiers: a linear model (logistic regression, LR) and a non-linear model (2-layer multi-layer perceptron, MLP). Then, the objective function for cognitive status classification is formulated as:

$$\min_{\theta_t} \frac{1}{M} \sum\nolimits_{i=1}^{M} \ell(f_t \circ f_{\text{FH}}(\mathbf{x}_i), \mathbf{y}_i^c),$$

where $\ell(\cdot)$ is the binary cross entropy loss. To achieve subject classification, we use a majority vote strategy so that if more than 50% of a subject's conversations are predicted as MCI by the conversation classifier, we classify that subject as MCI and NL otherwise. For both settings, we randomly sample 80% subjects as train subjects and the remaining subjects as test subjects. The conversations from training subjects are used to train the conversation classifier. The complete framework is illustrated in Figure 1.

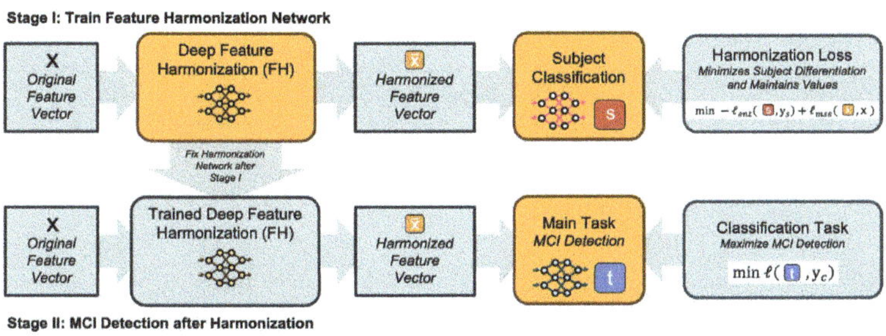

Fig. 1. The proposed subject harmonization process includes two stages. In the first stage, we train a deep harmonization network using an auxiliary subject classification task, which discourages differentiation among subjects and meanwhile retains the similarity between the original features and the harmonized ones. In the second stage, we fix the harmonization model and use the harmonized features to train the main learning task, i.e., the detection of MCI.

4. Experimental Results and Analysis

4.1. *Effectiveness and Generalizability of Subject Harmonization*

The design of harmonization is to remove the confounding factor of the variable of subjects. Therefore, we investigate the prediction power towards subjects using features before and after harmonization. The stronger the confounding variable, the better the features' prediction power differentiating subjects. A successful harmonization should greatly eliminate such prediction power.

In this experiment, conversations from individual subjects are assigned the same labels, while conversations from different subjects are assigned distinct labels. For example, all conversations from the first subject have the label 1, and all those from the second subject have the label 2. With a total of 74 subjects, we have 74 unique labels. We randomly split data (original or harmonized) into training and testing, with 80% of conversations for training and 20% for testing. We build a linear classifier (Logistic Regression) and a deep classifier (Multi-layer perceptron) using the training data and evaluate the performance in terms of accuracy using the data. For the harmonization network, we use a 3-layer Multi-layer Perceptron. We repeat the experiment for 100 random seeds, and report the average accuracy of predicting testing conversations' subject labels before and after harmonization in table 2. We use the same training and testing conversations for each random seed while evaluating before and after harmonization. We see a substantial decrease in subject classification performance in both models, showing the effectiveness of the harmonization design that removes the confounding variables' linear and non-linear effects.

We conduct a qualitative study that visualizes the distributions of the language markers before and after the subject harmonization in Figure 2. We use t-SNE[39] to plot the 99-dimensional language markers in a comprehensible 2-dimensional space, where conversations from the same subjects are assigned matching colors. From the visualization, we see that data points from the same subjects show a clear clustering structure of subjects, indicating subject

Table 2. Performance of subject classification tasks before and after subject harmonization.

Classifier	Before harmonization	After harmonization
Logistic Regression	0.921±0.007	0.221±0.012
Multi-layer Perceptron	0.905±0.007	0.219±0.038

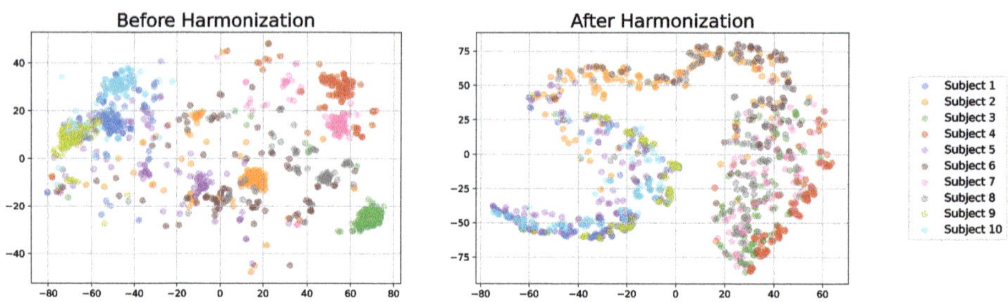

Fig. 2. The visualization of language markers extracted from conversations collected from 10 randomly selected subjects before and after subject harmonization. We see that a clear clustering structure exists before subject harmonization, which is successfully destroyed by the harmonization.

bias in the language markers. After the harmonization, such clustered structure is visually destroyed, showing the effectiveness of the purpose harmonization strategy.

4.2. *MCI Detection via Harmonized Language Markers*

We now investigate the predictive power of language markers in detecting MCI subjects. We compare a set of different harmonization approaches: a) generalized least squares,[21,38] commonly used for harmonizing linear effects and used age/gender/education as confounding variables; b) the proposed deep subject harmonization, which harmonizes against the subject variable but does not use demographic variables (age/gender/education); c) deep harmonization that does not use subject information and jointly harmonizes all demographic variables. d) deep harmonization approaches that harmonize only individual demographic variables.

When harmonizing demographic variables using a deep harmonization network, we construct category variables from age/gender/education (e.g., age between 75-79 as category 1, age between 80-84 as category 2) and train equation 1. We repeat the experiments for 100 random seeds and report the average and standard deviation of Area under the ROC curve (AUC), F1, Sensitivity, and Specificity on the test data in Table 3.

From the results, we find the following: 1) The non-linear model MLP using features from deep subject harmonization, which harmonizes the subject variable using a deep model, provides the best downstream classification performance on both conversation and subject predictions. 2) Both the linear and non-linear models benefit more from deep subject harmonization than generalized least squares. 3) For MLP, deep harmonization on demographic

Table 3. Performance of two cognitive status classification tasks over different harmonization methods.

Method for harmonization	Task Classifier	Performance metrics			
		AUC	F1	Sensitivity	Specificity
Conversation classification					
None	LR	0.583±0.098	0.557±0.092	0.570±0.123	0.557±0.101
	MLP	0.594±0.092	0.556±0.088	0.545±0.116	0.611±0.091
Generalized least squares[21]	LR	0.567±0.110	0.537±0.104	0.538±0.134	0.570±0.119
	MLP	0.545±0.109	0.522±0.103	0.516±0.132	0.574±0.125
Deep harmonization - subject (Proposed method)	LR	0.640±0.097	0.581±0.089	0.575±0.129	0.625±0.132
	MLP	0.646±0.092	0.558±0.101	0.541±0.136	0.640±0.126
Deep harmonization (- age & gender & education year)	MLP	0.527±0.120	0.517±0.119	0.593±0.227	0.427±0.235
Deep harmonization - age	MLP	0.596±0.107	0.538±0.101	0.535±0.166	0.608±0.178
Deep harmonization - gender	MLP	0.554±0.110	0.551±0.110	0.635±0.209	0.426±0.208
Deep harmonization - education year	MLP	0.611±0.102	0.589±0.080	0.654±0.141	0.477±0.165
Subject classification					
None	LR	0.591±0.124	0.579±0.126	0.593±0.166	0.568±0.169
	MLP	0.626±0.122	0.593±0.124	0.576±0.153	0.649±0.159
Generalized least squares[21]	LR	0.585±0.129	0.529±0.148	0.519±0.187	0.601±0.164
	MLP	0.568±0.122	0.568±0.138	0.565±0.175	0.605±0.175
Deep harmonization - subject (Proposed method)	LR	0.649±0.121	0.592±0.115	0.575±0.157	0.652±0.162
	MLP	0.657±0.113	0.571±0.118	0.546±0.152	0.655±0.152
Deep harmonization (- age & gender & education year)	MLP	0.538±0.148	0.539±0.165	0.637±0.272	0.381±0.282
Deep harmonization - age	MLP	0.614±0.122	0.577±0.133	0.585±0.205	0.603±0.217
Deep harmonization - gender	MLP	0.571±0.128	0.579±0.139	0.676±0.230	0.409±0.244
Deep harmonization - education year	MLP	0.639±0.122	0.632±0.091	0.736±0.159	0.417±0.218

Abbreviations: LR, Logistic Regression; MLP, Multi-layer Perceptron.

variables performs worse than generalized least squares, even though both jointly harmonize against all three demographic variables.

4.3. *Performance on Different Sub-Populations*

Table 4 presents the performance of conversation and subject classification on different sub-populations, i.e., different gender groups, education levels, and age groups. By zooming in on the performance of different sub-population groups, we want to inspect how the proposed subject harmonization impacts these groups, given that demographic variables are not used in the harmonization process. From the results, we see that the proposed subject harmonization consistently improved the performance of most groups, with the exception of 1) the higher educated group (Edu years 19-21), for both conversation and subject classification, and 2) minor performance drop in the Male group for the subject classification.

Table 4. Performance of two cognitive status classification tasks before and after the harmonization methods.

Groups	Performance compairsion					
	Before harmonization			After harmonization		
	AUC	Sensitivity	Specificity	AUC	Sensitivity	Specificity
Conversation classification						
Male	0.533±0.185	0.517±0.199	0.537±0.213	0.564±0.199	0.656±0.228	0.409±0.257
Female	0.621±0.112	0.554±0.140	0.641±0.103	0.673±0.106	0.475±0.181	0.728±0.143
Edu 12-15	0.483±0.162	0.529±0.182	0.447±0.165	0.618±0.186	0.586±0.205	0.599±0.234
Edu 16-18	0.621±0.163	0.490±0.185	0.715±0.110	0.668±0.146	0.452±0.241	0.735±0.160
Edu 19-21	0.857±0.096	0.790±0.182	0.743±0.161	0.732±0.323	0.647±0.418	0.498±0.257
Age 75-80	0.608±0.123	0.532±0.158	0.648±0.096	0.638±0.111	0.519±0.169	0.664±0.130
Age 81-87	0.500±0.231	0.483±0.224	0.529±0.317	0.517±0.309	0.456±0.263	0.512±0.369
Age 88-94	0.781±0.189	0.918±0.157	0.339±0.129	0.941±0.058	0.987±0.026	0.386±0.293
Subject classification						
Male	0.589±0.275	0.537±0.299	0.587±0.365	0.577±0.277	0.641±0.292	0.384±0.392
Female	0.653±0.152	0.600±0.184	0.665±0.187	0.691±0.158	0.491±0.204	0.751±0.184
Edu 12-15	0.480±0.211	0.530±0.226	0.377±0.291	0.624±0.215	0.601±0.218	0.603±0.247
Edu 16-18	0.694±0.241	0.549±0.327	0.828±0.199	0.699±0.228	0.445±0.295	0.756±0.221
Edu 19-21	1.000±0.000	0.929±0.258	0.921±0.260	0.754±0.395	0.607±0.457	0.508±0.445
Age 75-80	0.654±0.176	0.561±0.206	0.715±0.185	0.671±0.153	0.512±0.212	0.699±0.158
Age 81-87	0.515±0.309	0.501±0.331	0.569±0.431	0.541±0.379	0.474±0.320	0.536±0.444
Age 88-94	0.953±0.192	0.984±0.087	0.141±0.336	0.984±0.087	1.000±0.000	0.328±0.426

4.4. *Important Language Markers Before and After Harmonization*

In this section, we investigate the feature importance and compare the top language markers before and after harmonization. For linear models, feature importance can be directly derived from the model weights, and for non-linear MLP models used in this paper, we do not have such a straightforward way of getting them. We adopt commonly used permutation feature importance[40] to estimate the feature importance. We permute each feature's values and subsequently feed the modified dataset into our pipeline. After that, we derive the AUC score for both conversation and subject classification using this permutated dataset. The feature importance of a feature is then determined by computing the difference between the AUC values obtained from the original dataset and the permuted dataset. A larger decrease in AUC indicates higher importance of the respective feature in the classification model.

In table 5, we present the top 10 language features before and after the feature harmonization for both conversation and subject classification. We see that: 1) top features differ quite much before and after harmonization. Notably, we see "Nonfluencies" being the most important feature after harmonization, which better supports the pathology of dementia, where dementia (even at the preclinical stage) may impact a subject, making it harder to find the right words and therefore showing a higher number of nonfluencies during communication. 2)

Table 5. Top 10 language features before and after harmonization where the importance is w.r.t. the decreasing of AUC in both conversation classification and subject classification.

Before harmonization			After harmonization		
Feature name	Type	AUC drop	Feature name	Type	AUC drop
Conversation classification					
Negations	LIWC	0.02749	Nonfluencies	LIWC	0.00616
1st pers plural	LIWC	0.00587	Assent	LIWC	0.00471
Discrepancy	LIWC	0.00495	Insight	LIWC	0.00468
Assent	LIWC	0.00328	Affective processes	LIWC	0.00455
Family	LIWC	0.00325	T-unit per sentence	SC	0.00455
Tentative	LIWC	0.00324	3rd pers singular	LIWC	0.00439
Sexual	LIWC	0.00297	Causation	LIWC	0.00435
Auxiliary verbs	LIWC	0.00238	Certainty	LIWC	0.00418
Home	LIWC	0.00215	Mean length of sentence	SC	0.00414
Inhibition	LIWC	0.00204	Hear	LIWC	0.00395
Subject classification					
Negations	LIWC	0.03469	T-unit per sentence	SC	0.01203
Tentative	LIWC	0.00562	Mean length of sentence	SC	0.00969
Family	LIWC	0.00547	Negations	LIWC	0.00922
Textual lexical diversity	LD	0.00531	Clause	SC	0.00906
Home	LIWC	0.00438	Affective processes	LIWC	0.00859
Social processes	LIWC	0.00391	Causation	LIWC	0.00844
1st pers plural	LIWC	0.00359	Cognitive processes	LIWC	0.00828
Assent	LIWC	0.00344	Positive emotion	LIWC	0.00813
Personal pronouns	LIWC	0.00313	Inclusive	LIWC	0.00813
Discrepancy	LIWC	0.00313	Motion	LIWC	0.00750

Abbreviations: LIWC, Linguistic Inquiry and Word Count; SC, Syntactic Complexity; LD, Lexical Diversity.

more syntactic complexity features appear after harmonization for subject classification. The top features "T-unit per sentence" and "mean length of sentence" directly correlate to the language capability of constructing longer features.

5. Discussion

In this paper, we propose a subject harmonization algorithm to mitigate the distributional difference of digital biomarkers induced by subject variability. Our empirical results show that applying subject harmonization to language markers improves the performance of MCI detection. We show the effects of subject variability from a quantitative perspective using a subject prediction task, and also from a qualitative perspective from visible clusters in the visualization of language markers. Our experiments show that the proposed subject harmonization approach effectively mitigates the subject variability so that the harmonized data has much less power to differentiate among subjects. Meanwhile, we show that MCI detection models built from language markers harmonized by the proposed subject harmonization improve the predictive performance. The harmonization improves the AUC score of MCI prediction from 0.594 to 0.646 in conversation classification task and from 0.626 to 0.657 in subject classification task. We further investigated the sub-group performance of different age/gender/years of education, and we see that the performance of most groups have been improved.

Despite the improvement in prediction performance using language markers through the harmonization algorithm, future studies still need investigation. Firstly, the prediction performance from language markers is yet to be improved. A possible reason is the quality of the language markers and that we only used linguistic and syntactic information. We will study subject harmonization on additional feature variables, such as speech and video. Secondly, performing subject harmonization on demographic variables witnessed reduced predictive performance, indicating that the proposed deep harmonization network is currently not applicable to general harmonization usage. We plan to investigate theoretical relationship between the two harmonization types, and improve deep harmonization network to handle demographic variables. Thirdly, while we have successfully validated the positive impact of harmonization on language markers, it remains to confirm its efficacy on other data types. We plan to dedicate considerable time to applying the harmonization algorithm to different types of markers, such as clinical data or brain imaging data. This broader exploration will enable us to assess the generalizability and versatility of the harmonization technique across various data modalities, facilitating a more comprehensive understanding of its potential applications.

6. Acknowledgement

This material is based in part upon work supported by the National Science Foundation under Grant IIS-2212174, IIS-1749940, Office of Naval Research N00014-20-1-2382, and National Institute on Aging (NIA) RF1AG072449, R01AG051628, R01AG056102.

References

1. S. L. Murphy, K. D. Kochanek, J. Xu and E. Arias, Mortality in the united states, 2020 (2021).
2. B. Roark, M. Mitchell, J.-P. Hosom, K. Hollingshead and J. Kaye, Spoken language derived measures for detecting mild cognitive impairment, *IEEE Transactions on Audio, Speech, and Language Processing* **19**, 2081 (September 2011).
3. F. Tang, J. Chen, H. H. Dodge and J. Zhou, The joint effects of acoustic and linguistic markers for early identification of mild cognitive impairment, *Frontiers in digital health* **3**, p. 702772 (2022).
4. M. Asgari, J. Kaye and H. Dodge, Predicting mild cognitive impairment from spontaneous spoken utterances, *Alzheimer's & Dementia—Translational Research and Clinical Intervention* **3**, 219 (February 2017).
5. M. L. Manning, Improving clinical communication through structured conversation, *Nurs Econ* **24**, 268 (2006).
6. L. Chen, H. H. Dodge and M. Asgari, Topic-Based Measures of Conversation for Detecting Mild Cognitive Impairment, *Proc Conf Assoc Comput Linguist Meet* **2020**, 63 (Jul 2020).
7. S. Luz, F. Haider, S. de la Fuente, D. Fromm and B. MacWhinney, Alzheimer's dementia recognition through spontaneous speech: The ADReSS Challenge, in *Proceedings of INTERSPEECH 2020*, (Shanghai, China, 2020).

8. J. Li, Y. Rong, H. Meng, Z. Lu, T. Kwok and H. Cheng, Tatc: predicting alzheimer's disease with actigraphy data (2018).
9. X. Ouyang, Design and deployment of multi-modal federated learning systems for alzheimer's disease monitoring (2023).
10. J. Yang, K. Zhou, Y. Li and Z. Liu, Generalized out-of-distribution detection: A survey (2021).
11. J. Venugopalan, L. Tong, H. R. Hassanzadeh and M. D. Wang, Multimodal deep learning models for early detection of alzheimer's disease stage, *Scientific reports* **11**, p. 3254 (2021).
12. J. Zhou, L. Yuan, J. Liu and J. Ye, A multi-task learning formulation for predicting disease progression (2011).
13. J. Zhou, J. Liu, V. A. Narayan, J. Ye, A. D. N. Initiative *et al.*, Modeling disease progression via multi-task learning, *NeuroImage* **78**, 233 (2013).
14. Q. Wang, L. Zhan, P. M. Thompson, H. H. Dodge and J. Zhou, Discriminative fusion of multiple brain networks for early mild cognitive impairment detection (2016).
15. S. Gill, P. Mouches, S. Hu, D. Rajashekar, F. P. MacMaster, E. E. Smith, N. D. Forkert and Z. I. and, Using machine learning to predict dementia from neuropsychiatric symptom and neuroimaging data, *Journal of Alzheimer's Disease* **75**, 277 (May 2020).
16. Q. Wang, M. Sun, L. Zhan, P. Thompson, S. Ji and J. Zhou, Multi-modality disease modeling via collective deep matrix factorization (2017).
17. R.-H. Lin, C.-C. Wang and C.-W. Tung, A machine learning classifier for predicting stable MCI patients using gene biomarkers, *International Journal of Environmental Research and Public Health* **19**, p. 4839 (April 2022).
18. C. Yin, S. Li, W. Zhao and J. Feng, Brain imaging of mild cognitive impairment and alzheimer's disease, *Neural regeneration research* **8**, p. 435 (2013).
19. J. He, S. L. Baxter, J. Xu, J. Xu, X. Zhou and K. Zhang, The practical implementation of artificial intelligence technologies in medicine, *Nature medicine* **25**, 30 (2019).
20. E. J. Topol, High-performance medicine: the convergence of human and artificial intelligence, *Nature medicine* **25**, 44 (2019).
21. Q. Wang, L. Guo, P. M. Thompson, C. R. Jack Jr, H. Dodge, L. Zhan, J. Zhou, A. D. N. Initiative *et al.*, The added value of diffusion-weighted mri-derived structural connectome in evaluating mild cognitive impairment: A multi-cohort validation, *Journal of Alzheimer's Disease* **64**, 149 (2018).
22. E. Adeli, X. Li, D. Kwon, Y. Zhang and K. M. Pohl, Logistic regression confined by cardinality-constrained sample and feature selection, *IEEE transactions on pattern analysis and machine intelligence* **42**, 1713 (2019).
23. Q. Zhao, E. Adeli and K. M. Pohl, Training confounder-free deep learning models for medical applications, *Nature communications* **11**, p. 6010 (2020).
24. E. Adeli, Q. Zhao, A. Pfefferbaum, E. V. Sullivan, L. Fei-Fei, J. C. Niebles and K. M. Pohl, Representation learning with statistical independence to mitigate bias (2021).
25. M. Horry, S. Chakraborty, B. Pradhan, M. Paul, J. Zhu, H. W. Loh, P. D. Barua and U. R. Arharya, Debiasing pipeline improves deep learning model generalization for x-ray based lung nodule detection (2022).
26. A. Vento, Q. Zhao, R. Paul, K. M. Pohl and E. Adeli, A penalty approach for normalizing feature distributions to build confounder-free models (2022).
27. H. Namkoong and J. C. Duchi, Stochastic gradient methods for distributionally robust optimization with f-divergences, *Advances in neural information processing systems* **29** (2016).
28. T. Hashimoto, M. Srivastava, H. Namkoong and P. Liang, Fairness without demographics in repeated loss minimization (2018).
29. S. Jeong and H. Namkoong, Robust causal inference under covariate shift via worst-case subpopulation treatment effects (2020).

30. A. Zheng and A. Casari, *Feature engineering for machine learning: principles and techniques for data scientists* (" O'Reilly Media, Inc.", 2018).
31. The development and psychometric properties of liwc2015.
32. L. Ortega, Syntactic complexity in l2 writing: Progress and expansion, *Journal of Second Language Writing* **29**, 82 (September 2015).
33. X. Lu, Automatic analysis of syntactic complexity in second language writing, *International journal of corpus linguistics* **15**, 474 (2010).
34. M. M. Baese-Berk, S. Drake, K. Foster, D. yong Lee, C. Staggs and J. M. Wright, Lexical diversity, lexical sophistication, and predictability for speech in multiple listening conditions, *Frontiers in Psychology* **12** (June 2021).
35. W. Johnson, Studies in language behavior: A program of research, *Psychological Monographs* **56**, 1 (1944).
36. M. A. Covington and J. D. McFall, Cutting the gordian knot: The moving-average type–token ratio (mattr), *Journal of quantitative linguistics* **17**, 94 (2010).
37. P. M. McCarthy and S. Jarvis, Mtld, vocd-d, and hd-d: A validation study of sophisticated approaches to lexical diversity assessment, *Behavior research methods* **42**, 381 (2010).
38. R. McNamee, Regression modelling and other methods to control confounding, *Occupational and environmental medicine* **62**, 500 (2005).
39. L. van der Maaten and G. Hinton, Visualizing data using t-sne, *Journal of Machine Learning Research* **9**, 2579 (2008).
40. L. Breiman, Random forests, *Machine learning* **45**, 5 (2001).

Scalar-Function Causal Discovery for Generating Causal Hypotheses with Observational Wearable Device Data

Valeriya Rogovchenko, Austin Sibu, and Yang Ni†
Department of Statistics, Texas A&M University,
College Station, TX 77843, USA
†*E-mail: yni@stat.tamu.edu*

Digital health technologies such as wearable devices have transformed health data analytics, providing continuous, high-resolution functional data on various health metrics, thereby opening new avenues for innovative research. In this work, we introduce a new approach for generating causal hypotheses for a pair of a continuous functional variable (e.g., physical activities recorded over time) and a binary scalar variable (e.g., mobility condition indicator). Our method goes beyond traditional association-focused approaches and has the potential to reveal the underlying causal mechanism. We theoretically show that the proposed scalar-function causal model is identifiable with observational data alone. Our identifiability theory justifies the use of a simple yet principled algorithm to discern the causal relationship by comparing the likelihood functions of competing causal hypotheses. The robustness and applicability of our method are demonstrated through simulation studies and a real-world application using wearable device data from the National Health and Nutrition Examination Survey.

Keywords: Causal identifiability, digital health, NHANES, observational data, wearable device.

1. Introduction

The rise of wearable devices has revolutionized the way we collect and analyze health data, offering an unprecedented wealth of information about human health and behavior. These devices such as accelerometers and continuous glucose monitors allow for frequent measurement of various variables over time including physical activities, sleep patterns, electrocardiogram signals, and blood glucose levels. The availability of these measurements enables researchers to ask questions that previously could not be answered, e.g., how to quantify the effect of physical activities on all-cause mortality? In these types of scenarios, often, one variable (e.g., physical activities) is longitudinal/functional and the other (e.g., mortality) is a scalar. Thus, many statistical methods such as scalar-on-function regression models[10,17] have been successfully deployed to estimate the association of the scalar-function pair.

The focus of this paper is, however, different from the existing literature for modeling wearable device data. Instead of association, we investigate whether it is possible to discern the *causal* relationship between a scalar and a function. More specifically, we aim to identify which of the scalar-function pair is more likely to be the cause or effect given observational data alone.

We introduce a novel *scalar-function causal discovery* method to generate data-driven causal hypotheses. Revealing the causality underlying observed data can deepen our understanding of the physical mechanism involved in the data-generating process and potentially pave the way for better health interventions and policy-making.

The field of causal discovery has seen a significant surge in interest and development over recent years due to wide-ranging applicability across various domains.[4,5,12–14,16,22] While traditional causal discovery methods are typically tailored to handle either continuous or discrete variables exclusively, real-world scenarios are often far more complex. For example, in the fields of social and health sciences, data frequently comprise a mix of different types of variables, necessitating more versatile approaches.

In such scenarios, one may either discard discrete data or convert continuous data into a discrete form;[9,20] either way, a lot of information contained in the original data is lost. In light of these limitations, there have been some recent developments to discover causality for mixed data.[19,21] However, these methods have only been developed for scalar variables, which cannot be used for functional data. To deal with functional data, some very recent works[6,23] have been proposed, which, however, cannot accomodate scalar and/or discrete data. In summary, to the best of our knowledge, there are no existing methods that can identify causality between a continuous functional variable and a binary scalar variable.

This paper, therefore, aims to fill this critical gap in the causal discovery literature so that digital health researchers will have a powerful tool to identify causality in a wide range of observational wearable device data. Our approach is based on a probabilistic causal model that quantifies the likelihood of each possible causal direction (from function to scalar or from scalar to function). We theoretically establish the causal identifiability property of our model under common causal assumptions. Equipped with the identifiability property, we can simply identify causal directions based on likelihood functions.

We conduct simulation studies to assess the empirical identifiability of the proposed method. In addition, to validate our method in real-world scenarios, we present an application with two variables that have a clear causal relationship. Specifically, we consider mobility conditions and physical activities. Since it is clear that mobility issues may lead to reduced activities, we will test whether our method can correctly identify such causal relationship without prior knowledge using the National Health and Nutrition Examination Survey (NHANES) data.

The rest of the paper is organized in the following way. In Section 2, we describe the proposed scalar-function causal discovery model, theoretically prove that the causal relationship is identifiable, and develop a likelihood-based estimation procedure. In Section 3, we evaluate the proposed method through various simulations as well as a real wearable device dataset from NHANES, demonstrating its capability to correctly identify the true causal relationship. We conclude our paper with a brief discussion in Section 4.

2. Method

2.1. *Notations*

We use capital letters to denote random variables and small letters to denote their realized values. We use boldface to denote vectors or matrices and non-boldface to denote scalars. With a slight abuse of notation, we use $P(\cdot)$ to denote both probability mass and density functions, which can be understood from the context as it is determined by the type of the random variable under consideration. Let $\mathbb{M}^{n\times n}$ be the cone of $n \times n$ positive definite matrices.

2.2. *Causal Probability Model*

We are interested in identifying the causal relationship between two statistically dependent random variables: a random binary variable $Y \in \{0, 1\}$ and a random function measured on n time points $\boldsymbol{X} = (X(t_1), ..., X(t_n))^\top \in \mathbb{R}^n$. One can view these functional measurements as a finite realization of an infinite stochastic process $X(\cdot)$ such as the Gaussian process.[18]

We consider two competing causal hypotheses[a],

$$H_0 : \boldsymbol{X} \to Y \text{ or } \boldsymbol{X} \text{ causes } Y$$

$$\text{vs}$$

$$H_1 : Y \to \boldsymbol{X} \text{ or } Y \text{ causes } \boldsymbol{X}$$

Under each hypothesis, we will set up a probability model. Specifically, let $P_{\boldsymbol{X}\to Y}(\boldsymbol{X} = \boldsymbol{x}, Y = y)$ denote the probability model of H_0 and $P_{Y\to\boldsymbol{X}}(\boldsymbol{X} = \boldsymbol{x}, Y = y)$ denote the probability model of H_1. Using the probability chain rule, we have

$$\begin{aligned} P_{\boldsymbol{X}\to Y}(\boldsymbol{X} = \boldsymbol{x}, Y = y) &= P_{\boldsymbol{X}\to Y}(Y = y \mid \boldsymbol{X} = \boldsymbol{x}) \cdot P_{\boldsymbol{X}\to Y}(\boldsymbol{X} = \boldsymbol{x}), \\ P_{Y\to\boldsymbol{X}}(\boldsymbol{X} = \boldsymbol{x}, Y = y) &= P_{Y\to\boldsymbol{X}}(\boldsymbol{X} = \boldsymbol{x} \mid Y = y) \cdot P_{Y\to\boldsymbol{X}}(Y = y), \end{aligned} \tag{1}$$

where $P_{\boldsymbol{X}\to Y}(Y = y \mid \boldsymbol{X} = \boldsymbol{x})$ and $P_{\boldsymbol{X}\to Y}(\boldsymbol{X} = \boldsymbol{x})$ are respectively the conditional and marginal probability distributions under $H_0 : \boldsymbol{X} \to Y$ and similarly $P_{Y\to\boldsymbol{X}}(\boldsymbol{X} = \boldsymbol{x} \mid Y = y)$ and $P_{Y\to\boldsymbol{X}}(Y = y)$ are those under $H_1 : Y \to \boldsymbol{X}$. Next, we will discuss the choice of these four probability distributions.

For the marginal distribution of Y, we assume it to be a Bernoulli distribution with success probability $\rho \in (0, 1)$,

$$P_{Y\to\boldsymbol{X}}(Y = y) = \rho^y (1 - \rho)^{1-y}. \tag{2}$$

For the marginal distribution of $\boldsymbol{X}$, we assume it to be a multivariate Gaussian distribution with mean $\boldsymbol{\mu} \in \mathbb{R}^n$ and covariance matrix $\boldsymbol{\Sigma} \in \mathbb{M}^{n\times n}$,

$$P_{\boldsymbol{X}\to Y}(\boldsymbol{X} = \boldsymbol{x}) = \mathcal{N}(\boldsymbol{x} \mid \boldsymbol{\mu}, \boldsymbol{\Sigma}), \tag{3}$$

where $\mathcal{N}(\boldsymbol{x} \mid \cdot, \cdot)$ is the Gaussian probability density function evaluated at $\boldsymbol{x}$.

To model the conditional distribution of Y given $\boldsymbol{X}$, we adopt a linear logistic regression,

$$\log \frac{P_{\boldsymbol{X}\to Y}(Y = 1|\boldsymbol{X} = \boldsymbol{x})}{P_{\boldsymbol{X}\to Y}(Y = 0|\boldsymbol{X} = \boldsymbol{x})} = \alpha_0 + \boldsymbol{x}^\top \boldsymbol{\alpha}_1,$$

[a]Note that we are not performing null hypothesis testing. Our method is exploratory.

where α_0 is the intercept and $\boldsymbol{\alpha}_1 \neq \mathbf{0} \in \mathbb{R}^n$ are the slopes. That is, Y is conditionally Bernoulli,

$$P_{\boldsymbol{X}\to Y}(Y = y|\boldsymbol{X} = \boldsymbol{x}) = \phi_{\boldsymbol{x}}^y(1-\phi_{\boldsymbol{x}})^{1-y} \tag{4}$$

with the success probability depending on $\boldsymbol{X}$ through a sigmoid transformation,

$$\phi_{\boldsymbol{x}} = \frac{1}{1+e^{-\alpha_0-\boldsymbol{x}^\top\boldsymbol{\alpha}}}.$$

To specifiy $P_{Y\to\boldsymbol{X}}(\boldsymbol{X} \mid Y)$, we employ a multivariate linear regression model,

$$\boldsymbol{X} = \boldsymbol{\beta}_0 + Y\boldsymbol{\beta}_1 + \boldsymbol{\epsilon}$$

where $\boldsymbol{\beta}_0 \in \mathbb{R}^n$ are the intercepts, $\boldsymbol{\beta}_1 \neq \mathbf{0} \in \mathbb{R}^n$ are the slopes, and $\boldsymbol{\epsilon} \in \mathbb{R}^n$ are Gaussian errors with mean zero and covariance $\boldsymbol{\Omega}$. The multivariate linear regression model above implies,

$$P_{Y\to\boldsymbol{X}}(\boldsymbol{X} = \boldsymbol{x}|Y = y) = \mathcal{N}(\boldsymbol{x} \mid \boldsymbol{\theta}_y, \boldsymbol{\Omega}) \tag{5}$$

with $\boldsymbol{\theta}_y = \boldsymbol{\beta}_0 + y\boldsymbol{\beta}_1$.

Putting (1)-(5) together, we have

$$\begin{aligned} P_{\boldsymbol{X}\to Y}(\boldsymbol{X} = \boldsymbol{x}, Y = y) &= \phi_{\boldsymbol{x}}^y(1-\phi_{\boldsymbol{x}})^{1-y}\mathcal{N}(\boldsymbol{x} \mid \boldsymbol{\mu}, \boldsymbol{\Sigma}) \\ P_{Y\to\boldsymbol{X}}(\boldsymbol{X} = \boldsymbol{x}, Y = y) &= \mathcal{N}(\boldsymbol{x} \mid \boldsymbol{\theta}_y, \boldsymbol{\Omega})\rho^y(1-\rho)^{1-y} \end{aligned} \tag{6}$$

2.3. *Causal Identifiability*

Since we only have access to observational data, the two competing causal hypotheses may not be identifiable, i.e., $P_{\boldsymbol{X}\to Y}(\boldsymbol{X} = \boldsymbol{x}, Y = y) = P_{Y\to\boldsymbol{X}}(\boldsymbol{X} = \boldsymbol{x}, Y = y)$ for all $\boldsymbol{x}$ and y. For example, if both $\boldsymbol{X}$ and Y are Gaussian, they are not identifiable. Consequently, even with an infinite amount of data, one cannot tell these two causal models apart – clearly an undesirable feature. Fortunately, we will show, both theoretically and empirically, that the proposed model is identifiable.

Definition 1 (Causal Identifiability). We say H_0 and H_1 are identifiable if one cannot find any values of $\{\alpha_0, \boldsymbol{\alpha}, \boldsymbol{\mu}, \boldsymbol{\Sigma}\}$[b] and $\{\boldsymbol{\beta}_0, \boldsymbol{\beta}_1, \boldsymbol{\Omega}, \rho\}$[c] such that $P_{\boldsymbol{X}\to Y}(\boldsymbol{X} = \boldsymbol{x}, Y = y) = P_{Y\to\boldsymbol{X}}(\boldsymbol{X} = \boldsymbol{x}, Y = y)$ for all $\boldsymbol{x}$ and y.

Under the causal sufficiency assumption (i.e., there is no unmeasured confounder) commonly adopted in the literature,[2,4,11,13,15,22,23] we have the following identifiability theorem.

Theorem 1 (Causal Identifiability). *Assuming causal sufficiency, the causal hypotheses H_0 and H_1 are identifiable under model* (6).

Proof. We will show by contradiction. Suppose,

$$P_{\boldsymbol{X}\to Y}(\boldsymbol{X} = \boldsymbol{x}, Y = y \mid \alpha_0, \boldsymbol{\alpha}, \boldsymbol{\mu}, \boldsymbol{\Sigma}) = P_{Y\to\boldsymbol{X}}(\boldsymbol{X} = \boldsymbol{x}, Y = y \mid \boldsymbol{\beta}_0, \boldsymbol{\beta}_1, \boldsymbol{\Omega}, \rho) \tag{7}$$

[b] The parameters of $P_{\boldsymbol{X}\to Y}(\boldsymbol{X} = \boldsymbol{x}, Y = y)$
[c] The parameters of $P_{Y\to\boldsymbol{X}}(\boldsymbol{X} = \boldsymbol{x}, Y = y)$

for all $\boldsymbol{x} \in \mathbb{R}^n$ and $y \in \{0, 1\}$. Summing up both sides of (7) over y from 0 to 1, we have

$$\sum_{y=0}^{1} P_{\boldsymbol{X}\to Y}(Y = y \mid \boldsymbol{X} = \boldsymbol{x}, \alpha_0, \boldsymbol{\alpha}) P_{\boldsymbol{X}\to Y}(\boldsymbol{X} = \boldsymbol{x} \mid \boldsymbol{\mu}, \boldsymbol{\Sigma}) = \sum_{y=0}^{1} P_{Y\to\boldsymbol{X}}(\boldsymbol{X} = \boldsymbol{x} \mid Y = y, \boldsymbol{\beta}_0, \boldsymbol{\beta}_1, \boldsymbol{\Omega}) P_{Y\to\boldsymbol{X}}(Y = y \mid \rho) \tag{8}$$

The left-hand side of (8) is given by

$$\begin{aligned}
&\sum_{y=0}^{1} P_{\boldsymbol{X}\to Y}(Y = y \mid \boldsymbol{X} = \boldsymbol{x}, \alpha_0, \boldsymbol{\alpha}) P_{\boldsymbol{X}\to Y}(\boldsymbol{X} = \boldsymbol{x} \mid \boldsymbol{\mu}, \boldsymbol{\Sigma}) \\
&= P_{\boldsymbol{X}\to Y}(\boldsymbol{X} = \boldsymbol{x} \mid \boldsymbol{\mu}, \boldsymbol{\Sigma}) \sum_{y=0}^{1} P_{\boldsymbol{X}\to Y}(Y = y \mid \boldsymbol{X} = \boldsymbol{x}, \alpha_0, \boldsymbol{\alpha}) \\
&= P_{\boldsymbol{X}\to Y}(\boldsymbol{X} = \boldsymbol{x} \mid \boldsymbol{\mu}, \boldsymbol{\Sigma}) \\
&= \mathcal{N}(\boldsymbol{x} \mid \boldsymbol{\mu}, \boldsymbol{\Sigma}),
\end{aligned} \tag{9}$$

where the second equality is due to the law of total probability.

The right-hand side of (8) is given by

$$\begin{aligned}
&\sum_{y=0}^{1} P_{Y\to\boldsymbol{X}}(\boldsymbol{X} = \boldsymbol{x} \mid Y = y, \boldsymbol{\beta}_0, \boldsymbol{\beta}_1, \boldsymbol{\Omega}) P_{Y\to\boldsymbol{X}}(Y = y \mid \rho) \\
&= \rho \cdot P_{Y\to\boldsymbol{X}}(\boldsymbol{X} = \boldsymbol{x} \mid Y = 1, \boldsymbol{\beta}_0, \boldsymbol{\beta}_1, \boldsymbol{\Omega}) + (1-\rho) \cdot P_{Y\to\boldsymbol{X}}(\boldsymbol{X} = \boldsymbol{x} \mid Y = 0, \boldsymbol{\beta}_0, \boldsymbol{\beta}_1, \boldsymbol{\Omega}) \\
&= \rho \mathcal{N}(\boldsymbol{x} \mid \boldsymbol{\beta_0} + \boldsymbol{\beta_1}, \boldsymbol{\Omega}) + (1-\rho)\mathcal{N}(\boldsymbol{x} \mid \boldsymbol{\beta_0}, \boldsymbol{\Omega}).
\end{aligned} \tag{10}$$

Note that (9) is a Gaussian distribution whereas (10) is a mixture of Gaussian distribution. Therefore, for them to be equivalent, we must have $\rho = 0$, $\rho = 1$, or $\boldsymbol{\beta}_1 = \boldsymbol{0}$, which are degenerated cases (i.e., either Y is deterministically 0 or 1, or $\boldsymbol{X}$ and Y are independent). □

Although our theorem relies on the causal sufficiency assumption, the experiments in Section 3.1.3 empirically show that the proposed method is relatively robust to the presence of unmeasured confounders.

2.4. *Estimation*

Theorem 1 establishes a property of the probability model and therefore is a population-level result. It implies that for a large enough sample size, one can correctly identify the correct causal hypothesis even with observational data alone. For a finite sample, our identifiability result paves the way for a simple, yet useful, causal discovery algorithm based on the maximum likelihood estimation (MLE). We aim to determine whether $\boldsymbol{X}$ causes Y or vice versa by quantifying the respective likelihoods. Therefore, when provided with a dataset of N subjects, $(\boldsymbol{x}_1, y_1), \ldots, (\boldsymbol{x}_N, y_N)$, we conclude $H_0 : \boldsymbol{X} \to Y$ if

$$\max_{\alpha_0,\boldsymbol{\alpha},\boldsymbol{\mu},\boldsymbol{\Sigma}} \prod_{i=1}^{N} P_{\boldsymbol{X}\to Y}(\boldsymbol{X}=\boldsymbol{x}_i, Y=y_i \mid \alpha_0,\boldsymbol{\alpha},\boldsymbol{\mu},\boldsymbol{\Sigma}) > \max_{\boldsymbol{\beta}_0,\boldsymbol{\beta}_1,\boldsymbol{\Omega},\rho} \prod_{i=1}^{N} P_{Y\to\boldsymbol{X}}(\boldsymbol{X}=\boldsymbol{x}_i, Y=y_i \mid \boldsymbol{\beta}_0,\boldsymbol{\beta}_1,\boldsymbol{\Omega},\rho),$$

and $H_1 : Y \to \boldsymbol{X}$ otherwise. Note that the two competing hypotheses have the same model complexity (i.e., the same number of parameters) and hence a model complexity penalty, which is typically needed for model selection, is not necessary here. The factorized form of the proposed model (6) allows us to separately find the MLE of each of its four components using existing standard techniques.

However, we note that in our motivating application, $\boldsymbol{X}$ is high-dimensional (n = 1,440). For better statistical and computational efficiency, we choose to reduce its dimensionality before finding the MLE. Specifically, the functional principal component analysis (FPCA) is used, which can reduce the functional data into a few uncorrelated functional principal components (FPCs) that explain the most variation among all the functional bases. We decompose the covariance function of a stochastic process $X(\cdot)$ as,

$$Cov(X(s), X(t)) = \sum_{k=1}^{\infty} \lambda_k \psi_k(s)\psi_k(t),$$

where λ_k's are the nonnegative eigenvalues in descending order and $\psi_k(\cdot)$'s are the corresponding orthogonal eigenfunctions. By the Karhunen-Loève theorem,

$$X(t) = \mu(t) + \sum_{k=1}^{\infty} Z_k \psi_k(t),$$

where $\mu(t) = E[X(t)]$, $\{\psi_k(t)\}_{k=1}^{\infty}$ is referred to as the FPCs, and $\{Z_{ik}(t)\}_{k=1}^{\infty}$ denotes the corresponding FPC scores. In practice, we would choose the first $K \ll n$ FPC scores $\boldsymbol{Z} = (Z_1, \ldots, Z_K)^\top$ that explain 99% variance and replace $\boldsymbol{X}$ by $\boldsymbol{Z}$ in the proposed model when finding the MLEs.

Finally, to assess the uncertainty of our approach, we use the bootstrap[3] technique in our real data application. We first generate B bootstrap samples by resampling subjects with replacement. Each bootstrap sample has the same size as the original dataset. Then we apply our method to each bootstrap sample and record our choice between H_0 and H_1. The proportion of times that we choose H_0 or H_1 reflects our confidence toward each hypothesis.

3. Experiments

We first tested our model through various simulation scenarios on synthetic data where there is known ground truth. After confirming its effectiveness, we then applied our method to a real-world mobility-activity dataset, demonstrating its practical capability in generating plausible causal hypotheses.

3.1. *Simulations*

To assess the efficacy of the proposed model, we performed simulations on three different synthetic datasets including one with unmeasured confounders. Each simulation was repeated 500 times, measuring the accuracy by the frequency at which we correctly identified the true hypothesis. By considering varying sample sizes N between 50 and 200, we investigated the asymptotic behavior of our method. Furthermore, we examined the performance of the model under various signal strengths. For ease of exposition, $\boldsymbol{\delta}_i$ always denote the standard Gaussian white noises hereafter, i.e., $\boldsymbol{\delta}_i \sim \mathcal{N}(\mathbf{0}, \boldsymbol{I})$ where $\boldsymbol{I}$ is an identity matrix.

3.1.1. *Case 1: True Direction* $\boldsymbol{X} \to Y$

For each subject $i = 1, \ldots, N$, the functional data $\boldsymbol{x}_i$ were created by first sampling their mean $\boldsymbol{m}$ from a centered Gaussian process at $n = 30$ evenly spaced time points,

$$\boldsymbol{m} \sim \mathcal{GP}(0, \mathcal{K})$$

with the powered exponential covariance function,

$$\mathcal{K}(t, s) = \exp\{-|t - s|^{\kappa}\},$$

of which the power $\kappa = 1.9$, and then setting

$$\boldsymbol{x}_i = \boldsymbol{m} + \boldsymbol{\delta}_i.$$

We performed the FPCA[24] on $\boldsymbol{x}_1, \ldots, \boldsymbol{x}_N$ using the R package `fdapace`, and retained first K FPCs that explained 99% variance. We denote the standardized FPC scores by $\boldsymbol{z}_1, \ldots, \boldsymbol{z}_N$.

To create the causal dependency of y_i on $\boldsymbol{x}_i$ through $\boldsymbol{z}_i$, we generated y_i from a probit regression,

$$y_i = \begin{cases} 1 & \text{if} \quad y_i^* > 0 \\ 0 & \text{if} \quad y_i^* \leq 0 \end{cases},$$

where

$$y_i^* = 0.5 + \boldsymbol{z}_i^\top \boldsymbol{\gamma} + \epsilon_i$$

with $\epsilon_i \sim \mathcal{N}(0, 1)$. Here $\boldsymbol{\gamma} = (\gamma_1, \ldots, \gamma_K)^\top$ is the direct causal effect (signal), which will be varied at three levels: weak ($\gamma_k = \pm 1$), moderate ($\gamma_k = \pm 1.5$), and strong ($\gamma_k = \pm 3$).

The simulation results are reported in Table 1, showing an expected trend: the stronger the signal is, the more accurately the true causal direction can be discerned. Also, the accuracy approaches 100% as the sample size increases for the moderate and strong signal cases. Even with the weak signal, the accuracy was still good, around 90%. Note that for a non-identifiability model, the expected accuracy is 50%.

3.1.2. *Case 2: True Direction* $Y \to \boldsymbol{X}$

Exploring the reverse causal direction, we first generated the binary cause variable y_i from a Bernoulli distribution with a success probability of 0.5. Then we generated the functional effect variable,

$$\boldsymbol{x}_i = \boldsymbol{m}_{y_i} + \boldsymbol{\delta}_i,$$

where $\boldsymbol{m}_y \sim \mathcal{GP}(0, \mathcal{K}_y)$ for $y = 0, 1$ with the powered exponential covariance function $\mathcal{K}_y$ of which the power κ depends on y. Specifically, $\kappa = 1.9$ if $y = 1$, and $\kappa = 0.3$ (strong signal), 1.1 (moderate signal), or 1.7 (weak signal) if $y = 0$.

Table 1: Simulations. Accuracy of the proposed model in determining true causal directions in synthetic datasets over 500 simulations.

Case	Confounder	Signal	Sample size			
			50	100	150	200
$\boldsymbol{X} \to Y$	None	*weak*	92.8%	88.8%	90%	87.8%
		moderate	92.8%	97.2%	97.6%	99.6%
		strong	93.8%	98.6%	99%	99.8%
	Functional		94.6%	98.2%	99.2%	99.2%
	Binary		92.4%	99.2%	100%	100%
$Y \to \boldsymbol{X}$	None	*weak*	82.2%	89%	89.4%	93.6%
		moderate	83.4%	90.2%	96%	97.2%
		strong	97.8%	99%	99.4%	99.8%
	Functional		39.8%	61.6%	75%	83.8%
	Binary		63.6%	77%	85.2%	85.2%

As anticipated, our simulation results (Table 1) show that as the signal or the sample size increases, the accuracy approaches 100%.

3.1.3. *Case 3: Hidden Confounders*

The Simulation Cases 1&2 above demonstrate the validity of Theorem 1, i.e., causal directions can be identified even with observational data alone. We now empirically assess the robustness of the proposed method with respect to the violation of the causal sufficiency assumption, i.e., we test whether our method can still identify the correct causal direction in the presence of unmeasured confounders.

Our methodology hinges on determining the causality between two distinct types of variables, binary scalar and continuous functional. Thus, accordingly, we considered that the unobserved confounders, generically denoted by $\mathcal{C}$, are also either binary scalar or continuous functional. Consequently, we investigated four separate scenarios depicted in Fig. 1. We generated data from these four causal graphs and hid $\mathcal{C}$ from our method (i.e., only took $\boldsymbol{X}$ and Y as the inputs of our algorithm). As before, we recorded the frequency at which we correctly identified the causal direction between $\boldsymbol{X}$ and Y.

In Fig. 1 (a)&(b) where the confounder is binary, we generated the confounder c_i from a Bernoulli distribution with success probability 0.5. In Fig. 1 (a), the mean $\boldsymbol{m}_{c_i}$ of $\boldsymbol{x}_i$ was generated from a conditional Gaussian process $\boldsymbol{m}_c \sim \mathcal{GP}(0, \mathcal{K}_c)$ with the powered exponential covariance function $\mathcal{K}_c$ of which the power κ depends on c. Specifically, $\kappa = 1.9$ if $c = 1$ and $\kappa = 1.5$ if $c = 0$. Then as before, we set $\boldsymbol{x}_i = \boldsymbol{m}_{c_i} + \boldsymbol{\delta}_i$. Finally, we generated y_i from a probit regression model, $y_i = 1$ if $y_i^* > 0$ and $y_i = 0$ otherwise, where

$$y_i^* = 0.5 + 3 \cdot \boldsymbol{z}_i^\top \mathbf{1}_K + 3 \cdot c_i + \epsilon_i$$

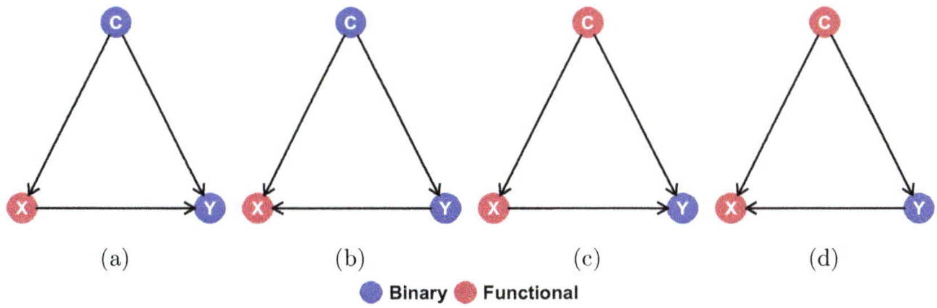

Fig. 1: Four confounding scenarios under consideration in Simulation Case 3.

with $\epsilon_i \sim \mathcal{N}(0,1)$, $\boldsymbol{z}_i$'s are the FPC scores of $\boldsymbol{x}_i$, and $\mathbf{1}_K = (1,\ldots,1)^\top$ is a vector of ones with length K.

In Fig. 1 (b), we generated y_i from a Bernoulli distribution with the success probability ρ_i dependent on c_i. More precisely, $\rho_i = 0.9$ if $c_i = 1$ and $\rho_i = 0.1$ if $c_i = 0$. Subsequently, the mean $\boldsymbol{m}_{c_i,y_i}$ of $\boldsymbol{x}_i$ was generated from a conditional Gaussian process $\boldsymbol{m}_{c,y} \sim \mathcal{GP}(0, \mathcal{K}_{c,y})$ with the powered exponential covariance function $\mathcal{K}_{c,y}$ of which the power κ depends on both c and y. To be specific, $\kappa = 1.9$ if $c = 1$ and $y = 1$, $\kappa = 0.5$ if $c = 1$ and $y = 0$, $\kappa = 1.0$ if $c = 0$ and $y = 1$, and $\kappa = 1.7$ if $c = 0$ and $y = 0$. Finally, we set $\boldsymbol{x}_i = \boldsymbol{m}_{c_i,y_i} + \boldsymbol{\delta}_i$.

In Fig. 1 (c)&(d), the functional confounder $\boldsymbol{c}_i$ was generated in the same way as $\boldsymbol{x}_i$ in Case 1 with $\kappa = 1.5$. Next, we performed the FPCA on $\boldsymbol{c}_1,\ldots,\boldsymbol{c}_N$ and retained the first J FPCs that explained 99% variance. We denote the standardized FPC scores by $\boldsymbol{d}_1,\ldots,\boldsymbol{d}_N$. In Fig. 1 (c), $\boldsymbol{m}$ was first generated from a centered Gaussian process $\boldsymbol{m} \sim \mathcal{GP}(0,\mathcal{K})$ with $\kappa = 1.9$. Then the dependence on the confounder was introduced by setting

$$\boldsymbol{x}_i = 0.5 + 5 \cdot \boldsymbol{m} + 5 \cdot \boldsymbol{c}_i + \boldsymbol{\delta}_i.$$

Finally, we generated y_i from a probit regression model, $y_i = 1$ if $y_i^* > 0$ and $y_i = 0$ otherwise, where

$$y_i^* = 0.5 + 5 \cdot \boldsymbol{z}_i^\top \mathbf{1}_K + 5 \cdot \boldsymbol{d}_i^\top \mathbf{1}_J + \epsilon_i$$

with $\epsilon_i \sim \mathcal{N}(0,1)$ and $\boldsymbol{z}_i$'s being the first K FPC scores of $\boldsymbol{x}_i$.

In Fig. 1 (d), we first generated y_i from a probit regression model, $y_i = 1$ if $y_i^* > 0$ and $y_i = 0$ otherwise, where

$$y_i^* = 0.5 + 3 \cdot \boldsymbol{d}_i^\top \mathbf{1}_J + \epsilon_i$$

with $\epsilon_i \sim \mathcal{N}(0,1)$. Then we generated mean processes $\boldsymbol{m}_{y_i}$ from a conditional Gaussian process $\boldsymbol{m}_y \sim \mathcal{GP}(0,\mathcal{K}_y)$. In this setting, the power κ of the powered exponential covariance function $\mathcal{K}_y$ depended on y: $\kappa = 1.9$ if $y = 1$, and $\kappa = 0.5$ if $y = 0$. To introduce the influence of the confounder, we defined

$$\boldsymbol{x}_i = \boldsymbol{m}_{y_i} + 3 \cdot \boldsymbol{c}_i + \boldsymbol{\delta}_i.$$

The results from these four confounding scenarios (Table 1) demonstrate the robustness of the proposed method. Particularly, as the sample size increased, our method achieved increasingly better accuracy and was significantly better than a random guess for large sample sizes.

3.2. *Real Data*

Next, we applied the proposed methodology to the data collected by the NHANES. This extensive study, conducted by the Centers for Disease Control, gathered a wide range of health and nutritional information about the U.S. population, including sociodemographic characteristics and various health conditions. To demonstrate the utility of the proposed method, we are particularly interested in two variables, physical activities $\boldsymbol{X}$ captured via hip-attached accelerometers and an indicator variable of mobility issues Y derived from self-reported household interview data. Given the logical assumption of $Y \to \boldsymbol{X}$ in this scenario, we aim to verify if our method can correctly identify this causal direction, primarily seeking to validate the effectiveness of our method in accurately determining causation from a known truth.

3.2.1. *Data Preprocessing*

Utilizing the NHANES dataset, we accessed activity data from hip-worn accelerometers during the 2003–2004 and 2005–2006 study waves. The magnitude of acceleration (movement "intensity") was captured using the ActiGraph AM-7164, delivering an objective measure of physical activity and bypassing the inconsistencies of self-reported data. Participants were instructed to wear the device for seven consecutive days, excluding swimming and bathing periods. The raw data were segmented into one-minute intervals or "epochs" with intensity readings accumulated per epoch and saved in long format (each row is a subject-minute).

The well-formatted data are contained in the R package `rnhanesdata`.[7,8] Following the preprocessing procedure in their paper,[8] we included individuals aged 50 to 85 and omitted non-compliant individuals who have excessive missing accelerometer data, leaving us with $N = 3,198$ subjects.

The activity data for each individual were aggregated over the 7-day period and transformed via $\log(1+x)$. This dataset is organized in a $7N \times 1440$ matrix, with one row designated for each subject-day across all NHANES waves, where 7 denotes the days each subject wore the accelerometer, and 1440 corresponds to the total number of minutes in a day.

The presence of any mobility issues was represented as a binary variable, categorized as either "No difficulty" or "Any difficulty," based on responses from the Physical Functioning questionnaire. Individuals were classified under "Any difficulty" if they reported challenges in climbing 10 stairs, walking a quarter mile, abstained from these activities, or required special walking equipment. Overall, there are 32.4% subjects in the sample who experience any mobility of movement problem.

3.2.2. *Results*

We generated $B = 100$ bootstrap samples and successfully identified the correct causal direction across all samples from comparing the maximized likelihoods of $Y \to \boldsymbol{X}$ and $\boldsymbol{X} \to Y$: the mobility issue Y unambiguously impacts an individual's level of physical activity $\boldsymbol{X}$ with high

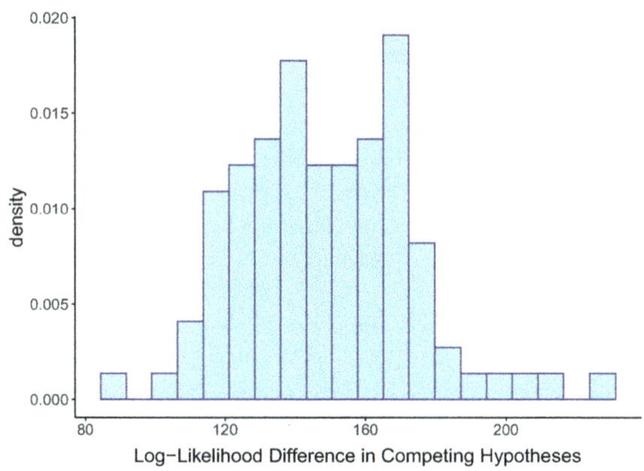

Fig. 2: Real data. Histogram depicting the maximized log-likelihood differences between the competing hypotheses $Y \to \boldsymbol{X}$ and $\boldsymbol{X} \to Y$.

confidence. As depicted in Fig. 2, the histogram illustrates the difference in the maximized log-likelihoods between these two competing hypotheses (the former minus the latter), which is noticeably bounded away from zero, meaning that $Y \to \boldsymbol{X}$ is far more likely than $\boldsymbol{X} \to Y$, which matches the presumed truth.

4. Discussion

In this paper, we have presented a new causal model for generating bivariate causal hypotheses with a continuous functional variable (e.g., physical activities) and a binary scalar variable (e.g., mobility issue indicator) in an exploratory fashion, which can provide insights as to which variable is more likely the cause. We theoretically proved that the underlying cause-effect relationship is identifiable with purely observational data under the causal sufficiency assumption. Empirically, we used a likelihood-based inference procedure and demonstrated the utility of the proposed method both under and beyond the causal sufficiency setting through simulation studies and a real-world wearable device application.

There are several areas where this paper could be strengthened and extended. First, our NHANES application has focused on physical activities and mobility issue because of their clear causal relationship. Having demonstrated it is possible to identify their causal relationship, we plan to analyze other variables in the data to generate causal hypotheses in an exploratory manner, which is an intended use of the proposed method.

Second, our identifiability theory operates under the assumption that there are no unmeasured confounders. Even though our empirical investigations have indicated a degree of robustness to the presence of confounders, a theoretical exploration of identifiability within this context would be interesting and particularly relevant in observational studies where the presence of unmeasured confounders is common.

Third, we have focused on the bivariate case and hence an extension to multivariate cases and leveraging additional publicly available datasets can considerably broaden the method's applicability. For example, the brain electroencephalogram dataset[1] comprises electroencephalogram signals collected over various trials with distinct stimuli for two groups - alcoholics and controls. By viewing the electroencephalogram signals as multivariate functional data, a recent paper[23] attempts to discern the causal relationships among these functions. The multivariate extension of our method could potentially enrich this research by providing additional insights into the causal relationships modified by the experimental groups by treating the group as a binary variable. Moreover, it should be relatively straightforward to extend our method to incorporate multiple categorical scalar variables.

Finally, a Bayesian inference approach could be adopted especially for multivariate cases where efficient searching strategies in the causal graph space are required. A Bayesian approach would make it easier to make finite-sample inferences with natural uncertainty quantification for complex causal graphs.

Acknowledgment

Ni's research was partially supported by NIH 1R01GM148974-01 and NSF DMS-2112943.

References

1. H. Begleiter. EEG Database. UCI Machine Learning Repository, 1999. DOI: https://doi.org/10.24432/C5TS3D.
2. J. Choi, R. Chapkin, and Y. Ni. Bayesian causal structural learning with zero-inflated poisson Bayesian networks. *Advances in neural information processing systems*, 33:5887–5897, 2020.
3. B. Efron. Bootstrap methods: another look at the jackknife. In *Breakthroughs in statistics: Methodology and distribution*, pages 569–593. Springer, 1992.
4. P. Hoyer, D. Janzing, J. M. Mooij, J. Peters, and B. Schölkopf. Nonlinear causal discovery with additive noise models. *Advances in neural information processing systems*, 21, 2008.
5. D. Janzing, J. Mooij, K. Zhang, J. Lemeire, J. Zscheischler, P. Daniušis, B. Steudel, and B. Schölkopf. Information-geometric approach to inferring causal directions. *Artificial Intelligence*, 182:1–31, 2012.
6. K.-Y. Lee and L. Li. Functional structural equation model. *Journal of the Royal Statistical Society Series B: Statistical Methodology*, 84(2):600–629, 2022.
7. A. Leroux. *rnhanesdata: NHANES Accelerometry Data Pipeline*, 2023. R package version 1.02.
8. A. Leroux, J. Di, E. Smirnova, E. J. Mcguffey, Q. Cao, E. Bayatmokhtari, L. Tabacu, V. Zipunnikov, J. K. Urbanek, and C. Crainiceanu. Organizing and analyzing the activity data in nhanes. *Statistics in biosciences*, 11:262–287, 2019.
9. S. Monti and G. F. Cooper. A multivariate discretization method for learning bayesian networks from mixed data. *arXiv preprint arXiv:1301.7403*, 2013.
10. J. S. Morris. Functional regression. *Annual Review of Statistics and Its Application*, 2:321–359, 2015.
11. Y. Ni. Bivariate causal discovery for categorical data via classification with optimal label permutation. *Advances in Neural Information Processing Systems*, 35:10837–10848, 2022.
12. J. Pearl et al. Models, reasoning and inference. *Cambridge, UK: CambridgeUniversityPress*, 19(2):3, 2000.
13. S. Shimizu, P. O. Hoyer, A. Hyvärinen, A. Kerminen, and M. Jordan. A linear non-gaussian acyclic model for causal discovery. *Journal of Machine Learning Research*, 7(10), 2006.

14. P. Spirtes, C. Glymour, and R. Scheines. *Causation, Prediction, and Search, 2nd Edition*, volume 1 of *MIT Press Books*. The MIT Press, December 2001.
15. P. Spirtes, C. N. Glymour, and R. Scheines. *Causation, prediction, and search*. MIT press, 2000.
16. O. Stegle, D. Janzing, K. Zhang, J. M. Mooij, and B. Schölkopf. Probabilistic latent variable models for distinguishing between cause and effect. *Advances in neural information processing systems*, 23, 2010.
17. C. D. Tekwe, R. S. Zoh, M. Yang, R. J. Carroll, G. Honvoh, D. B. Allison, M. Benden, and L. Xue. Instrumental variable approach to estimating the scalar-on-function regression model with measurement error with application to energy expenditure assessment in childhood obesity. *Statistics in medicine*, 38(20):3764–3781, 2019.
18. J.-L. Wang, J.-M. Chiou, and H.-G. Müller. Functional data analysis. *Annual Review of Statistics and Its Application*, 3(1):257–295, 2016.
19. W. Wenjuan, F. Lu, and L. Chunchen. Mixed causal structure discovery with application to prescriptive pricing. In *Proceedings of the 27th International Joint Conference on Artificial Intelligence*, pages 5126–5134, 2018.
20. M. Yamayoshi, J. Tsuchida, and H. Yadohisa. An estimation of causal structure based on latent lingam for mixed data. *Behaviormetrika*, 47:105–121, 2020.
21. Y. Zeng, S. Shimizu, H. Matsui, and F. Sun. Causal discovery for linear mixed data. In B. Schölkopf, C. Uhler, and K. Zhang, editors, *Proceedings of the First Conference on Causal Learning and Reasoning*, volume 177 of *Proceedings of Machine Learning Research*, pages 994–1009. PMLR, 11–13 Apr 2022.
22. K. Zhang and A. Hyvärinen. On the identifiability of the post-nonlinear causal model. 647–655, 2009.
23. F. Zhou, K. He, K. Wang, Y. Xu, and Y. Ni. Functional bayesian networks for discovering causality from multivariate functional data. *arXiv preprint arXiv:2210.12832*, 2022.
24. Y. Zhou, S. Bhattacharjee, C. Carroll, Y. Chen, X. Dai, J. Fan, A. Gajardo, P. Z. Hadjipantelis, K. Han, H. Ji, C. Zhu, H.-G. Müller, and J.-L. Wang. *fdapace: Functional Data Analysis and Empirical Dynamics*, 2022. R package version 0.5.9.

FedBrain: Federated Training of Graph Neural Networks for Connectome-based Brain Imaging Analysis

Yi Yang, Han Xie, Hejie Cui, Carl Yang[†]

Department of Computer Science, Emory University,
Atlanta, Georgia, USA
[†]*E-mail: j.carlyang@emory.edu*

Recent advancements in neuroimaging techniques have sparked a growing interest in understanding the complex interactions between anatomical regions of interest (ROIs), forming into brain networks that play a crucial role in various clinical tasks, such as neural pattern discovery and disorder diagnosis. In recent years, graph neural networks (GNNs) have emerged as powerful tools for analyzing network data. However, due to the complexity of data acquisition and regulatory restrictions, brain network studies remain limited in scale and are often confined to local institutions. These limitations greatly challenge GNN models to capture useful neural circuitry patterns and deliver robust downstream performance. As a distributed machine learning paradigm, federated learning (FL) provides a promising solution in addressing resource limitation and privacy concerns, by enabling collaborative learning across local institutions (*i.e.*, clients) without data sharing. While the data heterogeneity issues have been extensively studied in recent FL literature, cross-institutional brain network analysis presents unique data heterogeneity challenges, that is, the inconsistent ROI parcellation systems and varying predictive neural circuitry patterns across local neuroimaging studies. To this end, we propose FedBrain, a GNN-based personalized FL framework that takes into account the unique properties of brain network data. Specifically, we present a federated atlas mapping mechanism to overcome the feature and structure heterogeneity of brain networks arising from different ROI atlas systems, and a clustering approach guided by clinical prior knowledge to address varying predictive neural circuitry patterns regarding different patient groups, neuroimaging modalities and clinical outcomes. Compared to existing FL strategies, our approach demonstrates superior and more consistent performance, showcasing its strong potential and generalizability in cross-institutional connectome-based brain imaging analysis. The implementation is available here.

Keywords: Brain Connectome Analysis; Digital Health; Federated Learning

1. Introduction

In recent years, research in neuroscience has been driven to unravel the intricacies of the human brain and its connection to complex disorders such as bipolar disorder (BP) and Autism. Neuroimaging techniques, including fMRI and DTI, have emerged as crucial tools for facilitating the diagnosis of various diseases.[1] These techniques enable the construction of brain networks, which are essentially weighted connected graphs, where nodes represent anatomical regions of interest (ROIs) and edges represent their functional correlations or structural connections. By analyzing these networks, researchers gain valuable insights into the biological structures and functions of complex neural systems, aiding in the early detection of neurological disorders and advancing fundamental neuroscience research.

Graph Neural Networks (GNNs) have gained significant popularity in analyzing graph-structured data, demonstrating impressive performance across various domains like social networks, recommender systems, and gene/protein interactions.[2,3] In neuroscience, GNNs have been applied to brain network analysis, addressing tasks such as disease prediction and neural pattern discovery[4–9] However, deep learning models, including GNNs, heavily rely on large labeled datasets to obtain strong performance. Unfortunately, neuroimaging datasets are often relatively small due to the high complexity of data acquisition, preprocessing, and annotation, leading to significant model overfitting and limited generalization power.[10,11] For instance, the popular datasets for BP and HIV analysis consist of only a few dozen subjects,[12,13] making it particularly challenging for GNNs to effectively capture important neural circuitry patterns from the noisy networks. While there exist several relatively large multi-site neuroimaging studies, the these are still small compared to datasets in typical ML domains.[14]

Recently, federated learning (FL) has emerged as a promising solution to address the challenges of limited training data and computation resources in local studies.[15–17] FL operates by collaboratively training a centralized server model based on data privately stored by multiple local clients. The approach offers two notable advantages. First, it ensures privacy preservation since clients solely communicate model parameters with the server. Second, it facilitates knowledge generalization by client aggregatiion which can mitigate the overfitting issues typically associated with learning on small datasets. These aspects have contributed to the success of FL in various fields including healthcare applications[18] and graph learning.[19]

One significant challenge in FL is data heterogeneity, wherein the data distributions significantly differ across local data owners. Several FL algorithms[16,17] have been proposed to tackle the data heterogeneity challenge. However, these methods mostly focus on label distributions and fail to address the unique data heterogeneity scenarios in cross-institutional brain network analysis which can manifest in two key aspects. First, since network parcellation is traditionally an ad hoc process carried out by domain experts, it is difficult to assume or require all different institutions to conform to the same ROI atlas mapping systems when preprocessing their neuroimaging data. As a result, this leads to misalignment in network structures and ROI features across clients. Second, different institutions collect brain network data for different patient groups, with different neuroimaging techniques and towards different clinical purposes, which results in varying underlying predictive neural circuitry patterns.

In this work, we propose FedBrain, a personalized FL framework designed for GNN-based brain network analysis. Our framework comprises three key components: a GNN-based FL backbone, a federated atlas mapping mechanism, and a guided client clustering mechanism. To build our FL platform, we use the well-established `FedAvg` as a foundation, and our default GNN structure is an optimized GCN model.[4] To address the feature- and structure-wise heterogeneity issue due to potentially different atlas mapping systems used across local institutions, we introduce an autoencoder-based atlas mapping mechanism, which aims to project diverse ROI profiles onto a uniform sharable embedding space. To handle heterogeneous predictive neural circuitry patterns due to various neuroimaging modalities and clinical outcomes, we design a knowledge-guided client clustering mechanism by incorporating prior clinical knowledge into the dynamic clustering process of clients with similar data during FL.

To showcase the effectiveness of FedBrain on real-world datasets from different institutions, we conduct extensive empirical evaluations, comparing our framework to state-of-the-art methods. The results demonstrate that FedBrain outperforms the baselines across all clients, with a minimum relative gain of 21.36% in accuracy. Moreover, we conduct ablation studies and specific analyses on the proposed federated atlas mapping and guided clustering mechanisms to fully understand their contribution and robustness within the framework. The results confirmed the necessity of these components in improving overall model performance.

2. Related Work

GNNs for Brain Network Analysis. GNNs have gained significant attention for their effectiveness in analyzing graph-structured data,[20–22] with several pioneering models applied to brain network analysis. Notable examples include BrainGNN,[8] which uses ROI-aware graph convolutional and ROI-selection pooling layers to predict neurological biomarkers from fMRI data. Another approach, BrainNetCNN,[9] adopts a CNN framework with various convolutional filters designed to leverage the topological locality of structural brain networks. BrainNetTF[7] introduces a transformer architecture with an orthonormal clustering readout that considers ROI similarity within functional modules. Existing studies[5,23–25] have demonstrated GNNs can substantially improve performance in brain disorder predictions when sufficient data is available. However, the difficulty emerges when dealing with limited training samples in practical scenarios, especially for particular clinical studies.[26] This limitation hinders the full potential of GNNs for modeling brain network data, motivating designs capable of overcoming data scarcity and heterogeneity and improving performance in real clinical tasks.

FL on Graphs. FL has gained significant attention for collaboratively training deep learning models while preserving data privacy. Recently, it has been proven to be effective in the context of graphs. Some of the pioneering works have explored modeling clients as nodes in graphs,[27,28] and benchmark surveys[29] have contributed to the understanding of GNN-based FL across graphs in diverse data domains. FL on graphs can face a unique challenge, graph data heterogeneity. Some previous related works include FedCG[28] which addresses the challenge of statistical heterogeneity in FL by leveraging GNN models to extract interactions across domains; GCFL[30] which studies the specific graph-level heterogeneity across domains and proposes a dynamic clustered graph FL framework; and FedLit[31] which proposes a way to dynamically cluster the latent link types of graphs in FL to address the link-level heterogeneity across graphs. Nonetheless, the distinct ways in which heterogeneity manifests in brain network studies, such as the variance in parcellation systems and neural circuitry patterns, make most FL frameworks that emphasize generic graph structure learning inapplicable. While research on GNN-based FL for neuroimaging data has shown promise, existing techniques focus on privacy preservation[32] or domain adaptation.[33] These objectives inherently diverge from our approach, which aspires to bolster data alignment and augment client personalization.

3. The FedBrain Framework

3.1. *The FL Backbone*

The backbone FL structure of FEDBRAIN is based on federated averaging (`FedAvg`).[15] The essence is to aggregate the updated model parameters from local clients through a process of weighted averaging. These averaged parameters are then disseminated back to each client in the subsequent communication round. Specifically, when aggregating parameters, the server assigns a weight to each client in proportion to their respective sample size.

We utilize an optimized GCN[4] as backbone for both the server and client models. The ROI (*i.e.*, node) features are initialized with the connection profiles (*i.e.*, adjacency).[4] That is, the feature matrix $\boldsymbol{X}$ is equivalent to the adjacency $\boldsymbol{A}$ ($\boldsymbol{X} \equiv \boldsymbol{A}$), where $\boldsymbol{A}$ is parameterized by the node set $\mathcal{V} = \{v_n\}_{n=1}^{N}$ and the weighted edge set $\mathcal{E} = \mathcal{V} \times \mathcal{V}$.

3.2. *Federated Atlas Mapping*

Motivation. For brain network data, the ROI (*i.e.*, node) parcellation is determined by the brain atlas. Once a template is chosen, all brain networks within a dataset share the same ROI identities. However, in our cross-institutional setting, different institutions may utilize different parcellation systems. This leads to heterogeneity in both sizes and structures of the parcellated networks, as well as divergent meanings of ROI features (*i.e.*, connectivity profiles). While it is possible to manually convert between atlases, this process is laborious and requires extensive domain expertise. Therefore, we propose a data-driven transformation, as a preprocessing mechanism, that aims to align network features and structures across institutions, ensuring consistency in network dimensions and physical interpretations of features.

Autoencoder framework. To achieve uniform feature dimensions and network sizes, we employ a one-layer linear autoencoder (AE) to learn a dataset-specific projection. Given a target dimension M that is consistent across all datasets and an input feature $\boldsymbol{X} \in \mathbb{R}^{N \times N}$ ($N > M$), the objective is to learn a linear projection $\boldsymbol{W} \in \mathbb{R}^{N \times M}$, such that the projected representations preserve as much information as possible from the original features. The AE is optimized using the mean-squared-error (MSE) reconstruction objective, denoted as $\mathcal{L}_{rec} = (1/N)\|\boldsymbol{X} - \boldsymbol{X}\boldsymbol{W}\boldsymbol{W}^\top\|^2$. Intuitively, the projection $\boldsymbol{W}$ transforms initial features by applying a weighted linear combination on the original dimensions. Consequently, the columns of $\boldsymbol{W}$ learns to assign original dimensions into M groups. We exploit this concept to condense the network structure. To reduce the computational complexity, we formulate an assignment matrix $\boldsymbol{Z} \in \mathbb{R}^{N \times M}$ such that $\boldsymbol{Z}_{i,j} = \mathbb{1}[\boldsymbol{W}_{i,j} \in \arg\operatorname{top} k\,(\operatorname{col}_j(\boldsymbol{W}))]$. The matrix $\boldsymbol{Z}$ records the top-k greatest entries per each column in $\boldsymbol{W}$ and zeros out the rest. Ultimately, given a graph adjacency matrix $\boldsymbol{A}\,(\equiv \boldsymbol{X})$, we construct a compressed network $\boldsymbol{A}'$ by evaluating $\boldsymbol{A}' = \boldsymbol{Z}^\top \boldsymbol{A}\boldsymbol{Z}$.

Federated training. Apart from dataset-specific projections, aligning the physical interpretations of projected features across datasets is equally vital to mitigate structure- and feature-level heterogeneity. To achieve this, we leverage the FL approach to train the autoencoders with the intention of obtaining a global atlas projection. However, the architectural sizes of autoencoders across clients can vary due to the differing original data dimensions, which makes it challenging to communicate model parameters.

To address this issue, we propose a unified mapping method that aims to adapt the size of the global model to the varying dimensionality of each local dataset. Given a global projection $\boldsymbol{W}_G \in \mathbb{R}^{N_G \times M}$ based on the most detailed parcellation template with N_G defined ROIs, and a coarser template with N_L defined ROIs ($N_L < N_G$) employed for local data, our goal is to derive an assignment matrix $\boldsymbol{P}_L \in \mathbb{R}^{N_L \times N_G}$, which ensures the local projection $\boldsymbol{W}_L \in \mathbb{R}^{N_L \times M}$ is distributed through the mapping $\boldsymbol{W}_L = \boldsymbol{P}_L \boldsymbol{W}_G$. To achieve this, we leverage the 3D coordinates of the ROIs, denoted as $D_G \in \mathbb{R}^{N_G \times 3}$ for the global parcellation template and $D_L \in \mathbb{R}^{N_L \times 3}$ for the local template. We first calculate a distance matrix $\boldsymbol{S} \in \mathbb{R}^{N_L \times N_G}$, where $\boldsymbol{S}_{i,j} = d(\text{row}_i(D_L), \text{row}_j(D_G))$ represents the pairwise Euclidean distance between ROIs from the two templates. We then designate $\boldsymbol{P}_{L_{i,j}} = \mathbb{1}[\boldsymbol{S}_{i,j} = \arg\min(\text{col}_j(\boldsymbol{S}))]$. This implied that we only consider the minimum entry per each column of $\boldsymbol{S}$. Essentially, we enable $\boldsymbol{P}_L$ to learn a mapping that groups ROIs in the global template with those in the local template, based on their spatial proximity. During each communication round, clients start by downloading the server's parameter by applying the mapping $\boldsymbol{W}_L = \boldsymbol{P}_L \boldsymbol{W}_G$. Subsequently, each client sends their updated parameters back to the server, employing the inverse mapping $\boldsymbol{W}_L^* = \boldsymbol{P}_L^\top \boldsymbol{W}_L^*$.

3.3. *Guided Clustering*

Motivation. Beyond the discrepancies in network parcellation systems, another significant source of heterogeneity originates from the variability in predictive neural circuitry patterns, encompassing data modalities and clinical outcomes. These variances can result in a suboptimal adaptation of the generalized global model to specific local objectives. Therefore, our aim is to strike a balance between global generalization and local personalization. Moreover, as shown in Table 1, we notice that similar neural patterns are shared among certain client institution subgroups. This motivates us to integrate client clustering[30,34] into the FL process.

Clustered FL. When data distributions are similar among local clients, the average global model can achieve convergence for all local objectives. However, in instances of heterogeneity, the global model fails to adapt to local optimizations, resulting in stationary point convergence.[34] To mitigate stationary convergence, clients can be assigned to clusters with homogeneous data distributions, thereby initiating cluster-specific FL subroutines.

Constrained clustering. While gradient-based clustering effectively addresses the stationary point issue and improves performance over the basic `FedAvg`, the method is entirely data-driven, lacking consideration of shared clinical prior knowledge related to the neural circuitry patterns of each client. Consequently, heterogeneity may still exist within the formed clusters, necessitating further division of clusters. This often leads to the creation of singleton clusters, undermining the essence of collaborative learning. This phenomenon is demonstrated in Figure 2 (Section 4.4). Based on these observations, we propose a refined variant of the clustering method that incorporates shared prior knowledge to guide the clustering process. For instance, in terms of data modalities, it is intuitive to group clients with similar ROI connectivities and MRI data. Likewise, with regard to clinical outcomes, FL on a cluster level could benefit from learning similar objectives. To this end, we create must-links between pairs of clients that exhibit highly similar neural patterns and define cannot-links for those that don't. We introduce a weighted reward λ_{must} and penalty λ_{cannot} term, which are multiplied to the pairwise client similarity measure when creating must- and cannot-links.

4. Experiments

Datasets. We evaluate our framework using six real-world brain network datasets: BP,[12] HIV,[13] PPMI,[35] PNC,[36] ABIDE,[37] and ABCD.[38] We present key statistics for each dataset in Table 1. Among them, BP, HIV, and PPMI contain multiple data modalities. In light of this, we propose to employ every such modality to be learned on a separate FL client. Based on the available label information, we define two possible tasks – disease prediction (*i.e.*, patients *vs.* health controls) and gender prediction – both in the form of binary classification.

Table 1. Dataset statistics.

Dataset	Modality	Sample Size	Atlas	Network Size	Outcome	Class Number
BP	fMRI, DTI	97	Brodmann 82	82 × 82	Disease	2
HIV	fMRI, DTI	70	AAL 90	90 × 90	Disease	2
PPMI	PICo, Hough, FSL	754	Desikan-Killiany 84	84 × 84	Disease	2
PNC	fMRI	503	Power 264	264 × 264	Gender	2
ABIDE	fMRI	1009	Craddock 200	200 × 200	Disease	2
ABCD	fMRI	7901	HCP 360	360 × 360	Gender	2

Parameter setup. The downstream classifier consists of a single-layer MLP, and we use the negative log-likelihood measure as the optimization objective and accuracy as the evaluation metric. In the case of all FL baselines, a complete training procedure encompasses 80 communication rounds. For the `self-train` (*i.e.*, non-FL) baseline, each local model is trained for 80 epochs. Regarding FedBrain, we retain the top 3 entries in each column of the atlas mapping projection matrix for network transformation, and use the most detailed HCP 360 template to define the global model for our federated training of AEs.

Empirical analyses. The following sections are structured to assess (1) the performance of FedBrain in comparison to widely adopted FL frameworks, and (2) the contribution of the key components to the overall performance, supplemented by case studies.

4.1. *Overall performance comparison (RQ1)*

We present a comprehensive performance comparison in Table 2. We include the client (*i.e.*, dataset) name, along with its modality name if it contains multiple; average accuracy per each client; combined accuracy across all clients; and the minimum client-wise gain over the `self-train` baseline. To ensure fair comparisons, we apply the same GNN architecture and parameter setup to all methods. Our analysis reveals several key observations.

Firstly, FL baselines show significant improvement over `self-train`, with an average relative gain of 15.34% across all clients. Notably, clients with smaller sample sizes, like BP, HIV, and PNC, experience the most substantial performance enhancement, with an average relative gain of 19.31%. This highlights the valuable effect of collaborative learning and cross-institutional knowledge generalization in overcoming model overfitting on limited training

Table 2. Performance for each client is averaged from 10-fold cross-validation, the combined performance is averaged across all clients. We highlight the best in bold and the runner-up underlined.

Clients	BP-fMRI	BP-DTI	HIV-fMRI	HIV-DTI	PPMI-PICo	PPMI-Hough	PPMI-FSL	PNC	ABIDE	ABCD		
Accuracy	average										combine	min gain
`self-train`	0.5463(±0.019)	0.5012(±0.082)	0.5286(±0.035)	0.4571(±0.140)	0.6394(±0.034)	0.6570(±0.054)	0.6852(±0.041)	0.5034(±0.052)	0.5025(±0.007)	0.5342(±0.002)	0.5555(±0.073)	–
`FedAvg`	0.6037(±0.073)	0.5158(±0.013)	0.5457(±0.153)	0.5000(±0.078)	0.7925(±0.002)	0.7633(±0.031)	0.7925(±0.002)	0.5434(±0.008)	0.5044(±0.012)	0.5167(±0.017)	0.6078(±0.118)	-0.032
`FedProx`	0.6084(±0.117)	0.5853(±0.085)	0.6200(±0.132)	0.6029(±0.097)	0.7925(±0.002)	0.7536(±0.037)	0.7925(±0.002)	0.6057(±0.018)	0.5594(±0.003)	0.5700(±0.020)	0.6490(±0.088)	0.067
`SCAFFOLD`	0.5800(±0.120)	0.6400(±0.049)	0.6343(±0.070)	0.6629(±0.057)	0.7778(±0.000)	0.7944(±0.014)	0.7889(±0.014)	0.6015(±0.009)	0.5765(±0.090)	0.5980(±0.045)	0.6654(±0.084)	0.120
FedBrain	**0.7389(±0.066)**	**0.7500(±0.077)**	**0.7857(±0.071)**	**0.8143(±0.070)**	**0.8102(±0.010)**	**0.8102(±0.010)**	**0.8095(±0.010)**	**0.7275(±0.044)**	**0.6549(±0.034)**	**0.7033(±0.033)**	**0.7605(±0.052)**	**0.214**

resources. Moreover, FL training also results in slight performance improvements on larger datasets, such as PPMI, ABIDE, and ABCD, underscoring the positive impact of a global optimization scheme in enhancing local performance. However, it is worth noting that among the chosen FL baselines, there is a slightly increased performance variance across clients, mainly due to underlying heterogeneity arising from the unique characteristics of brain network data.

Secondly, among all the selected FL baselines, `SCAFFOLD` stands out as the top performer, exhibiting an impressive average gain of 5.89% over its competitors. This result highlights the robustness of `SCAFFOLD` in addressing client heterogeneity through controlled gradient correction. Additionally, along with `FedProx`, which is also capable of handling data and system heterogeneity, the performance variance is reduced compared to `FedAvg`. This further aligns with our motivation to develop a specialized solution for reducing brain network-specific heterogeneity, which is aimed to unleash the full potential of collaborative learning, reflected through enhanced performance across multiple datasets at greater consistency.

Lastly, FedBrain outperforms `SCAFFOLD` by a relative margin of 14.29%, while also significantly reducing performance variance across clients, indicating the value of tailoring FL approaches to consider the unique properties and characteristics of brain network data. Moreover, FedBrain demonstrates statistically significant improvements over the compared baselines, as validated by passing the paired t-test with $p = 0.05$ in comparison to all methods.

Table 3. Atlas mapping comparisons.

Accuracy	average	min gain
No Atlas Mapping	0.6845(±0.068)	–
Atlas Mapping	0.7246(±0.063)	0.0039
Federated Atlas Mapping	0.7605(±0.052)	0.0214

Table 4. Guided clustering comparisons.

Accuracy	average	min gain
No Clustering	0.6921(±0.071)	–
Non-guided Clustering	0.7231(±0.065)	0.0000
Guided Clustering	0.7605(±0.052)	0.0000

4.2. *Ablation studies (RQ2)*

We analyze the two key components of FedBrain: federated atlas mapping and guided clustering. To highlight the contribution of each, we keep the best configuration of one component fixed while evaluating the other. The results are presented in Table 3 and Table 4, where we present an averaged performance across all clients. Regarding the analysis for atlas mapping, we investigate its impact on overall performance both without the entire module and without federated training. When atlas mapping is not applied, we add a learnable linear projection head to the client's GNN model that is excluded from the FL process. In general, we make two main observations: **(1)** Ensuring consistency in feature and network dimensions reflects in a relative gain of 6.12% compared to the uncompressed baseline. **(2)** Aligning the physical meanings of projected features further boosts performance by 4.95%, showcasing its effectiveness in countering incongruous ROI parcellation systems.

Regarding client clustering, we compare two scenarios: without clustering and without shared prior knowledge guidance. Our key observations are as follows: **(1)** Personalizing client optimization through similarity-based clustering leads to a significant enhancement in downstream performance, with a relative margin of 4.48%. **(2)** By integrating clinical prior knowledge and constraints, we further enhance cluster-specific learning and knowledge generalization, resulting in a relative gain of 5.17% and a reduction in performance variance.

4.3. *Heterogeneity analysis of federated atlas mapping (RQ3)*

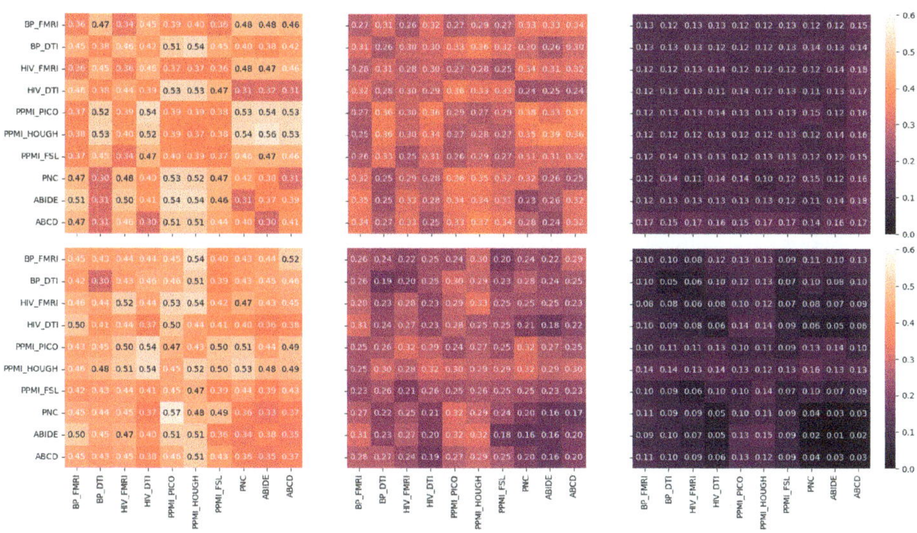

Fig. 1. Pairwise structure- (upper) and feature-level (lower) heterogeneity measures across all datasets compared on brain networks processed without atlas mapping (left), with atlas mapping but without federated training (mid), and full federated atlas mapping (right). The smaller the numeric measure, the less heterogeneity exists within the investigated pair.

To validate the contribution of the proposed federated atlas mapping in reducing structure- and feature-level heterogeneity, we employ two distinct quantitative metrics[30] to evaluate the averaged heterogeneity measure among brain networks across every pair of datasets. Firstly, regarding structure-level heterogeneity, we leverage the Anonymous Walk Embeddings (AWEs)[39] technique to generate representations for each brain network graph. We then calculate the Jensen-Shannon distance between every pair of AWE representations. Secondly, regarding feature-level heterogeneity, we analyze the empirical distribution of feature similarity between all pairs of linked nodes (ROIs) present in each graph. We then compute the Jensen-Shannon divergence between each pair of these distributions. We present our findings in Figure 1. Specifically, we compare the heterogeneity measures among brain networks and features processed under three scenarios: without federated atlas mapping, with atlas mapping but without federated training, and with full federated atlas mapping. Our observation suggests that atlas mapping along with federated training significantly reduces the level of heterogeneity across datasets in both network structures and ROI features.

In addition, we investigate the individual influence of the transformed network structure and ROI features on downstream performance. The summarized results can be found in Table 5. We observe that learning from either transformed network structures or ROI features leads to an average relative gain of 4.68% over the non-transformation baseline. The best performance is achieved when learning from both transformed structures and features, further validating the robustness of our design in reducing heterogeneity and enhancing task-wise performance simultaneously. Furthermore, we observe a significant reduction in time complexity when learned on transformed data. Given the original network and feature dimension N, a transformed dimension M ($M < N$), and a hidden size of F of the l-layer GNN model, the bounded complexity reduces from $O(l(N^2F + NF^2))$ to $O(l(M^2F + MF^2))$. Reflecting this to actual FL training with 80 communication rounds, the transformation reduces the time consumption from roughly 612 seconds to 266 seconds in completion time.

Table 5. Network transformation comparisons.

Transformation	average	min gain
None	0.6845(±0.068)	–
Structure	0.7042(±0.070)	-0.0126
Feature	0.7288(±0.060)	0.0357
Structure & Feature	0.7605(±0.052)	0.0417

Table 6. Cluster constraints comparisons.

Link	average	min gain
None	0.7231(±0.065)	–
Cannot	0.7337(±0.061)	0.0089
Must	0.7445(±0.057)	0.0148
Cannot & Must	0.7605(±0.052)	0.0235

4.4. *Clustering analysis of guided clustering (RQ4)*

We investigate the impact of the guided clustering approach on cluster formation. We focus on evaluating the effectiveness of this mechanism in grouping institutions (*i.e.*, clients) with similar neural circuitry patterns while also maintaining reasonable cluster sizes. We compare the outcomes with those obtained from the standard hierarchical clustering. We show a dendrogram visualization of the cluster results in Figure 2. Specifically, the linked branches depict the hierarchical relationships, with blue-colored lines representing singleton clusters, and other colors highlighting cluster assignments. Our observations indicate that incorporating clinical prior knowledge guidance substantially enhances the capability to identify and group clients with similar or near identical neural circuitry patterns. Our approach also avoids the produc-

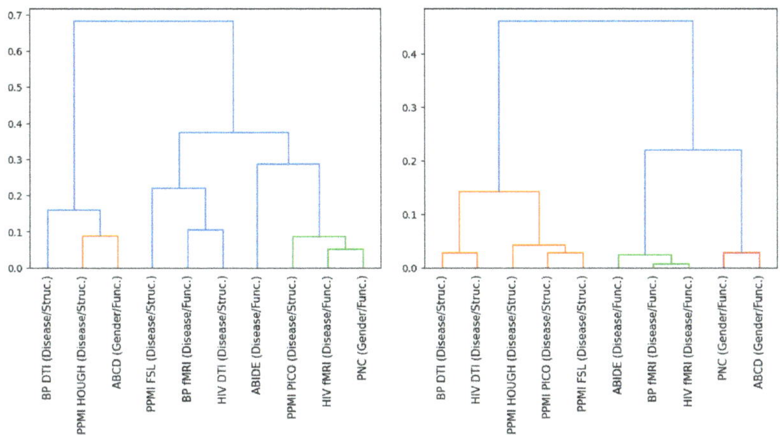

Fig. 2. Dendrogram visualization of cluster results from standard hierarchical clustering (left) and prior knowledge guided clustering (right). We list the client names alongside its clinical outcomes (*e.g.*, disease/gender) and data modalities (*e.g.*, functional/structural connectivities).

tion of singleton clusters, which were prominent when using the standard method.

Moreover, we study the impact on downstream performance when using clustering guidance that exclusively relies on either must- or cannot-link information. The results are presented in Table 6. We observe that sole cannot-link constraints lead to a relative gain of 1.47% over standard clustering. When guided by must-links alone, we achieve a further improvement of 1.53%, bringing the performance to within a mere 2.10% difference from considering both constraints. The findings suggest that must-link information plays a slightly more influential role in identifying similar neural circuitry patterns. On the other hand, cannot-link information proves valuable in averting additional intra-cluster heterogeneity, thereby reducing the likelihood of further cluster division and the formation of singleton clusters.

5. Conclusion

Cross-institutional brain network analysis has been a challenging task for conventional FL frameworks and GNN models. The presence of unique data heterogeneity, particularly in terms of inconsistent ROI parcellation systems and predictive neural circuitry patterns, poses a significant obstacle to effective collaborative training and knowledge generalization. To tackle these challenges, we propose FedBrain, a personalized GNN-based FL framework. Specifically, we leverage a data-driven atlas mapping mechanism to address the issue of incompatible ROI parcellation systems. Moreover, we incorporate clustered FL to enhance client personalization and integrate clinical prior knowledge to guide the clustering process. We conducted extensive experiments on multiple real-world brain network studies, demonstrating the superior performance of FedBrain compared to various state-of-the-art FL baselines.

We direct our future efforts to enhance FedBrain by addressing current limitations.

Firstly, we'll expand data considerations to include a wider array of atlas templates, clinical tasks, and clients with multi-modal data. Secondly, we'll optimize computational efficiency as the framework becomes more sophisticated. Thirdly, we'll delve into theoretical investigations to ensure strong privacy guarantees. Lastly, we plan to broaden empirical investigations by incorporating a broader set of data to validate the framework's robustness.

References

1. N. Yahata, J. Morimoto, R. Hashimoto, G. Lisi, K. Shibata, Y. Kawakubo, H. Kuwabara, M. Kuroda, T. Yamada, F. Megumi *et al.*, A small number of abnormal brain connections predicts adult autism spectrum disorder, in *Nat. Commun.*, (2016).
2. S. Wu, F. Sun, W. Zhang, X. Xie and B. Cui, Graph neural networks in recommender systems: a survey, in *ACM Comp. Surv.*, (2022).
3. W. Fan, Y. Ma, Q. Li, Y. He, E. Zhao, J. Tang and D. Yin, Graph neural networks for social recommendation, in *WWW*, (2019).
4. H. Cui, W. Dai, Y. Zhu, X. Kan, A. A. C. Gu, J. Lukemire, L. Zhan, L. He, Y. Guo and C. Yang, Braingb: A benchmark for brain network analysis with graph neural networks, in *IEEE TMI*, (2022).
5. X. Kan, H. Cui, J. Lukemire, Y. Guo and C. Yang, Fbnetgen: Task-aware gnn-based fmri analysis via functional brain network generation, in *MIDL*, (2022).
6. G. Luo, C. Li, H. Cui, L. Sun, L. He and C. Yang, Multi-view brain network analysis with cross-view missing network generation, in *IEEE BIBM*, (2022).
7. X. Kan, W. Dai, H. Cui, Z. Zhang, Y. Guo and C. Yang, Brain network transformer, in *NeurIPS*, (2022).
8. X. Li, Y. Zhou, N. Dvornek, M. Zhang, S. Gao, J. Zhuang, D. Scheinost, L. H. Staib, P. Ventola and J. S. Duncan, Braingnn: Interpretable brain graph neural network for fmri analysis, in *Medical Image Analysis*, (2021).
9. J. Kawahara, C. J. Brown, S. P. Miller, B. G. Booth, V. Chau, R. E. Grunau, J. G. Zwicker and G. Hamarneh, Brainnetcnn: Convolutional neural networks for brain networks; towards predicting neurodevelopment, in *NeuroImage*, (2017).
10. Y. Yang, H. Cui and C. Yang, Ptgb: Pre-train graph neural networks for brain network analysis, in *CHIL*, (2023).
11. Y. Yang, Y. Zhu, H. Cui, X. Kan, L. He, Y. Guo and C. Yang, Data-efficient brain connectome analysis via multi-task meta-learning, in *ACM SIGKDD*, (2022).
12. B. Cao, L. Zhan, X. Kong, P. S. Yu, N. Vizueta, L. L. Altshuler and A. D. Leow, Identification of discriminative subgraph patterns in fmri brain networks in bipolar affective disorder, in *Brain Informatics and Health*, (2015).
13. A. B. Ragin, H. Du, R. Ochs, Y. Wu, C. L. Sammet, A. Shoukry and L. G. Epstein, Structural brain alterations can be detected early in hiv infection, in *Neurology*, (2012).
14. R. A. Rossi and N. K. Ahmed, An interactive data repository with visual analytics, in *ACM SIGKDD Explorations Newsletter*, (2016).
15. B. McMahan, E. Moore, D. Ramage, S. Hampson and B. A. y Arcas, Communication-efficient learning of deep networks from decentralized data, in *PMLR*, (2017).
16. T. Li, A. K. Sahu, M. Zaheer, M. Sanjabi, A. Talwalkar and V. Smith, Federated optimization in heterogeneous networks, in *MLSys*, (2020).
17. S. P. Karimireddy, S. Kale, M. Mohri, S. Reddi, S. Stich and A. T. Suresh, Scaffold: Stochastic controlled averaging for federated learning, in *ICML*, (2020).
18. Q. Wu, X. Chen, Z. Zhou and J. Zhang, Fedhome: Cloud-edge based personalized federated learning for in-home health monitoring, in *IEEE TMC*, (2020).

19. M. Chen, W. Zhang, Z. Yuan, Y. Jia and H. Chen, Fede: Embedding knowledge graphs in federated setting, in *IJCKG*, (2021).
20. P. Veličković, G. Cucurull, A. Casanova, A. Romero, P. Lio and Y. Bengio, Graph attention networks, in *ICLR*, (2018).
21. K. Xu, W. Hu, J. Leskovec and S. Jegelka, How powerful are graph neural networks?, in *ICLR*, (2019).
22. T. N. Kipf and M. Welling, Semi-supervised classification with graph convolutional networks, in *ICLR*, (2017).
23. H. Cui, W. Dai, Y. Zhu, X. Li, L. He and C. Yang, Interpretable graph neural networks for connectome-based brain disorder analysis, in *MICCAI*, (2022).
24. Y. Zhu, H. Cui, L. He, L. Sun and C. Yang, Joint embedding of structural and functional brain networks with graph neural networks for mental illness diagnosis, in *IEEE EMBC*, (2022).
25. Y. Yu, X. Kan, H. Cui, R. Xu, Y. Zheng, X. Song, Y. Zhu, K. Zhang, R. Nabi, Y. Guo *et al.*, Learning task-aware effective brain connectivity for fmri analysis with graph neural networks, in *ISBI*, (2023).
26. R. Xu, Y. Yu, J. C. Ho and C. Yang, Weakly-supervised scientific document classification via retrieval-augmented multi-stage training, in *ACM SIGIR*, (2023).
27. A. Lalitha, O. C. Kilinc, T. Javidi and F. Koushanfar, Peer-to-peer federated learning on graphs, in *arXiv preprint*, (2019).
28. D. Caldarola, M. Mancini, F. Galasso, M. Ciccone, E. Rodolà and B. Caputo, Cluster-driven graph federated learning over multiple domains, in *CVPRW*, (2021).
29. C. He, K. Balasubramanian, E. Ceyani, C. Yang, H. Xie, L. Sun, L. He, L. Yang, S. Y. Philip, Y. Rong *et al.*, Fedgraphnn: A federated learning benchmark system for graph neural networks, in *ICLR-DPML*, (2021).
30. H. Xie, J. Ma, L. Xiong and C. Yang, Federated graph classification over non-iid graphs, in *NeurIPS*, (2021).
31. H. Xie, L. Xiong and C. Yang, Federated node classification over graphs with latent link-type heterogeneity, in *WWW*, (2023).
32. E. Darzidehkalani, M. Ghasemi-Rad and P. van Ooijen, Federated learning in medical imaging: part i: toward multicentral health care ecosystems, in *Journal of the American College of Radiology*, (2022).
33. X. Li, Y. Gu, N. Dvornek, L. H. Staib, P. Ventola and J. S. Duncan, Multi-site fmri analysis using privacy-preserving federated learning and domain adaptation: Abide results, in *Medical Image Analysis*, (2020).
34. F. Sattler, K.-R. Müller and W. Samek, Clustered federated learning: Model-agnostic distributed multitask optimization under privacy constraints, in *IEEE TNNLS*, (2020).
35. D. Aleksovski, D. Miljkovic, D. Bravi and A. Antonini, Disease progression in parkinson subtypes: the ppmi dataset, in *Neurol. Sci.*, (2018).
36. T. D. Satterthwaite, M. A. Elliott, K. Ruparel, J. Loughead, K. Prabhakaran, M. E. Calkins, R. Hopson, C. Jackson, J. Keefe, M. Riley *et al.*, Neuroimaging of the philadelphia neurodevelopmental cohort, in *Neuroimage*, (2014).
37. A. Di Martino, C.-G. Yan, Q. Li, E. Denio, F. X. Castellanos, K. Alaerts, J. S. Anderson, M. Assaf, S. Y. Bookheimer, M. Dapretto *et al.*, The autism brain imaging data exchange: towards a large-scale evaluation of the intrinsic brain architecture in autism, in *Molecular psychiatry*, (2014).
38. B. Casey, T. Cannonier, M. I. Conley, A. O. Cohen, D. M. Barch, M. M. Heitzeg, M. E. Soules, T. Teslovich, D. V. Dellarco, H. Garavan *et al.*, The adolescent brain cognitive development (abcd) study: imaging acquisition across 21 sites, in *Dev. Cogn. Neurosci.*, (2018).
39. S. Ivanov and E. Burnaev, Anonymous walk embeddings, in *ICML*, (2018).

Drug-repurposing and discovery in the era of "big" real-world data: how the incorporation of observational data, genetics, and other -omic technologies can move us forward

Megan M. Shuey
Division of Genetic Medicine, Department of Medicine, Vanderbilt Genetics Institute
Vanderbilt University Medical Center
2525 West End Ave. Ste 700, Nashville, TN, 37203, USA
Email: megan.m.shuey.1@vumc.org

Jacklyn N. Hellwege
Division of Genetic Medicine, Department of Medicine, Vanderbilt Genetics Institute
Vanderbilt University Medical Center
2525 West End Ave. Ste 700, Nashville, TN, 37203, USA
Email: jacklyn.hellwege@vumc.org

Nikhil Khankari
Division of Genetic Medicine, Department of Medicine, Vanderbilt Genetics Institute
Vanderbilt University Medical Center
2525 West End Ave. Ste 700, Nashville, TN, 37203, USA
Email: nikhil.khankari@vumc.org

Marijana Vujkovic
Division of Translational Medicine and Human Genetics
University of Pennsylvania Perelman School of Medicine
3400 Civic Center Blvd, Philadelphia, PA 19104, USA
Email: vujkovic@pennmedicine.upenn.edu

Todd L. Edwards
Division of Epidemiology, Vanderbilt Genetics Institute, Vanderbilt University Medical Center
2525 West End Ave. Ste 600, Nashville, TN, 37203, USA
Email: todd.l.edwards@vumc.org

This PSB 2024 session discusses the many broad biological, computational, and statistical approaches currently being used for therapeutic drug target identification and repurposing of existing treatments. Drug repurposing efforts have the potential to dramatically improve the treatment landscape by more rapidly identifying drug targets and alternative strategies for untreated or poorly managed diseases. The overarching theme for this session is the use and integration of real-world data to identify drug-disease pairs with potential therapeutic use. These drug-disease pairs may be identified through genomic, proteomic, biomarkers, protein interaction analyses, electronic health records, and chemical profiling. Taken together, this session combines novel applications of methods and innovative modeling strategies with diverse real-world data to suggest new pharmaceutical treatments for human diseases.

Keywords: drug repurposing, drug repositioning, -omics, machine learning, genetics, translational science

---

1. Drug development and repurposing

Drug discovery and development is a long and high-risk process with cumulative annual costs approaching $1 billion US dollars (Hinkson et al. 2020)·(Wouters et al. 2020), where over 85% of drug candidates will fail prior to completing clinical trials (Nielsch et al. 2016). Drug repurposing or repositioning of existing medications for new therapeutic uses can substantially reduce costs, time, and effort while providing additional treatment options to patients.

The increasing availability of large-scale electronical medical record (EMR) data, in combination with genomic, proteomic, molecular, and other biomedical data is enabling more cost-effective investigations of treatment response, adverse event profiling, and novel target identification. The use of "real-world data" presents a promising solution with the potential to dramatically reduce development time and cost. Furthermore, policy makers in the US and other countries are increasingly open to considering alternative sources of evidence beyond clinical trials in their decision-making processes. For instance, the 21st Century Cures Act encourages the use of real-world data to generate evidence of product effectiveness to help support approval of new indications for existing drugs (Dagenais et al. 2022).

Traditionally these sources encompass data from hospital or population-based health records, third-party health insurance claims, registries, and health surveys (Administration 2018). These data types are increasingly linked to novel types of biomedical data, such as genomic (or other "omic") data from large biobanks, biopsies, pathology tests, diagnostic imaging, and information related to social determinants of health (SDoH). Initiatives focused on data-driven drug repurposing, leveraging the expansive resources available within real-world data repositories have the potential to improve the efficiency of identifying potential treatments, while simultaneously reducing possible risks associated with drug development.

Computational approaches, such as machine learning, offer a powerful avenue to address specific challenges in drug development by harnessing the wealth of multi-dimensional data from various sources. For example, prior research has demonstrated the ability of machine learning algorithms to scan compound libraries to optimize the design of small molecules and evaluate molecular docking to estimate drug-target interactions, and use this to find repurposing targets for viral infections and cancer (Kumar et al. 2015; Mirza et al. 2016; Wang et al. 2017).

Phenotype-first approaches used in conjunction with machine learning are yet another example of identifying targets for drug repurposing. This methodology identifies a set of optimal treatment modalities using medical record history which has demonstrated increased efficacy of clinical conditions. This approach capitalizes on the growing availability of EMR data to evaluate acute and long-term therapeutic response based on individual-level, real-world clinical data. These deep-learning computational approaches can be coupled with genetics and advanced -omics data to elucidate the underlying mechanisms of disease. When combined with available drug-target datasets, this information can facilitate the identification of alternative treatment strategies (Allen et al. 2015; MacEachern and Forkert 2021; Wang et al. 2021; Xu et al. 2022).

Similarly, novel computational approaches that leverage genomic and transcriptomic methodologies, including but not limited to genome-wide association studies, genetically predicted gene expression analysis, and Mendelian randomization have the potential to identify and estimate the effect of drug repurposing on reducing risk of disease. These approaches are particularly appealing, given that drugs with genetic evidence from disease association studies have a two-fold higher likelihood of successfully reaching the market (Nelson et al. 2015; King et al. 2019).

The research team encompassing this panel has experience in developing such computational pipelines for identification of potential drug candidates for repurposing in diabetes treatment (Khankari et al. 2022; Shuey et al. 2022). This approach specifically uses a transcriptome-driven drug screening approach to identify candidate therapeutics. Subsequently, it validates these candidates through a two-step process by: 1) generating real-world evidence for drug efficacy using a self-controlled case series study design using large EMR datasets and quantifying changes in disease-associated biomarkers before and after treatment with identified candidates, and 2) generate genetic evidence for drug target efficacy for disease using the Mendelian randomization framework. We encourage participation in this series by other researchers who are involved in the development of strategies to aid in the identification and evaluation of drug repurposing opportunities.

2. Session contents

Here we describe briefly studies which will be presented during the session.

2.1. *List of topics captured in this session*

Our session includes presentations on the following diverse topics related to drug repurposing and discovery:

- Modeling of outcome risk based on medication exposure using propensity score matching
- Improved techniques for target identification from sequencing data
- Machine learning modeling of disease and protein interaction networks (???)
- High-throughput functional screening assays

2.2. *Systematic Estimation of Treatment Effect on Hospitalization Risk as a Drug Repurposing Screening Method*

In this manuscript, *Georgantas et al.* propose a simple, pragmatic screening approach for drug repurposing using real-world data. They incorporated time-to-event and propensity score matching with observational data from the UK Biobank to evaluate the roles of thousands of drug-disease pairs on hospitalization risk. This elegant use of high-dimensional real-world data suggests numerous repurposing opportunities for existing, commonly prescribed medications.

2.3. *Transcript-aware analysis of rare predicted loss-of-function variants in the UK Biobank elucidate new isoform-trait associations*

As whole exome and genome sequencing becomes more widely accessible, the ability to synthesize these results into meaningful discoveries is essential. Traditional burden testing approaches assume that all variants in a given gene have similar effects on gene function and fails to consider isoforms where this assumption is often violated. *Hoffing et al.* demonstrates how using transcript-specific annotations (rather than collapsed gene-based evaluations) to classify rare predicted loss-of-function (pLOF) mutations can dramatically impact effect estimates for rare variant association analyses. Their work links such pLOFs to tissue specificity, quantitative endophenotypes, and disease outcomes and has a distinct outcome for improving the outputs of such large-scale sequencing data for drug target identification. The results of this study have the potential to improve accuracy of rare variant-disease association studies that often serve to identify novel drug targets.

2.4. *Formulating new drug repurposing hypotheses using disease-specific hypergraphs*

In *Jain et al.*, the authors use disease-specific hypergraphs in which hyperedges of various lengths encode biological pathways to generate new repurposing targets which may be overlooked by classic knowledge graphs. These low-dimensional representations of drug-to-gene pathways are filtered to existing therapeutic approaches for Alzheimer's Disease and then evaluated using the multiscale interactome (MSI). Further, the seven targets not represented in MSI were evaluated by literature review, with many of these candidates having demonstrable impacts on brain development or disease processing that support a relationship with Alzheimer's Disease.

2.5. *Combined kinome inhibition states are predictive of cancer cell line sensitivity to kinase inhibitor combination therapies*

Kinase inhibitors are a staple in clinical oncology; however, monotherapy may lead to resistance in part due to compensation by other members of the kinase network or kinome. Combinatorial therapies have been suggested to combat this resistance. However, determining the best combination of kinase inhibitors is essential. To this end, *Joisa et al.* developed a high-throughput platform for evaluating combinatorial effects of multiple kinase inhibitors. By leveraging heterogenous data for the prediction of potential drug combination targets the authors identified the combination of MEK and PI3K inhibitors (Trametinib/Omipalisib). Their results are supported by this particular combination of inhibitors entering a recent phase 1 clinical trial which suggesting the potential for this method to identify other combinatorial therapies.

2.6. *The Human Protein Structure Targetome*

Ovanessians et al. utilized structure-based modeling of proteins for more than 20,000 human proteins curated from various protein databases to build a human "targetome". This approach was developed to prioritize protein-ligand pairs and accounts for the complexities of both protein structure and binding site affinities to prioritize drug targeting. The potential of this pipeline and strategies like this have the potential to advance drug design and development efforts by not only prioritizing candidates but informing various considerations in the drug development pipeline including competitive binding estimation.

2.7. *Modeling Path Importance for Effective Alzheimer's Disease Drug Repurposing*

The final manuscript in this session by *Xiang et al.* presents a modeling schema focused on building a large-scale protein-protein interaction network from various data sources. Their approach incorporated both available data about protein-protein interactions and existing drug-target interactions to develop a rich data resource for prioritization of biological systems, e.g. networks and pathways. Their models captured a network's rich topology and challenges the assumption that paths of equal length have equivalent importance in biological systems. Results were further supported by the prioritization of several drug candidates that are supported by previous publications and insurance claims data.

3. Conclusion

The authors in this session present six diverse papers that discuss methodologic improvements to guide potential drug discovery and repurposing. The session expands upon the application of commonly used techniques like improving prediction of loss-of-function mutations for target identification as well as modeling strategies using genetic data to evaluate medication exposure and outcomes. There is also a special emphasis on using machine learning techniques and available datasets to identify drug targets by considering disease, protein, and kinase interactions. We anticipate that these studies, results, and associated techniques can advance disease-specific target evaluation and drug repurposing.

Acknowledgements

JHN and MMS were supported by K12-HD043483. MV, TLE, and NK are supported by R01-DK134575.

References

Administration USFaD. 2018. Framework for FDA's Real World Evidence Program.
Allen BK, Mehta S, Ember SW, Schonbrunn E, Ayad N, Schurer SC. 2015. Large-Scale Computational Screening Identifies First in Class Multitarget Inhibitor of EGFR Kinase and BRD4. *Sci Rep* **5**: 16924.
Dagenais S, Russo L, Madsen A, Webster J, Becnel L. 2022. Use of Real-World Evidence to Drive Drug Development Strategy and Inform Clinical Trial Design. *Clin Pharmacol Ther* **111**: 77-89.
Hinkson IV, Madej B, Stahlberg EA. 2020. Accelerating Therapeutics for Opportunities in Medicine: A Paradigm Shift in Drug Discovery. *Front Pharmacol* **11**: 770.
Khankari NK, Keaton JM, Walker VM, Lee KM, Shuey MM, Clarke SL, Heberer KR, Miller DR, Reaven PD, Lynch JA et al. 2022. Using Mendelian randomisation to identify opportunities for type 2 diabetes prevention by repurposing medications used for lipid management. *EBioMedicine* **80**: 104038.

King EA, Davis JW, Degner JF. 2019. Are drug targets with genetic support twice as likely to be approved? Revised estimates of the impact of genetic support for drug mechanisms on the probability of drug approval. *PLoS Genet* **15**: e1008489.

Kumar V, Krishna S, Siddiqi MI. 2015. Virtual screening strategies: recent advances in the identification and design of anti-cancer agents. *Methods* **71**: 64-70.

MacEachern SJ, Forkert ND. 2021. Machine learning for precision medicine. *Genome* **64**: 416-425.

Mirza SB, Salmas RE, Fatmi MQ, Durdagi S. 2016. Virtual screening of eighteen million compounds against dengue virus: Combined molecular docking and molecular dynamics simulations study. *J Mol Graph Model* **66**: 99-107.

Nelson MR, Tipney H, Painter JL, Shen J, Nicoletti P, Shen Y, Floratos A, Sham PC, Li MJ, Wang J et al. 2015. The support of human genetic evidence for approved drug indications. *Nat Genet* **47**: 856-860.

Nielsch U, Fuhrmann U, Jaroch S. 2016. New Approaches to Drug Discovery. In *Handbook of Experimental Pharmacology,*, doi:10.1007/978-3-319-28914-4, pp. 1 online resource (VIII, 341 pages 100 illustrations, 363 illustrations in color. Springer International Publishing : Imprint: Springer,, Cham.

Shuey MM, Lee KM, Keaton J, Khankari NK, Breeyear JH, Walker VM, Miller DR, Heberer KR, Reaven PD, Clarke SL et al. 2022. A genetically supported drug repurposing pipeline for diabetes treatment using electronic health records. *medRxiv* doi:10.1101/2022.12.14.22283414: 2022.2012.2014.22283414.

Wang Y, Lin HQ, Wang P, Hu JS, Ip TM, Yang LM, Zheng YT, Chi-Cheong Wan D. 2017. Discovery of a Novel HIV-1 Integrase/p75 Interacting Inhibitor by Docking Screening, Biochemical Assay, and in Vitro Studies. *J Chem Inf Model* **57**: 2336-2343.

Wang Y, Yang Y, Chen S, Wang J. 2021. DeepDRK: a deep learning framework for drug repurposing through kernel-based multi-omics integration. *Brief Bioinform* **22**.

Wouters OJ, McKee M, Luyten J. 2020. Estimated Research and Development Investment Needed to Bring a New Medicine to Market, 2009-2018. *JAMA* **323**: 844-853.

Xu J, Mao C, Hou Y, Luo Y, Binder JL, Zhou Y, Bekris LM, Shin J, Hu M, Wang F et al. 2022. Interpretable deep learning translation of GWAS and multi-omics findings to identify pathobiology and drug repurposing in Alzheimer's disease. *Cell Rep* **41**: 111717.

Systematic Estimation of Treatment Effect on Hospitalization Risk as a Drug Repurposing Screening Method

Costa Georgantas[†1], Jaume Banus[1], Roger Hullin[2] and Jonas Richiardi[1]
[1]*Department of Radiology, Lausanne University Hospital and University of Lausanne, Lausanne, Switzerland*
[2]*Department of Cardiology, Lausanne University Hospital, Lausanne, Switzerland*
[†]*E-mail: costa.georgantas@chuv.ch*

Drug repurposing (DR) intends to identify new uses for approved medications outside their original indication. Computational methods for finding DR candidates usually rely on prior biological and chemical information on a specific drug or target but rarely utilize real-world observations. In this work, we propose a simple and effective systematic screening approach to measure medication impact on hospitalization risk based on large-scale observational data. We use common classification systems to group drugs and diseases into broader functional categories and test for non-zero effects in each drug-disease category pair. Treatment effects on the hospitalization risk of an individual disease are obtained by combining widely used methods for causal inference and time-to-event modelling. 6468 drug-disease pairs were tested using data from the UK Biobank, focusing on cardiovascular, metabolic, and respiratory diseases. We determined key parameters to reduce the number of spurious correlations and identified 7 statistically significant associations of reduced hospitalization risk after correcting for multiple testing. Some of these associations were already reported in other studies, including new potential applications for cardioselective beta-blockers and thiazides. We also found evidence for proton pump inhibitor side effects and multiple possible associations for anti-diabetic drugs. Our work demonstrates the applicability of the present screening approach and the utility of real-world data for identifying potential DR candidates.

Keywords: Drug repurposing; Propensity score matching; Cox regression; Real-world data

1. Introduction

Drug discovery is a rarely successful and extremely costly process that can span decades before commercialization. Drug repurposing (DR), or re-utilizing an existing medication for another use, has the potential to cut down the cost of development by a factor of 10.[1] DR is still dependent on clinical trial success and only approximately 30% of repurposed drugs go from phase I to market,[2] a process that can take multiple years. The majority of trials fail due to insufficient efficacy or the existence of other superior alternatives. Computational methods can reduce the chances of trial failure by selecting candidates that are likely to succeed and have already resulted in the identification of approved medications and promising candidates.[3,4]

A large number of computational DR approaches attempt to identify drug-disease associations by utilizing molecule structure, common pathways, or other known biological properties.[5] Signature matching and molecular docking use structural and chemical properties of molecules to identify similar drugs and therapeutic targets.[6] Other approaches use genome-wide summary statistics or biological pathway information to identify causal genes and new potential targets.[7] These methods generally attempt to use known information on the drug or disease in question to infer new treatment options.

Alternatively, electronic health records (EHRs) were used to identify potential alternative treatment targets based on documentation of side effects and clinical events.[8,9] Eguale et al.[10] used EHR data and Cox regression to associate off-label drug use with adverse drug events, Wu et al. have recently proposed another type of screening method using EHR records for the identification of drug-disease interactions.[11] Similarly, UK Biobank data has also been used to identify relations between treatment and phenotype, although these approaches generally focus on a specific phenotype and treatment pair. For instance, Ma et al.[12] used Cox regression in UK Biobank data to identify the benefits of glucosamine for type 2 diabetes. Pilling et al.[13] also used time-to-event modelling in UK Biobank to link lower vitamin D levels and hospitalization for delirium. Wu et al.[14] used PSM in UK Biobank for cost-benefit analysis of bariatric surgery.

Nevertheless, utilizing real-world data to isolate the effect of medication has proven challenging as this approach is highly prone to bias with the risk of creating spurious associations.[15] Indeed, in observational data, the characteristics of the treatment group are often very different from the average clinical study population. Propensity score matching (PSM)[16,17] is a statistical matching technique that attempts to associate subjects of the treatment group with similar subjects from the rest of the cohort to form a control group. Matched subjects have similar characteristics (as measured by selected covariates), limiting the impact of confounders in the estimation of the treatment effect. When time information of events is available, PSM can be combined with survival methods such as Cox regression[18] to estimate the relative risk between the treatment and control arms.[19]

In this work, we propose to model the risk of hospitalization w.r.t treatment for a large number of combinations of drugs and diseases. We effectively attempt to emulate thousands of clinical trials with hospitalization risk reduction as the endpoint. Our methodology is akin to genome-wide association studies (GWAS), in which a simple model is used to estimate the effect of a large number of loci in a hypothesis-free manner. As in GWAS, this form of drug-disease association study faces the risk of creating spurious relationships and requires further analysis, but can be seen as complementary to target-driven repurposing.[20] We applied our method to thousands of drug-disease pairs and showed that we can successfully re-identify associations that are already reported in UK Biobank, other observational cohorts, or controlled clinical trials. To the best of our knowledge, this is the first attempt to apply this type of systematic approach for treatment effect modelling.

2. Methods

2.1. *Dataset*

The UK Biobank[21] (UKBB) is a large observational dataset containing information on approximately 500K subjects over decades. During their initial visit to the UK Biobank assessment center, participants were interviewed about their medication use and completed a detailed questionnaire presenting questions on everyday habits, medical history, and mental health among others. A total of 1,233,630 treatments were reported, spanning 6745 different medications. Other biomarkers such as body mass index (BMI), blood pressure, and grip strength were also measured. Moreover, since the beginning of the study, more than 6 million hospitalization events were recorded in the form of an event date and a corresponding international classification of disease code (ICD10).

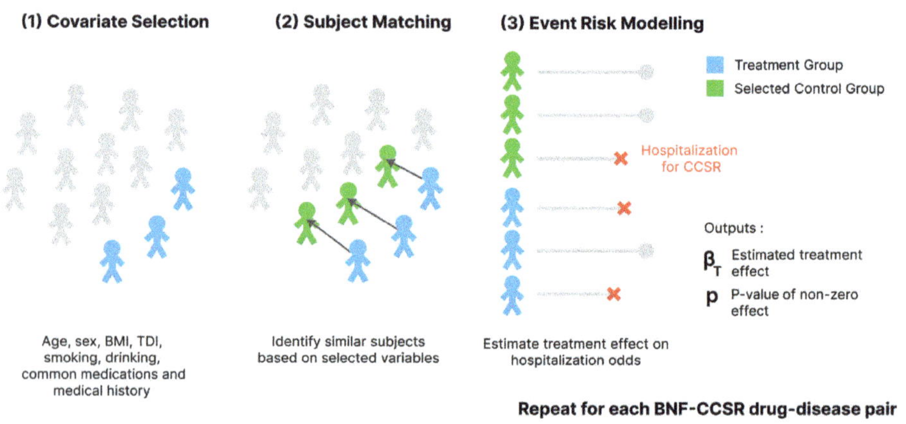

Fig. 1. Overview of the proposed method. It is composed of three main steps and is repeated for each BNF-CCSR drug-disease pair. (1) We use common comorbidities as covariates. Additionally, we included drugs and medical history, also respectively coded as BNF and CCSR, if they were present in more than 20% of the treatment group. We also remove subjects with a history of the CCSR code in question. (2) We use propensity score matching to find a similar non-treated subject for each subject of the treatment group, based on the selected covariates. (3) We use Cox regression, a proportional hazard ratio method, to estimate the treatment effect on hospitalization for the corresponding CCSR.

2.2. *Medication and Disease Selection*

ICD10 is a medical classification list from the World Health Organization used by many health organizations around the world that contains codes for over 70K diseases and symptoms. Some ICD10 codes represent similar phenotypes, for instance, I50.0, I50.1, and I50.9 correspond to congestive, left ventricular, and unspecified heart failure respectively. Using individual ICD10 codes for our analysis would be challenging due to the small number of events for each code, so we grouped them using Clinical Classifications Software Refined (CCSR)[22] v2023.1. CCSR is a classification system developed by the US Agency for Healthcare Research and Quality's Healthcare Cost and Utilization Project, that aggregates codes into clinically meaningful categories. We considered all CCSR categories spanning diseases of the circulatory system (CIR), endocrine, nutritional, and metabolic diseases (END) and symptoms, signs, and abnormal clinical and laboratory findings (SYM), as well as some others from diseases of the respiratory system (RSP), genitourinary system (GEN) and nervous system (NVS), totaling 77 phenotypes encompassing 3650 ICD10 codes.

Most of the medication types recorded in the UK Biobank dataset have very low frequency. Additionally, it is common for equivalent or similar active compounds to have different names, and no hierarchy is provided. To organize this data in a meaningful way, we mapped each medication to a corresponding British National Formulary (BNF) code. This code structure is used by the UK's National Health Service (NHS) to assign codes to drugs and chemicals and provides a fine-grained classification based on functionality. We used existing software[23] to map 3500 UKBB medications to 151 BNF codes. We only considered codes with at least 1000 subjects in the treatment group (power analysis with hazard ratio 0.6 and 80% power), resulting in 84 total BNF codes for analysis that include 93% of all reported medications in the UK Biobank during the first visit.

2.3. *Covariate and Subject Selection*

Medications (represented by BNF codes) and diseases (represented by CCSR codes) pairs were evaluated independently and the covariate and subject selection process was repeated for each pair. In total, we examined 6468 medication-disease pairs. We selected subjects from all available 500K participants who did not have a history of the CCSR code in question. Covariates can have a large impact on the estimated treatment effect and should be chosen carefully. In an attempt to be as general as possible, we used common demographics and risk factors: sex, age, BMI, Townsend deprivation index (TDI)[24] (related to poverty), smoking (current) and drinking habits (three times a week or more) as common covariates for all associations. For computational reasons, we capped the maximum number of subjects in the treatment group to 30,000, randomly sub-sampling when necessary.

To produce more precise matching and allow for more potential confounders, we also added medical history and medications as covariates if they were present in more than 20% of the treatment group. This percentage was evaluated for each individual drug-disease pair. Medical history was composed of both self-reported items and ICD10s prior to the first visit, grouped by CCSR coding. UKBB self-reported disease codes have their own representation, which were mapped to ICD10 and then to CCSR. Other medications were also selected by their corresponding BNF codes. This method of covariate selection has the advantage of being agnostic to the type of medication being considered. We used the same covariates for propensity score matching and Cox regression in all experiments.

2.4. *Propensity Score Matching and Pair Exclusion*

PSM consists in finding similar subjects in the control and treatment groups. This is done by fitting a logistic model and finding pairs of subjects that have the same probability of being in the same group. The unmatched subjects from the control group are then discarded. We used nearest neighbor distance as our matching method, and PSM was implemented in R using the matchit[25] package. PSM enables the estimation of the average effect of treatment in the treated individuals (ATT). In contrast to the average treatment effect (ATE), the ATT represents the effect of the drug on the treatment group, rather than the average population. As most drugs would not have any beneficial effect on a healthy population, we expect the effect of drugs for subjects that are already likely to be on treatment to be a more informative measure.

Despite still being widely used in retrospective studies, PSM has been criticized in the past[26] for potentially increasing imbalance between treatments and controls. However, this imbalance increase is only observed when groups are balanced initially, which was not the case in our experiments. Some alternatives to PSM, such as inverse probability of treatment weighting (IPTW) and Mahalanobis distance matching (MDM) were not considered due to their computational cost. In practice, we found PSM to produce balanced groups with minimal parameter tuning, and to be much more computationally efficient than other tested alternatives.

Additionally, unknown variables can bias the estimation of the treatment effect, to the point that the opposite effect can become statistically significant. This issue is not exclusive to PSM, and we observed that choosing the appropriate covariates was generally more impactful than the matching method itself. In some cases, the assignment of the treatment can deterministically depend on other variables, resulting in a lack of observation in the control group. Since PSM can introduce spurious relations between treatment and controls, careful interpretation of the treatment effects is always required. We report the number of balanced and unbalanced covariates for each pair in the summary statistics (mean standardized difference < 0.1).

We found that in some medication-disease pairs, some matched treatments and controls would be extremely dissimilar. Despite the large number of controls, it was simply not possible to match some subjects in the treatment groups in some cases. As an example, extremely morbidly obese subjects are almost always on the same medications. To address this issue, we computed the Huang distance[27] between each paired subject and discarded the pairs above an arbitrary threshold. The Huang distance was computed using both binary and normalized continuous covariates, treating CCSR history, sex, alcohol, and smoking habits as binary variables and the rest as continuous. In practice, we found only marginal improvements when excluding large-distance pairs.

2.5. *Cox Regression*

The Cox proportional hazard model[18] is a semi-parametric regression technique that estimates a relative hazard function, which represents a proportional risk of an event happening at time t.

The hazard function is of the form:

$$\lambda(t|\boldsymbol{X}_i) = \lambda_0(t)exp(\boldsymbol{X}_i \cdot \boldsymbol{\beta}) \tag{1}$$

where $\boldsymbol{X}_i$ represents the covariate vector for sample i, $\boldsymbol{\beta}$ are tunable regression coefficients also referred to as effect sizes and λ_0 is some common unknown hazard function that vanishes when estimating hazard ratios. Instead of binary categories, we considered a subject right-censored if no event had yet happened to that subject.

We fitted a separate model for each drug-disease pair. We considered all hospitalizations resulting in an ICD10 code contained in the CCSR category of choice as an event and only considered the first event if a subject had multiple events with the same CCSR code. Following the advice of Peter Austin,[19] we used a robust variance estimator and did not stratify on the matched sets. The output of the Cox regression is a treatment effect estimate β_T and a corresponding P-value for the null hypothesis of a zero effect for the drug-disease pair. The Cox regression was implemented in R, using the survival[28] package.

Using only the information from the assessment center, we could not consider how long subjects had been on treatment, neither how long they would stay on it, nor the medication dosage. We also could not measure if subjects changed treatment over time. To attempt to minimize the impact of some of these limitations, we only considered events that happened before a given number of years after the assessment center visit and varied this time event window to 1, 3, 5, and 10 years. We also experimented with maximum pair Huang distances of 1, 2, 3, and no cut-off. Finally, we evaluated the impact of including common medical history and/or medications in the treatment group as covariates. A graphical overview of the method is presented in Figure 1.

3. Results

We applied our method to 77 disease categories and 84 medication types, resulting in 6468 potential drug-disease associations. Our results are reported in Figure 2. When comparing negative and positive associations, we observed a clear bias towards unfavorable effects ($\beta_T > 0$, corresponding to increased risk of hospitalization and hazard ratio greater than 1) for all parameters, although some configurations are less biased than others. Since these medications have been thoroughly tested for safety and side effects, we expect this ratio to be more balanced. We attribute this imbalance in significant associations to a failure to find appropriate matches in PSM, making the control group systematically healthier than the treatments.

We found that the bias towards unfavorable effects did not vanish when reducing the pair Huang distance cut-off, implying that this discrepancy is due to non-observable variables. When inspecting significant associations, we found that drugs that were already used as a treatment for a CCSR category were consistently associated with a higher hospitalization risk for the same CCSR. Our explanation for this observation is that treatment was prescribed to high-risk subjects without any hospitalization event or self-report, making the treatment group inherently more at risk than matched controls.

As an example, anti-diabetic drugs (BNF 6.1.2) were consistently associated with a higher risk of diabetes (CCSR END002). This is likely due to the fact that we could not match for pre-diabetes effectively, and thus the treatment group was much more likely to end up diabetic than

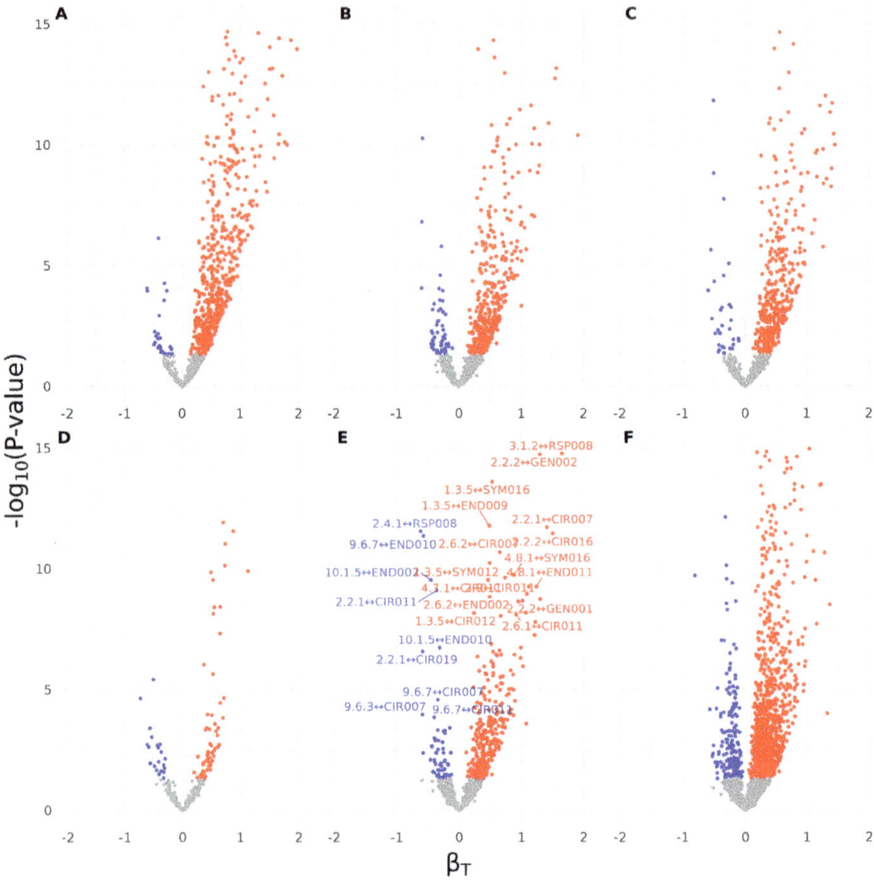

Fig. 2. Volcano plots of effect estimates for each drug-disease pair, spanning 84 medication categories and 77 phenotypes for multiple parameter choices. β_T: Cox regression coefficient; positive values indicate unfavorable effects (increased hazard ratio for hospitalization). Non-significant ($p > 0.05$) associations are reported in grey. Medications associated with a reduced or increased risk of hospitalization for the corresponding disease are reported in blue and red, respectively. **A**: Matching only with common covariates: sex, age, BMI, TDI, smoking, and drinking habits with an event time limit of 3 years. **B**: Matching with common covariates and medical history with an event time limit of 3 years. **C**: Matching with common covariates and other medications with an event time limit of 3 years. **D**: Matching with common covariates, medical history, and other medications with a maximum event time limit of 1 year post-visit. **E** : 3 year limit. **F** : 10 year limit.

the matched controls. Based on the previous results, we chose the combination of covariates, time, and Huang distance cut-offs that would result in the most balanced number of total associations (Maximum Huang distance of 3, Maximum time-to-event of 3 years, and including both medication and medical history). We used this configuration for all further analysis. Based on power analysis (60% hazard ratio and 80% power), we automatically discarded drug-disease pairs that included less than 100 events. We were able to estimate effects for 1013 pairs, and we used this number for correcting for multiple testing.

As we estimate the ATT, the correct interpretation of these measurements is that treated subjects would have a different risk of the corresponding CCSR code had they not taken the treatment, after correcting for all other known covariates. The root cause of this risk reduction cannot be inferred, and additional analysis is always required to determine the clinical relevance of this measured effect. Similarly, our method is also capable of measuring side effects that manifest as increased risks of hospitalization. We report all statistically significant effects after correcting for multiple testing (Bonferroni correction on the number of drug-disease pairs tested) in table 1, ordered by statistical significance. We expand further on each pairing in the next sub-sections.

Table 1. All statistically significant medications associated with a reduced risk of hospitalization for the corresponding disease ($p < 5 \cdot 10^{-5}$, after Bonferroni correction for multiple comparisons), ordered by P-value.

BNF	CCSR	Medication	Disease	Hazard Ratio	P-value
2.4.1	RSP008	Cardioselective Beta-blockers	COPD and bronchiectasis	0.54	2.9e-12
9.6.7	END010	Multivitamins	Disorders of lipid metabolism	0.56	4.5e-12
10.1.5	END002	Glucosamine	Diabetes mellitus	0.63	3e-10
2.2.1	CIR011	Thiazides	Coronary atherosclerosis	0.69	7.9e-10
10.1.5	END010	Glucosamine	Disorders of lipid metabolism	0.73	1.9e-07
2.2.1	CIR019	Thiazides	Heart failure	0.55	2.8e-07
9.6.7	CIR007	Multivitamins	Essential hypertension	0.7	2.7e-05

3.1. *Cardioselective Beta-blockers and COPD*

Beta-adrenergic blocking agents or beta-blockers (BNF 2.4) used for COPD (CCSR RSP008) constitute one of our most significant positive drug-disease pairs ($\beta_T = -0.56, p = 10^{-10}$) with a corresponding hazard ratio for hospitalization risk of 0.57. Historically, the use of beta-blockers was discouraged for COPD as non-selective beta-blockers can reduce lung function.[29] Nevertheless, several retrospective observational studies have shown that usage of beta-blockers can reduce mortality and other exacerbations in COPD.[30,31]

The beta-blocker BNF encoding does not separate between cardioselective and non-selective compounds, so we split this category into two, 2.4.1 and 2.4.2 for cardioselective and non-cardioselective beta-blockers respectively. We found a stronger effect and smaller P-value ($\beta_T = -0.62$, $p = 2.9 \cdot 10^{-12}$) for category 2.4.1 w.r.t 2.4 while results of non-selective beta-blockers were not significant; this corroborates recent observational findings.[32,33] Thus, our results agree with the consensus that cardioselective beta-blockers are not only safe for patients at risk of COPD but could also reduce their risk of hospitalization.[34] The effect of cardioselective beta-blockers for patients with COPD is the subject of an ongoing phase IV clinical trial (NCT03566667).[35]

3.2. *Glucosamine and Diabetes Mellitus*

Glucosamine is a widely used supplement for osteoarthritis that is often taken daily and has anti-inflammatory properties. While glucosamine has been shown to induce insulin resistance in rodents[36] this effect does not appear to be present in humans.[37] Nevertheless, similarly to our findings, another recent UK Biobank study also showed the potential of glucosamine for the prevention of diabetes.[12] Since glucosamine does not impact blood sugar levels, glucose tolerance, or insulin resistance, this effect is likely not direct. However, there is an established relation between inflammation and the occurrence of diabetes,[38] and even support for inflammatory pathways to be involved in its pathogenesis.[39] The anti-inflammatory properties of glucosamine and the reduction of symptoms of arthritis might explain its apparent benefits for diabetes. We also found a reduced risk of hospitalization for disorders of lipid metabolism, a CCSR category that includes different types of hypercholesterolemia and hyperlipidemia (corresponding to ICD10 E78). Thus, long-term glucosamine supplementation might be beneficial for the prevention of diabetes and other metabolic diseases.

3.3. *Multi-vitamin Supplementation*

There is mixed evidence for the benefits of multi-vitamin (MVM) supplementation for general health,[40,41] with a general consensus from clinical trials that MVM supplementation does not reduce CVD mortality. Recently, Che et al.[42] found that multivitamin/mineral supplementation was associated with a modest reduction in CVD events in the UK Biobank. In contrast, we find that MVM supplementation is associated with a substantial reduction in risks of disorders of lipid metabolism and essential hypertension.

We suspect that the average MVM user in UK Biobank is more health-conscious than their matched counterparts or has had MVM and other supplementations for a long time before their visit to the assessment center, thus biasing our estimates. Subjects were matched for their history of hypertension, use of non-opioid analgesics, lipid-regulating drugs, and glucosamine in addition to the common covariates. Adding other confounding variables such as diet and exercise might reduce the estimated effect of MVM, although we leave this analysis for future research.

3.4. *Thiazides and Heart Failure*

We report that thiazides, a family of diuretics, are associated with a reduced risk of hospitalization for coronary atherosclerosis and heart failure. This coincides with the results reported in previous studies such as the SPRINT clinical trial,[43] which showed the importance of intense systolic blood pressure management for the risk reduction of cardiovascular events. Additionally, more than half of heart failure cases have a history of hypertension.[44]

When inspecting the treatments and matched controls, we found that only approximately 5% of the control group was on some form of non-thiazide diuretic. The proportion of other blood pressure medications such as beta-blockers were otherwise similar. 98% of the treatment group had a history of hypertension, while the ratio for the control group was 96%. It is possible that the observed reduction in hospitalization risk might generalize to other types of diuretics.

Interestingly, we observed an opposite effect for loop diuretics (LD, BNF 2.2.2) and heart failure. As the number of subjects on thiazides was significantly larger than the LD group, the LD treatment group was matched with a large proportion of subjects on some other diuretic, which was not the case for thiazides. Furthermore, LD are also more likely to be already used for the management of heart failure, thus biasing our estimates.

Recent studies support the use of thiazides for the treatment of heart failure. Using data from the SPRINT study, Tsjimoto et al.[45] found that thiazides decreased the risk of events for heart failure in non-diabetics. In the CLOROTIC trial[46] the combination of thiazides and LD proved to be effective for the treatment of acute heart failure. Unfortunately, only approximately one hundred subjects used both thiazides and LD in our dataset, making the estimation of the effect of the combination of both treatments unfeasible. Nevertheless, our results underline the importance of hypertension management for the prevention of heart failure and the potential of thiazide diuretics.

3.5. *Other Associations*

We found 92 statistically significant associations (after Bonferroni correction) for medications that increase the risk of hospitalization ($\beta_T > 0$, $p < 5 \cdot 10^{-5}$). Four medication types included 52 of these associations, all of which are reported in Table 2. As previously explained, some of these associations are known to be spurious, for instance, aspirin (BNF 4.7.1) does not cause an increased risk of hospitalization for hypertension (CIR007). However, since aspirin is commonly prescribed to individuals at risk of hypertension and other diseases it is associated with the phenotype in our analysis. We observe a similar effect for loop diuretics and multiple cardiovascular diseases.

We also observed that proton pump inhibitors (PPIs, BNF 1.3.5) were associated with an increased risk for 23 diseases. We offer three potential explanations. 1) Subjects on PPIs have systematically poorer health than their matched counterparts, either due to unknown variables or scarcity of suitable matches in the control group. 2) PPIs are used in the treatment of multiple diseases in the list or other related comorbidities, thus biasing our estimates. 3) PPIs have measurable side effects and increase the risk of hospitalization for multiple diseases. Since PPIs are used for gastric acid-related disorders and have several known side-effects,[47,48] it is plausible for some of these associations to be causal. Further analysis would be required to estimate the causal effect of PPIs on these diseases, either by Mendelian randomization or a controlled study. We also come to a similar conclusion for anti-epileptic drugs (BNF 4.8.1), although the probability for these associations to be causal is lower.

Table 2. Medication associated with a higher risk of hospitalization ($\beta_T > 0$, $p < 5 \cdot 10^{-5}$) for the corresponding CCSRs, ordered by P-value from left to right. CIR: Diseases of the circulatory system; SYM: Symptoms, signs and abnormal clinical and laboratory findings, not elsewhere classified; RSP: Diseases of the respiratory system; GEN: Diseases of the genitourinary system; END: Endocrine, nutritional and metabolic diseases

Drug Category	CCSRs
Proton Pump Inhibitors	CIR007 (p =3.4e-36), SYM006, END010, CIR011, RSP008, SYM016, END009, SYM012, CIR012, END007, GEN003, GEN002, END011, SYM001, CIR031 GEN001 , CIR026, SYM010, END002, SYM013, SYM014, SYM017, CIR016
Loop Diuretics	GEN003 (p =4e-22), CIR019, GEN002, CIR016, END011, CIR003, SYM016 GEN001, CIR015, CIR011, END002, CIR031
Control of Epilepsy	SYM016 (p =1.8e-10), END011, SYM010, RSP008, CIR007, GEN003, SYM012 SYM015, SYM001
Non-opioid Analgesics	CIR007 (p =1e-26), END010, CIR011, RSP008, END002, CIR012, CIR026, END009

We also found 58 other risk-lowering associations ($\beta_T < 0$, $p < 0.05$) that were not statistically significant after correcting for multiple testing. Several of these associations were also reported in the literature and could have potential clinical applications. Anti-diabetic drugs (BNF 6.1.2), mostly composed of metformin (86% of treated subjects) and blood sugar lowering medications, were associated with reduced risk of 7 disease categories including conduction disorders ($p = 0.0023$), cardiac dysrhythmias ($p = 0.034$) and heart failure ($p = 0.047$). This corroborates the known cardiovascular benefits of metformin and other anti-diabetic drugs.[49–51]

4. Discussion

In this work, we proposed a purely phenotypic screening approach for drug repurposing that consists in systematically measuring medication effects on hospitalization risk from observational data. We showed that we could re-identify known repurposing candidates using simple extensively tested techniques for causal inference and time-to-event modelling. Grouping drugs and diseases by functionality allowed us to gather enough events to estimate potential effects while keeping fine-grained categories. We estimated the risk of hospitalization, making our method inherently preventive although some results could generalize to already hospitalized patients. While our results mostly corroborate known associations, the data for this study has been available for ten years and this method can be applied to new cohorts and treatments.

Due to the nature of the examined data, our study presents multiple limitations. The generally low frequency of events for each CCSR code made the estimation of most effects impossible. While more events could have been included by increasing the time event limit, this would have also introduced more spurious associations. Without utilizing general provider longitudinal data, we could not estimate the approximate dosage, length of treatment, or whether subjects swapped treatments after the first visit and we found a maximum time from visit of 3 years to be a good compromise. Medications were self-reported and no corresponding indication was provided. While matching for common medication has shown to produce less imbalance in general, it can also be counterproductive in cases where a single drug is used for multiple purposes and can result in inadequate matching. The quality of the matching itself is difficult to quantify as most of the bias comes from unmeasured variables, or due to irreconcilable differences between control and treatment groups.

Despite these limitations, biobanks have multiple advantages over typical EHR datasets. 1) All measurements were taken with the same methodology by a small number of assessment centers. 2) Measures such as BMI were taken at a single time point, making time-to-event analysis straightforward. In contrast, EHRs typically have a large portion of missing variables and information is spread over multiple records. 3) Subjects directly described in detail their medication intake and medical history. These variables would be more challenging to recover with EHR data and would likely be incomplete, as the subject history must be stitched up from past events. UK Biobank data allowed us to perform time-to-event analysis with relatively little pre-processing, and scaling up to thousands of tests was also straightforward to implement. To the best of our knowledge, we are the first to report associations for cardioselective beta-blockers, thiazides, and proton pump inhibitors in the UK Biobank.

Large-scale biobank data are a precious resource for understanding human health. While retrospective analysis is always biased and incomplete, it can be an effective tool to guide the design of future experiments that is complementary to other DR methods. Our proposed approach is especially effective at identifying repurposing candidates for preventive care of high-risk subjects. In the future, we plan on using longitudinal general provider prescription data to refine our estimates.

5. Code and Data Availability

Code used for the analysis and summary statistics for all drug-disease pairs in this manuscript is provided on a dedicated GitLab repository `https://gitlab.com/CGeorgantasCHUV/SYESTE`.

6. Acknowledgements

This research has been conducted using the UK Biobank resource under application number 80108, with funding from the Swiss National Science Foundation (Sinergia CRSII5_202276/1).

References

1. N. Nosengo, Can you teach old drugs new tricks?, *Nature* **534**, 314 (June 2016).
2. N. Krishnamurthy, A. A. Grimshaw, S. A. Axson, S. H. Choe and J. E. Miller, Drug repurposing: a systematic review on root causes, barriers and facilitators, *BMC Health Services Research* **22**, p. 970 (July 2022).
3. S. Pushpakom, F. Iorio, P. A. Eyers, K. J. Escott, S. Hopper, A. Wells, A. Doig, T. Guilliams, J. Latimer, C. McNamee, A. Norris, P. Sanseau, D. Cavalla and M. Pirmohamed, Drug repurposing: progress, challenges and recommendations, *Nature Reviews Drug Discovery* **18**, 41 (January 2019).

4. V. Parvathaneni, N. S. Kulkarni, A. Muth and V. Gupta, Drug repurposing: a promising tool to accelerate the drug discovery process, *Drug Discovery Today* **24**, 2076 (October 2019).
5. J. G. Moffat, F. Vincent, J. A. Lee, J. Eder and M. Prunotto, Opportunities and challenges in phenotypic drug discovery: an industry perspective, *Nature Reviews Drug Discovery* **16**, 531 (August 2017).
6. L. Pinzi and G. Rastelli, Molecular Docking: Shifting Paradigms in Drug Discovery, *International Journal of Molecular Sciences* **20**, p. 4331 (January 2019).
7. W. R. Reay and M. J. Cairns, Advancing the use of genome-wide association studies for drug repurposing, *Nature Reviews Genetics* **22**, 658 (October 2021).
8. H. Xu, M. C. Aldrich, Q. Chen, H. Liu, N. B. Peterson, Q. Dai, M. Levy, A. Shah, X. Han, X. Ruan, M. Jiang, Y. Li, J. S. Julien, J. Warner, C. Friedman, D. M. Roden and J. C. Denny, Validating drug repurposing signals using electronic health records: a case study of metformin associated with reduced cancer mortality, *Journal of the American Medical Informatics Association: JAMIA* **22**, 179 (January 2015).
9. M. Zhou, Q. Wang, C. Zheng, A. John Rush, N. D. Volkow and R. Xu, Drug repurposing for opioid use disorders: integration of computational prediction, clinical corroboration, and mechanism of action analyses, *Molecular Psychiatry* **26**, 5286 (September 2021).
10. T. Eguale, D. L. Buckeridge, A. Verma, N. E. Winslade, A. Benedetti, J. A. Hanley and R. Tamblyn, Association of Off-label Drug Use and Adverse Drug Events in an Adult Population, *JAMA internal medicine* **176**, 55 (January 2016).
11. P. Wu, S. D. Nelson, J. Zhao, C. A. Stone, Q. Feng, Q. Chen, E. A. Larson, B. Li, N. J. Cox, C. M. Stein, E. J. Phillips, D. M. Roden, J. C. Denny and W.-Q. Wei, DDIWAS: High-throughput electronic health record-based screening of drug-drug interactions, *Journal of the American Medical Informatics Association: JAMIA* **28**, 1421 (July 2021).
12. H. Ma, X. Li, T. Zhou, D. Sun, Z. Liang, Y. Li, Y. Heianza and L. Qi, Glucosamine Use, Inflammation, and Genetic Susceptibility, and Incidence of Type 2 Diabetes: A Prospective Study in UK Biobank, *Diabetes Care* **43**, 719 (April 2020).
13. L. C. Pilling, L. C. Jones, J. A. H. Masoli, J. Delgado, J. L. Atkins, J. Bowden, R. H. Fortinsky, G. A. Kuchel and D. Melzer, Low Vitamin D Levels and Risk of Incident Delirium in 351,000 Older UK Biobank Participants, *Journal of the American Geriatrics Society* **69**, 365 (February 2021).
14. T. Wu, K. B. Pouwels, R. Welbourn, S. Wordsworth, S. Kent and C. K. H. Wong, Does bariatric surgery reduce future hospital costs? A propensity score-matched analysis using UK Biobank Study data, *International Journal of Obesity* **45**, 2205 (October 2021).
15. Drug repurposing using real-world data, *Drug Discovery Today* **28**, p. 103422 (January 2023).
16. P. Rosenbaum and D. Rubin, The central role of the propensity score in observational studies for causal effects, *Biometrika* **70**, 41 (April 1983).
17. U. Benedetto, S. J. Head, G. D. Angelini and E. H. Blackstone, Statistical primer: propensity score matching and its alternatives, *European Journal of Cardio-Thoracic Surgery: Official Journal of the European Association for Cardio-Thoracic Surgery* **53**, 1112 (June 2018).
18. D. R. Cox, Regression Models and Life-Tables, *Journal of the Royal Statistical Society. Series B (Methodological)* **34**, 187 (1972).
19. P. C. Austin, The use of propensity score methods with survival or time-to-event outcomes: reporting measures of effect similar to those used in randomized experiments, *Statistics in Medicine* **33**, 1242 (March 2014).
20. A. G. Reaume, Drug repurposing through nonhypothesis driven phenotypic screening, *Drug Discovery Today: Therapeutic Strategies* **8**, 85 (December 2011).
21. C. Sudlow, J. Gallacher, N. Allen, V. Beral, P. Burton, J. Danesh, P. Downey, P. Elliott, J. Green, M. Landray, B. Liu, P. Matthews, G. Ong, J. Pell, A. Silman, A. Young, T. Sprosen, T. Peakman

and R. Collins, UK Biobank: An Open Access Resource for Identifying the Causes of a Wide Range of Complex Diseases of Middle and Old Age, *PLOS Medicine* **12**, p. e1001779 (March 2015).
22. Clinical Classifications Software Refined (CCSR) for ICD-10-CM Diagnoses.
23. A. Phil, UK Biobank Self Reported Medication Data parsing and matching.
24. B. Jarman, P. Townsend and V. Carstairs, Deprivation indices., *BMJ : British Medical Journal* **303**, p. 523 (August 1991).
25. D. Ho, K. Imai, G. King, E. Stuart, A. Whitworth and N. Greifer, MatchIt: Nonparametric Preprocessing for Parametric Causal Inference (June 2023).
26. G. King and R. Nielsen, Why Propensity Scores Should Not Be Used for Matching, *Political Analysis* **27**, 435 (October 2019).
27. Z. Huang, Clustering Large Data Sets With Mixed Numeric And Categorical Values1997.
28. T. M. Therneau, T. L. o. S.->. p. a. R. m. until 2009), A. Elizabeth and C. Cynthia, survival: Survival Analysis (March 2023).
29. W. MacNee, Beta-Blockers in COPD — A Controversy Resolved?, *New England Journal of Medicine* **381**, 2367 (December 2019).
30. F. H. Rutten, N. P. A. Zuithoff, E. Hak, D. E. Grobbee and A. W. Hoes, Beta-blockers may reduce mortality and risk of exacerbations in patients with chronic obstructive pulmonary disease, *Archives of Internal Medicine* **170**, 880 (May 2010).
31. P. M. Short, S. I. W. Lipworth, D. H. J. Elder, S. Schembri and B. J. Lipworth, Effect of beta blockers in treatment of chronic obstructive pulmonary disease: a retrospective cohort study, *BMJ (Clinical research ed.)* **342**, p. d2549 (May 2011).
32. Y.-L. Yang, Z.-J. Xiang, J.-H. Yang, W.-J. Wang, Z.-C. Xu and R.-L. Xiang, Association of β-blocker use with survival and pulmonary function in patients with chronic obstructive pulmonary and cardiovascular disease: a systematic review and meta-analysis, *European Heart Journal* **41**, 4415 (December 2020).
33. C.-M. Chung, M.-S. Lin, S.-T. Chang, P.-C. Wang, T.-Y. Yang and Y.-S. Lin, Cardioselective Versus Nonselective β-Blockers After Myocardial Infarction in Adults With Chronic Obstructive Pulmonary Disease, *Mayo Clinic Proceedings* **97**, 531 (March 2022).
34. B. Lipworth, J. Wedzicha, G. Devereux, J. Vestbo and M. T. Dransfield, Beta-blockers in COPD: time for reappraisal, *The European Respiratory Journal* **48**, 880 (September 2016).
35. J. Sundh, A. Magnuson, S. Montgomery, P. Andell, G. Rindler, O. Fröbert, M. Przybyszewska, A. Blomberg, M. Widmark, A. Palm, W. Greger, J. Ellingsen, L. Råhlén, T. Kipper, M. Hasselgren, C. Smith, F. Delijaj, K. Possler, J. Nilsson, N. Stenersen, H. Nguyen, D. Curiac, L. E. G. W. Vanfleteren, L. Johansson, F. Sjöberg, M. Ekström, J. S. Berglund, A. Lökke and the BRONCHIOLE investigators, Beta-blockeRs tO patieNts with CHronIc Obstructive puLmonary diseasE (BRONCHIOLE) – Study protocol from a randomized controlled trial, *Trials* **21**, p. 123 (January 2020).
36. L. Rossetti, M. Hawkins, W. Chen, J. Gindi and N. Barzilai, In vivo glucosamine infusion induces insulin resistance in normoglycemic but not in hyperglycemic conscious rats., *The Journal of Clinical Investigation* **96**, 132 (July 1995).
37. R. Muniyappa, R. J. Karne, G. Hall, S. K. Crandon, J. A. Bronstein, M. R. Ver, G. L. Hortin and M. J. Quon, Oral glucosamine for 6 weeks at standard doses does not cause or worsen insulin resistance or endothelial dysfunction in lean or obese subjects, *Diabetes* **55**, 3142 (November 2006).
38. K. E. Wellen and G. S. Hotamisligil, Inflammation, stress, and diabetes, *The Journal of Clinical Investigation* **115**, 1111 (May 2005).
39. Z. Tian, J. Mclaughlin, A. Verma, H. Chinoy and A. H. Heald, The relationship between rheumatoid arthritis and diabetes mellitus: a systematic review and meta-analysis, *Cardiovascular En-*

docrinology & Metabolism **10**, 125 (February 2021).
40. H.-Y. Huang, B. Caballero, S. Chang, A. Alberg, R. Semba, C. Schneyer, R. F. Wilson, T.-Y. Cheng, G. Prokopowicz, G. J. Barnes, J. Vassy and E. B. Bass, Multivitamin/mineral supplements and prevention of chronic disease, *Evidence Report/Technology Assessment* , 1 (May 2006).
41. J. Kim, J. Choi, S. Y. Kwon, J. W. McEvoy, M. J. Blaha, R. S. Blumenthal, E. Guallar, D. Zhao and E. D. Michos, Association of Multivitamin and Mineral Supplementation and Risk of Cardiovascular Disease: A Systematic Review and Meta-Analysis, *Circulation. Cardiovascular Quality and Outcomes* **11**, p. e004224 (July 2018).
42. B. Che, C. Zhong, R. Zhang, M. Wang, Y. Zhang and L. Han, Multivitamin/mineral supplementation and the risk of cardiovascular disease: a large prospective study using UK Biobank data, *European Journal of Nutrition* **61**, 2909 (September 2022).
43. The SPRINT Research Group, A Randomized Trial of Intensive versus Standard Blood-Pressure Control, *New England Journal of Medicine* **373**, 2103 (November 2015).
44. G. C. Oh and H.-J. Cho, Blood pressure and heart failure, *Clinical Hypertension* **26**, p. 1 (January 2020).
45. T. Tsujimoto and H. Kajio, Thiazide Use and Decreased Risk of Heart Failure in Nondiabetic Patients Receiving Intensive Blood Pressure Treatment, *Hypertension* **76**, 432 (August 2020).
46. J. C. Trulls, J. L. Morales-Rull, J. Casado, M. Carrera-Izquierdo, M. Snchez-Marteles, A. Conde-Martel, M. F. Dvila-Ramos, P. Llcer, P. Salamanca-Bautista, J. Prez-Silvestre, M. N. Plasn, J. M. Cerqueiro, P. Gil, F. Formiga, L. Manzano and CLOROTIC trial investigators, Combining loop with thiazide diuretics for decompensated heart failure: the CLOROTIC trial, *European Heart Journal* **44**, 411 (February 2023).
47. A. J. Schoenfeld and D. Grady, Adverse Effects Associated With Proton Pump Inhibitors, *JAMA Internal Medicine* **176**, 172 (February 2016).
48. C. E. Aubert, M. R. Blum, V. Gastens, O. Dalleur, F. Vaillant, E. Jennings, D. Aujesky, W. Thompson, T. Kool, C. Kramers, W. Knol, D. O'Mahony and N. Rodondi, Prescribing, deprescribing and potential adverse effects of proton pump inhibitors in older patients with multimorbidity: an observational study, *Canadian Medical Association Open Access Journal* **11**, E170 (January 2023).
49. G. Rena and C. C. Lang, Repurposing Metformin for Cardiovascular Disease, *Circulation* **137**, 422 (January 2018).
50. F. Luo, A. Das, J. Chen, P. Wu, X. Li and Z. Fang, Metformin in patients with and without diabetes: a paradigm shift in cardiovascular disease management, *Cardiovascular Diabetology* **18**, p. 54 (April 2019).
51. W. M. C. Top, A. Kooy and C. D. A. Stehouwer, Metformin: A Narrative Review of Its Potential Benefits for Cardiovascular Disease, Cancer and Dementia, *Pharmaceuticals* **15**, p. 312 (March 2022).

Transcript-aware analysis of rare predicted loss-of-function variants in the UK Biobank elucidate new isoform-trait associations

Rachel A. Hoffing, Aimee M. Deaton, Aaron M. Holleman, Lynne Krohn, Philip J. LoGerfo, Mollie E. Plekan, Sebastian Akle Serrano, Paul Nioi, Lucas D. Ward

Alnylam Pharmaceuticals
Cambridge, MA 02142, USA
Email: rhoffing@alnylam.com

A single gene can produce multiple transcripts with distinct molecular functions. Rare-variant association tests often aggregate all coding variants across individual genes, without accounting for the variants' presence or consequence in resulting transcript isoforms. To evaluate the utility of transcript-aware variant sets, rare predicted loss-of-function (pLOF) variants were aggregated for 17,035 protein-coding genes using 55,558 distinct transcript-specific variant sets. These sets were tested for their association with 728 circulating proteins and 188 quantitative phenotypes across 406,921 individuals in the UK Biobank. The transcript-specific approach resulted in larger estimated effects of pLOF variants decreasing serum *cis*-protein levels compared to the gene-based approach ($p_{binom} \leq 2x10^{-16}$). Additionally, 251 quantitative trait associations were identified as being significant using the transcript-specific approach but not the gene-based approach, including *PCSK5* transcript ENST00000376752 and standing height (transcript-specific statistic, P = $1.3x10^{-16}$, effect = 0.7 SD decrease; gene-based statistic, P = 0.02, effect = 0.05 SD decrease) and *LDLR* transcript ENST00000252444 and apolipoprotein B (transcript-specific statistic, P = $5.7x10^{-20}$, effect = 1.0 SD increase; gene-based statistic, P = $3.0x10^{-4}$, effect = 0.2 SD increase). This approach demonstrates the importance of considering the effect of pLOFs on specific transcript isoforms when performing rare-variant association studies.

Keywords: UK Biobank; rare variant; transcriptome; quantitative traits

1. Introduction

Alternative splicing allows for one gene to produce many transcript isoforms. When these isoforms differ in their coding sequence content, they can result in proteins with distinct molecular functions. Over 95% of protein-coding genes are alternatively spliced[1] which contributes to the large diversity of the human transcriptome and proteome. This process is instrumental in creating the complex and coordinated gene expression patterns that underlie all biological processes.

Alterations to the transcriptome by genetic variation is instrumental in driving differences in phenotypic expression. Many of these disruptions have been identified through genome-wide association studies (GWAS), which test the impact of common single nucleotide variants (SNVs) on phenotypes on a population scale. Studies such as these are critical in drug-discovery efforts, as they can be used to identify new therapeutic targets for disease. Additionally, lack of genetic validation of therapeutic hypotheses has been shown to reduce the likelihood of a successful clinical trial[2,3], suggesting that genetically validated targets are essential in the development process.

A large amount of phenotype heritability is not well captured through common-variant GWAS alone[4]. Rare, coding SNVs can be exceptionally disruptive to the transcriptome and have dramatic effects on phenotypes, even more so than common variants[5]. Rare variant association studies (RVAS) provide avenues to explain the "missing heritability" of traits[6], and provide a complementary approach to common-variant GWAS. Though, assessing genotype-phenotype associations with these low-frequency SNVs is difficult due to lack of sufficient sample size and statistical power.

One approach in ameliorating the statistical challenges of rare variant analysis is the aggregation of SNVs with similar predicted functional consequences, known as burden testing[7,8]. For example, an analysis may collect rare protein-truncating variants (PTVs), also known as predicted loss-of-function variants (pLOF), that are expected to result in non-functional gene products through nonsense-mediated decay[9]. These pLOF variants can then be aggregated and tested for their collective association with phenotypes of interest. This allows for an increase in statistical power and ability to detect genotype-phenotype associations which would otherwise be impossible at the level of single-variant tests.

However, burden testing assumes that all aggregated variants will have a similar effect on the function of the gene and, consequently, the associated phenotype. This assumption does not hold if a gene has multiple transcript isoforms with diverse downstream functions. For example, where a given SNV may encode a missense variant that is deleterious in some encoded protein isoforms but not others, or where an SNV may encode a variant of any function that overlaps some transcript isoforms but is not transcribed in others. The most common techniques for creating variant sets for burden testing consider the most deleterious consequence of an SNV across all documented isoforms. Subsequently, the expected impact of the SNV may be overestimated. Due to these challenges, we propose the inclusion of transcript-aware analyses when studying rare variants, in addition to the standard gene-based approach.

Our analysis uses whole-exome sequencing data from the UK Biobank to perform transcript-specific burden analyses on 406,921 individuals of European ancestry. Rare pLOFs were identified across 17,035 genes and aggregated by transcript, resulting in 55,558 unique, transcript-specific variant sets tested against the circulating levels of 728 *cis*-encoded proteins and 188 quantitative traits. The results of the transcript-specific burden tests were compared to the results from the maximally inclusive, standard, gene-based burden method.

2. Data

2.1. *UK Biobank*

The UK Biobank consists of approximately 500,000 volunteer participants, who were aged 40–69 years when recruited between 2006 and 2010[10,11]. Both array genotyping and whole-exome sequencing have been performed on most of these participants[12]. Data from genotyping, sequencing, questionnaires, primary care data, hospitalization data, cancer registry data, and death registry data were obtained through application number 26041. Proteomic profiling was also performed on a subset of participants through application number 65851[13]. Ethical oversight for the UK Biobank is provided by an Ethics and Governance Council which obtained informed consent from all participants to use these data for health-related research. Data management and analytics were performed using the REVEAL/SciDB translational analytics platform from Paradigm4.

2.2. *Variant calling and definition*

The source of genetic data for the main analysis was exome sequencing data. DNA from whole blood was extracted and sequenced by the Regeneron Genetics Center (RGC) using protocols previously described[14]. Of the variants called by RGC, additional quality-control filters were applied: Hardy-Weinberg equilibrium (among the European subpopulation, as defined by Pan-UKB[15]) $P > 1x10^{-10}$, and missingness across all individuals less than 2%. Variants were annotated using ENSEMBL Variant Effect Predictor (VEP)[16] version 109.3, using the LOFTEE plug-in version 1.0.4 to identify high-confidence predicted PTV variants[9] in protein-coding genes with minor allele frequency <1%. Bcftools was used to filter variants with genotype quality (GQ) > 20 and depth (DP) > 7 or 10 for SNPs and indels respectively. For the gene-based burden, variant effects were scored against all available transcripts in ENSEMBL, and the most severe predicted impact was retained. Variants were aggregated in each protein-coding gene as follows: pLOF variants were defined as "HC" (high confidence) from LOFTEE and their most severe consequence from VEP as "stop gained," "splice donor," "splice acceptor," or "frameshift." For the transcript-based burden, the consequence for each variant was assessed individually by transcript.

2.3. *Participant definition for overall analyses*

An initial round of quality control was performed by RGC, which removed subjects with evidence of contamination, discrepancies between chromosomal and reported sex, and high discordance between sequencing and genotyping array data. A European ancestry population was defined using data from the Pan-UKB Team[15], resulting in a set of 406,921 European ancestry individuals with exome sequencing data available. Two sets of genetic principal components (PCs) were defined, as described by Backman et al[17]: a set derived from common array variants, of which 10 were used, and a set derived from rare exome variants, of which 20 were used. Rare exome derived PCs were calculated by applying the following filters on variants on the autosomes: $MAF > 2.6x10^{-5}$ and < 0.01, Hardy-Weinberg equilibrium $P > 1x10^{-12}$, and genotype missingness < 2%. Regions of high LD were removed, and SNPs were pruned with PLINK's[18] indep-pairwise function, using a window-size of 1,000 base pairs, a step size of 100 base pairs, and an R^2 threshold of 0.1. Indels were removed, then R's Smart PCA was implemented to derive the PCs. Array derived PCs for the European subset were derived by imposing a MAF filter > 0.01 and INFO score = 1 before running Smart PCA.

2.4. *Phenotype sources*

The main source of phenotype data was from a release of structured data by the UK Biobank Data Showcase on December 22, 2022. We tested 188 quantitative phenotypes, including physical measures, blood counts, metabolomics, touchscreen questionnaire responses on family history, telomere length, and urine biochemistry. Quantitative traits were rank-based inverse normal transformed to have a mean of zero and a standard deviation of one.

2.5. *Tissue-expressed transcript isoforms*

GTEx version 8 bulk RNAseq data was aggregated across 54 tissue types from 948 donors. For each gene, expression was calculated across all tissues, identifying 145,219 transcripts with mean TPM expression > 0.

2.6. *Olink proteomics*

Characterization of 1,463 proteins across 54,306 individuals was undertaken by the UK Biobank Proteomics Project (UKB-PPP). Proteomic profiling was conducted across four panels utilizing the Olink Explore Assay. Sample collection, preparation, data pre-processing, and quality control is described in detail in Sun et al[13]. Quantified protein expression levels were rank-based inverse normal transformed to have a mean of zero and standard deviation of one.

3. Methods

3.1. *Transcript-specific variant set curation*

Rare, predicted loss-of-function (pLOF) variants sets (MAF < 1%) were created across 145,219 transcripts with mean TPM > 0 across all 53 GTEx tissue types. Overall, 72,769 transcripts had at least one overlapping rare pLOF variant. Identical variant sets that were representative of more than one transcript were combined into a single label, resulting in 55,558 unique transcript-specific variant sets across 17,035 genes.

3.2. *Whole-genome ridge regression analysis*

REGENIE v3.1.1[19] was used to perform a whole-genome ridge regression taking subject relatedness into account, while using a Firth approximation to estimate P values. For all quantitative traits, REGENIE was performed using an additive model across the entire European-ancestry population, including related individuals, controlling for age, sex, age^2, age x sex, age^2 x sex, 10 rare-variant derived principal components, and 20 common-variant derived principal components. For the Olink proteomics, batch numbers 1-7 were added as one-hot encoded covariates.

3.3. *Comparison of estimated effect sizes by approximating a binomial distribution*

The effect sizes across transcript and gene-based burden tests were compared in cases only where there was a significant association for a quantitative phenotype in both methods. Deviation from a binomial distribution was modeled using R's binom.test() to determine is the proportion of results with stronger associations in the transcript-based model differs from the null hypothesis.

3.4. *Binary case-control phenotype regression*

As a follow-up to the quantitative traits analysis, we tested a single binary phenotype, Alzheimer's disease, across multiple *TREM2* transcript-specific variant sets. Diagnoses were extracted from inpatient hospital diagnoses, the cancer and death registries, primary care, and self-reported data. We adjusted for age, sex, age^2, age x sex, age^2 x sex, 10 rare-variant derived principal components, 20 common-variant derived principal components, availability of primary care, and country of recruitment.

4. Results

4.1. *Transcript-specific variant sets show stronger associations with lower serum cis-protein levels*

To evaluate the validity of transcript-specific pLOF variant sets, they were first tested for their association with *cis*-encoded proteins. Variant sets with at least 10 carriers were tested across 728 circulating serum proteins in 47,297 individuals of European ancestry and compared to the gene-based approach. Several gene and transcript-specific variant sets were identical, and their removal resulted in 913 unique transcript variant sets tested across 432 serum protein levels. Among 580 results that were significant for both the transcript and gene-based burden approach, 75% (N = 437) had lower effect estimates on *cis*-serum proteins in the transcript-based burden (Figure 1), which is substantially greater than expected by chance ($p_{binom} \leq 2x10^{-16}$). Of the 437 transcript-based results with lower *cis*-protein effect estimates, 45 had non-overlapping 95% confidence intervals with the effect estimates of the gene-based approach.

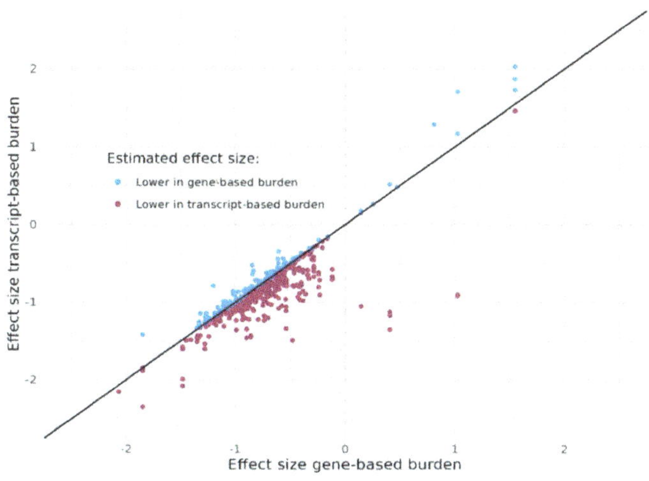

Figure 1. Comparison of estimated effect sizes on circulating serum proteins. Each dot represents an association of the transcript or gene-based burden with a *cis*-encoded protein.

4.2. *Some pLOF-cis protein associations are only detectable using transcript-specific variant sets*

The transcript-based burden on *cis*-proteins resulted in 35 associations across 21 loci that were non-significant in the gene-based burden. Of these associations only significant in the transcript-based burden, 22 associations across 12 loci had non-overlapping 95% confidence intervals with the gene-based approach (Table 1), and all of them had lower estimated effect sizes.

Gene/*cis*-protein	P value gene -based burden	Effect size gene -based burden	N carriers gene -based burden	95% CI gene -based burden	P value transcript -based burden	Effect size transcript -based burden	N carriers transcript-based burden	95% CI transcript-based burden	Transcripts
CD300LF	$5.3x10^{-1}$	-0.03	195	-0.1,0.1	$3.6x10^{-19}$	-1.0	38	-1.2,-0.8	ENST00000326165 ENST0000046491 ENST00000583937
CD84	$4.12x10^{-5}$	-0.2	51	-0.3,-0.1	$1.3x10^{-20}$	-1.0	10	-1.2,-0.8	ENST00000368048 ENST0000036805 ENST00000368054
CLEC10A	$3.0x10^{-4}$	-0.1	103	-0.2,-0.1	$7.1x10^{-18}$	-1.0	13	-1.2,-0.7	ENST00000416562 ENST00000571664
CPPED1	$4.6x10^{-2}$	-0.2	43	-0.5,0.01	$3.2x10^{-13}$	-1.6	13	-2.0,-1.1	ENST00000381774
MSR1	$1.4x10^{-2}$	-0.1	110	-0.2,-0.02	$5.0x10^{-13}$	-0.8	23	-1.0,-0.6	ENST0000026210
MSRA	$5.9x10^{-3}$	-0.2	93	-0.4,-0.1	$7.3x10^{-8}$	-1.3	13	-1.7,-0.8	ENST0000052824
NRP2	$1.8x10^{-1}$	0.1	32	-0.04,0.2	$2.6x10^{-14}$	-0.7	13	-0.9,-0.5	ENST00000357785
PLXNB2	$1.5x10^{-2}$	-0.1	85	-0.1,-0.01	$4.9x10^{-10}$	-0.4	23	-0.5,-0.3	ENST00000359337 ENST00000449103
SETMAR	$5.8x10^{-2}$	-0.1	133	-0.1,0.02	$1.2x10^{-09}$	-0.3	51	-0.4,-0.2	ENST00000425863
TREM2	$8.8x10^{-3}$	-0.2	66	-0.3,-0.04	$2.2x10^{-16}$	-1.2	17	-1.5,-0.9	ENST00000373113
TXNRD1	$5.4x10^{-5}$	-0.4	35	-0.6,-0.2	$1.3x10^{-9}$	-1.0	12	-1.3,-0.7	ENST0000050350 ENST00000524698 ENST0000052639 ENST0000052695 ENST0000052954
TYMP	$2.6x10^{-2}$	-0.2	54	-0.4,-0.02	$2.6x10^{-5}$	-0.9	13	-1.3,-0.5	ENST00000425169

Table 1. Transcript-specific results with significant association on circulating *cis*-proteins, and transcript-based burden 95% CI not-overlapping with gene-based burden 95% CI. Multiple transcripts listed when variant sets are identical.

From these data, we focused on *TREM2* as it has a known role in Alzheimer's disease (AD) risk. *TREM2* is primarily expressed in microglia, and rare loss-of-function mutations including the missense variant R47H have been shown to increase AD risk[20]. When testing *TREM2* transcript-specific pLOF variant sets (Figure 2), we observe more significant associations with larger reductions in serum TREM2 levels in the ENST00000338469 and ENST00000373113 models, compared to ENST00000373122 or the gene-based method (Table 2).

The primary variant that explains the difference in signal is rs538447052, a splice acceptor variant at the boundary of exon 4. The canonical transcript with the highest brain expression[21], ENST00000373113, and ENST00000338469, are both unaffected by rs538447052 as it functions there as an intron variant. By excluding rs538447052 from these variant sets, we see a much stronger association with decreasing serum TREM2.

Next, we tested the relationship between Alzheimer's disease and *TREM2* and its transcript isoforms. Our analysis is limited by a low number of affected carriers; however, we detect an enrichment of AD cases when using the more stringent *TREM2* transcript models, ENST00000373113 and ENST00000338469 (Table 2). This association is absent in the ENST00000373122 and gene-based models and is consistent with the weaker observed effects on serum TREM2.

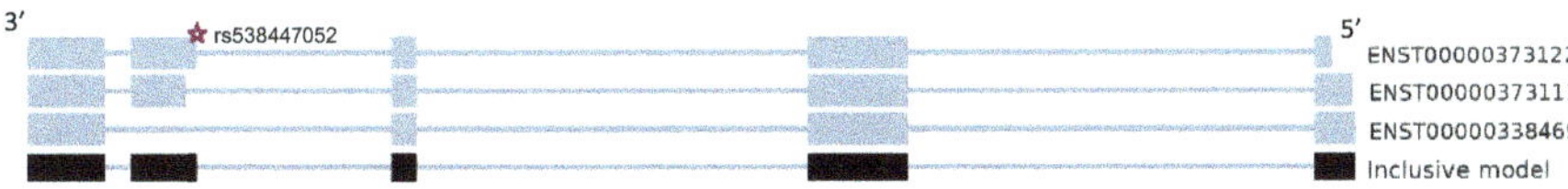

Figure 2. *TREM2* transcript models and the gene-based, inclusive model.

Transcripts	P value AD	Odds ratio AD	95% CI AD	N carriers AD	N carriers with AD	P value serum TREM2	Effect size serum TREM2	95% CI serum TREM2	N carriers serum TREM2
ENST00000338469	8.9×10^{-3}	10.3	2.6,40.9	48	2	7.2×10^{-16}	-1.4	-1.0,-1.7	12
ENST00000373113	1.2×10^{-2}	9.1	2.3,35.9	55	2	2.1×10^{-16}	-1.2	-1.4,-0.9	17
Inclusive model	9.6×10^{-2}	1.0	0.3,3.1	435	3	8.7×10^{-3}	-0.2	-0.3,0.0	66
ENST00000373122	9.8×10^{-2}	01.0	0.30,3.1	428	3	4.9×10^{-2}	-0.2	-0.3,0.0	61

Table 2. *TREM2* transcript-specific associations with AD and circulating TREM2 levels.

4.3. *Some pLOF-cis protein associations have opposite directions of effect in the transcript and gene-based models*

Most effect size estimates maintain their direction of effect when comparing the gene and transcript-based methods. However, six associations from three loci resulted in opposing estimated effect sizes (Table 3). In all six cases, the transcript-variant set of pLOFs associates with lower serum *cis*-protein levels, as expected, while the gene-based method associates with higher serum *cis*-protein levels.

Gene/*cis*-protein	P value gene -based burden	Effect size gene-based burden	95% CI gene-based burden	N carriers gene - based burden	P value transcript-based burden	Effect size transcript-based burden	95% CI transcript -based burden	N carriers transcript-based burden	Transcripts
BST1	8.8×10^{-69}	0.4	0.4,0.5	543	6.6×10^{-37}	-1.1	-1.3,-01.0	37	ENST00000265016; ENST00000382346
BST1	8.8×10^{-69}	0.4	0.4,0.5	543	1.1×10^{-36}	-1.2	-1.4,-1.0	34	ENST00000505785
BST1	8.8×10^{-69}	0.4	0.4,0.5	543	2.7×10^{-22}	-1.4	-1.6,-1.1	15	ENST00000514445
GPNMB	2.9×10^{-21}	0.2	0.1,0.3	446	1.1×10^{-212}	-1.1	-1.1,-1.0	93	ENST00000409458
HMOX2	2.2×10^{-21}	1.0	0.8,1.2	26	5.2×10^{-8}	-0.9	-1.2,-0.6	11	ENST00000570445; ENST00000575051
HMOX2	2.2×10^{-21}	1.0	0.8,1.2	26	1.4×10^{-7}	-0.9	-1.3,-0.6	10	ENST00000574466; ENST00000575129; ENST00000576827

Table 3. Significant transcript-based burden results with opposing effect sizes compared to the gene-based burden. Multiple transcripts are listed when the variant sets are identical.

The difference in variants captured by the *BST1, GPNMB,* and *HMOX2* gene-based and transcript-based variant sets are primarily attributable to variants missing from the terminal exon (Figure 3). The most significantly associated transcript variant set for each locus mainly exclude a single, frequent variant from the last exon, rs144539516, rs11537976, and rs11537976, respectively.

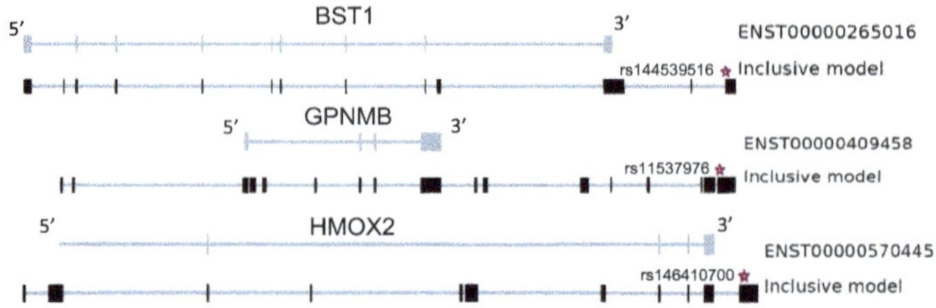

Figure 3: *BST1*, *GPNMB*, *HMOX2* inclusive gene-based model and representative transcript-based models.

Each of these excluded variants strongly associate with increased *cis*-serum protein levels when tested individually (Table 4). Rs144539516 and rs146410700 are 3' UTR variants in at least one transcript, which may affect the post-transcriptional stability of the RNA product. Rs146410700 also occasionally is identified as a missense variant in some transcripts and could influence protein stability, detectability, and post-translational regulation. Rs11537976 acts as a non-coding exon variant and may affect transcriptional regulation. In all instances, this provides an explanation for the unexpected gene-level association with increased protein.

Cis-protein	Rsid	P value	Effect size	95% CI	N carriers
GPNMB	rs11537976	1.5×10^{-233}	0.6	0.5, 0.6	318
BST1	rs144539516	7.6×10^{-102}	0.5	0.5, 0.6	506
HMOX2	rs146410700	1.2×10^{-91}	3.4	3.0, 3.7	11

Table 4: Single variant *cis*-protein association results for *BST1*, *GPNMB*, *HMOX2* variants rs11537976, rs144539516, and rs146410700

4.4. *Transcript-specific variant sets show stronger associations with quantitative traits*

Since the transcript-based variant sets show larger effects on circulating *cis*-proteins compared to the gene-based method, we next extended the analysis to quantitative traits. Transcript-specific pLOF variant sets with at least 10 carriers were tested for their association with 318 quantitative traits in 406,921 individuals of European ancestry and compared to the gene-based approach. After removing identical results between the transcript and gene-based approach, 6,981,491 transcript-trait and 2,740,011 gene-trait association tests were performed (Bonferroni corrected P value < 5.1×10^{-9}). Among 1,010 associations that were significant in both the transcript and gene-based approach, 73% (N = 745) had more extreme effect sizes in the transcript-specific approach (Figure 4), which is substantially larger than expected by chance ($p_{binom} \leq 2\times10^{-16}$). Of these, 75 had non overlapping 95% confidence intervals with the gene-based approach. Additionally, 46% of associations significant in both methods were more significant in the transcript-approach despite having a lower number of tested carriers in practically all instances.

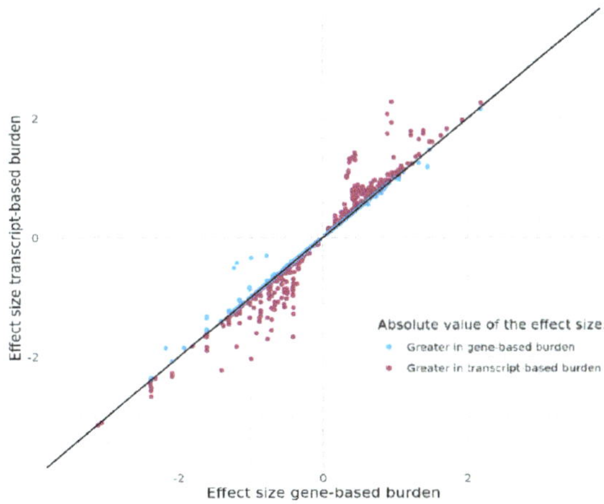

Figure 4. Comparison of estimated effect sizes on 188 quantitative traits for the transcript and gene-based burden. Each dot represents an association of the transcript or gene-based burden with a quantitative trait.

4.5. *Transcript-specific variant sets elucidate novel transcript-trait associations*

We identified 241 associations across 60 loci as being significant in the transcript-based approach but not in the gene-based burden. Of these, 56 transcript-trait associations had effect estimates with non-overlapping 95% confidence intervals with the gene-based burden (Table 5). These include *PCSK5* transcript ENST00000376752 and standing height (transcript-specific statistic, $P = 1.3x10^{-16}$, effect = 0.72 SD decrease; gene-based statistic, P = 0.02, effect = 0.05 SD decrease) and *LDLR* transcript ENST00000252444 and apolipoprotein B (transcript-specific statistic, $P = 5.7x10^{-20}$, effect = 1.0 SD increase; gene-based statistic, $P = 3.0x10^{-4}$, effect = 0.2 SD increase). These data reflect genotype-phenotype associations that would have been otherwise undetected if testing only the standard, gene-based burden.

Gene	Phenotype	P value gene-based burden	Effect size gene-based burden	N carriers gene-based burden	95% CI gene-based burden	P value transcript-based burden	Effect size transcript-based burden	N carriers transcript-based burden	95% CI transcript-based burden	Transcripts
EPB41	Reticulocyte percentage	$8.5x10^{-9}$	0.33	282	0.2,0.4	$6.9x10^{-22}$	1.1	77	0.8,1.3	ENST00000373800
LDLR	Apolipoprotein B	$3.0x10^{-4}$	0.17	425	0.1,0.3	$5.8x10^{-20}$	1.0	78	0.8,1.2	ENST00000252444
SCUBE3	Standing height	$1.2x10^{-4}$	-0.12	373	-0.2,-0.1	$1.4x10^{-18}$	-0.6	71	-0.8,-0.5	ENST00000274938
EPB41	Total bilirubin	$1.5x10^{-4}$	0.19	275	0.1,0.3	$8.3x10^{-17}$	0.8	74	0.6,1.0	ENST00000373800
PCSK5	Standing height	$1.9x10^{-2}$	-0.05	829	-0.1,-0.01	$1.3x10^{-16}$	-0.7	50	-0.9,-0.5	ENST00000376752
UGT1A9	Total bilirubin	$6.9x10^{-3}$	0.17	227	0.04,0.3	$8.7x10^{-16}$	0.4	355	0.3,0.5	ENST00000354728
TINF2	Telomere Length	$1.3x10^{-5}$	0.44	92	0.2,0.6	$1.1x10^{-15}$	2.1	14	1.5,2.6	ENST00000557921
PFKM	HbA1c	$4.8x10^{-8}$	-0.25	391	-0.3,-0.2	$1.4x10^{-15}$	-0.5	201	-0.6,-0.4	ENST00000549941
TTN	Systolic blood pressure	$8.6x10^{-9}$	-0.08	4430	-0.1,-0.1	$1.8x10^{-14}$	-0.2	1807	-0.2,-0.1	ENST00000359218

CHD2	Lymphocyte percentage	$1.4x10^{-5}$	0.50	67	0.3,0.7	$1.8x10^{-14}$	2.1	12	1.5,2.6	ENST00000394196
NF1	Lymphocyte percentage	$4.4x10^{-8}$	-0.29	312	-0.4,-0.19	$3.5x10^{-13}$	-1.4	25	-1.8,-1.0	ENST00000431387
IGF2BP2	Standing height	$2.5x10^{-8}$	-0.47	53	-0.6,-0.3	$4.0x10^{-13}$	-0.9	25	-1.1,-0.7	ENST00000346192; ENST00000382199
PKD1	Urate	$3.3x10^{-6}$	0.21	313	0.1,0.3	$4.5x10^{-13}$	0.6	97	0.4,0.8	ENST00000423118
CREB3L3	Apolipoprotein A	$6.5x10^{-4}$	-0.07	1709	-0.1,-0.03	$2.2x10^{-12}$	-0.4	205	-0.6,-0.3	ENST00000078445; ENST00000595923
CLEC11A	Standing height	$3.7x10^{-5}$	-0.02	13452	-0.03,-0.01	$3.5x10^{-12}$	-0.1	5719	-0.1,-0.04	ENST00000250340
TPM4	Thrombocyte volume	$9.1x10^{-6}$	0.62	42	0.3,0.9	$8.4x10^{-12}$	1.8	12	1.3,2.3	ENST00000586833
COL18A1	Apolipoprotein A	$6.6x10^{-4}$	-0.07	1876	-0.1,-0.03	$1.1x10^{-11}$	-0.2	620	-0.3,-0.2	ENST00000355480
PFKM	Pyruvate	$8.7x10^{-9}$	0.55	105	0.4,0.7	$1.4x10^{-11}$	1.2	30	0.9,1.6	ENST00000546465
RNF10	Reticulocyte count	$1.7x10^{-1}$	0.04	1296	-0.02,0.1	$3.2x10^{-11}$	0.7	76	0.5,0.9	ENST00000413266
ANK1	HbA1c	$6.4x10^{-3}$	-0.23	118	-0.4,-0.1	$8.3x10^{-11}$	-0.9	47	-1.1,-0.6	ENST00000520299
NF1	Standing height	$1.0x10^{-6}$	-0.17	322	-0.2,-0.1	$9.6x10^{-11}$	-0.8	27	-1.0,-0.5	ENST00000431387
MARCHF8	Erythrocyte distribution width	$1.1x10^{-4}$	0.17	456	0.1,0.3	$9.7x10^{-11}$	0.5	165	0.3,0.6	ENST00000319836; ENST00000395769
PRC1	Platelet crit	$1.7x10^{-5}$	-0.22	304	-0.3,-0.1	$1.8x10^{-10}$	-0.5	130	-0.7,-0.4	ENST00000442656
LARP1	Mean corpuscular hemoglobin	$1.7x10^{-2}$	-0.23	87	-0.4,-0.04	$8.9x10^{-10}$	-1.0	30	-1.3,-0.7	ENST00000518297
MARCHF8	Immature reticulocyte fraction	$7.7x10^{-6}$	0.21	441	0.1,0.3	$1.0x10^{-09}$	0.5	157	0.3,0.6	ENST00000319836; ENST00000395769
UGT1A8	Total bilirubin	$4.8x10^{-2}$	-0.18	99	-0.4,0	$1.4x10^{-09}$	0.4	227	0.3,0.5	ENST00000373450
CREB3L3	Triglycerides	$2.2x10^{-3}$	0.06	1880	0.02,0.1	$1.7x10^{-9}$	0.4	223	0.3,0.5	ENST00000078445; ENST00000595923
PTCH1	Standing height	$5.2x10^{-4}$	-0.16	167	-0.3,-0.1	$2.7x10^{-9}$	-0.4	87	-0.5,-0.3	ENST00000468211

Table 5. Transcript-specific results with significant quantitative traits associations, and 95% CI of effect size not-overlapping with the gene-based burden 95% CI. For loci with multiple significant results, or multiple highly correlated phenotypes, the result with the lowest P value is shown. Multiple transcripts are listed when the variant sets are identical.

4.6. *Transcript-specific variant sets limit pLOF variants in low expression exonic regions*

One method by which the transcript-aware variant sets improve burden testing is by excluding variants within weakly expressed exonic region. An example of this improvement can be shown with LDL cholesterol and the low-density lipoprotein receptor (*LDLR*). We evaluated seven distinct transcript-isoforms variant sets for their association with apolipoprotein B, the main protein found in LDL. All seven tested *LDLR* transcript sets were more statistically significant and had larger effect sizes as compared to the gene-based inclusive method (Table 6).

The best performing *LDLR* transcript, ENST00000252444, compared to the worst performing *LDLR* transcript, ENST00000557933, and the gene-based model, lacks pLOF variants primarily in two critical regions: the first exon and part of the penultimate exon, highlighted in pink (Figure 5).

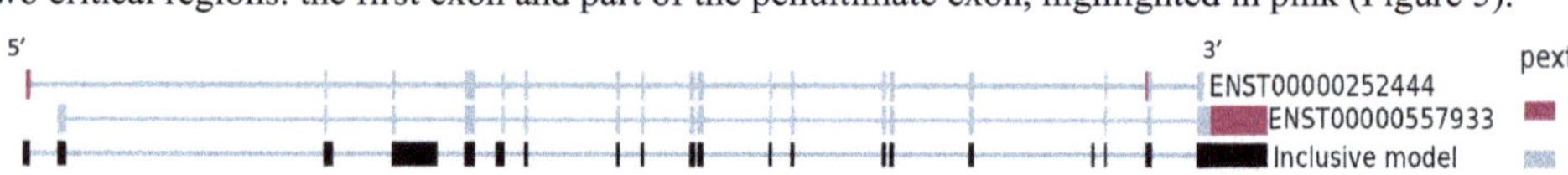

Figure 5. Two *LDLR* transcript models and the inclusive, gene-based model overlayed with pext = 0 regions in pink. No pLOF variants appear in the terminal exons of all three models.

In both regions, pext, or the proportion expressed across transcripts[22], is equal to 0, indicating that these regions have extremely low expression across all isoforms. All seven tested *LDLR* transcript-aware variant sets excluded some variants in the pext = 0 regions, and subsequently, resulted in an improved apolipoprotein B association compared to the gene-based method.

Gene	Phenotype	P value	Effect size	N carriers	95% CI	Transcripts	Median expression in all tissues (TPM)
LDLR	Apolipoprotein B	$5.8x10^{-20}$	1.0	78	0.8,1.2	ENST00000252444	7.8
LDLR	Apolipoprotein B	$6.7x10^{-18}$	0.9	84	0.7,1.1	ENST00000558518	0.5
LDLR	Apolipoprotein B	$3.2x10^{-16}$	1.0	67	0.7,1.2	ENST00000455727	0
LDLR	Apolipoprotein B	$1.0x10^{-14}$	0.9	65	0.7,1.2	ENST00000545707	0
LDLR	Apolipoprotein B	$4.4x10^{-13}$	0.9	56	0.7,1.2	ENST00000535915	0
LDLR	Apolipoprotein B	$1.6x10^{-12}$	0.6	118	0.5,0.8	ENST00000558013	0
LDLR	Apolipoprotein B	$1.7x10^{-9}$	0.5	124	0.4,0.7	ENST00000557933	0.1
LDLR	Apolipoprotein B	$3.8x10^{-4}$	0.2	287	0.1,0.3	Inclusive model	

Table 6. Comparison of *LDLR* transcript-based models and the inclusive, gene-based model on apolipoprotein B levels

4.7. *Transcript-specific variant sets exclude misannotated pLOF variants*

Additionally, the transcript-specific variant sets can improve association testing through the exclusion of misannotated variants. For example, polycysatin-1 (*PKD1)* is a well-characterized protein for its function in causing 85% of autosomal dominant polycystic kidney disease cases[23]. When damaged, the kidneys are unable to clear waste products like urea and creatinine which instead end up in high concentrations in the blood. Elevated serum urate is documented in rare-variant burden testing of *PKD1* pLOF variants[24]. Our results show an improved association of *PKD1* and urate using the transcript-based approach. When comparing the most significantly associated transcript variant set, ENST00000423118, and the gene-based burden, 12 variants are excluded. The most frequent among these is rs758337073, a *PKD1* variant labeled as "likely benign" by ClinVar[25]. Rs758337073 is a "stop gained" pLOF in ENST00000488185 and is subsequently designated as a pLOF in the gene-based method. However, rs758337073 is not considered a pLOF in 23/24 *PKD1* transcripts. ENST00000488185 has low overall expression, and zero expression in kidney cortex or medulla as shown by GTEx, indicating that this is likely a misannotated pLOF, and its inclusion in the gene-based method adds noise and dampens the *PKD1*-urate burden association (Figure 6).

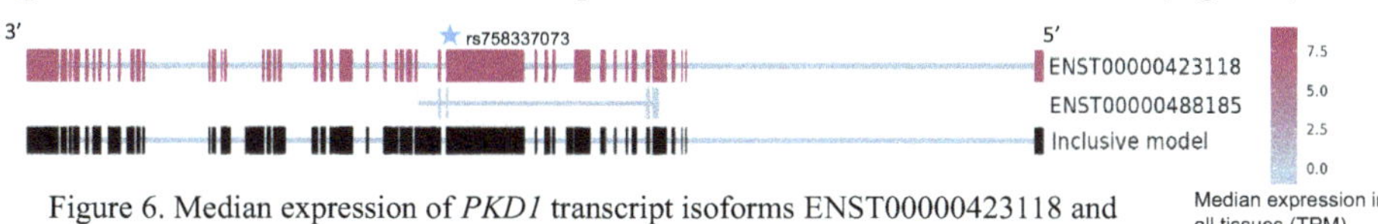

Figure 6. Median expression of *PKD1* transcript isoforms ENST00000423118 and ENST00000488185, and the inclusive, gene-based model.

5. Discussion

The drug discovery process is long, costly, and rarely ends in approval. Human genetic evidence provides an opportunity for novel target identification and validation for existing programs. Both common and rare variant genetic analyses have been shown to improv the chances of a successful clinical trial and form the basis of rational drug discovery and development[26].

Our analysis highlights the importance of incorporating transcript-aware analyses into RVAS. We find that a transcript-aware approach broadly leads to lower circulating levels of *cis*-proteins as compared to the gene-based method. Since we expect pLOFs to lead to nonsense-mediated decay, and a reduction of functional RNA and protein products, this indicates that the included variants are more likely to be functioning as true LOFs. This is also evident in quantitative-trait testing, where we observe increased absolute value of effect sizes for the isoform-specific variant sets. The transcript-level approach also identifies novel isoform-trait associations, and in rare cases, identifies associations with an opposite direction of effect as compared to the gene-based method, as is the case with *GPNMB*, *HMOX2*, and *BST1* and their proteins encoded in *cis*. These data indicate the potential for a transcript-aware approach to elucidate new genetically validated drug targets, some of which may be isoform-specific.

Previously published literature has highlighted the importance of considering transcript data in RVAS. Cummings et al. described variants overlapping low confidence transcripts as a main contributor to the false annotation of pLOF variants. To counteract this, the authors developed the "proportion expressed across transcripts" (pext) score which quantifies the expression of transcript isoforms and exons. When testing pLOF variants in low pext-scored regions, the authors reported effect sizes comparable to the inclusion of synonymous variants. However, testing pLOF variants in high pext-scored regions resulted in substantially larger effect sizes[22]. This is consistent with our results showing that the transcript-level burden leads to larger effect sizes, in some cases, like for *LDLR*, by excluding variants in low expression exonic regions.

Our approach is limited in several ways. By only using transcript isoforms detected in at least one of 53 GTEx tissues, we exclude transcripts that may be expressed in other tissue and cell types. For example, several quantitative ocular phenotypes were tested, but we did not utilize data on ocular transcript isoform expression. Additionally, our analysis was only conducted on European-ancestry individuals due to limited sample size of other ancestral groups; RVAS in other populations may yield additional associations.

One drawback to the transcript-aware approach is the reduction in sample size, as all isoform-aware variant sets are smaller than their gene-based counterparts. Additionally, a given gene can have multiple alternatively spliced, biologically relevant isoforms, where a pLOF variant in any number of those isoforms may lead to the same deleterious effect on a phenotype. In that case, testing a single transcript would not be a sufficient representation, and instead it would be better to use a more inclusive multi-transcript or gene-based approach.

It is possible to test all transcript-variant sets alongside the gene-based method, as we have done here. However, this leads to an exceptionally stringent P value threshold and many highly related experiments. We suggest a curated implementation of the transcript-approach by testing only specific transcripts chosen *a priori*, for example, only canonical transcripts, MANE-select transcripts[27] which intend to choose the most biologically relevant, representative isoform for each gene, or highly expressed transcript isoforms in relevant tissue types.

References

1. Pan, Q., Shai, O., Lee, L. J., Frey, B. J. & Blencowe, B. J. Deep surveying of alternative splicing complexity in the human transcriptome by high-throughput sequencing. *Nat Genet* **40**, 1413–1415 (2008).

2. King, E. A., Wade Davis, J. & Degner, J. F. Are drug targets with genetic support twice as likely to be approved? Revised estimates of the impact of genetic support for drug mechanisms on the probability of drug approval. *PLoS Genet* **15**, (2019).

3. Razuvayevskaya, O., Lopez, I., Dunham, I. & Ochoa, D. Why Clinical Trials Stop: The Role of Genetics. doi:10.1101/2023.02.07.23285407.

4. Manolio, T. A. *et al.* Finding the missing heritability of complex diseases. *Nature* vol. 461 747–753 Preprint at https://doi.org/10.1038/nature08494 (2009).

5. Lee, S., Abecasis, G. R., Boehnke, M. & Lin, X. Rare-variant association analysis: Study designs and statistical tests. *American Journal of Human Genetics* vol. 95 5–23 Preprint at https://doi.org/10.1016/j.ajhg.2014.06.009 (2014).

6. Wainschtein, P. *et al.* Assessing the contribution of rare variants to complex trait heritability from whole-genome sequence data. *Nat Genet* **54**, 263–273 (2022).

7. Li, B. & Leal, S. M. Methods for Detecting Associations with Rare Variants for Common Diseases: Application to Analysis of Sequence Data. *Am J Hum Genet* **83**, 311–321 (2008).

8. Madsen, B. E. & Browning, S. R. A groupwise association test for rare mutations using a weighted sum statistic. *PLoS Genet* **5**, (2009).

9. Karczewski, K. J. *et al.* The mutational constraint spectrum quantified from variation in 141,456 humans. *Nature* **581**, 434–443 (2020).

10. Sudlow, C. *et al.* UK Biobank: An Open Access Resource for Identifying the Causes of a Wide Range of Complex Diseases of Middle and Old Age. *PLoS Med* **12**, (2015).

11. Bycroft, C. *et al.* Genome-wide genetic data on ~500,000 UK Biobank participants. doi:10.1101/166298.

12. Szustakowski, J. D. *et al.* Advancing human genetics research and drug discovery through exome sequencing of the UK Biobank. *Nature Genetics* vol. 53 942–948 Preprint at https://doi.org/10.1038/s41588-021-00885-0 (2021).

13. Sun, B. B. *et al.* Genetic regulation of the human plasma proteome in 54,306 UK Biobank participants 2 3. *Population Analytics of Janssen Data Sciences* **20**,.

14. Van Hout, C. V. *et al.* Exome sequencing and characterization of 49,960 individuals in the UK Biobank. *Nature* **586**, 749–756 (2020).

15. Pan-UKB team. Pan UKB. https://pan.ukbb.broadinstitute.org (2020).

16. McLaren, W. *et al.* The Ensembl Variant Effect Predictor. *Genome Biol* **17**, (2016).

17. Backman, J. D. *et al.* Exome sequencing and analysis of 454,787 UK Biobank participants. *Nature* **599**, 628–634 (2021).

18. Purcell, S. *et al.* PLINK: A tool set for whole-genome association and population-based linkage analyses. *Am J Hum Genet* **81**, 559–575 (2007).

19. Mbatchou, J. *et al.* Computationally efficient whole-genome regression for quantitative and binary traits. *Nat Genet* **53**, 1097–1103 (2021).

20. Cheng-Hathaway, P. J. *et al.* The Trem2 R47H variant confers loss-of-function-like phenotypes in Alzheimer's disease. *Mol Neurodegener* **13**, (2018).

21. Moutinho, M. *et al.* TREM2 splice isoforms generate soluble TREM2 species that disrupt long-term potentiation. *Genome Med* **15**, (2023).

22. Cummings, B. B. *et al.* Transcript expression-aware annotation improves rare variant interpretation. *Nature* **581**, 452–458 (2020).

23. Peters, D. J. M. & Sandkuijl, L. A. Genetic Heterogeneity of Polycystic Kidney Disease in Europe1. in 128–139 doi:10.1159/000421651.

24. Karczewski, K. J. *et al.* Systematic single-variant and gene-based association testing of thousands of phenotypes in 394,841 UK Biobank exomes. *Cell Genomics* **2**, 100168 (2022).

25. Landrum, M. J. *et al.* ClinVar: Improving access to variant interpretations and supporting evidence. *Nucleic Acids Res* **46**, D1062–D1067 (2018).

26. Minikel, E. V., Painter, J. L., Dong, C. C. & Nelson, M. R. Refining the impact of genetic evidence on clinical success. doi:10.1101/2023.06.23.23291765.

27. Morales, J. *et al.* A joint NCBI and EMBL-EBI transcript set for clinical genomics and research. *Nature* (2022) doi:10.1038/s41586-022-04558-8.

Generating new drug repurposing hypotheses using disease-specific hypergraphs

Ayush Jain,[1,2,†] Marie-Laure Charpignon,[1] Irene Y. Chen,[3,4] Anthony Philippakis[1], Ahmed Alaa [3,4]
[1] *Broad Institute of MIT and Harvard, Cambridge, MA, USA*
[2] *Duke University, Durham, NC, USA*
[3] *UC Berkeley, Berkeley, CA, USA*
[4] *UC San Francisco, San Francisco, CA, USA*
† *Corresponding e-mail: a.jain@duke.edu*

The drug development pipeline for a new compound can last 10-20 years and cost over $10 billion. Drug repurposing offers a more time- and cost-effective alternative. Computational approaches based on network graph representations, comprising a mixture of disease nodes and their interactions, have recently yielded new drug repurposing hypotheses, including suitable candidates for COVID-19. However, these interactomes remain aggregate by design and often lack disease specificity. This dilution of information may affect the relevance of drug node embeddings to a particular disease, the resulting drug-disease and drug-drug similarity scores, and therefore our ability to identify new targets or drug synergies. To address this problem, we propose constructing and learning disease-specific hypergraphs in which hyperedges encode biological pathways of various lengths. We use a modified node2vec algorithm to generate pathway embeddings. We evaluate our hypergraph's ability to find repurposing targets for an incurable but prevalent disease, Alzheimer's disease (AD), and compare our ranked-ordered recommendations to those derived from a state-of-the-art knowledge graph, the multiscale interactome. Using our method, we successfully identified 7 promising repurposing candidates for AD that were ranked as unlikely repurposing targets by the multiscale interactome but for which the existing literature provides supporting evidence. Additionally, our drug repositioning suggestions are accompanied by explanations, eliciting plausible biological pathways. In the future, we plan on scaling our proposed method to 800+ diseases, combining single-disease hypergraphs into multi-disease hypergraphs to account for subpopulations with risk factors or encode a given patient's comorbidities to formulate personalized repurposing recommendations.

Supplementary materials and code: `https://github.com/ayujain04/psb_supplement`

Keywords: Hypergraphs, Precision Medicine, Drug Repurposing, Disease Specificity

1. Introduction

The development of new drugs can take more than 15 years, from the discovery and pre-clinical phase to review by regulatory agencies.[1] Hence, repurposing drugs previously approved by the Food and Drug Administration or European Medicines Agency serves as a convenient alternative since they are already known to be safe in human populations. From a research and development perspective, drug repurposing is a less risky enterprise. Indeed, following compound identification, repositioned drugs would generally hit the market in less than 10 years. Beyond time savings, this strategy brings significant cost savings, potentially reducing the average pharmaceutical pipeline's budget by over $5 billion compared to traditional drug development. To date, drug repurposing encompasses three main approaches: computational biomedicine,[2] biological experimentation, and their combination, e.g., through systems pharmacology.[3]

Computational approaches are both more time-effective and cost-effective than *in vitro* or *in vivo* biological experiments, which involve high-throughput screening or phenotypic screening based on animal and human models, respectively. Examples of available strategies include signature matching, genome-wide association studies, and the retrospective analysis of real-world clinical information.[4] Their use has been unlocked by the concurrent emergence of technical advances such as biological microarrays and the increase in data accessibility, as illustrated by the rapid growth of electronic health records and biobanks.[5]

Simultaneously, massive genomic databases and cell lines have yielded 20+ high-quality biological and biomedical knowledge graphs (KG) such as SPOKE[6] and PrimeKG[7] and aggregating platforms such as the KG-Hub[8] to ensure that the former can be shared and made interoperable for downstream graph machine learning tasks. Network-based methods for drug repurposing rely on the encoding of interactions between entities (i.e., drugs, diseases, proteins, biological functions) that can be heterogeneous (i.e., inhibition, binding). These representations can help address both predictive (e.g., polypharmacy side effects) and inferential (e.g., reasoning over causal pathways) questions. Prior graph representations such as the multi-scale interactome (MSI)[9] have proved useful in identifying agents that were previously repurposed and in formulating new potential drug repurposing candidates.

However, drug repurposing hypotheses output by algorithms or deep learning models deployed on KGs may appear as "black boxes." Yet structural and/or functional explanations are often desirable and necessary to understand the possible mechanisms of action underlying a predicted relationship between an existing drug and a disease – be it beneficial or detrimental. Further, KGs integrating various data sources are rarely disease-specific. Thus, they may result in overall drug similarities that do not hold for the pathology of interest or in spurious correlations. This concern is especially relevant for neurodegenerative diseases such as AD, given the presence of the blood-brain barrier[10] and differential gene expression levels and patterns in the brain – relative to other tissues – and across brain regions themselves.

Contributions. Hypergraphs have seen success in uncovering relationships in areas like marketing,[11] finance,[12] and computer vision.[13] Building upon this precedent in other disciplines, we propose disease-specific hypergraphs as the basis for data-driven drug repurposing. Importantly, hypergraphs allow encoding relationships among groups of nodes (i.e., hyperedges) rather than pairwise relationships (i.e., edges) only. In our study, hyperedges capture known biological pathways. First, we show that the properties of hypergraphs reflect relative disease complexity. Second, we transform disease-specific hypergraphs into weighted graphs where nodes encode biological pathways and weighted edges relate to the number of entities that they have in common (e.g., the number of shared genes or proteins). With the intent of encompassing disease specificity, we focus on pathways that start with a drug entity and end with the disease entity of interest, irrespective of their length. Using a modified node2vec[14] algorithm, we learn the disease-specific embeddings of each hyperedge. In particular, we use these low-dimensional representations to find original biological pathways that are highly similar to those whose starting entity is a drug currently prescribed to treat the disease or mitigate its progression. Then, we pool the top k or k% candidate biological pathways for various values of k and analyze the distributions of starting drug entities and middle gene entities. Such prevalences help gain a mechanistic understanding of promising drug classes and targets for repurposing. We illustrate our proposed method in the context of Alzheimer's disease (AD), a multi-factorial disease of aging that still has no cure despite recent progress.[15] We demonstrate that our proposed method outputs candidate biological pathways that are topologically non-obvious, i.e., they do not have any entities in common with the reference pathways involving currently prescribed drugs, besides the end disease entity. To assess the utility and complementary of learning disease-specific pathway embeddings, we contrast these non-obvious suggestions with those of the MSI, compare the corresponding rank orderings, and validate our findings by mining the biomedical literature to find supporting evidence. Our comparative analysis and publication search reveals that certain candidates that were highly ranked (i.e., in the top 10%) by our hypergraph-based learning approach for AD drug repurposing and had supporting evidence in the literature were missed by the MSI (i.e., in the bottom 33% across all drugs). Going forward, our proposed framework can be scaled to derive novel drug repurposing hypotheses for each of the 800+ major diseases currently registered on the KG-Hub[8] (i.e., excluding rare and orphan diseases).

2. Methods

In this section, we describe our proposed approach, which encompasses three main parts: hypergraph construction, pathway/hyperedge representation learning, comparative analysis with the MSI,[9] and mining of the biomedical literature to find supporting evidence.

2.1. *Hypergraph Construction*

We built disease-specific hypergraphs by querying the Hetionet[16] knowledge graph, which comprises 1,522 drugs, 5,734 side effects, and 137 diseases, to extract significant[a] biological pathways connecting each drug present in the KG to the disease of interest. Hetionet is an existing state-of-the-art knowledge graph that incorporates 11 node types (e.g., gene, symptom), allowing for vast heterogeneity in the node composition of "metapaths" going from a compound to a disease, which we sample from to create disease-specific hypergraphs (Figure 1a). We query the Hetionet to retrieve all paths starting at one of the 1,522 drugs and ending at a disease of choice. These paths are further grouped into metapath categories based on the type and order of nodes present within the path. For example, a metapath category could be "drug-gene1-gene2-targetDisease" or "drug-gene2-similarDisease-targetDisease." We

[a] A more detailed definition of significance follows.

use the direct weighted path count and adjusted p-value defined by the Hetionet to quantify the significance of a path, relative to others within its metapath category. We include the top 10% most significant paths within each category to create our induced disease-specific subgraph. We reasoned that selecting only the most significant pathways would help mitigate the resulting number of false positives among drug repurposing candidates. The largest connected component is treated as the subgraph of interest (Figure 1b); other components, generally much smaller in size, are ignored. All existing biological pathways in the resulting subgraph are explicitly unified as hyperedges, creating a disease-specific hypergraph (Figure 1c). Lastly, we transformed our disease-specific hypergraph into a disease-specific graph where the nodes now correspond to the biological pathway hyperedges that originally constituted the hypergraph. Two biological pathway nodes are connected if they share another element in their path besides the start entity (drug) and the end entity (disease). The edge weight is defined by the number of shared elements $\mathbf{w}$, normalized between 0 and 1 using min-max scaling (Figure 1d) to enable comparisons of graph structures across diseases.

2.2. *Biological Pathway Hyperedge Representation Learning*

Given a specific disease of interest, our study aimed to identify biological pathways analogous – in a learned distinct dimensional subspace – to those associated with drugs currently used to treat it. In particular, we conducted a case study on Alzheimer's disease and considered medications prescribed to alleviate the associated symptoms and behavioral complications. We focused primarily on three compounds: donepezil, galantamine, and memantine,[17–19] approved by the FDA in 1996, 2001, and 2003, respectively. Our approach is disease-agnostic and can be readily extended to other diseases than AD, upon the supply of a list of compounds currently used in clinical practice or previously suggested as repurposing candidates and provided access to adequate computing resources.

Our methodology involved initiating a random walk of fixed length L on the transformed, weighted graph G_w delineated in Figure 1 (d), commencing from any of the biological pathway nodes whose first path element was one of the drugs currently prescribed against the disease of interest. We accounted for the presence of weighted edges by sampling neighboring nodes proportionally to the strength of the connection. The random walker began at a selected node, then proceeded iteratively to an adjacent node chosen uniformly at random among possibly duplicated neighbors, and repeated this process for a predetermined number of steps. For each eligible starting node in our weighted graph G_w, a random walk was initiated, with a fixed length set at L=80. Each start node-specific random walk was replicated R=10 times, in light of the vast heterogeneity of node types and the resulting variation in feasible trajectories.

We denote by v_i the position of the random walker at iteration $i = 1$. At each iteration i of the random walk, the probability of transitioning from biological pathway x to biological pathway y is expressed as:

$$P(v_i = y|v_{i-1} = x) = \frac{\text{weight of edge between } x \text{ and } y}{\text{sum of weights of all edges leaving } x} \tag{1}$$

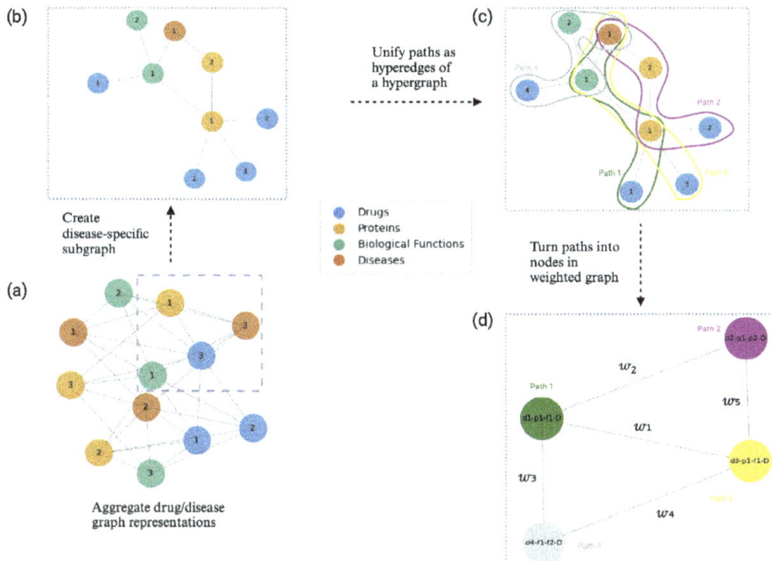

Fig. 1: Pipeline to derive disease-specific hypergraphs from existing KGs and learn contextual embeddings of biological pathways. (a) Full Hetionet graph with nodes of 11 types, including 1,522 drugs, 5,734 side effects, and 137 diseases. (b) Disease-specific subgraph, selecting only the biological pathways whose end node is the disease of interest. Of note, only the top 10% most significant pathways within each metapath category (as defined by their length and structure, e.g., drug-gene1-gene2-disease) are retained, based on a path importance score assigned by Hetionet.[16] (c) Disease-specific hypergraph unifying significant paths or hyperedges into a single structure. (d) Disease-specific weighted graph resulting from the transformation of the hypergraph described in (c). Each hyperedge in (c) becomes a node in (d) and nodes in (d) are connected if their corresponding biological pathways in (c) have at least another element in common, beyond the disease node. Each edge is assigned a weight **w**, corresponding to the number of elements common to the two biological pathways. Note that the weight does not include the disease node, which all pathways present in a given disease-specific hypergraph intersect at, by design; similarly, the compound node at the start of each reference biological pathway does not contribute to the weight either, continuing our focus on learning biological similarities between the drugs).

2.2.1. *Skip-Gram Model*

We interpreted the resulting random walks as sentences, utilizing the Word2Vec Skip-Gram model provided by gensim to develop node embeddings for each biological pathway.[14] This model predicts context words (nodes within the same walk) given a target word (a node).

Applied to the context of our disease-specific weighted graph, the embeddings of biological pathways learned through this process encapsulate the local neighborhood structure of the nodes and are subsequently used for our pathway similarity search.

The Skip-Gram model's objective is to devise word representations that effectively predict surrounding words in a sentence or document.[20] Formally stated, given a sequence of training words $w_1, w_2, ..., w_T$, the model aims to maximize the average log probability obtained via the chain rule:

$$\frac{1}{T}\sum_{t=1}^{T}\sum_{-k\leq j\leq k, j\neq 0} \log P(w_{t+j}|w_t) \tag{2}$$

where k denotes the size of the training context and T denotes the total number of training words. In linguistics, k often represents the typical length of a sentence; by analogy, in biology, it could encode the number of reactions occurring in cascade. Similarly, in linguistics, T often represents the size of the vocabulary, which can be language-specific; in biology, the total number of biological pathways involved is disease-specific. To alleviate the fact that pathway length can greatly vary, we guided the model to learn embeddings of fixed dimension $p = 64$ dimensions. We subsequently used cosine similarity as the metric to quantify similarity between any two biological pathways.

Our decision to utilize the Skip-Gram algorithm for learning embeddings was driven by our intent to infer semantic contextual relationships among biological pathways, given a specific disease. To learn the embeddings, we chose the random walk and skip-gram based approach to derive a first proof of concept of using a hypergraph structure and explicitly restricting it to a given disease.

2.3. *Methods for Evaluation*

Our approach to proposing repurposing hypotheses for a given disease of interest relies on identifying the top 10% of biological pathway embeddings having the highest cosine similarity with pathways initiating from any of the drugs known to mitigate or prevent disease progression and ending at the disease node. While pathways can include a variety of intermediary nodes (e.g., symptom, anatomical object, etc.), we selected exclusively those pathways with one or more gene intermediary nodes linking one of the 1,522 drug candidates to the considered disease. We reasoned that this feature would help focus our hypotheses on biologically plausible pathways and thus facilitate the interpretation of drug repurposing candidate rankings.

The data and methods of the multiscale interactome (MSI) were used as a baseline for comparative analysis. The MSI consists in a large biological KG with 1,566 drugs and 841 diseases and leverages a random walk approach to formulate repurposing hypotheses.

From our weighted graphs, we quantified the rank of each biological pathway in terms of its cosine similarity to a selected relevant pathway. We considered relevant pathways to be those whose starting drug entity was a drug currently indicated against the disease of interest. To obtain a single metric per drug, we either aggregated the cosine similarity scores of the pathways in which it was involved into an median value or used the pathway with the maximum similarity. Then, we used this summary metric to rank all considered drugs and contrast our own repurposing suggestions with those of the MSI. We used these metrics and the relative rankings of psychoanaleptics in the MSI vs. our hypergraph to compare their AUC.

From the MSI, we derived rankings of the most similar drug pairs based on the rankings of the drugs most similar to disease-specific drugs, based on the cosine similarity of their 64-dimensional embeddings. We also established a rank-ordered list of drugs most similar to the disease of interest (e.g., AD), given the cosine similarity between drug and disease embeddings. Notably, while the rank of a drug as derived from the MSI is the output of either a single similarity score (with the disease's embedding) or a couple of scores only (with the embeddings of known drugs against this disease), its rank as derived from our hypergraph-based approach is the output of a much larger double-averaging operation, across all pathways starting at the drug of interest and those starting at a drug already known to target this disease.

We aimed at uncovering any potential blind spots in the MSI that our methodology might successfully uncover. To this end, for drugs that our approach ranked among the top 10%, we retrieved the corresponding MSI-derived rankings for comparison. For each pair, we computed the absolute difference in rank. In addition, we computed an aggregate similarity metric between the MSI and our approach, defined as the size of the overlap between the sets of drugs appearing in the top 10% under each.

To further validate our methodology and the relevance of the resulting pathway embeddings, we undertook a deeper analysis of the drug repurposing suggestions that most differed between the MSI and our proposed method, based on the difference in ranks. In particular, we searched the biomedical literature for biological and/or clinical evidence about drug repurposing suggestions that fell within the bottom third of the MSI's rank-ordered list (i.e., compounds ranked 1,032 to 1,522) while being in the top 10% of ours.

3. Results

We conducted several experiments on our disease-specific hypergraph, using a sample of 18 prevalent and/or incurable diseases. First, we computed summary statistics about these 18 disease hypergraphs and formed clusters that reflect known disease complexity (Section 3.1) Second, we learned disease-specific embeddings of biological pathways on these hypergraphs to identify potential drug targets. To interpret our findings, we explored the distribution of genes involved in the resulting pathways (Section 3.2 and Figure 3). Our intent was to confirm some of the repurposing hypotheses that emerged from the MSI and to formulate new ones. Third, we reviewed the literature to gather information about targets overlooked by the MSI but documented in prior studies; we summarize our findings (Section 3.3). We also evaluated the AUC differences between the MSI and our Alzheimer's disease specific hypergraph (Appendix B in our supplement)

3.1. *Hypergraph Construction Underlines Known Disease Complexity*

Utilizing our method, we constructed 18 disease-specific hypergraphs (Figure 2). For each, we computed the number of hyperedges (biological pathways) and the number of weighted links among them; we mapped diseases on a scatterplot along these two dimensions. Using the k-means clustering algorithm (k=4), we grouped the 18 hypergraphs into four clusters. In addition to this visual representation, we quantified the number of protein nodes that each disease connects with in the MSI – a proxy to characterize disease complexity. Generally, more complex hypergraphs, with a larger number of hyperedges and links, were those of diseases involving a larger number of proteins. This suggests that the network properties of disease-specific hypergraphs could be leveraged to summarize their complexity and identify diseases that may be proximal based on their higher-order structure.The richness of the embeddings that we seek to learn will depend on the size of the underlying hypergraph; in particular, smaller hypergraphs may yield sparser embeddings. For instance, Figure 2 highlights a clear separation between chronic diseases such as Chronic Kidney Disease (CKD) and Coronary Heart Disease (CHD) and more complex diseases such as Rheumatoid Arthritis (RA) and Amyotrophic Lateral Sclerosis (ALS) involving auto-immune processes. Among the 18 hypergraphs, those of diseases in Cluster A boast more information to learn from and potentially uncover peripheral biological pathways of importance; this configuration prompted us to select one of the diseases in cluster A for our case study. Alzheimer's Disease (AD) was chosen due to its large and growing prevalence, as about 6.2 million Americans aged 65 and older are currently affected – a number which could rise to 13.8 million by 2060.[21]

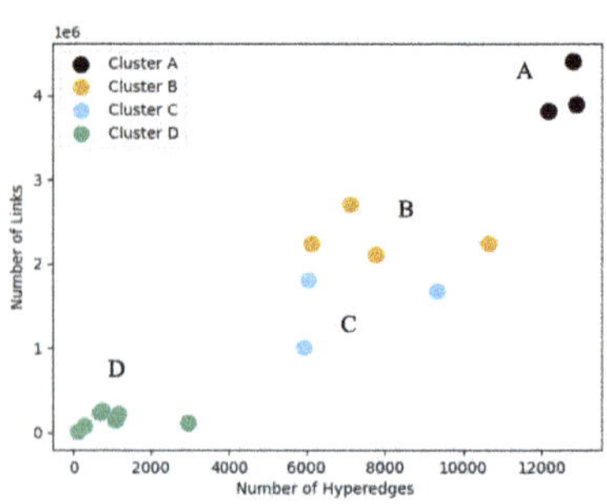

Cluster ID	Diseases Present	Proteins connected in the MSI
A	Alzheimer's Disease (1), Rheumatoid Arthritis (2), ALS (3)	(1) 73 (2) 162 (3) 49
B	Lupus (4), Psoriasis (5), Asthma (6), Hepatitis B (7)	(4) 60 (5) 52 (6) 92 (7) 6
C	Hypertension (8), Type 2 Diabetes (9), Type 1 Diabetes (10),	(8) 151 (9) 78 (10) 9
D	CHD (11), CKD (12), Diabetic Retinopathy (13), Renal Failure (14), Parkinson's Disease (15), AIDS (16), Atrial Fibrillation (17), Vascular Dementia (18)	(11) 15 (12) 39 (13) 13 (14) 13 (15) 39 (16) 1 (17) 34 (18) N/A

Fig. 2: The scatter plot represents four disease clusters, based on two structural attributes of their respective disease-specific hypergraphs, constructed as outlined in sections 1(b-c). The location of a given cluster indicates the complexity of the higher-order hypergraph structures and often reflects disease complexity. Diseases known to be highly complex (e.g., AD) are positioned in the top right corner; conversely, diseases deemed to be of lower complexity (e.g., renal failure) are situated in the bottom left corner. While disease complexity is primarily defined by the number of currently known biological pathways involved, we also provide the number of proteins to which each disease is directly connected (i.e., one-hop neighbors) in the MSI.[9]

3.2. *Hypergraph Representation Learning Identifies Repurposing Targets in Accordance with the MSI*

Both the MSI and our AD specific hypergraph shared a 50% overlap in drug categories for their repurposing suggestions: psycholeptics, psychoanaleptics, and drugs used in diabetes management, all of which are supported to have repurposing targets to AD in the literature.[22,23] The MSI also brought attention to drugs acting on the renin-angiotensin system, sex hormones, and other nervous system drugs, thereby grouping together common co-morbidity targets for AD treatment.[24–27] In contrast, our AD-specific hypergraph focused more on antineoplastic agents, cardiac therapy, and ophthalmologicals, each of which has been associated with AD in the literature, as well.[28–30]

Figure 3 shows how our hypergraphs can be used to add a layer of explainability to existing knowledge graphs. The figure also compares the number of gene targets that each of the suggestions from both our method and the MSI contained. After finding the top 10% of drugs most similar to AD in the MSI, we found all gene intermediary nodes in the paths starting at these drugs and ending at AD in our hypergraph. We then compared the makeup of these pathways to the makeup of the top 10% most similar pathways to those of donepezil, memantine, or galantamine (ranked by highest cosine similarity to any of the three drugs).

It is important to note the discrepancy in the number of paths considered when generating Figure 3(a)-(g) and Figure 3(h)-(n). The former, involving 926 paths, includes all paths in the AD hypergraph that start with a drug in the top 10% cosine similarity to AD in the MSI. Conversely, the latter, with 574 paths, only encompasses the top 10% of pathways that exhibit the highest cosine similarity to paths initiating from donepezil, memantine, or galantamine.

These findings underscore the efficacy of our disease-specific hypergraph approach in targeting drugs with pathways highly similar to those of known pertinent drugs when identifying potential candidates for repurposing. Moreover, these outcomes provide initial validation to our hypergraph representation learning method, which will be further discussed in the following subsection.

3.3. *Hypergraph Representation Learning Identifies Drug Repurposing Targets Discounted from the MSI but Present in Literature*

Hypergraph representation learning suggested 7 drug repurposing targets out of its top 30 (23%) that the MSI discounted (rank of $\geq$ 1032 in either column (2) or (3) of Table 1 in the supplement). The 7 drugs were eplerenone (diruetic), fosphenytoin (cardiac therapy), exemestane (endocrine therapy), eperisone (muscle relaxants), protriptyline (psychoanaleptics), ethotoin (antiepileptics), and pentamidine (antiprotozoals). 4 out of 7 (eplernone, pentamidine, exemestane, and protriptyline) of these have literature supporting their potential efficacy against AD. For the remaining 3 out of 7 (eperisone, ethotoin, and fosphenytoin), we explored tangential literature to evaluate the suggestion and/or looked upon the path that this druge headed in hopes of understanding why the prediction was made. Refer to Table 1 in the supplement for the exhaustive list of the top 30 repurposing suggestions based on pathway similarity to donepezil, memantine, or galantamine in the AD hypergraph.

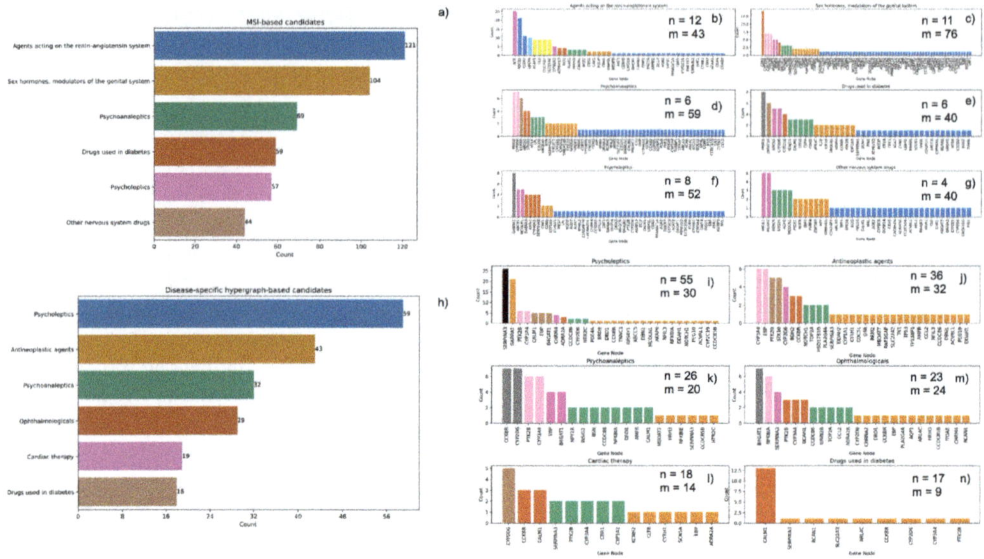

Fig. 3: (a)-(g) illustrate the number of gene targets within the paths of our AD hypergraph, originating from a drug node that ranks within the top 10% in terms of highest cosine similarity to the AD node in the MSI. (h)-(n) depict the number of gene targets in paths within our AD hypergraph that are within the top 10% in similarity to the pathways of donepezil, memantine, and galantamine, only considering paths with gene intermediary nodes.(a) presents the six categories with the most gene targets among the MSI's top 10% suggestions. Conversely, (h) displays the six categories with the most gene targets in the AD hypergraph's top 10% predictions based on similarity to donepezil, memantine, and galantamine. (b)-(g) further break down these top six categories from (a), demonstrating the count of each gene in the top 10% of predicted paths. Similarly, (i)-(n) break down the top six categories from (h), showing the count of each gene within the top 10% of paths similar to those of donepezil, memantine, and galantamine, as determined by cosine similarity. In these graphics, 'n' represents the number of unique drugs within this category, while 'm' signifies the number of unique gene targets in (b)-(g) and (i)-(n).

3.3.1. *Literature Review on 7 Targets Found by Hypegraph Representation Learning but Missed by MSI*

In this section, we delve into the literature that supports our hypotheses for drug repurposing. These drugs were identified as potential repurposing candidates, yet were overlooked by the MSI.

Eplernone has been observed to decrease brain damage, defined by cell death and cortical thinning, in a rat model.[31] Additionally, it is documented that eplernone enhances cognitive function in a mouse model of AD.[32] Another study reinforces these findings, illustrating that eplernone can mitigate cognitive deficits in the hippocampus of spontaneously hypertensive rats.[33] These outcomes coincide with the established correlation between hypertension and dementia/AD.[34] Lastly, an *in silico* pharmacological assessment of eplernone proposed that the drug holds potential in treating AD.[35]

Exemestane exhibits efficacy when AD patients are concurrently dealing with cancer, suggesting that women diagnosed with breast cancer who underwent treatment with tamoxifen or exemestane exhibited fewer instances of AD.[36] A subsequent study characterizes the relationship between AD and cancer, demonstrating that exemestane is proficient at managing cancer when it co-occurs with AD.[37] Additional research has suggested exemestane as a potential therapeutic for Parkinson's Disease (PD),[38] a neurodegenerative disorder associated with AD.[39]

Protriptyline was found to have the highest inhibitory activity among 140 FDA approved nervous system drugs against the three primary AD targets: AChE, BACE-1, and Aβ aggregation.[40,41] A study using an AD rat model concluded that protriptyline reduces oxidative damage and improves spatial memory in AD mice.[42]

Pentamidine, in a mouse model of AD, was found to inhibit Aβ-induced gliosis and neuroinflammation in AD mice.[43] However, to our knowledge, this is the only publication endorsing its use in alleviating AD, likely because pentamidine is unable to cross the blood-brain barrier. However, recent developments in nose-to-brain methods could surmount this obstacle.[10]

Ethotoin (antiepileptic), to our knowledge, doesn't have explicit literature connecting it to AD. There is, however, a study that warns that antiepileptics could escalate stroke risk in AD patients.[44] This elucidates a current limitation of our approach: we currently do not differentiate between positive and negative drug pathways to a disease.

Fosphenytoin's affect on AD is not specifically discussed in the literature. However, it is a prodrug of phenytoin,[45] which inhibits hippocampal tissue degradation and consequently the progression of AD.[46]

Eperisone lacks direct literature linking it to AD, to the best of our knowledge. When examining the pathway, starting at eperisone and ending at AD that was suggested to have a high similarity to galantamine (see Table 1. in the supplement) in our hypergraph, we find: Eperisone-Triporlidine-CYP2D6-AD. This pathway shows that eperisone was connected to AD by way of similarity to triporlidine. Triporlidine has been observed to enhance NREM sleep in AD patients.[47]

4. Conclusion and Future Directions

Our disease-specific hypergraphs have proven useful for clustering diseases based on their known complexity, identifying potential drug repurposing targets alongside existing methods, and discovering promising repurposing targets overlooked by state-of-the-art methods. We found that the disease hypergraphs formed four clear groups when comparing the number of hyperedges to the number of links between these hyperedges (see Figure 2). Additionally, in Figure 2, we see more complex hypergraphs correlated with more known disease complexity, which we assessed by counting the protein-disease connections in the MSI.

We also demonstrated the value of this method in generating drug repurposing suggestions for Alzheimer's disease (AD). We saw a significant overlap with the suggestions from the MSI when looking at the top 10% of suggestions from both methods, especially among drug categories with the highest number of gene targets in the pathways.

Among our top 30 repurposing suggestions for AD, ranked by pathway cosine similarity, we focused on pathways with one or more protein/gene intermediary nodes, hoping to keep our results grounded in biological relevance. Each suggestion comes with the drug pathway that supports its potential use in treating AD (see Table 1 in the supplement for the full list).

Additionally, our method also identified promising repurposing pathways for AD that the MSI overlooked. In fact, 7 out of our top 30 suggestions ranked in the lower third of the MSI's suggestions. We found supporting evidence for these suggestions in the scientific literature, both from studies that directly tested the drugs and from related research.

Looking ahead, we plan to enhance this method in several ways. We aim to refine our hypergraph construction by merging disease hypergraphs of co-occurring diseases such as AD, Type 2 Diabetes, and Hypertension. We will explore ways to improve our pathway embeddings and, crucially, we will look beyond the literature review to other forms of validation, including evidence from electronic health records and experimental studies.

Future research will explore the use of power iteration, page rank, and page rank with teleportation for learning additional sets of embeddings for each biological pathway-disease pair. We aim to compare the resulting outputs to our current pathway embeddings pairwise and to assess the sensitivity of the downstream similarity scores. To derive more robust drug repurposing candidates for a specific disease, several embeddings of biological pathways could be combined to minimize dependence on a particular algorithm or parameter set and instead maximize confidence, across representation learning approaches. Further, we can experiment with more specific designs of a true positive, perhaps using literature for or against a drug in the context of a disease of interest.

Additionally, we plan on doing more sensitivity analyses upon the comparison metric and learning method, editing parameters like the dimensions of the embedding vector (p), distance of each random walk (L), and amount of random walks taken per each node (R). We also plan on comparing our results to the hypergraphs encompassing more paths. Now that we have outlined a proof-of-concept for this design on the top 10% of paths ending at a disease of interest, we can explore how the suggestions compare to hypergraphs built with the top 20%, 25%, etc.

References

1. H. Xue, J. Li, H. Xie and Y. Wang, Review of Drug Repositioning Approaches and Resources, *International Journal of Biological Sciences* **14**, 1232 (July 2018).
2. T. N. Jarada, J. G. Rokne and R. Alhajj, A review of computational drug repositioning: strategies, approaches, opportunities, challenges, and directions, *Journal of Cheminformatics* **12**, p. 46 (July 2020).
3. S. Zhao and R. Iyengar, Systems pharmacology: network analysis to identify multiscale mechanisms of drug action, *Annual Review of Pharmacology and Toxicology* **52**, 505 (2012).
4. P. Wu, Q. Feng, V. E. Kerchberger, S. D. Nelson, Q. Chen, B. Li, T. L. Edwards, N. J. Cox, E. J.

Phillips, C. M. Stein, D. M. Roden, J. C. Denny and W.-Q. Wei, Integrating gene expression and clinical data to identify drug repurposing candidates for hyperlipidemia and hypertension, *Nature Communications* **13**, p. 46 (January 2022), Number: 1 Publisher: Nature Publishing Group.
5. G. S. Q. Tan, E. K. Sloan, P. Lambert, C. M. J. Kirkpatrick and J. Ilomäki, Drug repurposing using real-world data, *Drug Discovery Today* **28**, p. 103422 (January 2023).
6. J. H. Morris, K. Soman, R. E. Akbas, X. Zhou, B. Smith, E. C. Meng, C. C. Huang, G. Cerono, G. Schenk, A. Rizk-Jackson, A. Harroud, L. Sanders, S. V. Costes, K. Bharat, A. Chakraborty, A. R. Pico, T. Mardirossian, M. Keiser, A. Tang, J. Hardi, Y. Shi, M. Musen, S. Israni, S. Huang, P. W. Rose, C. A. Nelson and S. E. Baranzini, The scalable precision medicine open knowledge engine (SPOKE): a massive knowledge graph of biomedical information, *Bioinformatics* **39**, p. btad080 (02 2023).
7. P. Chandak, K. Huang and M. Zitnik, Building a knowledge graph to enable precision medicine, *Scientific Data* **10**, p. 67 (February 2023), Number: 1 Publisher: Nature Publishing Group.
8. KG-Hub—building and exchanging biological knowledge graphs | Bioinformatics | Oxford Academic.
9. C. Ruiz, M. Zitnik and J. Leskovec, Identification of disease treatment mechanisms through the multiscale interactome, *Nature Communications* **12**, p. 1796 (March 2021), Number: 1 Publisher: Nature Publishing Group.
10. F. Rinaldi, L. Seguella, S. Gigli, P. N. Hanieh, E. Del Favero, L. Cantù, M. Pesce, G. Sarnelli, C. Marianecci, G. Esposito and M. Carafa, inPentasomes: An innovative nose-to-brain pentamidine delivery blunts MPTP parkinsonism in mice, *Journal of Controlled Release* **294**, 17 (January 2019).
11. A. M A and A. Rajkumar, Hyper-IMRANK: Ranking-based Influence Maximization for Hypergraphs, in *5th Joint International Conference on Data Science & Management of Data (9th ACM IKDD CODS and 27th COMAD)*, CODS-COMAD 2022 (Association for Computing Machinery, New York, NY, USA, January 2022).
12. X. Ma, T. Zhao, Q. Guo, X. Li and C. Zhang, Fuzzy hypergraph network for recommending top-K profitable stocks, *Information Sciences* **613**, 239 (October 2022).
13. S. Bai, F. Zhang and P. H. S. Torr, Hypergraph convolution and hypergraph attention, *Pattern Recognition* **110**, p. 107637 (February 2021).
14. A. Grover and J. Leskovec, node2vec: Scalable Feature Learning for Networks (July 2016), arXiv:1607.00653 [cs, stat].
15. P. Scheltens, B. D. Strooper, M. Kivipelto, H. Holstege, G. Chételat, C. E. Teunissen, J. Cummings and W. M. v. d. Flier, Alzheimer's disease, *The Lancet* **397**, 1577 (April 2021), Publisher: Elsevier.
16. D. S. Himmelstein, M. Zietz, V. Rubinetti, K. Kloster, B. J. Heil, F. Alquaddoomi, D. Hu, D. N. Nicholson, Y. Hao, B. D. Sullivan, M. W. Nagle and C. S. Greene, Hetnet connectivity search provides rapid insights into how two biomedical entities are related (January 2023), Pages: 2023.01.05.522941 Section: New Results.
17. C.-C. Tan, J.-T. Yu, H.-F. Wang, M.-S. Tan, X.-F. Meng, C. Wang, T. Jiang, X.-C. Zhu and L. Tan, Efficacy and Safety of Donepezil, Galantamine, Rivastigmine, and Memantine for the Treatment of Alzheimer's Disease: A Systematic Review and Meta-Analysis, *Journal of Alzheimer's Disease* **41**, 615 (January 2014), Publisher: IOS Press.
18. M. Bond, G. Rogers, J. Peters, R. Anderson, M. Hoyle, A. Miners, T. Moxham, S. Davis, P. Thokala, A. Wailoo, M. Jeffreys and C. Hyde, The effectiveness and cost-effectiveness of donepezil, galantamine, rivastigmine and memantine for the treatment of Alzheimer's disease (review of Technology Appraisal No. 111): a systematic review and economic model., *Health technology assessment (Winchester, England)* **16**, 1 (2012), Number: 21 Publisher: NIHR Journals Library.

19. A. Burns, M. Rossor, J. Hecker, S. Gauthier, H. Petit, H.-J. Möller, S. Rogers and L. Friedhoff, The Effects of Donepezil in Alzheimer's Disease – Results from a Multinational Triall1, *Dementia and Geriatric Cognitive Disorders* **10**, 237 (May 1999).
20. T. Mikolov, I. Sutskever, K. Chen, G. Corrado and J. Dean, Distributed Representations of Words and Phrases and their Compositionality (October 2013), arXiv:1310.4546 [cs, stat].
21. 2021 Alzheimer's disease facts and figures, *Alzheimer's & Dementia: The Journal of the Alzheimer's Association* **17**, 327 (March 2021).
22. M. Citron, Alzheimer's disease: treatments in discovery and development, *Nature Neuroscience* **5**, 1055 (November 2002), Number: 11 Publisher: Nature Publishing Group.
23. M. M. Atef, N. M. El-Sayed, A. A. M. Ahmed and Y. M. Mostafa, Donepezil improves neuropathy through activation of AMPK signalling pathway in streptozotocin-induced diabetic mice, *Biochemical Pharmacology* **159**, 1 (January 2019).
24. R. Loera-Valencia, F. Eroli, S. Garcia-Ptacek and S. Maioli, Brain Renin–Angiotensin System as Novel and Potential Therapeutic Target for Alzheimer's Disease, *International Journal of Molecular Sciences* **22**, p. 10139 (January 2021), Number: 18 Publisher: Multidisciplinary Digital Publishing Institute.
25. IJMS | Free Full-Text | Brain Renin–Angiotensin System as Novel and Potential Therapeutic Target for Alzheimer's Disease.
26. L. Ghiadoni, Management of high blood pressure in type 2 diabetes: perindopril/indapamide fixed-dose combination and the ADVANCE trial [corrected], *Expert Opinion on Pharmacotherapy* **11**, 1647 (July 2010).
27. R. Li, J. Cui and Y. Shen, Brain sex matters: Estrogen in cognition and Alzheimer's disease, *Molecular and Cellular Endocrinology* **389**, 13 (May 2014).
28. M. Dahiya, A. Kumar, M. Yadav, P. Dhakla and S. Tushir, Therapeutic Targeting of Antineoplastic Drugs in Alzheimer's Disease: Discovered in Repurposed Agents, in *Drug Repurposing for Emerging Infectious Diseases and Cancer*, eds. R. C. Sobti, S. K. Lal and R. K. Goyal (Springer Nature, Singapore, 2023) pp. 329–345.
29. L. G. Howes, Cardiovascular Effects of Drugs Used to Treat Alzheimer's Disease, *Drug Safety* **37**, 391 (June 2014).
30. W. A. Fletcher, Ophthalmological aspects of Alzheimer's disease, *Current Opinion in Ophthalmology* **5**, p. 43 (December 1994).
31. X. Wang, Y. Zhu, S. Wang, Z. Wang, H. Sun, Y. He and W. Yao, Effects of eplerenone on cerebral aldosterone levels and brain lesions in spontaneously hypertensive rats, *Clinical and Experimental Hypertension* **42**, 531 (August 2020), Publisher: Taylor & Francis _eprint: https://doi.org/10.1080/10641963.2020.1723615.
32. L. Chen, R. Shi, X. She, C. Gu, L. Chong, L. Zhang and R. Li, Mineralocorticoid receptor antagonist-mediated cognitive improvement in a mouse model of Alzheimer's type: possible involvement of BDNF-H2S-Nrf2 signaling, *Fundamental & Clinical Pharmacology* **34**, 697 (2020), _eprint: https://onlinelibrary.wiley.com/doi/pdf/10.1111/fcp.12576.
33. Z. Lin, Y. Lu, S. Li, Y. Li, H. Li, L. Li and L. Wang, Effect of eplerenone on cognitive impairment in spontaneously hypertensive rats, *American Journal of Translational Research* **14**, 3864 (June 2022).
34. S. K. Raina, V. Chander, S. Raina, D. Kumar, A. Grover and A. Bhardwaj, Hypertension and diabetes as risk factors for dementia: A secondary post-hoc analysis from north-west India, *Annals of Indian Academy of Neurology* **18**, 63 (2015).
35. S. Hira, U. Saleem, F. Anwar, Z. Raza, A. U. Rehman and B. Ahmad, In Silico Study and Pharmacological Evaluation of Eplerinone as an Anti-Alzheimer's Drug in STZ-Induced Alzheimer's Disease Model, *ACS Omega* **5**, 13973 (June 2020), Publisher: American Chemical Society.
36. V. Das and M. Hajdúch, Randomizing for Alzheimer's disease drug trials should consider the

cancer history of participants, *Brain* , p. awad177 (May 2023).
37. J. Forés-Martos, C. Boullosa, D. Rodrigo-Domínguez, J. Sánchez-Valle, B. Suay-García, J. Climent, A. Falcó, A. Valencia, J. A. Puig-Butillé, S. Puig and R. Tabarés-Seisdedos, Transcriptomic and Genetic Associations between Alzheimer's Disease, Parkinson's Disease, and Cancer, *Cancers* **13**, p. 2990 (January 2021), Number: 12 Publisher: Multidisciplinary Digital Publishing Institute.
38. H. J. Son, S. H. Han, J. A. Lee, E. J. Shin and O. Hwang, Potential repositioning of exemestane as a neuroprotective agent for Parkinson's disease, *Free Radical Research* **51**, 633 (June 2017), Publisher: Taylor & Francis _eprint: https://doi.org/10.1080/10715762.2017.1353688.
39. N. R. Jabir, C. K. Firoz, S. S. Baeesa, G. M. Ashraf, S. Akhtar, W. Kamal, M. A. Kamal and S. Tabrez, Synopsis on the Linkage of Alzheimer's and Parkinson's Disease with Chronic Diseases, *CNS Neuroscience & Therapeutics* **21**, 1 (2015), _eprint: https://onlinelibrary.wiley.com/doi/pdf/10.1111/cns.12344.
40. S. B. Bansode, A. K. Jana, K. B. Batkulwar, S. D. Warkad, R. S. Joshi, N. Sengupta and M. J. Kulkarni, Molecular Investigations of Protriptyline as a Multi-Target Directed Ligand in Alzheimer's Disease, *PLOS ONE* **9**, p. e105196 (August 2014), Publisher: Public Library of Science.
41. Y. Wang, H. Wang and H.-z. Chen, AChE Inhibition-based Multi-target-directed Ligands, a Novel Pharmacological Approach for the Symptomatic and Disease-modifying Therapy of Alzheimer's Disease, *Current Neuropharmacology* **14**, 364 (May 2016).
42. V. Tiwari, A. Mishra, S. Singh, S. K. Mishra, K. K. Sahu, Parul, M. J. Kulkarni, R. Shukla and S. Shukla, Protriptyline improves spatial memory and reduces oxidative damage by regulating NFB-BDNF/CREB signaling axis in streptozotocin-induced rat model of Alzheimer's disease, *Brain Research* **1754**, p. 147261 (March 2021).
43. C. Cirillo, E. Capoccia, T. Iuvone, R. Cuomo, G. Sarnelli, L. Steardo and G. Esposito, S100B Inhibitor Pentamidine Attenuates Reactive Gliosis and Reduces Neuronal Loss in a Mouse Model of Alzheimer's Disease, *BioMed Research International* **2015**, p. e508342 (July 2015), Publisher: Hindawi.
44. T. Sarycheva, P. Lavikainen, H. Taipale, J. Tiihonen, A. Tanskanen, S. Hartikainen and A. Tolppanen, Antiepileptic Drug Use and the Risk of Stroke Among Community-Dwelling People With Alzheimer Disease: A Matched Cohort Study, *Journal of the American Heart Association* **7**, p. e009742 (September 2018), Publisher: American Heart Association.
45. S. M. Holliday, P. Benfield and G. L. Plosker, Fosphenytoin. Pharmacoeconomic implications of therapy, *PharmacoEconomics* **14**, 685 (December 1998).
46. V. Dhikav, Can phenytoin prevent Alzheimer's disease?, *Medical Hypotheses* **67**, 725 (2006).
47. A. Satpati, T. Neylan and L. T. Grinberg, Histaminergic neurotransmission in aging and Alzheimer's disease: A review of therapeutic opportunities and gaps, *Alzheimer's & Dementia: Translational Research & Clinical Interventions* **9**, p. e12379 (2023), _eprint: https://onlinelibrary.wiley.com/doi/pdf/10.1002/trc2.12379.

Combined kinome inhibition states are predictive of cancer cell line sensitivity to kinase inhibitor combination therapies

Chinmaya U. Joisa[1,2], Kevin A. Chen[3], Samantha Beville[2], Timothy Stuhlmiller[2], Matthew E. Berginski[2], Denis Okumu[2], Brian T. Golitz[2], Michael P. East[2], Gary L. Johnson[2], Shawn M. Gomez[1,2,*]

[1]*Joint Department of Biomedical Engineering, University of North Carolina at Chapel Hill, Chapel Hill, NC, USA and North Carolina State University, Raleigh, NC, USA*
[2]*Department of Pharmacology, University of North Carolina at Chapel Hill, Chapel Hill, NC, USA*
[3]*Department of Surgery, University of North Carolina at Chapel Hill, Chapel Hill, NC, USA*
* *Corresponding Author: Shawn M. Gomez (smgomez@unc.edu)*

Protein kinases are a primary focus in targeted therapy development for cancer, owing to their role as regulators in nearly all areas of cell life. Recent strategies targeting the kinome with combination therapies have shown promise, such as trametinib and dabrafenib in advanced melanoma, but empirical design for less characterized pathways remains a challenge. Computational combination screening is an attractive alternative, allowing in-silico filtering prior to experimental testing of drastically fewer leads, increasing efficiency and effectiveness of drug development pipelines. In this work, we generated combined kinome inhibition states of 40,000 kinase inhibitor combinations from kinobeads-based kinome profiling across 64 doses. We then integrated these with transcriptomics from CCLE to build machine learning models with elastic-net feature selection to predict cell line sensitivity across nine cancer types, with accuracy $R^2 \sim 0.75$-0.9. We then validated the model by using a PDX-derived TNBC cell line and saw good global accuracy ($R^2 \sim 0.7$) as well as high accuracy in predicting synergy using four popular metrics ($R^2 \sim 0.9$). Additionally, the model was able to predict a highly synergistic combination of trametinib and omipalisib for TNBC treatment, which incidentally was recently in phase I clinical trials. Our choice of tree-based models for greater interpretability allowed interrogation of highly predictive kinases in each cancer type, such as the MAPK, CDK, and STK kinases. Overall, these results suggest that kinome inhibition states of kinase inhibitor combinations are strongly predictive of cell line responses and have great potential for integration into computational drug screening pipelines. This approach may facilitate the identification of effective kinase inhibitor combinations and accelerate the development of novel cancer therapies, ultimately improving patient outcomes.

Keywords: Kinase signaling, precision medicine, systems biology, drug response prediction.

1. Introduction

Protein kinases, which serve as the primary conduits for information transfer within cells, are often implicated as key drivers in cancer development and have become a cornerstone in current targeted therapies [1]. The rapid expansion of kinase inhibitor therapies as an oncology drug class is exemplified by the FDA's approval of nearly 60 such therapies over the past 20 years [2]. Despite their initial promise, kinase-targeting monotherapies frequently give rise to resistance [3], in part due to the dynamic nature of the kinase network, i.e., the "kinome," which has been shown to reprogram and respond to the inhibition of single kinases by upregulating expression of partner pathways [4–6]. This necessitates the development of novel strategies to effectively target the kinome and harness the vast array of potential drug targets it offers.

One emerging strategy to counteract resistance involves the design of combination therapies, which perturb multiple targets with two or more drugs. These targets may be either known compensatory pathway partners, referred to as "horizontal pathway inhibition," or multiple targets within the same pathway, known as "vertical pathway inhibition" [7]. This approach has recently gained traction with the FDA approval of the combination of trametinib and dabrafenib for treating advanced melanoma [8]. This combination therapy "vertically" targets both BRAF and MEK within the RAF-MEK-ERK (MAPK) pathway, demonstrating the potential effectiveness of this strategy. However, this method of empirical design of combination therapies is not feasible for less characterized kinase pathways, and the sheer number of possible combinations of potential kinase targets (2^{500}) prevents brute-force screening or drug design.

To circumvent this issue, computational screening offers an appealing alternative, enabling the prediction of effective drug combinations in-silico prior to testing a reduced pool of potential combinations in-vitro. This method streamlines the drug development process, and when combined with patient-specific genomic profiling, can also enable personalized drug combination selection to potentially achieve resistance-proof responses in patients.

In recent years, a variety of computational approaches have been developed to predict combination therapy responses for cancer drug screening [9,10]. Most of these methods primarily rely on drug structure characteristics and cancer-specific baseline genomic profiling to predict effective drug combinations, spurred by advancements in the high-throughput acquisition of these data types. For example, a high-dimensional tensor-based modeling strategy used similar data and achieved impressive accuracy (Overall $R^2 \sim 0.8$) in predicting response to combination therapies, validated experimentally [11]. This approach and others employ intricate neural network architectures that, while capable of producing high performing models, can be challenging to interpret, posing a barrier to the broader understanding of their underlying mechanisms. Tree-based machine learning models on the other hand, although simpler and sometimes less powerful, are generally considered interpretable depending on the type of data fed to them [12]. Notably, drug-protein interactions, which are intuitively central to the process of phenotype reversal, have been relatively underexplored in these computational approaches. In part, the minimal amount of drug-target information leveraged in current response prediction efforts is because of the sheer amount of data generated by genomics and molecular fingerprinting, generating thousands of features for each measurement, while drug target data has been generally sparse with only a few annotated targets per drug. However, recent advances in technology to profile the interactions of clinical drugs with all the members of the kinome represent an unprecedented ability to measure drug-target information across ~500 proteins simultaneously in a quantitative manner [13,14]. The breadth, density, and ease of acquisition of this data, often measured at multiple dose points, is ideal for integration into machine learning models that can leverage diverse data types for drug response prediction.

Specifically, recent advances in proteomics techniques have facilitated the large-scale characterization of drug-kinase interactions, providing valuable information on the extent to which the entire kinome is inhibited by specific drugs or drug combinations. A landmark paper in 2017 used a mass spectrometry-based assay that used promiscuous kinase-binding compounds immobilized on beads to measure the binding competition between any given inhibitor and any given kinase (henceforth called the "kinobeads" assay) [15]. Using this assay, the kinome-wide binding profiles for ~230 clinical kinase inhibitors at eight doses each were elucidated using cancer cell lysates, forming the largest in-cell drug-target binding database publicly available at this time. The data generated from these assays allow interrogation of how clinical and investigational drugs interact with the entire kinome on an unprecedented scale. By analyzing the degree of inhibition of all kinases simultaneously for a given inhibitor, we can treat this as characterizing the degree of departure from the "baseline kinome state", thus moving through drug-induced alteration of multiple kinase activities to a new "kinome inhibition state". Given the degree to which modulation of the kinome alters cellular state and downstream behavior, these baseline kinome states and kinome inhibition states can be directly connected to various measured cellular phenotypes. We have recently demonstrated this idea by showing that kinome inhibition state is significantly predictive of cancer cell responses to kinase inhibitor monotherapies when integrated with cancer-specific information, such as baseline transcriptomics, using tree-based machine learning models [16].

In this work, we show that by combining the inhibition states of two kinase inhibitors, we can generate a hypothetical "combined" inhibition state for an untested inhibitor combination. In this manner, we can rationally use all combinatorial kinome inhibition states to sample all possible kinase target combinations, hypothetically including all pathway partners. By integrating these inhibition states with cancer-specific baseline transcriptomics, we demonstrate that the combined inhibition state can predict the sensitivity of cancer cell lines to inhibitor combination treatments from the NCI-ALMANAC dataset using interpretable machine learning models. We further validate these models experimentally by examining novel inhibitor combinations in a PDX-derived triple-negative breast cancer (TNBC) cell line. By focusing on dual-inhibitor drug-kinase interactions combined with cancer-specific baseline genomic profiling, we can enhance computation combination drug screening pipelines with combinatorial kinase targeting. Furthermore, this approach lays the foundation for the rational design and a priori prediction of combination kinase inhibitor treatments for patients with the potential to ultimately reduce single kinase inhibitor resistance acquisition by prior rational targeting of partner pathways and associated kinases.

2. Results

2.1. *Creating a Set of Combined Kinome Inhibition States Representing Current and Potential Kinase Inhibitor Combination Therapies*

In this work, we have focused on a specific set of 200 kinase inhibitors characterized using the kinobeads assay [15]. These inhibitors were profiled in-cell for their interactions with ~500 kinases and kinase-interacting proteins, across eight doses. From this data, as described previously (insert citation), we extracted monotherapy "kinome inhibition states", denoting the degree to which they inhibit each kinase in the kinome at eight doses on a scale of 0-1 (0 is complete inhibition and 1 is no inhibition of a given kinase). We next tested different methods to approximate the kinome inhibition state of a kinase inhibitor combination. Intuitively, this can be thought of as simply superimposing two individual monotherapy inhibition states, but for the few cases where different inhibitors target the same kinase, we found ways to accurately reflect the resulting effect on the kinome. Here, we tested combining monotherapy kinome inhibition state vectors through addition, multiplication, truncated multiplication (excluding kinase inhibition values >1). All three methods were compared for downstream model performance.

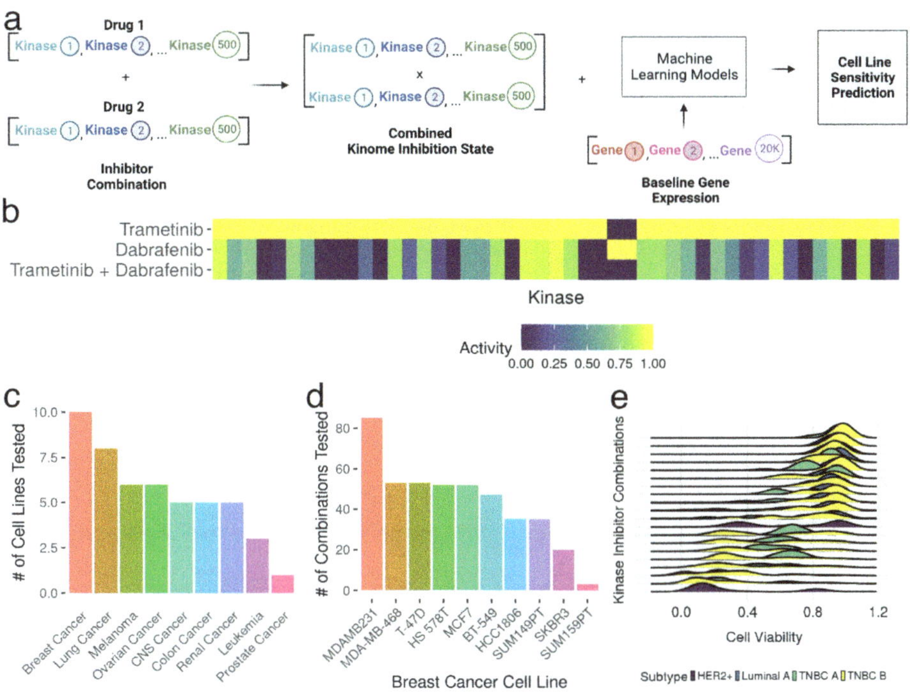

Figure 1. Kinome inhibition State Combination Modeling and Data Overview. (a) Schematic of modeling pipeline. (b) Heatmap showing the inhibition state of individual kinase inhibitors (row 1 and 2), and the hypothetical "combined" inhibition state for the two inhibitors (row 3) (c) Bar plot showing number of cell lines tested per cancer type in training data set (d) Bar plot showing number of unique combinations tested per cell line for the breast cancer subset of the training data set (e) Ridge plots showing cell viability (x-axis) variation for a random subset of different kinase inhibitor combinations (y-axis) in the NCI-ALMANAC data for breast cancer cell lines. Different breast cancer subtypes are represented with differing colors.

After combining the individual inhibition states, we were left with a dataset describing all possible pairwise combinations of ~220 kinase inhibitors. These ~45,000 combinations represent the kinome inhibition states of existing clinical therapies (example), therapies currently in clinical trials (example), as well as potential therapies. Together, they interrogate a search space that includes nearly every known kinase on the phylogenetic tree (Fig S1).

2.2. *Connecting Inhibited Kinome States with Cancer Cell Line Combination Sensitivities*

Next, we linked the data set describing kinase inhibitor combinations to their cell sensitivity phenotypes in the large-scale ALMANAC drug combination screen. The ALMANAC screen contains cell sensitivity data for 53 kinase inhibitor combinations, over ~200 unique dose combinations for 45 cell lines across 9 cancer types. Additionally, previous high-throughput combination screens conducted in our lab in breast cancer offered data for 56 inhibitor combinations in four cell lines. Ideally, we would like exact matches between the dose at which kinome inhibition state is profiled and the dose at which cell sensitivity was measured. However, there are very few exact matches between the datasets. To overcome this, we found the nearest dose (6 exact matches, 14 nearest matches at maximum differing by 1uM) at which kinome inhibition was profiled for each cell sensitivity measurement and connected the two datasets using these dose matches.

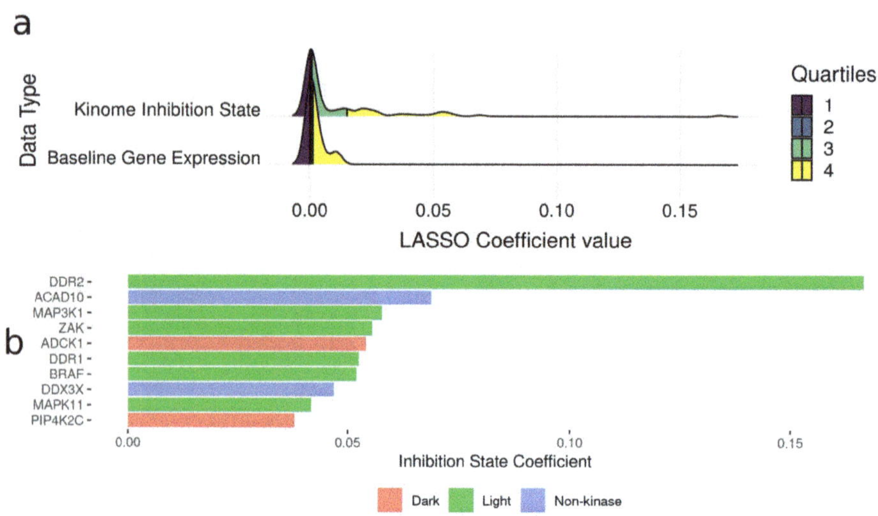

Figure 2. Feature Selection using an Elastic-net Regression Model against Cancer Cell Line Sensitivity. (a) Ridge plot showing the distribution of LASSO coefficient sizes as a metric for feature importance, for each feature type (b) Horizontal bar plot showing kinases with the largest elastic-net coefficient values, coloured by whether they are defined as "understudied" (Dark) or "well-characterized" (Light).

Additionally, we added cell line specific information to the dataset to complement the drug-specific kinome inhibition states. The CCLE database contains baseline transcriptomics data for ~1500 cancer cell lines, and almost all of the cell lines included in our data set were represented. Using this, we further added baseline gene expression into the dataset, now containing kinase inhibitor combinations, their inhibition state of the kinome, the cell line sensitivity to their treatment, as well as that cell line's baseline gene expression. In this way, the dataset connects the kinome inhibition states of inhibitor combinations to their cell sensitivity phenotypes.

The collected dataset represents a total of eight major cancer types, with the majority having ~7 cell lines represented each, while breast cancer had the most representation (11 cell lines). To ensure that the machine learning model downstream could find cancer-specific linkages between the kinome and cell sensitivity, we split the dataset into eight individual cancer type datasets and conducted all modeling on each data split in parallel.

2.3. *Elastic-Net Feature Selection Reveals Kinome Inhibition States to be Most Informative*

In our collected dataset, kinome inhibition states and baseline gene expression together represent ~20,000 variables or "features" that could affect the phenotype of cell sensitivity to kinase inhibitors. It is both practically prohibitive and ineffective to build models using all available features, and so keeping in mind computational efficiency we sought to filter down the dataset to include only the most informative features. To accomplish this "feature selection", we built our machine learning pipeline starting with an elastic-net regression [17] model built against the outcome of cell sensitivity. This generated coefficients for each feature, with the absolute value of the feature coefficient directly proportional to its predictive value for the outcome. We ensured non-informative features were not included in modeling by only considering features with non-zero coefficients. We fit the model on the entire dataset to visualize a snapshot of the feature coefficients globally. This revealed overwhelmingly larger coefficients for kinome inhibition states compared to baseline gene expression (Fig 2a), thus indicating that kinome inhibition states were globally more informative for cell sensitivity prediction compared to baseline gene expression.

For downstream model building, the data set was split into a training and testing set five times (five-fold cross validation). For the training set data to not have any influence on the test set (to prevent data leakage), the elastic net model is fit on only the training data, and features are selected within each fold. Parameters for the elastic net model and hyperparameters for the tested model types were also tuned this way.

2.4. *Machine Learning Models Can Predict Cancer Cell Line Sensitivity to Combination Therapies by Integrating Kinome Inhibition States and Baseline Transcriptomics*

After data set preparation and feature selection, we built machine learning models that can predict cell sensitivity to kinase inhibitor combinations. For each cancer type, three machine learning model types were tested: random forest, boosted trees (xgboost) and deep neural networks. Xgboost performed the best for all cancer types, with type-specific performance largely dependent on abundance of data in the training set (Fig 3b). The most abundant cancer type (breast) had the best performing model with an R^2 score of 0.93 (Fig 3b) while the lowest performing model was prostate cancer with $R^2 = 0.73$. Given our previous lab experience with breast cancer, we chose the breast cancer model for downstream experiments and validation.

Additionally, since the best-performing model was tree-based gradient boosting, we were able to further analyze the model using computed feature importance to find the most informative features in the data set based on the feature importance metric. Similar to the feature selection output, we saw much higher feature importance scores overall for kinome inhibition states when compared to baseline gene expression, and several kinases implicated in breast cancer dysfunction had high importance scores, such as MAP2K1/2 and EGFR(Fig. 3c).

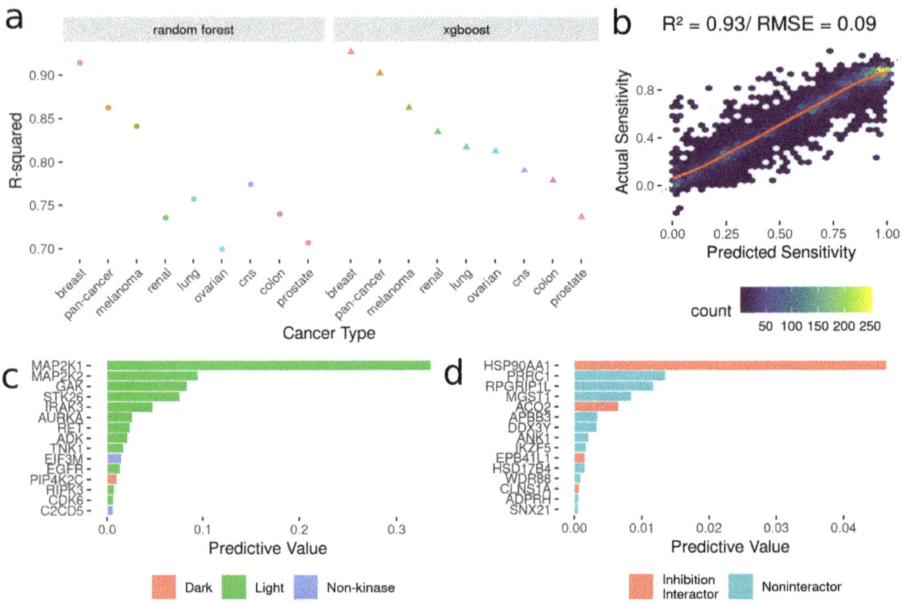

Figure 3. Development of Models to Predict Cancer Cell Line Sensitivities to Kinase Inhibitor Combination Therapies from Kinome Inhibition States. (a) Model performance metrics (R-squared) for Random Forest (dots) and XGBoost (triangles). (b) Scatter Plot of predicted sensitivity values from the best-performing model vs actual sensitivity values. The red line indicates a smooth fit through the data points. (c) Horizontal bar plot showing model importance of individual kinase inhibition states by importance values. (d) Horizontal bar plot showing model importance of individual baseline gene expression by importance values.

2.5. *Experimental Validation of Model Predictions in a PDX-Derived Triple Negative Breast Cancer Cell Line was Successful.*

We demonstrated that machine learning models using the kinome inhibition states of combination therapies along with cell-specific baseline gene expression could robustly predict cell sensitivity in multiple cancer types. However, to see if these predictive models could extend to real-world experiments, we experimentally validated 35 kinase inhibitor combinations in a PDX-tumor derived cell line(Fig 4A).

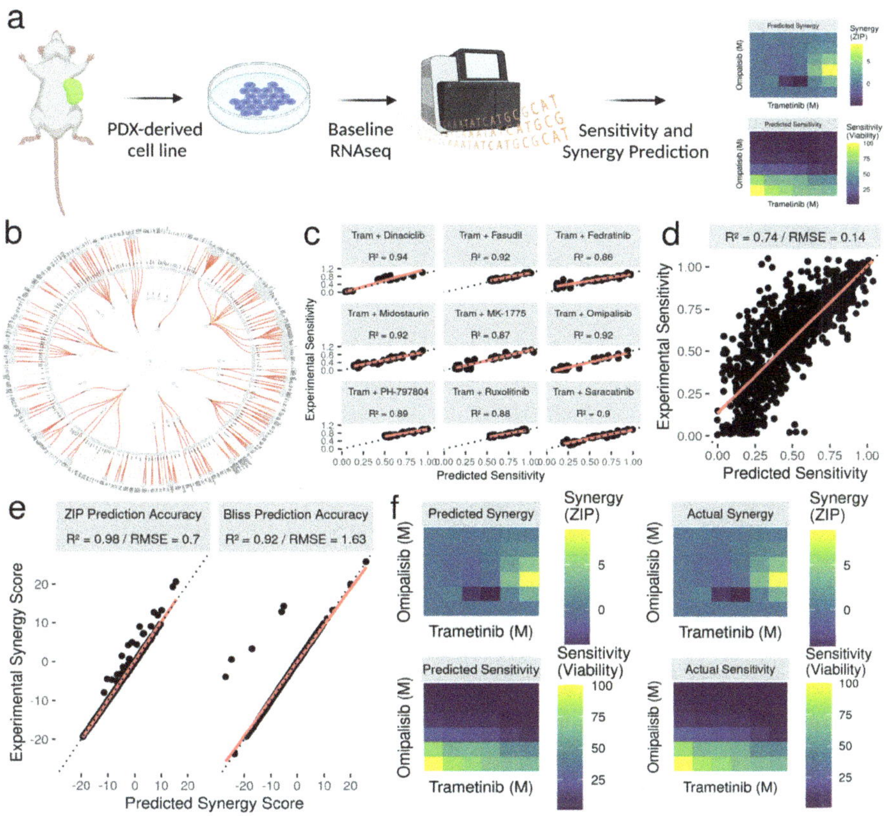

Figure 4. Experimental Validation of Model through a Trametinib Combination Screen in the WHIM12 Patient-Derived TNBC Cell Line. (a) Schematic showing experimental validation pipeline for the WHIM12 PDX-derived cell line. (b) Kinome phylogenetic map showing diversity of kinome targeted (red = inhibited by a validated kinase inhibitor combination). (c) Grid of scatter plots showing accuracy of top nine tested combinations. For all scatter plots, the dashed line indicates where perfect predictions would lie and the red line shows a linear fit through the data. Quantitative accuracy is represented by the R-squared score. (d) Scatter plot showing the global accuracy of model. (e) Grid of scatter plots showing accuracy of model predicted synergy scores compared to experimentally measured synergy scores across two metric types (ZIP, Bliss). (f) Grid of heatmap plots showing comparison of predicted vs experimentally measured sensitivity and synergy for the highly synergistic trametinib / omipalisib combination.

High-throughput cell line drug screens have been widely documented to suffer from a lack of reproducibility and poor translation to more complex samples like patient tumors. We sought to test whether our model of cell sensitivity in breast cancer, trained on 11 well-characterized immortalized cell lines, could effectively predict cell sensitivity in a PDX (Patient-Derived Xenograft) derived cell line. We chose the WHIM12 PDX-derived cell line, which was generated from a highly chemo-resistant TNBC tumor [18]. Previous experiments in the lab had conducted a drug combination screen in the WHIM12 cell line, out of which 35 kinase inhibitors were tested in combination with trametinib. Complementary baseline gene expression data was also generated through RNAseq. Using these in-house data, we were able to input the unseen WHIM12 gene expression into the trained model and predict the cell sensitivity outcomes of the conducted drug combination screen. We achieved robust prediction accuracy (Global $R^2 = 0.74$ / RMSE = 0.14) in predicting exact cell viability in response to treatment with 35 kinase inhibitor combinations, across 64 dose combinations (Fig 4c, d).

2.6. *Model Predictions Reveal Known Synergy in trametinib/omipalisib Combination*

The model predictions in the WHIM12 cell line were further interrogated for potential synergy. We generated synergy scores for all 35 combinations at each of the 64 dose points using the R package SynergyFinder [19] based on four different metrics: Zero-Interaction Potency [10] (ZIP), Bliss Independence [20], Highest Single-Agent (HSA), and Loewe Additivity [21]. Additionally, we generated similar synergy scores using the actual experimental data generated for validation as a comparison. We found a high degree of similarity (Global $R^2 \sim 0.94$/ RMSE ~ 0.5) between predicted and actual synergy, with trametinib + omipalisib as our most synergistic predicted combination, with a ZIP score of ~8 at certain dose combinations (Fig 4e, f). This is significant as the model predictions were in a TNBC PDX-derived line, and the trametinib/omipalisib combination represents the popular strategy of simultaneously targeting the MAPK and PI3K pathways[22].

3. Methods

Data Sources. The kinome profiling data set from the kinobeads assay was downloaded from the supplementary materials of Klaeger et al. 2017 [15]. For cancer cell line sensitivity to kinase inhibitor combinations, data was downloaded from (1) NCI-ALMANAC: cell sensitivity data was downloaded from the NCI wiki database (https://wiki.nci.nih.gov/display/NCIDTPdata/NCI-ALMANAC) and (2) Supplementary materials of previous lab combination screens published in Beville et al. 2019 [28] and Stuhlmiller et al. 2015 [29]. The CCLE gene expression set ("CCLE_expression.csv") was downloaded from the DepMap portal (https://depmap.org/portal/download/all/) to create the set of cancer cell lines and their gene expression characteristics. In-house baseline gene expression data for the PDX-derived WHIM12 line was downloaded from the GEO repository for the Zawitowski et al. paper[26] (GSE87424).

Data Preprocessing. The scripts implementing these descriptions are all available through github.

Klaeger et al. Kinobead Kinase Inhibition Profiles: As previously described [16], we read the values from the supplemental data table into R and produced a filtered list of kinase and kinase interactor relative intensity values. We imputed missing values with the default "no interaction" value of 1 and truncated likely outlier values to the 99.99 percentile (3.43).

Creating the Combination Inhibition State Data Set: To create a "combined" inhibition state of a given kinase inhibitor combination, we sought to superimpose the inhibition states of two individual states at specific doses. There were eight doses measured for each individual inhibitor, thus there were 64 possible combinations for each combination. We took the monotherapy kinome inhibition states from the Klaeger et al. set and computed a "combined" inhibition state for each kinase, based on three different combination schemes:

1. Simple Multiplicative: The simple conditional probability rule assumes two independent events (A and B in Eq. 1). Since the default "no interaction" inhibition value is 1, for kinases that are not targeted by both inhibitors simultaneously, the "combined" inhibition state (C') value is simply either one in monotherapy.
2. Truncated Multiplicative: A minority of measured kinase inhibition states (~1%) have values > 1 in the Klaeger et al. dataset, a possible artifact from the mass spectrometry measuring process. To avoid inflating those values, all >1 values were truncated at 1 and simple multiplication was performed as described above (Eq. 2).
3. Addition: All kinase inhibition states were inverted into "Percent Inhibition" values (A' and B'), where 0 denotes no inhibition and 100 denotes complete inhibition. Then, when two inhibition states were combined, they were added together and truncated at a max value of 100 (Eq. 3).

$$Eq. 1. \qquad C' = A * B$$

$$Eq. 2. \qquad C' = \min(1, A) * \min(1, B)$$

$$Eq. 3. \qquad C' = \min(100, \quad A' + B')$$

All three methods were tested in downstream modeling, resulting in minor variation. Truncated multiplied vectors were slightly more predictive (R^2 score of ~0.01 greater) so we used that scheme for all downstream modeling. In this way, we were able to compute hypothetical "combined" inhibition states for all possible combinations of ~220 inhibitors, altogether comprising ~2,000,000 combined inhibition states.

Dataset of Cancer Cell Line Sensitivity to Kinase Inhibitor Combinations: The cell sensitivity dataset from NCI-ALMANAC and previous lab publications were filtered to contain only kinase inhibitor small molecules, then summarized over replicates and converted to cell viability (1 = fully viable cell and 0 = full cell death). Relevant cancer types were annotated and individual cancer type datasets were subsetted for downstream cancer type-specific modeling.

Matching of Kinase Inhibitors between Inhibition State Dataset and Cell Line Sensitivity Dataset: The drug names from each dataset were read into R, and the package Webchem [30] was used to retrieve PubChem compound IDs (cid's). The two sets of drug names were then matched based on these reference IDs, with a total of ~100 matches between the two sets.

Baseline Gene Expression from CCLE: Data was preprocessed as described before [31] from the "CCLE_expression.csv" file. Cell line names were matched manually between CCLE and the NCI naming scheme. All cell lines represented in NCI-ALMANAC had a match in the CCLE database.

String: The STRING database [32] was processed as described previously [31] to annotate kinases and kinase interacting genes.

Modeling Techniques: To assess our models we used a random 5-fold cross validation strategy. We implemented Elastic-net regression using the glmnet engine [33] for the feature selection scheme [17], We compared the performance of three model types using this strategy: random forest using the ranger engine [34] and gradient boosting using the XGBoost engine [35]. Model performance was assessed by the R-squared value between predicted and actual outcome within the cross-validation scheme. For each model type and for the feature selection model, we tuned sets of 20 hyperparameters to find the best possible performer as follows: (a) Elastic-net: Penalty (0 - 0.1), Regularization (0.1-1) (b) Random Forest: Trees (100 - 2000) (c) XGBoost: Trees (100 - 1000), Tree Depth (4 - 30). After final model selection, we fit the model on the entire dataset and then made predictions on the experimental validation data.

All of the data and code written to support this paper is available through github (https://github.com/gomezlab/kinotype_combination_prediction).

Experimental Validation. 6x6 dose combination screens were performed in the WHIM12 cell line as described in Beville et al. 2019 [28]. Briefly, cells were seeded in 384-well plates and dosed with drug after 24h. The screening library was tested for growth inhibition alone or in combination with Trametinib across 6 doses: 10 nmol/L, 100 nmol/L, 300 nmol/L, 1 μmol/L, 3 μmol/L, and 10 μmol/L. 0.1% DMSO was included as the control for growth inhibition on each plate. Plates were incubated at 37°C for 96 hours and lysed using CellTiter-Glo Reagent (Promega, catalog. no. G7570). Luminescence was measured using a PHERAstar FS instrument and growth inhibition was calculated relative to DMSO-treated wells.

4. Discussion

Kinase inhibitors are one of the fastest growing drug classes for cancer therapy, with ~62 FDA approved in total against neoplasms [2]. With 500 potential druggable targets, there is significant interest in streamlining the kinase inhibitor screening process. We have previously introduced [16,23,24] the idea that the full spectrum of a given inhibitor's effect on the kinome as measured by recent advances in kinobead-competition/MS technology [15] can be represented as a "kinome inhibition state", i.e. a vector representing the effect of a given inhibitor on the kinome as a whole.

In this work, we have extended this idea to represent the kinome inhibition state of a combination of inhibitors, using a multiplicative probability model to "combine" the inhibition states of two given kinase inhibitors. By generating these "combined" inhibition states, we can vastly expand the search space targeted by inhibitor monotherapies, sampling all possible combinations of currently available therapies. To accomplish this, we used publicly available drug-kinome interaction data to generate snapshots of the combined effect of a combination therapy on the protein kinome. We then linked these kinome inhibition states of inhibitor combinations to cancer cell sensitivity phenotypes to combination treatment, creating a framework for predicting the efficacy of combination therapies in different cancer types.

We fit tree-based machine learning models on this linked data set to robustly predict precise cancer cell line sensitivity and synergy for untested kinase inhibitor combinations therapies and validate those predictions in complex patient derived samples. gradient-boosted tree models were highly accurate across cancer types (R^2 0.75-0.93), comparable to two recent neural-network driven attempts to predict cell line response to drug combinations [9,11]. We chose to validate our model predictions in the PDX-dervied WHIM12 line, reasoning that PDX-derived cell lines retain many of the molecular and genetic features of the xenografted original tumors. We were able to show that the models performed robustly on novel gene expression data (R^2 ~0.74) , representing its ability to extend to complex and clinical-adjacent samples compared to well-characterized cell line data.

One of the strengths of tree-based models is that they are considered to be interpretable through feature importance computation [12,25]. Using this, we were able to investigate the "black box" and query which specific kinase inhibition states and baseline genes were most predictive of cell sensitivity. We found that for the breast cancer model, the inhibition of the kinases MAP2K1/2 were the most informative by far. This is intuitive considering the most abundant kinase inhibitor in the dataset is the allosteric MEK inhibitor trametinib, but it must be noted that MEK inhibition is always only just one half of the kinome targeting in the combination. There has been increasing clinical interest recently in targeting the PI3K and MAPK pathways [22], and our lab has shown before that MEK1/2 inhibition in TNBC by trametinib induces widespread transcriptional adaptation, and that there is potential for clinical efficacy in complementary kinome targeting with trametinib [26]. Since our model's sensitivity predictions can effectively simultaneously predict synergy, our top synergy prediction for breast cancer according to the ZIP metric was trametinib and omipalisib, which we were able to validate experimentally in the WHIM12 line. This indicates that from the breast cancer screening data, the model was able to learn that targeting the complementary PI3K and MAPK pathways is effective and synergistic in TNBC.

Interestingly, the predicted high-synergy combination of trametinib/omipalisib was recently in phase I clinical trials for advanced solid tumors but failed due to patient intolerability [27]. This highlights some limitations of our modeling approach. Ideally, kinome inhibition state would be one of many different drug modalities included for response prediction, and we plan to further expand these models in the future by considering toxicity, drug structure and cancer-describing multi-omic data types not limited to baseline gene expression. Additionally, in this proof-of-concept study we utilized multiplicative probability models to generate the "combined" inhibition state of two inhibitors on the kinome, by assuming that the inhibition of a given kinase is mutually exclusive from that of other kinases. We know that kinases function physiologically as part of complex signaling networks, and their inhibition may have downstream effects on other kinases and signaling pathways. To address this limitation, future models will incorporate more biologically representative schemes to hypothesize combined kinome inhibition states.

In summary, through this work we demonstrate the development of a framework for predicting the efficacy of combination therapies in different cancer types using just kinome-drug interactions and baseline gene expression. We generated the combined "kinome inhibition state" and linked these states to cancer cell sensitivity phenotypes. First, we were able to show that a given combination therapy's cancer-agnostic interaction with the kinome was far more informative than baseline genomics in predicting downstream response. This is intuitive fundamentally, as drug-protein interactions are the primary means of drug effect on physiology, but this type of data is still underutilized in computational screening approaches. We then used machine learning models to predict cell line sensitivity and synergy for untested kinase inhibitor combination therapies and validated those predictions experimentally in complex patient derived samples.

Acknowledgements

We would like to thank UNC Research Computing for access to the computational resources necessary for this work. We would like to thank Michael P. East for his help with data compilation. This work was supported by grants through the National Institutes of Health (Grant #s CA274298, CA233811, CA238475, DK116204)

References

1. Kothari V, Wei I, Shankar S, Kalyana-Sundaram S, Wang L, Ma LW, et al. Outlier kinase expression by RNA sequencing as targets for precision therapy. Cancer Discov. 2013;3: 280–293.

2. Roskoski R Jr. Properties of FDA-approved small molecule protein kinase inhibitors: A 2023 update. Pharmacol Res. 2023;187: 106552.

3. Jiang L, Li L, Liu Y, Lu L, Zhan M, Yuan S, et al. Drug resistance mechanism of kinase inhibitors in the treatment of hepatocellular carcinoma. Front Pharmacol. 2023;14: 1097277.

4. Duncan JS, Whittle MC, Nakamura K, Abell AN, Midland AA, Zawistowski JS, et al. Dynamic reprogramming of the kinome in response to targeted MEK inhibition in triple-negative breast cancer. Cell. 2012;149: 307–321.

5. Engelman JA, Zejnullahu K, Mitsudomi T, Song Y, Hyland C, Park JO, et al. MET amplification leads to gefitinib resistance in lung cancer by activating ERBB3 signaling. Science. 2007;316: 1039–1043.

6. Chandarlapaty S. Negative feedback and adaptive resistance to the targeted therapy of cancer. Cancer Discov. 2012;2: 311–319.

7. Yesilkanal AE, Johnson GL, Ramos AF, Rosner MR. New strategies for targeting kinase networks in cancer. J Biol Chem. 2021;297: 101128.

8. Atkinson V, Sandhu S, Hospers G, Long GV, Aglietta M, Ferrucci PF, et al. Dabrafenib plus trametinib is effective in the treatment of BRAF V600-mutated metastatic melanoma patients: analysis of patients from the dabrafenib plus trametinib Named Patient Program (DESCRIBE II). Melanoma Res. 2020;30: 261–267.

9. Kuenzi BM, Park J, Fong SH, Sanchez KS, Lee J, Kreisberg JF, et al. Predicting Drug Response and Synergy Using a Deep Learning Model of Human Cancer Cells. Cancer Cell. 2020;38: 672-684.e6.

10. Yadav B, Wennerberg K, Aittokallio T, Tang J. Searching for Drug Synergy in Complex Dose-Response Landscapes Using an Interaction Potency Model. Comput Struct Biotechnol J. 2015;13: 504–513.

11. Julkunen H, Cichonska A, Gautam P, Szedmak S, Douat J, Pahikkala T, et al. Leveraging multi-way interactions for systematic prediction of pre-clinical drug combination effects. Nat Commun. 2020;11: 1–11.

12. Izza Y, Ignatiev A, Marques-Silva J. On Tackling Explanation Redundancy in Decision Trees. jair. 2022;75: 261–321bussy.

13. Cann ML, McDonald IM, East MP, Johnson GL, Graves LM. Measuring Kinase Activity-A Global Challenge. J Cell Biochem. 2017;118: 3595–3606.

14. Koleti A, Terryn R, Stathias V, Chung C, Cooper DJ, Turner JP, et al. Data Portal for the Library of Integrated Network-based Cellular Signatures (LINCS) program: integrated access to diverse large-scale cellular perturbation response data. Nucleic Acids Res. 2018;46: D558–D566.

15. Klaeger S, Heinzlmeir S, Wilhelm M, Polzer H, Vick B, Koenig P-A, et al. The target landscape of clinical kinase drugs. Science. 2017;358. doi:10.1126/science.aan4368

16. Berginski ME, Joisa CU, Golitz BT, Gomez SM. Kinome inhibition states and multiomics data enable prediction of cell viability in diverse cancer types. PLoS Comput Biol. 2023;19: e1010888.

17. Zou H, Hastie T. Regularization and Variable Selection via the Elastic Net. J R Stat Soc Series B Stat Methodol. 2005;67: 301–320.

18. Li S, Shen D, Shao J, Crowder R, Liu W, Prat A, et al. Endocrine-therapy-resistant ESR1 variants revealed by genomic characterization of breast-cancer-derived xenografts. Cell Rep. 2013;4: 1116–1130.

19. Ianevski A, Giri AK, Aittokallio T. SynergyFinder 2.0: visual analytics of multi-drug combination synergies. Nucleic Acids Res. 2020;48: W488–W493.

20. Bliss CI. THE TOXICITY OF POISONS APPLIED JOINTLY1. Ann Appl Biol. 1939;26: 585–615.

21. Chou TC, Talalay P. Quantitative analysis of dose-effect relationships: the combined effects of multiple drugs or enzyme inhibitors. Adv Enzyme Regul. 1984;22: 27–55.

22. Lee J, Liu H, Pearson T, Iwase T, Fuson J, Lalani AS, et al. PI3K and MAPK Pathways as Targets for Combination with the Pan-HER Irreversible Inhibitor Neratinib in HER2-Positive Breast Cancer and TNBC by Kinome RNAi Screening. Biomedicines. 2021;9. doi:10.3390/biomedicines9070740

23. Berginski ME, Jenner MR, Joisa CU, Herrera Loeza SG, Golitz BT, Lipner MB, et al. Kinome state is predictive of cell viability in pancreatic cancer tumor and stroma cell lines. bioRxiv. 2021. p. 2021.07.21.451515. doi:10.1101/2021.07.21.451515

24. Joisa CU, Chen KA, Berginski ME, Golitz BT, Jenner MR, Herrera Loeza SG, et al. Integrated Single-Dose Kinome Profiling Data is Predictive of Cancer Cell Line Sensitivity to Kinase Inhibitors. bioRxiv. 2022. p. 2022.12.06.519165. doi:10.1101/2022.12.06.519165

25. Lundberg SM, Erion G, Chen H, DeGrave A, Prutkin JM, Nair B, et al. Explainable AI for Trees: From Local Explanations to Global Understanding. arXiv [cs.LG]. 2019. Available: http://arxiv.org/abs/1905.04610

26. Zawistowski JS, Bevill SM, Goulet DR, Stuhlmiller TJ, Beltran AS, Olivares-Quintero JF, et al. Enhancer Remodeling during Adaptive Bypass to MEK Inhibition Is Attenuated by Pharmacologic Targeting of the P-TEFb Complex. Cancer Discov. 2017;7: 302–321.

27. Grilley-Olson JE, Bedard PL, Fasolo A, Cornfeld M, Cartee L, Razak ARA, et al. A phase Ib dose-escalation study of the MEK inhibitor trametinib in combination with the PI3K/mTOR inhibitor GSK2126458 in patients with advanced solid tumors. Invest New Drugs. 2016;34: 740–749.

28. Bevill SM, Olivares-Quintero JF, Sciaky N, Golitz BT, Singh D, Beltran AS, et al. GSK2801, a BAZ2/BRD9 Bromodomain Inhibitor, Synergizes with BET Inhibitors to Induce Apoptosis in Triple-Negative Breast Cancer. Mol Cancer Res. 2019;17: 1503–1518.

29. Stuhlmiller TJ, Miller SM, Zawistowski JS, Nakamura K, Beltran AS, Duncan JS, et al. Inhibition of Lapatinib-Induced Kinome Reprogramming in ERBB2-Positive Breast Cancer by Targeting BET Family Bromodomains. Cell Rep. 2015;11: 390–404.

30. Szöcs E, Stirling T, Scott ER, Scharmüller A, Schäfer RB. webchem: An R Package to Retrieve Chemical Information from the Web. J Stat Softw. 2020;93: 1–17.

31. Berginski ME, Joisa CU, Golitz BT, Gomez SM. Kinome Inhibition States and Multiomics Data Enable Prediction of Cell Viability in Diverse Cancer Types. bioRxiv. 2022. p. 2022.04.08.487646. doi:10.1101/2022.04.08.487646

32. Szklarczyk D, Gable AL, Nastou KC, Lyon D, Kirsch R, Pyysalo S, et al. The STRING database in 2021: customizable protein-protein networks, and functional characterization of user-uploaded gene/measurement sets. Nucleic Acids Res. 2021;49: D605–D612.

33. Friedman J, Hastie T, Tibshirani R. Regularization Paths for Generalized Linear Models via Coordinate Descent. J Stat Softw. 2010;33: 1–22.

34. Wright MN, Ziegler A. ranger: A Fast Implementation of Random Forests for High Dimensional Data in C++ and R. J Stat Softw. 2017;77: 1–17.

35. Chen T, Guestrin C. XGBoost: A Scalable Tree Boosting System. arXiv [cs.LG]. 2016. Available: http://arxiv.org/abs/1603.02754

Creation of a Curated Database of Experimentally Determined Human Protein Structures for the Identification of Its Targetome

Armand Ovanessians[1,2], Carson Snow[2], Thomas Jennewein[3], Susanta Sarkar[1,2], Gil Speyer[3] and Judith Klein-Seetharaman[†,1]

[1]*School of Molecular Sciences & College of Health Solutions, Arizona State University 850 N 5th Street, Phoenix, AZ 85012, USA*

[2]*Department of Physics, Colorado School of Mines 1500 Illinois St, Golden, CO 80401, USA*

[3]*Knowledge Enterprise, Arizona State University Tempe, AZ 85287, USA*

Emails: aovaness@asu.edu; cesnow@mines.edu; tjennewe@asu.edu; Susanta.Sarkar@asu.edu; speyer@asu.edu; † Corresponding author: Judith.Klein-Seetharaman@asu.edu

Abstract

Assembling an "integrated structural map of the human cell"[1] at atomic resolution will require a complete set of all human protein structures available for interaction with other biomolecules - the human protein structure targetome - and a pipeline of automated tools that allow quantitative analysis of millions of protein-ligand interactions. Toward this goal, we here describe the creation of a curated database of experimentally determined human protein structures. Starting with the sequences of 20,422 human proteins, we selected the most representative structure for each protein (if available) from the protein database (PDB), ranking structures by coverage of sequence by structure, depth (the difference between the final and initial residue number of each chain), resolution, and experimental method used to determine the structure. To enable expansion into an entire human targetome, we docked small molecule ligands to our curated set of protein structures. Using design constraints derived from comparing structure assembly and ligand docking results obtained with challenging protein examples, we here propose to combine this curated database of experimental structures with AlphaFold predictions[2] and multi-domain assembly using DEMO2[3] in the future. To demonstrate the utility of our curated database in identification of the human protein structure targetome, we used docking with AutoDock Vina[4] and created tools for automated analysis of affinity and binding site locations of the thousands of protein-ligand prediction results. The resulting human targetome, which can be updated and expanded with an evolving curated database and increasing numbers of ligands, is a valuable addition to the growing toolkit of structural bioinformatics.

Keywords: ligand binding; reverse molecular docking; high-performance computing

1. Introduction

The structures of proteins determine their ability to interact with other biomolecules, which is often at the heart of cellular functions and dysfunctions. Massive structural proteomics efforts have made large numbers of protein structures available in the protein databank.[5] While the coverage still falls short of completeness for any single organism, including human and other model organisms, let alone non-model organisms, the recent advent of molecular modeling approaches that rival experimental structure determination in accuracy in some cases,[2] now allows us to start imagining complete datasets of the entire structural proteome of an organism. Such datasets would allow us to start looking at the effects of natural and chemically synthesized small molecules in the context of all possible interactions. The availability of data and computing resources as well as development of new computational approaches are revolutionizing the field of drug discovery.[6] It is becoming increasingly clear that the traditional view of one drug-one protein target is too reductionist: Many successful drugs have multiple targets (for example, the popular anti-diabetic drug, metformin), and many metabolites do not only interact with the enzymes that use them to carry out chemical reactions but often thousands of other proteins.[7] Thus, target discovery is becoming increasingly important also for drug discovery, and reverse docking (i.e. binding of a given ligand to many proteins, as opposed to docking many ligands to one protein target) plays a major role in this field.[8] Looking at the entire set of human proteins that a ligand can potentially interact with - the human targetome - would allow us to answer fundamental questions about the functioning of cells while also improving drug discovery, drug repurposing and predictions of drug targets and toxicity. Finally, we may begin looking at complex mixtures of ligands with biological efficacy, such as natural extracts with positive health effects like lemon juice[9] and environmental pollutants such as asphalt,[10] comprised of thousands of individual compounds.[11]

Currently, docking and even reverse docking is carried out largely with limited subsets of protein structures[12],[13] To enable future systematic analysis of any biomolecular ligand with an organism's complete set of proteins, we describe an approach to create a database that contains a single representative of the optimal structure for each human protein. Our initial strategy is centered around devising a biologically pertinent methodology to rank experimentally derived protein structures as outlined in **Figure 1a**. We use the UniProt database[14] as our reference for all human protein sequences and retrieve the list of structure files from the protein databank.[5] To select the most representative structure, we adopted three key parameters for evaluation: coverage, depth, and resolution of the structures. "Coverage" refers to the count of residues in the protein's structure, indicating the structure's completeness. We prioritized this parameter due to its importance in understanding the overall integrity of a protein. Nevertheless, we encountered situations where a protein's structure, despite having less coverage, offered more meaningful insights due to its residue information being spread over a larger range of amino acids. To account for this, we introduced a novel metric, "depth", which calculates the discrepancy between the maximum and minimum residue numbers. After finally ranking by resolution, we obtained a list of 7606 unique human protein structure files, available on our GitHub page Here.

In the long term, we want to create a complete database to predict where and with what affinity different ligands bind to the human targetome. This will require automated tools to analyze the results obtained from docking ligands to human protein structures. It will also require supplementing experimentally determined structures with predicted structures. We here outline such methods and highlight design considerations using comparisons of known and predicted structures in general, and a specific challenging protein example, the insulin receptor (IR), in the context of structure assembly and ligand docking results. Based on this analysis, we here propose a pipeline that incorporates experimental structures, AlphaFold predictions,[2] multi-domain assembly using DEMO2,[3] docking with AutoDock Vina[4] and automated analysis of affinity and binding site location using the center of mass comparisons as well as Silhouette Score clustering optimization of predicted ligand volume overlap to classify binding pocket numbers and locations for a given protein-ligand pair, and across many proteins and many ligands. Our targetome-oriented, synergistic pipeline will augment protein structure and ligand interaction prediction practices. The current stage of implementation of this pipeline is the curated database of experimentally determined human protein structures, as well as the code used to create the database and to analyze the docking results, available here.

2. Materials and data sources

An initial naive download sourcing a spreadsheet listing experimental structures ignored specific chains and automatically chose the first in lists of multiple PDB codes for a given protein. This led to over 10% of the downloads being multiples of the same structures. In addition, these files would often have multiple models or chains, which either crashed the pre-processing codes due to inappropriate bounding box sizes or yielded huge search spaces that crashed the docking runs. The careful revision of the table –described in the following section– addressed most of these cases. Table 1 reflects the impact of these revisions, comparing the results of docking the ligand kaempferol against the full suite of downloaded structures. Out-of-memory and very large positive "overflow" affinity outputs indicated the two modes of run failure described above.

Table 1: Comparison of ligand kaempferol docking results from original naive scrape and then after table revision with specified chains following protocol shown in **Figure 1a**.

Statistic	Original dataset	Improved dataset
PDBs	6865	7529
Out of memory errors	288	0
Overflow affinities	399	244
Avg bounding box size	557279	212550

3. Methods

3.1. *Database Creation*

An overview of the database creation is shown in **Figure 1a**. First, we downloaded a comprehensive database comprising all 20,422 human protein sequences from the UniProt database.[14]

In the current implementation, we retained only those UniProt IDs with at least one experimental structure associated with it and a file deposited in the Protein Database (PDB).[5] This filtering criterion excluded 12,606 proteins, leaving 7,816 unique UniProt IDs in this subset, many of which were associated with multiple PDB files. To select the best representative structure, we defined several ranking criteria. Sometimes structures miss portions of the sequence, even if they were present during crystallization, often due to flexibility. This can be in loop regions, or at the ends. Often, specific domains have been chosen to represent a portion of the sequence. Because the structures of missing loop regions are typically ill defined, there is a benefit in having a larger stretch of the sequence covered, even if the total coverage is reduced by these missing loop regions. We wanted to have measures that capture both scenarios. Coverage refers to the total number of residues of a sequence that are associated with xyz coordinates in a sequence, while depth refers to the difference between the beginning and end of the structure, regardless of how many residues are missing in between. Moreover, for each PDB file corresponding to a UniProt ID, the scraper retrieved the resolution, the experimental method used (Electron Microscopy, X-ray crystallography, and NMR), and the chains of each PDB file. The latter was essential as a single PDB file can encapsulate multiple proteins. Thus, to compile the required information for this ranking, we designed a web scraper to extract content from the UniProt database.[14] Each DataFrame encompassed specific information for each protein structure, including:

(1) PDB ID
(2) Resolution
(3) Chains and their associated locations
(4) Experimental method used for structure determination
(5) Whether alpha carbons were the only present atom in the PDB file

While resolution and chain information was sourced directly from the UniProt database, coverage and depth information for each PDB file necessitated the scraping and local downloading of all PDB structures related to our 7,816 unique proteins from the RCSB PDB database.[5] 210 UniProt IDs lacked any PDB formatted structure available within the RCSB PDB[5] database, thereby reducing our working dataset to 7,606 unique proteins. Computing coverage involved iterating through the PDB file and enumerating the unique residues for each chain corresponding to the UniProt ID. Meanwhile, the depth metric was derived by calculating the difference between the final residue number and initial residue number of each chain within the associated PDB file. For example, if a PDB file started at residue 42 and ended at residue 200 the depth would be 158. In instances where multiple experimental methods for structure determination were utilized, we excluded NMR structures for a given UniProt ID because in protein NMR, there is no parameter identical to resolution,[15] complicating comparison with X-ray and cryo-EM structures. Ranking involved the following steps:

(1) Organize the DataFrame in a hierarchical manner based on the coverage, depth, and resolution of each PDB file.
(2) Purge structures that consist solely of alpha carbons provided that other structures are present.
(3) Implement the following decision-making rules iteratively until the top four structures remain unchanged:
    (a) If the coverage difference between a higher-ranked PDB file and a lower-ranked one falls within a +/- 20 amino acid range, assess the depth of the structures and adjust the ranking accordingly, favoring the structure with greater depth. This allows structures with missing residues in loop regions to be ranked highly.
    (b) In the case where the resolution of a higher-ranked PDB exceeds 4, rearrange the rows to rank the structures according to their resolutions in descending order. This rule balances coverage and resolution.

Upon securing a ranked list of PDB files for each UniProt ID, we extracted the highest-ranked PDB file for each respective UniProt ID and its associated chain/location information. For every top-rated PDB structure, all missing residues were obtained using the PDBParser package from the Biopython library.[16] Two UniProt IDs presented missing chain information and were subsequently excluded from our dataset, rendering us with 7,604 unique proteins as visualized in **Figure 1a**.

To obtain the AlphaFold complement of the experimentally known structures, we leveraged AlphaFold's API[2] to extract all associated AlphaFold models corresponding to the 7,604 UniProt IDs in our curated dataset. Using our top-ranked PDB file and the data of missing residue numbers for a specific UniProt ID, we computed AlphaFold's predicted confidence scores for both missing and present residues. Subsequently, we documented the AlphaFold residue confidence score for every residue, irrespective of its status (missing or present), in the highest-ranked PDB structure. We further computed the average AlphaFold confidence score for both missing and present residues in the top-ranked PDB structure for each UniProt ID as shown in **Figure 1a**.

3.2. *Multidomain Structure Prediction With DEMO2*

A protein structure dataset based on experimental structures is only limited by the availability of structural information for some parts of the sequence. Towards the aim of a complete human protein structure dataset, we will need to combine experimental data available for different parts of the sequence and/or integrate predictions of the missing parts. We evaluated the feasibility of using protein-protein docking to combine structural information from different sources into a complete model for a given UniProt sequence. We used DEMO2 software.[3] Neighboring domains were sequentially submitted to DEMO2 as pairwise structure files. For instance, in the case of the insulin receptor (IR), described in the results, the L1 and CR domains were initially introduced into DEMO2, followed by the insertion of CR and L2 domains. The output generated from both inputs was then transported into PyMol, where the structures were aligned based on the "common" domain – in this case, the CR domain.

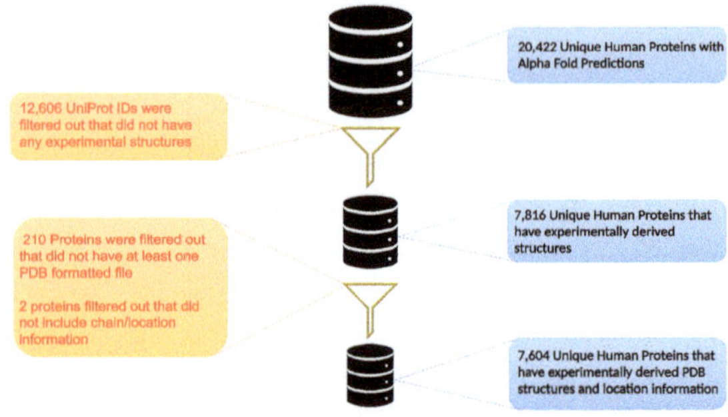

(a) Dataset Filtration

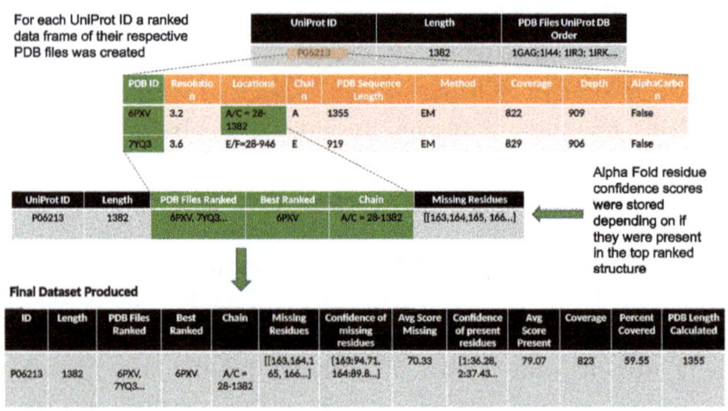

(b) Sequential steps undertaken to derive the final dataset

Fig. 1: Assembly and Composition of the Dataset.

This methodology was pursued iteratively until all desired domains were incorporated into the aligned structure.

3.3. *Analysis of Small Molecule Docking Positions*

3.3.1. *Prediction of Small Molecule Ligand Binding Sites with AutoDock Vina*

To identify putative ligand docking positions and quantify their relations to highly dense protein pocket regions, we utilized ligand-protein docking coordinates obtained from AutoDock Vina.[4] The table of structures was parsed for PDB code and specific chains. The PDB code was used to scrape from rcsb.org. The chain was subsequently used to excise the section of the PDB to use in the docking. To coordinate large-scale runs, individual AutoDock Vina scripts were automatically constructed, which employed PyMOL to determine the center of mass and bounding box for each protein, with these values stored in a configuration file. `reduce` and `prepare` scripts on protein and ligand pdbs preceded the docking run in the pipeline. These were sourced from the ADFR Suite of tools, although an updated, more robust, `reduce` script was later sourced from another repo (https://github.com/rlabduke/reduce).[17]

The AutoDock Vina code was run in batch mode using job array submissions to the SLURM scheduler on Arizona State University's Agave and Sol clusters.[18] Most jobs were completed using a single CPU and 4GB of RAM. **Figure 2** presents a logarithmic plot of runtimes (in seconds) versus ligand size (in atoms). The mean runtimes of these were strongly correlated ($\alpha = 0.746$) to number of atoms. As ligand size increases, the greater variation in runtime may be attributable to the number of flexible bonds or the total volume. To contrast, protein size in atoms and mean runtimes were uncorrelated. Cumulative runtime for a ligand across 7,527 proteins could take from hundreds to thousands of hours, but distribution across the 18,000 available cores on Sol dramatically reduced wall time. Outputs were stored in a directory structure with ligands at the top tier, each having several thousand protein directories containing affinity and output structure files for the top tier ligand.

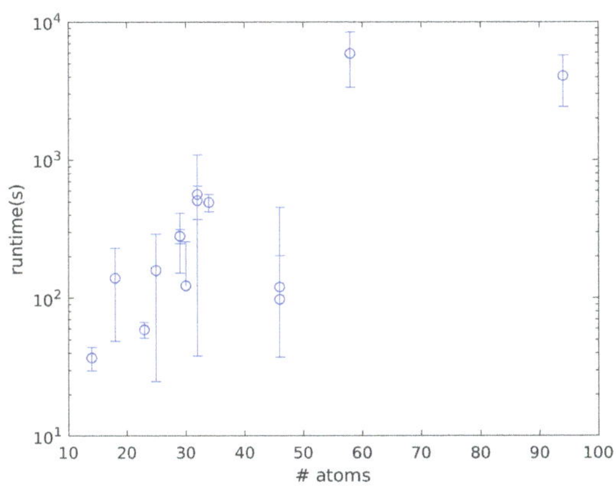

Fig. 2: Log plot of mean runtimes (in seconds) across 7,527 proteins versus ligand size (total atoms) While there was large variation in runtimes, indicated by error bars, the means were strongly correlated to ligand size.

3.3.2. *Point Cloud Clustering & Visualizations Created Using Delaunay Triangulation*

We analyzed the overlap of ligand docking positions using collections of three-dimensional point clouds that we rendered as surfaces by applying Delaunay triangulation. Delaunay triangulation is a useful method for plotting an arbitrary collection of coordinates as volumetric bodies. To further examine the spatial overlap of ligand-protein docking models for individual ligand-protein pairs, as well as the spatial overlap of docking positions for potentially competing ligands and their respective proteins, we deployed K-means clustering optimized using silhouette analysis. Silhouette analysis evaluates the density and separation between clusters, calculating a score by averaging the silhouette coefficient for each sample, which is computed as the difference between the mean intra-cluster distance and the mean nearest-cluster distance for each sample, normalized by the maximum value. The scores range between -1 and $+1$, where $+1$ indicates high separation of clusters and -1 indicates that the coordinates may have been assigned to the wrong cluster. By taking the highest-scoring configuration of clusters, we grouped ligand docking models into "locations" or "pockets."

As a metric for percent overlap of the volumetric surfaces rendered from the docking coordinates, we used **Equation 1**, where m is the number of models contained in an AutoDock Vina output file for a ligand-protein pair and k is the optimal number of clusters determined by the K-means algorithm. Fewer clusters result in a greater percent overlap, and in cases where the ratio of clusters to models is 1, the percent overlap is 0.

$$Percent\ Overlap(m, k) = (1 - \frac{k-1}{m-1}) * 100 \tag{1}$$

3.3.3. *Center of Mass*

PyMOL routines were employed for the center of mass calculations, which were used to prepare AutoDock Vina configuration scripts and in the post-processing of ligands for analysis.

4. Results and Discussion

4.1. *Human Protein Structure Database Creation*

There are 20,422 unique human protein sequences in UniProt,[14] out of which 7,816 have at least one PDB file associated with it.[5] A protein structure dataset based on experimental structures only is limited by the availability of structural information for some parts of the sequence. However, this number overestimates the availability of structural information because often only a single domain of a given human protein has been crystallized. The scale of this problem is highlighted in **Figure 3**, which compares the entire sequence lengths of the 20,422 human proteins to the coverage of sequences retrieved from the PDB. We can see that there is a drastic shift to a smaller number of amino acids covered in experimentally determined protein structures. Towards the aim of a complete human protein structure dataset, we will need to combine experimental data available for different parts of the sequence and/or integrate predictions of the missing parts. AlphaFold[2] provides a rich source of protein structure predictions that could be used, but we can see from **Figure 3** that the portions of sequences missing in existing protein structures are also the ones that it has least confidence in.

4.2. *Database Expansion Based on Multidomain Protein Interactions*

Ultimately, we wish to create a database of structures that covers the entire human proteome, and this will require inclusion of predictions. To illustrate the challenges and feasibility of expanding our dataset with AlphaFold predictions and/or by piecemealing domains of a given single UniProt ID for which domain structures have been determined independently in different experiments, we utilized the insulin receptor (IR) as a representative example. The IR is an important protein given its role in diabetes and the regulation of many cellular pathways, but it is also an experimentally challenging protein because it is a large, multimeric, multidomain, flexible membrane receptor. Thus, to this date, a full-length structure covering the entire

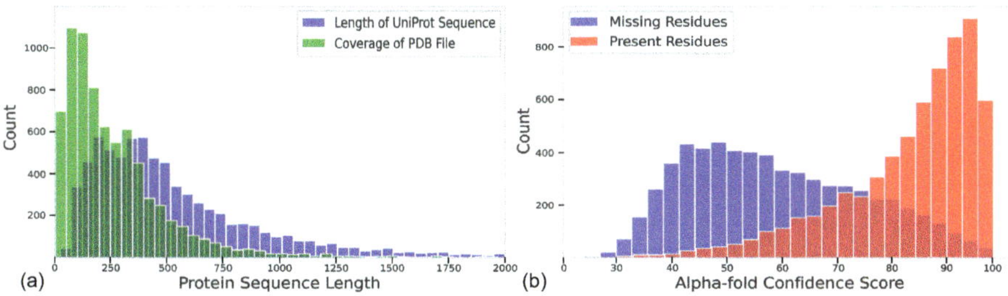

Fig. 3: **(a)** Distribution of protein length in UniProt in blue and the manually calculated coverage in green in the PDB. **(b)** AlphaFold's prediction confidence for amino acid residues, with missing residues represented in blue and present residues in red, in the context of the highest-ranked structure from the Protein Data Bank (PDB) taken from our dataset

UniProt sequence P06213 has yet to exist despite many efforts. Details of the different PDB files providing structural information and coverage for extracellular insulin binding domains, i.e., transmembrane and cytoplasmic kinase domains, have been reviewed.[19,20] 6PXV provides the most extensive coverage[21] representing the cryo-EM structure of the IR in complex with four insulin molecules. Although the full-length sequence was subjected to experimental analysis, structural data was only obtained for the extracellular domain.[21] Because the IR is a dimer, chains A and C in 6PXV are identical. Therefore, we focused our analysis solely on chain A. Initial steps involved utilizing PyMOL to visualize the distinctions between the experimentally derived structure of the IR and its predicted AlphaFold counterpart (AF-IR), depicted in **Figure 4**. Subsequently, we dissected both structures into their constituent domains: the leucine-rich repeat domains (L1-L2), a cysteine-rich region (CR), fibronectin type-III domains (FNIII-1-3), and the transmembrane domain (TM). Neighboring domains were sequentially inputted into DEMO2 (see Methods). For instance, the L1 and CR domains were initially introduced into DEMO2, followed by the insertion of CR and L2 domains. The output generated from both inputs was then transported into PyMol, where the structures were aligned based on the "common" domain - in this case, the CR domain. This methodology was pursued iteratively until all desired domains were incorporated into the aligned structure. We can see from **Figure 4** that DEMO2 not only reproduces the experimental cryo-EM structure as expected but also improves upon the initial AlphaFold prediction obtained when using the entire sequence. The integrated AlphaFold-IR structure portrayed in **Figure 4** is noticeably improved compared to AlphaFold's initial prediction. A significant portion of the error in both DEMO2 predicted structures **Figure 4** 6PXV and AF-IR structures can be attributed to an unconnected alpha helix from the FN3-2 domain.

4.3. *Database Expansion Based on Protein-Ligand Interactions*

Because our long-term goal is to view the human structure proteome as the targetome for small molecule ligands (and ultimately other biomolecules, but for now, we focus on small molecules), we used our protein structure datasets for docking more than 50 different ligands

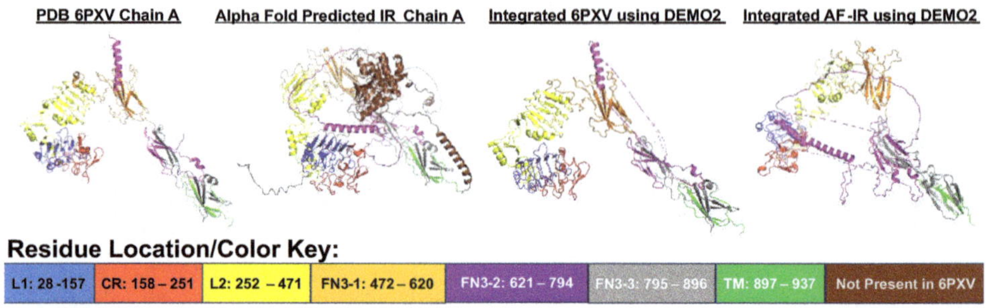

Fig. 4: Experimentally determined and predicted structures of IR.

of different sizes and physicochemical properties. We used AutoDock Vina (see Methods) and encountered a number of errors for the structures in our dataset, enumerated in Table 1.

4.4. *Automated Analysis of Ligand Prediction Results*

Even when looking at a single ligand, we now have thousands of AutoDock Vina prediction results. In the future, we plan to look at complex mixtures of ligands, which will result in even larger ligand docking datasets. Each AutoDock Vina result is a list of up to 9 docking poses for a given ligand-protein pair,[4] which vary by the details of the pose of the ligand based on bond rotations and interactions with different parts of the protein, resulting in different predicted locations and/or affinities. We know from many examples, that taking the best affinity prediction may miss biologically meaningful ligand binding pockets, which could in fact be representing allosteric and orthosteric pocket(s).[22,23,24] Furthermore, bond rotations in the ligand can result in drastic changes in predicted affinity, while the overall location of the binding pocket remains similar. To capture these insights on a large scale, we propose two approaches to automated analysis of the AutoDock Vina prediction based on the volume and center of mass of the ligands, respectively.

4.4.1. *Ligand-volume based binding pocket location analysis*

The development of a method to analyze AutoDock Vina prediction results by ligand volume overlap is shown in **Figure 5**. Volumetric analysis of four different ligand-protein pairs is shown to exemplify different common scenarios observed in AutoDock Vina predictions. An example of a low percent overlap in the volumetric surface plot for ligand 0A1 obtained from protein structure 3qtc, when docked to 1l9h (bovine rhodopsin, a G protein-coupled receptor), is shown in (a). We can see that the 9 predicted docking poses cluster into 5 easily distinguishable binding pockets. The opposite extreme is shown in (b), for ligand 00A obtained from PDB file 3cw8, docked to the same structure as in (a), 1l9h. All 9 docking poses are found in the same location, with 100 percent overlap. Other ligand-protein pairs show less clear results, for example, Benzo(a)pyrene (BaP), a hydrophobic ring structure ligand (c) and apigenin, a flavonoid ligand also with hydrophobic ring structures but with several oxygen-containing groups (d), when docked to the same protein (1ksg). Both ligands are of comparable size but different physicochemical properties, and both show overlap that is not easily distinguishable with this approach.

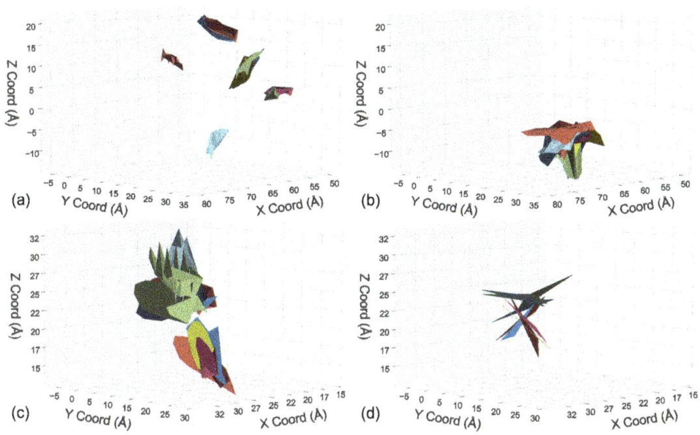

Fig. 5: Volumetric surface plots for different ligands or from original protein:docked protein pairs: (a) 0A1 3qtc:1l9h, (b) 00A 3cw8:1l9h, (c) Benzo(a)pyrene:1ksg, (d) apigenin:1ksg.

We clustered the volumetric overlap results using an optimized KMeans clustering algorithm (see Methods). The result is shown for the interaction of BaP with 1ksg in **Figure 6**a,b. We can see that we now obtain clear separation into two clusters, representing two distinct pockets in well-separated domains of the 1ksg protein structure, shown in **Figure 6**c.

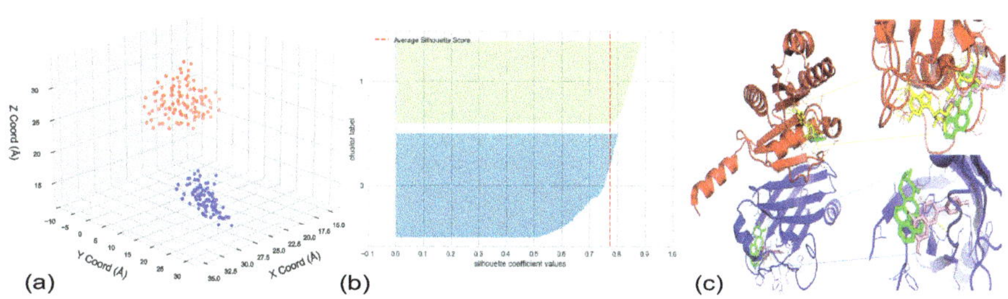

Fig. 6: Optimized Clustering Algorithm Deployed on BaP Ligand. (a) Number of Clusters = 2. Optimized using Silhouette Score. (b) BaP Models Percent Overlap = 87.5%. (c) Pymol representation of BaP, apigenin, and GTP in 1ksg structure.

4.4.2. *Ligand-center of mass based binding pocket location analysis*

A complementary approach to the volumetric overlap analysis is to reduce the complexity of ligand description to represent each pose by its center of mass. The result of this analysis for the same ligand:protein pair BaP:1ksg and apigenin:1ksg is shown in **Figure 7**. We can see that even in the lowest resolution representation of the ligand, where the coordinates of each atom in the molecule were collectively replaced with a single coordinate for the center of mass, the separation between pockets is not entirely clear. Furthermore, we can see that the known ligand binding pocket for the ligand that's actually bound to 1ksg, GTP, is located in the pocket on the top, which carries an overall lower predicted affinity than the regions on the right-hand side of the protein. To see how the pockets observed with these three ligands compare to a larger set of 50 ligands, we clustered the results using DBSCAN. They formed eight distinct clusters, with clear preferences for 4 of these pockets. The DBSCAN analysis was run over the entire set of proteins to create a distribution of cluster counts. From the left, this distribution sharply peaked at 7 clusters with a slowly decreasing long tail to the right.

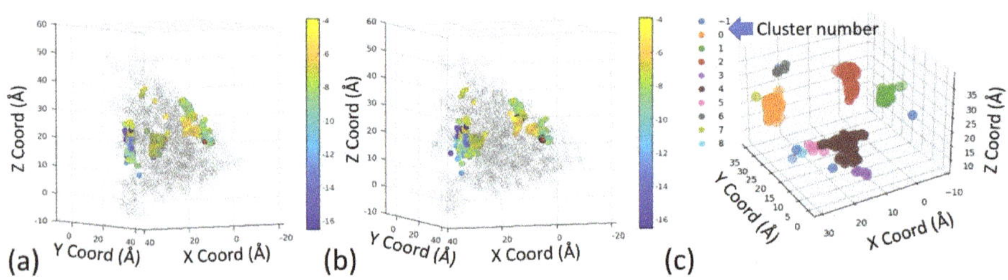

Fig. 7: Center of mass for apigenin ligand (a) and natural ligand BaP (b) docked to 1ksg. In (c), center of mass for 50 ligands are clustered with DBSCAN. Structure as in **Figure 6**c.

5. Conclusions and Future Work

In an era where assembling an "integrated structural map of the human cell"[1] at atomic resolution is no longer out of reach, cell structural bioinformatics will need to reconcile two extreme views of biomolecules inside cells: "selective" interaction of high-affinity ligands with single protein targets versus "everything binds to everything" the deciphering of which requires quantification of ligand and protein concentrations to determine chemical equilibria of binding. Our long-term goal is to assist this task and ongoing cell structural bioinformatics efforts by developing a human protein structure targetome database and a pipeline of automated tools that allow quantitative analysis of millions of protein-ligand interactions. Towards this goal, we present the docking of our current version of the human protein targetome to ligands using AutoDock Vina. We developed two complementary, automated analyses of affinity and binding site location using the center of mass comparisons, which can identify clusters at a coarse-grained level but ignores the size and shape of the ligands, as well as Silhouette Score clustering optimization of predicted ligand volume overlap, suitable for detailed analysis of ligand overlap when this level of detail is needed. In the future, we plan to use the human targetome and its ligand binding information to make predictions on the competition of ligands with different affinities to gain insights into challenging problems such as regulation of metabolic pathways, interactions with complex mixtures of nutrients and pollutants, and predicting off-target effects of drugs. With millions of known small molecules from natural sources and large numbers of ligands that can be synthesized in the laboratory, this pipeline will complement projects where experiments alone cannot reach the scale needed to gain biological insights.

Each iteration of the set of the structures comes with limitations. Our current dataset has the major limitation that it only represents a fraction (7606 of 20422 = 37%) of all human proteins. Currently, all structures are experimentally determined, while future iterations will also include predictions. To illustrate how predictions can be incorporated, we used an example, the insulin receptor, with sequential assembly of domains from N to C terminus. These strategies can be improved, for example, a sensitivity analysis for the sequence with which domains are assembled can be carried out. Other structure prediction and assembly strategies can be used that are specialized for the type of protein or domain or structural element, such as transmembrane helices. Users of the current and future protein structure datasets can further filter them if more uniform data are required or if the focus is on a given location, such as extracellular or a given subcellular compartment. Other limitations include the differences in quality of different structures, the lack of water molecules, ions and other solvents such as lipids, all known to be important contributors to ligand binding. This dataset can be subjected to future improvements in methods or filters as needed for a given use case.

The focus (and implementation status) of the current paper is the development of the curated database and tools for its analysis if it is used in target identification using tools such as AutoDock Vina.[4] The need to chose a method for docking of ligands presents another inherent limitation in this work. Autodock Vina,[4] for example, is very widely used and compares well with other methods,[25] but reverse docking in general suffers from large false positive rates due to limitations in scoring functions.[26] However, in most cases, a proper gold standard for target discovery is absent as it is typically unknown which proteins are true negatives (i.e., are not targets). The explosion in new computational methods using machine learning and artificial intelligence[6] can be used to replace or complement the reverse docking approach using Autodock Vina or related methods for example with state-of-the-art deep learning tools for ligand binding pocket predictions. The goal of the curated protein structure database described here was to improve coverage of the human structural proteome, while keeping the quality of the dataset as high as possible with state-of-the-art in data and tool availability to enable applications in cell structural bioinformatics.

6. Acknowledgment

S.K.S. was supported by the National Institute of General Medical Sciences of the National Institutes of Health under award number R01GM145210.

References

1. E. Lundberg, T. Ideker and A. Sali, Tools for assembling the cell: Towards the era of cell structural bioinformatics, https://psb.stanford.edu/workshop/tools/ (2023).
2. J. Jumper, R. Evans, A. Pritzel, T. Green, M. Figurnov, O. Ronneberger, K. Tunyasuvunakool, R. Bates, A. Žídek, A. Potapenko, A. Bridgland, C. Meyer, S. A. A. Kohl, A. J. Ballard, A. Cowie, B. Romera-Paredes, S. Nikolov, R. Jain, J. Adler, T. Back, S. Petersen, D. Reiman, E. Clancy, M. Zielinski, M. Steinegger, M. Pacholska, T. Berghammer, S. Bodenstein, D. Silver, O. Vinyals, A. W. Senior, K. Kavukcuoglu, P. Kohli and D. Hassabis, Highly accurate protein structure prediction with AlphaFold, *Nature* **596**, 583 (July 2021).
3. X. Zhou, C. Peng, W. Zheng, Y. Li, G. Zhang and Y. Zhang, DEMO2: Assemble multi-domain protein structures by coupling analogous template alignments with deep-learning inter-domain restraint prediction, *Nucleic Acids Research* **50**, W235 (May 2022).
4. J. Eberhardt, D. Santos-Martins, A. F. Tillack and S. Forli, AutoDock vina 1.2.0: New docking methods, expanded force field, and python bindings, *Journal of Chemical Information and Modeling* **61**, 3891 (July 2021).
5. H. Berman, J. Westbrook, Z. Feng, G. Gilliland, T. Bhat, H. Weissig, I. Shindyalov and P. Bourne, The protein data bank, *Nucleic Acids Research* **28**, 235 (2000).
6. A. V. Sadybekov and V. Katritch, Computational approaches streamlining drug discovery, *Nature* **616**, 673 (2023).
7. Y. Tian, N. Wan, H. Zhang, C. Shao, M. Ding, Q. Bao, H. Hu, H. Sun, C. Liu, K. Zhou, S. Chen, G. Wang, H. Ye and H. Hao, Chemoproteomic mapping of the glycolytic targetome in cancer cells, *Nat Chem Biol.* (2023).
8. A. Koutsoukas, B. Simms, J. Kirchmair, P. J. Bond, A. V. Whitmore, S. Zimmer, M. P. Young, J. L. Jenkins, M. Glick, R. C. Glen and A. Bender, From in silico target prediction to multi-target drug design: current databases, methods and applications, *J Proteomics* **74**, 2554 (2011).
9. S. Tejpal, A. Wemyss, C. Bastie and J. Klein-Seetharaman, Lemon extract reduces angiotensin converting enzyme (ace) expression and activity and increases insulin sensitivity and lipolysis in mouse adipocytes., *Nutrients* **12**, p. 2348 (2020).
10. E. Rozewski, O. Taqi, E. H. Fini, N. A. Lewinski and J. Klein-Seetharaman, Systems biology of asphalt pollutants and their human molecular targets, *Frontiers in Systems Biology* **2**, p. 928962 (2023).
11. P. Khare, J. Machesky, R. Soto, M. He, A. Presto and D. Gentner, Asphalt-related emissions are a major missing nontraditional source of secondary organic aerosol precursors, *Sci. Adv.* **6**, p. eabb9785 (2020).
12. S. Galati, M. D. Stefano, E. Martinelli, G. Poli and T. Tuccinardi, Recent advances in in silico target fishing, *Molecules* **26**, p. 5124 (2021).
13. H. Pérez-Sánchez, H. den Haan, J. Peña-García, J. Lozano-Sánchez, M. M. Moreno, A. Sánchez-Pérez, A. Muñoz, P. Ruiz-Espinosa, A. Pereira, A. Katsikoudi, J. G. Hernández, I. Stojanovic, A. Carretero and A. Tzakos, Dia-db: A database and web server for the prediction of diabetes drugs, *J Chem Inf Model* **60**, 4124 (2020).
14. A. Bateman, M.-J. Martin, S. Orchard, M. Magrane, S. Ahmad, E. Alpi, E. H. Bowler-Barnett, R. Britto, H. Bye-A-Jee, A. Cukura, P. Denny, T. Dogan, T. Ebenezer, J. Fan, P. Garmiri, L. J. da Costa Gonzales, E. Hatton-Ellis, A. Hussein, A. Ignatchenko, G. Insana, R. Ishtiaq, V. Joshi, D. Jyothi, S. Kandasaamy, A. Lock, A. Luciani, M. Lugaric, J. Luo, Y. Lussi, A. MacDougall, F. Madeira, M. Mahmoudy, A. Mishra, K. Moulang, A. Nightingale, S. Pundir, G. Qi, S. Raj, P. Raposo, D. L. Rice, R. Saidi, R. Santos, E. Speretta, J. Stephenson, P. Totoo, E. Turner, N. Tyagi, P. Vasudev, K. Warner, X. Watkins, R. Zaru, H. Zellner, A. J. Bridge, L. Aimo, G. Argoud-Puy, A. H. Auchincloss, K. B. Axelsen, P. Bansal, D. Baratin, T. M. B. Neto, M.-

C. Blatter, J. T. Bolleman, E. Boutet, L. Breuza, B. C. Gil, C. Casals-Casas, K. C. Echioukh, E. Coudert, B. Cuche, E. de Castro, A. Estreicher, M. L. Famiglietti, M. Feuermann, E. Gasteiger, P. Gaudet, S. Gehant, V. Gerritsen, A. Gos, N. Gruaz, C. Hulo, N. Hyka-Nouspikel, F. Jungo, A. Kerhornou, P. L. Mercier, D. Lieberherr, P. Masson, A. Morgat, V. Muthukrishnan, S. Paesano, I. Pedruzzi, S. Pilbout, L. Pourcel, S. Poux, M. Pozzato, M. Pruess, N. Redaschi, C. Rivoire, C. J. A. Sigrist, K. Sonesson, S. Sundaram, C. H. Wu, C. N. Arighi, L. Arminski, C. Chen, Y. Chen, H. Huang, K. Laiho, P. McGarvey, D. A. Natale, K. Ross, C. R. Vinayaka, Q. Wang, Y. Wang and J. Zhang, UniProt: the universal protein knowledgebase in 2023, *Nucleic Acids Research* **51**, D523 (November 2022).
15. M. Berjanskii, J. Zhou, Y. L. Y, G. Lin and D. W. DS, Resolution-by-proxy: a simple measure for assessing and comparing the overall quality of nmr protein structures, *J Biomol NMR.* **53**, 167 (2012).
16. P. J. A. Cock, T. Antao, J. T. Chang, B. A. Chapman, C. J. Cox, A. Dalke, I. Friedberg, T. Hamelryck, F. Kauff, B. Wilczynski and M. J. L. de Hoon, Biopython: freely available python tools for computational molecular biology and bioinformatics, *Bioinformatics* **25**, 1422 (2009).
17. Y. Zhang and M. F. Sanner, AutoDock CrankPep: combining folding and docking to predict protein–peptide complexes, *Bioinformatics* **35**, 5121 (06 2019).
18. D. M. e. a. Jennewein, The Sol Supercomputer at Arizona State University, in *Practice and Experience in Advanced Research Computing*, PEARC '23 (Association for Computing Machinery, New York, NY, USA, Jul 2023).
19. L. Ye, S. Maji, N. Sanghera, P. Gopalasingam, E. Gorbunov, S. Tarasov, O. Epstein and J. Klein-Seetharaman, Structure and dynamics of the insulin receptor: implications for receptor activation and drug discovery., *Drug Discov Today* **22**, 1092 (2017).
20. L. Kumar, W. Vizgaudis and J. Klein-Seetharaman, Structure-based survey of ligand binding in the human insulin receptor, *Br J Pharmacol* **179**, 3512 (2022).
21. E. Uchikawa, E. Choi, G. Shang, H. Yu and X. chen Bai, Activation mechanism of the insulin receptor revealed by cryo-EM structure of the fully liganded receptor–ligand complex, *eLife* **8** (August 2019).
22. C. Chu, J. Ji, R. Dagda, J. Jiang, Y. Tyurina, A. Kapralov, V. Tyurin, N. Yanamala, I. Shrivastava, D. Mohammadyani, K. Wang, J. Zhu, J. Klein-Seetharaman, K. Balasubramanian, A. Amoscato, G. Borisenko, Z. H. andn AM Gusdon, A. Cheikhi, E. Steer, R. Wang, C. Baty, S. Watkins, I. Bahar, H. Bayir and V. Kagan, Cardiolipin externalization to the outer mitochondrial membrane acts as an elimination signal for mitophagy in neuronal cells, *Nat Cell Biol.* **15**, 1197 (2013).
23. U. Schlattner, M. Tokarska-Schlattner, S. Ramirez, Y. Tyurina, A. Amoscato, D. Mohammadyani, Z. Huang, J. Jiang, N. Yanamala, A. Seffouh, M. Boissan, R. Epand, R. Epand, J. Klein-Seetharaman, M. Lacombe and V. Kagan, Dual function of mitochondrial nm23-h4 protein in phosphotransfer and intermembrane lipid transfer: a cardiolipin-dependent switch, *J Biol Chem.* **288**, 111 (2013).
24. N. Yanamala, E. Gardner, A. Riciutti and J. Klein-Seetharaman, The cytoplasmic rhodopsin-protein interface: potential for drug discovery, *Curr Drug Targets* **13**, 3 (2012).
25. V. T. Sabe, T. Ntombela, L. A. Jhamba, G. E. Maguire, T. Govender, T. Naicker and H. G. Kruger, Current trends in computer aided drug design and a highlight of drugs discovered via computational techniques: A review, *European Journal of Medicinal Chemistry* , p. 113705 (2021).
26. G. L. Warren, C. W. Andrews, A.-M. Capelli, B. Clarke, J. LaLonde, M. H. Lambert, M. Lindvall, N. Nevins, S. F. Semus, S. Senger, G. Tedesco, I. D. Wall, J. M. Woolven, C. E. Peishoff and M. S. Head, A critical assessment of docking programs and scoring functions, *J Med Chem* **49**, 5912 (2006).

Modeling Path Importance for Effective Alzheimer's Disease Drug Repurposing

Shunian Xiang[1*], Patrick J. Lawrence[1*], Bo Peng[2], ChienWei Chiang[1], PhD, Dokyoon Kim[3], PhD, Li Shen[3], PhD, and Xia Ning[1,2,4†], PhD

[1] *Biomedical Informatics Department, The Ohio State University, Columbus, OH 43210, USA*
[2] *Computer Science and Engineering Department, The Ohio State University, Columbus, OH 43210, USA*
[3] *Department of Biostatistics, Epidemiology, and Informatics, University of Pennsylvania, Philadelphia, PA 19104 USA*
[4] *Translational Data Analytics Institute, The Ohio State University, Columbus, OH 43210, USA*
[*] *Co-first author; authors contributed equally to this work*
[†] *E-mail: ning.104@osu.edu*

Recently, drug repurposing has emerged as an effective and resource-efficient paradigm for AD drug discovery. Among various methods for drug repurposing, network-based methods have shown promising results as they are capable of leveraging complex networks that integrate multiple interaction types, such as protein-protein interactions, to more effectively identify candidate drugs. However, existing approaches typically assume paths of the same length in the network have equal importance in identifying the therapeutic effect of drugs. Other domains have found that same length paths do not necessarily have the same importance. Thus, relying on this assumption may be deleterious to drug repurposing attempts. In this work, we propose MPI (Modeling Path Importance), a novel network-based method for AD drug repurposing. MPI is unique in that it prioritizes important paths via learned node embeddings, which can effectively capture a network's rich structural information. Thus, leveraging learned embeddings allows MPI to effectively differentiate the importance among paths. We evaluate MPI against a commonly used baseline method that identifies anti-AD drug candidates primarily based on the shortest paths between drugs and AD in the network. We observe that among the top-50 ranked drugs, MPI prioritizes 20.0% more drugs with anti-AD evidence compared to the baseline. Finally, Cox proportional-hazard models produced from insurance claims data aid us in identifying the use of etodolac, nicotine, and BBB-crossing ACE-INHs as having a reduced risk of AD, suggesting such drugs may be viable candidates for repurposing and should be explored further in future studies.

Keywords: Alzheimer's Disease; Drug Repurposing; Machine Learning.

1. Introduction

Alzheimer's Disease, denoted AD, is a progressive neurodegenerative disorder that accounts for 60%-70% of dementia cases and affects more than 50 million people worldwide today.[1,2] Given the large number of affected individuals and AD's life-threatening nature,[3] extensive resources have been dedicated to developing AD-modifying drugs. Since 2003, inefficacy or toxicity has accounted for a 95+% failure rate among candidates evaluated for AD treatment.[4,5] Furthermore, none of the current US Food and Drug Administration (FDA)-approved AD drugs are curative; they only slow disease progression. Because of the immense resources required to conduct clinical trials,[6] the numerous failed clinical trials have necessitated the development of a more resource-efficient method for AD drug discovery. In the last decade, the identification of new therapeutic indications for existing FDA-approved drugs, referred to as drug repurposing,[7] has emerged as an effective and resource-efficient paradigm for drug discovery.[8] This is an attractive option as the toxicity, pharmacokinetics, and pharmacodynamics of FDA-approved drugs have already been thoroughly investigated by previous clinical trials.[7,9]

Recently, the curation of comprehensive drug databases has enabled the development of computational methods for AD drug repurposing.[10–14] Among all the methods, network-based methods have shown promising results and emerged as a popular approach.[13,15,16] Network-based methods utilize comprehensive protein-protein, drug-target, and AD-protein interactions to effectively reveal potential therapeutic effects of drugs on AD. Though promising, existing methods[13] measure the therapeutic effects of drugs on AD primarily using count and length of the paths connecting drug nodes and the AD node in the network. Paths of the same length are considered equivalently effective at identifying the therapeutic effect of drugs by these methods. However, in other domains, paths of the same length have been shown to exhibit substantially different levels of importance.[17,18] As such, assuming equal length paths have equal importance could be detrimental to effective drug repurposing for AD.

In this work, we propose a novel method to conduct drug repurposing for AD, MPI (Modeling Path Importance), to address this limitation. Similar to existing methods,[13,14] MPI leverages the interactions between drugs and AD via proteins as indications of the potential therapeutic effects of drugs on AD. Based on the interactions, MPI introduces a scoring function to score and rank drugs for their anti-AD effectiveness. MPI is unique in that it learns node embeddings[19] and prioritizes important paths via these learned embeddings. Recent work[20] has shown that the learned node embeddings can effectively capture the rich structure information within a network. Thus, scoring paths using node embeddings allows MPI to utilize the network structure information to better prioritize paths for effective AD drug repurposing. Specifically, in this study, MPI leverages DeepWalk,[21] a widely used network learning approach, to generate node embeddings. Edges are scored using a normalized dot product between the learned node embeddings; paths and drugs are scored by multiplying individual edge scores. Note that because MPI serves as a general framework, other network learning approaches, such as Node2Vec[20] and graph neural networks,[22] could also be easily incorporated to generate node embeddings.

In this study, we construct a network to conduct drug repurposing for AD by combining protein-protein interactions (PPIs), drug-target interactions (DTIs), and AD-protein interactions (APIs) from multiple data sources. To investigate the effectiveness of MPI, we compare MPI against a commonly utilized network-based drug repurposing method for AD,[13,23] denoted as BSL, using our network. Our experimental results demonstrate that among the top-50 ranked drugs, MPI prioritizes 20% more drugs with anti-AD evidence compared to BSL. We examine published literature and analyze insurance claims meta data to evaluate the evidence of anti-AD activity among MPI's top prioritized candidates. The results of our evaluation find consensus between published experimental results and our own analysis for a few drug candidates. Notably, angiotensin converting enzyme inhibitors (ACE-INHs) represent a class of drugs that should be further explored for their anti-AD properties. Moreover, other drugs, such as nicotine, that enhance the brains response to acetylcholine and reduce cholinergic atrophy should be examined as well. Conversely, we find that, relative to other evaluated drugs, long-term use of trihexyphenidyl increases the risk of AD. This was corroborated by previously published *in vivo* experiments.[24] Finally, we find etodolac to confer the lowest risk of developing AD among all cyclooxygenase inhibitors (COX-INHs) in our network. Altogether, these findings suggest that MPI may be a viable option with respect to identifying repurposing candidates to treat AD.

2. Materials and Methods

2.1. *Network construction*

PPIs, DTIs and APIs have shown utility for AD drug repurposing.[13] As such, we construct our network using these interactions. Below, we describe our process for compiling the PPIs, DTIs and APIs used to construct our network from public data sources. In total, our network has 327,924, 2,854, and 230 edges corresponding to PPIs, DTIs, and APIs. These edges connect one AD node, 18,527 protein nodes, and 386 drug nodes.

2.1.1. *Protein-protein interactions (PPIs)*

Following Chen et al.,[13] we include a comprehensive list of human PPIs consisting of 327,924 interactions. This list aggregates a total of 21 bioinformatics and systems biology databases with combinations of five types of experimental evidence. We refer the audience of interest to Chen et al.[13] for a detailed description of the databases.

2.1.2. *Drug-target interactions (DTIs)*

We assemble drug-target interactions and bioactivity data from 4 commonly used databases (each downloaded in November 2022): the ChEMBL database[25] (v31), the binding database,[26] the therapeutic target database,[27] and the IUPHAR/BPS guide to pharmacology database.[28] We retain the drug-target interactions that satisfy the following inclusion criteria: 1) binding affinities, including K_i, K_d, IC_{50}, or EC_{50}, must be less than or equal to 10 μM; 2) protein targets and their respective proteins must have a unique UniProt[29] accession number; 3) protein targets must be marked as reviewed in the UniProt database; 4) protein targets must be present in *homo sapiens*.

Additionally, we retain drugs for which we have sufficient sample size to conduct quantitative analysis using MarketScan[30] insurance claims meta data (see Section 2.4). Specifically, included drugs have at least 100 patients with their first dose at least 2 years prior to an AD diagnosis (dx). Additionally, these drugs must have at least 15 patients who eventually received an AD dx. Applying these filters yielded 2,854 edges connecting 386 FDA-approved drugs to 548 protein targets.

2.1.3. AD-*protein interactions (APIs)*

The AD-associated proteins included in the network were identified from multiple sources. 54 β-amyloid-related proteins and 27 tauopathy-related proteins were obtained from Cheng et al..[13] The authors identified proteins that satisfied at least one of the following criteria: 1) the proteins are validated in large-scale amyloid or tauopathy genome-wide association studies; 2) *in vivo* experimental models exhibit evidence that knockdown or overexpression of the protein leads to AD-like amyloid or tau pathology. We also include 93 unique late-onset AD common risk proteins identified by 7 large-scale genetic studies .[31–37] We further incorporate a set of 118 AD-associated proteins introduced in at least 2 out of the 6 following databases (each was downloaded in November 2022): the online Mendelian inheritance in man database,[38] the comparative toxicogenomics database,[39] the HuGE navigator database,[40] the DisGeNET database[41] (v7.0), the ClinVar database[42] and the Open Targets database[43] (v22.09). In total, our network is comprised of 230 unique, AD-associated proteins. Each of the AD-associated proteins are connected to a single AD node with each edge between a protein and the AD node representing an API in our network.

2.2. *Modeling path importance for* AD *drug repurposing*

In this work, we denote the constructed network as G. Each node in G is denoted as v_i. Specifically, drug nodes, protein nodes and the AD node are v_i^d, v_i^g, and v_i^a , respectively. Note that the index, i, does not apply to the AD node as there are not multiple in our network. Each edge that connects node v_i to node v_j is denoted as e_{ij}. Each path is denoted as p_m, and the set of edges involved in a path is denoted $\mathbb{E}_{p_m}$. Below, we denote matrices, scalars and row vectors using uppercase, lowercase, and bold lowercase letters, respectively.

In MPI, we leveraged DeepWalk,[21] a widely used node embedding approach, to learn embeddings for each node in G. First, for each node v_i in the network, we conduct 256 random walks originating from this node, and terminating once the path length reaches 128. DeepWalk is then trained by sliding a window of length 10 over the generated paths. Nodes within the same window are forced to have similar embeddings following the objective function defined in the original paper.[21] Node embeddings for MPI are produced such that they have 128 dimensions.

After generating node embeddings, we score edge, e_{ij}, using a normalized dot product of the embedding of v_i^x (x=d, g or a) and v_j^y (y=d, g or a) as follows:

$$w_{ij} = \frac{\exp(\mathbf{v}_i^x \mathbf{v}_j^{y\top})}{\sum_{k \in \mathbb{V}} \exp(\mathbf{v}_i^x \mathbf{v}_k^{y\top})}, \tag{1}$$

where w_{ij} is the score of the edge e_{ij}; $\mathbf{v}_i^x$ and $\mathbf{v}_j^y$ is the learned embedding of node v_i^x and v_j^y, respectively; $\exp(\cdot)$ is the exponential function; and $\mathbb{V}$ is the set of all the nodes in the network. Note that, in Equation 1, only one of v_i^x and v_j^y could be the AD node. These edge scores are calculated with node embeddings which implicitly capture the rich structural information within the network. Thus, compared to existing methods, MPI can better leverage a network's structural information for AD drug repurposing. We calculate the score for each path by multiplying the scores of its individual edges as follows:

$$s_{p_m} = \prod_{e_{ij} \in \mathbb{E}_{p_m}} w_{ij}, \tag{2}$$

where s_{p_m} is the score of the path p_m; and $\mathbb{E}_p$ is the set of all the edges in the path p. The score for each drug (i.e., $s_{v_i^d}$) is then defined as the summation of the scores from all 3-hop or shorter paths that originate from the AD node and terminate at the drug node.

2.3. *Baseline method*

To evaluate the performance of MPI, we compare MPI against a network-based method recently developed by Cheng et al.,[13] denoted as BSL. BSL scores drugs based on the shortest distance between the drug targets and the AD-associated proteins (Section 2.1.3) in the network. Specifically, we denote $\mathbb{T}(i)$ as the set of protein targets associated with a given drug v_i^d, and denote $\mathbb{P}$ as the set of AD-associated proteins. The proximity between these two sets is calculated as the average shortest distance between elements in $\mathbb{T}(i)$ and $\mathbb{P}$ as follows:

$$r(\mathbb{T}(i), \mathbb{P}) = \frac{1}{|\mathbb{T}(i)| + |\mathbb{P}|} \left(\sum_{v_j \in \mathbb{T}(i)} \min_{v_k \in \mathbb{P}} d(v_j, v_k) + \sum_{v_k \in \mathbb{P}} \min_{v_j \in \mathbb{T}(i)} d(v_k, v_j) \right), \tag{3}$$

where $r(\mathbb{T}(i), \mathbb{P})$ is the proximity between these two sets; $|\mathbb{T}(i)|$ and $|\mathbb{P}|$ is the size of $\mathbb{T}(i)$ and $\mathbb{P}$, respectively; and $\min_{v_k \in \mathbb{P}} d(v_j, v_k)$ is the shortest distance between v_j and any elements in $\mathbb{P}$. Subsequently, we conduct a permutation test to assess the statistical significance of the calculated proximity. The resulting z-score from this test is used as the score of drug v_i.[13] In BSL, a lower drug score implies a higher potential for effective AD treatment.

2.4. *Validation using MarketScan database*

We use MarketScan medicare supplemental database from 2012–2021 to evaluate drug impact on AD onset via Cox proportional-hazard models.[30] The MarketScan database includes data for over 8 million unique individuals and is comprised of demographic information, administrative information, diagnoses, procedures, and pharmacy records. International Classification of Disease (ICD)-9/ICD-10 codes denote diagnoses and National Drug Codes (NDCs) record pharmacy claims. We use the ICD-9/ICD-10 codes listed in Supplementary Table S4♠ to define AD and comorbidities, which are included as covariates in Cox proportional-hazard models. We conduct our analysis over 1,632,218 unique individuals who were at least 65 years by 2022 and possessed a minimum of five years insurance enrollment prior of first AD diagnosis. Drugs from our constructed network are mapped to NDC codes by partial matching of generic names from MartketScan redbook. We only include patients who took or started taking a drug at least two years prior to AD diagnosis to mitigate the possibility that patients starting a drug already had AD given that AD is difficult to diagnosis.

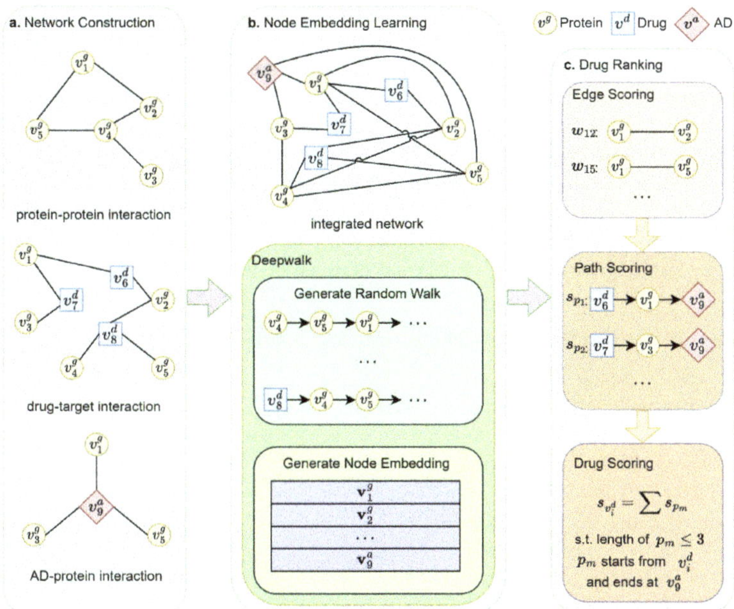

Fig. 1: Figure 1a shows the network construction process in MPI. Figure 1b shows the DeepWalk-based node embedding generation in MPI. Figure 1c shows the edge, path and drug scoring in MPI.

3. Results

3.1. MPI *for* AD *drug repurposing*

In this study, we curate a network consisting PPIs, DTIs and AGIs and propose a novel network-based method, MPI, for AD drug repurposing. We propose MPI with the following intuitions: 1) proteins that associated with AD are localized in the corresponding disease module within the comprehensive human PPI network; 2) the drug target(s) for a disease may also be targeted for other diseases (e.g., AD) owing to common functional targets and pathways elucidated by PPIs; 3) if a drug node is linked to the AD node through the paths of drug targets and AD-associated proteins in the PPI, the drug may have a treatment effect on AD.

We implement MPI using the following steps: 1) integrate AD-protein interactions, drug-target interactions and protein-protein interactions to generate a comprehensive network (Figure 1a), 2) employ DeepWalk to learn node embeddings which capture the structural information within the network (Figure 1b), and 3) score edges, paths and drugs based on the learned

♠Supplementary material and code can be found here: https://github.com/ninglab/MPI

embeddings to leverage the structural information for better AD drug repurposing (Figure 1c). Then we identify plausible treatment candidates from the top-ranked drugs using a literature search of the published evidence. We collected 327,924 PPIs from 21 bioinformatics and systems biology databases (Section 2.1.1). We also collected 2,854 DTIs from 4 commonly used databases (Section 2.1.2), and 230 comprehensive APIs from multiple resources (Section 2.1.3). By aggregating all the interactions, we construct a drug-protein-AD network comprised of 386 drug nodes, 18,527 protein nodes, 1 AD node, and 331,008 edges. More details about the network construction are available in Section 2. To the best of our knowledge, MPI is the first method which effectively repurposes drug candidates for AD treatment by prioritizing paths between drug nodes and the AD node using learned node embeddings.

3.2. *Comparing anti-*AD *evidence of* MPI*'s and* BSL*'s top-50 drugs*

We compare the top-50 drugs prioritized by MPI and BSL to evaluate their capacity for repurposing drugs to treat AD. Specifically, we score and rank all 386 drug nodes in our network using MPI and BSL. The complete rankings are reported in Supplementary Table S3. We then perform a literature search to evaluate the anti-AD evidence of the top-50 ranked drugs for both MPI and BSL. We define anti-AD evidence as any published experimental result(s), which demonstrate a drug either protects against the development of AD or ameliorates aberrant cellular phenotypes caused by AD. We present MPI's and BSL's top-10 drugs and their anti-AD evidence in Table 1 and Table 2, respectively. The complete rankings for the top-50 drugs and their anti-AD evidence is available in Supplementary Tables S1 and S2. Based on the significance of the anti-AD evidence, we categorized drugs into the following 6 types in decreasing order of significance: 1) drugs which are FDA-approved for AD treatment (approved); 2) drugs that have demonstrated anti-AD effects in completed clinical trials or are under investigation in AD clinical trials (clinical); 3) drugs which have demonstrated anti-AD effects in *in vivo* experiments (*in vivo*); 4) drugs which have

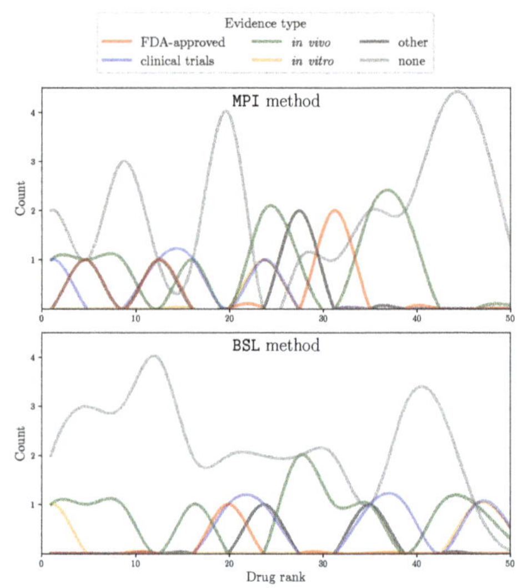

(a) Distribution of drug rank by evidence type for BSL and MPI methods.

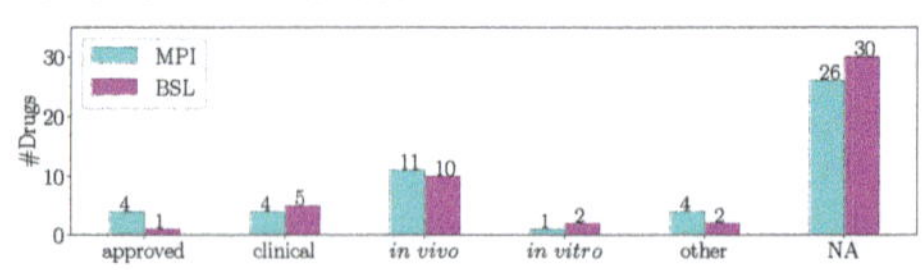

(b) Drug counts for each type of evidence among MPI's and BSL's top-50 drugs.

Fig. 2: Evaluation of drug rank distributions: MPI and BSL.

demonstrated anti-AD effects in *in vitro* experiments (*in vitro*); 5) drugs which show anti-AD effects in observational studies, cohort studies or analyses in insurance data (other); 6) drugs that either do not have the above 5 types of evidence or have been demonstrated ineffective or damaging for AD (NA). We present the distribution of the top-50 drugs from MPI and BSL over the different types of evidence in Figure 2a and the counts of each evidence type in Figure 2b. In Figure 2a, we observe more drugs with evidence ranked highly by MPI compared to BSL. This is supported by Figure 2b, which confirms that MPI identified more evidential anti-AD drugs compared to BSL in the top-50 ranked drugs. Specifically, among the top-50 ranked drugs, MPI prioritized 24 evidential anti-AD drugs while BSL only prioritized 20 evidential anti-AD drugs, demonstrating an improvement of 20%. Figures 2a and 2b also show MPI outperforms BSL in prioritizing drugs with significant evidence. MPI prioritizes all the 4 FDA-approved anti-AD drugs (e.g., galantamine, rivastigmine, donepezil and memantine) in our network among the top-50. In contrast, BSL prioritizes only a single FDA-approved anti-AD drug (donepezil) among the top-50.

We also observe in Table 1 and Table 2 that MPI is more effective than BSL at prioritizing anti-AD drugs among the very top (top-10) of the ranking list. That is, among the top-10 drugs, 6 drugs from MPI have anti-AD evidence including the FDA-approved AD drug galantamine, while only 4 drugs from BSL are evidential. As presented in Section 2, compared to BSL, MPI learns node embeddings to capture the rich structural information within the network, and leverage the structural information to better identify anti-AD drugs. The superior performance of MPI over BSL demonstrates the effectiveness of leveraging the network structural information to conduct repurposing to identify candidates for AD treatment. We also notice that both MPI and BSL prioritize 17 drugs in concordance within their top-50 drug lists. Among the 17 drugs, 5 drugs demonstrate anti-AD evidence: donepezil is an FDA-approved anti-AD drug; nicotine and rasagiline have clinical anti-AD evidence; and fluvoxamine and fluoxetine have *in vivo* anti-AD evidence. The drugs nicotine, rasagiline, fluvoxamine, and fluoxetine could be promising repurposing candidates. We leave the investigation of these drugs to future research.

3.3. *Identifying repurposing candidates with anti-*AD *activity*

In order to identify plausible candidates for repurposing, we produce Cox proportional-hazard models (see Section 2.4) to ascertain whether there is consensus between the MarketScan insurance data and the AD-related evidence we found for top ranked candidates prioritized by MPI. Specifically, we use hazard ratios (HR) to identify whether any evidential drug elicited reduced the risk of AD diagnosis among patients who took the drug compared against those that did not. We present each drug's HR with their significance levels in Supplementary Table S5b; the HR for each drug's covariates (sex, age, and additional common comorbidities) are reported in Supplementary Table S5c-t. A HR below 1 indicates that a drug has a protective effect, while a HR above 1 indicates that a drug has a damaging effect. Figure 3a plots Kaplan-Meier (KM) survival curves. These plots depict a patient's likelihood of being diagnosed with AD following long-term use of either an individual prescribed drug or a drug with a given mechanism of action (MOA). For MOAs, we group highly-prioritized drugs with published

Table 1: Top-10 Drugs from MPI

Drug	MOA	Indication	Anti-AD	Evidence
varenicline	AChR-Ag	smoking cessation	N	-
fosinopril	ACE-INH	hypertension	Y	*in vivo*[44]
nicotine	AChR-Ag	smoking cessation	Y	clinical[45]
nizatidine	histamine receptor antagonist	duodenal ulcer disease	N	-
piroxicam	COX-INH	osteoarthritis	Y	other[46,47]
meloxicam	COX-INH	osteoarthritis	Y	*in vivo*[48–50]
galantamine	AChE-INH	Alzheimer's disease	Y	approved
bromfenac	COX-INH	inflammation	N	-
etodolac	COX-INH	osteoarthritis	Y	*in vivo*[51]
pyridostigmine	AChE-INH	myasthenia gravis	N	-

In this table, the column "Drug" shows the identified top-10 ranked drugs; the column "MOA" shows the mechanism of action of each drug; the column "Indication" presents the indication of each drug; the column "Anti-AD " indicates if the drug has evidenced anti-AD effects; and the column "Evidence" presents the type of the evidence. In this table, ACE-INH represents the angiotensin converting enzyme inhibitor; COX-INH represents the cyclooxygenase inhibitor; AChE-INH represents the acetylcholinesterase inhibitor; and AChR-Ag represents the acetylcholine receptor agonist.

Table 2: Top-10 Drugs from BSL

Drug	MOA	Indication	Anti-AD	Evidence
tetracycline	bacterial 30S ribosomal subunit inhibitor	respiratory tract infections	Y	*in vitro*[52]
selegiline	monoamine oxidase inhibitor	Parkinson's Disease	N	-
ceftriaxone	bacterial cell wall synthesis inhibitor	gonorrhea	Y	*in vivo*[53]
ibuprofen	COX-INH	headache	N	-
levobunolol	adrenergic receptor antagonist	glaucoma	N	-
ketoprofen	COX-INH	rheumatoid arthritis	N	-
carbidopa	aromatic L-amino acid decarboxylase inhibitor	Parkinson's Disease	N	-
sulindac	COX-INH	osteoarthritis	Y	*in vivo*[54]
biotin	vitamin B	supplement	Y	*in vivo*[55]
lansoprazole	ATPase inhibitor	heartburn	N	-

In this table, the column "Drug" shows the identified top-10 ranked drugs; the column "MOA" shows the mechanism of action of each drug; the column "Indication" presents the indication of each drug; the column "Anti-AD " indicates if the drug has evidenced anti-AD effects; and the column "Evidence" presents the type of the evidence. In this table, ACE-INH represents the angiotensin converting enzyme inhibitor; COX-INH represents the cyclooxygenase inhibitor; AChE-INH represents the acetylcholinesterase inhibitor; and AChR-Ag represents the acetylcholine receptor agonist.

evidence of anti-AD activity (see Table 1). Bupropion (HR $= 1.04$; non-significant) was included as a negative control as clinical trials found the drug had no significant effect on cognition in AD patients.[56] Trihexyphenydil (HR $= 1.71$; $\alpha < 0.001$) was included as a positive control for damaging effects due to the evidence documented in Supplementary Table S1. The COX-INHs group includes the following drugs: piroxicam, meloxicam, etodolac, and flurbiprofen. The ACE-INHs group includes the following drugs: fosinopril, trandolapril, and lisinopril. Note that we only include blood brain barrier (BBB) crossing ACE-INHs in this group as non-BBB-crossing ACE-INHs have exhibited very limited effects on AD.[57] We also include time-to-event analysis for 4 of BSL's top prioritized drugs (See Supplementary Figure S1). Unlike MPI, we observe only one of BSL's drugs (sulindac) with reduced time-to-event compared to bupropion; however, this difference is not significant.

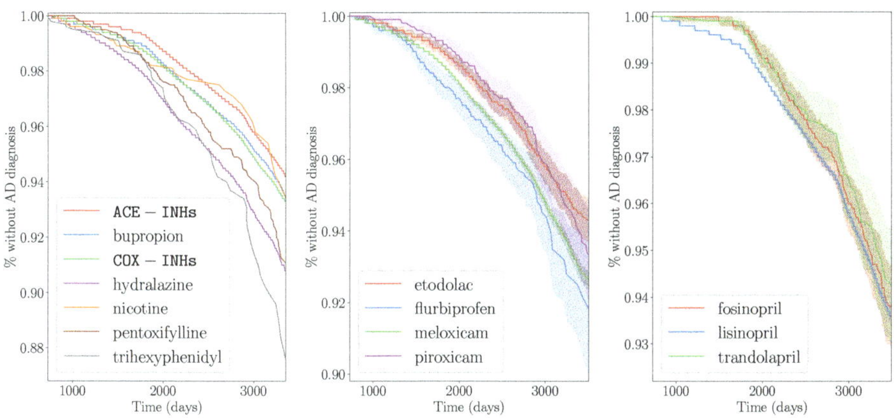

(a) Drugs and MOAs with published anti-AD evidence

(b) COX-INHs. Shaded regions represent 95% confidence intervals.

(c) ACE-INHs. Shaded regions represent 95% confidence intervals.

Fig. 3: Unadjusted Kaplan–Meier plots for cox proportional-hazard models

3.4. *Analyzing the MOAs of* MPI*'s top-50 drugs*

To identify groups of drugs whose anti-AD properties should be further examined and explored, we examine the top-50 drugs prioritized by MPI for any common MOAs. We find that COX-INHs and ACE-INHs are the most common MOAs prioritized by MPI. Both COX-INHs and ACE-INHs have published evidence of anti-AD activity. That said, experimental results suggest that long-term administration of COX-INHs may only have protective properties, reducing the risk of AD onset.[46] Moreover, meloxicam ($\mathrm{HR} = 0.86$; $\alpha < 0.05$), has even shown therapeutic potential, reversing cognitive decline via inhibition of neuronal apoptosis.[48,49] However, in Figure 3a, we observe that COX-INHs as a class do not yield reduced risk of AD compared to the negative control. That said, we find etodolac significantly reduces the risk of AD ($\mathrm{HR} = 0.78$; $\alpha < 0.001$) compared to other COX-INHs, including flurbiprofen ($\mathrm{HR} = 0.95$; non-significant) (Figure 3b). This suggests that only certain COX-INHs, such as etodolac, may elicit protective effects against AD onset. Importantly, this may be a result of differences in target as etodolac targets *COX2*, while flurbiprofen targets *COX1*. On the other hand, ACE-INHs were found to also protect against AD onset in Figure 3a. Specifically, we evaluate only ACE-INHs that cross the blood brain barrier (BBB) as previous insurance claims metadata analyses have indicated those that do not cross the BBB have no effect on AD.[57] To see if any of the BBB crossing ACE-INHs have a greater protective effect that others, we produce a KM plot for fosinopril, lisinopril, and trandolapril (Figure 3c). Unlike for COX-INHs, ACE-INHs do not elicit any significant by-drug difference in AD onset as illustrated in Figure 3c. While MPI prioritized four BBB crossing (BBBx) and four non-BBBx ACE-INHs in the top-50, the BBBx ACE-INHs had a lower average rank compared to the non-BBBx ACE-INHs (15 and 19, respectively).

Another important distinction between COX-INHs and ACE-INHs is that ACE-INHs have been shown to have some ameliorative potential; whereas, COX-INHs have only shown protective effects. In fact, fosninpril and lisinopril (ranked 2nd and 24th by MPI, respectively) was found to reduce cognitive decline in animal models of AD.[44,58] In Figure 3a, we find that BBBx ACE-INHs consistently exhibit decreased risk of AD relative to our negative control drug, bupropion. Additionally, there does not appear to be a significant difference between any of the BBBx ACE-INHs with respect to their protection against AD, indicating that they are possibly all viable candidates for repurposing. This is in agreement with other published evidence that has identified BBBx ACE-INHs as having protective effect on AD development. Interestingly, MPI prioritized 133.6% more COX-INHs and 700.9% more ACE-INHs than BSL in the top-50 from all such drugs in our network. MPI's ability to prioritize more drugs from MOAs with known anti-AD activity suggests that it may be a more viable option when identifying candidates for drug repurposing.

MPI also highly prioritizes drugs that increase the brain's response to acetylcholine, either by reducing its degradation (acetylcholinesterase inhibitors, AChE-INHs) or by stimulating its receptors (acetylcholine receptor agonists, AChR-Ags). This is important as acetylcholine's (ACh) synaptic bioavailability is an important contributor to AD progression. That is, there is evidence that cholenergic atrophy and ACh deficiency is linked with cognitive decline in AD patients.[59] Moveover, many of the current FDA-approved drugs indicated to slow AD progression target this mechanism of disease progression via AChE-INHs (e.g., donepezil, rivastigmine, and galantamine). AChR-Ags, also enhances ACh signaling. Such drugs, such as nicotine, accomplish this by increasing the response of ACh receptors located on the post-synaptic neuron. Interestingly, nicotine, was found to significantly improve cognition in patients with mild cognitive impairment, which is a precursor to AD.[45] We also find long-term nicotine use to have a protective effect ($\text{HR} = 0.532$; $\alpha < 0.001$), with respect to AD onset. In Figure 3a, we observe similar risk of developing AD to ACE-INHs after six to seven years. Conversely, we find evidence that long-term use of trihexyphenidyl, which reduces the activity of ACh receptors, is associated with AD-like neurodegeneration in rats.[24] This is corroborated by Figure 3a, where we observe the highest risk of AD elicited by trihexyphenidyl. More than eight years on trihexyphenidyl was associated with a substantial increase in the risk of AD relative to the other drugs evaluated in Figure 3a. These findings confirm that ACh signaling is closely linked with AD progression. As such, exploring other drugs and drug classes which either increase ACh synaptic bioavailability or enhance neuronal response to ACh should be further examined for anti-AD activity.

4. Discussion

In this work, we propose a novel network-based, AD-specific drug repurposing approach called MPI. MPI improves upon prior network-based methods by leveraging node embeddings learned via DeepWalk to prioritize AD-associated paths. Moreover, the use of learned embeddings allows MPI to more effectively capture a network's rich topology than previous approaches, such as BSL. In a direct comparison, we find that 20% more of MPI's highly prioritized drug candidates (top-50) have published anti-AD evidence compared to BSL's highly prioritized drug candidates. In addition to evidence in literature, we leverage insurance claims data to produce Cox proportional-hazard models. Among all the drugs we evaluate, these models identified BBBx ACE-INHs as having the lowest risk of AD. Similarly, etodolac was found to have the lowest risk of AD among the four COX-INHs we evaluated (Figure 3b), indicating that this drug in particular may have protective effect despite the class as a whole not exhibiting a significantly reduced risk of AD compared to our negative control (Figure 3a). Additionally, MPI highly prioritizes drugs that target the cholinergic system. Each of the approved AD drugs in our dataset that are also AChE-INHs are prioritized in the top-50 by MPI. MPI also highly prioritizes nicotine, an AChR-Ags. This prioritization is supported by both literature and our Cox models, which suggest nicotine is associated with reduced risk of AD. Altogether, the results presented in this work highlight etodolac, nicotine, and ACE-INHs as viable candidates for repurposing to treat AD and, as such, deserve further examination in future studies.

Despite its promising results, MPI exhibits a few limitations. The PPI network we construct is a simplification of molecular pathways. Like many other network-based approaches, MPI does not consider loops nor the directionality of PPI as these can be difficult for models to learn. In our context, this means that highly ranked candidates are only likely to be in close proximity to AD-related genes. To improve drug prioritization, models must be capable of identifying drugs that are both upstream of and in close proximity to these AD-related genes. In future studies, we will leverage directed interactions either by hard coding them or learning them. One way directionality might be learned is through the use of multi-omics data. Examining how changes to genomic and epigenomic profiles affect gene expression could facilitate learning where genes are in pathways. Furthermore, by leveraging multi-omics data, we may be able to provide more personalized drug recommendations.

5. Acknowledgments

This project was made possible, in part, by support from the National Institute of Aging grant no. 5R01AG071470. Any opinions, findings and conclusions or recommendations expressed in this paper are those of the authors and do not necessarily reflect the views of the funding agency.

References

1. M. V. F. Silva, C. d. M. G. Loures, L. C. V. Alves, L. C. de Souza, K. B. G. Borges and M. d. G. Carvalho, Alzheimer's disease: risk factors and potentially protective measures, Journal of biomedical science **26**, 1 (2019).
2. E. Passeri, K. Elkhoury, M. Morsink, K. Broersen, M. Linder, A. Tamayol, C. Malaplate,

F. T. Yen and E. Arab-Tehrany, Alzheimer's disease: Treatment strategies and their limitations, International journal of molecular sciences **23**, p. 13954 (2022).
3. T. Athar, K. Al Balushi and S. A. Khan, Recent advances on drug development and emerging therapeutic agents for alzheimer's disease, Molecular biology reports **48**, 5629 (2021).
4. T.-W. Yu, H.-Y. Lane and C.-H. Lin, Novel therapeutic approaches for alzheimer's disease: An updated review, International journal of molecular sciences **22**, p. 8208 (2021).
5. C. K. Kim, Y. R. Lee, L. Ong, M. Gold, A. Kalali and J. Sarkar, Alzheimer's disease: Key insights from two decades of clinical trial failures, Journal of Alzheimer's Disease **87**, 83 (2022).
6. J. L. Cummings, D. P. Goldman, N. R. Simmons-Stern and E. Ponton, The costs of developing treatments for alzheimer's disease: A retrospective exploration, Alzheimer's & Dementia **18**, 469 (2022).
7. S. Pushpakom, F. Iorio, P. A. Eyers, K. J. Escott, S. Hopper, A. Wells, A. Doig, T. Guilliams, J. Latimer, C. McNamee et al., Drug repurposing: progress, challenges and recommendations, Nature reviews Drug discovery **18**, 41 (2019).
8. P. Zhan, B. Yu and L. Ouyang, Drug repurposing: An effective strategy to accelerate contemporary drug discovery, Drug discovery today **27**, p. 1785 (2022).
9. C. G. Begley, M. Ashton, J. Baell, M. Bettess, M. P. Brown, B. Carter, W. N. Charman, C. Davis, S. Fisher, I. Frazer et al., Drug repurposing: Misconceptions, challenges, and opportunities for academic researchers, Science Translational Medicine **13**, p. eabd5524 (2021).
10. K. Park, A review of computational drug repurposing, Translational and clinical pharmacology **27**, 59 (2019).
11. D. N. Sosa and R. B. Altman, Contexts and contradictions: a roadmap for computational drug repurposing with knowledge inference, Briefings in Bioinformatics **23**, p. bbac268 (2022).
12. D. Morselli Gysi, Í. Do Valle, M. Zitnik, A. Ameli, X. Gan, O. Varol, S. D. Ghiassian, J. Patten, R. A. Davey, J. Loscalzo et al., Network medicine framework for identifying drug-repurposing opportunities for covid-19, Proceedings of the National Academy of Sciences **118**, p. e2025581118 (2021).
13. F. Cheng, R. J. Desai, D. E. Handy, R. Wang, S. Schneeweiss, A.-L. Barabasi and J. Loscalzo, Network-based approach to prediction and population-based validation of in silico drug repurposing, Nature communications **9**, p. 2691 (2018).
14. L. Cai, C. Lu, J. Xu, Y. Meng, P. Wang, X. Fu, X. Zeng and Y. Su, Drug repositioning based on the heterogeneous information fusion graph convolutional network, Briefings in bioinformatics **22**, p. bbab319 (2021).
15. K. Savva, M. Zachariou, M. M. Bourdakou, N. Dietis and G. M. Spyrou, Network-based stage-specific drug repurposing for alzheimer's disease, Computational and Structural Biotechnology Journal **20**, 1427 (2022).
16. J. Fang, P. Zhang, Q. Wang, Y. Zhou, C.-W. Chiang, R. Chen, B. Zhang, B. Li, S. J. Lewis, A. A. Pieper et al., Network-based translation of gwas findings to pathobiology and drug repurposing for alzheimer's disease, MedRxiv , 2020 (2020).
17. Y. Sun, J. Han, X. Yan, P. S. Yu and T. Wu, Pathsim: Meta path-based top-k similarity search in heterogeneous information networks, Proceedings of the VLDB Endowment **4**, 992 (2011).
18. P. Veličković, G. Cucurull, A. Casanova, A. Romero, P. Liò and Y. Bengio, Graph attention networks (2018).
19. S. Abu-El-Haija, B. Perozzi, R. Al-Rfou and A. A. Alemi, Watch your step: Learning node embeddings via graph attention, Advances in neural information processing systems **31** (2018).
20. A. Grover and J. Leskovec, Node2vec: Scalable feature learning for networks, p. 855–864 (2016).
21. B. Perozzi, R. Al-Rfou and S. Skiena, Deepwalk: Online learning of social representations, p. 701–710 (2014).
22. T. N. Kipf and M. Welling, Semi-supervised classification with graph convolutional networks,

arXiv preprint arXiv:1609.02907 (2016).
23. J. Fang, A. A. Pieper, R. Nussinov, G. Lee, L. Bekris, J. B. Leverenz, J. Cummings and F. Cheng, Harnessing endophenotypes and network medicine for alzheimer's drug repurposing, Medicinal research reviews **40**, 2386 (2020).
24. Y. Huang, Z. Zhao, X. Wei, Y. Zheng, J. Yu, J. Zheng and L. Wang, Long-term trihexyphenidyl exposure alters neuroimmune response and inflammation in aging rat: relevance to age and alzheimer's disease, Journal of Neuroinflammation **13**, 1 (2016).
25. D. Mendez, A. Gaulton, A. P. Bento, J. Chambers, M. De Veij, E. Félix, M. P. Magariños, J. F. Mosquera, P. Mutowo, M. Nowotka et al., Chembl: towards direct deposition of bioassay data, Nucleic acids research **47**, D930 (2019).
26. T. Liu, Y. Lin, X. Wen, R. N. Jorissen and M. K. Gilson, Bindingdb: a web-accessible database of experimentally determined protein–ligand binding affinities, Nucleic acids research **35**, D198 (2007).
27. X. Chen, Z. L. Ji and Y. Z. Chen, Ttd: therapeutic target database, Nucleic acids research **30**, 412 (2002).
28. S. D. Harding, J. F. Armstrong, E. Faccenda, C. Southan, S. P. Alexander, A. P. Davenport, A. J. Pawson, M. Spedding, J. A. Davies and NC-IUPHAR, The iuphar/bps guide to pharmacology in 2022: curating pharmacology for covid-19, malaria and antibacterials, Nucleic Acids Research **50**, D1282 (2022).
29. R. Apweiler, A. Bairoch, C. H. Wu, W. C. Barker, B. Boeckmann, S. Ferro, E. Gasteiger, H. Huang, R. Lopez, M. Magrane et al., Uniprot: the universal protein knowledgebase, Nucleic acids research **32**, D115 (2004).
30. Stanford Center for Population Health Sciences, MarketScan Medicare Supplemental (February 2023).
31. J.-C. Lambert, C. A. Ibrahim-Verbaas, D. Harold, A. C. Naj, R. Sims, C. Bellenguez, G. Jun, A. L. DeStefano, J. C. Bis, G. W. Beecham et al., Meta-analysis of 74,046 individuals identifies 11 new susceptibility loci for alzheimer's disease, Nature genetics **45**, 1452 (2013).
32. R. E. Marioni, S. E. Harris, Q. Zhang, A. F. McRae, S. P. Hagenaars, W. D. Hill, G. Davies, C. W. Ritchie, C. R. Gale, J. M. Starr et al., Gwas on family history of alzheimer's disease, Translational psychiatry **8**, p. 99 (2018).
33. I. E. Jansen, J. E. Savage, K. Watanabe, J. Bryois, D. M. Williams, S. Steinberg, J. Sealock, I. K. Karlsson, S. Hägg, L. Athanasiu et al., Genome-wide meta-analysis identifies new loci and functional pathways influencing alzheimer's disease risk, Nature genetics **51**, 404 (2019).
34. B. W. Kunkle, B. Grenier-Boley, R. Sims, J. C. Bis, V. Damotte, A. C. Naj, A. Boland, M. Vronskaya, S. J. Van Der Lee, A. Amlie-Wolf et al., Genetic meta-analysis of diagnosed alzheimer's disease identifies new risk loci and implicates aβ, tau, immunity and lipid processing, Nature genetics **51**, 414 (2019).
35. I. De Rojas, S. Moreno-Grau, N. Tesi, B. Grenier-Boley, V. Andrade, I. E. Jansen, N. L. Pedersen, N. Stringa, A. Zettergren, I. Hernández et al., Common variants in alzheimer's disease and risk stratification by polygenic risk scores, Nature communications **12**, p. 3417 (2021).
36. D. P. Wightman, I. E. Jansen, J. E. Savage, A. A. Shadrin, S. Bahrami, D. Holland, A. Rongve, S. Børte, B. S. Winsvold, O. K. Drange et al., A genome-wide association study with 1,126,563 individuals identifies new risk loci for alzheimer's disease, Nature genetics **53**, 1276 (2021).
37. C. Bellenguez, F. Küçükali, I. E. Jansen, L. Kleineidam, S. Moreno-Grau, N. Amin, A. C. Naj, R. Campos-Martin, B. Grenier-Boley, V. Andrade et al., New insights into the genetic etiology of alzheimer's disease and related dementias, Nature genetics **54**, 412 (2022).
38. A. Hamosh, A. F. Scott, J. S. Amberger, C. A. Bocchini and V. A. McKusick, Online mendelian inheritance in man (omim), a knowledgebase of human genes and genetic disorders, Nucleic acids research **33**, D514 (2005).

39. A. P. Davis, C. G. Murphy, C. A. Saraceni-Richards, M. C. Rosenstein, T. C. Wiegers and C. J. Mattingly, Comparative toxicogenomics database: a knowledgebase and discovery tool for chemical–gene–disease networks, Nucleic acids research **37**, D786 (2009).
40. W. Yu, M. Gwinn, M. Clyne, A. Yesupriya and M. J. Khoury, A navigator for human genome epidemiology, Nature genetics **40**, 124 (2008).
41. J. Piñero, À. Bravo, N. Queralt-Rosinach, A. Gutiérrez-Sacristán, J. Deu-Pons, E. Centeno, J. García-García, F. Sanz and L. I. Furlong, Disgenet: a comprehensive platform integrating information on human disease-associated genes and variants, Nucleic acids research , p. gkw943 (2016).
42. M. J. Landrum, J. M. Lee, M. Benson, G. R. Brown, C. Chao, S. Chitipiralla, B. Gu, J. Hart, D. Hoffman, W. Jang et al., Clinvar: improving access to variant interpretations and supporting evidence, Nucleic acids research **46**, D1062 (2018).
43. D. Ochoa, A. Hercules, M. Carmona, D. Suveges, J. Baker, C. Malangone, I. Lopez, A. Miranda, C. Cruz-Castillo, L. Fumis et al., The next-generation open targets platform: reimagined, redesigned, rebuilt, Nucleic acids research **51**, D1353 (2023).
44. D. Deb, K. Bairy, V. Nayak, M. Rao et al., Comparative effect of lisinopril and fosinopril in mitigating learning and memory deficit in scopolamine-induced amnesic rats, Advances in Pharmacological and Pharmaceutical Sciences **2015** (2015).
45. P. Newhouse, K. Kellar, P. Aisen, H. White, K. Wesnes, E. Coderre, A. Pfaff, H. Wilkins, D. Howard and E. Levin, Nicotine treatment of mild cognitive impairment: a 6-month double-blind pilot clinical trial, Neurology **78**, 91 (2012).
46. C. Zhang, Y. Wang, D. Wang, J. Zhang and F. Zhang, Nsaid exposure and risk of alzheimer's disease: an updated meta-analysis from cohort studies, Frontiers in aging neuroscience **10**, p. 83 (2018).
47. B. P. Imbimbo, V. Solfrizzi and F. Panza, Are nsaids useful to treat alzheimer's disease or mild cognitive impairment?, Frontiers in aging neuroscience **2**, p. 1517 (2010).
48. F. R. Ianiski, C. B. Alves, C. F. Ferreira, V. C. Rech, L. Savegnago, E. A. Wilhelm and C. Luchese, Meloxicam-loaded nanocapsules as an alternative to improve memory decline in an alzheimer's disease model in mice: involvement of na+, k+-atpase, Metabolic brain disease **31**, 793 (2016).
49. P. Guan, D. Zhu and P. Wang, Meloxicam inhibits apoptosis in neurons by deactivating tumor necrosis factor receptor superfamily member 25, leading to the decreased cleavage of dna fragmentation factor subunit α in alzheimer's disease, Molecular Neurobiology **60**, 395 (2023).
50. F. R. Ianiski, C. B. Alves, A. C. G. Souza, S. Pinton, S. S. Roman, C. R. Rhoden, M. P. Alves and C. Luchese, Protective effect of meloxicam-loaded nanocapsules against amyloid-β peptide-induced damage in mice, Behavioural Brain Research **230**, 100 (2012).
51. K. H. Elfakhri, I. M. Abdallah, A. D. Brannen and A. Kaddoumi, Multi-faceted therapeutic strategy for treatment of alzheimer's disease by concurrent administration of etodolac and α-tocopherol, Neurobiology of disease **125**, 123 (2019).
52. G. Forloni, L. Colombo, L. Girola, F. Tagliavini and M. Salmona, Anti-amyloidogenic activity of tetracyclines: studies in vitro, FEBS letters **487**, 404 (2001).
53. M. A. Tikhonova, T. G. Amstislavskaya, Y.-J. Ho, A. A. Akopyan, M. V. Tenditnik, M. V. Ovsyukova, A. A. Bashirzade, N. I. Dubrovina and L. I. Aftanas, Neuroprotective effects of ceftriaxone involve the reduction of aβ burden and neuroinflammatory response in a mouse model of alzheimer's disease, Frontiers in Neuroscience **15**, p. 736786 (2021).
54. J. P. Modi, H. Prentice and J.-Y. Wu, Sulindac for stroke treatment: neuroprotective mechanism and therapy, Neural Regeneration Research **9**, p. 2023 (2014).
55. K. M. Lohr, B. Frost, C. Scherzer and M. B. Feany, Biotin rescues mitochondrial dysfunction and neurotoxicity in a tauopathy model, Proceedings of the National Academy of Sciences **117**,

33608 (2020).
56. F. Maier, A. Spottke, J.-P. Bach, C. Bartels, K. Buerger, R. Dodel, A. Fellgiebel, K. Fliessbach, L. Frölich, L. Hausner et al., Bupropion for the treatment of apathy in alzheimer disease: a randomized clinical trial, JAMA network open **3**, e206027 (2020).
57. M. Ouk, C.-Y. Wu, J. S. Rabin, A. Jackson, J. D. Edwards, J. Ramirez, M. Masellis, R. H. Swartz, N. Herrmann, K. L. Lanctot et al., The use of angiotensin-converting enzyme inhibitors vs. angiotensin receptor blockers and cognitive decline in alzheimer's disease: the importance of blood-brain barrier penetration and apoe $\varepsilon 4$ carrier status, Alzheimer's Research & Therapy **13**, 1 (2021).
58. J. Thomas, H. Smith, C. A. Smith, L. Coward, G. Gorman, M. De Luca and P. Jumbo-Lucioni, The angiotensin-converting enzyme inhibitor lisinopril mitigates memory and motor deficits in a drosophila model of alzheimer's disease, Pathophysiology **28**, 307 (2021).
59. R. Knowles, Denitrification, Microbiological reviews **46**, 43 (1982).

Overcoming health disparities in precision medicine

Francisco M. De La Vega,[1] Kathleen C. Barnes,[2] Keolu Fox,[3] Alexander Ioannidis,[4] Eimear Kenny,[5] Rasika A. Mathias,[6] and Bogdan Pasaniuc.[7]

[1]Tempus Labs, Inc.,[2]Galatea Bio, Inc., [3]University of California San Diego, [4]Stanford University School of Medicine, [5]Ichan School of Medicine at Mount Sinai, [6]Johns Hopkins School of Medicine, [7]University of California Los Angeles.

1. Overview

Precision medicine and precision public health rely on the premise that determinants of disease incidence and differences in response to interventions can be identified, and their biology can be understood well enough for the development of individualized interventions that reduce the risk of disease and improve treatment. At the same time, well-documented racial and ethnic disparities exist throughout healthcare at the patient, provider, and healthcare system levels. These disparities are driven by a complex interplay among social, psychosocial, lifestyle, environmental, health system, and biological determinants of health (Freedman, et al. 2021). The aim of the PSB 2024 session "Overcoming health disparities in precision medicine" is to elicit the development of new methods and concepts than can be used in uncovering undetected biases, develop effective therapies and fair AI to improve precision healthcare and help reduce these disparities, and ultimately improve health equity.

2. Dealing with the lack of diversity in current research datasets

An overwhelming focus on individuals of European descent in past genomic studies, which account for 86% of all such research, has created inequities in precision medical insights and has limited scientific discovery (Fatumo et al. 2022). It is imperative to diversify genomic research data and to make investments aimed at understanding and eliminating these health inequalities.

In the meantime, methods that can use the currently available genomic and clinical data, which admittedly are lacking in diversity, to provide equitable prediction of phenotypes are needed. The paper by Comajoan Cara et al. (2024) in this proceedings introduces PopGenAdapt, a model that tackles the lack of diversity in genomic datasets by using semi-supervised domain adaptation techniques. The model effectively leverages labeled data from individuals of European ancestry and both labeled and unlabeled data from underrepresented populations. When tested in populations from Nigeria, Sri Lanka, and Hawaii, PopGenAdapt showed significant improvement in predicting disease outcomes compared to existing methods, highlighting its potential for more inclusive biomedical research.

On the other hand, the paper by Bonet et al. (2024) introduces a machine learning toolkit designed to directly improve the accuracy of genomic-based medical predictions for underrepresented populations. By employing techniques such as gradient boosting, ensembling, and population-conditional re-sampling techniques to address the lack of diversity, the method enhances phenotype prediction accuracy, achieving results comparable to those for well-represented European populations.

Ancestry can impact gene expression prediction methods as potentially undiscovered population variants and eQTNs affecting gene expression may exists in understudied populations. The paper by Mishra et al. introduces LA-GEM, a gene imputation model that incorporates local ancestry (LA) to improve gene expression predictions in African American populations. Tested on a cohort of 60 African American hepatocyte primary cultures, LA-GEM outperformed existing models like PrediXcan by reliably predicting the expression of unique genes critical to drug metabolism in this sample. The study highlights the value of leveraging local ancestry in gene imputation models for admixed populations to better understand disease susceptibility and drug response in all populations.

3. Development of fair machine learning algorithms

The development of fair algorithms and machine learning in healthcare is crucial for reducing health disparities, improving diagnostic accuracy, and building public trust. By minimizing biases, equitable healthcare and regulatory compliance is promoted, leading to more economically efficient systems.

One of the first steps to algorithmic equity is the thorough exploration of the input data to be used in their training for inherent and occult biases. The paper by Orlenko et al. (2024) provides examples of such necessary data exploration using cluster analysis to identify two distinct subgroups of elective spinal fusion patients based on insurance type. These findings reveal significant differences in characteristics and post-surgery outcomes related to socioeconomic and racial disparities. The aim is to inform the design of machine learning models to ensure fairness and minimize bias in healthcare predictions.

Methods designed to provide fair algorithmic predictions from the ground up are needed as well. Jun et al. (2024) use a fairness algorithm, Fairness-Aware Causal paThs (FACTS), to analyze nine years of electronic health records and social determinants of health to quantify disparities in MRSA infection outcomes. The study identified moderate disparities in age, gender, race, and income, revealing that comorbidities played a role in these disparities. Factors like kidney impairment and drug use affected racial disparity, while income and healthcare access affected gender disparity. The findings highlight the need for policies that address both clinical factors and social determinants to mitigate health disparities.

4. Race, genetic ancestry, and population structure

The persistent use of "race" and "ethnicity" in precision medicine, classifications rooted in perceived physical characteristics and cultural backgrounds, has generated substantial discussion (Nat. Acad. Sci. Eng. Med., 2023). However, for the purpose of examining health equity, these categories remain essential for identifying and addressing systemic disparities (Kahu et. Al. (2021). Established by the U.S. Office of Management and Budget in 1995, these classifications are integral for organizing data on social determinants of health and population demographics. They guide resource allocation, policymaking, and the development of culturally sensitive healthcare interventions.

The paper by Rhead et al. (2024) tackles the problem of missing disaggregated race and ethnicity data in real-world databases by introducing methods for imputing these categories using genetic ancestry from available genetic data. Analyzing data from over 100,000 cancer patients, ancestry-based machine learning methods were shown to outperform existing race imputation algorithms based on geolocation and surnames commonly used in administrative health data. The research offers a new way to improve real-world healthcare data for studying and ensuring healthcare equity and to enable its use in the development of diversity plans for clinical trials soon to be required per FDA guidance.

On the other hand, the study by Seagle et al. (2024) analyzes the genetic ancestry of 35,842 individuals over 100 birth years in the Southeastern United States, finding increasing levels of genetic admixture and heterozygosity in younger populations since 1990. This rise in diversity poses challenges to traditional genotype-phenotype relationship studies. The researchers explore the impact of increased admixture on health outcomes, discovering that greater genetic diversity was associated with protective effects against female reproductive disorders but elevated risks for diseases linked to autoimmune dysfunction. This highlights the influence of ancestral complexity on health disparities.

The social construct of race and ethnicity is far from precise, serving as a poor proxy for ancestry. In this vein, the study by Piekos et al. (2024) employs genetic ancestry rather than race to assess disease risk factors, leveraging data from the BioVU biobank. Researchers estimated six ancestry proportions and performed phenome-wide association studies, finding varying risks for conditions like 'Neoplasms' and 'Pregnancy Complications' based on different ancestries. The study also found that linear modeling was sufficient for assessing hypertension and atrial fibrillation risk in relation to ancestry, but not for renal failure, indicating the need for more complex models in certain cases.

5. Conclusion

The increased attention to social justice has emphasized the urgent need to tackle health disparities more effectively. Advanced computational and statistical approaches are essential for assessing and mitigating these disparities in healthcare. Their adoption is not just a technological advancement but also an ethical necessity for creating a healthcare environment that serves all communities effectively. We believe that the new methods in the collection of research papers accepted to this PSB 2024 proceedings can contribute to overcome disparities in precision medicine.

6. Acknowledgments

We thank the anonymous reviewers that helped in the peer review process of the submissions to this session.

References

Bonet D., Levin M., Mas Montserrat D. and Ioannidis A.G. (2024) Machine Learning Strategies for Improved Phenotype Prediction in Underrepresented Populations. In *Pacific Symposium on Biocomputing 2024.*

Comajoan Cara M., Mas Montserrat D., Ioannidis A.G. (2024) PopGenAdapt: Semi-Supervised Domain Adaptation for Genotype-to-Phenotype Prediction in Underrepresented Populations. In *Pac. Symp. on Biocomputing.*

Jun I., Ser S., Cohen S., Xu J., Lucero R.J., Bian J. and Prosperi M. (2024) Quantifying Health Outcome Disparity in Invasive Methicillin-Resistant Staphylococcus aureus Infection using Fairness Algorithms on Real-World Data. In *Pac. Symp. on Biocomputing.*

Fatumo S., Chikowore T., Choudhury A., Ayub M., Martin A.R., Kuchenbaecker K. (2022) A roadmap to increase diversity in genomic studies. *Nat Med.* 2022 Feb;28(2):243-250.

Freedman J.A., Abo M.A., Allen T.A., Piwarski S.A., Wegermann K., and Patierno S.R. (2021) Biological Aspects of Cancer Health Disparities. *Ann. Rev. Med.* 72:229-241

Kahu T.J., Ghazal Read, J. and Scheitler, A.J. (2021) The Critical Role of Racial/Ethnic Data Disaggregation for Health Equity. *Pop. Res. Pol. Rev.* 40:1-7.

Mishra M., Nahlawi L., Zhong Y., De T., Yang G., Alarcon C. and Perera M.A. (2024) LA-GEM: imputation of gene expression with incorporation of Local Ancestry. In *Pac. Symp. on Biocomputing.*

National Academies of Sciences, Engineering, and Medicine, Committee on the Use of Race, Ethnicity, and Ancestry as Population Descriptors in Genomics Research. *Using Population Descriptors in Genetics and Genomics Research: A New Framework for an Evolving Field.* (National Academies Press (US), (2023).

Orlenko A., Freda P.J., Ghosh A., Choi H., Matsumoto N., Bright T.J., Walker C.T., Obafemi-Ajayi T., and Moore J.H.. (2024) Cluster Analysis reveals Socioeconomic Disparities among Elective Spine Surgery Patients. In *Pac. Symp. on Biocomputing.*

Piekos J.A., Kim J., Keaton J.M, Hellwege J.N., Velez Edwards D.R., and Edwards T.L. (2024) Evaluating the Relationships Between Genetic Ancestry and the Clinical Phenome. In *Pac. Symp. on Biocomputing.*

Rhead B., Haffener P.E., Pouliot Y. and De La Vega F.M. (2024) Imputation of race and ethnicity categories using continental genetic ancestry from real- world genomic testing data. In *Pac. Symp. on Biocomputing.*

Seagle H.M., Mautz B.S., Hellwege J.N., Li C., Xu Y., Zhang S., Roden D.M., McGregor T.L., Velez Edwards D.R., Edwards T.L. (2024) Evidence of recent and ongoing admixture in the U.S. and influences on health and disparities. In *Pac. Symp. on Biocomputing.*

PopGenAdapt: Semi-Supervised Domain Adaptation for Genotype-to-Phenotype Prediction in Underrepresented Populations

Marçal Comajoan Cara[1,2], Daniel Mas Montserrat[1], Alexander G. Ioannidis[1,3,4]

[1]*Dept. of Biomedical Data Science, Stanford University School of Medicine, Stanford, CA, USA*
[2]*Dept. of Signal Theory & Communications, Universitat Politècnica de Catalunya, Barcelona, Spain*
[3]*Dept. of Biomolecular Engineering, University of California Santa Cruz, Santa Cruz, CA, USA*
[4]*Institute for Computational & Mathematical Engineering, Stanford University, Stanford, CA, USA*
e-mail: ioannidis@stanford.edu

The lack of diversity in genomic datasets, currently skewed towards individuals of European ancestry, presents a challenge in developing inclusive biomedical models. The scarcity of such data is particularly evident in labeled datasets that include genomic data linked to electronic health records. To address this gap, this paper presents PopGenAdapt, a genotype-to-phenotype prediction model which adopts semi-supervised domain adaptation (SSDA) techniques originally proposed for computer vision. PopGenAdapt is designed to leverage the substantial labeled data available from individuals of European ancestry, as well as the limited labeled and the larger amount of unlabeled data from currently underrepresented populations. The method is evaluated in underrepresented populations from Nigeria, Sri Lanka, and Hawaii for the prediction of several disease outcomes. The results suggest a significant improvement in the performance of genotype-to-phenotype models for these populations over state-of-the-art supervised learning methods, setting SSDA as a promising strategy for creating more inclusive machine learning models in biomedical research.

Our code is available at `https://github.com/AI-sandbox/PopGenAdapt`.

Keywords: phenotype prediction, semi-supervised, domain adaptation, underrepresented population

1. Introduction

Genomic data has become increasingly important for biomedical research, as it can reveal insights into the causes, diagnosis, prevention, and treatment of various diseases. However, the available data is predominantly from individuals of European ancestry, despite their making up only 16% of the global population. This disproportionate representation presents one of the major challenges in developing biomedical models and studies that can effectively generalize across diverse populations, posing the risk of exacerbating existing health disparities.[1] While widely adopted datasets such as the UK Biobank[2] provide rich phenotypic information from electronic health records, they lack diversity (see Fig. 1). On the other hand, highly diverse datasets, such as gnomAD,[3] lack phenotypic data, which makes them not directly usable to train supervised genotype-to-phenotype machine learning models, as phenotype labels for all the samples are required. New algorithmic solutions are needed in order to profit from all available data.

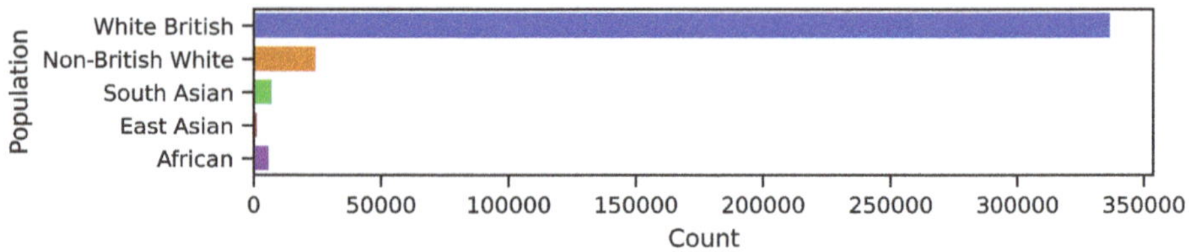

Fig. 1. Broad population counts in the UK Biobank.[2] Genetically inferred populations groups from the Global Biobank Engine.[4]

In this work, we propose PopGenAdapt, a semi-supervised domain adaptation (SSDA) method that can also exploit the available unlabeled data from underrepresented populations to improve the performance of phenotype prediction models. On the one hand, the semi-supervised nature of the proposed method makes possible the use of unlabeled data from underrepresented populations, as well as labeled data from large biobanks. On the other hand, the use of domain adaptation techniques makes it possible to still take advantage of the vast amount of data from individuals of European ancestry (the source domain), but to adapt the model predictions for a particular underrepresented population (the target domain). While SSDA has been previously applied to other types of data such as image and text, its application in genetics remains largely unexplored.

We adapt methods proposed for SSDA in computer vision for genotype-to-phenotype prediction and evaluate them in underrepresented population groups from Nigeria, Sri Lanka, and Hawaii. Our results predicting phenotypes including hypertension, diabetes, myxoedema, and asthma, demonstrate that SSDA can significantly enhance the performance of genotype-to-phenotype models in underrepresented populations, suggesting a promising direction for developing better machine learning models for diverse populations.

2. Background

2.1. *Genotype-to-Phenotype Prediction*

DNA is the hereditary material in humans and all living organisms, contributing to essential functions and appearance. While most positions in the DNA sequence are identical between individuals of the same species, some vary. Out of more than 3 billion positions, a typical human genome differs from the reference genetic sequence at 4 to 5 million sites ($\sim$1.5%).[5] In total, more than 600 million variable positions have been identified across different humans.[6] These variable positions are called single nucleotide polymorphisms (SNPs) and can be encoded as a ternary sequence, representing the counts of non-reference variants at each position, with 0 indicating that both maternal and paternal positions match the reference genome, 1 indicating that only maternal or paternal positions match, and 2 indicating that both are alternative variants.

Phenotypes are the observable characteristics of an organism that result from the interaction between its genotype (the genetic makeup determined by its DNA sequence) and the environment. These characteristics comprise physical and behavioral traits, as well as risk of developing certain diseases. Both the frequency distribution of genomic variants, and as a result, the distribution of phenotypes, vary across different populations. As a consequence, most studies developed for a particular population do not generalize well to other population groups.[1]

The goal of genotype-to-phenotype prediction is to use the genetic variation (SNP sequences) to estimate the phenotypes of an individual. Multiple machine learning models have been applied to solve this task, either using general-purpose methods like logistic regression, gradient boosting machines, or neural networks,[7,8] or through linear models specifically tailored to genetic data, such as PRS-CS,[9] SBayesR,[10] or snpnet.[11]

2.2. *Semi-Supervised Domain Adaptation*

Supervised learning is the framework most often adopted to train predictive models by using input samples and label pairs. However, in many real-world scenarios, such as in biomedical applications, obtaining labeled data can be challenging, involving time-consuming and expensive collection procedures. This limitation suggests the application of semi-supervised learning techniques, which can leverage both labeled and unlabeled data for training, providing better generalization than traditional supervised learning approaches.[12]

Both supervised and semi-supervised methods assume that the distribution of the training data (source domain) is the same as the one found during real-world deployment (target domain). However, this is not always the case, leading to distribution shifts that can drastically decrease the predictive performance. In order to address this shift, domain adaptation techniques have been proposed to properly adjust the models to bridge the gap between distributions and achieve accurate predictions in both the source and target domains.

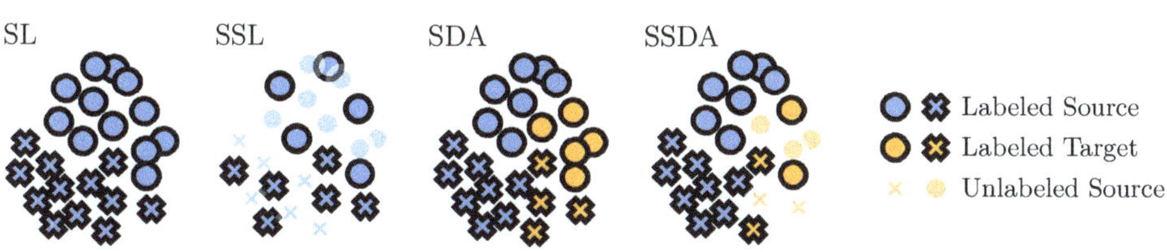

Fig. 2. Illustration of supervised learning (SL), semi-supervised learning (SSL), supervised domain adaptation (SDA), and semi-supervised domain adaptation (SSDA) in the case of binary classification. Circle and cross markers represent negative and positive classes, opaque and transparent markers represent labeled and unlabeled points, and blue and orange markers represent source and target domains, respectively.

Semi-supervised domain adaptation (SSDA) combines both semi-supervised learning and domain adaptation paradigms. The goal of SSDA is to leverage labeled data from a source domain, unlabeled data from the target domain, and a limited set of labeled data from the target domain, in order to obtain a machine learning model that achieves good performance within both domains.

In this paper, we adapt for genotype-to-phenotype prediction the state-of-the-art method of SSDA via Minimax Entropy (MME)[13] with Source Label Adaptation (SLA),[14] which was originally proposed in computer vision, considering different image domains, like photos, drawings, or paintings. Here, instead, we will consider different domains to be different populations.

2.2.1. *Minimax Entropy*

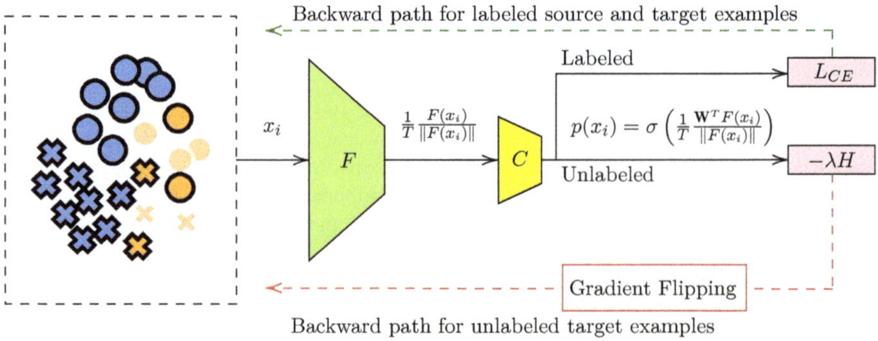

Fig. 3. Overview of the model architecture and minimax entropy proposed in Ref. 13.

Minimax Entropy (MME, Ref. 13, Fig. 3) proposes to use a neural network model consisting of a feature extractor F and a classifier C. At the output of F, ℓ_2 normalization and temperature scaling are applied, inspired by Ref. 15. In the original work, F is a pre-trained ResNet34,[16] an image classification network, and C is a single layer which takes $\frac{1}{T}\frac{F(x_i)}{\|F(x_i)\|}$ as input and outputs $g(x_i) = \sigma\left(\frac{1}{T}\frac{\mathbf{W}^T F(x_i)}{\|F(x_i)\|}\right)$. The weight vectors $\mathbf{W} = [w_1, ..., w_K]$ can be regarded as a representative point of each class k, or "prototype".

Both C and F are trained to classify labeled examples correctly by minimizing the cross-entropy loss L_{CE} on the labeled data, from both the source and target domains. However, to avoid overfitting on the source domain, which contains a larger amount of samples, as well as to take advantage of the unlabeled target data, it has been proposed to use an adversarial regularization term, the Minimax Entropy. MME is formulated as adversarial training between F and C, in which F is trained to minimize the conditional entropy H of the neural network predictions from unlabeled target data $p(x_t)$, whereas C is trained to maximize the entropy of the predictions $p(x_t)$. This adversarial learning forces F to learn discriminative features, while C estimates domain-invariant prototypes reducing the overfitting to the source domain. The overall adversarial learning objective functions are:

$$\hat{\theta}_F = \underset{\theta_F}{\operatorname{argmin}}\, L_{CE} + \lambda H \tag{1}$$

$$\hat{\theta}_C = \underset{\theta_C}{\operatorname{argmin}}\, L_{CE} - \lambda H \tag{2}$$

where λ is a hyperparameter to control the tradeoff between classification on labeled examples and the minimax entropy training. To simplify the training process, MME makes use of a gradient reversal layer[17] to flip the gradient between C and F with respect to H, allowing to perform the minimax training with a single forward and backward pass.

2.2.2. *Source Label Adaptation*

Source Label Adaptation (SLA, Ref. 14) is a framework that considers source data as a noisily-labeled version of the target data and gradually adapts the source labels to the target space. Specifically, inspired by Refs. 18,19, for each source point x_i^s, one constructs a modified source label $\tilde{y}_i^s$ by combining, with a tradeoff ratio α, the original source label y_i^s and the prediction of a source label adaptation model p:

$$\tilde{y}_i^s = (1-\alpha)y_i^s + \alpha p(x_i^s) \tag{3}$$

Note that p cannot be the current unadapted model g as it would overfit to the source data due to the larger number of samples, resulting in almost no effect. Thus, it has been proposed to train on the target domain data. However, to avoid simple memorization of the target data due to the low number of labeled samples available, it has been proposed to use a prototypical network (protonet),[20] a model for few-shot learning. Given a feature extractor F, the prototype of class k is defined as the center of features with the same class:

$$c_k = \frac{1}{N_k}\sum_{i=1}^{N} 1_{\{y_i=k\}} F(x_i) \tag{4}$$

Then, a protonet produces a distribution over classes for a query point x_i based on a softmax with temperature τ over the Euclidean distances to the prototypes in the embedding space:

$$p(x_i)_k = \frac{\exp(-d(F(x_i), c_k)\tau)}{\sum_{k'} \exp(-d(F(x_i), c_{k'})\tau)} \tag{5}$$

Moreover, in Ref. 14 it is proposed to derive the prototypes using the unlabeled data available by using, for each unlabeled target instance x_i^u, pseudo labels $\tilde{y}_i^u$ computed by the current model g:

$$\tilde{y}_i^u = \underset{k}{\operatorname{argmax}}\, g(x_i^u)_k \tag{6}$$

Using these pseudo labels, we can get pseudo centers by Eq. 4, and further define with them a Protonet with Pseudo Centers (PPC) by Eq. 5. Next, the PPC is applied to Eq. 3 to compute the modified source labels $\tilde{y}_i^s$ for each source instance x_i^s. Finally, the real source labels y_i^s are replaced by the cleaned source labels $\tilde{y}_i^s$ in the computation of the cross-entropy for the labeled source part of the whole dataset. The loss for labeled target data can still be a standard cross-entropy loss. Other loss terms can still be included, like the minimax entropy proposed in Ref. 13.

In practice, the SLA framework is only applied after W warmup steps in which the model is trained normally with the original source labels to obtain an initial robust model, and then the pseudo labels are only recomputed every I steps for efficiency. Since the SLA[14] paradigm of considering the source labels as noisy from the target domain viewpoint and cleaning them is orthogonal to the ideas in MME,[13] both approaches can be combined to get superior results. We refer to this combination as MME-SLA.

2.3. *Semi-Supervised Learning and Domain Adaptation for Genotype-to-Phenotype Prediction*

Semi-supervised learning techniques have been previously applied in genotype-to-phenotype prediction. For example, Ref. 21 proposed a method to predict the residual feed intake in dairy cattle using both labeled and unlabeled samples. However, the samples are assumed to be from the same domain, so the method would still have the problem of not generalizing to other populations.

Likewise, domain adaptation techniques have also been applied in genotype-to-phenotype prediction. For instance, Refs. 22–24, proposed several transfer learning techniques to also improve prediction performance for underrepresented populations. However, the proposed approaches cannot utilize unlabeled samples, thereby still grappling with the scarcity of labeled data from underrepresented populations. Consequently, the achieved performance improvement remains limited. To our knowledge, this is the first work to combine both approaches by applying semi-supervised domain adaptation for genotype-to-phenotype prediction.

3. Method

3.1. *Data*

We apply these methods to predict multiple disease outcomes, including hypertension, diabetes, myxoedema, and asthma, for individuals from populations underrepresented in commonly used datasets, including Nigeria, Sri Lanka, and Hawaii, available in the UK Biobank[2] and the PAGE study.[25] In order to have meaningful results, we limit the phenotypes to these four, as they have a high enough case count within the three target populations. For each phenotype, we use balanced data from white British individuals as the source domain, obtained by removing samples from the majority class. Then, for each phenotype and target population, we use the labeled source domain data as well as labeled and unlabeled data from the target domain. To test the method's efficacy, we use a subset of labeled data exclusive to the target underrepresented population.

To establish which samples constitute the target population domain, we propose two approaches, which show two different ways in which we can take advantage of the availability of datasets, even when the labeled data in the target domain is very scarce. The first approach is adopted for the Nigerian and Sri Lankan populations, and the second one for the Hawaiian population.

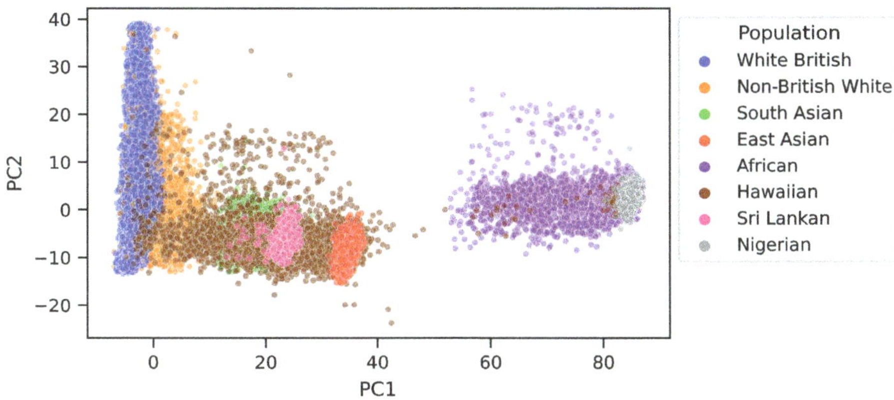

Fig. 4. Two-dimensional PCA projection of the samples in the UK Biobank and the Hawaiian dataset. The PCA was fitted with only the samples from the UK Biobank. Note that all samples marked as Sri Lankan fall within the South Asian genetic ancestry cluster, and all the Nigerian ones fall within the African cluster.

The first approach only uses data from the UK Biobank.[2] To establish which samples constitute the target population, we combine the genetically inferred ancestry available from the Global Biobank Engine[4] (white British, non-British white, South Asian, East Asian, or African) and the country of birth reported in the UK Biobank.[2] We use both fields because the inferred genetic ancestry provides a continental-level description, encompassing many regions within each label. On the other hand, the country of birth alone is not representative of the ancestry composition within the UK Biobank due to high selection bias, as the samples were collected in assessment centers in the United Kingdom, so many individuals in the data born outside the United Kingdom are still of English genetic ancestry. By filtering both by inferred population group and country of birth, we ensure that the definition of the target domain is precise.

In particular, for the case of Nigeria, we only keep the samples that are of African genetic ancestry and born in Nigeria, and for the case of Sri Lanka, the samples that are of South Asian genetic ancestry and born in Sri Lanka. This results in a total of 852 samples for the Nigeria group and 535 samples for the Sri Lanka group. Once we have the samples from the target domain, since the UK Biobank has phenotype labels for all the samples, we artificially unlabel half of them for the purpose of evaluating the proposed method. For training, we use all the unlabeled samples plus only 10 labeled samples from the target domain, 5 negative and 5 positive, alongside all the labeled samples from the source domain. The rest of the labeled individuals of the target domain are split into two equal parts using stratified sampling to create the validation and test sets. Note that we can only use labeled data for validation and testing.

The second approach to define the target domain shows how additional unlabeled datasets can be employed. To achieve this, in addition to the UK Biobank,[2] we use a dataset of SNP sequences (without phenotype labels) of 5,862 Native Hawaiian individuals from the PAGE study.[25] In this setting, we only have unlabeled data from the target population. Note that we cannot use the country of birth field, as the people born in Hawaii are labeled as born in the USA. To have labeled data in the target domain, we propose to use the nearest neighbor of each sample from the Hawaiian dataset within the UK Biobank, excluding the white British individuals to avoid having repeated samples in both the source and target domains. For efficiency, we compute the distances between samples on the first 50 principal components, instead of using the raw SNP sequences. After removing duplicated individuals that are the nearest neighbor to more than one sample from the Hawaiian dataset, we obtained 1,689 labeled samples. While it is unlikely that the UK Biobank contains this many individuals of Hawaiian ancestry, the closer distribution of these samples to the Native Hawaiian population makes the domain more apt to model them than using samples of predominantly European ancestry.

The second approach to defining the target domain is less accurate than the first one, as it includes samples from other similar populations. However, it has the advantage that it results in a larger number of samples, which can be helpful for unbalanced phenotypes with a low positive case count, and to counteract the effect of having a noisier target domain definition. In this scenario, we use 50% of the labeled target samples for the training set, 25% for the validation set, and 25% for the test set. Note that the unlabeled samples used for training are the ones from the Hawaiian dataset from the PAGE study.[25]

Table 1. Size of sets used for training and evaluation for each population. Note that a combination of white British as the source domain plus another population as the target domain is always used.

Population	Training labeled	Training unlabeled	Val. + Test labeled	Total
White British	*	0	0	*
Nigeria	10	213	106 + 107	852
Sri Lanka	10	134	67 + 67	535
Hawaii	822	5,862	412 + 413	7,507

*White British set size depends on class balancing, but in all cases is >40,000.

We use the variants that are both in the UK Biobank data and the Hawaiian dataset, resulting in 83,362 overlapping SNPs. Note that we decided to not impute the SNPs outside the intersection to avoid introducing a bias. Most algorithms to perform statistical imputation are based on the available samples, and since in this scenario most of them are not from the underrepresented population, the imputation could result in incorrect values.

Table 2. Case counts for each disease and population (only considering labeled samples).

Population	Hypertension	Myxoedema	Diabetes	Asthma
White British*	114,687 (50.00%)	21,471 (50.00%)	23,099 (50.00%)	45,192 (50.00%)
Nigeria	105 (47.08%)	11 (4.93%)	38 (17.04%)	21 (9.41%)
Sri Lanka	60 (41.66%)	18 (12.50%)	41 (28.47%)	29 (20.13%)
Hawaii	535 (31.68%)	69 (4.09%)	153 (9,05%)	197 (11.66%)

*White British phenotypes are balanced by undersampling the negative class.

3.2. *Model*

We adopt the MME-SLA[13,14] method originally proposed for classification tasks in computer vision for genotype-to-phenotype prediction by replacing the ResNet34[16] backbone model used in the original works with a multi-layer perceptron (MLP). Specifically, we use a 4-layer MLP with GELU activations,[26] layer normalization,[27] and a residual connection[16] between the output of the first layer and the input of the last one. The choice of activation and the use of layer normalization and a residual connection is commonly adopted in modern architectures such as Transformers[28] and has been proven to help improve the performance of the models, as well as their stability during training. The initial layer of the network takes an input size corresponding to the number of SNPs and reduces it to a hidden size of 256. Subsequently, the two middle layers maintain the same input and output dimensions of 256. Next, before the last layer, ℓ_2 normalization and temperature scaling with $T = 0.05$ is applied, as proposed in Refs. 13,15. Lastly, the last layer, which acts as the classifier, produces an output size equivalent to the number of classes. We call the complete model PopGenAdapt.

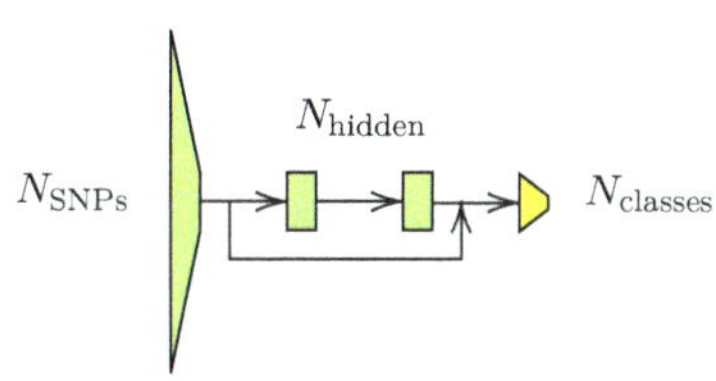

Fig. 5. Diagram of the backbone MLP model for PopGenAdapt.

The backbone MLP model without the MME-SLA components is also used as the baseline model to compare how applying SSDA improves against a typical supervised learning approach.

We train the baseline and PopGenAdapt models for each combination of target population and phenotype using a batch size of 64, the AdamW optimizer[29] with weight decay of 0.01, and the same learning rate scheduler used in Ref. 14. We use randomized hyperparameter search to tune several hyperparameters. For the baseline method, we only tune the learning rate. For PopGenAdapt, we tune both the learning rate and the MME-SLA[13,14] hyperparameters (λ, α,

τ, W, and I). Table 3 shows the hyperparameter space from which the values were sampled. Our experiments showed that the resulting performance is highly sensitive to the choice of hyperparameters, as also pointed out in the paper which introduced SLA.[14] We select the model with the best validation AUROC on the target domain and perform the final testing on a separate hold-out test set also using AUROC on the target domain.

Table 3. Definition of the distribution of the hyperparameter space.

Hyperparameter	Probability distribution
Learning rate	$\text{LogUniform}(10^{-5}, 10^{-2})$
MME tradeoff λ	$\text{Uniform}(0, 1)$
SLA mix ratio α	$\text{Uniform}(0, 1)$
SLA temperature τ	$\text{Uniform}(0, 1)$
SLA warmup W	UniformChoice({100, 500, 1000, 2000, 5000})
SLA update interval I	UniformChoice({5, 10, 100, 500, 1000, 2000, 5000})

Note that while PopGenAdapt employs both the labeled and unlabeled samples, the baselines are trained on the subset that is labeled, as it has no way of using the unlabeled samples.

The training and inference was performed with an NVIDIA GeForce GTX 1080 Ti GPU (11 GB), and took between 10 and 50 minutes, depending on the number of samples and the hyperparameters, for each configuration.

4. Results

We compare PopGenAdapt with the baseline model consisting only of the backbone MLP (MLP Base), as well as with the state-of-the-art genotype-to-phenotype snpnet[11] model, and PRS-CSx,[30] which is an extension of PRS-CS[9] to improve polygenic prediction in ancestrally diverse populations. Note that since snpnet and PRS-CSx are supervised models, like in the case of the baseline model, they can not exploit the unlabeled samples.

We show the results obtained for each of the four phenotypes on the three tested target underrepresented populations in Tables 4–6.

PopGenAdapt outperforms snpnet, PRS-CSx, and the baseline model on average and in the majority of evaluated scenarios. Moreover, we observe that snpnet, PRS-CSx, and the baseline model obtain in multiple cases an AUROC below 0.5, indicating a predictive performance worse than the one obtained by random guessing. We note that this does not happen in any of the experimented cases for PopGenAdapt. Considering that snpnet and the MLP baseline methods do not perform any type of domain adaptation, it makes sense for this to happen, as the models are tested on a domain that differs from the one in which most of the training samples are.

We hypothesize that a possible reason for the poor performance of snpnet on non-European populations is due to the use of the lasso in the method, which performs SNP selection, thus

Table 4. AUROC for the Nigerian population.

Method	Hypertension	Myxoedema	Diabetes	Asthma	Average
snpnet[11]	0.4647	0.7949	0.4699	0.4646	0.5485
PRS-CSx[30]	0.3884	0.1827	**0.6046**	0.4306	0.4016
MLP Base	**0.5488**	0.4423	0.5376	0.4886	0.5043
PopGenAdapt	0.5088	**0.8109**	0.5516	**0.5619**	**0.6083**

Table 5. AUROC for the Sri Lankan population.

Method	Hypertension	Myxoedema	Diabetes	Asthma	Average
snpnet[11]	0.4500	0.4631	0.4603	0.5871	0.4901
PRS-CSx[30]	0.4852	0.4863	0.3991	0.5379	0.4771
MLP Base	**0.5898**	0.5137	0.5952	0.5939	0.5731
PopGenAdapt	0.5778	**0.5902**	**0.6723**	**0.6091**	**0.6123**

Table 6. AUROC for the Hawaiian population.

Method	Hypertension	Myxoedema	Diabetes	Asthma	Average
snpnet[11]	0.6132	**0.6148**	0.5235	0.4857	0.5593
PRS-CSx[30]	0.5423	0.5881	0.4577	0.5087	0.5242
MLP Base	0.6104	0.5162	0.5041	0.5666	0.5493
PopGenAdapt	**0.6135**	0.5556	**0.5791**	**0.5811**	**0.5823**

excluding completely some variants. Possibly, as the training data is mostly from the source domain, many of the SNPs excluded are the ones that remain useful for making the prediction on the target population. All this reflects the usefulness and need for domain adaptation techniques to be used when the data of the target domain is limited, like in the case of underrepresented populations.

Furthermore, PRS-CSx has poor performance in most settings. We believe that the small number of samples of the target population still had an effect on this case, reflecting the value of incorporating unlabeled samples when the labeled data is scarce. Another possible limitation that could result in bad performance is the use of a relatively small number of SNPs, although this is shared across all four methods.

Finally, we also observe that the supervised methods suffer less in the Hawaiian dataset, probably due to the higher number of labeled samples for training that are used in this scenario and the fact that the Hawaiian target domain is less precise, resulting in less advantage for domain adaptation. The target domain, in this case, is less precise due to the nearest neighbor approach used to establish the labeled samples, as well as due to the fact that Pacific Islanders are admixed populations, resulting in more variability across the samples, as can be observed in Fig. 4.

5. Conclusion

In this work, we presented PopGenAdapt, a model that applies semi-supervised domain adaptation techniques for genotype-to-phenotype prediction. We also proposed two approaches to set the target domain samples and evaluated the model to predict several disease outcomes in three different underrepresented populations. The results show that by using SSDA on underrepresented populations, the prediction performance can be improved over state-of-the-art supervised methods. Consequently, we show SSDA is a promising technique to help overcome health disparities in precision medicine by exploiting the availability of unlabeled data from underrepresented populations while still taking advantage of the greater magnitude of labeled data available from populations of European ancestry.

Nonetheless, there are still some limitations and avenues for future work. Due to the limited data on the underrepresented population we had available from the UK Biobank,[2] we did not study the influence the ratio of labeled and unlabeled samples could have on the attained performance, as using more samples for training would have left too few for validation and testing. Moreover, the scalability of the method to a larger number of SNPs also remains to be assessed. Further work on the approach could also include the possibility of learning from GWAS summary statistics instead of the SNP sequences or to also support continuous phenotypes apart from categorical ones. Possibly, there is also room for improvement on the base model used, as more powerful deep learning architectures could be evaluated. Furthermore, considering the integration of PopGenAdapt on emerging paradigms such as federated learning or differential privacy[31] could further enhance the applicability of the method in biomedical research and healthcare.

Acknowledgments

We would like to thank Mark A. Penjueli for data pre-processing of the PAGE study dataset. The computing for this project was performed on the Sherlock cluster from Stanford University and on the TSC CALCULA cluster from the Department of Signal Theory and Communications (TSC) of the Polytechnic University of Catalonia (UPC). We would like to thank both institutions for supplying computational resources and support that contributed to these results. This work was partially supported by NIH under award R01HG010140 and by a grant from the Stanford Institute for Human-Centered Artificial Intelligence (HAI). This research has been conducted using the UK Biobank Resource under Application Number 24983.

References

1. A. R. Martin, M. Kanai, Y. Kamatani, Y. Okada, B. M. Neale and M. J. Daly, Clinical use of current polygenic risk scores may exacerbate health disparities, *Nature Genetics* **51** (April 2019).

2. C. Bycroft *et al.*, The UK Biobank resource with deep phenotyping and genomic data, *Nature* **562** (October 2018).
3. K. J. Karczewski *et al.*, The mutational constraint spectrum quantified from variation in 141,456 humans, *Nature* **581** (May 2020).
4. N. Sinnott-Armstrong, Y. Tanigawa, D. Amar, N. Mars, C. Benner, M. Aguirre, G. R. Venkataraman, M. Wainberg, H. M. Ollila, T. Kiiskinen, A. S. Havulinna, J. P. Pirruccello, J. Qian, A. Shcherbina, F. Rodriguez, T. L. Assimes, V. Agarwala, R. Tibshirani, T. Hastie, S. Ripatti, J. K. Pritchard, M. J. Daly, M. A. Rivas and FinnGen, Genetics of 35 blood and urine biomarkers in the UK Biobank, *Nature Genetics* **53** (February 2021).
5. The 1000 Genomes Project Consortium, A global reference for human genetic variation, *Nature* **526** (October 2015).
6. B. V. Halldorsson *et al.*, The sequences of 150,119 genomes in the UK Biobank, *Nature* **607** (July 2022).
7. E. R. Bartusiak, M. Barrabés, A. Rymbekova, J. Gimbernat-Mayol, C. López, L. Barberis, D. Mas Montserrat, X. Giró-I-Nieto and A. G. Ioannidis, Predicting Dog Phenotypes from Genotypes, in *2022 44th Annual International Conference of the IEEE Engineering in Medicine and Biology Society (EMBC)*, (IEEE, September 2022).
8. M. John, F. Haselbeck, R. Dass, C. Malisi, P. Ricca, C. Dreischer, S. J. Schultheiss and D. G. Grimm, A comparison of classical and machine learning-based phenotype prediction methods on simulated data and three plant species, *Frontiers in Plant Science* **13** (November 2022).
9. T. Ge, C.-Y. Chen, Y. Ni, Y.-C. A. Feng and J. W. Smoller, Polygenic prediction via Bayesian regression and continuous shrinkage priors, *Nature Communications* **10** (April 2019).
10. L. R. Lloyd-Jones, J. Zeng, J. Sidorenko, L. Yengo, G. Moser, K. E. Kemper, H. Wang, Z. Zheng, R. Magi, T. Esko, A. Metspalu, N. R. Wray, M. E. Goddard, J. Yang and P. M. Visscher, Improved polygenic prediction by Bayesian multiple regression on summary statistics, *Nature Communications* **10** (November 2019).
11. J. Qian, Y. Tanigawa, W. Du, M. Aguirre, C. Chang, R. Tibshirani, M. A. Rivas and T. Hastie, A fast and scalable framework for large-scale and ultrahigh-dimensional sparse regression with application to the UK Biobank, *PLOS Genetics* **16** (October 2020).
12. Y. Ouali, C. Hudelot and M. Tami, An Overview of Deep Semi-Supervised Learning (July 2020).
13. K. Saito, D. Kim, S. Sclaroff, T. Darrell and K. Saenko, Semi-supervised domain adaptation via minimax entropy, in *Proceedings of the IEEE/CVF International Conference on Computer Vision (ICCV)*, (IEEE, October 2019).
14. Y.-C. Yu and H.-T. Lin, Semi-supervised domain adaptation with source label adaptation, in *Proceedings of the IEEE/CVF Conference on Computer Vision and Pattern Recognition (CVPR)*, (IEEE, June 2023).
15. W.-Y. Chen, Y.-C. Liu, Z. Kira, Y.-C. F. Wang and J.-B. Huang, A Closer Look at Few-shot Classification, in *International Conference on Learning Representations*, (OpenReview, 2019).
16. K. He, X. Zhang, S. Ren and J. Sun, Deep Residual Learning for Image Recognition, in *Proceedings of the IEEE Conference on Computer Vision and Pattern Recognition (CVPR)*, (IEEE, June 2016).
17. Y. Ganin and V. Lempitsky, Unsupervised Domain Adaptation by Backpropagation, in *Proceedings of the 32nd International Conference on Machine Learning*, (PMLR, July 2015).
18. X. Wang, Y. Hua, E. Kodirov, S. S. Mukherjee, D. A. Clifton and N. M. Robertson, ProSelfLC: Progressive Self Label Correction Towards A Low-Temperature Entropy State (2022).
19. S. Reed, H. Lee, D. Anguelov, C. Szegedy, D. Erhan and A. Rabinovich, Training Deep Neural Networks on Noisy Labels with Bootstrapping (2015).
20. J. Snell, K. Swersky and R. Zemel, Prototypical Networks for Few-shot Learning, in *Advances in Neural Information Processing Systems*, (Curran Associates, Inc., 2017).
21. C. Yao, X. Zhu and K. A. Weigel, Semi-supervised learning for genomic prediction of novel traits

with small reference populations: an application to residual feed intake in dairy cattle, *Genetics Selection Evolution* **48** (November 2016).

22. D. M. Reyes, A. Bose, E. Karavani and L. Parida, *FairPRS: adjusting for admixed populations in polygenic risk scores using invariant risk minimization*, in *Biocomputing 2023*, (World Scientific Publishing Company, November 2022).
23. T. Gu, Y. Han and R. Duan, *A transfer learning approach based on random forest with application to breast cancer prediction in underrepresented populations*, in *Biocomputing 2023*, (World Scientific Publishing Company, November 2022).
24. M. Muneeb, S. Feng and A. Henschel, Transfer learning for genotype–phenotype prediction using deep learning models, *BMC Bioinformatics* **23** (November 2022).
25. G. L. Wojcik *et al.*, Genetic analyses of diverse populations improves discovery for complex traits, *Nature* **570** (June 2019).
26. D. Hendrycks and K. Gimpel, Gaussian Error Linear Units (GELUs) (2016).
27. J. L. Ba, J. R. Kiros and G. E. Hinton, Layer Normalization (July 2016).
28. K. Han, A. Xiao, E. Wu, J. Guo, C. XU and Y. Wang, Transformer in transformer, in *Advances in Neural Information Processing Systems*, eds. M. Ranzato, A. Beygelzimer, Y. Dauphin, P. Liang and J. W. Vaughan (Curran Associates, Inc., 2021).
29. I. Loshchilov and F. Hutter, Decoupled Weight Decay Regularization, in *International Conference on Learning Representations*, (OpenReview, 2019).
30. Y. Ruan *et al.*, Improving polygenic prediction in ancestrally diverse populations, *Nature Genetics* **54** (May 2022).
31. M. Joshi, A. Pal and M. Sankarasubbu, Federated Learning for Healthcare Domain - Pipeline, Applications and Challenges, **3** (Nov 2022).

LA-GEM: imputation of gene expression with incorporation of Local Ancestry

Mrinal Mishra†, Layan Nahlawi†, Yizhen Zhong, Tanima De, Guang Yang, Cristina Alarcon and Minoli A. Perera

Department of Pharmacology, Center for Pharmacogenomics, Feinberg School of Medicine, Northwestern University
Chicago, Illinois, USA
Email: minoli.perera@northwestern.edu

Gene imputation and TWAS have become a staple in the genomics medicine discovery space; helping to identify genes whose regulation effects may contribute to disease susceptibility. However, the cohorts on which these methods are built are overwhelmingly of European Ancestry. This means that the unique regulatory variation that exist in non-European populations, specifically African Ancestry populations, may not be included in the current models. Moreover, African Americans are an admixed population, with a mix of European and African segments within their genome. No gene imputation model thus far has incorporated the effect of local ancestry (LA) on gene expression imputation. As such, we created LA-GEM which was trained and tested on a cohort of 60 African American hepatocyte primary cultures. Uniquely, LA-GEM include local ancestry inference in its prediction of gene expression. We compared the performance of LA-GEM to PrediXcan trained the same dataset (with no inclusion of local ancestry) We were able to reliably predict the expression of 2559 genes (1326 in LA-GEM and 1236 in PrediXcan). Of these, 546 genes were unique to LA-GEM, including the *CYP3A5* gene which is critical to drug metabolism. We conducted TWAS analysis on two African American clinical cohorts with pharmacogenomics phenotypic information to identity novel gene associations. In our IWPC warfarin cohort, we identified 17 transcriptome-wide significant hits. No gene reached are prespecified significance level in the clopidogrel cohort. We did see suggestive association with *RAS3A* to P2RY12 Reactivity Units (PRU), a clinical measure of response to anti-platelet therapy. This method demonstrated the need for the incorporation of LA into study in admixed populations.

Keywords: Local Ancestry, Gene Expression Model, LA-GEM, PrediXcan, Gene Imputation, Population-specific Genetic Variations, Admixed Populations, Ancestry-specific Gene Associations

1. Introduction

It is widely acknowledged that large-scale genetic studies investigating human diseases have often failed to encompass the extensive diversity seen in global populations, as they primarily focus on individuals of European descent.[1]This insufficiency of ethnic diversity in such studies limits our understanding of the genetic underpinnings of human diseases and intensifies health disparities. Moreover, the paucity of ethnic diversity in human genomics research could lead to a potentially hazardous deficiency, or even errors, in our capacity to apply genetic research findings to clinical procedures or public health policies.

† Contributed equally to the work.

PrediXcan is one of the first and most popular methods used to predict gene expression levels in different tissues or cell types for use in transcriptome-wide association studies (TWAS).[2] The method leverages large publicly available multi-omic datasets that includes paired single nucleotide polymorphism (SNP) data and gene expression data from multiple individuals and tissues.[2-3] By training a predictive model on these reference datasets, PrediXcan can predict the expression levels of a given gene in a new individual, based on that person's genetic variation. Outside data can be trained through various available methods.[4,5] There are various extensions to PrediXcan that have been developed which extend this method to multi-tissue TWAS and causal gene prioritization.[5-9]

In any association studies, undetected population stratification can lead to false-positive. Therefore, it is critical to implement appropriate correction to adjust these effects.[10] One such measure, used in genome-wide association studies (GWAS), is the inclusion of principal components (PCs), with the first few PCs estimating global ancestry (GA) in the cohort. GA is largely directed by demographic history of the population. However, for admixed population the effects of nearby SNPs or epigenetic changes has been shown to have a significant effect of gene expression[11]. Thus, local ancestry may be an important consideration in gene expression prediction. Here we have incorporated LA as predictor in PrediXcan framework to assess the if including this variable in the African American population resulting in the improved predictability of the gene models.

2. Methods

In this paper, we propose a modification to PrediXcan method titled LA-GEM (Local Ancestry based Gene Expression prediction Model) to incorporate local ancestry predictors (LA) along with cis region genetic variants in the development of gene expression prediction models. We have used our African American multi-omic hepatocyte dataset (N = 60) to create gene expression prediction models, however this method can be used on any multi-omic data from an admixed cohort in which local ancestry inference is available.

2.1. *Primary Hepatocyte Cohort*

Sixty-three African Ancestry (AA) primary human hepatocyte (PHHs) cultures were acquired. AA PHHs were either purchased from commercial companies (BioIVT, TRL/Lonza, Life technologies, Corning, and Xenotech), or isolated in-house from cadaveric livers. Livers with active cancer or a history of hepatocarcinoma were excluded from the study. To account for differences in PHH sourcing, transcriptomic data went through additional QC measures (i.e., PC visualization, batch correction) to ensure any differences from source and isolation method were corrected. PHHs were isolated from cadaveric livers using a modified two-step collagenase perfusion procedure previously described in Park et. al.[12] Only hepatocyte cultures with RNA Integrity Number (RIN) over 8 and with sufficient RNA to conduct NGS were used in the study.

2.2. *Genotyping, quality control and imputation*

DNA was obtained from around 1 million cells of each PHH culture using the Gentra Puregene (Qiagen) kit following manufacturer's protocol. All extracted DNA samples were barcoded for genotyping. Illumina Infinium Multi-Ethnic Global Kit was used for SNP genotyping and standard genotyping protocol was followed. SNPs were filtered out before imputation based on following criterion: (1) SNPs present on the sex and mitochondrial chromosomes. They were filtered out as they could alter the minor allele frequency (MAF) values (2) SNPs having A/T or C/G as it may introduce flip-strand issues. (3) SNPs with low genotype quality (call rate < 0.95).

Using PLINK[9], individuals with discordant sex information were identified using the sex check function and duplicates or related individuals were identified using the identity-by-descent (IBD) method. An IBD cutoff score of 0.125 was used, indicating third-degree relatedness or closer. No samples were removed after these QC steps. SNPs with MAF<0.05 were removed. Patient ancestries were confirmed using a principal component analysis (PCA) plot of linkage disequilibrium (LD) pruned genotype data. LD pruning was conducted to identify the principal dimensions of genetic variation between samples. Samples that did not cluster along the spectrum for AA within this PCA plot of raw genotype data were removed.[11] One individual was excluded after sample and genotyping QC analysis, leaving 62 individuals.

Genotypes were imputed by the TOPMed imputation server (version 1.6.6)12-14 using the TOPMed r2 reference panel, GRCh38/hg38 array build, and 0.3 estimated r2 (rsq) filter threshold. Post-imputation QC includes removal of SNPs with poor imputation quality scores (<0.8), failed Hardy-Weinberg equilibrium tests ($p < 0.00001$), and low MAFs (<0.05). This resulted in a total of 5,189,820 SNPs included for model building.

2.3. *Local ancestry inference*

LA was inferred using RFMix (v.1.5.4). RFMix takes as input a set of reference panels (populations with known ancestry) and a set of test individuals, and uses a hidden Markov model to infer the most likely ancestry of each segment of the test individuals' genomes. The output of RFMix is a set of probabilities for each test individual, indicating the likelihood that a specific haplotype segment comes from one of the reference populations.[13] In this analysis we use Yoruba (African Ancestry) and American white (CEU – European Ancestry) as our refences populations.

2.4. *RNA-sequencing and Quality Control*

Total RNA was extracted from each PHH culture three days after plating using the Qiagen RNeasy Plus mini kit. Samples with an RNA integrity number (RIN) less than 8 were removed from analysis. This resulted in the removal of 2 samples leaving 60 individuals at the end. Libraries were prepared for sequencing using the TruSeq RNA Sample Prep Kit, Set A (Illumina) per manufacturer's protocol. The cDNA libraries were prepared and sequenced using either HiSeq2500 (Illumina) or HiSeq4000 (Illumina) instruments by the University of Chicago's Functional Genomics core, producing single-end 50bp reads with approximately 50 million reads per sample. As two instruments were used in this study, we were cognizant of potential batch effect and incorporated methods for correction as previously described.[14]

2.5. *Gene Expression Quantification*

Gene expression was quantified using a collapsed gene model following the GTEx isoform collapsing procedure[15]. To evaluate gene-level expression, reads were mapped to genes referenced with GENCODE(v.25) using RNA-SeQC. HTSeq supplied raw counts for gene expression analysis using Bioconductor package DESeq2(v1.20.0). Counts were normalized by regularized log transformation, batch correction was performed using ComBat-Seq[14], and PCA was performed using DESeq2.

Gene expression was normalized by trimmed means of M-values normalization method (TMM) implemented in edgeR.[16] Transcripts per million (TPM) was calculated by first normalizing counts by gene length and then by read depth.[17] Gene expression values were filtered based on expression thresholds < 0.1 TPM in at least 20% of samples and ≤ 6 reads in at least 20% of samples. The expression values for each gene were normalized across samples with inverse normal transformation. To account for unmeasured confounding variables in transcriptome data, we used probabilistic estimation of expression residuals (PEER).[18]

2.6. *LA-GEM Framework*

LA-GEM consists of mainly three steps:

For gene expression prediction, a linear model was trained using reference panel that includes genotype, LA predictor, interaction predictor (interaction between genotype and LA predictor) and corresponding expression data[2,19] using the following training model equation[19]:

$$y_g \sim \sum_{a,b,c} w_a S_a + w_b A_b + w_c I_c + \varepsilon \tag{1}$$

where w_a, w_b and w_c are the regression parameter needed to be trained, S = $(S_1, , S_2, \ldots, S_a)$ is the genotype data in the cis region of interest, A = $(A_1, A_2, \ldots, A_b)$ is the local ancestry predictors for all SNP positions in the cis region and I = $(I_1, I_2, \ldots, I_c)$ is the Interaction predictor (I = S x A).

Genetically regulated gene expressions are then determined using the above model for new dataset that include combination of genotype and local ancestry information using the following equation:

$$\hat{y}_g \sim \sum_{d,e,f} w_d S_d + w_e A_e + w_f I_f \tag{2}$$

Estimated genetically regulated gene expressions ŷ_g is then associated to the phenotype using the following equation:

$$Z \sim \hat{y}_g + \varepsilon \tag{3}$$

LA-GEM prediction models were trained on 60 African American PHH samples followed by 5-fold cross-validation. Gene models with an average correlation $\rho \geq 0.1$ and $P < 0.05$ between predicted and observed Expression were deemed well predicted.

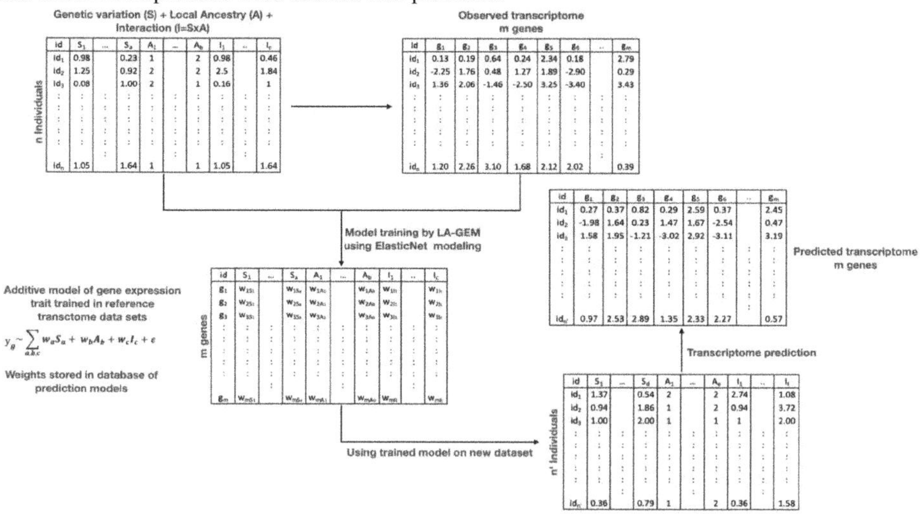

Fig. 1. Flowchart showing LA-GEM workflow.

2.7. ***TWAS association using LA-GEM gene imputation.***

As a proof of concept, we use LA-GEM to impute hepatic gene expression in two clinical cohort to identify novel gene associations to drug response. As the expression of hepatic genes are especially important in platelet function and drug metabolism, we imputed gene expression of 1323 genes which were then used in the TWAS conducted using PrediXcan.[2] We prespecified a TWAS p-value of 3.8×10^{-5} as significant (0.05/1323).

2.7.1. *African American warfarin Cohort*

Through the International Warfarin Pharmacogenomics consortium (IWPC) we collect information from 340 African American patients on warfarin as well as 199 African Americans who were part of the University of Alabama Birmingham Warfarin cohort assess though dbGAP (phs000708.v1.p1). Briefly, clinical and demographic data on stable warfarin dose was collected, defined as the dose of warfarin needed to elicit and INR within therapeutic rage (2-3) for three consecutive clinical visits as previously described.[20]

2.7.2. *ACCOuNT Clopidogrel cohort*

Through the ACCOuNT Consortium[21] we recruited 180 African Americans on the anti-platelet drug, clopidogrel. All subjects included in the TWAS had a biomarker measure of clopidogrel response, P2Y12 Reactivity Units (PRU). All subjects were on 75 mg of clopidogrel for at least 15 days with inclusion and exclusion criteria as described previously.[21]

2.8. *Log Ratio of Interaction Predictors*

To quantify the relative influence of interaction predictors in our LA-GEM model, we calculated a Log Ratio for each gene using the formula:

$$\text{Log Ratio} = \log_2(\text{Count of Interaction Predictors} + 1) - \log_2(\text{Count of SNP Dosage Predictors} + 1)$$

A positive Log Ratio indicates that a gene relies more heavily on interaction predictors, while a negative value suggests greater reliance on genetic dosage predictors.

2.9. *Code Availability*

The LA-GEM model was implemented in R and employs SNP-based local ancestry calculated using RFMix version 1.5.4. The source code is publicly available and can be accessed at https://github.com/pereralab/LA-GEM.

3. Results

We built two gene expression prediction models, LA-GEM and PrediXcan (using AA PHH as training). We assessed predictive performance using five-fold cross-validation (R2 of model performance). We found that LA-GEM was able to impute 1323 genes at a rho>0.1, p-value ≤ 0.05 (Average rho = 0.397) as compared to 1236 genes imputed well using the PrediXcan model (Average rho = 0.403) in the same dataset without LA (Fig. 2). The average number of predictors for LA-GEM is shown in Table 1.

Table 1 – Summary table showing total number of Predictable genes and number of different Predictors used to train the model.

	LA-GEM
Number of Predictable genes	1323
Number of Predictors	71702
Number of SNP Dosage Predictor	46028
Number of Interaction Predictors (L.A X SNP Dosage)	25674

Fig. 2. Venn diagram showing number of predictable genes in each of the model.

3.1. *Gene list enrichment analysis of predictable genes*

KEGG Pathway enrichment analysis (Statistical overrepresentation test) was performed using g:Profiler[28] for predictable genes (1323 genes) obtained from LA-GEM. The analysis yielded significant enrichments for several pathways as shown in Fig, 3, notably those linked to pharmacogenomics. Among these, three pathways were found to be prominently enriched: "Metabolism of xenobiotics by cytochrome P450" (KEGG:00980) with a fold enrichment of 3.37 and an adjusted p-value of 0.00285, "Drug metabolism - cytochrome P450" (KEGG:00982) with a fold enrichment of 3.18 and an adjusted p-value of 0.01097, and "Drug metabolism - other enzymes" (KEGG:00983) with a fold enrichment of 2.74 and an adjusted p-value of 0.04196.

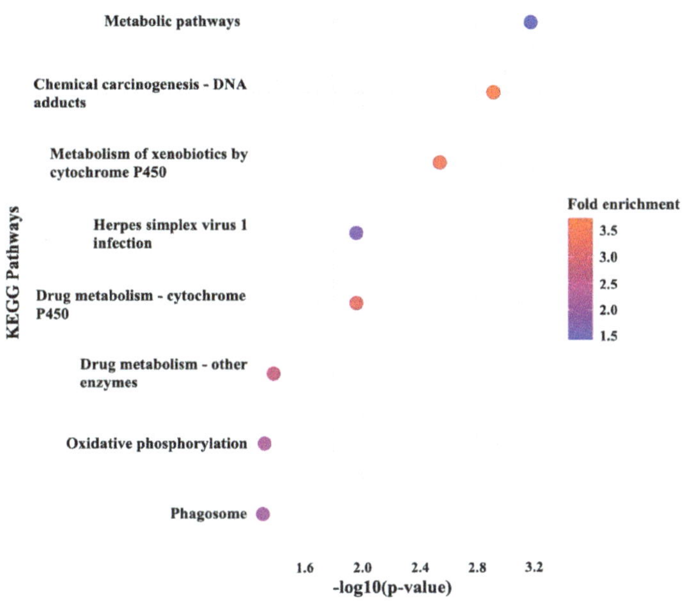

Fig. 3. Gene set enrichment of 1323 predictable genes obtained from LA-GEM. Y-axis show categories with their corresponding -$\log_{10}$(p-value) in the X-axis. Color shows the fold enrichment value for each of the processes.

3.2. *Genes unique to LA-GEM*

Among the 1323 predictable genes, 546 genes were found to be unique to LA-GEM model which were not reported by PrediXcan model as significant (Fig. 2). Out of the 546 unique genes, 2 genes (*MME* and *LRRC37A2*) were found to be strongly associated with global West African ancestry as previously reported[12]. In addition, *CYP3A5 CYP1A1, CYP4F2, CBR1,* and *UGT2A1* was also among the genes unique to LA-GEM which is known to show significant variability in level of expression between population of different ancestry and are important to drug response[22].

3.3. *Genes unique to PrediXcan*

Among the 1323 predictable genes, 459 genes were found to only in the PrediXcan model (Fig. 2). Out of the 459 genes, 6 genes (*DHODH*, *SNAI1*, *RBBP9*, *ENSG00000271239*, *NPR2*, and *SLC39A11*) were found to be strongly associated with global West African ancestry as previously reported.[12]

3.4. *Genes common to LA-GEM and PrediXcan*

Among the 1323 predictable genes, 777 genes were found to be well imputed by both models. Out of these 777 genes, 4 genes (*CDK18*, *GREM2*, *COL26A1* and *MMP20-AS1*) were found to be strongly associated with West African ancestry as previously reported.[12] The rho average for *CDK18* and *GREM2* were higher in LA-GEM (0.48 versus 0.33 and 0.28 versus 0.26, respectively) but the inverse was true for *COL26A1* and *MMP20-AS* (0.39 versus 0.44 and 0.32 versus 0.54 respectively) The rho average for these genes were evenly distributed around the diagonal (Fig. 4), suggesting one model did not outperform the other in these commonly imputed genes. For genes that were unique to the PrediXcan model, the average difference in rho between models was 0.42. For those gene that were uniquely to LA-GEM the average difference in Rho was 0.46. However, these differences in prediction accuracy were not a significant difference between the two groups of genes ($p = 0.07$).

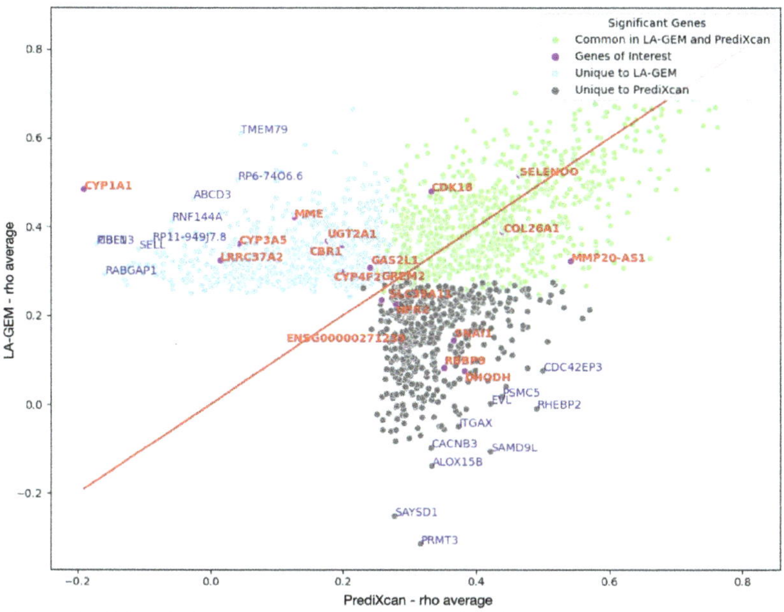

Fig. 4. Correlation plot between rho-averages of gene well predicted with LA-GEM and PrediXcan models. Top 10 genes showing the maximum rho-average difference between methods are labelled in dark blue color. Red line shows perfect correlation. Well predicted genes unique to LA-GEM model are shown in light blue. Well predicted genes unique to PrediXcan model are shown in grey. Well predicted genes common between LA-GEM and PrediXcan model are shown in light green. Genes of interest with pharmacogenomic relevance or which are associated with West African ancestry are shown in violet and are labelled in red.

3.5. *Differential Role of Interaction Predictors in LA-GEM and PrediXcan Models*

In the process of model training for LA-GEM, we observed differences in the role played by the type of predictors, especially interaction predictors, in model efficacy. Among the 546 genes uniquely imputed by the LA-GEM model, 137 genes (or approximately 25% of these significant genes) exhibited a positive Log Ratio of the Count of Interaction predictors. This observation underscores the relevance of interaction predictors as significant contributors in the unique imputation capability of the LA-GEM model.

In contrast, among the 777 genes that were common between LA-GEM and PrediXcan, only 119 genes (approximately 15% of these significant genes) had a positive Log Ratio of the Count of Interaction predictors. This relatively lower proportion suggests that the common genes might rely less on interaction predictors in the LA-GEM model than the genes unique to it.

The detailed distribution of the Log Ratio of the Count of Interaction predictors for these gene sets is depicted in Fig 5. This difference in the involvement of interaction predictors between genes unique to LA-GEM and those common with PrediXcan provides further insight into the distinguishing features of these models.

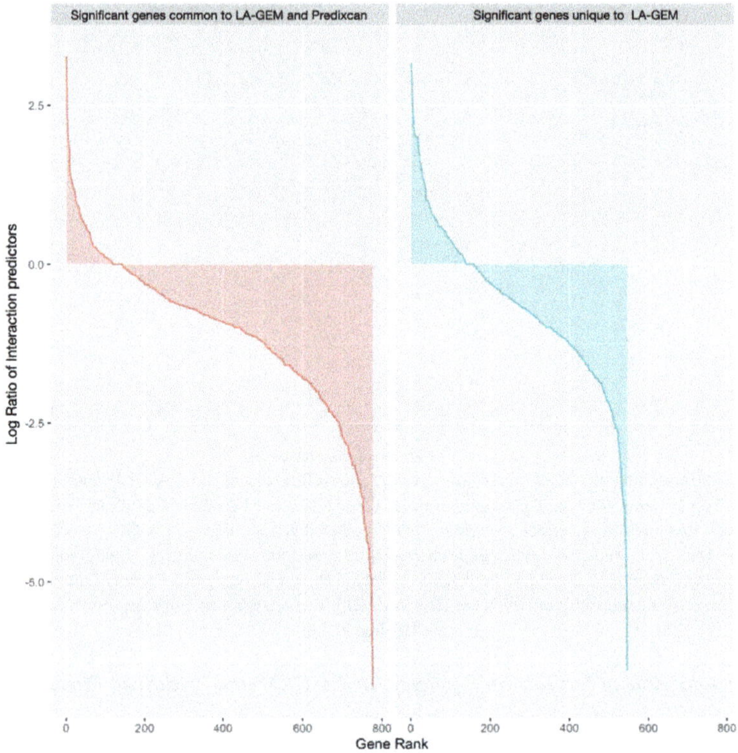

Fig. 5. Distribution of Positive Log Ratios of Count of Interaction Predictors in Genes Unique to LA-GEM and Common to LA-GEM and PrediXcan

3.6. *TWAS association to warfarin dose*

Using the IWPC warfarin cohort we imputed hepatocyte gene expression (restricted to those genes that were well imputed – N = 1325) and conducted a TWAS. The top associations are shown in the Manhattan plot (Fig. 6). Bonferroni corrected significant associations were found with 17 genes. No association was seen with known warfarin genes *VKORC1*, or *CYP2C9* as these gene were not well imputed in our models.

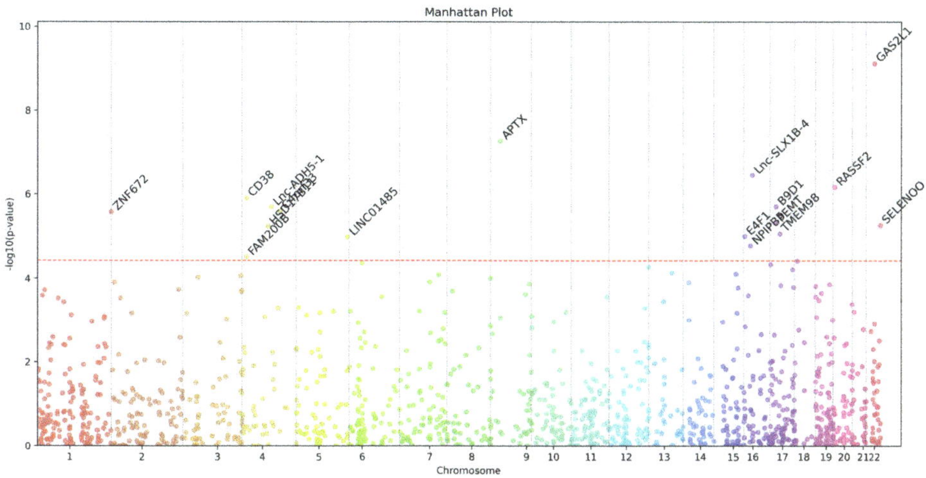

Fig. 6. Manhattan plot of TWAS results. The figure shows the association of imputed gene expression to stable warfarin dose in the IWPC cohort. The x-axis show the relative genomic position of each gene tested (N = 1323) and the y-axis show the Log(10) p-value. The red dashed line marked the threshold of significance for this study.

3.7. *TWAS association to PRU in patient taking clopidogrel.*

Using the ACCOuNT cohort, we imputed the hepatic gene expression in 180 African American patients on clopidogrel. We found no transcriptome-wide significant associations. However, one top association showed *RASA3* gene expression associated with increased PRU (p = 0.0014, Beta = 0.61). This gene has known association to platelet aggregation.[29]

4. Discussion

This study introduces a novel computational model, LA-GEM, designed to enhance gene expression prediction by integrating local ancestry (LA) predictors with cis-regional genetic variants. The development and deployment of such a model emerge from the understanding that complex trait prediction may be augmented by considering population-specific genetic variations. In many traditional models, such as PrediXcan, the unique genetic contributions of LA are not considered, potentially leading to overlooked associations.[2]

Our findings revealed that LA-GEM improved gene expression prediction compared to PrediXcan in some genes, suggesting that the inclusion of LA predictors can effectively supplement traditional cis-regional genetic variants. This improvement was demonstrated by the imputation of 1323 genes at a rho>0.1, p-value ≤ 0.05 by LA-GEM, compared to 1236 genes imputed by PrediXcan without considering LA.

Beyond these numbers, our study unveiled a set of 546 genes uniquely predicted by LA-GEM and 777 genes common in both LA-GEM and PrediXcan AA model. Out of 1323, 6 genes (*MME* and *LRRC37A2*, *CDK18*, *GREM2*, *COL26A1* and *ENSG00000281655*) were previously found to be associated with global West African ancestry and exhibited significant differential expression when compared to individuals of European descent.[12] These genes are not only statistically significant but also relevant to pharmacogenomics. For instance, *GREM2*, a gene involved in developmental processes[23,] is also associated with allopurinol efficacy[24]. and *MME*, implicated in neuropeptide degradation[25] and associated with ACE inhibitor-induced cough[26], were amongst the uniquely predicted genes. Lastly variants in *COL26A1* have been associated to Aspirin-intolerant asthma.[27]

Importantly, this study highlights the valuable implications of integrating LA predictors in gene expression models for drug response studies. By significantly predicting genes such as *CYP3A5, CYP1A1, CYP4F2, CBR1,* and *UGT2A1* - well-known contributors to drug metabolism and disease progression[30-33] - our model may aid in TWAS studies of inter-individual variations in drug responses and adverse drug reactions in African Americans. A particular emphasis should be placed on *CYP3A5*. This gene has been widely recognized for variability between different ethnic groups. The splice variant *CYP3A5*3*, associated with reduced enzyme activity, is less frequent in African populations, resulting in a functional enzyme in African populations. As most European carry the *CYP3A5*3,* the effect of this enzyme on drug response is not well accounted for in studies of European individuals. *CYP3A5* is thought to contribute to drug efficacy and toxicity, including responses to immunosuppressants such as tacrolimus.[34-35]

We applied LA-GEM to the African American warfarin and clopidogrel cohorts, demonstrating its utility in clinical studies. The warfarin cohort revealed 17 genes with significant associations with warfarin dose requirement, providing novel potential genetic influencers of warfarin dosage response beyond the well-known *VKORC1* and CYP2C9 genes[36-37]. The most significate TWAS hit was *GAS2L1 (*associated with increased warfarin dose requirement, $p = 7.7\times10^{-10}$), which has previously been associated with thrombocytopenia in women.[38] Also, the gene *SELENOO* on chromosome 22 showed association to decrease warfarin dose requirement ($p = 5.5\times10^{-6}$). A previous study in Sub-Saharan Africans found variants near this gene associated to increase R-6 Hydroxy-warfarin metabolite measurement.[39]

In the ACCOuNT clopidogrel cohort, we discovered an association between *RASA3* gene expression and increase P2Y12 Reactivity Units (PRU) level. While the most notable role of *RASA3* involves platelet function and hemostasis[29], this gene's function is not limited to platelets and the bloodstream. It is broadly expressed in many tissues, including the brain, lungs, and kidneys, suggesting it might have additional roles outside of platelets. In cancer biology, the Ras and RAP GTPases regulated by *RASA3* are often involved in tumorigenesis. For instance, inactivation of GAPs (like *RASA3*) can lead to overactive Ras signaling, which can contribute to the development of cancer.[40] This gene has also been associated to pulmonary hypertension in Sickle Cell Disease.[41]

In terms of computational efficiency, LA-GEM and PrediXcan showed similar performance during the model training phase. Specifically, for our limited dataset of 60 hepatocyte samples, both models completed the training within a time frame of approximately 2 to 3 hours. It's worth noting that the computational time is expected to scale linearly with the size of the sample pool, thus offering scalability as more comprehensive datasets become available.

Several innovative methods have set the stage in ancestry inform gene expression prediction. Notable among these are METRO[42], which enhances transcriptome-wide association studies (TWAS) through a likelihood-based inference framework, and MATS[43], which jointly analyzes samples from multiple populations to account for ancestral heterogeneity in gene expression effects. Additionally, a study by Lauren et al.[44] addressed the genetic architecture of gene expression across diverse populations, emphasizing the necessity for diverse population sampling in genomics. Despite their valuable contributions, none of these methods utilize SNP-based local ancestry as an intrinsic part of their predictive models. Our approach, LA-GEM, distinctively integrates SNP-based local ancestry predictors along with cis-regional variants to make more nuanced gene expression predictions. This unique aspect of LA-GEM not only adds a new layer of granularity to the existing methodologies but also paves the way for future explorations in this growing field.

While our findings are promising, there are several limitations to our study. First, we constructed the LA-GEM models with a limited cohort of 60 hepatocyte cultures. This is reflective of the overall lack of comprehensive multi-omics data in the African American population. With greater amounts of data on which to build these models, we will be better able to predict tissue specific patterns in the under-represented populations. This is also evident by the much greater number of well imputed gene available for the GTEx liver model (N = 3356) which is built on 153 liver samples. It should be noted that only 12 of these sample have any African Ancestry. Second, it is clear that there are still genes that are better predicted without the addition of LA. This suggests that to comprehensively use TWAS in African American population may require both LA-aware as well as traditional gene imputation methods. Lastly, the validation of LA-GEM in other tissues and larger cohorts remains a crucial next step. Ultimately, the incorporation of LA predictors can contribute significantly to personalized medicine, paving the way for treatments and interventions more attuned to a unique admixed genetic background of African Americans.

In conclusion, our study underscores the need for inclusion of LA in genomic methods. LA-GEM serves as a valuable tool in this endeavor, providing novel insights into the genomic architecture of complex traits in multiethnic populations, and highlighting the importance of considering local ancestry when predicting gene expression. The potential to uncover novel ancestry-specific gene associations can revolutionize our understanding of the interplay between genetics, disease, and therapeutic responses.

5. Acknowledgment

This work was made possible for through the following grants R01MD009217 (NIH, NIMHD), and R21HG011695 (NIH, NHGRI)

References

1. Popejoy, A. B., & Fullerton, S. M. (2016). Genomics is failing on diversity. Nature, 538(7624), 161–164. https://doi.org/10.1038/538161a
2. Gamazon ER, Wheeler HE, Shah KP, Mozaffari SV, Aquino-Michaels K, Carroll RJ, Eyler AE, Denny JC; GTEx Consortium; Nicolae DL, Cox NJ, Im HK. A gene-based association method for mapping traits using reference transcriptome data. Nat Genet. 2015 Sep;47(9):1091-8. doi: 10.1038/ng.3367. Epub 2015 Aug 10. PMID: 26258848; PMCID: PMC4552594.
3. Mikhaylova AV, Thornton TA. Accuracy of Gene Expression Prediction From Genotype Data With PrediXcan Varies Across and Within Continental Populations. Front Genet. 2019 Apr 3;10:261. doi: 10.3389/fgene.2019.00261. PMID: 31001318; PMCID: PMC6456650.
4. Xu Z, Wu C, Wei P, Pan W. A Powerful Framework for Integrating eQTL and GWAS Summary Data. Genetics. 2017 Nov;207(3):893-902. doi: 10.1534/genetics.117.300270. Epub 2017 Sep 11. PMID: 28893853; PMCID: PMC5676241.
5. Gusev A, Ko A, Shi H, Bhatia G, Chung W, Penninx BW, Jansen R, de Geus EJ, Boomsma DI, Wright FA, Sullivan PF, Nikkola E, Alvarez M, Civelek M, Lusis AJ, Lehtimäki T, Raitoharju E, Kähönen M, Seppälä I, Raitakari OT, Kuusisto J, Laakso M, Price AL, Pajukanta P, Pasaniuc B. Integrative approaches for large-scale transcriptome-wide association studies. Nat Genet. 2016 Mar;48(3):245-52. doi: 10.1038/ng.3506. Epub 2016 Feb 8. PMID: 26854917; PMCID: PMC4767558.
6. Hu Y, Li M, Lu Q, Weng H, Wang J, Zekavat SM, Yu Z, Li B, Gu J, Muchnik S, Shi Y, Kunkle BW, Mukherjee S, Natarajan P, Naj A, Kuzma A, Zhao Y, Crane PK; Alzheimer's Disease Genetics Consortium,; Lu H, Zhao H. A statistical framework for cross-tissue transcriptome-wide association analysis. Nat Genet. 2019 Mar;51(3):568-576. doi: 10.1038/s41588-019-0345-7. Epub 2019 Feb 25. PMID: 30804563; PMCID: PMC6788740.
7. Wainberg M, Sinnott-Armstrong N, Mancuso N, Barbeira AN, Knowles DA, Golan D, Ermel R, Ruusalepp A, Quertermous T, Hao K, Björkegren JLM, Im HK, Pasaniuc B, Rivas MA, Kundaje A. Opportunities, and challenges for transcriptome-wide association studies. Nat Genet. 2019 Apr;51(4):592-599. doi: 10.1038/s41588-019-0385-z. Epub 2019 Mar 29. PMID: 30926968; PMCID: PMC6777347.
8. Mancuso N, Shi H, Goddard P, Kichaev G, Gusev A, Pasaniuc B. Integrating Gene Expression with Summary Association Statistics to Identify Genes Associated with 30 Complex Traits. Am J Hum Genet. 2017 Mar 2;100(3):473-487. doi: 10.1016/j.ajhg.2017.01.031. Epub 2017 Feb 23. PMID: 28238358; PMCID: PMC5339290.
9. Purcell S, Neale B, Todd-Brown K, Thomas L, Ferreira MA, Bender D, Maller J, Sklar P, de Bakker PI, Daly MJ, Sham PC. PLINK: a tool set for whole-genome association and population-based linkage analyses. Am J Hum Genet. 2007 Sep;81(3):559-75. doi: 10.1086/519795. Epub 2007 Jul 25. PMID: 17701901; PMCID: PMC1950838.
10. Kang SJ, Larkin EK, Song Y, Barnholtz-Sloan J, Baechle D, Feng T, Zhu X. Assessing the impact of global versus local ancestry in association studies. BMC Proc. 2009 Dec 15;3 Suppl 7(Suppl 7):S107. doi: 10.1186/1753-6561-3-s7-s107. PMID: 20017971; PMCID: PMC2795878.

11. Zhong Y, De T, Alarcon C, Park CS, Lec B, Perera MA. Discovery of novel hepatocyte eQTLs in African Americans. PLoS Genet. 2020 Apr 20;16(4):e1008662. doi: 10.1371/journal.pgen.1008662. PMID: 32310939; PMCID: PMC7192504.
12. Park CS, De T, Xu Y, Zhong Y, Smithberger E, Alarcon C, Gamazon ER, Perera MA. Hepatocyte gene expression and DNA methylation as ancestry-dependent mechanisms in African Americans. NPJ Genom Med. 2019 Nov 25;4:29. doi: 10.1038/s41525-019-0102-y. PMID: 31798965; PMCID: PMC6877651.
13. Maples BK, Gravel S, Kenny EE, Bustamante CD. RFMix: a discriminative modeling approach for rapid and robust local-ancestry inference. Am J Hum Genet. 2013 Aug 8;93(2):278-88. doi: 10.1016/j.ajhg.2013.06.020. Epub 2013 Aug 1. PMID: 23910464; PMCID: PMC3738819.
14. Zhang, Y., Parmigiani, G., & Johnson, W. E. (2020). ComBat-seq: batch effect adjustment for RNA-seq count data. NAR genomics and bioinformatics, 2(3), lqaa078. https://doi.org/10.1093/nargab/lqaa078
15. GTEx Consortium, Laboratory, Data Analysis &Coordinating Center (LDACC)—Analysis Working Group, Statistical Methods groups—Analysis Working Group, Enhancing GTEx (eGTEx) groups, NIH Common Fund, NIH/NCI, NIH/NHGRI, NIH/NIMH, NIH/NIDA, Biospecimen Collection Source Site—NDRI, Biospecimen Collection Source Site—RPCI, Biospecimen Core Resource—VARI, Brain Bank Repository—University of Miami Brain Endowment Bank, Leidos Biomedical—Project Management, ELSI Study, Genome Browser Data Integration &Visualization—EBI, Genome Browser Data Integration &Visualization—UCSC Genomics Institute, University of California Santa Cruz, Lead analysts:, Laboratory, Data Analysis &Coordinating Center (LDACC):, NIH program management:, ... Montgomery, S. B. (2017). Genetic effects on gene expression across human tissues. Nature, 550(7675), 204–213. https://doi.org/10.1038/nature24277.1993).
16. Robinson MD, McCarthy DJ, Smyth GK. edgeR: a Bioconductor package for differential expression analysis of digital gene expression data. Bioinformatics. 2010 Jan 1;26(1):139-40. doi: 10.1093/bioinformatics/btp616. Epub 2009 Nov 11. PMID: 19910308; PMCID: PMC2796818.
17. Zhao S, Ye Z, Stanton R. Misuse of RPKM or TPM normalization when comparing across samples and sequencing protocols. RNA. 2020 Aug;26(8):903-909. doi: 10.1261/rna.074922.120. Epub 2020 Apr 13. PMID: 32284352; PMCID: PMC7373998.
18. Stegle O, Parts L, Piipari M, Winn J, Durbin R. Using probabilistic estimation of expression residuals (PEER) to obtain increased power and interpretability of gene expression analyses. Nat Protoc. 2012 Feb 16;7(3):500-7. doi: 10.1038/nprot.2011.457. PMID: 22343431; PMCID: PMC3398141.
19. Zou H, & Hastie T. Regularization and Variable Selection via the Elastic Net. Journal of the Royal Statistical Society, Series B (Statistical Methodology). 2005 67(2), 301–320. http://www.jstor.org/stable/3647580
20. Perera MA, Cavallari LH, Limdi NA, Gamazon ER, Konkashbaev A, Daneshjou R, Pluzhnikov A, Crawford DC, Wang J, Liu N, Tatonetti N, Bourgeois S, Takahashi H, Bradford Y, Burkley BM, Desnick RJ, Halperin JL, Khalifa SI, Langaee TY, Lubitz SA, Nutescu EA, Oetjens M, Shahin MH, Patel SR, Sagreiya H, Tector M, Weck KE, Rieder MJ, Scott SA, Wu AH, Burmester JK, Wadelius M, Deloukas P, Wagner MJ, Mushiroda T, Kubo M, Roden DM, Cox NJ, Altman RB, Klein TE, Nakamura Y, Johnson JA. Genetic

variants associated with warfarin dose in African-American individuals: a genome-wide association study. Lancet. 2013 Aug 31;382(9894):790-6. doi: 10.1016/S0140-6736(13)60681-9. Epub 2013 Jun 5. PMID: 23755828; PMCID: PMC3759580.

21. Friedman PN, Shaazuddin M, Gong L, Grossman RL, Harralson AF, Klein TE, Lee NH, Miller DC, Nutescu EA, O'Brien TJ, O'Donnell PH, O'Leary KJ, Tuck M, Meltzer DO, Perera MA. The ACCOuNT Consortium: A Model for the Discovery, Translation, and Implementation of Precision Medicine in African Americans. Clin Transl Sci. 2019 May;12(3):209-217. doi: 10.1111/cts.12608. Epub 2019 Feb 12. PMID: 30592548; PMCID: PMC6510376.
22. Galaviz-Hernández C, Lazalde-Ramos BP, Lares-Assef I, Macías-Salas A, Ortega-Chavez MA, Rangel-Villalobos H, Sosa-Macías M. Influence of Genetic Admixture Components on CYP3A5*3 Allele-Associated Hypertension in Amerindian Populations From Northwest Mexico. Front Pharmacol. 2020 May 11;11:638. doi: 10.3389/fphar.2020.00638. PMID: 32477124; PMCID: PMC7232668.
23. Kosinski C, Li VS, Chan AS, Zhang J, Ho C, Tsui WY, Chan TL, Mifflin RC, Powell DW, Yuen ST, Leung SY, Chen X. Gene expression patterns of human colon tops and basal crypts and BMP antagonists as intestinal stem cell niche factors. Proc Natl Acad Sci U S A. 2007 Sep 25;104(39):15418-23. doi: 10.1073/pnas.0707210104. Epub 2007 Sep 19. PMID: 17881565; PMCID: PMC2000506.
24. Brackman DJ, Yee SW, Enogieru OJ, Shaffer C, Ranatunga D, Denny JC, Wei WQ, Kamatani Y, Kubo M, Roden DM, Jorgenson E, Giacomini KM. Genome-Wide Association and Functional Studies Reveal Novel Pharmacological Mechanisms for Allopurinol. Clin Pharmacol Ther. 2019 Sep;106(3):623-631. doi: 10.1002/cpt.1439. Epub 2019 May 23. PMID: 30924126; PMCID: PMC6941886.
25. Roques BP, Noble F, Daugé V, Fournié-Zaluski MC, Beaumont A. Neutral endopeptidase 24.11: structure, inhibition, and experimental and clinical pharmacology. Pharmacol Rev. 1993 Mar;45(1):87-146. PMID: 8475170.
26. Morice AH, Fontana GA, Sovijarvi AR, Pistolesi M, Chung KF, Widdicombe J, O'Connell F, Geppetti P, Gronke L, De Jongste J, Belvisi M, Dicpinigaitis P, Fischer A, McGarvey L, Fokkens WJ, Kastelik J; ERS Task Force. The diagnosis and management of chronic cough. Eur Respir J. 2004 Sep;24(3):481-92. doi: 10.1183/09031936.04.00027804. PMID: 15358710.
27. Pasaje CF, Kim JH, Park BL, Cheong HS, Kim MK, Choi IS, Cho SH, Hong CS, Lee YW, Lee JY, Koh IS, Park TJ, Lee JS, Kim Y, Bae JS, Park CS, Shin HD. A possible association of EMID2 polymorphisms with aspirin hypersensitivity in asthma. Immunogenetics. 2011 Jan;63(1):13-21. doi: 10.1007/s00251-010-0490-8. Epub 2010 Nov 18. PMID: 21086123.
28. Raudvere U, Kolberg L, Kuzmin I, Arak T, Adler P, Peterson H, Vilo J. g:Profiler: a web server for functional enrichment analysis and conversions of gene lists (2019 update). Nucleic Acids Res. 2019 Jul 2;47(W1):W191-W198. doi: 10.1093/nar/gkz369. PMID: 31066453; PMCID: PMC6602461.
29. Stefanini L, Paul DS, Robledo RF, Chan ER, Getz TM, Campbell RA, Kechele DO, Casari C, Piatt R, Caron KM, Mackman N, Weyrich AS, Parrott MC, Boulaftali Y, Adams MD, Peters LL, Bergmeier W. RASA3 is a critical inhibitor of RAP1-dependent platelet activation. J Clin Invest. 2015 Apr;125(4):1419-32. doi: 10.1172/JCI77993. Epub 2015 Feb 23. PMID: 25705885; PMCID: PMC4396462.

30. Zanger UM, Schwab M. Cytochrome P450 enzymes in drug metabolism: regulation of gene expression, enzyme activities, and impact of genetic variation. Pharmacol Ther. 2013 Apr;138(1):103-41. doi: 10.1016/j.pharmthera.2012.12.007. Epub 2013 Jan 16. PMID: 23333322.
31. Caldwell MD, Awad T, Johnson JA, Gage BF, Falkowski M, Gardina P, Hubbard J, Turpaz Y, Langaee TY, Eby C, King CR, Brower A, Schmelzer JR, Glurich I, Vidaillet HJ, Yale SH, Qi Zhang K, Berg RL, Burmester JK. CYP4F2 genetic variant alters required warfarin dose. Blood. 2008 Apr 15;111(8):4106-12. doi: 10.1182/blood-2007-11-122010. Epub 2008 Feb 4. PMID: 18250228; PMCID: PMC2288721.
32. Lal S, Sandanaraj E, Wong ZW, Ang PC, Wong NS, Lee EJ, Chowbay B. CBR1 and CBR3 pharmacogenetics and their influence on doxorubicin disposition in Asian breast cancer patients. Cancer Sci. 2008 Oct;99(10):2045-54. doi: 10.1111/j.1349-7006.2008.00903.x. PMID: 19016765.
33. Court MH, Hao Q, Krishnaswamy S, Bekaii-Saab T, Al-Rohaimi A, von Moltke LL, Greenblatt DJ. UDP-glucuronosyltransferase (UGT) 2B15 pharmacogenetics: UGT2B15 D85Y genotype and gender are major determinants of oxazepam glucuronidation by human liver. J Pharmacol Exp Ther. 2004 Aug;310(2):656-65. doi: 10.1124/jpet.104.067660. Epub 2004 Mar 25. PMID: 15044558.
34. Staatz CE, Tett SE. Clinical pharmacokinetics and pharmacodynamics of tacrolimus in solid organ transplantation. Clin Pharmacokinet. 2004;43(10):623-53. doi: 10.2165/00003088-200443100-00001. PMID: 15244495.
35. Birdwell KA, Decker B, Barbarino JM, Peterson JF, Stein CM, Sadee W, Wang D, Vinks AA, He Y, Swen JJ, Leeder JS, van Schaik R, Thummel KE, Klein TE, Caudle KE, MacPhee IA. Clinical Pharmacogenetics Implementation Consortium (CPIC) Guidelines for CYP3A5 Genotype and Tacrolimus Dosing. Clin Pharmacol Ther. 2015 Jul;98(1):19-24. doi: 10.1002/cpt.113. Epub 2015 Jun 3. PMID: 25801146; PMCID: PMC4481158.
36. Johnson JA, Gong L, Whirl-Carrillo M, Gage BF, Scott SA, Stein CM, Anderson JL, Kimmel SE, Lee MT, Pirmohamed M, Wadelius M, Klein TE, Altman RB; Clinical Pharmacogenetics Implementation Consortium. Clinical Pharmacogenetics Implementation Consortium Guidelines for CYP2C9 and VKORC1 genotypes and warfarin dosing. Clin Pharmacol Ther. 2011 Oct;90(4):625-9. doi: 10.1038/clpt.2011.185. Epub 2011 Sep 7. PMID: 21900891; PMCID: PMC3187550.
37. Wadelius M, Chen LY, Downes K, Ghori J, Hunt S, Eriksson N, Wallerman O, Melhus H, Wadelius C, Bentley D, Deloukas P. Common VKORC1 and GGCX polymorphisms associated with warfarin dose. Pharmacogenomics J. 2005;5(4):262-70. doi: 10.1038/sj.tpj.6500313. PMID: 15883587.
38. Gnatenko DV, Zhu W, Xu X, Samuel ET, Monaghan M, Zarrabi MH, Kim C, Dhundale A, Bahou WF. Class prediction models of thrombocytosis using genetic biomarkers. Blood. 2010 Jan 7;115(1):7-14. doi: 10.1182/blood-2009-05-224477. Epub 2009 Sep 22. PMID: 19773543; PMCID: PMC2803693.
39. Asiimwe IG, Blockman M, Cohen K, Cupido C, Hutchinson C, Jacobson B, Lamorde M, Morgan J, Mouton JP, Nakagaayi D, Okello E, Schapkaitz E, Sekaggya-Wiltshire C, Semakula JR, Waitt C, Zhang EJ, Jorgensen AL, Pirmohamed M. A genome-wide association study of plasma concentrations of warfarin enantiomers and metabolites in sub-

Saharan black-African patients. Front Pharmacol. 2022 Sep 23;13:967082. doi: 10.3389/fphar.2022.967082. PMID: 36210801; PMCID: PMC9537548.

40. Vigil D, Cherfils J, Rossman KL, Der CJ. Ras superfamily GEFs and GAPs: validated and tractable targets for cancer therapy? Nat Rev Cancer. 2010 Dec;10(12):842-57. doi: 10.1038/nrc2960. Epub 2010 Nov 24. PMID: 21102635; PMCID: PMC3124093.
41. Prohaska CC, Zhang X, Schwantes-An TL, Stearman RS, Hooker S, Kittles RA, Aldred MA, Lutz KA, Pauciulo MW, Nichols WC, Desai AA, Gordeuk VR, Machado RF. RASA3 is a candidate gene in sickle cell disease-associated pulmonary hypertension and pulmonary arterial hypertension. Pulm Circ. 2023 Apr 1;13(2):e12227. doi: 10.1002/pul2.12227. PMID: 37101805; PMCID: PMC10124178.
42. Li Z, Zhao W, Shang L, Mosley TH, Kardia SLR, Smith JA, Zhou X. METRO: Multi-ancestry transcriptome-wide association studies for powerful gene-trait association detection. Am J Hum Genet. 2022 May 5;109(5):783-801. doi: 10.1016/j.ajhg.2022.03.003. Epub 2022 Mar 24. PMID: 35334221; PMCID: PMC9118130.
43. Knutson KA, Pan W. MATS: a novel multi-ancestry transcriptome-wide association study to account for heterogeneity in the effects of cis-regulated gene expression on complex traits. Hum Mol Genet. 2023 Apr 6;32(8):1237-1251. doi: 10.1093/hmg/ddac247. PMID: 36179104; PMCID: PMC10077507.
44. Mogil LS, Andaleon A, Badalamenti A, Dickinson SP, Guo X, Rotter JI, Johnson WC, Im HK, Liu Y, Wheeler HE. Genetic architecture of gene expression traits across diverse populations. PLoS Genet. 2018 Aug 10;14(8):e1007586. doi: 10.1371/journal.pgen.1007586. PMID: 30096133; PMCID: PMC6105030.

Cluster Analysis reveals Socioeconomic Disparities among Elective Spine Surgery Patients

Alena Orlenko*[1], Philip J. Freda*[1], Attri Ghosh[1], Hyunjun Choi[1], Nicholas Matsumoto[1], Tiffani J. Bright[1], Corey T. Walker[2], Tayo Obafemi-Ajayi[3], Jason H. Moore[1]

[1]*Department of Computational Biomedicine*
[2]*Department of Neurosurgery*
Cedars-Sinai Medical Center, Los Angeles, California, USA

[3]*Engineering Program*
Missouri State University, Springfield, Missouri, USA

This work demonstrates the use of cluster analysis in detecting fair and unbiased novel discoveries. Given a sample population of elective spinal fusion patients, we identify two overarching subgroups driven by insurance type. The Medicare group, associated with lower socioeconomic status, exhibited an over-representation of negative risk factors. The findings provide a compelling depiction of the interwoven socioeconomic and racial disparities present within the healthcare system, highlighting their consequential effects on health inequalities. The results are intended to guide design of fair and precise machine learning models based on intentional integration of population stratification.

Keywords: clustering; fairness; equity; explainability; feature importance; informatics.

1. Introduction

Advances in machine learning (ML) technologies paralleled with increased clinically relevant data availability have led to major progress in precision medicine over the past decade.[1] Data-driven solutions, particularly ML methods, are becoming integral to personalized predictive medicine as they can inform clinical decision support systems, generate accurate patient risk stratification models, and contribute to intelligent guideline development using high-dimensional complex medical data.[2] Indeed, ML-based approaches have generated robust predictive models in the diagnoses of several diseases such as cardiovascular diseases,[3] type II diabetes,[4] and early-stage Alzheimer's disease[5] and for post-surgical outcomes and treatment response in several procedures including cardiac surgery[6] and spinal surgeries.[2,7,8] Thus, clinicians can utilize this information to evaluate risk of poor diagnoses and adverse outcomes, assisting clinical decision making by providing personalized assessments of the benefits and consequences related to undergoing or delaying invasive procedures.

The rates of spine surgery, an invasive procedure, have been steadily increasing over the past few decades.[9] With the proportion of the elderly population projected to dramatically

*These authors contributed equally to the paper.

This work was supported by National Institutes of Health (USA) grant R01 LM010098.

increase in the coming years, utilization of spinal procedures is expected to follow as degenerative spine conditions become more prevalent.[10] Spinal fusions generally require extensive muscle dissection and reconstruction of the spinal column, which typically necessitates significant post-operative opioid consumption and comes with considerable post-operative risks.[11] With the potential for long recovery periods and the risk of the development of opioid dependency as a result of these surgeries, outcome prediction in spinal fusion surgeries has become an important area of research. To accurately predict outcomes, it is crucial to consider patient diversity, which stems from various sources, including but not limited to biological, societal, environmental, and psychosocial factors.[12] These sources of diversity can result in significantly different outcomes, ultimately affecting a patient's long-term quality of life after surgery.

For data-driven predictive models to become widely and safely adopted in clinical settings, key research challenges still remain to be resolved. These include assessing clinical heterogeneity and avoiding bias in decision-making. Complex ML algorithms have an inherent tendency for biased decisions that disproportionately impact underrepresented demographic groups leading to possible discriminatory outcomes.[13] This concern is frequently overlooked in study design, resulting in unequal treatment of minority individuals.[14] We seek to examine the intricate heterogeneity in clinical data to identify any differential patient subgroups, if present. This will enable us to mitigate bias in the ML decision-making for clinical systems.

Cluster analysis has been applied in a wide range of applications as an exploratory tool to enhance knowledge discovery.[15–17] It can help by identifying more homogeneous subgroups for effective ML models. The goal is to detect and characterize novel sub-types that exhibit differing clinical patterns and/or outcome trajectories that may benefit from different treatment options. Ultimately, the validity of any sub-grouping paradigm depends on whether the resulting sub-groups uncover/expose some biologic or genetic variation, which can be used to predict prognoses, recurrent risks, or treatment responses. However, most of the approaches employ a single clustering algorithm with limited explainability.[15–17] To overcome these limitations, we introduce a novel clustering framework to examine and characterize a cohort of patients that have undergone elective spinal fusion surgery at Cedars-Sinai Medical Center.

2. Data

The dataset consists of electronic health records (EHR) of 5,214 elective spinal fusion (ESF) surgery procedures derived from 4,930 patients (ages 18-85) at the authors' single institution from 2013 to 2022. Only patients who survived after surgery, with two or fewer procedures are included. If the second procedure was conducted within seven days of the first, the most recent is retained. Patients with a second procedure conducted after seven days but less than a year apart are excluded. Forty-five features from the patient's health records were selected and integrated in the cluster analysis. These features span baseline characteristics/demographics, pre-surgery clinical labs, vitals, medication lists, past medical history, post-operative care, and social status, as guided by domain expert (C.T.W.).

The race feature consolidates both self-reported race and ethnicity information. Self-reported ethnicity of "Hispanic", regardless of race, is represented as "Hispanic". Race designation of "Asian" or "Native Hawaiian or other Pacific Islander" are categorized as

"Asian/Pacific Islander". "Native American or Alaska Native", "Other", "Patient declined", "Unknown", or missing are all consolidated as "Other". Social status features include insurance type, marital status, smoking, and alcohol use. Patients with commercial or private insurance are grouped as "commercial" while Medicare, California's Medicaid program (Medi-Cal), or all other government insurance are categorized as "medicare". Vitals features include systolic blood pressure (SBP), body mass index (BMI) and pain score. We include the most frequently used lab value results from the EHR that had less than 50% of missing data (11: hemoglobin, white blood cell (WBC) count, red blood cell (RBC) count, platelet count, potassium, sodium, chloride, blood urea nitrogen (BUN), creatinine, calcium, and blood type) . Selected post-operative care features are discharge disposition, length of stay, and readmission status. Past medical history (PMH) features (yes/no) are derived by aggregating the ICD codes relevant to specific conditions of interest (metabolic, anxiety, chronic pain, mood, headache, nicotine, other psychiatric, opioid substance use disorder (SUD), alcohol SUD, cannabis SUD, and other SUD). Medication list features are derived based on usage of medications under 7 broad categories, as defined by the domain expert. These include muscle relaxers, non-opioid analgesic, psychiatric, sleep, medication-assisted treatment, gabapentinoids, and "other".

The summary of the baseline characteristics is presented in Table 1. For a complete list of the medications that map to each medication feature as well as the ICD codes that map to each PMH feature, see supplementary file [a]. Data request approved by the Cedars-Sinai Honest Enterprise Research Brokers (HERB) committee. This research study was carried out under the guidelines and approval of the Cedars-Sinai Institutional Review Board.

Table 1. Demographic summary of elective spinal fusion surgery patient sample (n = 5,214).

Characteristic	Distribution
Age	median: 67 range: 18 - 85; 65+: 57.59%
Gender	Male: 46.47% Female: 53.53%
Race	White: 75.66%, Hispanic: 10.32%, Black/African-American: 6.75%, Asian: 3.55%, Other: 3.72%
Insurance type	Medicare: 45.09% (65+: 88.74%) (Medicare: 96.85%, Medi-Cal: 0.02%, Other government: 0.01%) Commercial: 53.93% (65+: 31.80%), No Insurance: 0.98%
Marital status	Single: 17.97%, Married: 63.57%, Divorced: 9.51%, Widowed: 6.23%, Significant other: 2.51%, Unknown: 0.21%

3. Methods

To ensure a fair and unbiased model, we propose a robust automated system that integrates multiple clustering algorithms, ensemble internal validation metrics, automated ML (autoML)-driven explainability, and post-hoc univariate statistical analysis.

The data curation steps involve the detection of erroneous, non-biologically plausible values, and/or outliers. Domain expert guidance in conjunction with outlier analyses are applied to ensure mitigation of potential bias and possible human data entry errors. These values are dropped and imputed, rather than dropping the entire sample. Missing values are imputed using the multivariate feature imputation (*IterativeImputer* method in Python).[18] All 45 fea-

[a]Supplementary information is available at: https://github.com/EpistasisLab/PSB2024_spine/

tures are not highly intercorrelated as evident from passing the correlation filter analysis using the Pearson and Spearman rank correlations (≤ 0.85).

We perform an automated clustering method that incorporates hyperparameter sampling across various algorithms that permutes the distance type (Euclidean, Manhattan), and number of clusters (k=[2:10]), when applicable. It exploits five individual algorithms (Spectral, Agglomerative, k-means, Birch, and Gaussian mixture).[19] We also conduct an ensemble clustering model that leverages these individual methods using the mixture model consensus metric in *OpenEnsemble*.[20,21] Our model includes TooManyCells (TMC) spectral hierarchical clustering method,[22] for a total of seven methods with 68 permutations. To integrate TMC into the automated clustering pipeline, we implement an extension that aggregates cluster labels with multiple terminal cluster nodes starting at the root node. The depth of the tree partition serves as a TMC hyper-parameter. The optimal clustering output is determined using the ensemble internal validation metric model introduced by Nguyen et al..[23] The model assigns a final score based on a consensus of five metrics (Calinski-Harabasz, Davies-Bouldin, Silhouette score, $\mathcal{I}$, and Xie-Benie).[24] Each metric ranks its top 15 results and sets the remainder to zero. The ensemble model assigns a final overall rank score to each clustering outcome based on the weighted sum of the individual ranking assignment of each metric.

Key novelty of our clustering framework is that we utilize a model-agnostic approach to evaluate the feature importance and assess which key discriminant features are driving cluster separation with an autoML tool, TPOT.[25] TPOT evaluates the informative contributions of features to clustering results by predicting cluster labels with each feature independently. In contrast to the current state-of-the-art methods for evaluating feature importance (such as SHapley Additive exPlanation,[26] Permutation feature importance, Gini impurity in Random Forest[27]), TPOT overcomes the single model limitation as it searches and optimizes across multiple ML algorithms. For each feature, we run the TPOT optimization (across 13 different classifiers configuration), and extract the best-performing model performance as the feature importance metric. This provides insight into the key discriminant input features and guides the next steps of analysis. Visualization of results is performed using ISOMAP[28] and TMC dendograms. Code for all the methods are available at (https://github.com/EpistasisLab/PSB2024_spine).

Univariate global statistical tests are conducted, as post-hoc analyses, to assess which features exhibit differences among the cluster groups. The method of analysis differs depending on the measurement scale of the feature. Features with significant test results suggest utility in clustering. For continuous features, we test for normality using Shapiro-Wilk tests. All features are non-normally distributed. Thus, we employ non-parametric Mann-Whitney tests (or Kruskal-Wallis tests in case of multiple groups). For categorical and binomial features, we use Chi-square tests of independence. The resulting p-values of these tests are corrected for multiple testing using the Benjamini-Hochberg procedure.

4. Results

4.1. *Entire ESF sample is stratified by socioeconomic factor of insurance.*

Upon evaluating ensemble clustering on the overall cohort of 5,214 surgeries, Table 2A shows k-means with two clusters consistently outperforms other methods across internal validation

Table 2. Top ranked results for 1st and 2nd order clustering based ensemble validation rank scores.

Output [Cluster sizes]	CH (rank)	Db (rank)	I (rank)	Sil (rank)	Xb (rank)	Overall Rank
A. Clustering on entire cohort						
*k*means-2 [2852, 2362]	550.01 (15)	3.03 (14)	0.619 (15)	0.098 (15)	2.348 (15)	74
GaussianMixture-2 [2732, 2482]	549.56 (14)	3.03 (15)	0.619 (14)	0.097 (14)	2.348 (14)	71
Spectral (euclidean)-2 [2863, 2351]	533.62 (13)	3.08 (11)	0.597 (13)	0.095 (13)	2.436 (13)	63
B. 2nd order clustering on C1 group						
*k*means-2 [1872, 980]	202.13 (15)	3.57 (0)	0.398 (15)	0.089 (14)	3.180 (3)	47
Spectral (manhattan)-2 [1931, 921]	168.17 (13)	3.86 (0)	0.341 (14)	0.081 (13)	3.705 (0)	40
Mixture model-2 [1638, 1214]	168.23 (14)	4.06 (0)	0.305 (13)	0.074 (12)	4.142 (0)	39
C. 2nd order clustering on C2 group						
*k*means-2 [1476, 886]	204.92 (15)	3.27 (11)	0.503 (15)	0.094 (15)	2.700 (15)	69
GaussianMixture-2 [1474, 888]	204.88 (14)	3.27 (10)	0.503 (14)	0.094 (14)	2.702 (14)	64
Mixture model-2 [1473, 889]	195.48 (13)	3.36 (1)	0.480 (13)	0.094 (13)	2.834 (13)	54

metrics. Top ranking methods (*k*-means, Gaussian Mixture, spectral, TooManyCells) return similar 2-cluster partitions and display high consistency as top performers across all five metrics. Subsequent analyses are conducted on the *k*-means-2 result (C1 and C2). The visualization of the subgroups is shown using both ISOMAP (Figure 1(a)) and TMC (Figure 1(b)). Note: TMC performs its embedded technique (spectral hierarchical clustering) prior to visualization, hence, not representing C1 and C2 separation exactly. TPOT feature importance analysis reveals that insurance type, a potential socioeconomic factor, is most important to cluster separation explainability (100% balanced accuracy (B-Acc.)). Age, discharge disposition, and PMH metabolic are of less importance (79.1%, 64.2%, 62.7% B-Acc. respectively). Mapping the insurance type label with TMC dendrograms confirms this as well (Figure 1). Cluster C1 consists of all patients with "commercial insurance" and 40 of "no insurance" while C2 has all patients on "medicare insurance" and 11 with "no insurance".

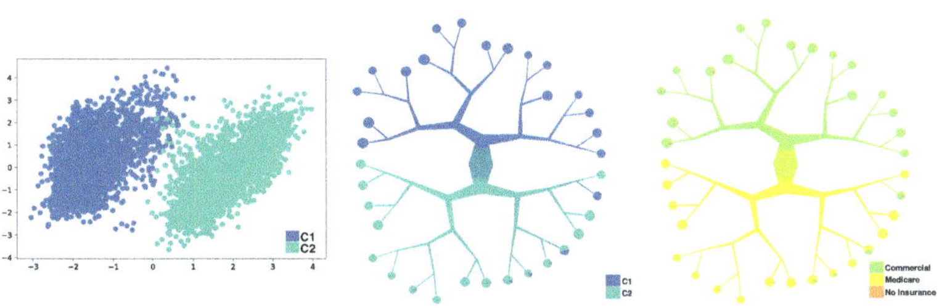

(a) ISOMAP representation (b) C1 & C2 using TMC (c) TMC with insurance

Fig. 1. Visualization of *k*-means-2 results on entire cohort.

Age is a determinant for medicare eligibility (65+) in the USA. Thus, we conduct univariate statistical analyses between C1 and C2 (insurance-driven clusters) as well as between and within age-stratified subgroups. Figure 2 illustrates the experimental design of these analyses. The pairwise comparisons are conducted as follows: Exp 1: C1 vs. C2; Exp 2: 65+ subgroups of C1 and C2 i.e., C1 $\geq$ 65 vs. C2 $\geq$ 65; Exp 3: 65- subgroups of C1 and C2 i.e., C1 < 65 vs. C2 < 65; Exp 4: within C1: C1 $\geq$ 65 vs. C1 < 65; Exp 5: within C2: C2 $\geq$ 65 vs. C2 < 65.

4.1.1. *Univariate analysis reveals health disparities associated with insurance types.*

Figure 3 summarizes the key features that differ significantly at the entire cohort level between C1 and C2 and when age-stratified (Exp 1, 2, and 3). Nine features display age-independence as they are statistically different across all three comparisons (Figure 3a). These are race, marital status, discharge disposition, hemoglobin, platelet count, RBC count, potassium, and two PMH features (metabolic and anxiety). We also observe that there are some features that are not different between C1 and C2 (Exp 1), but do exhibit significant differences within the 65- comparisons (Exp 3) (Figure 3b). These features (PMH features of pain score, other psychiatric disorders, nicotine use, headache, other SUDs, and use of non-opioid analgesics) imply some possible health disparities between the two socioeconomic driven groups after accounting for the age factor. (Note, an additional significant feature, PMH of other SUD, isn't shown in the figure, as it affects less than 5% of the overall population.) There are no features that are significant only between 65+ subgroups (Exp 2) and not at the entire cohort level (Exp 1). See Supplementary file [b] for complete details of all the pairwise comparisons.

The analysis also reveals some features that are significant across all three comparisons (Exp 1, 2 and 3), which are also significant within C1 and C2 when stratified by age (Exp 4 and 5). These include race, platelet count, RBC count, marital status, discharge disposition, and PMH features of metabolic and anxiety (see Supplementary file[b]). Features such as hemoglobin are significant within C1 (Exp 4) but not C2 (Exp 5). All PMH features are significantly different within C2 age-stratified groups. Overall, negative health factors, such as lower hemoglobin, RBC, platelet count, potassium levels, and higher incidence of metabolic disease and anxiety are associated with C2, indicating socioeconomic health disparities.

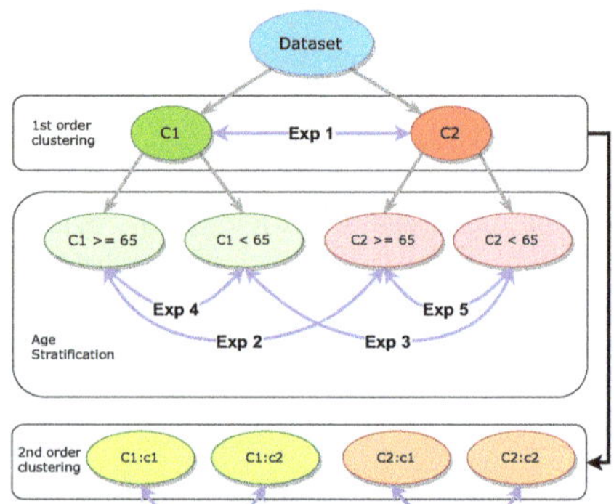

Fig. 2. Experimental design of cluster analyses and pairwise comparisons.

4.1.2. *Adverse outcomes are disproportionately observed in minority racial groups.*

From the pairwise comparisons (Exps 1-5), race is consistently significant. The 65- population in C2 had a larger proportion of non-white patients (60% compared to 73% in C1), with the disparity being most prominent in the Black/African-American demographic with a wide

[b]Supplementary information is available at: https://github.com/EpistasisLab/PSB2024_spine/

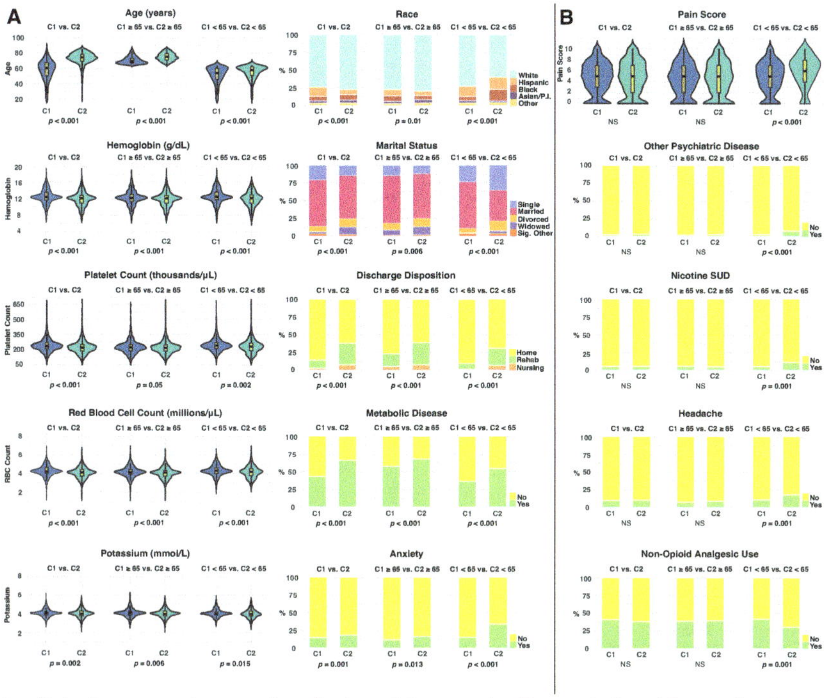

Fig. 3. Pairwise comparison results of selected features for Exp 1, 2, & 3 (C1 vs. C2; C1 $\geq$ 65 vs. C2 $\geq$ 65; C1 $<$ 65 vs. C2 $<$ 65) significant across all in (A), and only for Exp 3 in (B).

percentage gap of 16% vs 5.7% (Figure 3). Given a predominantly White cohort, it is important to highlight that complex ML models may inadvertently neglect pattern associations within minority classes. We recognize the importance of deeper exploration into race since our clustering model could potentially marginalize significant patterns linked to minority groups. This section further examines race-related differentiation at both cohort and cluster levels.

We observe significant differences for post-operative care outcomes (discharge disposition, length of hospital stay (LOS), and readmission rate) between race groups in multiple comparisons (Figure 4). At the entire cohort level, Blacks exhibit a higher proportion of adverse outcomes in all scenarios (see Figure 4). The "Other" group (Native American or Alaskan Native, Other, patient declined, and unknown) also demonstrates increased rates of adverse outcomes for discharge disposition and LOS. We subsequently examine the cluster and age-stratified groups to identify whether the adverse outcome over-representation in Blacks and "Other" remain independent of insurance and age. Likewise, for readmission rate and discharge disposition, the higher adverse outcome effect remains significant in C2, specifically in the 65+ subgroup. However, LOS is independent of race in C2 as adverse outcomes become

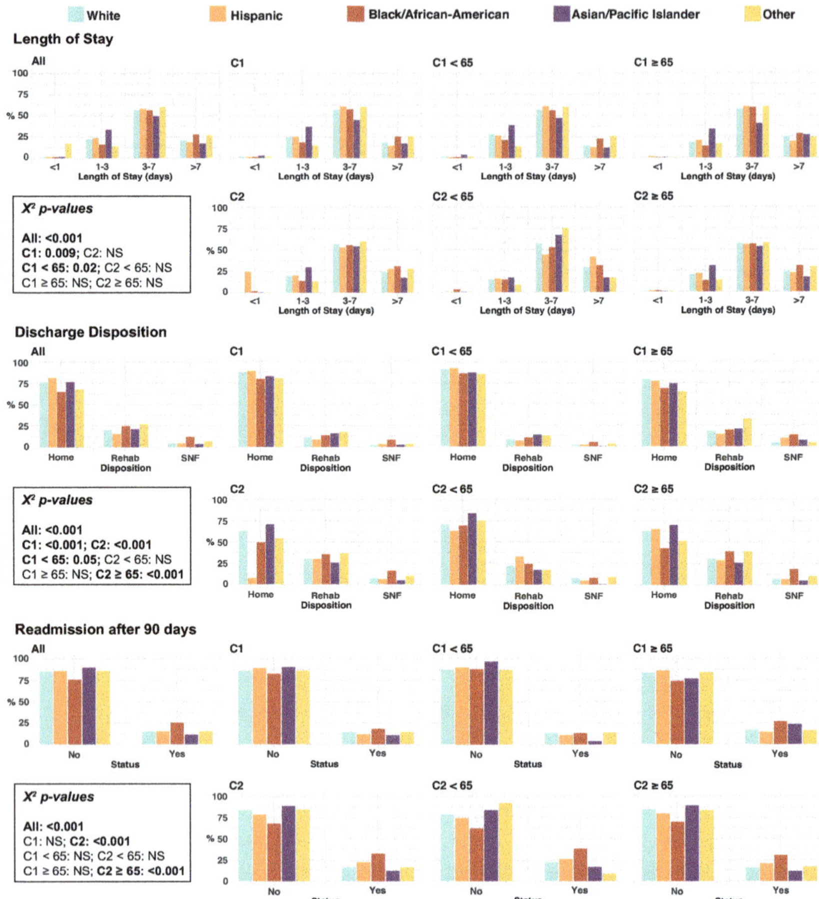

Fig. 4. Pairwise comparisons of clinical outcomes across subgroups by race.

more prominent for all groups, likely denoting a combined effect of socioeconomic disparities and advanced age. Race appears to also be an important factor in C1 with Blacks and "Other" having higher LOS (> 7 days) and discharges to other than home compared to other groups. These results, although limited due to small non-white sample sizes, indicate that race is an important discriminant of health outcomes for ESF surgery.

4.2. *Second-order clustering reveals clinical and demographic heterogeneity*

Given the overwhelmingly distinct clusters driven by socioeconomic factors, we reiterate the automated clustering on C1 and C2 separately to further examine the insurance-associated heterogeneity. This is denoted as second-order clustering (Exp 6 and 7) in Figure 2.

The top-ranking clustering results for both experiments are illustrated in Table 2B and C. We observe that in both instances, the k-means-2 result is the most optimal method. For C2, all the high-ranking algorithms unanimously identified 2-cluster solutions with minor size distribution differences. In contrast for C1, though the 2-cluster solution is the best method overall, there is more variance among the metrics. Visual inspection of ISOMAP decomposition and TMC dendrograms with cluster labels confirm that C2 clusters (C2:c1 and C2:c2) display more separation compared to C1 (C1:c1 and C1:c2) (Figures 5 and 6).

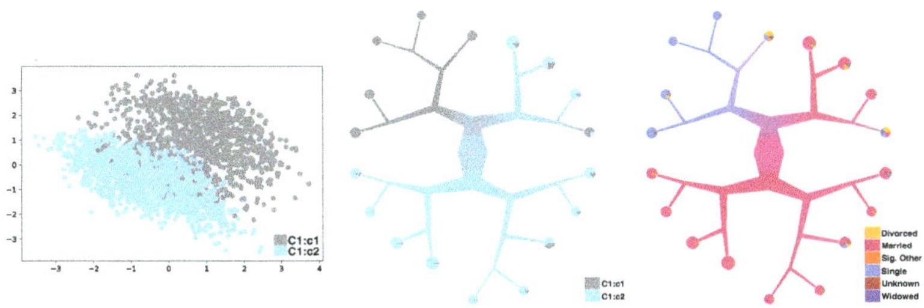

(a) ISOMAP representation (b) TMC with $k$means-2 (c) Marital status TMC

Fig. 5. Optimal clustering result on C1 subgroup:kmeans-2 optimal result.

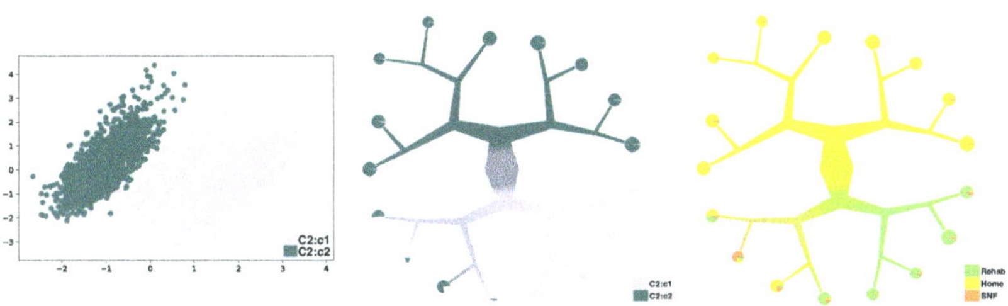

(a) ISOMAP representation (b) TMC with $k$means-2 (c) Discharge TMC

Fig. 6. Optimal clustering result on C2 subgroup: $k$means-2 optimal result.

TPOT feature importance analysis identifies marital status as highly discriminant for C1:c1 and C1:c2 groups, and discharge disposition for C2:c1 and C2:c2, both with 100% B-Acc. For C1, Age trails with 57.4% B-Acc. For C2, LOS, hemoglobin, and readmit predicts label with 64.2%, 61.8%, and 60.9% B-Acc. respectively. This is illustrated using the TMC dendrograms overlaid with the discriminant features in Figures 5(c) and 6(c). C1:c2 consists entirely of all married patients while C1:c1 contains all others. In C2, the two clusters (C2:c1 and C2:c2) are stratified primarily by discharge disposition. C2:c1 (n = 886) consists mainly of patients discharged to rehab and skilled-nursing facilities (SNF) while C2:c2 (n = 1,476) is comprised of almost all home discharge patients (99.86%). We also observe that the second-

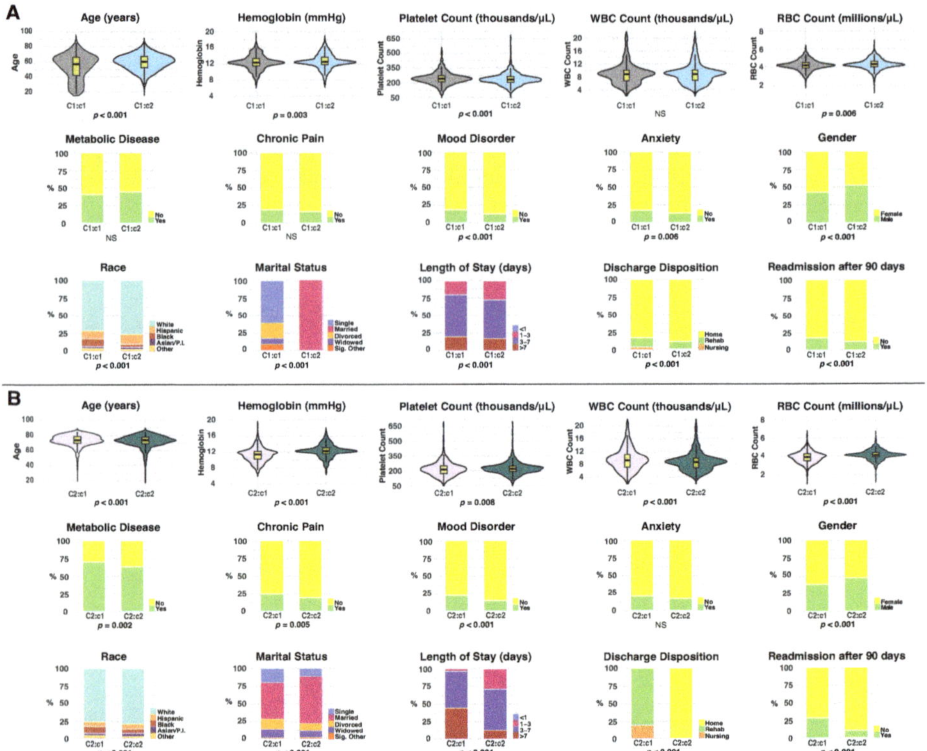

Fig. 7. Selected significant features from univariate analysis of pairwise comparisons on second-order clustering on (A) cluster C1 (Exp 6) and (B) cluster C2 (Exp 7).

order clustering yields subgroups of disproportionate sizes (large vs. small) compared to the first-order clustering.

From univariate analysis results (Figure 7), statistically significant differences are observed for both comparisons (Exp 6 and 7) for age, race, gender, discharge disposition, readmission, LOS, platelet count, RBC count, hemoglobin, BUN, creatinine, chloride, calcium, sodium, and PMH features of anxiety and mood of which selected features are illustrated in 7. Overall, we observe that C2 displays a higher level of complexity and divergence. The features that drive the C2:c1 vs. C2:c2 divergence are LOS $>$ 7 days (44% vs 13%), readmission rate (29% vs 11%), and lower median hemoglobin values (11.4 vs. 12.3) (Figure 7B).

5. Discussion

In this study, we elaborate our commitment towards constructing equitable and unbiased ML models. Our initial intention was the development of a predictive model specific to elective spine fusion surgery, however, during the course of our investigation, we identified the necessity for deeper understanding of potential disparities present within our dataset to more accurately address clinical inquiries. The manifestation of bias within ML algorithms through data sources has been substantially highlighted in prior literature.[13,29,30] To combat this, we employ a robust automated multiple clustering approach to scrutinize our dataset for potential bias factors, prior to developing an ML model. Investigation of subpopulation structure in clinical cohorts is an important area of research and has significant implications for patient care and treatment. However, the methodologies used in most studies[15–17] are limited in that they usually implement a single clustering technique without conducting exploratory investigations of their results, potentially overlooking components driving heterogeneity. Our framework addresses these shortcomings by employing automated cluster analysis with hyperparameter tuning and a multi-metric performance score. The framework, enhanced by autoML-driven feature importance estimation along with univariate analysis, allowed us to uncover and explain drivers of population divergence. We demonstrate its capabilities in uncovering inherent patterns of heterogeneity in patients undergoing ESF, an invasive medical procedure that is associated with risks of many adverse outcomes.[11]

The cluster analysis uncovers two diverse subgroups (C1 and C2), each exhibiting unique characteristics, driven mainly by socioeconomic factors (insurance type and race). It is important to note that the entire ESF sample is almost evenly split between insurance types (54% commercial insurance). This indicates increasing equity of access as patients with medicare coverage have historically experienced limited access to certain medical procedures, including elective spinal fusion.[10] However, disheartening but not surprising, is the observed significant health disparities in the cohort driven by socioeconomic factors. Similarly, there are several recent studies[31,32] highlighting that racial minorities, and those with lower socioeconomic status, are at higher risk of adverse outcomes. The C2 subgroup contains all medicare insurance patients and is characterized by an increased proportion of minority groups compared to C1, though the overall sample is primarily White (Table 1). C2 patients have higher occurrences of non-home discharge dispositions, clinically remarkable past medical histories, especially with respect to metabolic-related diseases and anxiety, as well as clinical lab values associated with poor prognoses (Figure 3). In particular, the under 65 C2 patients (266) have significantly higher pain scores and a higher prevalence of nicotine substance abuse, headaches, other psychiatric disorders, and conditions already noted (metabolic and anxiety). These characteristics are not surprising, however, what is notable is that the socioeconomic factor of insurance overwhelms the clustering results, compelling us to adjust for it prior to characterizing the underlying heterogeneity with second-order clustering on C1 and C2 separately.

Both C1 and C2 contain one of two sub-clusters that are smaller and associated with poor health outcomes (C1:c1 and C2:c1). Interestingly, C1 is stratified by marital status with C1:c2 consisting of all married patients while C1:c1, its adverse outcome subcluster, is made up of all other marital status groups (Figure 5(c)). C2 is stratified by discharge disposition. C2:c1, its adverse outcome group, consists of almost all non-home discharged patients (99.86%) (Figure 6(c)). Despite the unique characteristics that differentiate the adverse outcome subclusters (C1:c1 and C2:c1), they share striking similarities as both are comprised of patients presenting suboptimal values of numerous labs, PMH of mood disorder, poor outcomes (LOS, discharge, and readmission), and higher proportions of minority patients (Figure 7). Though similarities exist, the proportions of patients with negative indicators of health (and their magnitudes) are greater in C2:c1 compared to C1:c1 (Figure 7). This is also true for race as C2:c1 has a higher proportion of minority patients. This aligns with the validation metrics analysis (Table 2) which indicates more separation in C2 compared to C1. In addition, C2:c1 has significantly suboptimal WBC count, PMH of metabolic and chronic pain, and use of gabapentin while C1:c1 has more prominence of PMH of anxiety, alcohol, other psychiatric disorders, nicotine use, and other SUDs (Figure 7). We acknowledge that these characteristics are probably due to a combination of social, environmental, and biological factors. However, interestingly, overall better prognoses are strongly associated with "married" status (Figure 7A).

The conspicuous racial partitioning observed at both levels of clustering highlights the importance of conducting thorough exploratory analysis and incorporation of fair algorithms in ML. The race-stratified analysis further validates findings on existing socioeconomic disparities within the ESF sample, especially for post-surgery event outcomes (Figure 4). All relatively poor outcome subgroups (C2 as a whole, under 65 age-stratified cohort in C2, C1:c1, and C2:c1) have significantly more minority patients (Figures 3,7). Interestingly, the over-representation of Blacks and "Other" are similar in both C1:c1 and C2:c1 (Blacks: $\approx$10% and "Other": $\approx$4.5% (which includes Native Americans)). This is concerning given the overall low percentage of Blacks (6.75%) and "Other" (3.72%) in the entire sample. Note that "Other" also includes self-reported race entries of "Other", "patient declined", and "Unknown", which are often associated with privacy, self-identity/profiling, and trust concerns.[33] Constructing "Other" with Native-Americans, Alaskan Natives, and individuals with no reported race is not optimal and was done due to small sample sizes. Nevertheless, identifying higher proportions of these individuals in the adverse risk clusters is likely driven by cumulative disparity factors associated with these groups. These implications are important as identifying patients with needs for specialized care could lead to substantial improvements in clinical outcomes.

Complex pattern recognition models can sometimes overlook minority groups due to imbalanced data, potentially leading to biased results and unfair outcomes.[13] Here, we showcase a framework that mitigates these issues by incorporating information about heterogeneous subgroups into the clinical risk score model. With thorough evaluation and validation, our discovery from clustering results has the potential to be actionable in clinical settings, allowing diverse groups of patients and clinicians to receive more precise estimates of treatment success and risk of developing adverse effects. This approach can be transferred to other domains that require clinical decision support. Moreover, as we observe racial and socioeconomic indicators playing key roles in explaining disproportional adverse effect distribution, it is important to continue advocating for more fair healthcare policies, especially for preventative care access. By identifying socioeconomic status and race as significant determinants of health outcomes, our two-tier approach averts a potential scenario of introducing health disparities due to algorithmic bias. We are enthusiastic about the development and deployment of our methodology in predictive modeling in clinical settings to assist surgeons and patients in real-time decision-making regarding the most efficacious ESF surgery options. These clusters could be utilized in a sampling scheme to mitigate bias in ML models aimed at predicting outcomes, by incorporating feature engineering based on the cluster labels into the model as well as exploring risk score ML models with discovered population stratification. This study presents a compelling illustration of the heterogeneity within the healthcare system and underscores the need for personalized medicine as a strategic approach to enhance healthcare and reduce health disparities. Therefore, we strongly advocate for others to employ a similar rigorous approach to data integration in order to better comprehend potential biases.

References

1. P. Rajpurkar, E. Chen, O. Banerjee and E. J. Topol, AI in health and medicine, *Nat Med* **28**, 31 (Jan 2022).
2. A. Goyal, C. Ngufor, P. Kerezoudis, B. McCutcheon, C. Storlie and M. Bydon, Can machine learning algorithms accurately predict discharge to nonhome facility and early unplanned readmissions following spinal fusion? analysis of a national surgical registry: Presented at the 2019 aans/cns section on disorders of the spine and peripheral nerves, *Journal of Neurosurgery: Spine* **31**, 568 (2019).
3. C. Krittanawong, H. U. H. Virk, S. Bangalore, Z. Wang, K. W. Johnson, R. Pinotti, H. Zhang, S. Kaplin, B. Narasimhan, T. Kitai *et al.*, Machine learning prediction in cardiovascular diseases: a meta-analysis, *Scientific reports* **10**, p. 16057 (2020).
4. U. Ahmed, G. F. Issa, M. A. Khan, S. Aftab, M. F. Khan, R. A. Said, T. M. Ghazal and M. Ahmad, Prediction of diabetes empowered with fused machine learning, *IEEE Access* **10**, 8529 (2022).
5. C. Kavitha, V. Mani, S. Srividhya, O. I. Khalaf and C. A. Tavera Romero, Early-stage alzheimer's disease prediction using machine learning models, *Frontiers in public health* **10**, p. 853294 (2022).
6. P.-Y. Tseng, Y.-T. Chen, C.-H. Wang, K.-M. Chiu, Y.-S. Peng, S.-P. Hsu, K.-L. Chen, C.-Y. Yang and O. K.-S. Lee, Prediction of the development of acute kidney injury following cardiac surgery by machine learning, *Critical care* **24**, 1 (2020).
7. B. M. Stopa, F. C. Robertson, A. V. Karhade, M. Chua, M. L. Broekman, J. H. Schwab, T. R. Smith and W. B. Gormley, Predicting nonroutine discharge after elective spine surgery: external validation of machine learning algorithms: Presented at the 2019 aans/cns joint section on disorders of the spine and peripheral nerves, *Journal of Neurosurgery: Spine* **31**, 742 (2019).
8. A. Siccoli, M. P. de Wispelaere, M. L. Schröder and V. E. Staartjes, Machine learning–based preoperative predictive analytics for lumbar spinal stenosis, *Neurosurgical Focus* **46**, p. E5 (2019).
9. S. S. Rajaee, H. W. Bae, L. E. Kanim and R. B. Delamarter, Spinal fusion in the united states: analysis of trends from 1998 to 2008, *Spine* **37**, 67 (2012).
10. D. Badin, C. Ortiz-Babilonia, F. N. Musharbash and A. Jain, Disparities in elective spine surgery for medicaid beneficiaries: a systematic review, *Global Spine Journal* **13**, 534 (2023).
11. S. J. S. Bajwa and R. Haldar, Pain management following spinal surgeries: an appraisal of the available options, *Journal of craniovertebral junction & spine* **6**, p. 105 (2015).
12. A. Finkelstein, M. Gentzkow and H. Williams, Sources of geographic variation in health care: Evidence from patient migration, *The quarterly journal of economics* **131**, 1681 (2016).
13. T. P. Pagano, R. B. Loureiro, F. V. N. Lisboa, G. O. R. Cruz, R. M. Peixoto, G. A. d. S. Guimarães, L. L. d. Santos, M. M. Araujo, M. Cruz, E. L. S. de Oliveira *et al.*, Bias and unfairness in machine learning models: a systematic literature review, *arXiv preprint arXiv:2202.08176* (2022).
14. Z. Obermeyer, B. Powers, C. Vogeli and S. Mullainathan, Dissecting racial bias in an algorithm used to manage the health of populations, *Science* **366**, 447 (Oct 2019).

15. T. Ahmad, M. J. Pencina, P. J. Schulte, E. O'Brien, D. J. Whellan, I. L. a, D. W. Kitzman, K. L. Lee, C. M. O'Connor and G. M. Felker, Clinical implications of chronic heart failure phenotypes defined by cluster analysis, *J Am Coll Cardiol* **64**, 1765 (Oct 2014).
16. L. Marisa, A. s, A. Duval, J. Selves, M. P. Gaub, L. Vescovo, M. C. Etienne-Grimaldi, R. Schiappa, D. Guenot, M. Ayadi, S. Kirzin, M. Chazal, J. F. jou, D. Benchimol, A. Berger, A. Lagarde, E. Pencreach, F. Piard, D. Elias, Y. Parc, S. Olschwang, G. Milano, P. Laurent-Puig and V. Boige, Gene expression classification of colon cancer into molecular subtypes: characterization, validation, and prognostic value, *PLoS Med* **10**, p. e1001453 (2013).
17. K. A. e. a. Hoadley, Cell-of-Origin Patterns Dominate the Molecular Classification of 10,000 Tumors from 33 Types of Cancer, *Cell* **173**, 291 (Apr 2018).
18. S. van Buuren and K. Groothuis-Oudshoorn, mice: Multivariate imputation by chained equations in R, *Journal of Statistical Software* **45**, 1 (2011).
19. K. Al-Jabery, T. Obafemi-Ajayi, G. Olbricht and D. Wunsch, *"Computational Learning Approaches to Data Analytics in Biomedical Applications"* (Academic Press, 2019).
20. T. Ronan, S. Anastasio, Z. Qi, R. Sloutsky, K. M. Naegle and P. H. S. V. Tavares, Openensembles: a python resource for ensemble clustering, *The Journal of Machine Learning Research* **19**, 956 (2018).
21. D. Yeboah, L. Steinmeister, D. B. Hier, B. Hadi, D. C. Wunsch, G. R. Olbricht and T. Obafemi-Ajayi, An explainable and statistically validated ensemble clustering model applied to the identification of traumatic brain injury subgroups, *IEEE Access* **8**, 180690 (2020).
22. G. W. Schwartz, Y. Zhou, J. Petrovic, M. Fasolino, L. Xu, S. M. Shaffer, W. S. Pear, G. Vahedi and R. B. Faryabi, TooManyCells identifies and visualizes relationships of single-cell clades, *Nat Methods* **17**, 405 (Apr 2020).
23. T. en, K. Nowell, K. E. Bodner and T. Obafemi-Ajayi, Ensemble validation paradigm for intelligent data analysis in autism spectrum disorders, in *2018 IEEE Conference on Computational Intelligence in Bioinformatics and Computational Biology (CIBCB)*, 2018.
24. Y. Liu, Z. Li, H. Xiong, X. Gao and J. Wu, Understanding of internal clustering validation measures, in *2010 IEEE international conference on data mining*, 2010.
25. R. S. Olson, R. J. Urbanowicz, P. C. Andrews, N. A. Lavender, L. C. Kidd and J. H. Moore, Automating biomedical data science through tree-based pipeline optimization, in *Applications of Evolutionary Computation*, eds. G. Squillero and P. Burelli (Springer International Publishing, Cham, 2016).
26. S. M. Lundberg, G. Erion, H. Chen, A. DeGrave, J. M. Prutkin, B. Nair, R. Katz, J. Himmelfarb, N. Bansal and S.-I. Lee, From local explanations to global understanding with explainable ai for trees, *Nature Machine Intelligence* **2**, 2522 (2020).
27. L. Breiman, Random forests, *Machine learning* **45**, 5 (2001).
28. J. B. Tenenbaum, V. de Silva and J. C. Langford, A global geometric framework for nonlinear dimensionality reduction, *Science* **290**, 2319 (Dec 2000).
29. W. Sun, O. Nasraoui and P. Shafto, Evolution and impact of bias in human and machine learning algorithm interaction, *Plos one* **15**, p. e0235502 (2020).
30. N. Mehrabi, F. Morstatter, N. Saxena, K. Lerman and A. Galstyan, A survey on bias and fairness in machine learning, *ACM computing surveys (CSUR)* **54**, 1 (2021).
31. L. Wang, N. A. Berger, D. C. Kaelber, P. B. Davis, N. D. Volkow and R. Xu, Covid infection rates, clinical outcomes, and racial/ethnic and gender disparities before and after omicron emerged in the us, *medRxiv* (2022).
32. D. Quan, L. Luna Wong, A. Shallal, R. Madan, A. Hamdan, H. Ahdi, A. Daneshvar, M. Mahajan, M. Nasereldin, M. Van Harn *et al.*, Impact of race and socioeconomic status on outcomes in patients hospitalized with covid-19, *Journal of general internal medicine* **36**, 1302 (2021).

33. S. J. Hong, B. Drake, M. Goodman and K. A. Kaphingst, Race, trust in doctors, privacy concerns, and consent preferences for biobanks, *Health communication* **35**, 1219 (2020).

Evidence of recent and ongoing admixture in the U.S. and influences on health and disparities

Hannah M. Seagle[1], Jacklyn N. Hellwege[1,2], Brian S. Mautz[2,3], Chun Li[4], Yaomin Xu[5], Siwei Zhang[5], Dan M. Roden[1,3,6,7], Tracy L. McGregor[8], Digna R. Velez Edwards[1,6,9], Todd L. Edwards[1,3]*

[1]Vanderbilt Genetics Institute, [2]Department of Medicine, [5]Department of Biostatistics, [6]Department of Biomedical Informatics, [7]Department of Pharmacology and Biomedical Informatics, [8]Department of Pediatrics, [9]Department of Obstetrics and Gynecology, Vanderbilt University School of Medicine, Nashville, TN 37203, USA

[3]Janssen Pharmaceutical Companies of Johnson & Johnson, Spring House, PA 19477, USA

[4]Department of Population and Quantitative Health Sciences, Case Western Reserve University, Cleveland, OH 44106, USA

**Email: todd.l.edwards@vumc.org*

Many researchers in genetics and social science incorporate information about race in their work. However, migrations (historical and forced) and social mobility have brought formerly separated populations of humans together, creating younger generations of individuals who have more complex and diverse ancestry and race profiles than older age groups. Here, we sought to better understand how temporal changes in genetic admixture influence levels of heterozygosity and impact health outcomes. We evaluated variation in genetic ancestry over 100 birth years in a cohort of 35,842 individuals with electronic health record (EHR) information in the Southeastern United States. Using the software STRUCTURE, we analyzed 2,678 ancestrally informative markers relative to three ancestral clusters (African, East Asian, and European) and observed rising levels of admixture for all clinically-defined race groups since 1990. Most race groups also exhibited increases in heterozygosity and long-range linkage disequilibrium over time, further supporting the finding of increasing admixture in young individuals in our cohort. These data are consistent with United States Census information from broader geographic areas and highlight the changing demography of the population. This increased diversity challenges classic approaches to studies of genotype-phenotype relationships which motivated us to explore the relationship between heterozygosity and disease diagnosis. Using a phenome-wide association study approach, we explored the relationship between admixture and disease risk and found that increased admixture resulted in protective associations with female reproductive disorders and increased risk for diseases with links to autoimmune dysfunction. These data suggest that tendencies in the United States population are increasing ancestral complexity over time. Further, these observations imply that, because both prevalence and severity of many diseases vary by race groups, complexity of ancestral origins influences health and disparities.

Keywords: Disparities; Electronic Health Records; Health Outcomes; Admixture.

1. Introduction

Genetic admixture has previously been used to identify geographic variability and historical migration patterns across several human populations[1-11] and to investigate the genetic basis of diseases[12-14]. Two studies have shown temporal increases in heterozygosity due to urbanization, one in a Croatian population[15] and one in a U.S. population of European ancestry[16]. However, these studies of admixture have not connected migratory or urbanization patterns to health outcomes. One study that performed a meta-analysis on populations of individuals with European American and African American ancestry found a positive association between levels of heterozygosity and mortality in humans[17]. However, this study investigated only a single outcome and omitted the impact of temporal trends in admixture and heterozygosity on epidemiological outcomes. Further, these studies have not explored variability in admixture with respect to age or generational trends.

Understanding temporal changes in ancestry and heterozygosity has important implications for individual- and population-level health in humans that remain unexplored. Using human population genetic data to study the connection between ancestry, heterozygosity, and health is ideal due to the substantial number of individuals with genetic data linked to electronic health records (EHR)[18,19]. Further, many diseases and their etiologies recorded in EHRs are known in detail and are well classified, facilitating the estimation of the relationship between heterozygosity and disease risks.

In our cohort of 35,842 individuals from the Southeastern U.S., we investigated temporal changes and variance of admixture by age with de-identified information from the EHR on race, ethnicity, and year of birth linked to genotype data from the Illumina HumanExome array in Vanderbilt University Medical Center's biorepository resource (BioVU)[18]. In addition, we used a phenome-wide association study (PheWAS[20]) to connect genetic data with the clinical phenome capturing clinical disease outcomes in BioVU. This approach allowed us to investigate the relationship between increased ancestral complexity and disease risk. Our study provides important insights into the changing landscape of genetic admixture in a clinical context.

2. Methods

2.1. *Study Population*

Individuals were selected from the BioVU DNA repository which links clinical data from de-identified electronic medical records to DNA samples obtained from patients at Vanderbilt University Medical Center (VUMC)[18]. Each individual's race was designated in the Electronic Health Record (EHR) as either White, Black, Asian, Pacific Islander, American Indian/Alaska Native, or declined/unknown, and an ethnicity of Hispanic/Latino, Not Hispanic/Latino, or declined/unknown. BioVU also contains third-party designated race, which is a good predictor of genetically estimated ancestry in this database[21]. This study of de-identified data was determined to be non-human subject research by the institutional review board (IRB) of Vanderbilt University, Nashville, TN.

2.2. *DNA Extraction and Genotyping*

All DNA samples were isolated from whole blood using the Autopure LS system (QIAGEN Inc., Valencia, CA). Genomic DNA was quantitated via an ND-8000 spectrophotometer and DNA quality was evaluated via gel electrophoresis. Individuals were genotyped using the Illumina Infinium HumanExome Array [12v1-1] (Illumina Inc., San Diego, CA). The data were processed for genotype calling using Illumina's Genome Studio (Illumina Inc., San Diego, CA).

2.3. *Genotyping Quality Control*

Data on 240,117 SNPs and 35,842 individuals (16,289 males and 19,552 females) were available prior to implementation of quality control (QC) measures. No individuals were excluded for low genotyping efficiency (<98%). 6,599 SNPs were excluded for low genotyping efficiency (<98%) and 71,667 SNPs were monomorphic. Twenty-six individuals (14 EHR males and 12 EHR females) were excluded for inconsistent genetic and database sex. After QC, 163,135 SNPs remained for analyses in 35,456 individuals. No SNPs were removed for deviations from Hardy-Weinberg equilibrium.

2.4. *Quantification and Statistical Analyses*

Descriptive statistics on demographic and clinical characteristics were expressed as means with standard deviation or median with interquartile range for continuous covariates and as frequencies or proportions for categorical data using SPSS statistical software (IBM Corporation, Armonk, NY) (Table 1).

Table 1. Summary of demographic characteristics of study individuals

Race*	White	Black	Hispanic/Latino	Asian	Other/Unknown
N (%)	28,723 (80.1)	4,129 (11.5)	550 (1.5)	270 (0.75)	2,170 (2.8)
Male %	46.9%	38.9%	43.6%	42.5%	39.6%
Birth Year					
Mean (SD)	1957 (24.1)	1968 (26.0)	1976 (25.6)	1959 (19.8)	1955 (20.2)
Median (IQR)	1951 (1938-1971)	1964 (1948-1995)	1977 (1957-2000)	1958 (1944-1972)	1953 (1940-1967)
Range	1905-2012	1908-2011	1918-2012	1915-2010	1906-2012

*Non-overlapping categories

A subset of 2,678 ancestry-informative markers (AIMs) were selected for subsequent analysis. We chose AIMs from the ExomeChip selected to have strong differences between African and European ancestry populations as well as between Asian and European ancestry populations. AIMs were used instead of pruned SNP data due to the particular composition of the ExomeChip platform, which was designed with a panel of AIMs to enable evaluation of ancestry.

2.4.1. *Principal component analysis*

EIGENSTRAT v6.0.1 software was used to conduct principal component analysis (PCA) to estimate continuous axes of ancestry from AIMs in all populations together[22]. SPSS was used to create plots of individuals, stratified by birth year which demonstrate trends in changing demography in individuals over time as shown in Fig. 1.

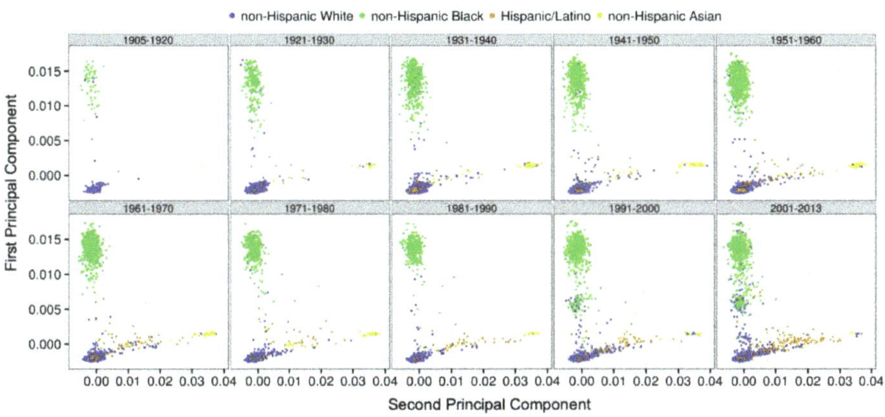

Fig. 1. Principal components plots of study individuals by decade of birth between 1905-2013. Principal components were calculated in EIGENSTRAT and anchored to populations in 1000 Genomes. Sample size information can be found in Table 1.

2.4.2. *STRUCTURE analysis*

STRUCTURE software v2.3.3[23,24] was used to quantify ancestry in combined study and 1000 Genomes Project Phase 3 individuals using the AIMs[22]. We estimated proportions of ancestry assuming ancestral clusters (*K*) ranging from one to 16, where 16 is the number of sub-populations in the 1000 Genomes Phase 3 data plus two. We assumed unlinked SNPs and used 5,000 iterations of burn-in and 10,000 iterations for analysis without providing population information to the software. We observed that the –log-likelihood of the data given *K* did not vary significantly for *K*'s greater than three and observed that *K*'s greater than three primarily subdivided the European populations (data not shown). The three STRUCTURE clusters corresponded to African, Asian, and European ancestry based on comparisons to the 1000 Genomes reference data (data not shown).

Each of the three derived proportions of continental ancestry from STRUCTURE were regressed onto birth year using generalized additive models with integrated smoothness estimation (GAM)[25] implemented in the R package mgcv in all study individuals (Fig. 2), Non-Hispanic Whites, Non-Hispanic Blacks, Hispanics/Latinos, and Non-Hispanic Asians (data not shown). We derived admixture proportion, α, for the i-th individual using the formula $\alpha_i = 1 -$ maximum(%European, %African, %East Asian). We regressed α_i onto birth year using generalized additive models with integrated smoothness estimation for all study individuals, Non-Hispanic White, Non-Hispanic Black, Hispanic/Latino, and Non-Hispanic Asian (Fig. 3).

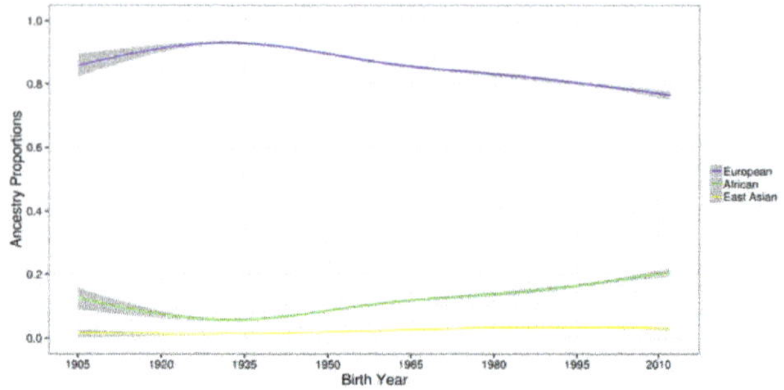

Fig. 2. Proportions of ancestry derived by STRUCTURE analysis of AIMS for study individuals plotted against birth year. Shaded regions represent 95% confidence intervals.

It has been previously shown that when parental populations stop contributing to admixture, that the variance of admixture proportions decreases rapidly, and when parental populations continue to contribute to the admixture, the variance of admixture proportions increases over time[26]. To test the null hypothesis that the observed levels of admixture in our data were not due to ongoing and increasing rates of admixture, we analyzed the association between the variance of α and birth year. We used the software package MVtest and modeled the admixture proportion variance as a log-linear function of birth year and five principal components of ancestry using estimating equations.

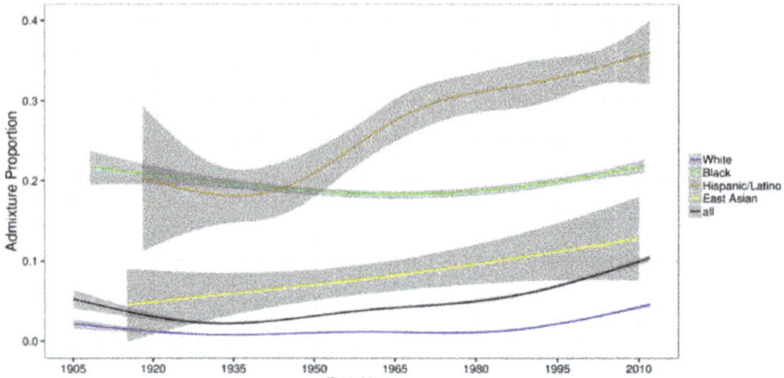

Fig. 3. Admixture proportion plotted against birth year. Admixture proportion is defined as 1 minus the maximum ancestry proportion. The smoothing curves are obtained using the generalized additive model method (gam) with a cubic spline basis implemented in R package mgcv and plotted using R package ggplot2. Sample size information can be found in Table 1. Shaded regions represent 95% confidence intervals. EHR-designated race categories are non-overlapping.

2.5. *Analysis of Admixture Proportion Variance Over Time*

We modeled admixture proportion mean and variance simultaneously as functions of birth year and covariates. Specifically, let m_i be the mean and σ_i^2 be the variance of the trait Y_i for the i-th individual. We model them as

$$m_i = \beta_0 + \beta_g G_i + \sum_{j=1}^{p} \beta_j X_{ij} \quad (1)$$

and

$$\ln(\sigma^2{}_i) = \gamma_0 + \gamma_g G_i + \sum_{j=1}^{p} \gamma_j X_{ij}\,. \quad (2)$$

where G_i is the birth year or variable of interest for the i-th individual and $X_{i1}, \ldots, X_{ip}$ are p covariates. In this model the variance is monotonic with respect to birth year, an assumption that holds in most circumstances. The parameters are estimated simultaneously. This framework allows for testing of the null hypothesis of no effect on mean, variance, or both for any term in the model. These correspond to a mean test with null H_0: $\beta_g = 0$, a variance test with H_0: $\gamma_g = 0$, both having one degree of freedom (DF), and a 2-DF test with H_0: $\beta_g = 0, \gamma_g = 0$.

2.6. *Model Fitting with Estimating Equations*

The parameters are estimated through the estimating equations approach, which does not require a full specification of the outcome distribution, but only a few constraints for the parameters of interest. These constraints are often written as equations, and the parameter estimates can be obtained by solving the equations. The asymptotic distribution for the parameter estimates can be derived[27]. Specifically, suppose the random variable has mean $m_i = \beta_0 + \beta_g G_i + \sum_{j=1}^{p} \beta_j X_{ij}$, and log-variance $\ln(\sigma^2{}_i) = \gamma_0 + \gamma_g G_i + \sum_{j=1}^{p} \gamma_j X_{ij}$. There are $k = 2(p+2)$ parameters, which can be written as a vector, $\theta = (\beta, \gamma)$, where $\beta = (\beta_0, \beta_g, \beta_1, \ldots, \beta_p)$ and $\gamma = \gamma_0, \gamma_g, \gamma_1, \ldots, \gamma_p)$. Let y_i and $x_i = (1, g, x_{i1}, \ldots, x_{ip})^T$ be the observed values for subject i. If we had assumed normality for the outcome, the log-likelihood for the observation i would have been

$$l_i(\theta) = -\frac{1}{2}\log(2\pi) - \frac{1}{2}\gamma' x_i - \frac{(y_1 - \beta' x_i)^2}{2\exp(\gamma' x_i)'} \quad (3)$$

for which the partial derivatives with respect to the parameters θ is a k-vector,

$$\Psi_i(\theta) = \begin{pmatrix} \frac{\partial l_i}{\partial \beta} \\ \frac{\partial l_i}{\partial \gamma} \end{pmatrix} = \begin{pmatrix} \frac{y_i - \beta' x_i}{\exp(\gamma' x_i)} \\ \frac{1}{2}\left[\frac{(y_i - \beta' x_i)^2}{\exp(\gamma' x_i)} - 1\right] x_i \end{pmatrix}, \quad (4)$$

and maximum likelihood estimates of the parameters could have been obtained by solving the k equations $\sum_{i=1}^{n} \Psi_i(\theta) = 0$. This motivated us to use these k equations,

$$\sum_{i=1}^{n} \Psi_i(\theta) = 0, \quad (5)$$

as the starting point for our estimating equations approach to obtain parameter estimates $\hat{\theta} = (\hat{\beta}, \hat{\gamma})$. If normality holds, then $\hat{\theta}$ are the maximum likelihood estimates. Note that although the estimating equations were motivated by the Gaussian likelihood, one can always start from these equations to obtain $\hat{\beta}$ and $\hat{\gamma}$, whether normality holds or not, and proceed with statistical inference using the M-estimation theory[27]. This is a major advantage for using estimating equations. The partial derivative of $\Psi_i(\theta)$ is a $k \times k$ matrix, denoted as $\psi_i(\theta)$. Using the M-estimation theory, we have

$$\sqrt{n}(\hat{\theta} - \theta)^d \rightarrow N(0, V), \quad (6)$$

where the $k \times k$ covariance matrix V can be estimated as $A^{-1}B(A^{-1})^T$ with $A = -\frac{1}{n}\sum_{i=1}^{n} \psi_i(\hat{\theta})$

$$\text{and } B = \frac{1}{n}\sum_{i=1}^{n} \Psi_i(\hat{\theta})\Psi_i(\hat{\theta})^T.$$

If our interest is on the effect of G, the asymptotic result for the joint distribution for the parameter estimates $\hat{\beta}_g$ and $\hat{\gamma}_g$ is

$$\sqrt{n}\left[\begin{pmatrix}\hat{\beta}_g \\ \hat{\gamma}_g\end{pmatrix} - \begin{pmatrix}\beta_g \\ \gamma_g\end{pmatrix}\right] \xrightarrow{d} N(0, V_2), \quad (7)$$

where V_2 is the corresponding 2×2 submatrix of V, with diagonal values denoted as $\widehat{\sigma_{\beta}^2}_g$ and $\widehat{\sigma_{\gamma}^2}_g$, respectively. A mean test (H_0: $\beta_g = 0$) can be performed by comparing $\sqrt{n}\beta_g$ with $N(0, \widehat{\sigma_{\beta}^2}_g)$, and similarly, a variance test (H_0: $\gamma_g = 0$) by comparing $\sqrt{n}\hat{\gamma}_g$ with $N(0, \widehat{\sigma_{\gamma}^2}_g)$. A 2-DF joint test ($H_0$: $\beta_g = 0, \gamma_g = 0$) can be performed by comparing $n(\hat{\beta}_g, \hat{\gamma}_g)V_2^{-1}\begin{pmatrix}\hat{\beta}_g \\ \hat{\gamma}_g\end{pmatrix}$ with a chi-squared distribution with two degrees of freedom. MVtest software for genetic analysis of SNP data or general analysis of variables is freely available at https://github.com/edwards-lab/MVtest.

2.7. *Heterozygosity Analysis*

Standardized measures of heterozygosity among the AIMs were calculated to evaluate trends in heterozygosity over time relative to expectations. We first estimated the expected number of heterozygous genotypes in an individual in the k-th subpopulation as

$$H_k = \sum_i 2p_{ik}q_{ik} \quad (8)$$

where the sum is over all SNPs in our analysis, and p_{ik} and $q_{ik} = 1-p_{ik}$ are the allele frequencies for the i-th SNP. Hardy-Weinberg equilibrium was assumed. Then for every individual j in the k-th subpopulation, we standardized the observed number of heterozygous genotypes, O_{kj}, by comparing it with the expected number H_k:

$$(O_{kj} - H_k)/H_k \quad (9)$$

Standardized heterozygosity was regressed onto birth year using GAM for all study individuals, and the results were plotted for Non-Hispanic White, Non-Hispanic Black, Hispanic/Latino, and Non-Hispanic Asian.

2.8. *Analysis of Long-Range Linkage Disequilibrium*

To evaluate the presence of admixture long-range linkage disequilibrium (LRLD), pairwise linkage disequilibrium (LD) D' statistics were calculated for all pairs of common (MAF > 0.05) SNPs within 10 megabases (Mb) using Haploview software[28]. D' statistics were regressed onto physical distance between SNPs using generalized additive models with integrated smoothness estimation for distances in the interval from 9-10 Mb for each birth decade (Fig. 4).

2.9. *United States Census Data Analysis*

We downloaded the 1% representative sample of individual-level response to the American Community Survey from the Integrated Public Use Microdata Series (IPUMS) (IPUMS USA, Minneapolis, MN). We regressed the number of major race groups claimed by individuals onto their reported birth year using generalized additive models with integrate d smoothness estimation
and frequency weights provided by IPUMS for TN, the South East Central census region, and the entire U.S. (Fig. 5). For individual groups, such as "White" and "Black or African American", we plotted all individuals who responded affirmatively to those items; thereby, the samples for the individual race group plots in Fig. 5 are not independent and overlap at observation where participants claim two or more race groups.

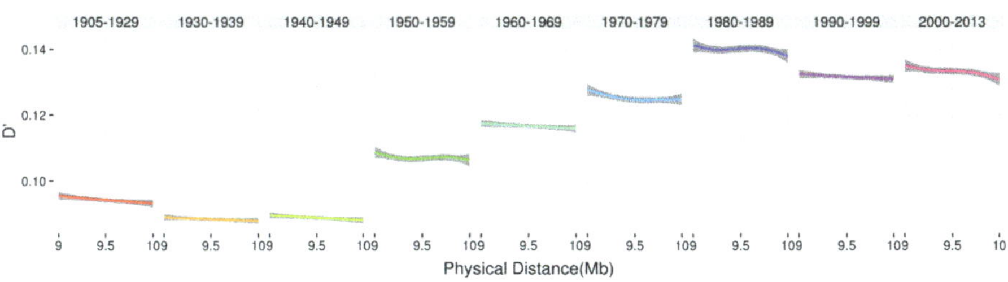

Fig. 4. Pairwise D' for SNPs between 9-10 Mb for all common SNPs on the exome array in all study individuals stratified by intervals of birth decade. Shaded regions represent 95% confidence intervals.

2.10. *Phenotype Classification*

Each individual was classified according to 1,645 phenotypes based on the International Classification of Disease, Ninth Revision, Clinical Modification (ICD9) Codes[20]. Our classification strategy includes all ICD9 features except for procedures. Additionally, the system is hierarchical such that disease subtypes are also classified, such as cardiac arrhythmias are the parent to atrial fibrillation and atrial flutter. Additional phenotypes that are not represented directly in the ICD9 hierarchy are also included, such as inflammatory bowel disease as the parent for Crohn's disease and ulcerative colitis. Diagnoses that were not possible for an individual were set to missing, such as pregnancy for biological males, or prostate disease for biological females. Detailed feature of all phenotype algorithms used are available from: http://phewascatalog.org.

2.11. *Clinical Outcomes*

For each phenotype, we regressed the binary outcome onto the standardized heterozygosity from equation 9 above, adjusted for birth year and the top 5 principal components of ancestry using logistic regression. We limited analysis to outcomes with 40 or more cases and individuals with at least 2 ICD9 codes. We determined the threshold for statistical significance by Bonferroni correction for the number of analyses where the model converged.

3. Results

We evaluated 2,678 ancestry informative markers (AIMs) from genetic data in Vanderbilt University's BioVU. These AIMs were from ExomeChip data in a cohort of 28,723 White, 4,129 Black, 550 Hispanic/Latino, and 270 Asian individuals, based on EHR-third party race designation. The demographic characteristics of study individuals are presented in Table 1.

3.1. *Analysis of Temporal Trends in Genetic Admixture*

After combining our data with the 1000 Genomes as a reference group[29], we calculated principal components to identify patterns of ancestry in each individual. We used the ancestral classifications to test for temporal trends in mean and variance in admixture proportion.

Analysis of temporal trends in genetic admixture showed an increase in ancestral diversity over time. Plots of the first two principal components demonstrated a distinct pattern change in younger generations. To assess the trend of increasing admixture in younger individuals, we calculated the admixture proportion, defined as (1-Predominant fraction of ancestry) for each EHR-designated race (Fig. 3). The admixture proportion consistently increased with younger ages in Asian and Hispanic/Latino groups. In White individuals, the level of admixture remained stable until approximately 1990 when it began to increase notably. Black individuals presented with a decrease in admixture

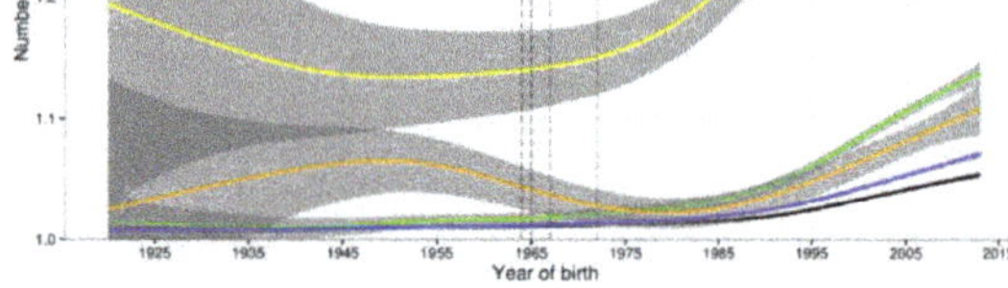

Fig. 5. Plot of individual-level 2013 U.S. Census data for the East South Central Division. Sample size information can be found in Table 1. Shaded regions represent 95% confidence intervals. Race groups are non-overlapping.

proportion in those born in the early- to mid-20th century, but an increase after the 1980's. In all age groups, each recorded race had a small number of individuals who plotted outside of the expected clusters. In later birth years, an increased number of individuals midway between the European and African clusters appear, creating a new cluster representing Black-White biracial children (Fig. 1). Stratifying these plots by EHR-designated race revealed that individuals in this ancestral cluster identify as both White and Black (data not shown). In addition to the clear biracial cluster apparent in the principal component plots, the overall proportions of ancestry across all recorded race groups exhibit increasing admixture in younger individuals. Additionally, a significant increase in the variance of admixture proportions over time was observed (variance coefficient = 0.0193 $\pm$ 0.0009 [SE], p-value $< 1.44\times10^{-100}$), indicating that there is a linear increase in variance of the admixture proportion of 0.0193 with every birth year. This finding is consistent with recent and ongoing admixture[26].

We detected similar patterns in three additional sources. First, we compared the rate of heterozygosity to birth year to assess the extent of isolate breaking in our data over time. This occurs when genetically distinct populations reproduce resulting in a temporary excess of heterozygosity. We standardized individual heterozygosity estimates by comparing this estimate with the expected heterozygosity for the EHR-designated race of the individual as described in the methods. Analysis of the standardized heterozygosity by birth year and stated race strongly supports the finding of increasing ancestral diversity in younger individuals (Fig. 6). The timing of inflection for increased standardized heterozygosity varied between race groups, but the data indicated that ancestral diversity has accelerated rapidly in Asian, non-Hispanic Black, and non-Hispanic White cohorts since approximately 1980, while Hispanic/Latino groups have exhibited a relatively steady rate of increasing diversity since the 1940s. This finding reflects the increasing number of children born to biological parents of predominantly different ancestral backgrounds over the past few decades.

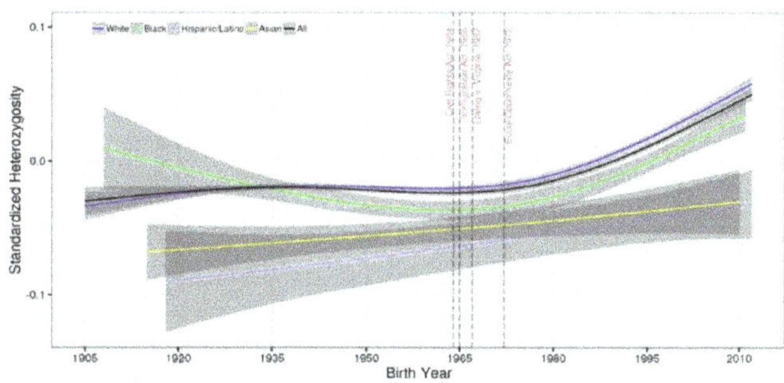

Fig. 6. Standardized heterozygosity plotted against birth year for non-overlapping EHR-designated race groups. Sample size information can be found in Table 1. Shaded regions represent 95% confidence intervals.

Second, because recent admixture leads to increased LRLD, we verified patterns by estimating pairwise LRLD in our dataset. Using common single nucleotide polymorphisms (SNPs) (minor allele frequency [MAF] > 5%) from the genotyping array, we calculated pairwise D' using Haploview[28] and plotted against physical distance for all pairs of SNPs between 9-10 megabases of each other (Fig. 4). These results show a small drop in LRLD in individuals born before the 1950s, followed by significant steady increases in the 1950s through the 1980s and fluctuation at higher levels in the 1990s to 2010s. This finding is consistent with the results of the admixture proportion analysis (Fig. 3), where admixture proportions decreased from the 1910s to the 1940s, and then steadily increased thereafter.

Third, we estimated changes in the number of races indicated in self-reported ancestry in the 2013 American Communities Survey. Respondents were instructed to select Hispanic/Latino/Spanish status and all applicable races for each individual in the household[30]. We analyzed the average number of race categories selected for each individual by birth year and stratified these results within race categories for Tennessee, the East South-Central Region which includes Tennessee, Alabama, Kentucky, and Mississippi (Fig. 5). The results suggest that younger individuals are more likely to indicate multiple races. The inflection point appears to have been earlier in the Asian race category but is demonstrated in all race groups by the mid-1960s in the entire U.S. sample. These data mirror the findings of our cohort genetic analyses.

3.2. *Changes in Health Diagnoses with Admixture*

To investigate the possible impact of increased ancestral complexity on human health and disparities, we evaluated the association between individual heterozygosity and disease diagnoses using a phenome-wide association study approach (PheWAS[20]). Increasing genetic admixture resulted in fewer diagnoses of female reproductive traits across all data (Table 2). These results remain statistically significant after correction for multiple tests. Phenotype codes for "disorders of menstruation and other abnormal bleeding from female genital tract" and "irregular menstrual cycle/bleeding" were significantly associated with protection by increasing heterozygosity (p-value = 7.21×10^{-6} and 4.37×10^{-5}, respectively; Table 2). Other protective findings were also gynecological in nature, including cervical cancer/dysplasia and abnormal Papanicolaou smear results. Significant phenotypes in adults were predominantly detected for biological females. Outside of genitourinary findings, other nominally significant associations (Bonferroni significant $< p \leq 0.05$) show increased risk with genetic admixture and include atopic dermatitis, AV Block, obstructive asthma, and Sicca syndrome.

Table 2. Results from the phenome-wide association study of heterozygosity and clinical outcomes for full sample.

PheCode	Phenotype	P-value	OR (95% Confidence Interval)
626	Disorders of menstruation and other abnormal bleeding from female genital tract	7.21×10^{-6}	0.37 (0.24 – 0.57)
626.1	Irregular menstrual cycle/bleeding	4.37×10^{-5}	0.37 (0.23 – 0.60)
939	Atopic/contact dermatitis due to other or unspecified	2.86×10^{-4}	1.82 (1.32 – 2.52)
180	Cervical cancer and dysplasia	4.02×10^{-4}	0.19 (0.08 – 0.48)
792.1	Papanicolaou smear of cervix or vagina with atypical squamous cells	5.36×10^{-4}	0.21 (0.09 – 0.51)
180.3	Cervical intraepithelial neoplasia [CIN] [Cervical dysplasia]	6.24×10^{-4}	0.15 (0.05 – 0.45)
426.2	Atrioventricular [AV] block	7.62×10^{-4}	2.97 (1.58 – 5.61)
495.11	Chronic obstructive asthma with exacerbation	8.13×10^{-4}	4.63 (1.89 – 11.34)

4. Discussion

Mitigating racial disparities in health is a significant challenge for precision medicine. Some of these population-level health differences may be caused by phenotypic variability associated with ancestral genetic backgrounds. Admixture introduces additional complexity to genetic studies of health disparities and the effects of historical, ongoing, and increasing admixture on population-level health are not well understood. This study evaluates the level of admixture over time and the relationship between ancestral diversity and population health from a clinical perspective.

In our Southeastern United States cohort of 35,842 individuals, we found that for individuals with an EHR race designation of White, the mean proportion of European ancestry decreased from 98% to 92% after the 1990s as the proportion of African ancestry increased to 6%. For individuals designated as Black in the EHR, the mean African proportion decreased by 3% after the 1990s. The European ancestry proportion in the EHR-designated Hispanic/Latino group decreased by 15% after the 1980s (data not shown). Comparing these changes to historical socio-cultural shifts in our cohort's geographic region provides context for these results. In the Southeastern U.S., laws and policies enforced segregation of populations of European and African ancestry. Consistent with these socio-cultural boundaries, there is little change in admixture through the 1960s. Additionally, despite legal rulings and socio-cultural transformation, there remained a very slow increase in admixture and heterozygosity for an additional 20-30 years, followed by a sharp increase over the next few decades.

Our results, qualitatively mirrored in the 2013 American Communities Survey, show that younger individuals are more likely to have greater ancestral diversity than older age groups. The results of our long-range disequilibrium (LRLD) analyses support this notion, with LRLD consistently increasing for individuals born between 1990 and 2010. It is important to note, however, that other possible sources of LRLD (e.g. drift, epistatic selection) cannot be necessarily ruled out, although they seem unlikely given the recent nature of admixture and formation of LRLD. This increase in ancestral heterogeneity of the younger population may also lend itself to more powerful admixture mapping projects in populations not traditionally considered for these types of studies.

Further, we show that changes in population genetic parameters have important consequences for individual and population-level health. Several statistically significant ($p < 5\times10^{-5}$) associations of genetic diversity with adult female genitourinary diagnosis codes (e.g. irregular menstrual cycle/bleeding, cervical cancer/dysplasia) were observed. These novel findings linking admixture to protection from menstruation and gynecological abnormalities suggest that ancestral diversity may decrease risk of disorders that could affect reproduction. Further, the changes in reproductive diagnoses were detected predominantly for biological females, suggesting a potential sex-specific population clinical response to changes in admixture. However, the sex-specific response we detected could also be a result of differences in treatment for reproductive health. For example, male reproductive traits, such as sperm quality, may not be routinely checked and reported as with female reproductive parameters.

Other patterns emerge when considering nominally significant ($p < 0.05$) PheWAS results. Several of these diagnoses increase risk with increasing genetic diversity. Importantly, each of these diseases have at least suggestive links to autoimmune dysregulation, including atopic dermatitis[31], AV block[32,33], asthma[34,35], and Sicca/Sjögren syndrome[36]. These patterns suggest a connection between increased heterozygosity and increased activity in the immune system. Because our results show continued increase in genetic admixture over time, it is possible that there will be increases in prevalence of these types of diseases with time as well. Future research should address these immunity-disease relationships with respect to admixture to determine the validity and consistency of these patterns.

The present study has several limitations that warrant consideration. First, the use of EHR data may have high levels of missingness and can introduce inherent selection bias due to patients seeking care at tertiary care centers. Furthermore, given the constraints imposed by our limited sample size and the unavailability of comprehensive reference data for Hispanic/Latino and Native American populations, we were unable to estimate Native American ancestry in this study. Therefore, to provide more robust insights into individuals who identify as Hispanic/Latino and/or Native American, it is necessary to independently validate these results using larger datasets with more diverse reference data.

The concept of race was utilized in this study to reflect demographic dynamics in our cohort's geographic region and to investigate changes to admixture and heterozygosity within these groups. Although the concept of race is a construct with social underpinnings and has limited biological meaning[37], race is often captured in the clinical setting and is the basis for some clinical decision making. It is important to consider the changing implications of classifying individuals by race given the trend of increasing genetic diversity observed in this work and others[38]. As prevalence of many diseases and some drug efficacies vary by race, understanding race-associated factors in patients with complex ancestries may be increasingly important for effective delivery of precision medical care.

5. Acknowledgements

The dataset used in the analyses described were obtained from Vanderbilt University Medical Center's BioVU which is supported by institutional funding and by the Vanderbilt CTSA grant UL1 TR000445 from NCATS/NIH. This work was supported by the National Institutes of Health (grant numbers RC2GM092618, U01HG004603, K23HG000001, R25/T32CA160056, R01LM010685). H. Seagle was funded by T32 GM145734-01 (PI: David Samuels). The authors thank Dr. Joshua C. Denny, MD, MS for his contributions to the manuscript.

References

1. Bryc, K., Durand, E. Y., Macpherson, J. M., Reich, D., & Mountain, J. L. The genetic ancestry of African Americans, Latinos, and European Americans across the United States. *Am J Hum Genet* **96**, 37-53 (2015).
2. Baharian, S. et al. The Great Migration and African-American Genomic Diversity. *PLOS Genet* **12**, e1006059 (2016).

3. Byrc, K. et al. Colloquium paper: genome-wide patterns of population structure and admixture among Hispanic/Latino populations. *Proc Natl Acad Sci U S A* **107**, 8954–8961 (2010).
4. Han, E. et al. Clustering of 770,000 genomes reveals post-colonial population structure of North America. *Nat Commun* **8**, 14238 (2017).
5. Monero-Estrada, A. et al. Human genetics. The genetics of Mexico recapitulates Native American substructure and affects biomedical traits. *Science.* **344**, 1280–1285 (2014).
6. Tishkoff, S.A. et al. The genetic structure and history of Africans and African Americans. *Science* **344**, 1035-1044 (2009).
7. Wang, C. et al. Genome-wide analysis of runs of homozygosity identifies new susceptibility regions of lung cancer in Han Chinese. *J Biomed Res* **27**, 208-214 (2013).
8. Lao, O. et al. Correlation between genetic and geographic structure in Europe. *Curr Biol* **18**, 1241-8 (2008).
9. Leslie, S. et al. The fine-scale genetic structure of the British population. *Nature* **519**, 309-14 (2015).
10. Novembre, J. et al Genes mirror geography within Europe. *Nature **456**, 98-101 (2008).*
11. Pena, S.D. et al. The genomic ancestry of individuals from different geographical regions of Brazil is more uniform than expected. *PLoS One* **6**, e17063 (2011).
12. Franceschini, N. et al. Genome-wide association analysis of blood pressure traits in African-Ancestry Individuals reveals common associated genes in African and Non-African populations. *Am J Hum Genet* **93**, 545-554 (2013).
13. Kato, N. et al. Trans-ancestry genome-wise association study identifies 12 genetic loci influencing blood pressure and implicates a role for DNA methylation. *Nat Genet* **47**, 1282-1293 (2015).
14. Reich, D., & Patterson, N. Will admixture mapping work to find disease genes? *Philo Trans R Soc Lon B Biol Sci* **60**, 1605-1607 (2005).
15. Rudan, I. et al. Quantifying the increase in average human heterozygosity due to urbanization. *Eur J Hum Genet* **16**, 1097-1102 (2008).
16. Nalls, M., A. et al. Measures of autozygosity in decline: globalization, urbanization, and its implications for medical genetics. *PLoS Genet* **5**, e1000415 (2009).
17. Bihlmeyer, N. A. et al. Genetic diversity is a predictor of mortality in humans. *BMC Genet* **15**, 159 (2014).
18. Roden, D.M. et al. Development of a large-scale de-identified DNA biobank to enable personalized medicine. *Clin Pharmacol Ther* **84**, 362-369 (2008).
19. Sudlow, C. et al. UK Biobank: an open access resource for identifying the causes of a wide range of complex diseases of middle and old age. *PLoS Med* **12**, e1001779 (2015).
20. Denny, J.C. et al. Systematic comparison of phenome-wide association study of electronic medical record data and genome-wide association study data. *Nat Biotechnol* **31**, 1102-1111 (2013).
21. Dumitrescu, L. et al. Assessing the accuracy of observer-reported ancestry in a biorepository linked to electronic medical records. *Genet Med* **12**, 648-650 (2010).

22. Price, A.L. et al. Principal components analysis corrects for stratification in genome-wide association studies. *Nat Genet* **38**, 904-9 (2006).
23. Falush, D., Stephens, M. & Pritchard, J.K. Inference of population structure using multilocus genotype data: linked loci and correlated allele frequencies. *Genetics* **164**, 1567-87 (2003).
24. Pritchard, J.K., Stephens, M. & Donnelly, P. Inference of population structure using multilocus genotype data. *Genetics* **155**, 945-59 (2000).
25. Wood, S.N. Fast stable restricted maximum likelihood and marginal likelihood estimation of semiparametric generalized linear models. *J R Statist Soc B* **73**, 3-36 (2011).
26. Verdu, P., & Rosenberg, N.A. A general mechanistic model for admixture histories of hybrid populations. *Genetics* **189**, 1413-1426 (2011).
27. Stefanski, L.A. & Boos, D.D. The Calculus of M-Estimation. *Am Stat* **56**, 29-38 (2022).
28. Barrett, J.C., Fry, B., Maller, J., & Dally M.J. Haploview: analysis and visualization of LD maps. *Bioinformatics* **21**, 263-265 (2005).
29. Durbin, R.M. et al. A map of human genome variation from population-scale sequencing. *Nature* **467**, 1061-1073 (2010).
30. Ruggles, S. Big microdata for population research. *Demography* **51**, 287-297 (2014).
31. Mittermann, I. et al. Autoimmunity and atopic dermatitis. *Curr Opin Allergy Clin Immunol* **4**, 367-371 (2004).
32. Buyon, J.P. et al. Autoimmune-associated congenital heart block: demographics, mortality, morbidity, and recurrence rates obtained from a national neonatal lupus registry. *J Am Coll Cardiol* **31**, 1658-1666 (1998).
33. Villuendas, R. et al. Autoimmunity and atrioventricular block of unknown etiology in adults. *J Am Coll Cardiol* **63**, 1335-1336 (2014).
34. Barnes, P.J. Immunology of asthma and chronic obstructive pulmonary disease. *Nat Rev Immunol* **8**, 183-192 (2008).
35. Tedeschi, A., and Asero, R. Asthma and autoimmunity: a complex but intriguing relation. *Expert Rev Clin Immunol* **4**, 767-776 (2008).
36. Brito-Zerón, P., Izmirly, P.M., Ramos-Casals, M., Buyon, J.P., and Khamashta, M.A. The clinical spectrum of autoimmune congenital heart block. *Nat Rev Rheumatol* **11**, 301-312 (2016).
37. Maglo, K.N., Mersha, T.B., and Martin, L.J. Population genomics and the statistical values of race: an interdisciplinary perspective on the biological classification of human populations and implications for clinical genetic epidemiological research. *Front Genet* **7**, 22 (2016).
38. Wendt, F.R. et al. Modeling the longitudinal changes of ancestry diversity in the Million Veteran Program. *Hum Genomics* **17**, 46 (2023).

EVALUATING THE RELATIONSHIPS BETWEEN GENETIC ANCESTRY AND THE CLINICAL PHENOME

Jacqueline A. Piekos[1-3^] and Jeewoo Kim,[1-3†] Jacob M. Keaton,[4] Jacklyn N. Hellwege,[1,5‡] Todd L. Edwards[1,5] and Digna R. Velez Edwards[1-3,5]

1. Vanderbilt Genetics Institute, Vanderbilt University, Nashville, Tennessee 37203; 2. Department of Obstetrics and Gynecology, Vanderbilt University Medical Center Nashville, Tennessee 37232; 3. Department of Biomedical Informatics, Vanderbilt University Medical Center Nashville, Tennessee 37232; 4. National Human Genome Research Institute, National Institutes of Health, Bethesda, MD 20892; 5. Department of Medicine, Vanderbilt University Medical Center Nashville, Tennessee 37203

Corresponding authors: todd.l.edwards@vumc.org, digna.r.velez.edwards@vumc.org

Abstract

There is a desire in research to move away from the concept of race as a clinical factor because it is a societal construct used as an imprecise proxy for geographic ancestry. In this study, we leverage the biobank from Vanderbilt University Medical Center, BioVU, to investigate relationships between genetic ancestry proportion and the clinical phenome. For all samples in BioVU, we calculated six ancestry proportions based on 1000 Genomes references: eastern African (EAFR), western African (WAFR), northern European (NEUR), southern European (SEUR), eastern Asian (EAS), and southern Asian (SAS). From PheWAS, we found phecode categories significantly enriched neoplasms for EAFR, WAFR, and SEUR, and pregnancy complication in SEUR, NEUR, SAS, and EAS ($p < 0.003$). We then selected phenotypes hypertension (HTN) and atrial fibrillation (AFib) to further investigate the relationships between these phenotypes and EAFR, WAFR, SEUR, and NEUR using logistic regression modeling and non-linear restricted cubic spline modeling (RCS). For EAS and SAS, we chose renal failure (RF) for further modeling. The relationships between HTN and AFib and the ancestries EAFR, WAFR, and SEUR were best fit by the linear model (beta $p < 1 \times 10^{-4}$ for all) while the relationships with NEUR were best fit with RCS (HTN ANOVA $p = 0.001$, AFib ANOVA $p < 1 \times 10^{-4}$). For RF, the relationship with SAS was best fit with a linear model (beta $p < 1 \times 10^{-4}$) while RCS model was a better fit for EAS (ANOVA $p < 1 \times 10^{-4}$). In this study, we identify relationships between genetic ancestry and phenotypes that are best fit with non-linear modeling techniques. The assumption of linearity for regression modeling is integral for proper fitting of a model and there is no knowing a priori to modeling if the relationship is truly linear.

Keywords: genetic ancestry, health disparities, PheWAS, linear modeling

---

*Vanderbilt University Medical Center's BioVU is supporting by institutional funding, 1S10RR025141-01 and by the CTSA grant UL1TR000445 from NCATS/NIH.
^Work partially supported by T32GM080178
†Work supported by T32GM007347 and TL1TR002244
‡Work Partially supported by K12 HD043483

1. Introduction

Race is a social construct that is an imprecise way to classify groups prevalence of heritable risk factors, therefore there is a growing consensus in clinical and population research to move away from the use of race in the context of disease risk. Some racial disparities in health condition risks documented in the epidemiological literature may be due to non-biological differences between racial groups.[1] Geographic or genetic ancestry has been proposed as a more precise approach to capture differences in disease etiology that may be due to acquired biological differences in human populations. We hypothesize that when populations have evolutionarily adapted to a specific environment encounter different circumstances, disease risks can be influenced, and disparities can arise when compared to a population that is in evolutionary equilibrium with that environment. If this hypothesis is true, then this relationship would be detectable as an association between genetically inferred proportions of ancestry and disease risk. Improved understanding of how different geographic ancestries are responding to modern environments, nutrition, and behavioral lifestyles could help us understand genetic causes of diseases and improve healthcare.

Current approaches to precision medicine focus on a patient's clinical history and are often combined with known genetic risk factors, such as causal monogenic variants and more recently polygenic risk scores. Over the last several decades, race has been incorporated into clinical risk prediction models for several conditions when racial differences have been observed in disease prevalence, particularly for estimating drug responses. Race has also been used for medical tools such as calibrating eGFR measures for assessment of kidney disease risk. However, multiple studies have shown that administratively determined race or self-reported race are imprecise estimates of an individual's genetic ancestry, and thus use of race in modeling is a flawed approach.[2,3] Imprecise racial/ancestral identification may lead to lack of response to a personalized treatment plan that depends on a strong assumption of race capturing biological differences. Furthermore, recent work by several groups have shown that for some diseases genetic ancestry (global ancestry)[4] may directly interact with a patient's clinical characteristics to modify risk for disease and that this interaction varies at specific points in their genome (local ancestry).[5-7]

Within this study we leverage the rich phenotypic information available from Vanderbilt University Medical Center's (VUMC) biobank, BioVU, to evaluate the relationship between global geographic ancestry and the clinical phenome using phenome wide association study (PheWAS). From PheWAS results, we sought to identify enriched phenotype categories for ancestry groups and selected phenotypes within them for additional modeling. Selected phenotypes were then modeled using logistic regression and restricted cubic splines (RCS) to further investigate the relationship between phenotype and ancestry group. Studies usually make the strong assumption that the relationship between genetic ancestry and disease risk is linear. We chose to explore if fitting a non-linear model better described the relationship.

2. Methods

2.1. *Study Population*

The BioVU DNA Repository is a de-identified database of electronic health records (EHR) that are linked to patient DNA samples at VUMC. A detailed description of the database and how it is maintained has been published elsewhere.[8] BioVU participant DNA samples were genotyped on a custom Illumina Multi-Ethnic Genotyping Array (MEGA-ex; Illumina Inc., San Diego, CA, USA).

Quality control included excluding samples or variants with missingness rates above 2%, excluded if consent had been revoked, sample was duplicated, or failed sex concordance checks. Imputation was performed on the Michigan Imputation Server v1.2.410 using Minimac4[9] and the Haplotype Reference Consortium (HRC) panel v1.1.[10]

2.2. *Ancestry Estimations of BioVU Participants*

Estimation of ancestry proportion for BioVU participants based upon 1000 Genomes reference data has been described elsewhere.[11] In brief, the 1000 Genome populations were grouped into six super-population by geographic ancestry of east African (EAFR), west African (WAFR), southern European (SEUR), northern European (NEUR), east Asian (EAS), and south Asian (SAS) as described in Keaton, et. al 2021[12] using ADMIXTURE.[13] The six ancestry groups were projected onto BioVU to determine proportion of the six ancestries for all samples. Ancestry proportion of samples in the cohort was visualized by plotting subjects along the x-axis and their corresponding stacked ancestry proportions on the y-axis. Subjects were sorted by increasing SEUR ancestry.

2.3. *Ancestry Phenome Wide Association Study*

We conducted hypothesis-free PheWAS analyses of evaluating phecodes in the phenome with each of the six ancestries. Each ancestry was used as the main predictor in separate analysis, adjusted for age, sex, and body mass index (BMI). PheWAS was performed with the R package 'PheWAS' version 2.[14] 1,875 clinical disease phenotypes called phecodes from Phecode Map 1.2 were evaluated.[15] A p-value of 2.7×10^{-5} was the threshold for significance to correct for multiple testing (Bonferroni correction of 0.05/1,875 phecodes tested).

2.3.1. *Hypergeometric Testing of Enrichment*

Post PheWAS, phecodes were mapped to phenotypes and the phenotypes were grouped into sixteen categories from the phecodes map. We then conducted hypergeometric testing for enrichment for each phecode category within each ancestry PheWAS result. The hypergeometric distribution function HYPGEOM.DIST from excel was used to calculate fold change and significance level for each category. Threshold for significance was 0.003 to correct for multiple testing (Bonferroni correction of 0.05/16 phecode groups tested). Hypergeometric testing results were visualized by plotting the -log(p-value) of enrichment for each category as a function of fold change. Phecode categories pregnancy complication and neoplasms were visualized by graphing each phecode in the categories by -log(p-value) as a function of effect size. Plots were made with R 4.2.2.[16]

2.3.2. *Selection of Phecodes for Modeling*

In PheWAS results, we looked for phecodes that differed in relationship between EAS and SAS, and between EAFR, WAFR and NEUR, SEUR. Renal failure (RF) was selected for further modeling in EAS and SAS. The pre-made phecode categories do not always capture all relevant codes to a certain system. To focus more on the cardiac system, we extracted phenotypes using the key terms "hypertens", "heart", "card", "valv", "fibril", "coronary", and "angina." After manual review, we excluded codes pertaining to "poisoning by agents primarily affecting the cardiovascular system" and "heartburn". Selected cardiac phecodes were visualized by plotting the -log(p-value) of the phecodes as a function of effect size using R 4.2.2. From this cardiac systems plot, we selected phenotypes hypertension (HTN) and atrial fibrillation (AFib) for further modeling with EAFR, WAFR, NEUR, and SEUR.

2.4. *Logistic Regression Modeling of Select Phecodes*

Selected phenotypes were modeled as logistic regression and RCS using the R package "rms" version 6.2-0.[17] Each ancestry was used as the main predictor in separate models. Phenotypes were modeled as a function of ancestry proportion (ANC) using (Eq. 1) for logistic regression.

$$P\{Y = 1|X\} = \beta_0 + \beta_{ANC}X_{ANC} + \beta_{age}X_{age} + \beta_{sex}\beta_{sex} + \beta_{BMI}X_{BMI} \tag{1}$$

Odds ratios (OR) and confidence intervals (CI) calculated for each ancestry from logistic regression are given for a 10% increase in ancestry proportion. Phenotypes were modeled as a function of ancestry proportion using (Eq. 2) for RCS with three knots (a,b,c).

$$P\{Y = 1|X\} = \beta_0 + \beta_{ANC}X_{ANC} + \beta_{age}X_{age} + \beta_{sex}\beta_{sex} + \beta_{BMI}X_{BMI} + \beta_a(X_{ANC} + a)^3 + \beta_b(X_{ANC} + b)^3 + \beta_c(X_{ANC} + c)^3 \tag{2}$$

Knot positions were determined by default "rms" placement. Odds ratios for RCS were calculated using integrated "rms" functions for a quartile increase in ancestry from the 25th to 50th percentile and for the 50th to 75th percentile. Significance threshold for ANOVA tests of significant model improvement with RCS over linear was 0.004 (Bonferroni correction of 0.05/12 [six ancestries * two models]).

3. Results

3.1. *Genetic Ancestry of BioVU Participants*

There were 71,140 participants from BioVU, 59.06% of which were female, the average age was 54.09 (SD = 18.15), and the average BMI was 29.03 (SD = 7.27). (Table 1) Ancestry proportions for all individuals in BioVU are visualized in Figure 1. From the six ancestry proportions calculated, the ancestry group SEUR represented the largest proportion of genetic ancestry with a population average of 60.9%, followed by NEUR with 22.4%, WAFR with 6.41%, EAFR with 7.07%, SAS with 1.40%, and EAS with 1.76%. (Table 1)

3.2. *PheWAS Summarized with Hypergeometric Testing*

There were 404 phecodes significantly associated with EAFR, 396 with WAFR, 414 with SEUR, 150 with NEUR, 68 with SAS, and 74 with EAS. (Table 2) Hypergeometric testing of phecode categories identified enriched and de-enriched categories of phecodes. (Figure 2A) EAFR, WAFR and SEUR were de-enriched for 'injuries and poisonings' and 'musculoskeletal' and enriched for 'neoplasms.'

Table 1. Population characteristics of the BioVU cohort.

	Mean (SD) or N (%)
Age (years)	54.08 (18.15)
BMI (kg/m^2)	29.03 (7.27)
Sex (Females)	42016 (59%)
NEUR	22.43 (9.72)
SEUR	60.93 (23.29)
EAFR	7.07 (15.4)
WAFR	6.41 (14.09)
EAS	1.76 (9.6)
SAS	1.4 (6.05)

Kg: kilogram; m: meters

Phecodes significant within neoplasms showed opposite directions of effect for NEUR and SEUR groups versus WAFR and EAFR groups. (Figure 2B) Codes representing skin cancer and other skin neoplasms increased in odds with increasing NEUR and SEUR ancestry proportion but decreased in odds with increasing WAFR and EAFR ancestry proportion. Conversely uterine leiomyoma had increased odds with increased EAFR and WAFR ancestry proportion and decreased odds with increased SEUR and NEUR ancestry proportion. (Figure 2B) EAFR was additionally enriched for 'genitourinary.' 'Pregnancy complications' was enriched in NEUR, SEUR, EAS, and SAS. When investigated further, it was revealed the significant phecodes in the category were almost all in the decreased direction for NEUR and SEUR and increased direction for EAFR, WAFR, EAS, and SAS. (Figure 2C)

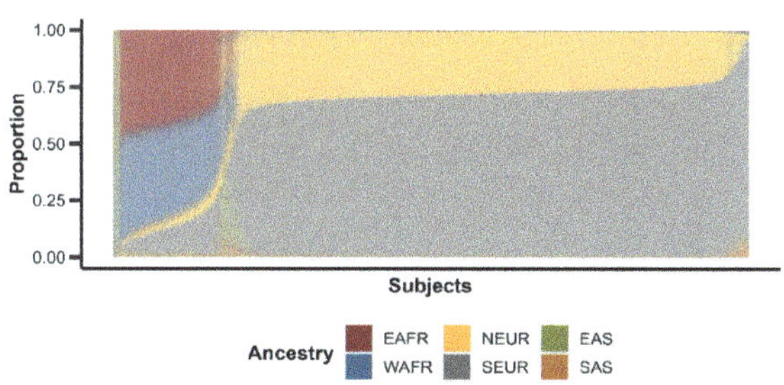

Figure 1. Structure plot of the genetic ancestry make up of BioVU Participants. Subjects are aligned on the x-axis by proportion of SEUR. NEUR: northern European; SEUR: southern European; EAFR: eastern African; WAFR: western African; EAS: eastern Asian; SAS: southern Asian ancestry.

3.3. *Modeling Ancestry Proportion*

We identified 103 phecodes that included cardiac keyword/phrases. The most significant phecodes were phecodes representing hypertension and its consequences. Increasing EAFR and WAFR ancestry proportion increases odds for the phecodes and increasing SEUR and NEUR ancestry proportion decreases odds for the conditions. Phecodes involving atrial fibrillation and related codes

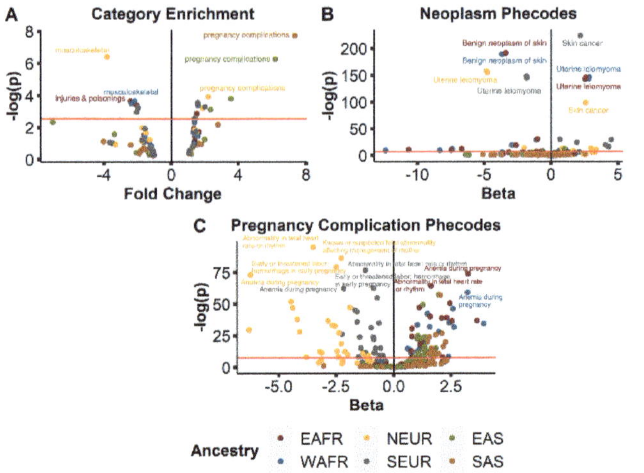

Figure 2. Volcano plots of fold change from hypergeometric testing or ancestry coefficient from PheWAS plotted against the negative log transformed p-value for A) Phecode categories B) neoplasm and C) pregnancy complications. Created with BioRender.com

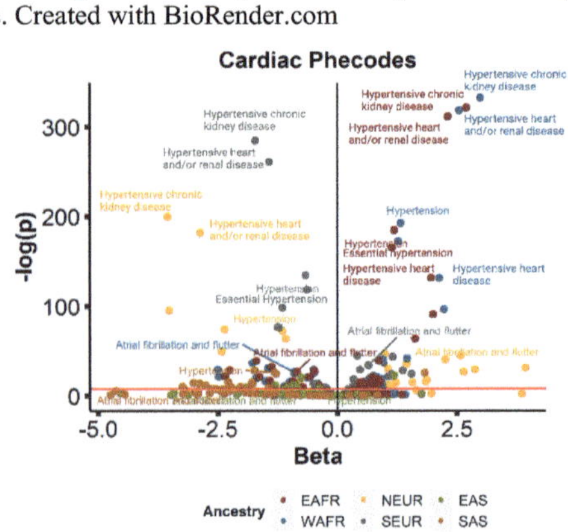

Figure 3. Volcano plot of selected phecodes related to the cardiac system. Coefficient of phecode from PheWAS is on the x-axis and the y-axis is negative log transformation of p-value. Created with BioRender.com

were significantly associated with the same ancestries, but in opposite directions: increasing NEUR and SEUR increase odds while EAFR and WAFR decrease odds. (Figure 3)

Table 2. Significant results from hypergeometric testing of phecode categories for each ancestry. Positive values indicated an enrichment of significant phecodes within that category while negative values indicate de-enrichment. Significance level is 0.003.

Ancestry ~ N Significant Codes	Fold Change	P-value
EAFR ~ 404		
genitourinary	1.39	0.003
injuries & poisonings	-2.4	2.24×10^{-4}
musculoskeletal	-2.05	3.89×10^{-4}
neoplasms	1.51	0.001
WAFR ~ 396		
injuries & poisonings	2.35	3.35×10^{-4}
musculoskeletal	-2.17	2.31×10^{-4}
neoplasms	1.57	4.13×10^{-4}
SEUR ~ 414		
injuries & poisonings	-2.05	0.001
musculoskeletal	-1.96	5.77×10^{-4}
neoplasms	1.58	2.83×10^{-4}
pregnancy complications	2	5.06×10^{-4}
NEUR ~ 150		
infectious diseases	1.87	6.38×10^{-4}
musculoskeletal	-3.83	4.17×10^{-4}
pregnancy complications	2.2	1.27×10^{-4}
SAS ~ 68		
pregnancy complications	7.32	1.97×10^{-8}
EAS ~ 74		
digestive	2.34	7.63×10^{-4}
mental disorders	3.56	1.57×10^{-4}
pregnancy complications	6.16	5.57×10^{-7}

We then investigated phecodes 401 'hypertension' (HTN) and 427.2 'atrial fibrillation' (AFib) with modeling in EAFR, WAFR, NEUR, and SEUR. (Figure 4A) When modeled linearly, each ancestry was associated with HTN and AFib ($p < 0.003$). (Table 3) When HTN and AFib were modeled using RCS, the ANOVA test revealed adding the complexity of non-linearity did significantly improve the model for NEUR ($p = 0.001$, $p < 1 \times 10^{-4}$ respectively) but not for EAFR, WAFR, and NEUR ($p > 0.003$). (Figure 4) Increasing ancestry proportion by 10% in the linear model gave an OR for HTN of 2.29 (95% CI: 2.11 - 2.48) for EAFR, 2.73 (95% CI: 2.48 - 3.01) for WAFR, 0.27 (95% CI: 0.22 - 0.33) for NEUR, and 0.73 (95% CI: 0.70 - 0.75) for SEUR, visualized in the top row panels of Figure 4A. For AFib, a 10% increase in ancestry proportion yields ORs of 0.58 (95% CI: 0.49 - 0.68) for EAFR, 0.53 (95% CI: 0.44 - 0.63) for WAFR, 4.39 (95% CI: 3.07 -

6.27) for NEUR, and 1.31 (95% CI: 1.22 - 1.40) for SEUR when modeled linearly and is visualized in the third row of panels in Figure 4A. Only NEUR had significant ANOVA p-values for the RCS models in both HTN and AFib. Increasing NEUR ancestry in RCS modeling of HTN from 25th to 50th percentile in NEUR ancestry proportion gave an OR of 0.96 (95% CI: 0.94 - 0.98) and the 50th to 75th percentile increase gave an OR of 0.99 (95% CI: 0.98 - 1.01). In RCS modeling of AFib, increase from 25th to 50th percentile in NEUR ancestry yielded an OR of 1.02 (95% CI: 0.99 - 1.06) and the 50th to 75th percentile increase yielded an OR of 0.98 (95% CI: 0.96 - 1.01). (Table 3) RCS models for HTN and AFib are visualized in the second and fourth row of panels in Figure 4A, respectively.

Table 3. Results of logistic regression and restricted cubic spline modeling for hypertension and atrial fibrillation in northern European, southern European, west African, and east African ancestry; and renal failure in eastern Asian and southern Asian ancestry.

	Logistic Regression		Restricted Cubic Spline		
	OR* (95% CI)	**P-value**	**OR± (95% CI)**	**OR ‡ (95% CI)**	**ANOVA P-value**
Atrial Fibrillation					
SEUR	1.31 (1.22-1.40)	$<1x10^{-4}$	1.00 (0.98-1.03)	1.00 (0.98-1.02)	0.06
NEUR	4.39 (3.07-6.27)	$<1x10^{-4}$	1.02 (0.99-1.06)	0.98 (0.96-1.01)	$<1x10^{-4}$
EAFR	0.58 (0.49-0.68)	$<1x10^{-4}$	0.99 (0.99-1.00)	0.97 (0.96-0.99)	0.01
WAFR	0.53 (0.44-0.63)	$<1x10^{-4}$	0.99 (0.99-1.00)	0.98 (0.96-0.99)	0.02
Hypertension					
SEUR	0.72 (0.70-0.75)	$<1x10^{-4}$	0.98 (0.96-0.99)	0.99 (0.98-1.00)	0.25
NEUR	0.27 (0.22-0.33)	$<1x10^{-4}$	0.96 (0.94-0.98)	0.99 (0.98-1.01)	0.001
EAFR	2.29 (2.11-2.48)	$<1x10^{-4}$	1.00 (0.999-1.003)	1.003 (.996-1.01)	0.30
WAFR	2.73 (2.48-3.01)	$<1x10^{-4}$	1.00 (0.998-1.002)	1.00 (0.99-1.01)	0.11
Renal Failure					
EAS	0.96 (0.73-1.26)	0.78	1.09 (1.08-1.11)	1.18 (1.15-1.21)	$<1x10^{-4}$
SAS	0.15 (0.06-0.37)	$<1x10^{-4}$	1.00 (0.98-1.03)	1.00 (0.98-1.03)	0.41

*Odds ratio given for 10% increase of ancestry proportion
±Odds ratio given for 25th to 50th percentile of ancestry proportion
‡ Odds ratio given for 50th to 75th percentile of ancestry proportion

In PheWAS results, phecode 585 ‘renal failure’ showed different relationships with EAS and SAS ancestry proportion; RF was significantly associated with SAS, but not for EAS. (Table 3) When modeled linearly, SAS ancestry proportion was significantly associated with RF ($p < 1x10^{-4}$)

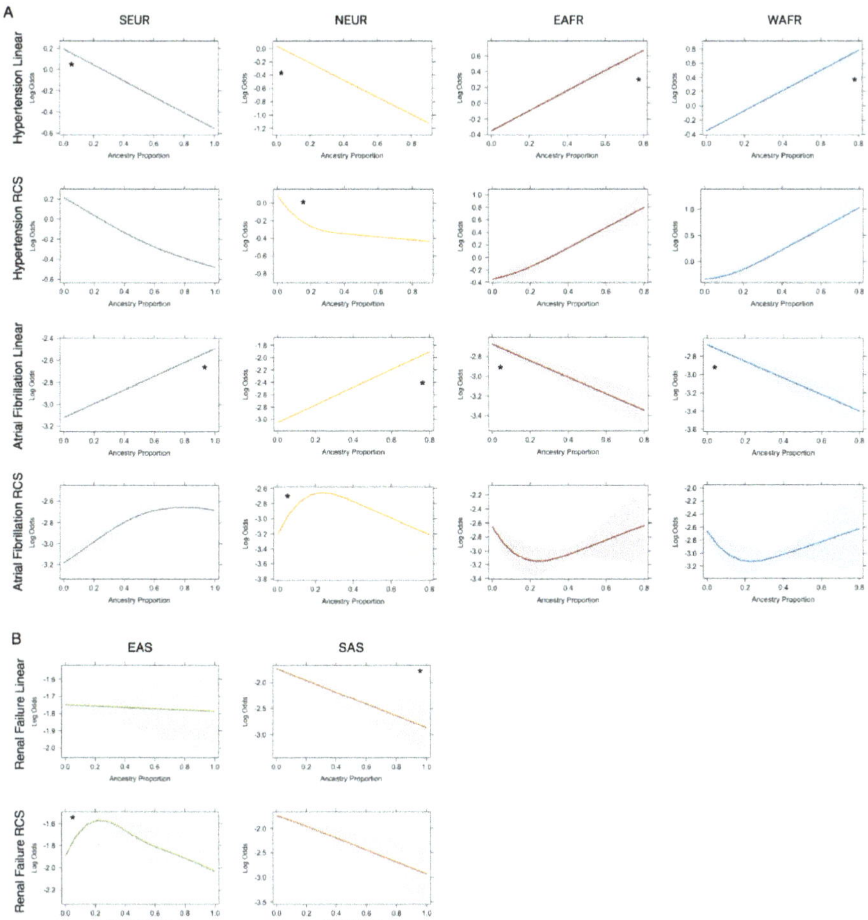

Figure 4. Linear modeling using logistic regression and restricted cubic spline (RCS) modeling of select phenotypes. A) Hypertension and atrial fibrillation risk models for SEUR, NEUR, EAFR, WAFR. B) Renal failure models for EAS and SAS. Log odds of outcome was graphed as a function of ancestry proportion adjusted for age, sex, and BMI. * = significant model. Created with BioRender.com.

but adding non-linear complexity did not significantly improve the model ($p = 0.41$). (Figure 4B) EAS was not significantly associated with RF when modeled linearly ($p = 0.78$). Modeling with the non-linear RCS revealed a significant relationship between RF and EAS ancestry proportion ($p < 1x10^{-4}$). (Figure 4B) For SAS, a 10 % increase in ancestry proportion had an OR of 0.15 (95% CI:

0.06 - 0.37) when modeled linearly. In RCS modeling, increasing from the 25th to 50th percentile of EAS ancestry proportion increases odds for RF by 1.09 (95% CI: 1.08 – 1.11) and increasing from the 50^{h} to 75th percentile increases odds by 1.18 (95% CI: 1.15 – 1.21). (Table 3)

4. Discussion

We present an evaluation of the relationships between genetic ancestry proportions and the clinical phenome of the BioVU cohort. Our analyses revealed significantly enriched and de-enriched phecode categories for each ancestry group studied. We further evaluated the relationship between genetic ancestry and risk for HTN, AFib, and RF using linear and non-linear modeling methods.

4.1. *Relationships Between Ancestry and the Clinical Phenome*

Phecode categories that were de-enriched for PheWAS associations were 'injuries and poisonings' and 'musculoskeletal' for EAFR, WAFR, SEUR and EAFR, WAFR, SEUR, NEUR respectively. Both categories represent codes that are not conditions typically considered heritable. 'Injuries and poisonings' category comprises codes related to non-pathologic fractures, trauma injuries, and poisonings, all events caused by environment. Phecodes in musculoskeletal involve injuries or deformities of joints, bones, and muscles acquired from usage of the body. One specific phenotype to mention in this category is osteoporosis, where increasing NEUR and SEUR ancestry increased risk for codes relating to osteoporosis (phecodes 743, 743.1, 743.11) and spine curvature (737, 737.3), while the same codes have a protective effect with increasing EAFR and WAFR ancestry. Studies have shown increased bone mineral density and lower rates of osteoporosis associated in Black women compared to non-Hispanic White women.[18] Our genetic ancestry study findings support this previously observed epidemiological relationship.

In the 'neoplasm' category, many of the phecodes were in the risk direction for SEUR and NEUR ancestries and in the protective direction for WAFR and EAFR. The top significant neoplasm codes refer to skin cancer and other neoplasms of skin. The biological relationship between geographic ancestry and skin cancer has been well documented; populations in equatorial regions produce more melanin to protect against DNA damage from UV radiation while populations out towards the poles have evolved to produce less melanin due to less UV exposure.[19,20] It is possible that individuals of European genetic ancestry migrated away from the environments where they adapted to be at equilibrium and are now in new environments they are at disequilibrium with.[21]

One of the few exceptions to the pattern seen in 'neoplasms' were the phecodes 218 and 218.1, representing 'uterine leiomyoma' (or fibroids). Increasing EAFR and WAFR ancestries increases odds for fibroids while increasing SEUR and NEUR ancestries was protective against fibroids. This relationship pattern is consistent with previous epidemiology literature. Black women have been found to develop fibroids at younger ages, were more likely to have a clinical diagnosis, and to have had a hysterectomy from fibroids.[22] The overall odds of developing fibroids by age 50 were 2.9 times higher among Black women compared to White women.[22] Due to the significant racial disparities that exist for fibroids, it has been hypothesized that there is a genetic component to the condition, with a heritability estimate of ~30%.[23] Previous genetic studies have found African genetic ancestry proportion to be associated with fibroids diagnosis[12] and multiple fibroids.[24] Our study further supports the theory that African genetic ancestry may explain a portion of the risk for fibroids.

The pregnancy complication category was significantly enriched in NEUR, SEUR, EAS, and SAS. Within the category, significantly associated phecodes were all in the protective direction for NEUR and SEUR and in the risk direction for EAS and SAS. Racial disparities in maternal health outcomes have been well documented for White and Black women, with Black women having significantly higher adverse maternal outcomes compared to White women.[25] There have been many external factors posited for why Black women in US experience pregnancy complications and maternal mortality at much higher rates.[26] Trends in pregnancy complications for Asian women are less well-documented. A study of fertility treatment outcomes in Asian American women found decreased success of treatment in the forms of lower pregnancy rates and live births.[27] Using genetic ancestry proportions as a study variable may help to fill in some of the missing epidemiological gaps that still are pervasive in historically under-represented racial groups.

4.2. *Modeling Ancestry Proportion Linearly and non-Linearly*

From the phecodes grouped into the cardiac category, we saw a striking pattern. Several phecodes representing HTN and hypertensive disorders and consequences were found to be at increased risk in EAFR and WAFR and decreased risk in NEUR and SEUR. An opposite trend was seen for phecodes representing AFib and related codes; SEUR and NEUR were at increased risk while EAFR and WAFR were at decreased risk. This pattern follows what has been reported in literature.[28-30] Our study shows the trends we see for HTN and AFib are due in part to genetic ancestry.

While plenty of studies have focused on external causes and contributions to the higher prevalence of HTN in Black individuals,[31] it is known to be heritable.[32] A small (N = 998), previous study evaluated the relationship between African genetic ancestry proportion in self-identified Black individuals and hypertension and found the highest quartile of African genetic ancestry proportion had 8% higher prevalence than the lowest quartile.[33] Marden et al. used African genetic ancestry proportion to tease apart the contributions of genetics and socioeconomic status to HTN prevalence and found that their accounted socioeconomic factors only explained one-third of the difference in prevalence measured.[33] We as well sought to use genetic ancestry to determine its contribution to HTN disease risk as it helps to avoid confounders. The previous study and ours have both found African genetic ancestry to be associated with HTN risk and prevalence.

Within our evaluation of RF, we found linear modeling to be sufficient to model the relationship with SAS ancestry. EAS ancestry was not significantly associated with RF in PheWAS or when modeled individually linearly. Allowing for flexibility with non-linear RCS modeling revealed a relationship between EAS ancestry and RF. Only with the RCS model were we able to detect an OR of 1.18 (95% CI: 1.15-1.21) with an increase from 50th to 75th percentile of EAS ancestry. EAS was the ancestry group with the most skewed data density of the six groups, with the 3rd quartile ancestry proportion value being just 0.45% and one of our smallest sub-sample sizes with 760 self-identified individuals. The RCS model may have performed better due to being able to compensate for the skewness of data. Many wide-scale analyses perform only linear modeling which may not detect relationships, as seen for RF in EAS ancestry PheWAS. The risk trends for EAS and RF from RCS modeling have been reported previously in literature. Higher rates of end stage renal disease and increased risk of projected kidney failure have been observed in Far East, Southeast Asia, and Indian populations as compared to White populations.[34,35] The linear model for SAS and RCS model for EAS recapitulate these findings. Assuming linear relationships between genetics and disease may cause associations to be missed, highlighting the need to consider non-linear modeling methods such as RCS.

4.3. *Considerations and Strengths*

While our study found some phenotype relationships that were consistent with epidemiology studies based on self-identified race, we did not evaluate the potential contribution of proportions of admixture on disease risk. On average, people had an admixture proportion of 0.33 (+/- 0.12) amongst the 6 super populations we determined with the more granular division of Southern and Northern European, Eastern and Western African, and East and South Asian. Within our cohort, those who self-identified as non-Hispanic Black had on average 78.6% African ancestry (EAFR + WAFR), 19.4% European ancestry (SEUR + NEUR), and 1.99% Asian ancestry (EAS + SAS). Those who self-identified as non-Hispanic White had on average 6.85% African ancestry, 98.0% European ancestry, and 1.26% Asian ancestry. Our study was limited in its ability to test more admixed populations where these methods may be more useful in identifying phenotypes associated with genetic ancestry.

We only used one ancestry as a predictor variable per model. Different geographic ancestries may interact differently, and this study does not account for various combinations of genetic ancestry proportions. Further investigation is needed to understand how the different genetic ancestries interact with each other and modify risk. A potential limitation of our study is the way in which some phenotypes may be diagnosed. Some phenotypes such as chronic kidney disease rely on algorithms that use self-reported race as a criterion to determine diagnosis, for example estimated glomerular filtration rate (eGFR) algorithms have historically used race as a coefficient in the equation for measuring eGFR levels which may bias diagnoses across racial and ethnic groups.[36]

In this study, we identified hundreds of traits in the clinical phenome that are associated with ancestry proportion. From our selected studies of enriched phecode categories and modeling of HTN and AFib, we observed many relationships between ancestry and phecodes that matched the epidemiology literature between self-identified race and traits. We used RCS to model a significant relationship between RF and EAS ancestry, one that was not originally identified from linear modeling. We highlighted a few phenotypes in this paper as an exploratory investigation into the potential of RCS modeling for ancestry proportion and disease risk.

Most traditional epidemiology literature notes the shortcomings of their studies revolve around using the societal construct of race, a lack of healthcare access for underrepresented groups and low-income individuals, and external environmental factors. Adjusting for race to better account for these factors like socioeconomic status or systemic discrimination in addition to using genetic ancestry proportion, which capture heritable contributions, may provide more comprehensive models. Future work controlling for genetic ancestry that demonstrates significant associations with race would highlight systemic factors affecting outcomes that are not captured by ancestry alone. In addition to utilizing genetic ancestry, we show how alternative modeling methods can be useful especially in a case of an underrepresented ancestry group where linear models may not be as successful to describe more complicated associations. Our study displays how genetic ancestry can be leveraged in furtherance of studying disease risk where traditional epidemiological studies have fallen short.

Acknowledgements

We would like to thank the participants of BioVU for their enrollments.

References

1. Yudell, M., Roberts, D., DeSalle, R., and Tishkoff, S. (2016). Taking race out of human genetics. Science (American Association for the Advancement of Science) *351*, 564-565. 10.1126/science.aac4951.
2. Vyas, D.A., Eisenstein, L.G., and Jones, D.S. (2020). Hidden in Plain Sight — Reconsidering the Use of Race Correction in Clinical Algorithms. New England Journal of Medicine *383*, 874-882. 10.1056/NEJMms2004740.
3. Paulus, J.K., Wessler, B.S., Lundquist, C.M., and Kent, D.M. (2018). Effects of Race Are Rarely Included in Clinical Prediction Models for Cardiovascular Disease. J Gen Intern Med *33*, 1429-1430. 10.1007/s11606-018-4475-x.
4. Borrell, L.N., Elhawary, J.R., Fuentes-Afflick, E., Witonsky, J., Bhakta, N., Wu, A.H.B., Bibbins-Domingo, K., Rodríguez-Santana, J.R., Lenoir, M.A., Gavin, J.R., et al. (2021). Race and Genetic Ancestry in Medicine — A Time for Reckoning with Racism. The New England journal of medicine *384*, 474-480. 10.1056/NEJMms2029562.
5. Kumar, R., Nguyen Ea Fau - Roth, L.A., Roth La Fau - Oh, S.S., Oh Ss Fau - Gignoux, C.R., Gignoux Cr Fau - Huntsman, S., Huntsman S Fau - Eng, C., Eng C Fau - Moreno-Estrada, A., Moreno-Estrada A Fau - Sandoval, K., Sandoval K Fau - Peñaloza-Espinosa, R.I., Peñaloza-Espinosa Ri Fau - López-López, M., et al. (2013). Factors associated with degree of atopy in Latino children in a nationwide pediatric sample: the Genes-environments and Admixture in Latino Asthmatics (GALA II) study. J Allergy Clin Immunol *132*, 896-905. 10.1016/j.jaci.2013.02.046.
6. Neophytou, A.M., White, M.J., Oh, S.S., Thakur, N., Galanter, J.M., Nishimura, K.K., Pino-Yanes, M., Torgerson, D.G., Gignoux, C.R., Eng, C., et al. (2016). Air Pollution and Lung Function in Minority Youth with Asthma in the GALA II (Genes-Environments and Admixture in Latino Americans) and SAGE II (Study of African Americans, Asthma, Genes, and Environments) Studies. Am J Respir Crit Care Med *193*, 1271-1280. 10.1164/rccm.201508-1706OC.
7. Choudhury, A., Aron, S., Botigué, L.R., Sengupta, D., Botha, G., Bensellak, T., Wells, G., Kumuthini, J., Shriner, D., and Fakim, Y.J. (2020). High-depth African genomes inform human migration and health. Nature *586*, 741-748.
8. Roden, D.M., Pulley Jm Fau - Basford, M.A., Basford Ma Fau - Bernard, G.R., Bernard Gr Fau - Clayton, E.W., Clayton Ew Fau - Balser, J.R., Balser Jr Fau - Masys, D.R., and Masys, D.R. (2008). Development of a large-scale de-identified DNA biobank to enable personalized medicine. Clin Pharmacol Ther *84*, 362-369. 10.1038/clpt.2008.89.
9. Das, S., Forer, L., Schönherr, S., Sidore, C., Locke, A.E., Kwong, A., Vrieze, S.I., Chew, E.Y., Levy, S., McGue, M., et al. (2016). Next-generation genotype imputation service and methods. Neat Genet *48*, 1284-1287. 10.1038/ng.3656.
10. McCarthy, S.A.-O., Das, S.A.-O., Kretzschmar, W.A.-O., Delaneau, O., Wood, A.R., Teumer, A., Kang, H.M., Fuchsberger, C., Danecek, P., Sharp, K., et al. (2016). A reference panel of 64,976 haplotypes for genotype imputation. Neat Genet *48*, 1279-1283. 10.1038/ng.3643.
11. Edwards, T.L., Giri, A., Hellwege, J.N., Hartmann, K.E., Stewart, E.A., Jeff, J.M., Bray, M.J., Pendergrass, S.A., Torstenson, E.S., Keaton, J.M., et al. (2019). A Trans-Ethnic Genome-Wide Association Study of Uterine Fibroids. Front Genet *10*, 511. 10.3389/fgene.2019.00511.
12. Keaton, J.M., Jasper, E.A., Hellwege, J.N., Jones, S.H., Torstenson, E.S., Edwards, T.L., and Velez Edwards, D.R. (2021). Evidence that geographic variation in genetic ancestry associates with uterine fibroids. Human genetics *140*, 1433-1440.
13. Alexander, D.H., and Lange, K. (2011). Enhancements to the ADMIXTURE algorithm for individual ancestry estimation. BMC bioinformatics *12*, 1-6.
14. Denny, J.C., Ritchie, M.D., Basford, M.A., Pulley, J.M., Bastarache, L., Brown-Gentry, K., Wang, D., Masys, D.R., Roden, D.M., and Crawford, D.C. (2010). PheWAS: demonstrating the feasibility of a phenome-wide scan to discover gene-disease associations. Bioinformatics *26*, 1205-1210. 10.1093/bioinformatics/btq126.

15. Wu, P., Gifford, A., Meng, X., Li, X., Campbell, H., Varley, T., Zhao, J., Carroll, R., Bastarache, L., and Denny, J.C. (2018). Developing and Evaluating Mappings of ICD-10 and ICD-10-CM codes to Phecodes. BioRxiv, 462077.
16. Team, R.C. (2013). R: A language and environment for statistical computing.
17. Harrell, F.E. (2023). rms: Regression Modeling Strategies (CRAN).
18. Barrett-Connor, E., Siris, E.S., Wehren, L.E., Miller, P.D., Abbott, T.A., Berger, M.L., Santora, A.C., and Sherwood, L.M. (2005). Osteoporosis and fracture risk in women of different ethnic groups. Journal of bone and mineral research *20*, 185-194.
19. Agbai, O.N., Buster, K., Sanchez, M., Hernandez, C., Kundu, R.V., Chiu, M., Roberts, W.E., Draelos, Z.D., Bhushan, R., and Taylor, S.C. (2014). Skin cancer and photoprotection in people of color: a review and recommendations for physicians and the public. Journal of the American Academy of Dermatology *70*, 748-762.
20. Avise, J.C., and Ayala, F.J. (2010). Human skin pigmentation as an adaptation to UV radiation. In In the Light of Evolution: Volume IV: The Human Condition, (National Academies Press (US)).
21. Asadi, L.K., Khalili, A., and Wang, S.Q. (2023). The sociological basis of the skin cancer epidemic. International Journal of Dermatology *62*, 169-176.
22. Baird, D.D., Dunson, D.B., Hill, M.C., Cousins, D., and Schectman, J.M. (2003). High cumulative incidence of uterine leiomyoma in black and white women: Ultrasound evidence. American Journal of Obstetrics and Gynecology. 10.1067/mob.2003.99.
23. Bray, M.J., Davis, L.K., Torstenson, E.S., Jones, S.H., Edwards, T.L., and Velez Edwards, D.R. (2019). Estimating uterine fibroid SNP-based heritability in European American women with imaging-confirmed fibroids. Human heredity *84*, 73-81.
24. Bray, M.J., Edwards, T.L., Wellons, M.F., Jones, S.H., Hartmann, K.E., and Velez Edwards, D.R. (2017). Admixture mapping of uterine fibroid size and number in African American women. Fertility and Sterility *108*, 1034-1042.e1026. https://doi.org/10.1016/j.fertnstert.2017.09.018.
25. MacDorman, M.F., Thoma, M., Declcerq, E., and Howell, E.A. (2021). Racial and ethnic disparities in maternal mortality in the United States using enhanced vital records, 2016–2017. American journal of public health *111*, 1673-1681.
26. Saluja, B., and Bryant, Z. (2021). How implicit bias contributes to racial disparities in maternal morbidity and mortality in the United States. Journal of women's health *30*, 270-273.
27. Vu, M.H., Nguyen, A.-T.A., and Alur-Gupta, S. (2022). Asian Americans and infertility: genetic susceptibilities, sociocultural stigma, and access to care. F&S Reports *3*, 40-45.
28. Zilbermint, M., Hannah-Shmouni, F., and Stratakis, C.A. (2019). Genetics of hypertension in African Americans and others of African descent. International journal of molecular sciences *20*, 1081.
29. Keaton, J.M., Hellwege, J.N., Giri, A., Torstenson, E.S., Kovesdy, C.P., Sun, Y.V., Wilson, P.W., O'Donnell, C.J., Edwards, T.L., and Hung, A.M. (2021). Associations of biogeographic ancestry with hypertension traits. Journal of hypertension *39*, 633.
30. Marcus, G.M., Alonso, A., Peralta, C.A., Lettre, G., Vittinghoff, E., Lubitz, S.A., Fox, E.R., Levitzky, Y.S., Mehra, R., and Kerr, K.F. (2010). European ancestry as a risk factor for atrial fibrillation in African Americans. Circulation *122*, 2009-2015.
31. Usher, T., Gaskin, D.J., Bower, K., Rohde, C., and Thorpe Jr, R.J. (2018). Residential segregation and hypertension prevalence in black and white older adults. Journal of Applied Gerontology *37*, 177-202.
32. Kolifarhood, G., Daneshpour, M., Hadaegh, F., Sabour, S., Mozafar Saadati, H., Akbar Haghdoust, A., Akbarzadeh, M., Sedaghati-Khayat, B., and Khosravi, N. (2019). Heritability of blood pressure traits in diverse populations: a systematic review and meta-analysis. Journal of human hypertension *33*, 775-785.
33. Marden, J.R., Walter, S., Kaufman, J.S., and Glymour, M.M. (2016). African ancestry, social factors, and hypertension among non-Hispanic Blacks in the Health and Retirement Study. Biodemography and social biology *62*, 19-35.
34. Derose, S.F., Rutkowski, M.P., Crooks, P.W., Shi, J.M., Wang, J.Q., Kalantar-Zadeh, K., Kovesdy, C.P., Levin, N.W., and Jacobsen, S.J. (2013). Racial differences in estimated GFR decline, ESRD, and mortality in an integrated health system. American journal of kidney diseases *62*, 236-244.
35. Kataoka-Yahiro, M., Davis, J., Gandhi, K., Rhee, C.M., and Page, V. (2019). Asian Americans & chronic kidney disease in a nationally representative cohort. BMC nephrology *20*, 1-10.

36. Uppal, P., Golden, B.L., Panicker, A., Khan, O.A., and Burday, M.J. (2022). The Case Against Race-Based GFR. Delaware Journal of Public Health *8*, 86.

Machine Learning Strategies for Improved Phenotype Prediction in Underrepresented Populations

David Bonet[1,2], May Levin[1], Daniel Mas Montserrat[1], and Alexander G. Ioannidis[1,3]
[1] *Stanford University, Stanford, CA, US*
[2] *Universitat Politècnica de Catalunya, Barcelona, Spain*
[3] *University of California Santa Cruz, Santa Cruz, CA, US*
E-mail: ioannidis@stanford.edu

Precision medicine models often perform better for populations of European ancestry due to the over-representation of this group in the genomic datasets and large-scale biobanks from which the models are constructed. As a result, prediction models may misrepresent or provide less accurate treatment recommendations for underrepresented populations, contributing to health disparities. This study introduces an adaptable machine learning toolkit that integrates multiple existing methodologies and novel techniques to enhance the prediction accuracy for underrepresented populations in genomic datasets. By leveraging machine learning techniques, including gradient boosting and automated methods, coupled with novel population-conditional re-sampling techniques, our method significantly improves the phenotypic prediction from single nucleotide polymorphism (SNP) data for diverse populations. We evaluate our approach using the UK Biobank, which is composed primarily of British individuals with European ancestry, and a minority representation of groups with Asian and African ancestry. Performance metrics demonstrate substantial improvements in phenotype prediction for underrepresented groups, achieving prediction accuracy comparable to that of the majority group. This approach represents a significant step towards improving prediction accuracy amidst current dataset diversity challenges. By integrating a tailored pipeline, our approach fosters more equitable validity and utility of statistical genetics methods, paving the way for more inclusive models and outcomes.

Keywords: Genetics; Precision Medicine; Machine Learning; Phenotype Prediction; Bioinformatics.

1. Introduction

In recent years, genome-wide association studies (GWAS) have provided many insights into the genetic basis of complex traits and diseases. However, these findings predominantly benefit populations of European descent due to their over-representation in genomic datasets. Individuals with Asian, African, and other ancestries only represent a small fraction of the available datasets.[1] Although individuals of European descent constitute ~79% of GWAS participants,[2] they account for less than a quarter of the global population. This disproportionate representation creates a limitation in precision medicine, because statistical models built to infer disease risks or health-related traits can perform poorly for individuals from populations that were underrepresented when creating the model, exacerbating health disparities. Despite initiatives to include a broader range of populations in genetic studies and biobanks,[3–7] the proportion of non-European individuals in GWAS studies has stagnated in the last decade.[2,8] This imbalance has a direct impact on Polygenic Risk Score (PRS) prediction for underrepresented populations,[9] making clinical applications based on PRS significantly more accurate for individuals of European descent, but less effective for other populations.[10–12] This disparity has raised ethical concerns within the scientific and clinical community.[1,3,8,13] While most studies only use European individuals and European-derived statistics to build predictive models,[8,11,14] recent studies have explored including non-European training data in PRS construction, but this has only proven effective when a large number of training samples of non-European target populations are available.[15]

Phenotype prediction utilizes genetic information to forecast an organism's observable characteristics, known as phenotypes. These traits can range from disease susceptibility to other attributes, enabling personalized treatments based on individual genetic profiles. Machine learning (ML) and deep learning (DL) models used to predict phenotype and population structure from genomic data[14,16–20] are similarly negatively impacted by imbalanced datasets. Vokinger et al.[21] highlighted the presence of bias in ML-based medicine prediction pipelines. Specifically, they revealed how a naive application of simple ML methods can showcase an overall good performance, yet still produce biased predictions favoring the majority population at the cost of lower accuracy for underrepresented groups. Efforts to mitigate this bias exist, such as Afrose et al.[22] who created a double prioritized bias correction technique that involves training customized prediction models for specific subpopulations. However, this approach is limited to binary classification tasks and is not generalizable to other prediction problems.

Conventionally, the statistical methods that are applied for genomic prediction problems linearly combine the effects of different genetic variants on an individual's risk of disease. Some of the most widely used regression models include Lasso,[23] a linear method with ℓ_1 penalty, Elastic net[24] with ℓ_1 and ℓ_2 penalty, and efficient implementations of both.[14] Although being the routine choice in most studies, linear models are not able to capture non-linear genetic interactions that can contribute to a phenotype.[25] The ability of non-linear predictive models to capture genetic interactions could help improve performance generalization across populations.[26,27] Neural networks, a complex non-linear method, have recently gained traction in computational biology,[28,29] but require vast amounts of data for training. Large-scale biobanks, such as the UK Biobank,[30] provide such expansive datasets. However, the small proportion of samples from minority populations hinders robust generalization across different genetic backgrounds. In contrast, gradient boosting (GB) algorithms,[31] such as eXtreme Gradient Boosting (XGBoost)[32] and LightGBM,[33] have frequently demonstrated superior performance for tabular data and small-sized datasets,[34,35] and have already been explored in biological studies for tasks such as local ancestry inference,[36] protein-protein interactions,[37] and drug-gene interactions.[38] In the realm of genotype-to-phenotype prediction, recent research has also highlighted the potential benefits of using such nonlinear predictive models.[39,40]

In this paper, we aim to improve phenotype prediction for diverse and underrepresented populations. We propose a more inclusive genomic research approach that uses multi-ancestry data together with advanced machine learning techniques to boost the predictability of complex traits across a broader range of populations. Our method leverages several machine learning techniques such as boosting, and ensembling, and we propose population-conditional weighting and re-sampling techniques to generate more accurate models for underrepresented populations without requiring large sample sizes of non-European training data. Fig. 1 illustrates the workflow of our approach, starting with the formation of the data set through the application of various machine learning techniques and data de-biasing methods. We compare our approach with state-of-the-art statistical genetics models on the UK Biobank, conducting a systematic evaluation across 12 phenotypes in European (British), African, East Asian, and South Asian individuals. Given that the majority population is of European descent, we observe a large gap in phenotype prediction accuracy for minority populations when using classical linear methods. This disparity only grows when European-only data is used to train any of the prediction models. We demonstrate how the application of our method helps narrow this accuracy gap, balance the performance across populations, and obtain state-of-the-art phenotype prediction results for multi-ancestry datasets.

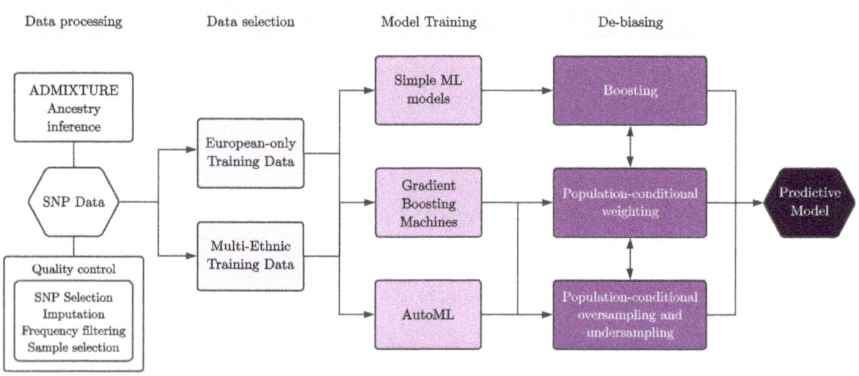

Fig. 1. A schematic representation of our predictive modeling pipeline, starting from the initial data ingestion to the application of various ML methods and de-biasing techniques.

2. Methods

2.1. *Dataset preparation*

We utilize a dataset extracted from the UK Biobank[30] that includes European (British), South Asian, African, and East Asian individuals (see Fig. 2). We use the pre-computed population labels from the Global Biobank Engine (GBE),[41] inferred based on genetic clustering with ADMIXTURE software[42] results, which provides a maximum likelihood estimation of an in-

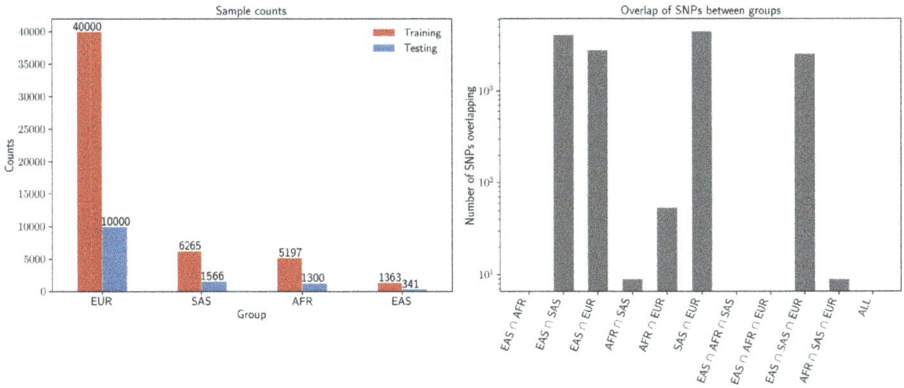

Fig. 2. EUR - European British, SAS - South Asian, AFR - African, EAS - East Asian (Left) Sample counts per group in the training and testing set. (Right) Percentage of SNP overlap between the selected sets of SNPs per group using the MAF filter.

dividual's genetic ancestry clustering from multilocus genotype datasets.

Single nucleotide polymorphism (SNP) sequences are encoded using a ternary system, where at each genomic position, an individual i has $n_{ij} \in \{0, 1, 2\}$ copies of the minority SNP j. To address high dimensionality and retain the most informative SNPs, we apply a SNP selection process. Minor allele frequency (MAF) filtering is applied with a 1.25% threshold, keeping a set $\mathcal{S}_p$ of 10000 SNPs for each population $p \in P$, such that $|\mathcal{S}_p| = 10000$. After SNP selection for each population, we computed the union of these sets. It is important to note that not all sets necessarily overlap with every other set. The union is represented by:

$$\mathcal{S}_{\text{union}} = \bigcup_{p \in P} \mathcal{S}_p \tag{1}$$

This resulted in a unified set of SNPs where $|\mathcal{S}_{\text{union}}| = 31153$, which is then used for all individuals, creating a dataset of 66032 individuals and 31153 features. Fig. 2 shows the intersection size of the sets of selected SNPs $\mathcal{S}_p$ for all intersections of populations $p \in P$. We observe that the highest overlap is between South Asian, East Asian, and European populations, while the selected set of SNPs for the African population has practically no overlap with the others. Any subsequent missing SNPs within the samples underwent mode imputation to ensure data completeness.

To conduct our experiments, we study a set of phenotypes included in the GBE,[41] listed in Table 1. Details regarding the correspondence of the GBE to the UK Biobank can be found in the GBE paper. We selected the available phenotypes with minimal missing data for the minority populations, and that also showed good predictive performance from genotype features.[43] We analyze both binary phenotypes (absence or presence of the phenotype) and continuous phenotypes to evaluate model performance across both classification and regression tasks.

Additionally, we ensured there is no missing data and filter samples that have missing phenotypic information. The dataset is partitioned into a training set and a testing set, comprising 80% and 20% of the data, respectively. We applied stratified sampling, ensuring the proportion of samples from each population closely mirrors their proportion in the overall dataset.

Table 1. We present results on 12 phenotypes, 10 continuous and 2 binary ones.

Variable	**Type**	**Variable**	**Type**
Standing height	Continuous	Weight	Continuous
Ankle spacing width	Continuous	Impedance of whole body	Continuous
HDL cholesterol	Continuous	Apolipoprotein A	Continuous
Urate	Continuous	Total bilirubin	Continuous
Plateletcrit	Continuous	Red blood cell (erythrocyte) count	Continuous
Diabetes	Binary	Atrial fibrillation	Binary

2.2. *Algorithmic models*

We explore a wide range of machine learning methods to improve phenotype prediction on underrepresented populations. Some algorithms serve as standalone models, capable of making predictions without supplementary techniques. Other algorithms we describe in this section, such as boosting, are techniques that can be used to further improve the performance of a base machine learning model. Finally, we explore complex machine learning systems that combine multiple models and automate the process of machine learning.

Linear models We include the Least Absolute Shrinkage and Selection Operator (Lasso),[23] a linear regression method that performs variable selection (i.e., identifies the most important predictors) and regularization, which prevents overfitting by constraining the model parameters. It does this by imposing an ℓ_1 penalty, effectively reducing some coefficients to zero. We also use Elastic Net,[24] a regularized method that combines ℓ_1 and ℓ_2 penalties, allowing coefficient shrinkage and feature selection.

Boosting We consider boosting,[44] a powerful ensemble machine learning technique that constructs a strong predictive model by combining multiple *weak learners*—simple models— that are trained sequentially. In each iteration of the boosting process, a new weak learner is trained giving more importance to the instances that were poorly predicted by the previous models, meaning the model attempts to correct the errors of its predecessors. This procedure is repeated sequentially, with each new model targeting the instances where the combined ensemble has performed the worst. The final model is a weighted combination of all the weak models, which often yields a strong predictive performance by aggregating the strengths of all individual models. Decision trees are the most common type of weak learners used in boosting algorithms. However, we also study how boosting can help improve predictive performance when traditional linear methods used in the field, such as Elastic Net, are used as weak learners.

Gradient boosting machines A specific implementation of the boosting techniques are gradient boosting machines (GBM). The key idea behind GBMs is the use of the gradient descent algorithm to minimize a loss function, which quantifies how well the model predicts the target variable. In each iteration, rather than directly focusing on the poorly predicted instances, a new decision tree is fit to the negative gradient (residuals) of the loss function with respect to the prediction of the ensemble model from the previous stage. This new decision tree provides a direction in which the prediction should be adjusted to minimize the loss function. The predictions are then updated by taking a step in this direction. Extreme Gradient Boosting (XGBoost)[32] and LightGBM[33] are two optimized implementations of GBMs that have gained significant popularity due to their efficiency and performance. XGBoost offers several advanced features such as regularized boosting, handling of missing values, and tree-pruning that makes it faster and more robust. LightGBM also offers high performance and efficiency but is particularly notable for its effectiveness with large datasets and high-dimensional data, due to its innovative histogram-based algorithm that reduces memory usage and speeds up training.

AutoML Automated Machine Learning (AutoML)[45] refers to the automated process of end-to-end model development, encompassing steps from feature engineering to model selection, hyperparameter tuning, and model evaluation. AutoML methods have been developed to streamline the machine learning pipeline while reducing time and expertise required to develop effective predictive models. In particular, we consider AutoGluon[46] (AG), a state-of-the-art AutoML framework known for its robust performance, efficiency and ease of use. AutoGluon automatically trains and optimizes multiple models such as neural networks, nearest neighbors, linear models, and gradient boosting machines, combining them into a stacked ensemble.

2.3. *Population-conditional re-sampling solutions*

We introduce a set of population-conditional re-sampling techniques to address population imbalance in datasets. These techniques serve as auxiliary methods designed to reduce model bias towards the majority population and can be integrated with any predictive model. While we focus on human populations in this work, these techniques can also be applied to any data where samples can be grouped into different populations, groups, or categories. Moreover, they are suitable for tasks beyond single-target classification, such as regression, and they can also be extended to multi-output tasks.

Population-conditional oversampling and undersampling We modify the traditional oversampling and undersampling techniques used in imbalanced classification tasks, and adapt them to address imbalances at the population level, regardless of the target variables (both categorical and continuous). We organize the training dataset as $\boldsymbol{X} \in \mathbb{R}^{N \times d}$ such that each row represents an individual, and the target variable or variables are concatenated to the rest of the input features as the final attributes. The population label is then used as a downstream label $\boldsymbol{y}' \in \mathbb{R}^N$ for the oversampling or undersampling rule, originally designed to work with single-target imbalanced classification datasets, such that the "minority" samples are those pertaining to the populations with lowest representation in the dataset. After this procedure, we discard the population labels and split the columns of the re-sampled training dataset as features and targets and fit the prediction models.

We explore population-conditional random oversampling (OS) by picking samples at random with replacement from the minority populations. We also adapt the Synthetic Minority Over-sampling Technique (SMOTE),[47] which is commonly used to address class imbalances by generating synthetic samples. Our modification enables us to synthetically increase the number of instances from the minority populations in the training set. Note that in the case of regression tasks, our approach differs from existing SMOTE variations for regression,[48,49] which involve identifying "minority" samples based on the distribution of the target values rather than external categorical labels associated with the samples. Finally, we also consider adapting the SMOTE-Edited Nearest Neighbours (SMOTE-ENN) algorithm,[50] a method that combines both oversampling and undersampling techniques. Our proposed population-conditional variation can also be applied to any other re-sampling technique originally designed to address class imbalance in classification problems.

Population-conditional weighting In a similar fashion, traditional class-based sample weighting techniques for class imbalance give more importance to underrepresented classes in the target variable. In contrast, we propose to emphasize the individual instances from underrepresented populations given the population labels each sample has assigned, regardless of their target variable. We calculate N_p, the size (i.e. number of samples) of each population $p \in P$ in the training set, and assign a weight $w_p = \frac{N}{N_p}$ to each sample corresponding to population p, inversely proportional to the size of its population, where N is the total size of the training dataset.

2.4. *Evaluation setup*

For training, data is either filtered to only contain European ancestry individuals, mirroring the typical bias seen in many genetic studies, or contain the complete, multi-ethnic dataset that includes individuals from underrepresented populations. The testing data is fixed and contains samples from each population group, allowing the assessment and model performance comparison across each population in all the experiments. Model hyperparameters are adjusted by 5-fold cross validation, with hyperparameter configurations drawn from comprehensive search spaces until 1000 configurations are explored or a search budget of 120 hours is reached. Then, the model is fitted on the full training set with the chosen hyperparameter configuration, and evaluated on the held out test set (20% of the data).

Predictive performance is evaluated using the coefficient of determination (R^2) for regression tasks, and the Area Under the Receiver Operating Characteristic Curve (ROC AUC) for classification tasks. R^2 represents the proportion of variance in the predicted phenotype that is explained by the genotype, and its value lies between 0 and 1. An R^2 nearing 1 signifies the model's high accuracy in phenotype prediction using the given genetic data. In contrast, values approaching 0 highlight the model's limited predictive capability. ROC AUC measures the model's ability to distinguish between the positive and negative classes. The value ranges from 0 to 1, with 0.5 indicating performance equivalent to random chance, and values approaching 1 indicating high predictive accuracy.

3. Results

3.1. *Continuous phenotypes*

We first analyze the use of multi-ethnic data and the predictive performance of several linear and non-linear models, including Lasso, Elastic Net, LightGBM, and XGBoost, for the 10 continuous phenotypes described in Table 1. Fig. 3 shows the increase in R^2 when training the models with multi-ethnic data, compared to training with only with European individuals on a linear model (Lasso), which is the common practice in the field. Note that relative performance (ratio) cannot be computed per population, as the baseline model obtains an R^2 of 0 for some population groups when predicting some of the phenotypes. We observe that prediction performance significantly improves across all populations and methods when including multi-ethnic data in training. Specifically, the gradient boosting method LightGBM is the model that obtains the highest boost in predictive performance consistently across all ancestry backgrounds, including European and underrepresented ones.

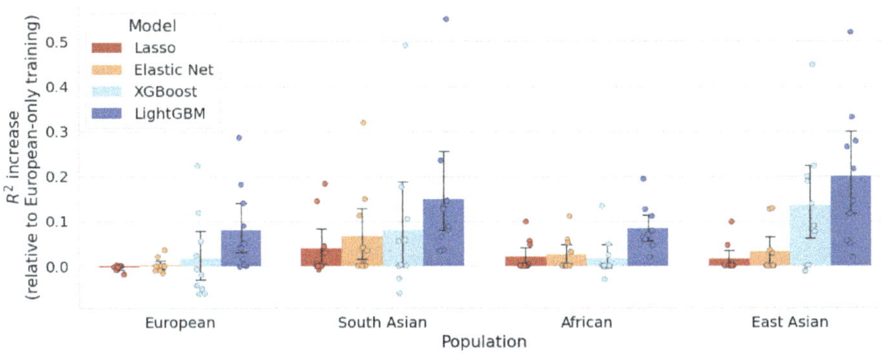

Fig. 3. Aggregated results of increase in R^2 for the 10 continuous phenotypes, with a 95% confidence interval, comparing the scores for models trained on multi-ethnic data (including populations underrepresented in the UK Biobank) versus models trained exclusively on the British-with-European-ancestry population.

In an effort to gain deeper insights into how various methodologies can influence a phenotype, we focus on the Standing Height phenotype. Fig. 4 shows our experiments on different models and techniques, with the complexity of machine learning techniques increasing from left to right. Our experiments begin with Elastic Net (EN), starting from a simple linear model trained on individuals of European descent. We then include multi-ethnic data and introduce population-conditional weighting during training. Subsequently, we explore creating an ensemble of Elastic Nets using boosting. As a more complex boosting algorithm, we include LightGBM, followed by AutoGluon, an AutoML method that trains multiple ML models to form a stacked ensemble, including LightGBM as one of its members.

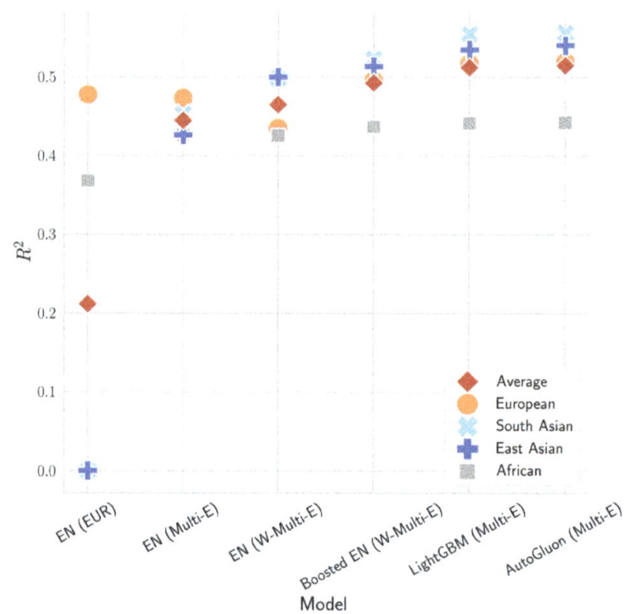

Fig. 4. Comparison of R^2 scores across diverse populations for the Standing Height phenotype. "EN" represents Elastic Net. The population used for training is provided in parenthesis, with "EUR" signifying European-only training data, and "Multi-E" indicating the use of multi-ethnic data. The symbol "W" marks the application of population-conditional sample weighting.

We note incremental performance for all populations, starting with Elastic Net which yields an R^2 of 0 for South Asian and East Asian individuals when trained solely on European data. Introducing multi-ethnic data leads to significant R^2 improvements, narrowing the performance gap between populations. Moreover, population-conditional weighting boosts performance for underrepresented groups. Finally, non-linear methods like LightGBM and AutoGluon have proven especially effective for the European, South Asian and East Asian populations. Gains are more modest for the African samples due to the higher genetic variation

Table 2. R^2 results for standing height. All the proposed population-conditional (PC) re-sampling methods use multi-ethnic training data. EN: Elastic Net, AG: AutoGluon.

Population	Training	Lasso	EN	Boosted EN	LightGBM	XGBoost	AG
European	European-only	0.508	0.477	0.506	**0.520**	**0.520**	**0.520**
	Multi-ethnic	0.497	0.473	0.503	0.517	0.517	0.519
	PC-Random OS	0.451	0.435	0.492	0.503	0.501	0.510
	PC-SMOTE	0.465	0.422	0.499	0.513	0.505	0.508
	PC-SMOTE-ENN	0.189	0	0.132	0.319	0.388	0.372
	PC-Weighted	0.452	0.435	0.496	0.506	0.501	0.513
South Asian	European-only	0	0	0	0	0	0
	Multi-ethnic	0.342	0.452	0.460	0.554	0.547	0.554
	PC-Random OS	0.506	0.499	0.523	0.542	0.549	0.557
	PC-SMOTE	0.486	0.480	0.509	0.552	0.533	0.541
	PC-SMOTE-ENN	0.520	0.467	0.525	0.544	0.548	0.553
	PC-Weighted	0.506	0.498	0.523	0.537	0.543	**0.563**
African	European-only	0.374	0.368	0.372	0.373	0.355	0.351
	Multi-ethnic	0.442	0.427	0.440	0.441	0.437	**0.443**
	PC-Random OS	0.442	0.426	0.439	0.386	0.429	0.439
	PC-SMOTE	0.434	0.411	0.421	0.400	0.433	0.414
	PC-SMOTE-ENN	**0.443**	0.397	0.427	0.401	0.431	0.418
	PC-Weighted	0.442	0.426	0.437	0.406	0.423	0.442
East Asian	European-only	0	0	0	0	0	0
	Multi-ethnic	0.174	0.426	0.413	0.535	0.513	0.534
	PC-Random OS	0.487	0.500	0.511	0.536	0.540	0.548
	PC-SMOTE	0.466	0.479	0.497	0.525	0.526	0.547
	PC-SMOTE-ENN	0.490	0.479	0.502	0.534	0.533	0.547
	PC-Weighted	0.487	0.500	0.513	0.511	0.524	**0.552**

within this group, making phenotype prediction a more challenging task. Models trained on multi-ethnic datasets can still struggle to capture the intricate relationships between genotype and phenotype specific to African populations. As we integrated increasingly complex and de-biasing techniques, we observed an overall improvement in R^2, underscoring that non-linear models, multi-ethnic data, and de-biasing techniques collectively drive enhanced results.

Table 2 provides a comprehensive comparison of various models in predicting standing height across different ancestry groups using diverse training techniques. For the individuals of European descent, training with either European-only or multi-ethnic data showcased similar results, with LightGBM, XGBoost, and AutoGluon emerging as top performers. In contrast, for the South Asian and East Asian groups, introducing multi-ethnic data and applying the proposed population-conditional re-sampling significantly improves predictive performance. The best results in the Asian groups are obtained applying the population-conditional sample weighting with AutoGluon. For the African group, top performance was observed not only with AutoGluon trained on multi-ethnic data but also with the Lasso combined with the population-conditional SMOTE-ENN. This finding underscores the importance of not only model choice but also nuanced training strategies, especially for diverse groups.

3.2. *Binary phenotypes*

We extend our experiments to classification models to observe if they follow similar trends as the regression results presented above. Table 3 showcases the ROC AUC results for two binary phenotypes (diabetes and atrial fibrillation). For both phenotypes, AutoGluon frequently achieves the highest ROC AUC scores, followed by LightGBM, outperforming the linear models. Particularly, the population-conditional weighted training improves model outcomes for the underrepresented groups when using multi-ethnic data.

Table 3. Performance of various models and training techniques in predicting binary phenotypes (Diabetes and Atrial Fibrillation), as measured by ROC AUC scores per group. The proposed population-conditional (PC) method uses multi-ethnic training data.

Phenotype	Population	Training	Lasso	Elastic Net	LightGBM	AutoGluon
Diabetes	European	European-only	0.520	0.530	0.585	0.604
		Multi-ethnic	0.501	0.508	0.606	**0.616**
		PC-Weighted	0.494	0.495	0.562	0.610
	South Asian	European-only	0.546	0.547	0.550	0.570
		Multi-ethnic	0.535	0.535	0.539	0.562
		PC-Weighted	0.528	0.533	0.563	**0.586**
	African	European-only	0.508	0.527	0.483	0.509
		Multi-ethnic	0.527	0.533	0.507	0.494
		PC-Weighted	0.516	0.516	**0.543**	0.493
	East Asian	European-only	0.391	0.409	0.480	0.552
		Multi-ethnic	0.421	0.385	0.554	0.579
		PC-Weighted	0.400	0.429	**0.638**	0.558
Atrial fibrillation	European	European-only	0.537	0.537	0.591	0.625
		Multi-ethnic	0.538	0.539	0.594	0.624
		PC-Weighted	0.538	0.537	0.609	**0.629**
	South Asian	European-only	0.504	0.485	0.562	0.513
		Multi-ethnic	0.478	0.498	0.547	0.548
		PC-Weighted	0.479	0.501	0.487	**0.586**
	African	European-only	0.544	0.521	0.559	**0.665**
		Multi-ethnic	0.554	0.532	0.523	0.592
		PC-Weighted	0.550	0.509	0.499	0.566
	East Asian	European-only	0.350	0.424	**0.596**	0.405
		Multi-ethnic	0.313	0.397	0.507	0.459
		PC-Weighted	0.322	0.424	0.542	0.374

4. Conclusions and Future Work

Our results advocate for the implementation of non-linear and ensemble methods, particularly LightGBM and AutoGluon, combined with the proposed population-conditional techniques to enhance genotype-to-phenotype prediction tasks for populations underrepresented in existing datasets. Strategies such as boosting and population-conditional sample weighting and re-sampling proved to be influential additions in order to better generalize across population and improve prediction accuracy. These methods were effective for both continuous and binary phenotypes, demonstrating their applicability for both regression and classification models.

Our study illustrates the use of methodological advancements to enhance prediction accuracy in the face of a lack of diverse genetic datasets. While the ideal solution would simply be the inclusion of more representative datasets, this is not an accurate reflection of the current data landscape. As such, we recommend for our models and techniques to be implemented when researchers are dealing with datasets of biased representation, especially in genetics. Using these methods should be a priority in situations demanding equitable outcomes, such as in clinical studies.

Failure to address these disparities could engender biases in precision medicine, which might unfavorably impact underrepresented populations. While our study addressed twelve phenotypes, expanding this focus to include other disease phenotypes in future research could yield a deeper understanding of genetic influences on disease. Although AutoGluon includes simple neural network models, future work could delve into a broader spectrum of deep learning architectures, including convolutional layers and attention mechanisms.

The moderate improvement in the African population compared to the Asian groups when applying multi-ethnic training and population-conditional re-sampling can be attributed to the inherent genetic diversity present within the African group, as the SNPs selected for this study are predominantly enriched for representation in Eurasian populations. For future work, a more refined SNP selection tailored for more diverse ancestral backgrounds could potentially enhance the predictive performance and rectify this limitation. A deeper investigation into linkage disequilibrium among SNPs could also optimize the SNP selection process by minimizing redundancies. Although models studied are able to capture genotype-phenotype relationships, covariates, particularly genetic principal components, could allow for a more accurate accounting of the underlying population structure. Incorporating advanced explainable ML techniques[51] alongside further analysis of covariates can elucidate the underlying mechanisms through which non-linear relationships boost predictive performance, offering a clearer insight into genotype-phenotype mappings. These approaches could refine model performance and enhance prediction accuracy across different ancestry backgrounds.

Given the prevalent bias in many clinical and genetic datasets,[10] underrepresented populations are often overlooked, with potentially grave implications for health outcomes. This issue is especially pertinent in an era where precision health methods and AI algorithms are becoming increasingly prominent. Thus, implementing strategies such as those presented in our study could considerably enhance the equability and effectiveness of precision medicine for underrepresented groups.

Acknowledgments

This work was partially supported by NIH under award R01HG010140 and by a grant from the Stanford Institute for Human-Centered Artificial Intelligence (HAI). This research has been conducted using the UK Biobank Resource under Application Number 24983.

References

1. A. B. Popejoy and S. M. Fullerton, Genomics is failing on diversity, *Nature* **538**, 161 (2016).
2. J. MacArthur, E. Bowler, M. Cerezo, L. Gil, P. Hall, E. Hastings, H. Junkins, A. McMahon, A. Milano, J. Morales *et al.*, The new nhgri-ebi catalog of published genome-wide association studies (gwas catalog), *Nucleic acids research* **45**, D896 (2017).
3. C. D. Bustamante, F. M. De La Vega and E. G. Burchard, Genomics for the world, *Nature* **475**, 163 (2011).
4. P. Song, A. Gupta, I. Y. Goon, M. Hasan, S. Mahmood, R. Pradeepa, S. Siddiqui, G. S. Frost, D. Kusuma, M. Miraldo *et al.*, Data resource profile: understanding the patterns and determinants of health in south asians—the south asia biobank, *International Journal of Epidemiology* **50**, 717 (2021).
5. L. Cheng, C. Shi, X. Wang, Q. Li, Q. Wan, Z. Yan and Y. Zhang, Chinese biobanks: present and future, *Genetics Research* **95**, 157 (2013).
6. S. Lee, P. E. Jung and Y. Lee, Publicly-funded biobanks and networks in east asia, *SpringerPlus* **5**, 1 (2016).
7. G. Wojcik, M. Graff, K. Nishimura *et al.*, Genetic analyses of diverse populations improves discovery for complex traits, *Nature* **570**, 514 (2019).
8. A. R. Martin, M. Kanai, Y. Kamatani, Y. Okada, B. M. Neale and M. J. Daly, Clinical use of current polygenic risk scores may exacerbate health disparities, *Nature genetics* **51**, 584 (2019).
9. N. R. Wray, M. E. Goddard and P. M. Visscher, Prediction of individual genetic risk to disease from genome-wide association studies, *Genome research* **17**, 1520 (2007).
10. A. R. Martin, C. R. Gignoux, R. K. Walters, G. L. Wojcik, B. M. Neale, S. Gravel, M. J. Daly, C. D. Bustamante and E. E. Kenny, Human demographic history impacts genetic risk prediction across diverse populations, *The American Journal of Human Genetics* **100**, 635 (2017), PMID: 28366442; PMCID: PMC5384097.
11. L. Duncan, H. Shen, B. Gelaye, J. Meijsen, K. Ressler, M. Feldman, R. Peterson and B. Domingue, Analysis of polygenic risk score usage and performance in diverse human populations, *Nature communications* **10**, p. 3328 (2019).
12. A. D'Amour, K. Heller, D. Moldovan, B. Adlam, B. Alipanahi, A. Beutel, C. Chen, J. Deaton, J. Eisenstein, M. D. Hoffman *et al.*, Underspecification presents challenges for credibility in modern machine learning, *The Journal of Machine Learning Research* **23**, 10237 (2022).
13. A. C. Need and D. B. Goldstein, Next generation disparities in human genomics: concerns and remedies, *Trends in Genetics* **25**, 489 (2009).
14. J. Qian, Y. Tanigawa, W. Du, M. Aguirre, C. Chang, R. Tibshirani, M. A. Rivas and T. Hastie, A fast and scalable framework for large-scale and ultrahigh-dimensional sparse regression with application to the uk biobank, *PLoS genetics* **16**, p. e1009141 (2020).
15. O. Weissbrod, M. Kanai, H. Shi, S. Gazal, W. J. Peyrot, A. V. Khera, Y. Okada, A. R. Martin, H. K. Finucane *et al.*, Leveraging fine-mapping and multipopulation training data to improve cross-population polygenic risk scores, *Nature Genetics* **54**, 450 (2022).
16. A. Dominguez Mantes, D. Mas Montserrat, C. D. Bustamante, X. Giró-i Nieto and A. G. Ioannidis, Neural admixture for rapid genomic clustering, *Nature Computational Science*, 1 (2023).
17. D. M. Montserrat, C. Bustamante and A. Ioannidis, Lai-net: Local-ancestry inference with neural

networks, in *ICASSP 2020-2020 IEEE International Conference on Acoustics, Speech and Signal Processing (ICASSP)*,
18. J. Gimbernat-Mayol, A. Dominguez Mantes, C. D. Bustamante, D. Mas Montserrat and A. G. Ioannidis, Archetypal analysis for population genetics, *PLoS Computational Biology* **18**, p. e1010301 (2022).
19. B. Oriol Sabat, D. Mas Montserrat, X. Giro-i Nieto and A. G. Ioannidis, Salai-net: species-agnostic local ancestry inference network, *Bioinformatics* **38**, ii27 (2022).
20. E. R. Bartusiak, M. Barrabés, A. Rymbekova, J. Gimbernat-Mayol, C. López, L. Barberis, D. M. Montserrat, X. Giró-i Nieto and A. G. Ioannidis, Predicting dog phenotypes from genotypes, in *2022 44th Annual International Conference of the IEEE Engineering in Medicine & Biology Society (EMBC)*,
21. K. Vokinger, S. Feuerriegel and A. Kesselheim, Mitigating bias in machine learning for medicine, *Communications Medicine* **1** (2021).
22. S. Afrose, W. Song, C. Nemeroff *et al.*, Subpopulation-specific machine learning prognosis for underrepresented patients with double prioritized bias correction, *Communications Medicine* **2**, p. 111 (2022).
23. R. Tibshirani, Regression shrinkage and selection via the lasso, *Journal of the Royal Statistical Society: Series B (Methodological)* **58**, 267 (1996).
24. H. Zou and T. Hastie, Regularization and variable selection via the elastic net, *Journal of the Royal Statistical Society: Series B (Statistical Methodology)* **67**, 301 (2005).
25. T. F. Mackay, Epistasis and quantitative traits: using model organisms to study gene–gene interactions, *Nature Reviews Genetics* **15**, 22 (2014).
26. Z. Dai, N. Long and W. Huang, Influence of genetic interactions on polygenic prediction, *G3: Genes, Genomes, Genetics* **10**, 109 (2020).
27. F. Morgante, W. Huang, C. Maltecca and T. F. Mackay, Effect of genetic architecture on the prediction accuracy of quantitative traits in samples of unrelated individuals, *Heredity* **120**, 500 (2018).
28. J. Jumper, R. Evans, A. Pritzel, T. Green, M. Figurnov, O. Ronneberger, K. Tunyasuvunakool, R. Bates, A. Žídek, A. Potapenko *et al.*, Highly accurate protein structure prediction with alphafold, *Nature* **596**, 583 (2021).
29. N. Sapoval, A. Aghazadeh, M. G. Nute, D. A. Antunes, A. Balaji, R. Baraniuk, C. Barberan, R. Dannenfelser, C. Dun, M. Edrisi *et al.*, Current progress and open challenges for applying deep learning across the biosciences, *Nature Communications* **13**, p. 1728 (2022).
30. C. Sudlow, J. Gallacher, N. Allen, V. Beral, P. Burton, J. Danesh, P. Downey, P. Elliott, J. Green, M. Landray *et al.*, Uk biobank: an open access resource for identifying the causes of a wide range of complex diseases of middle and old age, *PLoS medicine* **12**, p. e1001779 (2015).
31. J. H. Friedman, Greedy function approximation: a gradient boosting machine, *Annals of statistics*, 1189 (2001).
32. T. Chen and C. Guestrin, Xgboost: A scalable tree boosting system, in *Proceedings of the 22nd acm sigkdd international conference on knowledge discovery and data mining*,
33. G. Ke, Q. Meng, T. Finley, T. Wang, W. Chen, W. Ma, Q. Ye and T.-Y. Liu, Lightgbm: A highly efficient gradient boosting decision tree, in *Advances in Neural Information Processing Systems*, eds. I. Guyon, U. V. Luxburg, S. Bengio, H. Wallach, R. Fergus, S. Vishwanathan and R. Garnett (Curran Associates, Inc., 2017).
34. R. Shwartz-Ziv and A. Armon, Tabular data: Deep learning is not all you need, *Inf. Fusion* **81**, p. 84–90 (may 2022).
35. L. Grinsztajn, E. Oyallon and G. Varoquaux, Why do tree-based models still outperform deep learning on typical tabular data?, in *Thirty-sixth Conference on Neural Information Processing Systems Datasets and Benchmarks Track*,

36. A. Kumar, D. M. Montserrat, C. Bustamante and A. Ioannidis, Xgmix: Local-ancestry inference with stacked xgboost, *BioRxiv*, 2020 (2020).
37. J. C. Beltran, P. Valdez and P. Naval, Predicting protein-protein interactions based on biological information using extreme gradient boosting, in *2019 IEEE Conference on Computational Intelligence in Bioinformatics and Computational Biology (CIBCB)*,
38. P. Xuan, C. Sun, T. Zhang, Y. Ye, T. Shen and Y. Dong, Gradient boosting decision tree-based method for predicting interactions between target genes and drugs, *Frontiers in genetics* **10**, p. 459 (2019).
39. N. F. Grinberg, O. I. Orhobor and R. D. King, An evaluation of machine-learning for predicting phenotype: studies in yeast, rice, and wheat, *Machine Learning* **109**, 251 (2020).
40. A. Medvedev, S. M. Sharma, E. Tsatsorin, E. Nabieva and D. Yarotsky, Human genotype-to-phenotype predictions: Boosting accuracy with nonlinear models, *PloS one* **17**, p. e0273293 (2022).
41. G. McInnes, Y. Tanigawa, C. DeBoever, A. Lavertu, J. E. Olivieri, M. Aguirre and M. A. Rivas, Global biobank engine: enabling genotype-phenotype browsing for biobank summary statistics, *Bioinformatics* **35**, 2495 (2019).
42. D. H. Alexander, J. Novembre and K. Lange, Fast model-based estimation of ancestry in unrelated individuals, *Genome research* **19**, 1655 (2009).
43. Y. Tanigawa, J. Qian, G. Venkataraman, J. M. Justesen, R. Li, R. Tibshirani, T. Hastie and M. A. Rivas, Significant sparse polygenic risk scores across 813 traits in uk biobank, *PLoS Genetics* **18**, p. e1010105 (2022).
44. Y. Freund, R. E. Schapire *et al.*, Experiments with a new boosting algorithm
45. F. Hutter, L. Kotthoff and J. Vanschoren, *Automated machine learning: methods, systems, challenges* (Springer Nature, 2019).
46. N. Erickson, J. Mueller, A. Shirkov, H. Zhang, P. Larroy, M. Li and A. Smola, Autogluon: An automated machine learning framework (2020).
47. N. V. Chawla, K. W. Bowyer, L. O. Hall and W. P. Kegelmeyer, Smote: synthetic minority over-sampling technique, *Journal of artificial intelligence research* **16**, 321 (2002).
48. L. Torgo, R. P. Ribeiro, B. Pfahringer and P. Branco, Smote for regression, in *Portuguese conference on artificial intelligence*,
49. P. Branco, L. Torgo and R. P. Ribeiro, Smogn: a pre-processing approach for imbalanced regression, in *First international workshop on learning with imbalanced domains: Theory and applications*,
50. G. E. Batista, R. C. Prati and M. C. Monard, A study of the behavior of several methods for balancing machine learning training data, *ACM SIGKDD explorations newsletter* **6**, 20 (2004).
51. V. Belle and I. Papantonis, Principles and practice of explainable machine learning, *Frontiers in big Data*, p. 39 (2021).

Quantifying Health Outcome Disparity in Invasive Methicillin-Resistant *Staphylococcus aureus* Infection using Fairness Algorithms on Real-World Data[†]

Inyoung Jun[1], Sarah E. Ser[1], Scott A. Cohen[1], Jie Xu[2], Robert J. Lucero[3], Jiang Bian[2], and Mattia Prosperi[1*]
[1]*Department of Epidemiology, University of Florida*
Gainesville, FL 32611, USA
[2]*Department of Health Outcomes & Biomedical Informatics, University of Florida*
Gainesville, FL 32611, USA
[3]*School of Nursing, University of California, Los Angeles*
Los Angeles, CA 90095, USA
[*]*Email: m.prosperi@ufl.edu*

This study quantifies health outcome disparities in invasive Methicillin-Resistant *Staphylococcus aureus* (MRSA) infections by leveraging a novel artificial intelligence (AI) fairness algorithm, the Fairness-Aware Causal paThs (FACTS) decomposition, and applying it to real-world electronic health record (EHR) data. We spatiotemporally linked 9 years of EHRs from a large healthcare provider in Florida, USA, with contextual social determinants of health (SDoH). We first created a causal structure graph connecting SDoH with individual clinical measurements before/upon diagnosis of invasive MRSA infection, treatments, side effects, and outcomes; then, we applied FACTS to quantify outcome potential disparities of different causal pathways including SDoH, clinical and demographic variables. We found moderate disparity with respect to demographics and SDoH, and all the top ranked pathways that led to outcome disparities in age, gender, race, and income, included comorbidity. Prior kidney impairment, vancomycin use, and timing were associated with racial disparity, while income, rurality, and available healthcare facilities contributed to gender disparity. From an intervention standpoint, our results highlight the necessity of devising policies that consider both clinical factors and SDoH. In conclusion, this work demonstrates a practical utility of fairness AI methods in public health settings.

Keywords: AI fairness; Methicillin-resistant *Staphylococcus aureus*; Health outcome disparity

1. Introduction

Invasive Methicillin-Resistant *Staphylococcus aureus* (MRSA) infections pose a significant public health concern. According to the Centers for Disease Control and Prevention (CDC), MRSA infections account for a substantial proportion of healthcare-associated infections, affecting both inpatient and outpatient settings[1]. These infections, characterized by resistance to all beta-lactam antibiotics, have been associated with increased morbidity, mortality, and healthcare costs.

It is widely recognized that socioeconomic and demographic factors influence transmission and care outcomes of infectious diseases, including MRSA. For example, See et al. (2017) shed light on the complex interplay between race, socioeconomic factors, and MRSA infections[2]. Gualandi et al.

[†] Work partially supported by grant NIH NIAID 1R01AI141810; NIH NIA R33AG062884;

(2018) analyzed surveillance data in the USA from nine US states (20+ million people) and found that the risk of MRSA infection in African Americans was double the risk in other racial/ethnic groups, even when rates were decreasing[3]. Nonetheless, Mohnasky et al. (2021) found that in a prospective cohort of individuals seen in a single, large US medical center for over 20 years, social disparity in MRSA outcomes was explained by differences in comorbidities between racial/ethnic groups[4]. Thus, contrasts among studies could be explained by population selection and modeling choices. Many studies on quantification of health outcome disparity within invasive MRSA infections have been associational in nature, and further research is necessary to deconstruct and understand the underlying causal mechanisms driving such disparity in order to identify potential avenues for intervention. Such advancement can help develop effective strategies to mitigate the impact of the disease which target not only the majority of the population, but also specific subpopulations that might be more vulnerable, e.g., the elderly or ethnic/racial minorities.

To address the aforementioned challenges, we employ a recently developed artificial intelligence (AI) fairness algorithm, the Fairness-Aware Causal paThs (FACTS) decomposition[5]. FACTS is able to decompose disparity of an outcome measure with respect to a variable of interest into multiple causal pathways, and to quantify the relative contribution of each path. We apply FACTS on large real world electronic health record (EHR) data collated over 9 years from a large healthcare provider in Florida, USA, linked with contextual social determinants of health (SDoH).

2. Materials and Methods

2.1. *Ethics Statement*

This study obtained approval from the Institutional Review Board (#IRB201900652) of the University of Florida (UF). The authors strictly adhere to the research integrity and ethical principles outlined in the Declaration of Helsinki.

2.2. *Data Source*

We analyzed deidentified EHR data from the UF Health's Integrated Data Repository (IDR, https://idr.ufhealth.org/), which includes two primary hospitals in Gainesville and Jacksonville, and several other outpatient clinics in Florida. The IDR-EHR data includes patients' demographics, residence (here masked into 3-digit zip codes), laboratory tests (encoded with Logical Observation Identifiers Names and Codes, LOINC), drug prescriptions (RxNorm terminology), clinical procedures and diagnoses (International Classification of Disease, ICD 9th and 10th revision). In this study all ICD-10 codes were converted to ICD-9 format following General Equivalence Mappings guideline of Centers for Medicare & Medicaid Services[6] since the sample predominantly consisted of ICD-9 codes. Data requests can be directed to IDR (https://idr.ufhealth.org/research-services/) in compliance with institutional, state and US Federal policies; authors are willing to share study procedures for reproducing results.

We linked individual patient records to the county-level social determinants of health (SDoH) variables using multiple external sources. SDoH variables used in this work were: Median Household Income[7]; Rurality (urban or rural based on the Federal Bureau of Investigation

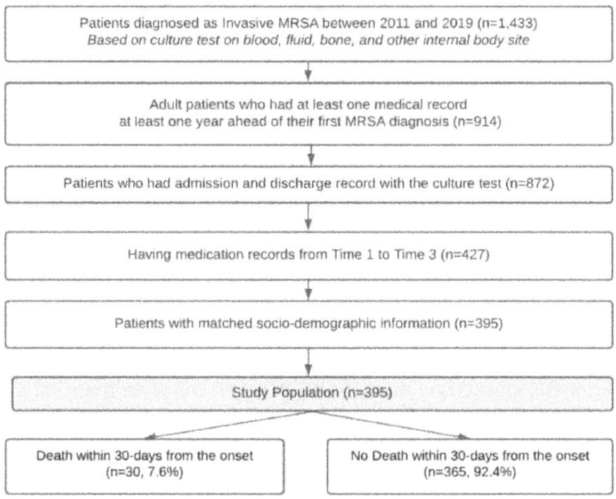

Fig. 1. Flowchart of Study Population

metropolitan criteria)[8]; Health Insurance Coverage (proportion of residents of uninsured populations)[9], and Access to Healthcare Facilities (proximity and availability of healthcare facilities such as hospitals, clinics, and primary care providers in the area; number of hospital beds per 100,000 population)[10].

2.3. *Study Design, Study Population, Variables*

We included adults aged 18 years and older at the time of diagnosis of invasive MRSA (ICD-9-CM: 041.11) at UF Health between January 1, 2011, and July 1, 2019. To ensure comprehensive medical information availability, patients without complete sociodemographic information and without a prior medical record from at least one year before their first invasive MRSA diagnosis were excluded. Excluding these patients mitigated potential bias arising from missing comprehensive past medical information. To follow up comprehensive antibiotic treatment, we defined three time points for each patient's antibiotic treatment. Time 1 was defined as the empiric treatment stage which the patient will receive without any test results confirmed when they got infected. Time 2 was defined as the time when their initial antibiotic susceptibility testing was revealed, and Time 3 was 7 days from time 2 (reflecting patient's latest clinical progression). A detailed clinical justification of the choice of the three time points is given in a prior work [11]. Individuals who were missing antibiotic treatment history for those three time points were dropped from the study. **Fig. 1** provides an overview of the inclusion criteria cascade.

The study's index/baseline date was set corresponding to the first invasive MRSA diagnosis, and the outcome was 30-day mortality. The patients' variables at index date included age, gender (male vs. female), race (African American vs. White), Charlson's comorbidity index (CCI), history of antibiotic usage, prior history of kidney impairment, types of infection (i.e., bloodstream infection or not), severity of infection (i.e., transfer to intensive care unit, ICU, or not), and the SDoH panel. Additional variables upon admission included the treatment course (i.e., whether they received vancomycin or not at each time point), and side effects (i.e., nephrotoxicity developed after the initial treatment).

2.4. *Causal Assumptions and FACTS*

Using literature search and authors' consensus, we created a partially directed acyclic graph (pDAG) connecting SDoH with individual clinical measurements before/upon diagnosis of invasive MRSA infection, treatments, side effects, and outcomes. Double-edged arrows might represent unmeasured confounding between two variables (e.g., income and rurality).

Each arrow in the pDAG is supported by at least one finding from our literature search. Race was associated with previous vancomycin use[12,13], types of infection[14], severity of infection[14], prior kidney impairment[15], prior drug resistance[2,16], income[17], and chronic comorbidities[18–23]. Income and health insurance were linked[24,25]. Sex was associated with health insurance[26], income[27], and chronic comorbidities[28–30]. Income and rurality were linked[31]. Rurality was also associated with access to healthcare facilities[32] , and healthcare facilities were associated with chronic comorbidities[33–35] . Age was associated with health insurance coverage[25] and chronic comorbidities[21,36,37]. Previous vancomycin use was associated with prior drug resistance[38] and vancomycin at Time 1[39,40]. Type

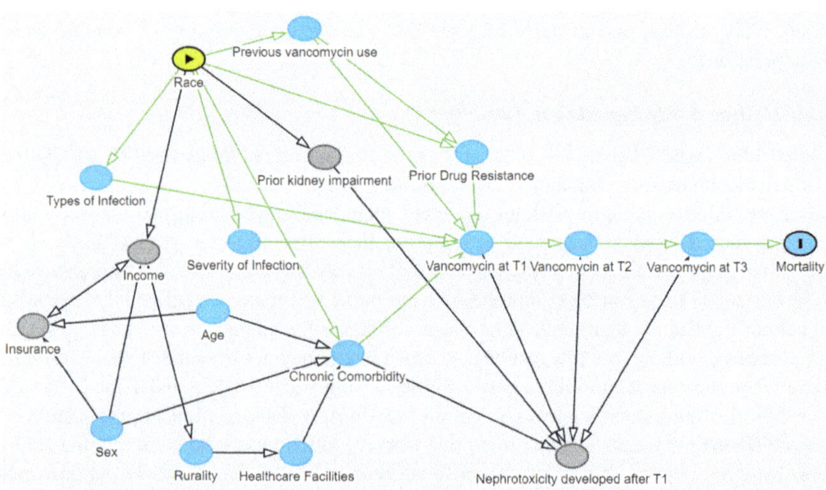

Fig. 2. Partially directed acyclic graph representing the causal relationships among clinical, sociodemographic variables, and MRSA 30-day mortality (race is displayed as the exposure variable).

of infection[41], severity of infection[40,41] , chronic comorbidities[39] , and prior drug resistance[39] were also associated with vancomycin at Time 1. Vancomycin at T1 was associated with vancomycin at Time 2, and vancomycin at Time 2 was associated with vancomycin at Time 3[42]. Chronic comorbidities and prior kidney impairment were associated with nephrotoxicity which developed after Time 1[43]. Vancomycin at Time 2 and Vancomycin at Time 3 were also linked with nephrotoxicity which developed after Time 1[43–45]. Vancomycin at Time 3 was associated with mortality[46].

The final pDAG is provided in **Fig. 2**. We selected race, income, gender, and age as exposure variables. The pDAG was used to calculate an adjustment set to identify the effect of the exposures with respect to MRSA outcome, quantifying the potential disparity in terms of odds ratios using a main-effects logistic regression. After this analysis, we applied the FACTS on our pDAG using the same exposures[5]. In detail, FACTS builds a prediction model of the outcome using all variables (through the XGBoost algorithm), then uses a given pDAG and a 'sensitive' attribute of interest (i.e., exposure, like gender or race) to calculate the contribution to outcome disparity for all paths involving such sensitive attribute. Finally, it ranks and outputs the most important paths.

2.5. *Software*

We conducted our analyses in R (https://www.r-project.org/), using the libraries 'tidyverse'[47] and 'data.table'[48] for data preprocessing, 'comorbidity'[49] for calculating Charlson's comorbidity index (CCI), and 'tidycensus'[50] for extracting the US Census Bureau's data APIs. The DAG and the adjustment sets were done with dagitty (https://www.dagitty.net/). For the FACTS analysis, we applied Python based on the code available at: https://github.com/weishenpan15/FACTS.

3. Results

3.1. *Characteristics of the Study Population*

We identified 1,433 individuals admitted to the hospital and diagnosed with an invasive MRSA infection between 2011 and 2019, based on the bio/tissue-sample source and the culture test (i.e., blood, fluid, bone, and other internal body site). After matching with socio-demographic information based on the three digits zip-code information of each patient, and after filtering based on all inclusion criteria as described in the methods, a total of 395 patients constituted the final study sample **(Fig. 1**).

Table 1 describes the baseline characteristics of the sample. Patients were 55.2 years old on average. The percentage of females was 50.1%, and 39.0% of the population was African American. In terms of county-level SDoH, the average number of healthcare facilities was 189, 50.9% of individuals lived in urban areas, the median household income was $48,300, and 21.8% was the prevalence of being uninsured. The mean Charlson's comorbidity index was 6.40. The prevalence of patients who were administered vancomycin before this invasive MRSA infection was 68.6%, and for 35.9% of patients, there was record of multiple drug resistance (MDR). Eighty percent of patients had a bloodstream infection, and 52.2% of patients were transferred to the ICU. The percentages of vancomycin usage at each time point were 97%, 79%, and 66.1% respectively. While

Table 1. Variable Characteristics of the Study Population (N=395)

Variables	Measure; Mean (SD), Median [Min, Max] or N (%)
Individual-level EHR variables	
Age	55.2 (16.4), 56 [19, 96]
Age – 65+ years old	121 (30.6%)
Gender – Female	198 (50.1%)
Race – African American	154 (39.0%)
Charlson's comorbidity index (CCI)	6.40 (3.92), 6 [0,20]
History of antibiotic usage (Vancomycin)	271 (68.6%)
Prior history of kidney impairment	211 (53.4%)
Types of Infection - Bloodstream	316 (80.0%)
Severity of Infection – ICU stay	206 (52.2%)
Prior Drug Resistance	142 (35.9%)
Nephrotoxicity developed	40 (10.1%)
Vancomycin use at Time1	383 (97.0%)
Vancomycin use at Time2	312 (79.0%)
Vancomycin use at Time3	261 (66.1%)
County-level sociodemographic variables	
Number of Healthcare Facilities (number of beds/100,000)	189 (30.3), 192 [107, 367]
Area – Urban	201 (50.9%)
Median household Income	$48,300 (7,900), $44,700 [$39,500, $67,400]
Insurance coverage (% of Uninsured)	21.8 (4.17), 21.9 [13.8, 35.7]

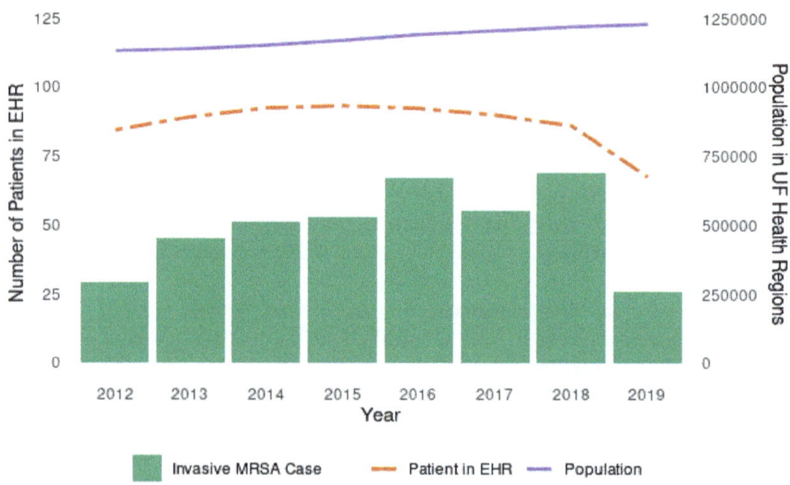

Fig. 3. Annual trends in invasive MRSA cases, EHR patients, and local population

being treated, 10.1% of patients developed nephrotoxicity, and 53.4% had prior history of kidney impairment.

In **Fig. 3**, we described the annual trend of the number of patients with invasive MRSA infection in our study sample. Additionally, we depicted the number of bacterial infection patients registered in our EHR system and the population of two Florida counties where large hospitals of the UF Health Network are situated (i.e., Alachua county and Duval county). The combined population of both counties exhibited an increasing trend over the years, while the number of invasive MRSA cases fluctuated annually. For the invasive MRSA cases, the data only covers half of 2019, from January to July.

3.2. *Quantification of Health Outcome Disparity*

We estimated the total and direct effects of age, race, income, and gender on the 30-day mortality outcome. For each of the exposure variables, we report the odds ratios (OR) and 95% confidence intervals (CI) obtained by fitting a logistic regression model with the adjustment set variables identified through the pDAG (**Table 2**). Income showed the strongest association with outcome disparity (total effect OR 0.44, 95% CI 0.17-0.99), followed by age, gender, and race. All effect estimates, except that of income, included OR=1 in the 95%CI.

We then ran the FACTS algorithm on the same set of exposures (**Table 3**). Of note, FACTS needs all binary variables, so we split the numeric variables based on their median. Overall, all paths showed absolute low weights, close to zero, for both accuracy and disparity metrics. Results did not change when including only clinical variables or clinical and SDoH variables in the pDAG and associated paths. There were no relevant paths detected for income. Comorbidity was detected as a disparity path for age, race, and gender. Antibiotic use, timing, and renal toxicity were relevant with respect to race, while income, rurality and number of healthcare facilities were relevant for gender disparity.

Table 2. Total and direct effects of age, race, income, and gender on to risk of 30-day mortality in invasive MRSA Infection.

Sensitive Variable	Model	Odds Ratio (95% CI)
Age (65+ years old =1 vs. younger=0)	Outcome ~ Age	2.11 (0.98, 4.48)
	Outcome ~ Age + Chronic Comorbidity + Previous Vancomycin Use + Prior Drug Resistance + Severity of Infection + Types of Infection	1.64 (0.74, 3.60)
Race (African American=1 vs. White=0)	Outcome ~ Race	0.77 (0.34, 1.65)
	Outcome ~ Race + Types of Infection + Severity of Infection + Previous Vancomycin Use + Prior Drug Resistance + Chronic Comorbidity	0.58 (0.24, 1.30)
Income (Below median=1 vs. Upper=0)	Outcome ~ Income	0.44 (0.17, 0.99)
	Outcome ~ Income + Chronic Comorbidity + Race + Rurality + Gender	0.74 (0.24, 2.14)
Gender (Female=1 vs. Male=1)	Outcome ~ Gender	1.15 (0.54, 2.45)
	Outcome ~ Gender + Chronic Comorbidity + Previous Vancomycin Use + Prior Drug Resistance + Severity of Infection + Types of Infection	1.24 (0.57,2.72)

Table 3. FACTS decomposition of disparity in 30-day mortality from invasive MRSA infection, with respect to age, race, income, and gender

Sensitive Variable	Clinical-only			Clinical + SDoH		
	Disparity Path	Disparity	Accuracy	Disparity Path	Disparity	Accuracy
Age	Comorbidity	-0.01162	0.00840	-	-	-
Race	Comorbidity	-0.05714	0.03361	Comorbidity	-0.05428	0.02857
	Prior Kidney Impairment	0.02571	-0.01512	Prior Vancomycin Use	-0.01142	-0.00672
	Nephrotoxicity developed after Time 1 ↔Vancomyc in use at Time 2 ↔ Vancomycin use at Time 3	0.01428	-0.00840	Prior Kidney Impairment	0.00857	-0.00504
Income	-	-	-	-	-	-
Gender	Comorbidity	-0.05084	0.02521	Insurance	-0.05101	0.02857
	-	-	-	Income→ Rurality	0.01333	-0.00672
	-	-	-	Income→ Rurality → Facility	0.00265	0.02857

4. Discussion

We deconstructed sociodemographic disparity in 30-day mortality among invasive MRSA infections, using EHR data and fairness AI methods. Upon explicit expert-derived causal assumptions, we found moderates to strong effects of age, gender, race, and income on mortality, although the 95% CIs included no difference in risk among groups. Our fairness analysis, conducted using the FACTS algorithm, revealed that comorbidity status was the most significant contributor to outcome disparity across age, race, and gender, while no distinct paths could be found for income. For race, antibiotic usage, timing, and prior kidney impairment contributed to disparity, while SDoH contributed to outcome disparity among genders. Age and income are well-known risk factors for mortality, and confirming their effects was clearly expected. Prior kidney impairment, identified through pre-infection creatinine levels, could contribute to the observed differences in invasive MRSA mortality rates between racial groups. Kidney impairment significantly influences the clinical management of MRSA infections in hospitals. Beyond its effect on the immune response, renal impairment also complicates the choice and dosage of anti-MRSA antibiotics that can be safely administered. For instance, vancomycin, the most commonly used antibiotic agent for treating these infections and a known iatrogenic cause of acute kidney injury, necessitates close monitoring and dosage adjustments based on renal function[42] . Further, while creatinine levels served as an indicator of renal function in our analysis, clinical teams during the study period were likely assessing for renal impairment using creatinine-based equations that vary by race, e.g., estimated glomerular filtration rate (eGFR), and changing accordingly the medical management of the patient. As a result, use of eGFR in clinical practice could have confounded the disparity paths that we decomposed[51]. Compared to other studies, our findings are consistent with recent literature that analyzed individuals diagnosed with *S. aureus* bacteremia, reporting no differences in mortality between racial groups[4,52]. However, it has to be noted that study populations are heterogeneous and demographic groups exhibit strong differences in risk factors. We found that variations in mortality rates are partially attributable to the burden of underlying comorbidities, therapeutic choices, and SDoH that differ among ages, incomes, genders, and races. Of note, our results align with another recent analysis of EHR in Florida that quantified the effect of demographics and SDoH on outcome disparity in urinary tract infections (UTIs), where comorbidity, number of healthcare facilities, income and insurance were also found to be involved in disparity paths with respect to race.[11]

The decision to use FACTS for this study was driven by the algorithm's emphasis on causal pathways which account for both directed and undirected arrows between variables in partially directed acyclic graphs (pDAGs). While statistical-based algorithms focus on assessing whether all groups have the same metric for outcomes, causal-based fairness is more concerned with analyzing the presence of causal effects of a sensitive attribute on outcomes, including path-specific fairness.[53] Although studies exist that focus on path-specific effects [54–57], the FACTS algorithm introduced a novel approach. The algorithm concentrates not only on causal paths but also on uncovering overlooked sources of disparity that may contribute to model disparity. The capability of the FACTS algorithm to consider undirected relationships among risk factors in pDAGs is pivotal, especially when relationships are unclear. Therefore, public health researchers could benefit directly from using advanced algorithm (i.e., FACTS) by quantifying the unknown weights of factors to the model disparity.

Our study has several limitations. Firstly, our causal assumptions may be incorrect, and we did not account for unmeasured confounders in our models. Despite researcher's best efforts to define a DAG, it remains a challenge in real-life situations to accurately represent all variable relationships. This brings up concerns of incorrect assumptions and the potential for reverse causality. However, both FACTS and the generalized adjustment criterion can work with partial DAGs, which can mitigate some of these issues. Another recommended approach is to estimate effects using multiple DAGs, each incorporating different assumptions. Secondly, due to our strict inclusion criteria, only one third of patients who potentially had an invasive MRSA infection were included, which likely introduced selection bias (i.e., exclusion bias). Exclusion bias arises when particular members of a population are excluded from a study due to criteria set by researchers. The patients who were not included in our study constituted about two-thirds of the patients with invasive MRSA; these patients were mainly excluded due to a lack of sequential antibiotic prescription records. The excluded patients might have exhibited different disparity pathways if sufficient information had been available to conduct such disparity analyses. However, by starting from what is available to us, discovered pathways by FACTS could provide initial inferences about the invasive MRSA population. These inferences could be further refined as principles for secondary data collection become more standardized in research and therefore minimizing missing information. Therefore, despite the constraint of not encompassing every patient in our EHR, our study can still offer valuable and profound insights. We aimed to identify diverse causal pathways of disparity and by meticulously delineating our cohort definition in this preliminary analysis. This approach was intended to curtail information bias and mitigate the impact of missing data. By sharing our method transparently, we seek to contribute meaningful insights, informed by a clear and comprehensive understanding of the available data, that can elucidate the disparity pathways prevalent in the broader invasive MRSA population. Thirdly, given the sample size and observed effect sizes, type II errors were also likely. Fourthly, the current release of the FACTS algorithm is capable of handling only binary variables; it is anticipated that future versions of the algorithm will expand its capabilities.

5. Conclusion

In conclusion, this work demonstrates the practical utility of fairness AI methods in public health settings. The FACTS framework can be useful to explore intervention strategies for optimizing health outcomes among different sociodemographic groups using actionable variables in the causal pathways, e.g., reducing rates of comorbidities in vulnerable populations, and equalizing SDoH. For future studies, it is paramount to relax the population selection constraints, and to explore multiple different causal assumptions to reduce residual bias.

References

1. Centers for Disease Control and Prevention (U.S.). *Antibiotic Resistance Threats in the United States, 2019.* Centers for Disease Control and Prevention (U.S.); 2019. doi:10.15620/cdc:82532

2. See I, Wesson P, Gualandi N, et al. Socioeconomic Factors Explain Racial Disparities in Invasive Community-Associated Methicillin-Resistant Staphylococcus aureus Disease Rates. *Clin Infect Dis*. 2017;64(5):597-604. doi:10.1093/cid/ciw808

3. Gualandi N, Mu Y, Bamberg WM, et al. Racial Disparities in Invasive Methicillin-resistant Staphylococcus aureus Infections, 2005–2014. *Clin Infect Dis*. 2018;67(8):1175-1181. doi:10.1093/cid/ciy277

4. Mohnasky MC, Park L, Eichenberger E, et al. 1373. Racial Disparities in Clinical Characteristics and Outcomes for Methicillin Susceptible and Methicillin-Resistant Staphylococcus aureus Bacteremia. *Open Forum Infectious Diseases*. 2021;8(Supplement_1):S773. doi:10.1093/ofid/ofab466.1565

5. Pan W, Cui S, Bian J, Zhang C, Wang F. Explaining Algorithmic Fairness Through Fairness-Aware Causal Path Decomposition. *Proceedings of the 27th ACM SIGKDD Conference on Knowledge Discovery & Data Mining*. Published online 2021.

6. ICD-9-CM to and from ICD-10-CM and ICD-10-PCS Crosswalk or General Equivalence Mappings. NBER. Accessed July 31, 2023. https://www.nber.org/research/data/icd-9-cm-and-icd-10-cm-and-icd-10-pcs-crosswalk-or-general-equivalence-mappings

7. Bureau UC. American Community Survey Data. Census.gov. Accessed October 21, 2022. https://www.census.gov/programs-surveys/acs/data.html

8. Florida. FBI. Accessed July 29, 2023. https://ucr.fbi.gov/crime-in-the-u.s/2018/crime-in-the-u.s.-2018/tables/table-10/table-10-state-cuts/florida.xls

9. How Healthy is your County? | County Health Rankings. County Health Rankings & Roadmaps. Accessed January 18, 2023. https://www.countyhealthrankings.org/county-health-rankings-roadmaps

10. Hospital Beds - Florida Health CHARTS - Florida Department of Health. Accessed October 21, 2022. https://www.flhealthcharts.gov/ChartsReports/rdPage.aspx?rdReport=NonVitalIndNoGrp.Dataviewer&cid=0313

11. Jun I, Cohen SA, Ser SE, et al. Optimizing Dynamic Antibiotic Treatment Strategies against Invasive Methicillin-Resistant Staphylococcus Aureus Infections using Causal Survival Forests and G-Formula on Statewide Electronic Health Record Data. In: Le T, Li J, Ness R, et al., eds. *Proceedings of The KDD'23 Workshop on Causal Discovery, Prediction and Decision*. Vol 218. Proceedings of Machine Learning Research. PMLR; 2023:98-115. https://proceedings.mlr.press/v218/inyoung23a.html

12. Olesen SW, Grad YH. Racial/Ethnic Disparities in Antimicrobial Drug Use, United States, 2014-2015. *Emerg Infect Dis*. 2018;24(11):2126-2128. doi:10.3201/eid2411.180762

13. Young EH, Strey KA, Lee GC, et al. National Disparities in Antibiotic Prescribing by Race, Ethnicity, Age Group, and Sex in United States Ambulatory Care Visits, 2009 to 2016. *Antibiotics*. 2023;12(1). doi:10.3390/antibiotics12010051

14. Rybak MJ, Zasowski EJ, Jorgensen SCJ, et al. Risk Factors for Bloodstream Infections Among an Urban Population with Skin and Soft Tissue Infections: A Retrospective Unmatched Case-Control Study. *Infect Dis Ther*. 2019;8(1):75-85. doi:10.1007/s40121-018-0227-9

15. Eneanya ND, Boulware LE, Tsai J, et al. Health inequities and the inappropriate use of race in nephrology. *Nature Reviews Nephrology*. 2022;18(2):84-94. doi:10.1038/s41581-021-00501-8

16. Denoble A, Reid HW, Krischak M, et al. Bad bugs: antibiotic-resistant bacteriuria in pregnancy and risk of pyelonephritis. *Am J Obstet Gynecol MFM*. 2022;4(2):100540. doi:10.1016/j.ajogmf.2021.100540

17. Racial disparities in income and poverty remain largely unchanged amid strong income growth in 2019. Economic Policy Institute. Accessed August 1, 2023. https://www.epi.org/blog/racial-disparities-in-income-and-poverty-remain-largely-unchanged-amid-strong-income-growth-in-2019/

18. Cancer Disparities - Cancer Stat Facts. SEER. Accessed January 22, 2023. https://seer.cancer.gov/statfacts/html/.html

19. Cancer Disparities - NCI. Published August 4, 2016. Accessed January 22, 2023. https://www.cancer.gov/about-cancer/understanding/disparities

20. Cheng YJ, Kanaya AM, Araneta MRG, et al. Prevalence of Diabetes by Race and Ethnicity in the United States, 2011-2016. *JAMA*. 2019;322(24):2389-2398. doi:10.1001/jama.2019.19365

21. Deshpande AD, Harris-Hayes M, Schootman M. Epidemiology of diabetes and diabetes-related complications. *Phys Ther*. 2008;88(11):1254-1264. doi:10.2522/ptj.20080020

22. Ward E, Jemal A, Cokkinides V, et al. Cancer disparities by race/ethnicity and socioeconomic status. *CA Cancer J Clin*. 2004;54(2):78-93. doi:10.3322/canjclin.54.2.78

23. Zavala VA, Bracci PM, Carethers JM, et al. Cancer health disparities in racial/ethnic minorities in the United States. *Br J Cancer*. 2021;124(2):315-332. doi:10.1038/s41416-020-01038-6

24. Lee DC, Liang H, Shi L. The convergence of racial and income disparities in health insurance coverage in the United States. *International Journal for Equity in Health*. 2021;20(1):96. doi:10.1186/s12939-021-01436-z

25. Keisler-Starkey K, Bunch LN. Health Insurance Coverage in the United States: 2020.

26. Dec 21 P, 2022. Women's Health Insurance Coverage. KFF. Published December 21, 2022. Accessed January 23, 2023. https://www.kff.org/womens-health-policy/fact-sheet/womens-health-insurance-coverage/

27. Barroso A, Brown A. Gender pay gap in U.S. held steady in 2020. Pew Research Center. Accessed January 22, 2023. https://www.pewresearch.org/fact-tank/2021/05/25/gender-pay-gap-facts/

28. O'Neil A, Scovelle AJ, Milner AJ, Kavanagh A. Gender/Sex as a Social Determinant of Cardiovascular Risk. *Circulation*. 2018;137(8):854-864. doi:10.1161/CIRCULATIONAHA.117.028595

29. Mosca L, Barrett-Connor E, Wenger NK. Sex/gender differences in cardiovascular disease prevention: what a difference a decade makes. *Circulation*. 2011;124(19):2145-2154. doi:10.1161/CIRCULATIONAHA.110.968792

30. Ntritsos G, Franek J, Belbasis L, et al. Gender-specific estimates of COPD prevalence: a systematic review and meta-analysis. *Int J Chron Obstruct Pulmon Dis*. 2018;13:1507-1514. doi:10.2147/COPD.S146390

31. Data show U.S. poverty rates in 2019 higher in rural areas than in urban for racial/ethnic groups. Accessed January 22, 2023. http://www.ers.usda.gov/data-products/chart-gallery/gallery/chart-detail/?chartId=101903

32. Johnston KJ, Wen H, Kotwal A, Joynt Maddox KE. Comparing Preventable Acute Care Use of Rural Versus Urban Americans: an Observational Study of National Rates During 2008-2017. *J Gen Intern Med*. 2021;36(12):3728-3736. doi:10.1007/s11606-020-06532-4

33. Siminoff LA. Access and equity to cancer care in the USA: a review and assessment. *Postgraduate Medical Journal*. 2005;81(961):674-679. doi:10.1136/pgmj.2005.032813

34. Yee AM, Mazumder PK, Dong F, Neeki MM. Impact of Healthcare Access Disparities on Initial Diagnosis of Breast Cancer in the Emergency Department. *Cureus*. 2020;12(8):e10027. doi:10.7759/cureus.10027

35. Mannoh I, Hussien M, Commodore-Mensah Y, Michos ED. Impact of social determinants of health on cardiovascular disease prevention. *Curr Opin Cardiol*. 2021;36(5):572-579. doi:10.1097/HCO.0000000000000893

36. Forouhi NG, Wareham NJ. Epidemiology of diabetes. *Medicine (Abingdon)*. 2014;42(12):698-702. doi:10.1016/j.mpmed.2014.09.007

37. Risk Factors: Age - NCI. Published April 29, 2015. Accessed January 22, 2023. https://www.cancer.gov/about-cancer/causes-prevention/risk/age

38. Amberpet R, Sistla S, Parija SC, Thabah MM. Screening for Intestinal Colonization with Vancomycin Resistant Enterococci and Associated Risk Factors among Patients Admitted to an Adult Intensive Care Unit of a Large Teaching Hospital. *J Clin Diagn Res*. 2016;10(9):DC06-DC09. doi:10.7860/JCDR/2016/20562.8418

39. Leekha S, Terrell CL, Edson RS. General Principles of Antimicrobial Therapy. *Mayo Clin Proc*. 2011;86(2):156-167. doi:10.4065/mcp.2010.0639

40. Karam G, Chastre J, Wilcox MH, Vincent JL. Antibiotic strategies in the era of multidrug resistance. *Critical Care*. 2016;20(1):136. doi:10.1186/s13054-016-1320-7

41. VanEperen AS, Segreti J. Empirical therapy in Methicillin-resistant Staphylococcus Aureus infections: An Up-To-Date approach. *J Infect Chemother*. 2016;22(6):351-359. doi:10.1016/j.jiac.2016.02.012

42. Jeffres MN. The Whole Price of Vancomycin: Toxicities, Troughs, and Time. *Drugs*. 2017;77(11):1143-1154. doi:10.1007/s40265-017-0764-7

43. Filippone E, Kraft W, Farber J. The Nephrotoxicity of Vancomycin. *Clin Pharmacol Ther*. 2017;102(3):459-469. doi:10.1002/cpt.726

44. Liu J, Tong SYC, Davis JS, Rhodes NJ, Scheetz MH, CAMERA2 Study Group. Vancomycin Exposure and Acute Kidney Injury Outcome: A Snapshot From the CAMERA2 Study. *Open Forum Infectious Diseases*. 2020;7(12):ofaa538. doi:10.1093/ofid/ofaa538

45. Sawada A, Kawanishi K, Morikawa S, et al. Biopsy-proven vancomycin-induced acute kidney injury: a case report and literature review. *BMC Nephrol*. 2018;19:72. doi:10.1186/s12882-018-0845-1

46. Schweizer ML, Richardson K, Vaughan Sarrazin MS, et al. Comparative Effectiveness of Switching to Daptomycin Versus Remaining on Vancomycin Among Patients With Methicillin-resistant Staphylococcus aureus (MRSA) Bloodstream Infections. *Clinical Infectious Diseases*. 2021;72(Supplement_1):S68-S73. doi:10.1093/cid/ciaa1572

47. Wickham H, Averick M, Bryan J, et al. Welcome to the Tidyverse. *Journal of Open Source Software*. 2019;4(43):1686. doi:10.21105/joss.01686

48. Dowle M, Srinivasan A. *Data.Table: Extension of `data.Frame`*.; 2023.

49. Gasparini A. comorbidity: An R package for computing comorbidity scores. *Journal of Open Source Software*. 2018;3(23):648. doi:10.21105/joss.00648

50. Walker K, Herman M. *Tidycensus: Load US Census Boundary and Attribute Data as "tidyverse" and 'Sf'-Ready Data Frames*.; 2023. https://walker-data.com/tidycensus/

51. Williams WW, Hogan JW, Ingelfinger JR. Time to Eliminate Health Care Disparities in the Estimation of Kidney Function. *New England Journal of Medicine*. 2021;385(19):1804-1806. doi:10.1056/NEJMe2114918

52. Ruffin F, Dagher M, Park LP, et al. Black and White Patients With Staphylococcus aureus Bacteremia Have Similar Outcomes but Different Risk Factors. *Clinical Infectious Diseases*. 2023;76(7):1260-1265. doi:10.1093/cid/ciac893

53. Wang X, Zhang Y, Zhu R. A brief review on algorithmic fairness. *MSE*. 2022;1(1):7. doi:10.1007/s44176-022-00006-z

54. Chiappa S. Path-Specific Counterfactual Fairness. *AAAI*. 2019;33(01):7801-7808. doi:10.1609/aaai.v33i01.33017801

55. Kusner MJ, Loftus J, Russell C, Silva R. Counterfactual Fairness. In: Guyon I, Luxburg UV, Bengio S, et al., eds. *Advances in Neural Information Processing Systems*. Vol 30. Curran Associates, Inc.; 2017. https://proceedings.neurips.cc/paper_files/paper/2017/file/a486cd07e4ac3d270571622f4f316ec5-Paper.pdf

56. Nabi R, Shpitser I. Fair Inference on Outcomes. *AAAI*. 2018;32(1). doi:10.1609/aaai.v32i1.11553

57. Wu Y, Zhang L, Wu X, Tong H. PC-Fairness: A Unified Framework for Measuring Causality-based Fairness. In: Wallach H, Larochelle H, Beygelzimer A, Alché-Buc F d', Fox E, Garnett R, eds. *Advances in Neural Information Processing Systems*. Vol 32. Curran Associates, Inc.; 2019. https://proceedings.neurips.cc/paper_files/paper/2019/file/44a2e0804995faf8d2e3b084a1e2db1d-Paper.pdf

Imputation of race and ethnicity categories using genetic ancestry from real-world genomic testing data

Brooke Rhead*, Paige E. Haffener*, Yannick Pouliot, and Francisco M. De La Vega†
Tempus Labs, Inc.
Chicago, IL, 60654, USA

The incompleteness of race and ethnicity information in real-world data (RWD) hampers its utility in promoting healthcare equity. This study introduces two methods—one heuristic and the other machine learning-based—to impute race and ethnicity from genetic ancestry using tumor profiling data. Analyzing de-identified data from over 100,000 cancer patients sequenced with the Tempus xT panel, we demonstrate that both methods outperform existing geolocation and surname-based methods, with the machine learning approach achieving high recall (range: 0.859-0.993) and precision (range: 0.932-0.981) across four mutually exclusive race and ethnicity categories. This work presents a novel pathway to enhance RWD utility in studying racial disparities in healthcare.

Keywords: Race; ethnicity; ancestry; imputation, disparities, equity, real-world data.

1. Introduction

Real-world data (RWD) offers insights into disease etiology, therapy outcomes, and racial disparities in healthcare.[1,2] However, its utility in improving healthcare equity is limited by the significant sparsity of race and ethnicity data. This gap, attributable to factors such as lack of capture, data loss during transfer and de-identification,[3,4] and shortcomings in electronic health record integrations,[5] leads to reliance on limited, potentially biased datasets that may result in poorly generalizable results and biased disease outcome predictors.[4]

Several remediation strategies have been proposed, including improving data collection, conducting complete case analysis, modeling missingness in analyses, supplementing with additional data, and employing imputation methodologies.[5] Existing imputation methods, many of which leverage census data based on geolocation and correlations between people's surnames and their self-reported race and ethnicity,[6,7] achieve moderate accuracy and require access to protected health information (PHI), limiting their applicability.[8,9]

Molecular tumor profiling, an assay used in support of therapy decisions in cancer patients, is often accompanied by a wealth of multimodal RWD that, once de-identified, can be harnessed for research.[10] This can include clinical metadata, imaging, and molecular data, such as DNA variants on a set of cancer related genes and transcript sequences from different patient tissues.[11]

Inferring genetic ancestry, or more accurately, genetic similarity to reference populations,[12] from molecular testing sequencing data, offers a potential solution to the challenge of missingness in race and ethnicity data. The granularity of such inferences is contingent on the availability of allele frequency data across samples from reference populations, with the most common level of genetic

* Joint first authorship.
† Correspondence: francisco.delavega@tempus.com

ancestry inference being at super-population level categories, as described by the 1000 Genomes Project.[13] Although genetic ancestry is not equivalent to race or ethnicity, a strong correlation between these two concepts has been observed among US populations.[14,15] We propose to leverage this correlation and the genetic information available in molecular testing RWD using two methods — one heuristic and the other based on machine learning — to impute mutually exclusive race and ethnicity categories from genetic ancestry. Here, we benchmark these methods and find they outperform previously reported race and ethnicity imputation methods, with a machine learning-based method providing the most accurate imputation.

2. Methods

The categorizations of race and ethnicity in this study adhere to the standards developed by the US Office of Management and Budget,[16] which are also used in the US census. These standards are based on two self-reported questions: a) Race (American Indian or Alaska Native, Asian, Black or African American, Native Hawaiian or Other Pacific Islander, and White); and b) Ethnicity (Hispanic or Latino and Not Hispanic or Latino). However, these categories present analytical challenges due to the orthogonal race and ethnicity questions, and it is often more practical to consolidate answers to these two questions into non-overlapping classes,[17] defined in this study as: Hispanic or Latino, non-Hispanic (NH) Asian, NH Black, and NH White, with the other races having insufficient numbers at the moment to develop reliable models in our source data. This consolidation allows for a more streamlined and comprehensive analysis of race and ethnicity in the context of RWD.

2.1. *Data*

Genomic and clinical data from patients of multiple cancer diagnoses was obtained from the Tempus database. The selected cohort consisted of 132,523 de-identified records of patients whose tissues were sequenced with the Tempus xT next-generation sequencing (NGS) panel (596-648 genes, v2-v4, tumor-normal matched when tissue available)[11,18] from 2018 to 2022. These records had been previously de-identified for other studies and passed minimal data quality filters. A total of 33,232 records had populated race, ethnicity, and geolocation data and belonged to one of the four non-overlapping race and ethnicity categories that we imputed: 4,357 Hispanic or Latino, 1,258 NH Asian, 3,120 NH Black, and 24,497 NH White. Race and ethnicity information in the Tempus database is obtained from a combination of electronic health record integrations and data abstraction from clinical documents and can be self-declared by patients or observed by practitioners. Information could be missing because there was no attempt to collect it, because patients or practitioners abstained from answering, or because it was not captured in the Tempus database. Analyses were performed using de-identified data under human subject research exemption granted by Advarra, Inc. Institutional Review Board, protocol Pro00042950.

2.2. *Determination of genetic ancestry*

We estimated genetic ancestry proportions using a re-implementation of the ADMIXTURE supervised global genetic ancestry estimation algorithm.[19] This approach calculated the proportions of ancestries for five super-populations—Africa (AFR), the Americas (AMR), East Asia (EAS), Europe (EUR), and South Asia (SAS)—using a previously published bespoke set of 654 ancestry informative markers (AIMs).[20] Briefly, AIMs were selected from single-nucleotide variants present in the reference samples that intersect with the targeted regions of the Tempus xT NGS assay, are not protein-changing, and are present at significantly different frequencies across the reference populations.[21] We sourced reference allele frequency data for these AIMs from the 1,000 Genomes Project,[13] the Human Genome Diversity Project,[22] and the Simons Genome Diversity Project databases.[23] In the case of the AMR super-population, we excluded the 1,000 Genomes Project's admixed "AMR" population and only included allele frequencies for Native American individuals available in the other sources. To evaluate the accuracy of our methods, we compared our global ancestry proportion estimates on whole-genome sequencing data from the Pan-Cancer Analysis of Whole Genomes Project,[24] with published global ancestry proportions determined by summing genome-wide local ancestry segments derived using the RFMix method.[25] This comparison yielded an average mean squared error, normalized to the sum of population proportions present in the dataset, of 0.12. The Tempus xT assay utilizes matched normal tissue when available (present for 51% of the study cohort) to classify variants as either germline or somatic, but germline variants can still be inferred in the absence of normal tissue.[11] The genetic ancestry proportion estimation method utilizes variant calls from normal tissue or those deemed to be germline. To assess performance when no matched normal tissue is available, we estimated proportions from both the tumor sample and the matched normal sample for a subset of patients (N = 3,358) and found that the five estimated proportions were highly concordant, with Pearson's correlation coefficient ranging from 0.9977 to 0.9999.[20]

2.3. *Benchmarking and performance metrics for race and ethnicity category imputation*

We relied on our cohort's stated race and ethnicity data as available in the Tempus database as our ground truth. To assess the performance of imputation methods, we employed a range of accuracy measures specific to each predicted race or ethnicity category. *Recall*, also called *sensitivity* or *true positive rate,*[26] measures the proportion of individuals correctly assigned to a category among all individuals truly in that category. *Precision*, or *positive predictive value*,[26] is the fraction of relevant instances among the retrieved instances, i.e., the proportion of correctly assigned individuals among all those assigned to a category. The *F1-score* is the harmonic mean of precision and recall, providing a balance between these two metrics. We also evaluated several measures of overall accuracy. *Cohen's kappa*[27] is a measure of agreement between predicted and true categories, accounting for the possibility of agreement occurring by chance. The *correct rate*, or *accuracy,* measures the proportion of all predictions that are correct.[26] *Log loss* quantifies the difference between predicted probabilities of belonging to a class and the true value (0 or 1) of belonging to that class, with lower log loss indicating better model performance[28]. The area under the receiver operating characteristic curve, or *AUC*, is a measure of model performance based on sensitivity and specificity across all classification thresholds and thus is not sensitive to any specific chosen threshold. *prAUC* is an analogous measure based on precision and recall. A predicted probability threshold of >0.5 was used for all metrics that rely on a single classification for each subject.

In addition to these common measures, we also utilized metrics proposed by Elliot *et al.*[6] The *weighted error* compares the true prevalence of race/ethnicity in the validation dataset to the predicted prevalence, providing an indication of the overall error rate. The *weighted correlation* measures the weighted average correlation (calculated using vectors of indicators) between true race and ethnicity and imputed category for each of the four categories, with weights equal to true prevalence. Together, these metrics offer a comprehensive evaluation of the performance of our imputation methods.

2.4. *Heuristic imputation of race and ethnicity*

We initially imputed mutually exclusive race and ethnicity categories from genetic ancestry proportions using a set of heuristics (Table 1) in part derived from admixture proportions reported in the literature for Black and Hispanic or Latino groups in the United States.[15] We defined four categories: Hispanic or Latino, NH Asian, NH Black, and NH White. Patients who did not fit the categories defined by these heuristics were labeled "complex." This latter category could be considered a no-call, as patients classified as such are typically excluded from any downstream analyses, and for comparison with other methods described below.

Table 1. Race and ethnicity imputation heuristics from genetic ancestry. Super-population codes: AFR, Africa; AMR, Americas; EAS, East Asia; EUR, Europe; SAS, South Asia.

Imputed category	Super-population genetic ancestry thresholds
Hispanic or Latino	>10% AMR and >70% combined AMR, EUR, and AFR
NH Asian	>70% combined EAS and SAS
NH Black	>20% AFR, <10% AMR, and >70% combined AFR and EUR
NH White	>80% EUR and <10% AMR
Complex	Remaining patients not meeting above thresholds

2.5. *Machine learning imputation of race and ethnicity*

We also developed machine learning (ML)-based imputation methods, wherein an ML algorithm is trained to classify subjects into race and ethnicity categories based on genetic ancestry and other inputs. For all models, a single train+test and validation set was assembled from the 33,232 patient records with stated race and ethnicity that fit our imputation categories and with available home address 3-digit ZIP code. Features used by these models included genetic ancestry proportions for AFR, AMR, EAS, EUR, and SAS; US census division of patient's home state (nine geographic groupings of states defined by the US Census Bureau: Pacific, Mountain, West North Central, West South Central, East North Central, East South Central, South Atlantic, Middle Atlantic, and New England); and "demographic proportions," i.e., proportions of Hispanic or Latino, NH Asian, NH Black, and NH White residing in each patient's three-digit ZIP code tabulation area (ZCTA), as available from the 2021 5-year American Community Survey and mapped to three-digit ZIP codes using UDS Mapper.[29] We split the train+test and validation sets 90/10 while maintaining the US census division proportions in each set to ensure that the sets were aligned well for populations whose genetic ancestry proportions vary by U.S. geography, e.g., Hispanic or Latino.[15] We evaluated models using three groups of features: 1) *ML-ancestry*: genetic ancestry proportions only; 2) *ML-ancestry+geolocation*: genetic ancestry proportions and US census divisions; 3) *ML-ancestry+demographics*: genetic ancestry proportions and demographic proportions.

We implemented all machine learning models in R using the caret package (v 6.0.94).[30] A number of models based on supervised training algorithms were evaluated, including models based on the random forest (method="rf") and gradient boosting (method="gbm") algorithms. We ultimately chose a boosted logistic regression algorithm (method="LogitBoost",[28] presented here) as it provided the ability to make no-call assignments and applied a probabilistic threshold in classification. Boosted logistic regression is a supervised machine learning algorithm that utilizes negative log-likelihood as a cost function. It iteratively builds decision trees to classify subjects, where each iteration is trained on a sample (with replacement) of the data in which subjects who were incorrectly classified in the previous round are more frequently sampled. The final classifier consists of a weighted combination of decision trees, where trees with lower log loss have more weight, and it returns the probabilities of belonging to each category for each subject. We chose to assign "No Call" to any subject with all probabilities ≤0.5. All models were trained using 10-fold cross validation. Grid expansion was performed to evaluate boosting iterations from 1 to 100 in intervals of 10. The optimal number of iterations and the final model were selected based on the lowest log loss value.

3. Results

3.1. *Comparison of performance of race and ethnicity imputation methods*

Table 2 summarizes the overall performance of the heuristic assignment method and each of the ML models. The ML model that utilized combined genetic ancestry proportions and demographic proportions (the proportions of the population in a patient's three-digit ZCTA belonging to Hispanic or Latino, NH Asian, NH Black, and NH White) achieved the best mean F1-score (0.957), Cohen's kappa (0.936), correct rate (0.974), log loss (0.122), AUC (0.982), and prAUC (0.946) whereas the heuristic method performed the worst by most metrics: mean F1-score 0.939, Cohen's kappa 0.903, correct rate 0.959, weighted correlation 0.876, and weighted error 0.009. The ML model that solely considered genetic ancestry proportions achieved the best weighted correlation (0.930) and weighted error (0.007), whereas the ML model that included geolocation in the form of the US Census district of a patient's home address state had intermediate performance by most metrics.

Table 2. Overall performance of race and ethnicity imputation methods for the validation set (N=3,319). Metrics that rely on a single classification threshold used a predicted probability of ≥0.5 for computation. Refer to section **2.5** for ML method descriptions. Best performing metric indicated with bold.

Imputation Method	Mean F1-Score	Cohen's Kappa	Correct Rate	Weighted Correlation	Weighted Error	Log Loss	AUC	prAUC
Heuristic	0.939	0.903	0.959	0.876	0.009	-	-	-
ML-ancestry	0.954	0.934	0.973	**0.930**	**0.007**	0.127	0.980	0.930
ML-ancestry+geolocation	0.955	0.935	0.973	0.926	0.009	0.131	0.979	0.898
ML-ancestry+demographics	**0.957**	**0.936**	**0.974**	0.928	0.013	**0.122**	**0.982**	**0.946**

When evaluating performance by category, we found that recall, precision, and F1-score were all at or above 0.932 for the NH Asian, NH Black and NH White categories (Table 3). Performance of all imputation methods was worst for the Hispanic or Latino category, with recall ranging from 0.859-0.887, precision from 0.833-0.964, and F1-score from 0.859-0.909.

Table 3. Performance of race and ethnicity imputation methods on validation set (N=3,319) per classification category. Refer to section **2.5** for ML method descriptions. Best performing metric for each category indicated with bold.

		Classification Category, N			
Metric	Imputation Method	Hispanic or Latino, 435	NH Asian, 130	NH Black, 301	NH White, 2,463
Recall	Heuristic	**0.887**	**0.983**	0.983	0.966
	ML-ancestry	0.876	0.962	0.993	0.987
	ML-ancestry+geolocation	0.877	0.969	0.983	0.988
	ML-ancestry+demographics	0.859	0.976	**0.993**	**0.990**
Precision	Heuristic	0.833	**0.935**	0.942	**0.985**
	ML-ancestry	0.938	0.933	0.967	0.981
	ML-ancestry+geolocation	0.941	0.932	**0.969**	0.981
	ML-ancestry+demographics	**0.964**	0.932	0.968	0.978
F1-Score	Heuristic	0.859	**0.959**	0.962	0.976
	ML-ancestry	0.906	0.947	**0.980**	**0.984**
	ML-ancestry+geolocation	0.908	0.950	0.976	**0.984**
	ML-ancestry+demographics	**0.909**	0.954	**0.980**	**0.984**

3.2. *Performance of heuristic method*

Perhaps unsurprisingly, the heuristic method for assigning race and ethnicity categories based on genetic ancestry proportions alone underperformed by all measures as compared to the ML models (cf. Table 3). For the Hispanic or Latino category (the most difficult to predict using the selected features), the heuristic method did have the highest recall (0.887), but this was achieved at the cost of low precision (0.833), also reflected in this method obtaining the lowest F1-score (0.859) for that category. The heuristic method did achieve the highest recall, precision, and F1-score for the NH Asian category. Overall, although the heuristic method did not perform as well as the ML method, its performance was not far behind, achieving an overall correct classification rate of ~96% compared to ~97% for the ML models. The no-call rate (i.e., patients assigned to the "complex" category) was 2.5%.

3.3. *Performance of ML-ancestry boosted logistic regression model*

We found that the boosted logistic regression model that utilized only genetic ancestry proportions improved upon the heuristic method for all overall performance metrics, with an overall correct classification rate of 97.3%. It had lower recall (0.876) but higher precision (0.938) for the Hispanic or Latino category than the heuristic method. The model had a recall of 0.962-0.993 for the three non-Hispanic categories, indicating that it correctly identifies the vast majority of patients in those categories and is usually correct in its predictions, with precision ranging from 0.933-0.981. The no-call rate was very low at 0.7%.

3.4. *Performance of ML models including geolocation and demographics*

Adding geolocation or demographic composition obtained from patients' home address ZCTA areas to the genetic ancestry proportions (*ML-ancestry+geolocation* and *ML-ancestry+demographics*) slightly improved model performance according to most metrics, yielding a correct classification rate of 97.3% and 97.4%, respectively. The *ML-ancestry+demographics* model had the best overall performance by all metrics except the less commonly used weighted metrics, which emphasize performance according to the true prevalence of each race and ethnicity category in the validation dataset. Individual category performance metrics followed a similar pattern to that of the *ML-ancestry* model. Notably, the *ML-ancestry+geolocation* model had the best precision for the Hispanic or Latino category (0.964), which may be desirable for use cases where correct predictions of this category are valued over high recall. The no-call rate was 1.1% for *ML-ancestry+geolocation* and 1.0% for *ML-ancestry+demographics*.

3.5. *Reclassification of stated race and ethnicity categories by imputation*

We selected the *ML-ancestry* model for further characterization because of its minimal input needs by applying it to the entire labeled dataset, regardless of whether geolocation data was available (N=35,229). The resulting confusion matrix (Table 4) compares the imputed categories with the stated race and ethnicity from the Tempus database, including the rate of no-calls and the number and fraction of misclassified records for each stated category. The confusion matrix for the validation dataset mirrors this table in terms of percentages (data not shown).

Table 4. Confusion matrix comparing imputed race and ethnicity category to stated category for the *ML-ancestry* model on all labeled data, including records without geolocation information (N=35,229). Percentage of each stated category and numbers of patients (in parentheses) are indicated in each cell. Total percentage and number of misclassified patients for each stated category is given in the last row.

	Stated category			
Imputed category	Hispanic or Latino	NH Asian	NH Black	NH White
Hispanic or Latino	**82.3% (4,059)**	0.2% (2)	0.5% (18)	0.8% (195)
NH Asian	0.8% (39)	**96.5% (1,285)**	0.2% (6)	0.2% (57)
NH Black	1.8% (91)	0.1% (1)	**97.7% (3,231)**	0.2% (49)
NH White	11.4% (560)	2.5% (33)	0.6% (19)	**98.5% (25,266)**
No Call	3.7% (180)	0.8% (10)	1.0% (32)	0.4% (96)
Misclassified	14.0% (690)	2.7% (36)	1.3% (43)	1.2% (301)

Additionally, Figure 1 provides a visual representation of the allocation of patients from their stated race and ethnicity to their imputed categories through a flow diagram.

The confusion matrix further indicates that the Hispanic or Latino category experienced the highest rates of no-calls (3.7%) and misclassifications (14.0%), whereas the NH White category had the lowest (0.4% and 1.2%, respectively). The flow diagram in Figure 1 illustrates that most patients were assigned to their stated category, with the majority of misclassifications occurring between Hispanic or Latino and NH White categories. Nevertheless, the overall misclassification rate of this model was very low at 0.9%.

Fig. 1. Flow diagram showing the relationship between stated (left) and imputed (right) race and ethnicity.

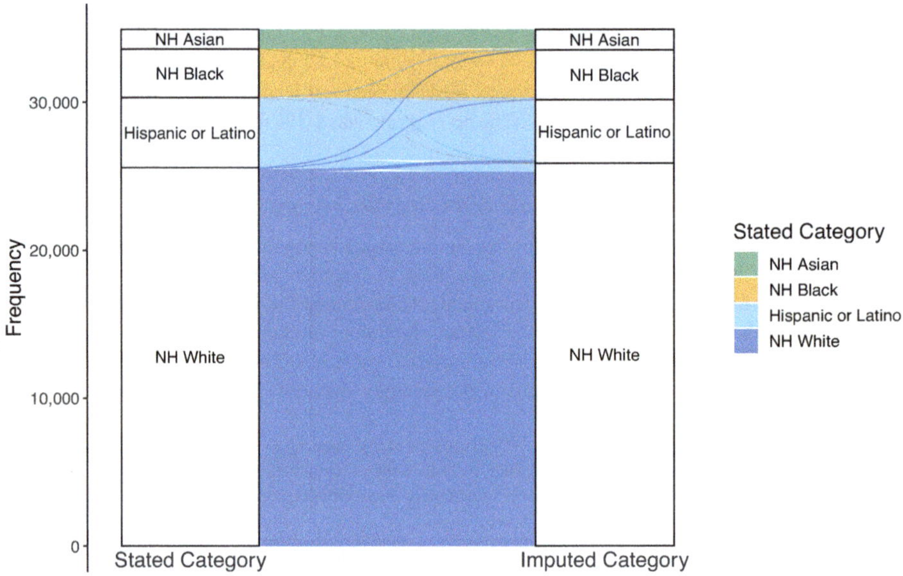

categories with the *ML-ancestry* model in all labeled data, including records without geolocation information and excluding no-calls (N=34,911).

3.6. *Distribution of race and ethnicity categories imputed on unlabeled patients*

We also imputed race and ethnicity categories using the *ML-ancestry* model for all patients in the cohort (N=132,523) and examined the distribution of availability of race and ethnicity labels across categories (Figure 2). A total of 35,229 patients belonged to one of the four imputation categories according to their stated race and ethnicity data ("labeled"). There were 62,674 patients with no available race or ethnicity data at all ("unlabeled"), and an additional 34,620 with only partial information, i.e., either stated race or stated ethnicity (or both) were available, but patients did not fall into one of the four imputation categories, most frequently because ethnicity was unavailable ("partially labeled"). Imputed categories had comparable levels of unlabeled data, with the No Call and NH Asian categories having the most (53% and 52%, respectively) and NH Black having the least (44%). The Hispanic or Latino category had the highest level of labeled data by far (40%) due to the definition of that category only requiring a stated ethnicity of "Hispanic or Latino" and allowing stated race to be any value, including a missing value. The remaining categories had 22-26% labeled data. We observed that about half of each of the NH Asian, NH Black, and NH White imputed categories had records with a concordant stated race but a missing ethnicity (data not shown).

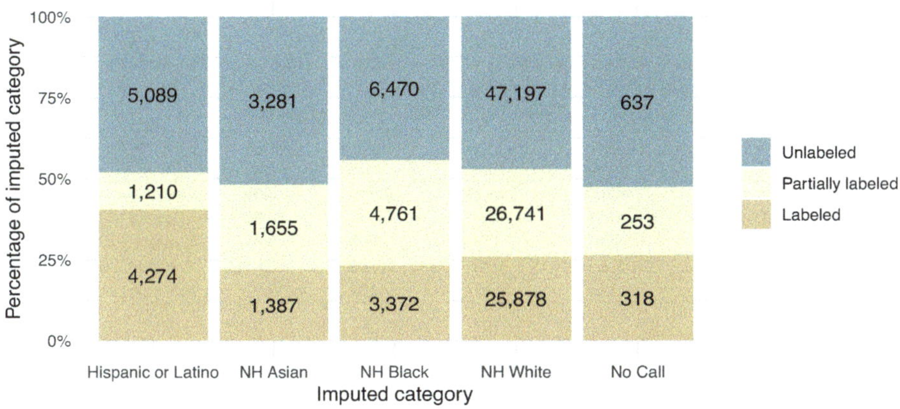

Fig. 2. Counts of patients in the full dataset (N=132,523) by label availability status and race and ethnicity category as imputed using the *ML-ancestry* model. Labeled = stated race and ethnicity are available, and a patient falls into one of: Hispanic or Latino, NH Asian, NH Black, or NH White based on this information. Unlabeled = neither stated race nor ethnicity is available. Partially labeled = either stated race or ethnicity is available, but the patient cannot be placed in one of the four listed categories.

3.7. *Analysis of potential biases*

The dataset used to develop our ML models is heavily imbalanced, with the largest group of patients (~74%) having a stated category of NH White, and the smallest group (~4%) having a stated category of NH Asian, potentially leading to overfitting to the majority category and biasing model performance. To address these potential problems, we evaluated additional models beyond those discussed here, wherein each model was trained in the same way except that each train+test set was downsampled to require an equal number of patients in each category, matching that of the category with the smallest number of patients. However, the downsampled models exhibited worse overall performance by all of our metrics and, within each category, had lower F1-scores (data not shown). Additionally, the performance metrics computed on the train+test sets during cross-validation were only slightly better than those computed on the validation set, alleviating concerns of overfitting. Importantly, the performance metrics we considered included metrics broken down by classification category to enable evaluation of whether any particular category was underperforming relative to the others. We also considered metrics that are suited to imbalanced data, such as Cohen's kappa.

4. Discussion

Although a direct comparison of our methods with other imputation methods was not possible due to the absence of PHI (such as surnames or addresses) in our de-identified dataset, we compared our performance to that reported in the literature. Our models consistently and substantially outperformed these prior methods.[6–9,31] E.g., the weighted correlation of our *ML-ancestry* model was 24-33 percentage points better, while its weighted error was an order of magnitude lower than other methods (Table 5).

Table 5. Comparison of performance metrics of *ML-ancestry* and other imputation methods based on metrics reported in the literature. Best performing metric indicated with bold. BISG = Bayesian Improved Surname and Geocoding;[6] CTBF = CT-based full;[9] CTBR = CT-based reduced.[9]

Imputation Method	Cohen's Kappa	Correct Rate	Weighted Correlation	Weighted Error	Reference
ML-ancestry	**0.934**	**0.973**	**0.930**	**0.007**	This study
BISG	0.58	0.78	0.597	0.089	Xue et al, 2019a[9]
CTBF	0.67	0.81	0.668	0.048	Xue et al, 2019a[9]
CTBR	0.65	0.81	0.595	0.051	Xue et al, 2019a[9]
Random Forest	0.67	0.807	0.672	0.025	Xue et al, 2019b[8]

In our study, the category with the lowest recall was Hispanic or Latino, ranging from 86-89%. This category also had the highest level of no-calls (3.4% vs. ≤1%). Prior methods report an even more pronounced drop in performance for this category.[6–9,31] However, the *ML-ancestry+demographics* model provided the best precision (96%) at a good recall rate (86%), while the Heuristic method provided the best recall (89%) but at a significantly lower level of precision (83%). Although the intended use of the imputation may dictate the best trade-off, we believe that precision is the most important feature as minimization of misclassified subjects is generally more desirable. The drop in performance in the Hispanic or Latino category may be due to the fact that self-affiliation with this category corresponds more with culture and language than with genetic similarity,[14] with levels of admixture within this group varying widely depending on country of origin and among the coasts of the US.[15]

As with all RWD analyses, our work has potential limitations. Differences between patients with complete vs. incomplete stated race and ethnicity could affect model training and therefore imputation performance. The unequal distribution of imputed categories in labeled and unlabeled data suggests that there are indeed some slight differences in the composition of patients who lack race and ethnicity data, with imputed NH Asian category most likely to be missing this information, but therefore also most able to benefit from imputation. Given the limited numbers of American Indian or Alaska Native and Native Hawaiian or Other Pacific Islander individuals in our dataset as well as the insufficient public allele frequency information from these groups, we are unable to develop models to impute those categories, meaning those individuals will be misclassified, typically as Hispanic or Latino and NH Asian, respectively. As the Tempus database grows and additional AIM allele frequencies become available, our model could be retrained to enable classification using these additional categories. While the performance of our models on populations outside the US is unknown, or indeed with differently ascertained population samples, our results suggest that retraining with additional data pertaining to those populations could yield similar performance in other settings.

When developing our race imputation methods, we adhered to established recommendations for ethical imputation.[32] We audited input data for bias, scrutinized methodological choices for potential bias introduction, rigorously assessed the accuracy of the imputed data, and our aims are to use this data to study or reduce disparities. Our adherence to these guidelines underscores our commitment to the responsible use of race imputation in promoting equity in healthcare.

5. Conclusions

Addressing racial disparities is pivotal to advancing equity in precision medicine. However, the frequent unavailability of data disaggregated by race and ethnicity in RWD can lead to biased outcome predictors,[34] inadequate representation in clinical trials,[33] and poorly targeted policies, potentially exacerbating disparities.[34] While the ultimate goal is to have complete self-reported data for optimal race and ethnicity information, our study highlights the efficacy of using genetic ancestry data to impute these categories in a de-identified setting, mitigating the challenge of data sparsity for these data in RWD from US populations. Our approach could allow more accurate identification of racial disparities in certain healthcare settings where genetic data are available, contributing to the development of fair artificial intelligence predictors and more targeted and equitable healthcare interventions.

6. Acknowledgments

We appreciate the feedback received on this work from Funmi Olopade (University of Chicago), John Carpten (City of Hope), Jose Trevino (Virginia Commonwealth University), Carlos D. Bustamante (Stanford University), Joel Dudley, Calvin Chao, Ezra Cohen and Kate Sasser (Tempus Labs). We also acknowledge Frank Nothaft, Rafael Esleyer, Nick Riggan, and Arvind Prasad (Tempus Labs) for their invaluable assistance in procuring data needed for this work. We thank Vanessa Nepomuceno from the Tempus Publications team for copyediting the manuscript.

References

1. Dang, A. Real-World Evidence: A Primer. *Pharm. Med.* **37**, 25–36 (2023).
2. Resnic, F. S. & Matheny, M. E. Medical Devices in the Real World. *N. Engl. J. Med.* **378**, 595–597 (2018).
3. Studna, A. Executive Roundtable: The Rise of RWD in Clinical Research. *Applied Clinical Trials.* 28 September 2023, https://www.appliedclinicaltrialsonline.com/view/executive-roundtable-the-rise-of-rwd-in-clinical-research (2023).
4. Cullen, M. R. *et al.* A framework for setting enrollment goals to ensure participant diversity in sponsored clinical trials in the United States. *Contemp. Clin. Trials* **129**, 107184 (2023).

5. Cabreros, I., Agniel, D., Martino, S. C., Damberg, C. L. & Elliott, M. N. Predicting Race And Ethnicity To Ensure Equitable Algorithms For Health Care Decision Making. *Health Aff.* **41**, 1153–1159 (2022).
6. Elliott, M. N., Fremont, A., Morrison, P. A., Pantoja, P. & Lurie, N. A New Method for Estimating Race/Ethnicity and Associated Disparities Where Administrative Records Lack Self-Reported Race/Ethnicity. *Health Serv. Res.* **43**, 1722–1736 (2008).
7. Derose, S. F., Contreras, R., Coleman, K. J., Koebnick, C. & Jacobsen, S. J. Race and Ethnicity Data Quality and Imputation Using U.S. Census Data in an Integrated Health System. *Med Care Res Rev* **70**, 330–345 (2012).
8. Xue, Y., Harel, O. & Aseltine, R. Comparison of Imputation Methods for Race and Ethnic Information in Administrative Health Data. *2019 13th Int Conf Sampl Theory Appl Sampta* **00**, 1–4 (2019).
9. Xue, Y., Harel, O. & Aseltine, R. H. Imputing race and ethnic information in administrative health data. *Health Serv. Res.* **54**, 957–963 (2019).
10. Walther, Z. & Sklar, J. Molecular Tumor Profiling for Prediction of Response to Anticancer Therapies. *Cancer J.* **17**, 71–79 (2011).
11. Beaubier, N. *et al.* Integrated genomic profiling expands clinical options for patients with cancer. *Nat Biotechnol* **37**, 1351–1360 (2019).
12. National Academies of Sciences, Engineering, and Medicine, Committee on the Use of Race, Ethnicity, and Ancestry as Population Descriptors in Genomics Research. *Using Population Descriptors in Genetics and Genomics Research: A New Framework for an Evolving Field.* (National Academies Press (US), (2023).
13. Auton, A. *et al.* A global reference for human genetic variation. *Nature* **526**, 68 74 (2015).
14. Lu, C., Ahmed, R., Lamri, A. & Anand, S. S. Use of race, ethnicity, and ancestry data in health research. *Plos Global Public Health* **2**, e0001060 (2022).
15. Bryc, K., Durand, E. Y., Macpherson, J. M., Reich, D. & Mountain, J. L. The Genetic Ancestry of African Americans, Latinos, and European Americans across the United States. *Am J Hum Genet.* **96**, 37 53 (2015).
16. Budget, O. of M. and. Standards for the classification of federal data on race and ethnicity. *Fed. Reg.* **62**, 58782 (1997).
17. Flanagin, A., Frey, T., Christiansen, S. L. & Committee, A. M. of S. Updated Guidance on the Reporting of Race and Ethnicity in Medical and Science Journals. *JAMA* **326**, 621–627 (2021).
18. Beaubier, N. *et al.* Clinical validation of the Tempus xO assay. *Oncotarget* **9**, 25826-25832 (2018).
19. Alexander, D. H., Novembre, J. & Lange, K. Fast model-based estimation of ancestry in unrelated individuals. *Genome Res* **19**, 1655-1664 (2009).
20. Miyashita, M. *et al.* Molecular profiling of a real-world breast cancer cohort with genetically inferred ancestries reveals actionable tumor biology differences between European ancestry and African ancestry patient populations. *Breast Cancer Res.* **25**, 58 (2023).
21. Kosoy, R. *et al.* Ancestry informative marker sets for determining continental origin and admixture proportions in common populations in America. *Hum Mut.* **30**, 69–78 (2009).

22. Bergström, A. *et al.* Insights into human genetic variation and population history from 929 diverse genomes. *Science* **367**, 6484:eaay5012 (2020).
23. Mallick, S. *et al.* The Simons Genome Diversity Project: 300 genomes from 142 diverse populations. *Nature* **538**, 201–206 (2016).
24. Campbell, P. J. *et al.* Pan-cancer analysis of whole genomes. *Nature* **578**, 82–93 (2020).
25. Maples, B. K., Gravel, S., Kenny, E. E. & Bustamante, C. D. RFMix: A Discriminative Modeling Approach for Rapid and Robust Local-Ancestry Inference. *Am J Hum Genet.* **8**, 93(2), 278-88 (2013).
26. Pepe, M. S. *The Statistical Evaluation of Medical Tests for Classification and Prediction.* Oxford University Press (2004).
27. Cohen, J. A Coefficient of Agreement for Nominal Scales. *Educ. Psychol. Meas.* **20**, 37–46 (1960).
28. Dettling, M. & Bühlmann, P. Boosting for tumor classification with gene expression data. *Bioinformatics* **19**, 1061–1069 (2003).
29. UDS Mapper: Zip code to ZCTA crosswalk. 28 September 2023, https://udsmapper.org/zip-code-to-zcta-crosswalk/.
30. caret: Classification and Regression Training. 28 September 2023, https://cran.r-project.org/web/packages/caret/index.html.
31. Grundmeier, R. W. *et al.* Imputing Missing Race/Ethnicity in Pediatric Electronic Health Records: Reducing Bias with Use of U.S. Census Location and Surname Data. *Health Serv Res* **50**, 946–960 (2015).
32. Brown, K. S., Ford, L., Ashley, S., Stern, A. & Narayanan, A. Ethics and Empathy in Using Imputation to Disaggregate Data for Racial Equity: Recommendations and Standards Guide. Urban Institute research report (2021).
33. Loree, J. M. *et al.* Disparity of Race Reporting and Representation in Clinical Trials Leading to Cancer Drug Approvals From 2008 to 2018. *JAMA Oncol* **5**, e191870 (2019).
34. Obermeyer, Z., Powers, B., Vogeli, C. & Mullainathan, S. Dissecting racial bias in an algorithm used to manage the health of populations. *Science* **366**, 447–453 (2019).

Precision Medicine: Innovative methods for advanced understanding of molecular underpinnings of disease

Yana Bromberg†
Department of Biology, Emory University, 1510 Clifton Rd
Department of Computer Science, Emory University, 400 Dowman Dr.
Atlanta, GA 30317, USA
Email: yana.bromberg@emory.edu

Hannah Carter†
Department of Medicine, University of California San Diego, 9500 Gilman Dr.
La Jolla, CA 92093, USA
Email: hkcarter@health.ucsd.edu

Steven E. Brenner
Department of Plant & Microbial Biology, University of California Berkeley, 461 Koshland Hall
Berkeley, California 94720-3102, USA
Email: brenner@compbio.berkeley.edu

Precision medicine, also often referred to as personalized medicine, targets the development of treatments and preventative measures specific to the individual's genomic signatures, lifestyle, and environmental conditions. The series of Precision Medicine sessions in PSB has continuously highlighted the advances in this field. Our 2024 collection of manuscripts showcases algorithmic advances that integrate data from distinct modalities and introduce innovative approaches to extract new, medically relevant information from existing data. These evolving technology and analytical methods promise to bring closer the goals of precision medicine to improve health and increase lifespan.

1. Introduction

Precision medicine involves tailoring medical decisions and treatments to individual patients in a data-driven manner. The accumulation of medically-relevant and, particularly, molecular data has uncovered the potential for mechanistic insight into disease processes facilitating clinical decision making. Advances in genomic techniques, e.g. spatial transcriptomics and single cell analysis, have further enabled identification of the genetic biomarkers of patient drug responses, susceptibility to diseases, and other medically-relevant outcomes. At the same time, the enormous scale of this data has stimulated use of novel computational methods, resulting in, e.g., the recent explosion in deep learning-based, biological and medical data analysis techniques.

† Corresponding Authors

While the concept of personalized medicine stretches back nearly two decades – just slightly longer than our PSB session – the implementation of precision medicine in practice (still) remains in its early stages. Novel technologies require new integrative approaches to advance the state of the art in this field. In our 2024 session we feature work from researchers across diverse domains, who integrate various omics data to provide valuable insights into disease mechanisms, diagnosis, and treatment. In this collection, we explore their cutting-edge advancements in more detail.

2. Session Contributions

2.1. *Transcriptome and Histopathology Integration*

A number of studies submitted to our session focused on integrating spatial transcriptomics and histopathology data and demonstrating the potential of this combination to enhance our understanding of tumor biology. Song et al enriched their transcriptome-driven findings by incorporating morphological features extracted from histopathology images to enable a comprehensive analysis of tumor architecture via feature clustering. Azher et al employed contrastive learning and Graph Convolutional Neural Networks (GCN) to predict disease stage, lymph node metastasis, and survival prognosis in cancer patients. Meanwhile, Srinivasan et al developed a transformer-based model to shed light on the molecular pathways involved in skin aging due to light exposure. Their findings not only contribute to our understanding of their chosen conditions but also demonstrate the potential of their approaches for studying other diseases.

2.2. *Spatial Proteomics: Revealing Tissue Microenvironments.*

Wu et al introduced innovative methods to analyze tissue microenvironments at high resolution using spatial proteomics. By measuring inferred protein polarity, they identified distinct subpopulations of immune cells within tumors, shedding light on potential markers of better prognosis. This approach holds promise for identifying patients who may respond favorably to specific treatments.

2.3. *Microbiome Analysis: A Closer Look at Gut Health.*

Sapoval et al proposed a novel metagenomic analysis pipeline that bypasses the need for genome assembly, allowing for direct comparisons between patients and healthy controls. This reference-free approach is particularly valuable for studying conditions like myalgic encephalomyelitis/chronic fatigue syndrome (ME/CFS), where the gut microbiome plays a crucial role. Understanding dysbiosis at this level can help identify potential disease markers and therapeutic targets.

2.4. *Polygenic Risk Scores: Bridging Genomics and Disease.*

Cardone et al examined the role of a lymphocyte count PRS (polygenic risk score) in predicting CD4 T-cell recovery in individuals with HIV undergoing anti-retroviral therapy. While their findings indicated limited PRS impact compared to clinical factors, they underscore the importance of considering multiple variables in precision medicine studies. Kember et al focused on improving PRS accuracy for cardiometabolic traits; their findings emphasize the need for implementing separate scoring mechanisms for diverse ancestries.

2.5. *Integrative Methods for Clustering, Meta-analysis, Deconvolution, and Network Rewiring.*

Numerous contributions aimed at enhancing integrative methods for meta-analysis, subtype detection, cell type deconvolution, and network rewiring. Zhang et al introduced nSEA, an algorithm for unsupervised clustering of low grade Gliomas, uncovering a novel subtype with clinical implications. Huang et al proposed a multi-modal clustering approach that combines various data types to cluster tumor samples across different cancer types, offering a more holistic perspective on cancer classification.

In the realm of meta-analysis, Fukutani et al overcame batch effects in gene expression data, highlighting the importance of robust analytical techniques in large-scale studies. Sufriyana et al employed data-driven ontology inference to uncover novel gene sets relevant to subtypes of preeclampsia, showcasing the power of meta-analysis in identifying novel biological processes.

Deconvolution, a vital tool for deciphering cellular composition from omics data, faced challenges in understanding nascent RNA. Maas et al introduced an adaptation for nascent RNA sequencing, addressing the nuances of this emerging field.

Finally, Dannenfelser et al explored how alternative splicing rewires protein interaction networks in cancer. Their Splitpea method provides insights into the complex interplay between alternative splicing and disease, offering a novel perspective on cancer biology.

References

1. Azher, Z., Fatemi, M., Lu, Y., Srinivasan, G., Diallo A., Christensen, B., Salas, L., Kolling IV, F., Perreard, L., Palisoul, S., Vaickus, L., Levy, J. (2024) Spatial Omics Driven Crossmodal Pretraining Applied to Graph-based Deep Learning for Cancer Pathology Analysis. *Pacific Symposium on Biocomputing 2024.*
2. Dannenfelser, R., Yao, V., (2024) Splitpea: uncovering cancer patient-specific protein interaction network rewiring. *Pacific Symposium on Biocomputing 2024.*
3. Fukutani, K.F., Hampton, T.H., Bobak, C.A., MacKenzie, T.A. , Stanton, B.A. (2024) Application of Quantile Discretization and Bayesian Network Analysis to Publicly Available Cystic Fibrosis Data Sets. *Pacific Symposium on Biocomputing 2024.*
4. Maas, Z., Sigauke, R., Dowell, R. (2024) Deconvolution of Nascent Sequencing Data Using Transcriptional Regulatory Elements. *Pacific Symposium on Biocomputing 2024.*
5. Sapoval, N., Tanevski, M., Treangen, T.J. (2024) KombOver: Efficient k-core and K-truss based characterization of chronic disease impact on the human gut microbiome. *Pacific Symposium on Biocomputing 2024.*
6. Song, J., Ramaswamy, V.G., Lamstein, J., Webb, M., Zada, G., Finkbeiner, S., Craig, D.W. (2024) Enhancing Spatial Transcriptomics Analysis by Integrating Image-Aware Deep Learning Methods. *Pacific Symposium on Biocomputing 2024.*
7. Srinivasan, G., Davis, M., LeBoeuf, M., Fatemi, M., Azher, Z., Lu, Y., Diallo, A., Saldias Montivero, M., Kolling IV, F., Perrard, L., Salas, L., Christensen, B., Palisoul, S., Tsongalis, G. Vaickus, L., Preum, S. Levy, J. (2024) Potential to Enhance Large Scale Molecular Assessments of Skin Photoaging through Virtual Inference of Spatial Transcriptomics from Routine Staining. *Pacific Symposium on Biocomputing 2024.*
8. Sufriyana, H., Wu, Y., Su, E.C.Y. (2024) Transcriptome and interactome connecting endometrial-decidua-placental origin of preeclampsia subtypes: A preliminary study. *Pacific Symposium on Biocomputing 2024.*

9. Wu, E., Wu, Z., Mayer, A.T., Trevino, A.E., Zou, J. (2024) Polarity measurements from spatial proteomics imaging suggest immune cell engagement. *Pacific Symposium on Biocomputing 2024.*
10. Zhang, Z., Wang, C., Zhao, Z., Yi, Z., Durmaz, A., Yu, J., Bebek, G. (2024) nSEA: n-Node Subnetwork Enumeration Algorithm Identifies Lower Grade Glioma Subtypes with Altered Subnetworks and Distinct Prognostics. *Pacific Symposium on Biocomputing 2024.*
11. Cardone, K.M., Dudek, S., Keat, K., Bradford, Y., Cindi, Z., Daar, E.S., Gulick, R., Riddler, S.A., Lennox, J.L., Sinxadi, P., Haas, D.W., Ritchie, M.D. (2024) Lymphocyte Count Derived Polygenic Score and Interindividual Variability in CD4 T-cell Recovery in Response to Antiretroviral Therapy. *Pacific Symposium on Biocomputing 2024.*
12. Huang, C., Kuan, P.F. (2024) intCC: An efficient weighted integrative consensus clustering of multimodal data. *Pacific Symposium on Biocomputing 2024.*
13. Kember, R.L., Verma, A., Verma, S.S., Xiao, B., Lucas, A., Kripke, C.M., Judy, R., Chen, J., Damrauer, S.M., Rader, D.J., Ritchie, M.D. (2024) Polygenic risk scores for cardiometabolic traits demonstrate importance of ancestry for predictive precision medicine. *Pacific Symposium on Biocomputing 2024.*

Enhancing Spatial Transcriptomics Analysis by Integrating Image-Aware Deep Learning Methods

Jiarong Song[1,4], Josh Lamstein[2], Vivek Gopal Ramaswamy[2], Michelle Webb[4], Gabriel Zada[3], Steven Finkbeiner[3], and David W. Craig[1,4,5†]

[1]Department of Integrated Translational Sciences; City of Hope, Duarte, CA 91010, USA
[2]Center for Systems and Therapeutics, Gladstone Institutes, San Francisco, CA 94158, USA
[3]USC Radiosurgery Center, Keck School of Medicine of USC, CA 91008, USA
[4]Dept of Translational Genomics, Keck School of Medicine of USC, CA 91008, USA
[5]Departments of Neurology and Physiology, UCSF, San Francisco CA 91458
[†]Email: dacraig@coh.org

Spatial transcriptomics (ST) represents a pivotal advancement in biomedical research, enabling the transcriptional profiling of cells within their morphological context and providing a pivotal tool for understanding spatial heterogeneity in cancer tissues. However, current analytical approaches, akin to single-cell analysis, largely depend on gene expression, underutilizing the rich morphological information inherent in the tissue. We present a novel method integrating spatial transcriptomics and histopathological image data to better capture biologically meaningful patterns in patient data, focusing on aggressive cancer types such as glioblastoma and triple-negative breast cancer. We used a ResNet-based deep learning model to extract key morphological features from high-resolution whole-slide histology images. Spot-level PCA-reduced vectors of both the ResNet-50 analysis of the histological image and the spatial gene expression data were used in Louvain clustering to enable image-aware feature discovery. Assessment of features from image-aware clustering successfully pinpointed key biological features identified by manual histopathology, such as for regions of fibrosis and necrosis, as well as improved edge definition in EGFR-rich areas. Importantly, our combinatorial approach revealed crucial characteristics seen in histopathology that gene-expression-only analysis had missed.

Supplemental Material:
https://github.com/davcraig75/song_psb2014/blob/main/SupplementaryData.pdf

Keywords: Spatial transcriptomics; Deep learning; Image-aware clustering

1. Introduction

Mapping the spatial organization of genes and cells in tissues is the foundation for understanding higher-level molecular and cellular processes driving disease pathogenesis. In the past decade, paradigm-shifting approaches such as single-cell RNA-seq (scRNA) have provided unprecedented insights into cellular populations. More recently, Spatial Transcriptomics (ST) methods have emerged (e.g., Visium ST), providing a view of cellular RNA expression and disease pathology in ~~the context of neighboring cel~~ls and structures[1,2]. Instead of measuring a single bulk transcriptome from a tissue section, ST obtains thousands of transcriptomes across a tissue section at spatially distinct spots, where each spot covers a few cells with Visium[3], or provides sub-cellular data as with

MERFISH[4] and Xenium. These emerging ST technologies have unlocked unprecedented possibilities for exploring the transcriptomic architecture of multicellular organisms, revealing intricate cellular heterogeneity in diverse tissues and disease states[5].

However, analytical methodologies that do not take into account spatial context or the underlying histopathology frequently limit the full potential of these potent technologies. In the case of ST analysis methods, many are extensions of earlier strategies that lack direct incorporation of spatial information between spots or not to directly leverage the underlying imaging data of protein and cellular structure. For example, current clustering techniques, such as Louvain and k-means clustering provided by Seurat[6], primarily focus on gene expression[7], often neglecting spatial context and the potential complementary information that can be gleaned from tissue morphology. This incomplete fusion of transcriptomic and morphological data limits our ability to fully understand the cellular ecosystem within tissues, particularly in cancer related states. In particular groups, efforts are being made to integrate imaging data in order to better capture the richness of information embodied in high resolution H&E images[7–10].

In our exploration of the utility of advanced analytical methods for spatial transcriptomics, we developed a novel approach we termed "stMIC" (Spatial Transcriptomics and Morphological Integrated Clustering). Central to stMIC's design is the incorporation of a form of Convolutional Neural Network (CNN) deep learning, specifically the Residual Network-50 or ResNet-50, which is characterized by its versatility and effectiveness across a wide array of applications[3,11]. The Resnet-50 architecture uses the concept of residual connections, and, with its existing prior training, is highly effective for image classification, object detection, and image segmentation[12]. To underline the clinical implications of our study, we deployed our method on previously histopathological assessed disease specimens, with a particular focus on aggressive malignancies such as glioblastoma and triple-negative breast cancer.

2. Method

2.1. *Visium Spatial Gene Expression Assay, Sequencing, and Preprocessing*

Freshly frozen, OCT-embedded tissues were cryosectioned and mounted on Visium spatial gene expression slides (10x Genomics, #1000184), which contain four 6.5 mm * 6.5 mm capture areas comprising 5,000 barcoded spatial features each. Hematoxylin and eosin (H&E) staining was applied, and microscopic images were obtained subsequently with a Zeiss Axioscan2 microscope using a 10x objective. After staining, tissues underwent a permeabilization process to facilitate RNA binding to the slide surface, which was determined using the Spatial Tissue Optimization procedure (10x Genomics, #1000193). The on-slide cDNA synthesized from immobilized RNA was used to generate sequencing libraries, which were paired-end sequenced on an Illumina NovaSeq 6000 instrument to produce a minimum of ~250 million read pairs per sample[13]. The Triple-Negative Breast Cancer (TNBC) sample utilized in our study was originally characterized using Spatial Transcriptomics (ST) in the work of Bassiouni et al[13].

The Space Ranger pipelines (version 1.1.0; 10x Genomics) were employed to preprocess the sequencing data. Demultiplexing of BCL data and conversion to FASTQ format were accomplished using the *spaceranger mkfastq* pipeline. Further, the spaceranger count pipeline enabled read alignment to the human reference genome GRCh38, UMI counting, and the generation of feature-spot matrices corresponding to the microscopic tissue image. This pipeline also provided automatic tissue detection and fiducial alignment based on the image. Raw gene expression data underwent Counts Per Million (CPM) normalization, and subsequent log transformation, followed by scaling the data to zero mean and unit variance. All these steps were completed using Scanpy (version:1.6.0).[14]

2.2. *Histopathological Image Annotation and Evaluation*

Frozen tissues from each block selected for study were stained with H&E. Images derived from the Visium slides were examined and annotated by a pathologist utilizing Adobe Photoshop software[13]. Regions characterized by blood vessel, necrosis, dense immune cell infiltrates, or stromal fibrosis were indicated when applicable.

2.3. *Histological Image Segmentation and Patch Selection*

The primary stage of image preprocessing entailed segmenting whole-slide H&E histological images from patient samples into smaller patches. Patches were chosen such that each entirely encompassed corresponding spots under the tissue, as indicated in the second column of the "tissue_positions_list.csv" if it is 1. For each spot S_i, a corresponding patch was defined, with the geographic center of the patch aligned with the center of the spot, denoted by coordinates (x_i,y_i) from the last two columns in "tissue_positions_list.csv" file. The patch dimensions were such that both the height and width were equivalent to the diameter of the spot, d, from scalefactors_json.json. Thus, the boundaries of the patch were formally determined by the following coordinates: the upper boundary at ($x_i+d/2$), the lower boundary at ($x_i-d/2$), the left boundary at ($y_i-d/2$) , and the right boundary at ($y_i+d/2$).

2.4. *Feature Extraction Using ResNet-50*

We implemented a convolutional neural network (CNN) model for feature extraction from each patch. We utilized a pretrained ResNet-50 model (Tensorflow version: 2.6.0) trained on the ImageNet dataset for optimal performance in our task. Specifically, it was employed with its top fully-connected layer excluded, selecting "avg" pooling mode for feature extraction (Tensorflow version: 2.6.0), and everything else was default setting. The segmented histology image patches, read in using OpenCV (version: 4.5.3) and resized to (224,224,3), served as inputs. The ResNet-50 model subsequently outputted a 2048-dimensional feature array that represented the patch's morphological features. Feature standardization was achieved using StandardScaler from sklearn (version 0.22.1), which removes the mean and scales to unit variance.

2.5. *Integration of Matrices and Clustering*

Both the normalized gene expression matrix and morphological feature matrix underwent Principal Component Analysis (PCA), separately processed through the top 10 principal components. The resulting matrices were concatenated based on the corresponding barcode of the spots. Clustering was accomplished using the Louvain modularity optimization algorithm with a resolution range of 1.5-1.9 and k-neighbors set at 39, implemented in Orange Data Mining (version

3.30.1). This configuration yielded a stable set of clusters, paralleling the cluster numbers obtained through the current analytical approach solely based on gene expression.

2.6. *Evaluation Measures*

The performance of our method was evaluated using multiple validation metrics. The Adjusted Rand Index (ARI) served as the initial metric, quantifying the similarity between clustering assignments relative to the pathologist's annotation[15]. Gene set enrichment analysis (GSEA) was then performed at the cluster level via the Broad Institute's GenePattern software (RRID:SCR_003199). Utilizing the FindAllMarkers feature of Seurat (version: 4.3.0.1), differentially expressed genes within chosen clusters were identified via Wilcoxon rank sum testing.

Gene lists from each selected cluster were subjected to Pre Ranked GSEA, contrasting against chosen gene sets (H: hallmark gene sets[16], C2:CP:KEGG[17], C4[18], C7: immunesigdb[19]), with permutations set at 1000 and the collapse dataset selected as "Remap_only". The final step involved visualizing spatial expression patterns for genes of interest using the SpatialPlot feature in Seurat.

2.7. *Implementation*

stMIC has been developed with Python 3.7 as a user-friendly pipeline. Setting up and tutorials are described in the stMIC GitHub page: https://github.com/USCDTG/stMIC.

2.8. *Supplemental Material*

Supplemental Material referred to in the paper may be found at the following URL: https://github.com/davcraig75/song_psb2014/blob/main/SupplementaryData.pdf

3. Results

3.1. *Pipeline*

Our primary goal was to enhance identification of biological features from 10X Visium ST by incorporating deep learning analysis. We first show the default approach in Figure 1A, where "spot-level" normalized gene expression obtained via the Space Ranger pipeline is first reduced by principal component analysis (PCA) from (spots x genes) to (spots x M) where M is 10. This is frequently followed by graph-based clustering using a sparse nearest neighbor graph, followed by Louvain Modularity Optimization to identify highly-connected modules in the graph. We note that other clustering methods, such as K-means, are used and are presenting the default clustering method of Space Ranger.

Our stMIC pipeline is shown in Figure 1B and includes partitioning or splitting the ST histological image into segmented tissue spots, followed by feature extraction in the ResNet-50 model on the underlying image for each spot. Thus, if there were 4,096 passing spots, we would have the same number of images. Computationally, this step took approximately 2.5-5 minutes per whole slide image on a machine with an NVIDIA Tesla T4 GPU with 16GB of VRAM, depending on the number of spots under tissue. The gene expression data was simultaneously processed with the Space Ranger pipelines, turning raw sequencing data into normalized feature-spot matrices within 3 hours on a standard bioinformatics workstation with 32 CPU cores.

Following dimensionality reduction of both morphological and gene expression data via Principal Component Analysis (PCA), we concatenated and performed clustering using the Louvain algorithm, an operation that took roughly 1-2 minutes on the same workstation. To evaluate our method's performance, we used multiple validation metrics, including the Adjusted Rand Index (ARI) and Gene Set Enrichment Analysis (GSEA).

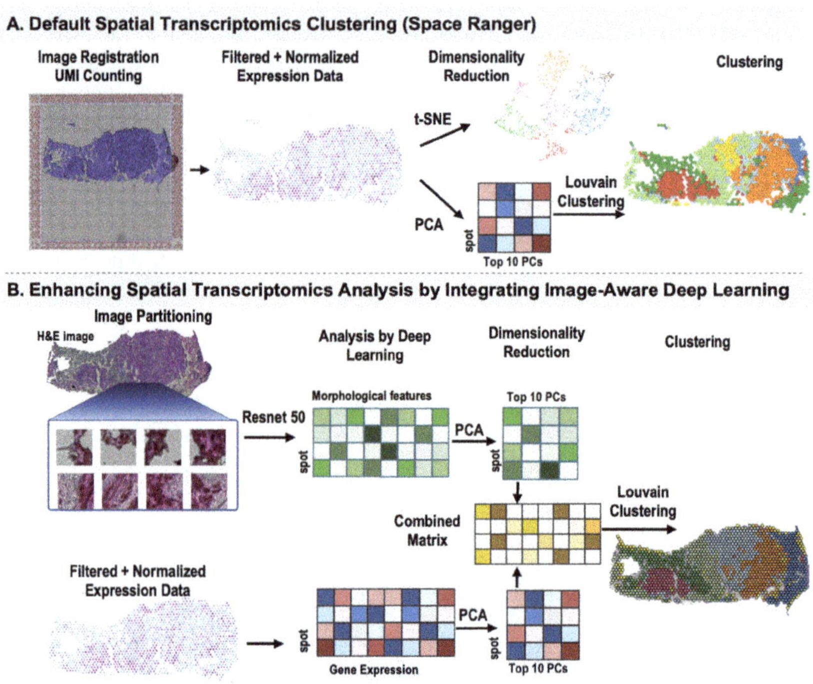

Fig. 1. (A) Default spatial transcriptomics clustering approach, e.g. used in 10X Space Ranger Pipeline. (B) stMIC: ST analysis integrating image-aware deep-learning analysis. High-resolution histology images undergo an initial cropping process into smaller patches, driven by the location and size of spots within the tissue. Subsequently, these patches are introduced to a deep learning model, ResNet-50, resulting in the production of a morphological feature matrix. Principal Component Analysis (PCA) is applied to this matrix, from which the top 10 principal components (PCs) are selected. A parallel procedure is enacted on the gene expression matrix derived from the spatial transcriptomics dataset. The reduced matrices from both the morphological and gene expression data are then concatenated at each spot to form a unified matrix.

3.2. *Application to human glioblastoma spatial transcriptomics data*

In the exploration of the pretrained ResNet-50 model's proficiency, we started with a representative glioblastoma sample, FFD1. This sample encompasses 3,594 spots and 33,538 genes, obtained from the 10x Genomics Visium platform. The analysis commences by contrasting the clustering outcomes from both the Louvain methodology, which is only dependent on gene expression (GEBC) (Fig. 2B), and the approach leveraging H&E histology image feature extraction via ResNet-50 (Fig. 2C). For enhanced visual interpretability, each cluster is assigned a unique color.

Further exploration of the three most significantly differentially expressed genes (DEGs) within this cluster, *HBB*, *HBA1*, and *HBA2*, revealed their critical role in blood biochemistry. *HBB* encodes the beta-globin protein, while *HBA1* and *HBA2* code for the alpha-globin protein, forming essential components of hemoglobin[20]. Predominantly, the expression of these genes was concentrated within the region defined as cluster 10 (Fig. 2D), affirming the blood vessel identity of this cluster.

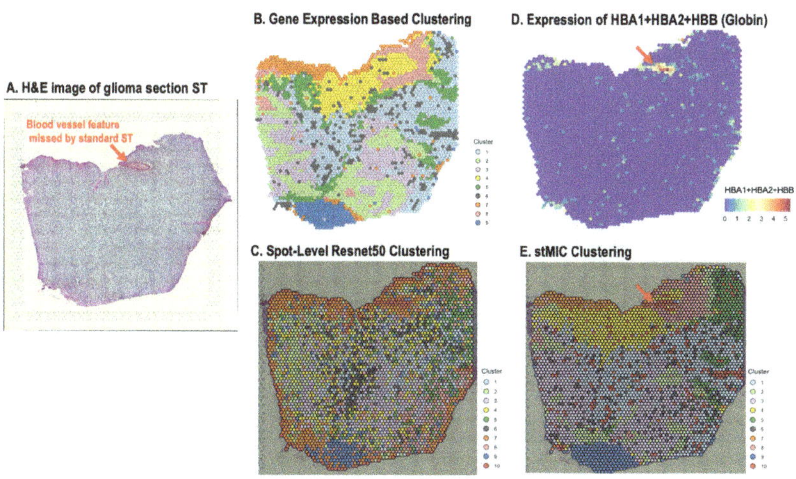

Fig. 2. Comparative analysis of clustering methods and spatial expression patterns of FFD1. (A) Haematoxylin and eosin (H&E) staining image of Glioblastoma sample. (B) Spatial domains identified by gene-expression-based Louvain clustering method. (C) Spatial domains identified by ResNet-50 feature extraction method. (D) Spatial distribution of the top upregulated genes in cluster 10 (HBA1, HBB, HBA2). (E) Spatial domains identified by stMIC method.

These findings attest to the remarkable capability of ResNet-50 in uncovering areas that remain undetected by the gene-expression-based methodology, which relies strictly on gene expression. These advantages provide a solid foundation for developing an even more robust and comprehensive analysis method, stimulating the formulation of an integrated approach. Motivated by this, we proceeded to implement an advanced strategy that harnesses both histological and gene expression information. This integrative method not only enhances our ability to discern various tissue regions, but also seeks to offer a more nuanced, multidimensional view of the complex histopathological landscape. The harmonization of image-derived and transcriptomics data enables us to move beyond the limitations of each individual data type, allowing a more holistic exploration of biological phenomena at the tissue level. Importantly, stMIC strategy not only successfully pinpoints the blood vessel area, but also enhances edge definition in this key area (Fig. 2E), thus increasing precision in key region detection. These results highlight the potential of the integrated image-aware method methodology for providing comprehensive and accurate histopathological profiling.

We set the number of clusters at ten for all methods to compare the clustering results of SpaGCN, SpaCell, and stLearn. SpaCell, which employs an autoencoder for dimension reduction, failed to detect the EGFR-rich region (not detailed in the main results) and did not align closely with the pathologist-annotated blood vessel region as depicted in Supplementary Fig. 1D. Notably, both SpaGCN and stLearn in their default implementation were also unable to identify the blood vessel region, as shown in Supplementary Fig 1.E&F. Still any interpretation of features missed or seen should be taken with caution since these types of features were not part of their development.

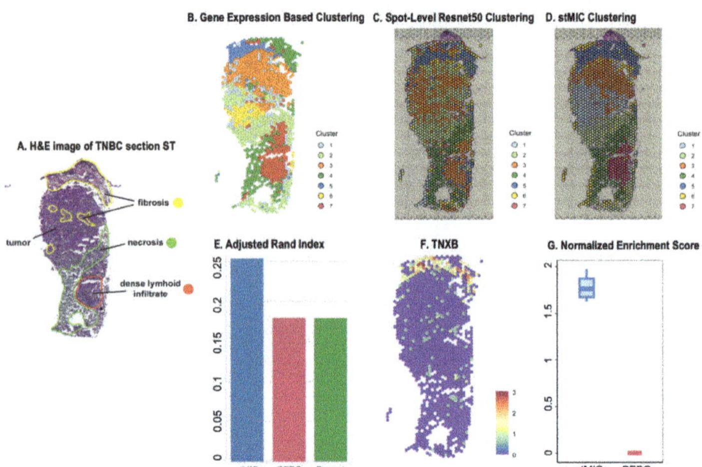

Fig. 3. Comparative analysis of clustering methods and spatial expression patterns of Slide 120D. (A) Histology and manually annotated structure for slide 120D. (B) Spatial domains identified by gene expression-based clustering method. (C) Spatial domains identified by ResNet-50 feature extraction method. (D) Spatial domains identified by stMIC method. (E) Adjusted Rand Index (ARI) in stMIC, gene expression based (GEBC), and ResNet clustering methods determined sections against the ground truth labels (pathologist annotation). (F) Spatial expression of the fibrosis marker gene. (G) boxplot of GSEA for cluster 4 of sample 120D against selected gene sets. NES = Normalized enrichment score.

3.3. *Application to human triple-negative breast cancer spatial transcriptomics data*

In our evaluation of the triple-negative breast cancer sample, termed Slide 120D, ResNet enhanced the performance in distinguishing between fibrotic and necrotic regions, a result that is corroborated by pathologist annotations (Fig. 3A,C). The gene expression-based method could not differentiate these distinct regions, splitting the fibrotic region into two clusters (Fig. 3B). Using an approach previously applied to a glioblastoma sample, we employed our integrated image-aware method, stMIC. Consequently, our stMIC method not only improved clustering of the fibrotic and necrotic regions but also significantly enhanced the accuracy in identifying the dense lymphoid infiltrate region (Fig. 3D). To assess and compare the performance of two clustering methods, the Adjusted Rand Index (ARI) was employed as an evaluation metric. The ARI is a widely used measure that quantifies the similarity between two clustering assignments by comparing their agreement with respect to the ground truth, in this case, the pathologist's annotation. Clustering based on the integration of morphological features with gene expression was more consistent with the regional annotations obtained by pathologists (ARI = 0.2536) compared to gene expression-based clustering method (ARI = 0.1796) and ResNet-50 feature extraction clustering (ARI = 0.1793) (Fig. 3E). To discern each method's proficiency in detecting intricate tissue structures, we observed stLearn's clustering closely mirrored that of the gene expression-based Louvain clustering. Both stLearn and SpaGCN does not segregate the fibrosis and necrosis regions (Supplementary Fig 2.E&F). Specifically, SpaCell managed to identify almost the entire fibrosis region but did not fully differentiate between these two crucial areas (Supplementary Fig. 2D). The Adjusted Rand Index (ARI) is 0.173 for SpaGCN, 0.151 for stLearn, and 0.125 for SpaCell (Supplementary Fig 2H). At some level, these results must be taken with caution since these tools had not been evaluated or designed for these types of pathologies.

From a biological point of view, improved clustering is supported by the fact that the key marker tenascin-XB (TNXB) of fibrosis shows high differential expression in cluster 4 (Fig. 3D&F). Tenascin-XB, a key component of the extracellular matrix, has been linked to tissue remodeling and fibrosis, and is often upregulated in fibrotic tissues[21] . Its elevated expression and localization within the fibrotic region not only strengthen the characterization of this region but also implies the ongoing process of tissue remodeling - a common event in fibrosis[22]. This discovery again emphasizes the capacity of our integrated method to reveal crucial biological elements and events, contributing to a more comprehensive understanding of tumor progression.

Furthermore, stMIC approach proved better in capturing biologically relevant features within the detected fibrotic region compared to the gene expression-based clustering method. Gene set enrichment analysis revealed significant distinctions between the two methods (Fig. 3G). Only results with a false discovery rate (FDR) < 0.05 are displayed. While no significant pathways were detected using the expression-based method, our integrated approach identified four significantly enriched pathways. These pathways were indicative of active immunity, antigen processing and presentation, as well as humoral immune and inflammatory responses, all with Normalized Enrichment Scores (NES) greater than 1.5 (Supplementary Fig. 2G). These findings demonstrate the profound immune involvement within the fibrotic area, further underscoring the added value of stMIC in capturing these nuanced dynamics.

In another triple-negative breast cancer sample, 094D, in a horizontal comparison of results from the gene expression based, ResNet-50, and stMIC methods with annotation, stMIC approach unveiled critical biological phenomena (Fig. 4A-D). Notably, it achieved higher accuracy in capturing the middle right fibrotic region, a conclusion that was further corroborated by the higher Adjusted Rand Index (ARI) in the stMIC (ARI = 0.1479) vs. gene-expression alone (0.1111) or ResNet feature clustering (0.1051). In a comparison between the clustering results of stMIC and spaCell, Figure 4G indicates SpaCell did not differentiate between these sub-clusters (Supplementary Fig. 3D). While both spaGCN and stLearn could differentiate these sub-clusters((Supplementary Fig 3.E&F), their performance was not markedly superior to stMIC. When measured against pathologist annotations using the ARI), SpaGCN, stLearn and SpaCell recorded ARIs of 0.098, 0.127 and 0.091, respectively (Supplementary Fig 3H).

In-depth analysis of marker genes, facilitated by the stMIC approach, revealed notable features absent from gene-expression-only clustering. Specifically, our approach discerned that what was identified as cluster 1 in gene-expression-only clustering comprised two distinct, biologically relevant clusters. One of these exhibited increased expression of hypoxia markers CA9/NDRG1, while the other was characterized by the presence of IFIT1, a marker indicative of an active interferon response (Fig. 4 F-H).

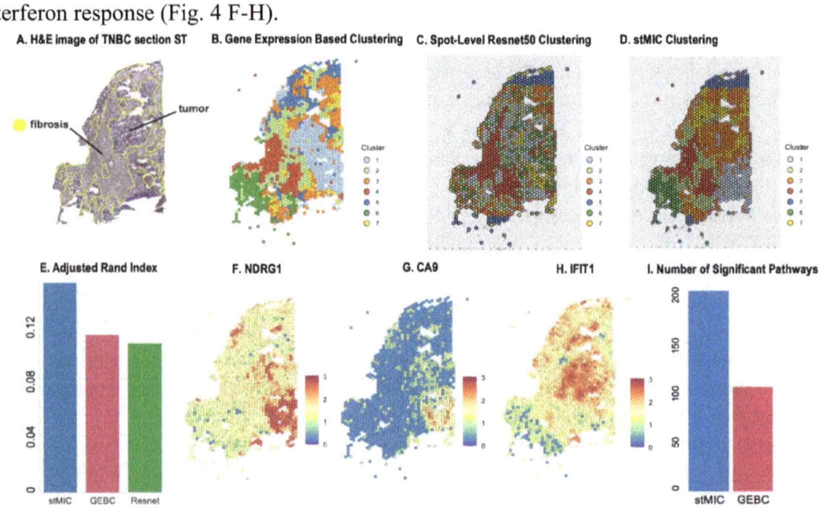

Fig. 4. Comparative analysis of clustering methods and spatial expression patterns of Slide 94D. (A) Histology and manually annotated structure for slide 94D. (B) Spatial domains identified by gene-expression-based Louvain clustering method. (C) Spatial domains identified by ResNet-50 feature extraction method. (D) Spatial domains identified by stMIC method. (E) Adjusted Rand Index (ARI) in stMIC, gene-expression (graph-based), and ResNet-50 clustering methods determined sections against the ground truth labels (pathologist annotation). (F-G) Spatial expression of hypoxia marker genes. (H) Spatial expression of hypoxia marker gene. (I) Boxplot of GSEA for cluster 4 of sample 94D against selected gene sets.

The genes CA9 (Carbonic Anhydrase IX) and NDRG1 (N-Myc Downstream Regulated 1) (Fig. 4F&G) are known to be upregulated under hypoxic conditions, which often occur in solid tumors such as breast cancer due to inadequate oxygen supply[23,24]. This upregulation is a response to the low oxygen tension in an attempt to adapt to the harsh microenvironment. Hypoxia within tumors is associated with increased invasiveness, resistance to therapy, and a poor prognosis, thus indicating a potentially more aggressive disease state within this cluster[25].

In addition to these hypoxia markers, stMIC discerns a separate region with high expression of IFIT1 (Fig. 4H). This interferon-induced protein has been associated with active interferon signaling and an ongoing immune response. Active interferon signaling can have complex implications in cancer, possessing both tumor-suppressive and tumor-promoting properties[26]. Meanwhile, an ongoing immune response may reflect the immune system's attempts to counteract tumor progression or could even suggest the shaping influence of the immune system on the tumor's behavior[27]. Beside this, *IFIT1* has been associated with chemotherapy response in breast cancer[28], suggesting that this region might be more susceptible to chemotherapy.

In fact, previous studies have validated the relevance of these two distinct clusters, but this was determined only through a comprehensive joint expression analysis spanning 28 diverse samples[13]. Thus, while expression-analysis does identify these sub-clusters, it is reliant on the comparative context gleaned from multiple, external samples. Furthermore, our stMIC approach has identified significantly more enriched pathways in cluster 4 than what the gene-expression-based method alone was able to capture (Fig.4 I). The gene set enrichment score of top five Hallmark pathways are further compared between gene expression based vs. stMIC clustering method (Supplementary Fig.3E). Only results with a false discovery rate (FDR) < 0.05 are displayed. This difference underscores the potential of the integrated image-aware approach to provide a richer, more nuanced analysis of gene expression within specific tissue regions. In conclusion, our integrated image-aware approach has provided new insights into the spatial heterogeneity of the 094D triple-negative breast cancer sample, unraveling the complexities of hypoxia and immune responses within the tumor environment. The incorporation of morphological information in gene expression analysis can enhance the resolution of tumor substructure identification and pave the way for a more nuanced understanding of tumor biology, with potential implications for treatment strategies.

4. Discussion

Our results showed how integrating a naïve pre-trained ResNet-50 into Spatial Transcriptomics workflows identified features missed from standard ST gene-expression analysis. Second, our results show an improved ability to recapitulate features identified by a pathologist.

An image-aware, integrated approach can identify previously overlooked or undervalued features. The most prominent feature identified from the combined approach was the glioma's blood vessel and vascularization features. These features are unmistakable in the histology image. However, only three genes (*HBA*, *HBA1*, and *HBA2*) within this biological feature have significant expression within the gene expression data. While this is expected from globin-producing erythrocytes, these three genes did not contribute significantly to the first 10 PCs used in gene expression clustering. In practice, given that hallmarks of the disease involve post-translational modifications, many types of features would be less evident by gene expression alone. The ability to highlight these features underscores the power of stMIC approach that draws from the strengths of deep learning and the rich data afforded by spatial transcriptomics.

Clear identification of novel features and improved clustering is encouraging, given that there are several areas where the approaches used here could be improved or optimized. First, we utilized a naive, pre-trained ResNet-50 model. Training approaches could lead to improved clustering. However, histopathological image data often have well-described biases that limit transferability [29]. Indeed, some groups have identified approaches for internal training, as in the case of the tool stMIC, which shows promising early results in disease relevant contexts[30]. Thus, the value seen even in this naive model is notable, as improvement was seen without further training.

Our approach differs significantly from RESEPT[31], stLearn[32], and SpaCell[9], which are well established pipelines using the ResNet-50 model in spatial transcriptomics analysis. RESEPT uses a graph autoencoder to embed ST gene expression data into a three-dimensional representation and maps this to an RGB image for visual analysis using the ResNet101 deep learning model. Unlike RESEPT, which does not consider H&E image data, our approach integrates both spatial gene expression and H&E image data to capture comprehensive morphological and transcriptional details, offering the potential for richer biological insights. Like stLearn, our pipeline utilizes a pre-trained ResNet-50 model, but we diverge by integrating the image data with gene-expression clustering, rather than just normalization, offering a richer exploration of tissue features. By comparison, SpaCell also uses a ResNet-50 model for feature extraction and integrates imaging with transcriptomics; they differ by using autoencoders to reduce features and mainly focusing on classification/prediction.

Within this work, we leveraged pathology annotated samples to assess performance, and very importantly, while some methods employ models on non-pathological systems, such as Spatial Transcriptomics (ST) on a mouse cortex or similar non-disease settings, their utility can be limited in capturing the full spectrum of cellular interactions in aggressive human diseases like cancer. Instead, our approach prioritizes pathological samples from patients, given their inherent complexity and heterogeneity. By targeting unique pathological landmarks and intricacies, our method seeks to illuminate tumor progression and identify key biological elements and events. While we can see improved results using pathology annotation as a benchmark, it has limitations. First, these samples were predominantly done on fresh-frozen tissue, whereas pathologists typically prefer annotation on FFPE fixed tissues. With the emergence of spatial transcriptomics approaches that work with FFPE tissue, these ideally annotated histology images will become more available; however, FFPE usually comes with lower quality ST due to degraded RNA.

A deeper dive into understanding the exact mechanisms by which ResNet-50 outperforms solely gene-expression-based approaches would be beneficial. Unraveling the strengths and limitations of image-aware deep learning model in the context of spatial transcriptomics can provide a foundation for optimizing or even designing new architectures that can more comprehensively capture pathology-related features. We recognize this as a crucial next step and are planning further investigations to elucidate the specific advantages of these models and to guide the development of even more effective methodologies.

Some advanced deep learning architectures have emerged that could potentially enhance the methods presented in this study. The Vision Transformer (ViT)[33], for instance, processes images by segmenting them into fixed-size patches and then leverages the Transformer architecture, offering the potential to extract more nuanced spatial relationships crucial for spatial transcriptomics. Similarly, the MLP-Mixer[34], through its unique mixing of tokens with multilayer perceptrons, and the Swin Transformer[35], with its shifted windows approach, can be particularly advantageous for capturing intricate spatial hierarchies and features from histology images. Our study predominantly utilized the ResNet-50 model for its proven efficacy in image analysis. However, integrating recent architectures like ViT, MLP-Mixer, and Swin Transformer might allow for a more comprehensive feature extraction, bridging image-based nuances with spatial transcriptomic insights.

Several potential extensions and improvements could be made to the presented approaches. Consideration could be given to incorporating other types of omics data to enrich the data pool further. Refinement of the deep learning model could enhance performance by integrating new layers or algorithms. A promising avenue to explore is the incorporation of spatial distances between spots into the analysis, offering a more nuanced understanding of cellular organization.

A broader sample set and more detailed annotations would certainly strengthen the robustness of our model. The requirement for specialized technical expertise to operate and interpret results is another hurdle that needs to be overcome. Furthermore, validating our methodology using a more comprehensive range of patient-derived data is necessary to ascertain the model's clinical relevance and translational applicability.

Beyond cancer, the potential application of our approach to other diseases warrants further exploration. With the in-depth view of the tumor microenvironment that our method provides, we foresee a crucial role for it in the realm of personalized medicine, particularly given advances in immune-oncology treatments. As our understanding of cellular heterogeneity within tissues becomes increasingly nuanced, an image-informed ST approach will likely serve as a powerful tool in understanding the role of the immune microenvironment. This work underscores the exciting potential of spatial transcriptomics and deep learning in shaping the future of understanding disease heterogeneity.

Availability of data and materials.

The stMIC code has been developed with Python 3.7 as a user-friendly pipeline. Code, setting up and tutorials are described in the stMIC GitHub page: https://github.com/USCDTG/stMIC.

Supplemental Material referred to in the paper may be found at the following URL: https://github.com/davcraig75/song_psb2014/blob/main/SupplementaryData.pdf

References

1. Rao, A., Barkley, D., França, G. S. & Yanai, I. Exploring tissue architecture using spatial transcriptomics. Nature 596, 211–220 (2021).
2. Asp, M., Bergenstråhle, J. & Lundeberg, J. Spatially Resolved Transcriptomes-Next Generation Tools for Tissue Exploration. Bioessays 42, e1900221 (2020).
3. Vickovic, S. et al. High-definition spatial transcriptomics for in situ tissue profiling. Nat. Methods 16, 987–990 (2019).
4. Chen, K. H., Boettiger, A. N., Moffitt, J. R., Wang, S. & Zhuang, X. RNA imaging. Spatially resolved, highly multiplexed RNA profiling in single cells. Science 348, aaa6090 (2015).
5. Li, Q., Zhang, X. & Ke, R. Spatial Transcriptomics for Tumor Heterogeneity Analysis. Front. Genet. 13, 906158 (2022).
6. Stuart, T. et al. Comprehensive Integration of Single-Cell Data. Cell 177, 1888–1902.e21 (2019).
7. Edsgärd, D., Johnsson, P. & Sandberg, R. Identification of spatial expression trends in single-cell gene expression data. Nat. Methods 15, 339–342 (2018).
8. Hu, J. et al. SpaGCN: Integrating gene expression, spatial location and histology to identify spatial domains and spatially variable genes by graph convolutional network. Nat. Methods 18, 1342–1351 (2021).
9. Tan, X., Su, A., Tran, M. & Nguyen, Q. SpaCell: integrating tissue morphology and spatial gene expression to predict disease cells. Bioinformatics 36, 2293–2294 (2020).
10. Ifvarsson, Falk, Vidman & Thorén. [Social Welfare Department's general advice on prevention, diagnosis and treatment of eye infections in newborn infants]. Jordemodern 99, 398–403 (1986).
11. Website. https://doi.org/10.48550/arXiv.1512.03385 doi:10.48550/arXiv.1512.03385.
12. Kutluer, N., Solmaz, O. A., Yamacli, V., Eristi, B. & Eristi, H. Classification of breast tumors by using a novel approach based on deep learning methods and feature selection. Breast Cancer Res. Treat. 200, 183–192 (2023).
13. Bassiouni, R. et al. Spatial Transcriptomic Analysis of a Diverse Patient Cohort Reveals a Conserved Architecture in Triple-Negative Breast Cancer. Cancer Res. 83, 34–48 (2023).
14. Wolf, F. A., Angerer, P. & Theis, F. J. SCANPY: large-scale single-cell gene expression data analysis. Genome Biol. 19, 15 (2018).
15. Steinley, D. Properties of the Hubert-Arabie adjusted Rand index. Psychol. Methods 9, 386–396 (2004).
16. Liberzon, A. et al. The Molecular Signatures Database (MSigDB) hallmark gene set collection. Cell Syst 1, 417–425 (2015).
17. Kanehisa, M. & Goto, S. KEGG: kyoto encyclopedia of genes and genomes. Nucleic Acids Res. 28, 27–30 (2000).
18. Segal, E., Friedman, N., Koller, D. & Regev, A. A module map showing conditional activity of expression modules in cancer. Nat. Genet. 36, 1090–1098 (2004).
19. Godec, J. et al. Compendium of Immune Signatures Identifies Conserved and Species-Specific Biology in Response to Inflammation. Immunity 44, 194–206 (2016).

20. Richter, F., Meurers, B. H., Zhu, C., Medvedeva, V. P. & Chesselet, M.-F. Neurons express hemoglobin alpha- and beta-chains in rat and human brains. J. Comp. Neurol. 515, 538–547 (2009).
21. Cohen, C. et al. The roles of Tenascin C and Fibronectin 1 in adhesive capsulitis: a pilot gene expression study. Clinics 71, 325–331 (2016).
22. Caja, L. et al. TGF-β and the Tissue Microenvironment: Relevance in Fibrosis and Cancer. Int. J. Mol. Sci. 19, (2018).
23. Kuhlenbäumer, G., Stögbauer, F., Ringelstein, E. B. & Young, P. Hereditary Peripheral Neuropathies. (Springer Science & Business Media, 2005).
24. Shamis, S. A. K., Edwards, J. & McMillan, D. C. The relationship between carbonic anhydrase IX (CAIX) and patient survival in breast cancer: systematic review and meta-analysis. Diagn. Pathol. 18, 46 (2023).
25. Jing, X. et al. Role of hypoxia in cancer therapy by regulating the tumor microenvironment. Mol. Cancer 18, 157 (2019).
26. Minn, A. J. Interferons and the Immunogenic Effects of Cancer Therapy. Trends Immunol. 36, 725–737 (2015).
27. Hiam-Galvez, K. J., Allen, B. M. & Spitzer, M. H. Systemic immunity in cancer. Nat. Rev. Cancer 21, 345–359 (2021).
28. Weichselbaum, R. R. et al. An interferon-related gene signature for DNA damage resistance is a predictive marker for chemotherapy and radiation for breast cancer. Proc. Natl. Acad. Sci. U. S. A. 105, 18490–18495 (2008).
29. Hägele, M. et al. Resolving challenges in deep learning-based analyses of histopathological images using explanation methods. Sci. Rep. 10, 6423 (2020).
30. Zuo, C. et al. Elucidating tumor heterogeneity from spatially resolved transcriptomics data by multi-view graph collaborative learning. Nat. Commun. 13, 5962 (2022).
31. Chang, Y. et al. Define and visualize pathological architectures of human tissues from spatially resolved transcriptomics using deep learning. Comput. Struct. Biotechnol. J. 20, 4600–4617 (2022).
32. Pham, D. et al. stLearn: integrating spatial location, tissue morphology and gene expression to find cell types, cell-cell interactions and spatial trajectories within undissociated tissues. bioRxiv 2020.05.31.125658 (2020) doi:10.1101/2020.05.31.125658.
33. Dosovitskiy, Alexey, et al. An image is worth 16x16 words: Transformers for image recognition at scale. arXiv preprint arXiv:2010.11929 (2020).
34. Tolstikhin, Ilya O., et al. Mlp-mixer: An all-mlp architecture for vision. Advances in neural information processing systems 34 (2021): 24261-24272.
35. Liu, Ze, et al. Swin transformer: Hierarchical vision transformer using shifted windows. Proceedings of the IEEE/CVF international conference on computer vision. 2021.

Spatial Omics Driven Crossmodal Pretraining Applied to Graph-based Deep Learning for Cancer Pathology Analysis

Zarif L. Azher
Thomas Jefferson High School for Science and Technology
Alexandria, VA 22312, USA
Email: 2024zazher@tjhsst.edu

Michael Fatemi
University of Virginia, Department of Computer Science
Charlottesville, VA 22904, USA
Email: myfatemi04@gmail.com

Yunrui Lu, Gokul Srinivasan, Alos B. Diallo
EDIT, Department of Pathology and Laboratory Medicine, Dartmouth Hitchcock Medical Center
Lebanon, NH 03756, USA
Email: yunrui.lu@dartmouth.edu, gokulsrin@gmail.com, alos.b.diallo.gr@dartmouth.edu

Brock C. Christensen, Lucas A. Salas
Department of Epidemiology, Geisel School of Medicine at Dartmouth
Lebanon, NH 03756, USA
Email: brock.c.christensen@dartmouth.edu , lucas.a.salas@dartmouth.edu

Fred W. Kolling IV, Laurent Perreard
Genomics Shared Resource, Dartmouth Cancer Center
Lebanon, NH 03756, USA
Email: fred.w.kolling.iv@dartmouth.edu, laurent.perreard@dartmouth.edu

Scott M. Palisoul, Louis J. Vaickus, Joshua J. Levy*†
EDIT, Department of Pathology and Laboratory Medicine, Dartmouth Hitchcock Medical Center
Lebanon, NH 03756, USA
Email: scott.m.palisoul@hitchcock.org, louis.j.vaickus@hitchcock.org, joshua.j.levy@dartmouth.edu

Graph-based deep learning has shown great promise in cancer histopathology image analysis by contextualizing complex morphology and structure across whole slide images to make high quality downstream outcome predictions (ex: prognostication). These methods rely on informative representations (i.e., embeddings) of image patches comprising larger slides, which are used as node attributes in slide graphs. Spatial omics data, including spatial transcriptomics, is a novel paradigm offering a wealth of detailed information. Pairing this data with corresponding histological imaging

* To whom correspondence should be addressed.
† Work supported by grants P20GM130454, P20GM104416 to JL and K08CA267096 to LV.

localized at 50-micron resolution, may facilitate the development of algorithms which better appreciate the morphological and molecular underpinnings of carcinogenesis. Here, we explore the utility of leveraging spatial transcriptomics data with a contrastive crossmodal pretraining mechanism to generate deep learning models that can extract molecular and histological information for graph-based learning tasks. Performance on cancer staging, lymph node metastasis prediction, survival prediction, and tissue clustering analyses indicate that the proposed methods bring improvement to graph based deep learning models for histopathological slides compared to leveraging histological information from existing schemes, demonstrating the promise of mining spatial omics data to enhance deep learning for pathology workflows.

Keywords: spatial omics, transcriptomics, deep learning, graphs, cancer, colon cancer.

1. Introduction

1.1. *Deep Learning for Pathology*

In recent years, countless studies have demonstrated the potential for deep learning algorithms to solve challenging biomedical tasks, thereby improving risk stratification and alleviating the potential for clinical burnout by making tedious and unreliable tasks faster and more quantitative, potentially leading to improved patient health outcomes [1]. These algorithms are formulated on computational heuristics – specifically, machine learning -- which can make sense of many complex data types through the dynamic derivation of relevant patterns and features [2–4]. Analysis of pathology data, including whole slide imaging (WSI) – microscopic images of patient tissue – is an emerging application in this space, as WSIs are routinely collected and used for patient monitoring, diagnosis, and prognostication. Existing works have shown that specially designed deep learning algorithms, inspired by processes of the central nervous system, may be able to automate or assist in these tasks [5]. Most deep neural networks study small micromorphological changes given the enormity of these gigapixel images. Graph convolutional networks (GCNs), however, are a promising method in this domain, as they can effectively model macro and micro architectural features present across WSI in a human-interpretable manner [6]. Generally, these methods split WSI into patches (i.e., more manageable subimages), extract numeric representations (i.e., "embeddings") from each patch using a predetermined algorithm, and construct a graph where the nodes are given patch embeddings and edges are formed based on spatial adjacency [7–9]. Such methods have been applied for tumor stage prediction [9], survival analysis [8], and derive numerical representations of WSI that can be combined with other omics and imaging modalities [7].

The optimal algorithm used to extract node features is an area of ongoing research, though many works presently use a ResNet convolutional neural network (CNN) pretrained on the ImageNet database [10] for this task [8,11,12]. It has become increasingly common to additionally train these CNNs on various image tasks orthogonal to the task at hand to prepopulate an information registry of features which will ultimately improve predictive performance in other settings; these techniques are known as pretraining. Recently, self-supervised techniques have emerged as promising pretraining methodologies, where images are compared from several different vantage points without being explicitly labeled. Cross-modal pretraining has recently been highlighted as a common self-supervised method by leveraging complementary "paired" information across multiple input data types (e.g., images and text) which can improve the representation of all involved modalities. Here, we investigate the utility of using spatial omics data, which is paired at 50-micron resolution to the histological information, to pretrain an encoder model for these patches, to demonstrate the power of leveraging spatial omics for deep learning-based pathology methods which are particularly suited for analysis using graph neural networks (GNNs).

1.2. *Spatial Omics*

Omics data – such as gene expression quantification and DNA methylation – have traditionally been collected on a bulk scale where measurements are taken across an entire sample or tissue section. Recent advancements in technology have allowed for collection on a more granular scale, such as the single cell level, or across specific spots/regions in a slide sample [13]. Prior studies have demonstrated that deep learning through specialized architectures like GCNs can mine spatial omics data to build a more comprehensive understanding of spatial cellular heterogeneity, especially as it pertains to how the tumor microenvironment can facilitate/inhibit further disease progression [14,15]. Notably, this type of data is not yet commonly available at large scale due to the prohibitive cost of these assays as well as batch effects and selection of limited slide area, meaning that methods which can learn from spatial omics data and effectively transfer this knowledge to improve other tasks may be valuable. Zeng et al [15] previously developed a model which utilized contrastive learning to mine a shared representation between image patches and corresponding spatial transcriptomics; however, their investigation centered on driving improved understanding on gene domains, rather than attempting to leverage the method to enhance downstream clinical outcome modeling in situations where only WSI – and no ST data – is available.

1.3. *Contributions*

We hypothesize that additional biological information can be learnt from spatially resolved transcriptomics data that may prove relevant for enhancing prediction models across a range of histological analyses. Existing works applying GCNs for WSI analysis have not yet leveraged spatial omics data to enhance modeling across orthogonal tasks. In part, this is because the quality of histological slides for spatially co-registered omics data has been limited as the standard Visium spatial transcriptomics (ST) workflow featured manual staining and low-resolution imaging– this information does not readily transfer to prediction models on higher resolution histological slides. Now, with the development of assays such as the CytAssist which permit the use of sophisticated laboratory processing (i.e., autostaining and 40X imaging prior to Visium profiling), the quality of slides has remarkably increased and allows for training image models that may more readily transfer to related domains. Here, we assess the ability of spatially resolved omics data to enhance predictions on a range of different histological assessment tasks by presenting an initial evaluation of a crossmodal pretraining mechanism using matched WSI and spatial omics measurements as means to encode biological information within WSI graphs to apply in scenarios where spatial omics data is not available. We compare this method against other common pretraining schemes on downstream predictive analyses (staging, lymph node metastasis, survival prognostication) of WSI, as well as explore generated image patch embeddings. Accurate methods for these downstream predictive tasks may enable more personalized patient treatments. In this study, we expect developed models which can mine for spatial molecular information to outperform the compared approaches on these tasks. We aim to demonstrate the potential benefits of utilizing spatial omics – spatial transcriptomics, in particular – methods to enhance deep learning-driven pathology analysis.

2. Methods

2.1. *Data Collection and Preprocessing*

Visium spatial transcriptomics data matched with WSI was collected from four colorectal cancer patients from the Dartmouth Hitchcock Medical Center, to serve as a training dataset for the crossmodal patch embedding method. This process was conducted through the 10x Genomics Visium spatial transcriptomics workflow, featuring H&E staining, followed by mRNA profiling and whole slide imaging. Spatial transcriptomics data were filtered to include the top 1000 most variable genes across slides identified by SpatialDE [16]. Separately, 708 WSIs were collected from colorectal cancer patients from the Dartmouth Hitchcock Medical Center, for whom, histological stage annotations were available. Finally, WSIs were obtained for a cohort of 350 colorectal cancer patients from The Cancer Genome Atlas (TCGA) for whom survival information and lymph node metastasis information was available. All WSIs were stain normalized using the Macenko [17] method. Collected WSIs were split into non overlapping 224 x 224 patches via the PathflowAI Python package [18], whose embeddings served as node attributes in a graph. We compared several methods described below to encode information for these patches, which is the main focus of this study. Nodes were connected with edges based on spatial adjacency using the *knn_graph* (k-nearest neighbor) method from the torch_cluster Python package, with k=16. Patients from the in-house dataset and TCGA were separately partitioned into training, validation, and testing sets using a random 80/10/10 split. The collected datasets and the downstream tasks they were used on, are summarized below:

1. **Visium spatial transcriptomics slides (n=4; 20,000 spots/patches; Co-Registered Spatial Transcriptomics, H&E WSI):** to pretrain contrastive crossmodal model
2. **Dartmouth Hitchcock Medical Center (n=708 H&E WSI):** used for histological stage prediction and clustering analysis
3. **TCGA Cohort (n=350 H&E WSI):** used for lymph node metastasis prediction, survival prognostication, and tumor infiltrating lymphocyte (TIL) alignment analysis

All analyses were conducted on a machine using a single Nvidia Tesla v100 GPU with 32 gigabytes of VRAM, and 100 gigabytes of RAM.

2.2. *Patch Level Pretraining Methods*

Three embedding production methods were compared for the 224x224 patches used as nodes of the graphs representing WSI.

2.2.1. *ImageNet-Pretrained ResNet18*

A ResNet18 CNN model pre trained on the ImageNet dataset (commonly used for embedding histopathology patches) was accessed using the *torchvision* Python package (https://github.com/pytorch/vision). The model was truncated through the penultimate layer, to extract length 512 vectors/embeddings for each input patch.

2.2.2. *Ciga Self Supervised Histopathology Pretrained ResNet18*

A separate ResNet18 CNN model pretrained using a self-supervised learning (SSL) SimCLR [19] contrastive procedure on histpathological imaging datasets was similarly accessed and truncated through the penultimate layer to extract length 512 embeddings for all patches. In summary, SimCLR employs an objective function that encourages similarity between embeddings from augmented (i.e., "corrupted") views of the same image, while penalizing based on dissimilarity between views from different images. This model was made publicly available by Ciga et al [20], and has been previously shown to outperform the aforementioned ImageNet-pretrained model on a variety of downstream modeling tasks.

2.2.3. *Spatial Omics-driven Crossmodal Pretrained Encoder*

A contrastive cross-modal model encoding image patches and spatial transcriptomic profiles was created, similar to the model implemented by Zeng et al [21]. Input images patches of size 224x224 were encoded into embeddings of size 512 units, using the feature extraction portion of a CNN initialized with weights initialized from the ResNet model trained by Ciga et al. Spatial transcriptomics profiles containing expression of the most spatially variable 1000 genes across Visium slides, selected to avoid overfitting on genes with imprecise expression, were encoded with three standard fully connected (FC) layers of size 512. The embeddings from co-registered patches from each modality (ST, WSI) were passed through a common projection layer of size 512, to output a single embedding per modality (ie; one vector of length 512 which describes an image patch, and one of length 512 which describes the corresponding gene expression). Crossmodal and unimodal contrastive penalties are applied using the SimCLR loss function [19]; during training, several augmentation strategies were applied to both the image patches and corresponding transcriptomic profiles to generate "corrupted" representations of each data type as means for comparison. Transcriptomic profiles were randomly masked and corrupted with noise with 30% probability. Images were augmented using a series of random flips, color jitter transforms, random grayscaling, random rotation, and random image solarization. Both the original and augmented image patches and transcriptomics profiles were encoded using the aforementioned neural network layers. The loss mechanism penalizes the model based on the difference between the embeddings from the original and augmented data from each modality. A crossmodal loss is used to maximize the similarity between the corrupted image and transcriptomic embeddings from the same patch. These three loss functions (augmented image to image, augmented transcriptomics to transcriptomics, augmented image to augmented transcriptomics) were summed to optimize the crossmodal contrastive model.

This model was trained for 150 epochs with a batch size of 8 and a learning rate of 0.00001. Visium sections corresponding to six patients were partitioned into the training set, and tissue sections from two patients were partitioned to the validation set. Validation set loss was used to inform selection of the top model, following training. The RELU activation function was applied to outputs of every layer. The image encoder pretrained using the spatially co-registered transcriptomics information and the subsequent projection head were retained for subsequent analysis, and were used to embed image patches which GNN models were to operate on. The remaining layers of this pretrained model were not utilized. The usage of this image encoder derived using this training protocol for other ancillary tasks is the primary focus of this study, compared to the other image encoders (weights from ImageNet, Ciga et al.). This model is further described along with data collection procedure, in **Figure 1**.

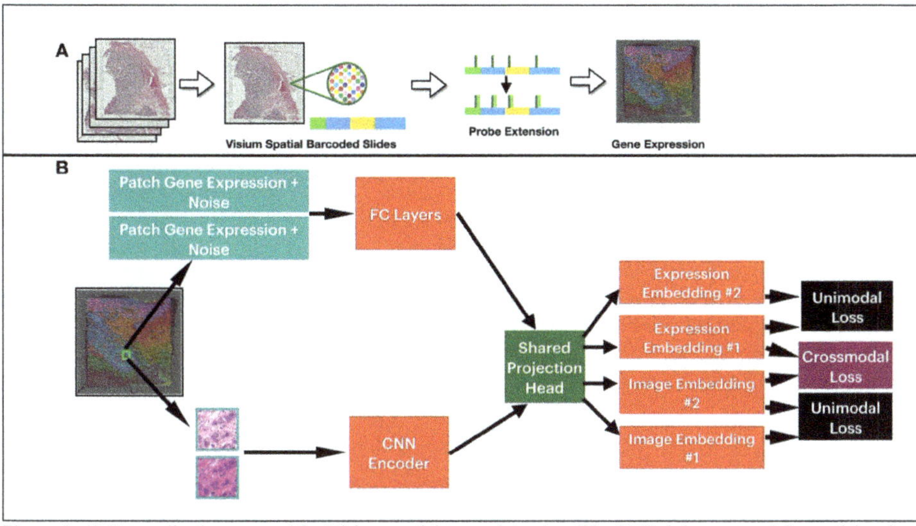

Figure 1: A) Data collection protocol for Visium spatial transcriptomics slide. **B)** Training protocol for spatial omics-driven crossmodal contrastive model; two views are generated per modality, per patch; each view is passed through the corresponding branch of the crossmodal model; embeddings are transformed using a shared projection head; unimodal and crossmodal contrastive losses are applied to output embeddings.

2.3. *Downstream Outcome Prediction*

We sought to understand whether CNN encoders, pretrained on co-registered spatial transcriptomics data, could enhance the predictions on a range of different GCN tasks. A graph convolutional network was constructed to take an input graph of nodes represented by length 512 embeddings, followed by three GCNConv graph convolutional layers [22] to contextualize and aggregate embeddings into length 128, with SAGEPooling pooling [23] layers (ie: 30% of patches retained, for subsequent layers; SAGEPooling stochastically samples higher-order neighborhoods of patches) placed after each convolutional layer. These pooling layers learn to downsample graphs, to push the model to learn focused information relevant to the training task. Graph embeddings were aggregated using global mean pooling after each SAGEPooling layer. These embeddings were combined using the JumpingKnowledge mechanism, resulting in a single vector of length 128 to represent the entire input graph/WSI. Finally, two fully connected layers were applied to this embedding, followed by a single output layer. The model (**Figure 2**) was applied to the following prognostication-focused experiments/outputs to assess patch encoding mechanisms:

2.3.1. *Histological Stage Prediction*

The in-house dataset was used to train and assess model capability to predict dichotomized tumor histological stage (T-stage; signifies depth of invasion) - low (stage 0, stage 1, stage 2) or high (stage 3, stage 4). A sigmoid function was applied to the output of the final layer in the GCN, and model training was supervised using a binary crossentropy loss function.

2.3.2. *Survival Prognostication*

The TCGA dataset was used to train and evaluate GCNs to assess for time to death using hazard predictions, indicating the real-time risk of death. Model training was supervised using a standard Cox loss, which considers the predicted risk, patient censor status, and duration (either days to death or days to last follow up). This setup entails the proportional hazards assumption, that predictors have a constant hazard ratio (i.e., relative risk between two patient groups) over time.

All GCN models were trained for up to 30 epochs, using a learning rate of 0.001 and batch size 8. Top model checkpoints were selected for evaluation following training, based on validation set loss. GCN models were implemented using the Pytorch Geometric [24] Python package. Three separate GCN models were trained for each prediction task - one for each patch embedding mechanism. Stage prediction and lymph node metastasis models were evaluated on held-out test sets using F1-score and area under the curve (AUC), while C-index was used to evaluate prognostication models. These metrics are reported using 95% confidence interval derived from 1000 sample non-parametric bootstrapping procedures.

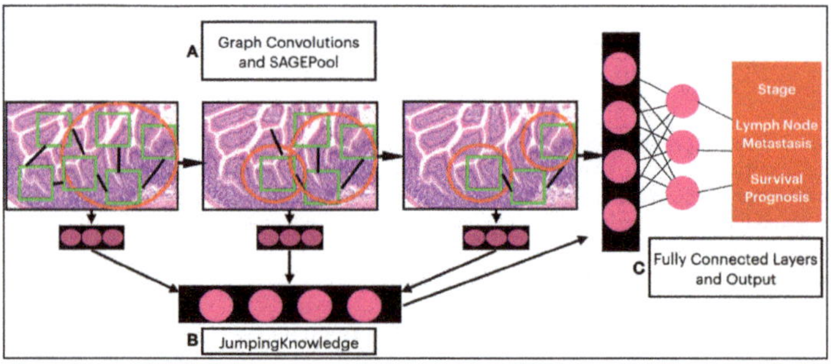

Figure 2: Overview of generalized GCN for downstream outcome modeling; initial patch embeddings vary across experimentation. A) Graph convolution layers contextualize each node embedding; after each such layer, SAGEPool operators aggregate nodes/patch embeddings, removing up to 70% of them, to only retain informative ones. B) A JumpingKnowledge scheme aggregates embeddings across graphs to create a single embedding for the image. C) The image embedding is used to make downstream predictions.

2.4. *Embedding Clustering Quality Analysis*

The ability of patch embeddings to capture morphological and molecular heterogeneity across slides was assessed across embedding methods, using an unsupervised clustering approach and the in-house dataset. For each WSI in the dataset, KMeans clustering (k=5; chosen via coarse optimization to ensure stability when run numerous times) was applied to the patch embeddings derived by each pretraining method (standard ResNet, Ciga et al, spatial pretrained) to elucidate sub-groups of patches implicitly captured by the representations. Clusters were plotted across slides to visually ensure that they represented different morphologies and structures within slides. Subsequently, the Calinski-Harabanz (CH) index [25] and the Davies-Bouldin (DB) index [26] were computed for the clustering result for each pretraining strategy. The ANOVA-based CH score assesses the density and separation of clusters, with a higher value indicating greater density within clusters and separation among different clusters. Similarly, the DB index measures the ratio between within-cluster and cross-cluster separation. Thus, superior patch embeddings should result in a relatively high CH index and low DB index. The per-WSI scores were used to calculate average CH index and DB score at a 95% confidence interval, for each pretraining method.

2.5. *TIL-based Model Interpretation*

Previous research has demonstrated the importance of tumor infiltrating lymphocytes (TILs) and the tumor microenvironment on the progression of colon cancer [27]. We sought to demonstrate the interpretability of GCN models developed here using the TCGA dataset, by comparing regions of WSI given high attention with previously published predicted TIL maps [28] for corresponding slides. Patches deemed important by GCN models trained on lymph node metastasis prediction were determined by extracting patches remaining in WSI graphs following the final pooling layer; for a given patch, being left in its graph by a GCN model following three pooling layers, indicates its significance to the model. The coordinates of these patches were compared to those describing the locations of predicted TILs via Wald Wolfowitz testing [29], where the null hypothesis would indicate high overlap between these two sets of coordinates. Accordingly, Wald Wolfowitz testing was used to calculate a test statistic per slide per GCN model trained with each patch embedding method– negative values of this test statistics, W, represents the localization of TILs. Spearman's rank correlation coefficients (alpha p-value = 0.05) were calculated to evaluate the relationship between the test statistic (W), and predicted hazard. A negative correlation coefficient would suggest a statistically significant association between predicted hazard and TIL spatial localization, following biological knowledge holding that TILs help inhibit colon cancer proliferation and migration [30]. Test statistics were further dichotomized to indicate presence/lack of TIL localization, to compare these relationships across the GCN model using embeddings derived from the Ciga et al method, versus the model using spatially pretrained embeddings.

3. Results[†]

3.1. *Quantitative Predictive Analysis*

Held out testing-set performance for GCNs trained to predict stage, lymph node metastasis, and survival prognosis, are presented in **Table 1**; models which used patch embeddings derived from the spatial omics-driven mechanism outperformed those using the compared methods for all three experiments.

Table 1: Test set performance metrics (95% confidence interval) of GCNs trained using various patch embedding mechanisms, for binary stage prediction, lymph node metastasis prediction, and survival prognostication.

Task	Measure	ImageNet ResNet	Ciga et al ResNet	Spatial Pretrained
Stage Prediction	**AUC**	0.935 ± 0.003	0.948 ± 0.002	**0.981 ± 0.001**
	F1-Score	0.863 ± 0.004	0.858 ± 0.004	**0.878 ± 0.004**
Lymph Node Metastasis	**AUC**	0.651 ± 0.004	0.612 ± 0.004	**0.708 ± 0.003**
	F1-Score	0.560 ± 0.002	0.630 ± 0.003	**0.671 ± 0.005**
Survival Prognostication	**C-index**	0.597 ± 0.003	0.582 ± 0.002	**0.638 ± 0.002**

† Supplementary materials can be found at the following DOI: https://doi.org/10.5281/zenodo.8197573.

For the classification experiments, models using embeddings derived from the spatial omics-driven mechanism outperformed those which used embeddings from the ImageNet-trained ResNet18 CNN by an average of 6.98% measured by AUC, and outperformed models using embeddings derived from the ResNet18 pretrained by Ciga et al, by average of 9.47%. GCNs using spatial omics-driven embeddings (C-index 0.638) also outperformed ImageNet-trained ResNet18 embeddings (C-index 0.597) and embeddings derived from the model trained by Ciga et al (C-index 0.582).

3.2. *Clustering Evaluation*

A KMeans clustering approach paired with CH index and DB index calculation was employed to compare the abilities of these different pretraining approaches to elucidate molecular and morphological heterogeneity across slides; the results of this analysis are presented in **Table 2**. An example visualization including regions of a slide assigned to clusters indicating by different coloring, is presented in **Figure 3**; additional examples are available in Supplementary Figures S2 and S3.

Embeddings from the contrastive crossmodal spatial model resulted in a significantly higher CH index and lower DB index, versus both the ImageNet-pretrained ResNet and the ResNet trained on histopathology datasets via self-supervised learning by Ciga et al.

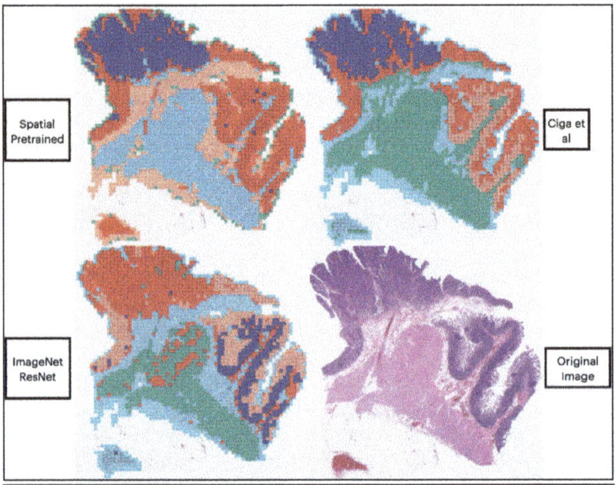

Figure 3: Example visualization of clustering of embeddings derived using various methods, for a single WSI.

Table 2: Clustering quality metrics calculated across embedding methods.

Measure	ImageNet ResNet	Ciga et al ResNet	Spatial Pretrained
Calinski-Harabanz Index	643.76 ± 17.51	786.70 ± 20.40	**2605.68 ± 70.66**
Davies-Bouldin Index	1.90 ± 0.01	1.719 ± 0.01	**0.975 ± 0.01**

3.3. *Model Interpretation*

Spearman's correlation coefficient values testing the relationship between lymph node metastasis risk predicted by GCN models using various patch embedding mechanisms and TIL localization elucidated via Wald Wolfowitz testing, are presented in **Table 3** along with corresponding p-values, suggesting both the Ciga and spatial pretrained models were able to derive TIL-associated embeddings related to instantaneous hazards. Boxplot visualizations comparison of predicted model risk and dichotomized TIL alignment are presented in **Supplementary Figure S4**.

Table 3: Spearman's correlation coefficient values for TIL localization versus predicted lymph node metastasis risk, across GCN models using various patch embedding methods.

	ImageNet ResNet	**Ciga et al ResNet**	**Spatial Pretrained**
Spearman's Coefficient	-0.061	-0.426	-0.218
Spearman's P-value	0.2693	2.2e-16	7.74e-5

4. Discussion and Conclusion

This is the first study which aims to determine whether leveraging spatial omics data to pretrain image patch encoders using a cross modal contrastive mechanism can improve downstream performance in graph convolutional networks, which may improve automated cancer patient analysis. While most prior research leveraged a GCN to integrate spatially localized omics with imaging for spot-level spatial transcriptomics enhancement or histological feature extraction tied to bulk transcriptional characteristics, our approach discerns spatial transcriptomics features from standalone slides. Recognizing the inaccessibility of spatial transcriptomics data, we employed transfer learning to apply extracted spatial transcriptomics features to a diverse range of subsequent tasks. We compared spatial omics-driven embeddings against those extracted from a standard ResNet18 CNN pretrained on the ImageNet dataset, and a ResNet18 pretrained using self-supervised learning on histopathology datasets. GCN models trained and evaluated using the spatially enhanced embeddings outperformed those using the baseline embedding methods on three downstream tasks – stage prediction, lymph node metastasis prediction, and prognostication. This suggests that incorporating spatial transcriptomics information into the pretraining process of image patch encoders, enhances the quality of learned representations, beyond what is extracted from state-of-the-art techniques which use solely images for patch encoding pretraining.

Additional quantitative analysis from clustering patch embeddings indicates that the models leveraging spatially-pretrained embeddings were superior at capturing distinct heterogeneities across slides, versus models using patch embeddings from existing strategies. Thus, we expect future applications of the developed spatial pretraining method for patch embeddings, to improve the performance of workflows aiming to capture tissue heterogeneity, including tumor subcompartment segmentation.

Furthermore, Wald Wolfowitz testing paired with Spearman's correlation coefficients, suggests that GCN models using embeddings from the spatial pretraining method and the Ciga et al method, learned to highlight TILs to contextualize prognostic assessment of cancerous tissue when considering lymph node metastatic potential, particularly in patients whom the models understood to be at lower risk. The Spearman's coefficient value for the GCN model using ImageNet ResNet patch representations was markedly closer to 0 versus the other two methods, indicating far weaker correlation in this relationship. Interestingly, the magnitude of the coefficient for the GCN model using the Ciga et al embeddings was nearly double that of the spatially pretrained embeddings, indicating that the Ciga et al method may induce greater tendency to turn to TILs for understanding patient profiles. Though this does not indicate greater predictive power among models, that such nuances can be extrapolated related to model reasoning, demonstrates the interpretability of graph-based modeling for cancer histopathology, and further emphasizes the importance of enhancing the ability of such methods.

Overall, our results indicate that spatial omics data can be effectively mined in a crossmodal fashion, to improve existing image-based deep learning workflows to analyze cancer histopathology; this also adds to the growing body of literature [31–33] which reflects the importance of enhancing pretraining mechanisms as a basis of improving deep learning models for cancer histopathology. Notably, ours is the first study to mine spatial omics data in the pretraining process to enhance the capability of such image-based models, while others have focused on mechanisms which use solely imaging. Several AI methods also exist to integrate spatial transcriptomics with histology through contrastive learning to improve the identification of spatial domains. This work differs from prior approaches as it aims to improve the extraction of imaging information on held-out tissue slides from which Visium spatial transcriptomics assaying has not been done, training with paired imaging and spatial expression data to enhance this capability.

A key limitation of this study is the relatively small dataset used to pretrain the spatially-enhanced crossmodal contrastive model; spatial transcriptomics data was only generated for 4 total slides due to high resource and time costs and the limited size of the tissue placement area on Visium slides. Furthermore, coarse hyperparameter search was used to select GCN architecture parameters, as a detailed experiment here was beyond the scope of this study. It should be noted that optimization of the convolutional neural network and GCN parameters can be done end-to-end, i.e., simultaneously, which can improve predictive results– as will incorporating additional varied histologies and tumor characteristics, improved specimen processing/imaging using the CytAssist and commensurate hardware to fit larger models. Future works will seek to use larger cohorts to pretrain the spatial model to improve quality of extracted embeddings. Additionally, the embeddings from the spatially enhanced model can be evaluated for use in applications other than GCNs, such as Transformer networks – which have become popular in cancer histopathology in recent years [34,35] – histology image search, and multimodal data integration.

5. Acknowledgements and Location of Supplementary Material

The results published here are in part based on data generated by the TCGA Research Network: https://cancer.gov/tcga. The authors acknowledge the support of the Center for Clinical Genomics and Advanced Technology in the Department of Pathology and Laboratory Medicine of the Dartmouth Hitchcock Health System which includes the Pathology Shared Resource, at the Dartmouth Cancer Center with NCI Cancer Center Support Grant 5P30 CA023108-37. Spatial transcriptomics assays were carried out in the Genomics and Molecular Biology Shared Resource (GMBSR) at Dartmouth which is supported by NCI Cancer Center Support Grant 5P30CA023108 and NIH S10 (1S10OD030242) awards. Spatial studies were conducted through the Dartmouth Center for Quantitative Biology in collaboration with the GMBSR with support from NIGMS (P20GM130454) and NIH S10 (S10OD025235) awards. Supplementary materials can be found at the following DOI: https://doi.org/10.5281/zenodo.8197573. Code for primary model implementation can be found at the following Github repository: https://github.com/zarif101/histopath_spatial_omics_pretrain

References

1. Egger, J. *et al.* Medical deep learning—A systematic meta-review. *Computer Methods and Programs in Biomedicine* **221**, 106874 (2022).
2. Cao, C. *et al.* Deep Learning and Its Applications in Biomedicine. *Genomics, Proteomics & Bioinformatics* **16**, 17–32 (2018).
3. Shamshirband, S., Fathi, M., Dehzangi, A., Chronopoulos, A. T. & Alinejad-Rokny, H. A review on deep learning approaches in healthcare systems: Taxonomies, challenges, and open issues. *Journal of Biomedical Informatics* **113**, 103627 (2021).
4. Van Der Laak, J., Litjens, G. & Ciompi, F. Deep learning in histopathology: the path to the clinic. *Nat Med* **27**, 775–784 (2021).
5. Dimitriou, N., Arandjelović, O. & Caie, P. D. Deep Learning for Whole Slide Image Analysis: An Overview. *Front. Med.* **6**, 264 (2019).
6. Ahmedt-Aristizabal, D., Armin, M. A., Denman, S., Fookes, C. & Petersson, L. A survey on graph-based deep learning for computational histopathology. *Computerized Medical Imaging and Graphics* **95**, 102027 (2022).
7. Azher, Z. L., Vaickus, L. J., Salas, L. A., Christensen, B. C. & Levy, J. J. Development of biologically interpretable multimodal deep learning model for cancer prognosis prediction. in *Proceedings of the 37th ACM/SIGAPP Symposium on Applied Computing* 636–644 (ACM, 2022). doi:10.1145/3477314.3507032.
8. Chen, R. J. *et al.* Whole Slide Images are 2D Point Clouds: Context-Aware Survival Prediction Using Patch-Based Graph Convolutional Networks. in *Medical Image Computing and Computer Assisted Intervention – MICCAI 2021* (eds. De Bruijne, M. et al.) vol. 12908 339–349 (Springer International Publishing, 2021).
9. Levy, J., Haudenschild, C., Barwick, C., Christensen, B. & Vaickus, L. Topological Feature Extraction and Visualization of Whole Slide Images using Graph Neural Networks. *Pac Symp Biocomput* **26**, 285–296 (2021).
10. Deng, J. *et al.* ImageNet: A large-scale hierarchical image database. in *2009 IEEE Conference on Computer Vision and Pattern Recognition* 248–255 (IEEE, 2009). doi:10.1109/CVPR.2009.5206848.
11. Guan, Y. *et al.* Node-aligned Graph Convolutional Network for Whole-slide Image Representation and Classification. in *2022 IEEE/CVF Conference on Computer Vision and Pattern Recognition (CVPR)* 18791–18801 (IEEE, 2022). doi:10.1109/CVPR52688.2022.01825.
12. Wu, W., Liu, X., Hamilton, R. B., Suriawinata, A. A. & Hassanpour, S. Graph Convolutional Neural Networks for Histologic Classification of Pancreatic Cancer. *Archives of Pathology & Laboratory Medicine* (2023) doi:10.5858/arpa.2022-0035-OA.
13. Wu, Y., Cheng, Y., Wang, X., Fan, J. & Gao, Q. Spatial omics: Navigating to the golden era of cancer research. *Clinical & Translational Med* **12**, (2022).
14. Biancalani, T. *et al.* Deep learning and alignment of spatially resolved single-cell transcriptomes with Tangram. *Nat Methods* **18**, 1352–1362 (2021).
15. Zeng, Z., Li, Y., Li, Y. & Luo, Y. Statistical and machine learning methods for spatially resolved transcriptomics data analysis. *Genome Biol* **23**, 83 (2022).
16. Svensson, V., Teichmann, S. A. & Stegle, O. SpatialDE: identification of spatially variable genes. *Nat Methods* **15**, 343–346 (2018).

17. Macenko, M. *et al.* A method for normalizing histology slides for quantitative analysis. in *2009 IEEE International Symposium on Biomedical Imaging: From Nano to Macro* 1107–1110 (IEEE, 2009). doi:10.1109/ISBI.2009.5193250.
18. Levy, J. J., Salas, L. A., Christensen, B. C., Sriharan, A. & Vaickus, L. J. PathFlowAI: A High-Throughput Workflow for Preprocessing, Deep Learning and Interpretation in Digital Pathology. *Pac Symp Biocomput* **25**, 403–414 (2020).
19. Chen, T., Kornblith, S., Norouzi, M. & Hinton, G. A Simple Framework for Contrastive Learning of Visual Representations. (2020) doi:10.48550/ARXIV.2002.05709.
20. Ciga, O., Xu, T. & Martel, A. L. Self supervised contrastive learning for digital histopathology. *Machine Learning with Applications* **7**, 100198 (2022).
21. Zeng, Y. *et al.* Identifying spatial domain by adapting transcriptomics with histology through contrastive learning. *Briefings in Bioinformatics* **24**, bbad048 (2023).
22. Kipf, T. N. & Welling, M. Semi-Supervised Classification with Graph Convolutional Networks. (2016) doi:10.48550/ARXIV.1609.02907.
23. Lee, J., Lee, I. & Kang, J. Self-Attention Graph Pooling. (2019) doi:10.48550/ARXIV.1904.08082.
24. Fey, M. & Lenssen, J. E. Fast Graph Representation Learning with PyTorch Geometric. (2019) doi:10.48550/ARXIV.1903.02428.
25. Calinski, T. & Harabasz, J. A dendrite method for cluster analysis. *Comm. in Stats. - Theory & Methods* **3**, 1–27 (1974).
26. Davies, D. L. & Bouldin, D. W. A Cluster Separation Measure. *IEEE Trans. Pattern Anal. Mach. Intell.* **PAMI-1**, 224–227 (1979).
27. Bai, Z. *et al.* Tumor-Infiltrating Lymphocytes in Colorectal Cancer: The Fundamental Indication and Application on Immunotherapy. *Front. Immunol.* **12**, 808964 (2022).
28. Saltz, J. *et al.* Spatial Organization and Molecular Correlation of Tumor-Infiltrating Lymphocytes Using Deep Learning on Pathology Images. *Cell Rep* **23**, 181-193.e7 (2018).
29. Magel, R. C. & Wibowo, S. H. Comparing the Powers of the Wald-Wolfowitz and Kolmogorov-Smirnov Tests. *Biom. J.* **39**, 665–675 (1997).
30. Idos, G. E. *et al.* The Prognostic Implications of Tumor Infiltrating Lymphocytes in Colorectal Cancer: A Systematic Review and Meta-Analysis. *Sci Rep* **10**, 3360 (2020).
31. Azher, Z. L. *et al.* Assessment of emerging pretraining strategies in interpretable multimodal deep learning for cancer prognostication. *BioData Mining* **16**, 23 (2023).
32. Cheerla, A. & Gevaert, O. Deep learning with multimodal representation for pancancer prognosis prediction. *Bioinformatics* **35**, i446–i454 (2019).
33. Schirris, Y., Gavves, E., Nederlof, I., Horlings, H. M. & Teuwen, J. DeepSMILE: Contrastive self-supervised pre-training benefits MSI and HRD classification directly from H&E whole-slide images in colorectal and breast cancer. *Medical Image Analysis* **79**, 102464 (2022).
34. Chen, R. J. & Krishnan, R. G. Self-Supervised Vision Transformers Learn Visual Concepts in Histopathology. (2022) doi:10.48550/ARXIV.2203.00585.
35. Li, Z. *et al.* Vision transformer-based weakly supervised histopathological image analysis of primary brain tumors. *iScience* **26**, 105872 (2023).

Potential to Enhance Large Scale Molecular Assessments of Skin Photoaging through Virtual Inference of Spatial Transcriptomics from Routine Staining

Gokul Srinivasan
Department of Pathology and Laboratory Medicine, Dartmouth Hitchcock Medical Center
Lebanon, NH 03756, USA
Email: gokul.srinivasan.23@dartmouth.edu

Matthew J. Davis, Matthew R. LeBoeuf
Department of Dermatology, Dartmouth-Hitchcock Medical Center
Lebanon, NH 03756, USA
Email: Matthew.J.Davis@hitchcock.edu, Matthew.Leboeuf@hitchcock.org

Michael Fatemi, Zarif L. Azher, Yunrui Lu, Alos B. Diallo, Marietta K. Saldias Montivero
Department of Pathology and Laboratory Medicine, Dartmouth Hitchcock Medical Center
Lebanon, NH 03756, USA
Email: myfatemi04@gmail.com, zarif.azher@gmail.com, Yunrui.Lu@dartmouth.edu, Alos.B.Diallo.GR@dartmouth.edu, Marietta.K.Saldias.Montivero.GR@dartmouth.edu

Fred W. Kolling IV, Laurent Perrard
Genomics Shared Resource, Dartmouth Cancer Center
Lebanon, NH 03756, USA
Email: Fred.W.Kolling.IV@dartmouth.edu, Laurent.Perreard@dartmouth.edu

Lucas A. Salas, Brock C. Christensen, Thomas J. Palys, Margaret R. Karagas
Department of Epidemiology, Geisel School of Medicine at Dartmouth
Lebanon, NH 03756, USA
Email: Lucas.A.Salas@dartmouth.edu, Brock.C.Christensen@dartmouth.edu, Thomas.J.Palys@dartmouth.edu, Margaret.R.Karagas@Dartmouth.edu

Scott M. Palisoul, Gregory J. Tsongalis, Louis J. Vaickus
Department of Pathology and Laboratory Medicine, Dartmouth Hitchcock Medical Center
Lebanon, NH 03756, USA
Email: Scott.M.Palisoul@hitchcock.org, gregory.j.tsongalis@hitchcock.org, louis.j.vaickus@hitchcock.org

Sarah M. Preum
Department of Computer Science, Dartmouth College
Hanover, NH 03766, USA
Email: Sarah.Masud.Preum@dartmouth.edu

Joshua J. Levy*†
Department of Pathology and Laboratory Medicine, Dartmouth Hitchcock Medical Center
Lebanon, NH 03756, USA
Email: joshua.j.levy@dartmouth.edu

The advent of spatial transcriptomics technologies has heralded a renaissance in research to advance our understanding of the spatial cellular and transcriptional heterogeneity within tissues. Spatial transcriptomics allows investigation of the interplay between cells, molecular pathways, and the surrounding tissue architecture and can help elucidate developmental trajectories, disease pathogenesis, and various niches in the tumor microenvironment. Photoaging is the histological and molecular skin damage resulting from chronic/acute sun exposure and is a major risk factor for skin cancer. Spatial transcriptomics technologies hold promise for improving the reliability of evaluating photoaging and developing new therapeutics. Challenges to current methods include limited focus on dermal elastosis variations and reliance on self-reported measures, which can introduce subjectivity and inconsistency. Spatial transcriptomics offers an opportunity to assess photoaging objectively and reproducibly in studies of carcinogenesis and discern the effectiveness of therapies that intervene in photoaging and preventing cancer. Evaluation of distinct histological architectures using highly-multiplexed spatial technologies can identify specific cell lineages that have been understudied due to their location beyond the depth of UV penetration. However, the cost and inter-patient variability using state-of-the-art assays such as the 10x Genomics Spatial Transcriptomics assays limits the scope and scale of large-scale molecular epidemiologic studies. Here, we investigate the inference of spatial transcriptomics information from routine hematoxylin and eosin-stained (H&E) tissue slides. We employed the Visium CytAssist spatial transcriptomics assay to analyze over 18,000 genes at a 50-micron resolution for four patients from a cohort of 261 skin specimens collected adjacent to surgical resection sites for basal cell and squamous cell keratinocyte tumors. The spatial transcriptomics data was co-registered with 40x resolution whole slide imaging (WSI) information. We developed machine learning models that achieved a macro-averaged median AUC and F1 score of 0.80 and 0.61 and Spearman coefficient of 0.60 in inferring transcriptomic profiles across the slides, and accurately captured biological pathways across various tissue architectures.

Keywords: Deep Learning, Machine Learning, Spatial Transcriptomics, Skin Photoaging.

1. Introduction

Spatial transcriptomics is an innovative and rapidly evolving field in biomedical research that combines the power of genomics and spatial mapping techniques to gain insights into the spatial organization of gene expression within complex tissues, such as the skin. By providing a detailed view of gene expression patterns in relation to cellular and tissue architecture, spatial transcriptomics has quickly become a valuable tool for biomedical research, including dermatological research.

The skin is the largest organ in the body, composed of multiple cell types that each play a crucial role in maintaining its structure and function. Though traditional genomic analysis techniques, such as bulk-RNA sequencing (RNA-seq), and disaggregated techniques, such as single cell RNA sequencing (scRNA-seq), have provided valuable information about cellular heterogeneity and disease progression, they lack the ability to assess localized gene expression patterns that may relate

* To whom correspondence should be addressed.

† Work supported by grants P20GM130454, P20GM104416 to JL and K08CA267096 to LV.

with cell-cell interactions and architecture to support tissue function. Spatial transcriptomics approaches uniquely allow researchers to examine gene expression patterns within their anatomical and histological context, enabling a deeper understanding of the underlying molecular mechanisms driving skin biology, carcinogenesis, and disease progression.

An important potential application of spatial transcriptomics in dermatology is to advance the emerging study of skin aging.[1] The skin serves as a barrier between the environment and the body where it is exposed to near-constant insults, including ultraviolet radiation (UVR), mechanical stress, and toxicants.[2] These exposures, along with genetic influences, combine to induce skin damage, reduced function, and, ultimately, a characteristic loss of elasticity of the skin largely reflecting degradation of the collagen matrix.[3] More recently, Zou et al. created a single-cell transcriptomic atlas of human skin aging using eyelid tissue and identified cell-type-specific associations with human skin aging.[1] Further characterization of cellular changes that incorporate spatial information in skin can inform therapeutic strategies and interventions to combat age-related skin alterations and disease.

Currently, spatial transcriptomics technologies at whole transcriptomic-level multiplexing are incredibly costly and prone to several sources of variation (e.g., within/between-subject variation), limiting broad application. Recently, deep learning models have been proposed as a cost-saving alternative to predict spatial gene expression from routine tissue stains.[4–6] For instance, the DeepSpaCE approach includes convolutional neural networks (CNNs) for spatial gene cluster and gene expression prediction in human breast cancer tissue sections.[4] Another modeling paradigm aimed at predicting spatial gene expression across breast and cutaneous tumor data used a mix of transformer and graph neural network-based approaches.[6] In addition, the performance of several different modeling approaches for spatial gene expression prediction in tissue was recently compared using stage-III (pT3) colorectal tumors. Though these studies demonstrate the potential to infer spatial expression patterns using histomorphological data, several crucial questions remained unanswered, including the applicability of these methods to non-cancerous tissue sections, to other biological domains (e.g., dermatology), as well as the extent to which prediction modeling can preserve salient biological pathways and relationships required for downstream analysis on larger cohorts.

In this pilot study, we develop and validate a deep learning method for the prediction of spatial gene expression across spatially variable genes in routine H&E-stained skin tissue. Predictions can be used to create synthetic multidimensional tissue maps—similar to those produced through spatial transcriptomic profiling—for tissues without corresponding spatial transcriptomics data. Use of deep learning models promises to reduce the cost and time associated with spatial transcriptomics data acquisition for dermatological applications, greatly expanding access to the technology and its range of unique insights. Interrogating pathways associated with spatially inferred genes can advance our knowledge of skin biology, improve diagnostic tools, and pave the way for more personalized treatment strategies.

2. Methods and Materials

In this work, we attempt to predict the spatial gene expression of Visium spatial transcriptomics spots distributed across 40X magnification H&E slides. To this end, we use the following methods:

1. **Data collection and annotation:** Acquired H&E whole slide images (WSI), and spatially registered Visium CytAssist assayed spatial transcriptomics slides from 4 human cheek skin tissue samples collected from sites histologically adjacent to basal cell carcinoma (BCC) and

squamous cell carcinoma (SCC) during skin cancer removal surgery. These samples were then graded by dermatologists for their solar elastosis status (two mild, two severe). Additionally, dermatologists annotated regions corresponding to distinct histological entities (e.g., epidermis, eccrine glands, hair follicles, sebaceous glands, and vascular/endothelial infrastructure). Instances of actinic keratoses were documented from a larger cohort.

2. **Preprocessing:** Preprocess gene expression and WSI subarrays to capture spatially variable genes and genetically dense regions of tissue.
3. **Model development:** Configure the SWIN-T transformer to perform two distinct modeling tasks, binary (dichotomized expression) and continuous gene expression prediction, on the 1000 most spatially variable genes.
4. **Leave one-patient-out cross-validation:** Evaluation on held-out slides/patients as a measure of external applicability.
5. **Recover spatial biology inferences:** Model performance was further measured using: 1) pathway analysis for high performing genes, 2) topological consistency between ground truth and predicted expression, and 3) the ability to recapitulate genes and pathways associated with distinct histological structures.

Each of these steps will be further detailed in the ensuing sections.

2.1. *Data Collection*

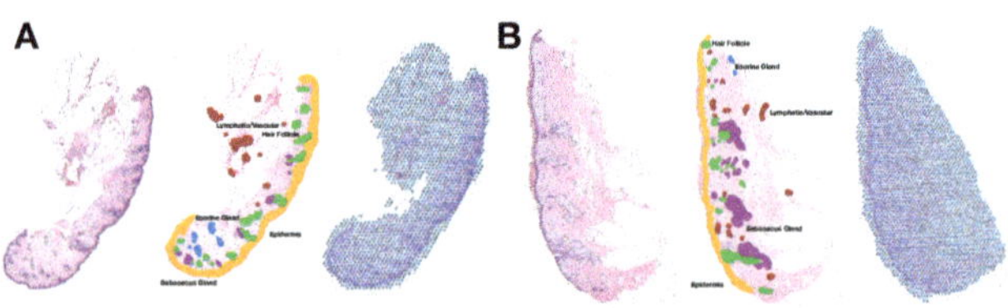

Figure 1: Cohort Description. 261 WSIs were scanned. Four of these slides underwent further spatial transcriptomics profiling and were annotated for distinct histological architectures. Two are shown here. From left to right, the WSI, histological annotations, and Visium spatial transcriptomics spot array for **(A)** sample 14 and **(B)** sample 167.

Four specimens were collected for profiling from a cohort of 261 tissue samples obtained in a single site Mohs micrographic surgery (MMS) clinic between March 1st 2022, and October 10th 2022. The samples were mostly from the head and neck, and all from sites histologically adjacent to either basal cell carcinoma or squamous cell carcinoma, as confirmed by histologic analysis of frozen section slides. The tissue was removed as part of standard surgical practice as Burow's triangle flaps for skin grafting/reconstruction. Triangles are normally discarded—two triangles were collected per patient, in some cases bisected. One triangle underwent formalin fixation while the other triangle was frozen. Formalin-fixed specimens were breadloafed, encased in paraffin-embedded tissue blocks, and sectioned and stained for hematoxylin and eosin (H&E) using Autostainers for subsequent imaging at 40x resolution (0.25 micron/pixel) using Aperio GT450 scanners. Tissue slides were transported to the Genomics core, where after tissue decoverslipping, the Visium CytAssist device was used to transfer transcriptomic probes from the original glass slides to 11mmx11mm capture areas on Visium slides. Sections from two patients were placed into each capture area to conserve costs and separated during the analysis stage. Whole transcriptomic profiling was accomplished after mRNA permeabilization, poly(A) capture, and probe hybridization. The eosin stain for the tissue sections were imaged using CytAssist, which were then co-registered to the original 40X whole slide images (WSI). Given the limited sample size due to the spatial transcriptomics assay costs, four specimens were selected, representing cheek tissue from four females, two with mild elastosis (participant #178 and #14, ages 24 and 76 respectively), two with severe elastosis (participant #167 and #107, ages 55 and 84) respectively.

2.2. *Preprocessing*

Prior to processing, Visium spatial transcriptomics profiles for samples contained 18,085 genes measured across several thousand locations throughout each slide. Each profile was then subjected to preliminary filtering, where genes and spots were filtered according to their abundance (i.e., cells with less than 500 genes, genes expressed in less than 3 cells, and cells with more than 15% mitochondrial gene expression were filtered out). After filtering out the regions lacking tissue using a custom annotation tool augmented by the SAM, the total number of Visium spots per slide reached 2561, 3279, 3547, and 1737, each sampled in a honeycomb formation. Each Visium spot covers a circular capture area with a diameter of 50-micron (~200 pixels) at 40x magnification. After sequencing, we used the SpaceRanger package to preprocess the Visium reads into gene count matrices.

Every whole slide image (WSIs) used for the Visium assay captures an area (size of capture area– 11 × 5.5 mm– half the capture area per patient) that spans tens of thousands of pixels along each dimension. Accordingly, to make the prediction task computationally tractable, we subdivided every WSI into square 512 by 512-pixel image patches (i.e., subarrays) centered on each Visium spot. The gene expression of the central 50-micron Visium spots were aligned to each image patch. Data present within the image patch but falling outside the capture area of the Visium spot were considered to have less direct relevance to the cells being assayed. Spots were additionally annotated based on the aforementioned tissue histological structure using the Annotorious OpenSeadragon plugin.

2.3. *Model Development*

2.3.1. *Inference Targets*

As predicting all of the genes assayed is computationally intractable, we used the SpatialDE library to select the top 1000 genes based on their mean spatial variance (MSV) across all slides (i.e., selected genes that exhibited the greatest spatial variation across the 4 slides). We then tested the capacity of our models to predict both dichotomized and log gene expression for all 1000 genes.

More specifically, in the dichotomized prediction task, patches were classified as having a "high" or "low" gene expression for each gene if the expression of the gene at that patch location was greater or lower than its mean gene expression across all other Visium spots in the corresponding WSI. This approach follows existing work detailed in Fatemi et al.[5] For this task, models were trained using a binary cross entropy loss function.

In the continuous expression task, by contrast, models were trained to predict the log-pseudocount log(1+counts) gene expression for each gene within the corresponding image patch region. For this task, loss was calculated using the mean squared error, which was found to be comparable to modeling counts using the zero-inflated negative binomial distribution.[5]

2.3.2. *Modeling Approach*

Previous work has established the importance of spatial and neighborhood context information in both the dichotomized and continuous gene expression tasks.[5] In this study, we leveraged the SWIN-T vision transformer, a hierarchical transformer that has gained repute for building hierarchical feature maps by iteratively merging information from nearby image patches in deeper layers.[7] Transformers divide images into smaller subimages, and numerical descriptors are extracted for each subimage using convolutional filters, along with information on the relative positioning of the subimages. Self-attention mechanisms are used to route information across the image based the relevance of one subregion of the image to another. In both the dichotomized and continuous expression tasks, the output layer of the base SWIN-T model was modified. In particular, the output layer was expanded to consist of two feed-forward layers of sizes 768 and 2000, chosen through coarse experimentation to maximize model performance. Both the dichotomized and continuous expression models yielded predictions for the 1000 most spatially variable genes.

2.3.3. *Data Augmentation and Hyperparameter Selection*

To improve the robustness and generalizability of these models to varied histological contexts, all images in the training set were subject to a series of data augmentation transformations implemented using the Albumentations package.[8] Images were first resized to 448 by 448 pixels in size, the input dimensionality for the SWIN-T model. Horizontal flips and random brightness contrast were then performed with probabilities 0.5 and 0.2, respectively. A shift, scale, and rotate transformation was also applied to every image with a probability of 0.3. A shifting limit of 0.1, a scaling limit of 0.1, and a rotation limit of 30 were used here. Additionally, random rectangular areas of the images were erased– a maximum of 8 holes were produced per image, each hole obscuring at most 16 by 16 pixels.

Hyperparameters were obtained for both the dichotomized and continuous expression models via a coarse hyperparameter grid search. For the dichotomized models, optimal performance was observed while using a batch size, learning rate, and training length of 64, $0.5\times10{-}6$, and 20 epochs. Whereas for the continuous models, optimal performance was observed while using a batch size, learning rate, and training length of 64, $0.33 \times 10{-}6$, and 20 epochs. The Lion optimizer was used in both cases.[9]

2.4. *Cross Validation*

Model performance was measured via leave-one-patient-out cross-validation (LOOCV). In this procedure, three of the four Visium spatial transcriptomics samples were used for training and validation, while the remaining sample was used for testing. This procedure was repeated four times to account for all possible training/testing combinations. Reported performance metrics for each gene are the macro-averaged (across slides, weighting each slide equally) median (across genes) area under the receiver operating characteristic curve (AUROC) and F1 score (F1) statistics for the dichotomized task, and correlation coefficients (Spearman coefficient) to compare true versus predicted pseudocounts– log(1+counts)– for the continuous task. Macro performance statistics underwent 1,000 sample nonparametric bootstrapping across the Visium spots to yield 95% confidence intervals.

2.5. *Biological Salience*

To assess for the model's ability to capture meaningful biological information from tissue histology, model predictions were also scrutinized for their ability to 1) recapitulate a range of biologically salient pathways, 2) maintain the shape and spatial signature of ground truth spatial gene expression data in lower dimensional space (i.e., preserves key relationships; spots cluster similarly) using the aligned-UMAP procedure, and 3) facilitate the inference of biologically salient features, such as histological markers. These tasks are detailed below.

2.5.1. *Pathway Analyses*

Given the nature of histomorphological data, it is unreasonable to expect that every gene can be predicted from tissue histology alone. Accordingly, we sought to determine the biological pathways associated with sets of differentially performing genes to answer understand what biological properties make a gene amenable to prediction. We utilized the GO Biological Process 2023 database through the EnrichR package, to perform a pathway analysis on predicted genes stratified by decile after ranking genes based on predictive performance.[10,11] The top 3 pathways were selected for each decile—from 90th to 100th decile (i.e., top performing genes) to the 0th to 10th device (i.e., worst performing genes)—based on their combined score (i.e., magnitude of representation and statistical significance). Detected pathways were also filtered by tissue specificity (i.e., could reasonably be involved with the skin).

We further sought to identify whether the gene signatures correspondent to different histological architectures was congruent between true and predicted expression. First, the top 100 most differentially expressed genes were found using the Wilcoxon rank-sum test in a one vs. rest fashion for each tissue architecture (e.g., follicles versus non-follicular structures) using both predicted and ground truth data. A pathway analysis using GO Biological Process 2023 database through the EnrichR package was then performed for the top true and predicted differentially expressed genes for each architecture. The top 10 pathways were selected by combined score for each histological category. Detected pathways were compared between ground truth and predicted gene expression for each sample under the hypothesis that similar pathways should be associated with the same architectures.

2.5.2. *Similar Clustering of Visium Spots and Consistent Topology via Aligned-UMAP*

Model predictions were further assessed for their ability recapitulate the topology (i.e., relationships between spots) of ground truth Visium spatial gene expression data within a lower dimensional space. This was accomplished through the comparison of Uniform Manifold Approximation and Projection (UMAP) embeddings (i.e., numerical representations that could be plotted in a 2D scatterplot; closer points share similar expression/biological relevance) for the ground truth and predicted expression profiles (on held-out slides) extracted using the SWIN-T model. Ground truth and predicted gene expression profiles for each slide were co-projected to a lower dimensional space using the Aligned-UMAP procedure to preserve the relative orientation and alignment between spots to enable comparison between the approaches. Each Visium spot from the WSI was plotted as 2D scatterplot point and colored according to its gene expression profile as dictated by the Leiden clustering algorithm. In other words, ground truth Visium spots sharing similar transcriptional information are grouped to the same Leiden cluster, while genetically dissimilar spots are grouped to different Leiden clusters. These ground truth cluster assignments were overlaid on the scatterplots for the predicted expression patterns. It is expected that the relative positioning between the clusters would be preserved in the 2D scatterplot for the predicted expression, which would measure the extent to which model predictions recapitulated patterns associated with distinct histological regions of each WSI.

Aside from overlaying the original ground truth clusters, predicted expression profiles were also separately clustered through the Leiden algorithm, yielding a separate set of cluster assignments for the same Visium spots. These assignments were compared to the ground truth Visium spatial transcriptomics profiles' clusters. Similar clustering assignments would provide further evidence for greater correspondence between transcriptional data and information derived from the histology.

3. Results[†]

3.1. *Prediction of Spatial Transcriptomic Patterns from Histology*

In the dichotomized prediction task, the SWIN-T vision transformer model achieved a macro-averaged (i.e., across genes) median AUC and F1 score of 0.80 and 0.61, respectively, across the testing sets (**Supplementary Table 1**). The model performed best on genes *ADIPOQ*, *PLIN1*, and *PKP3* (involved in fatty acid metabolism[12], triacylglycerol storage[13], and desmosome function and stability[14], respectively) , and worst on genes *ANKRD35*, *ALAS1*, and *MIA* (of which the latter two are known to be involved in heme biosynthesis[15] and melanocyte migration[16], respectively) . Dichotomized model predictions for genes *ADIPOQ*, *PLIN1*, and *PKP3* are visualized across sample #14 and #178 (**Figure 2**), demonstrating spatial concordance between true and predicted expression. In the continuous prediction task, models achieved a macro-averaged median Spearman coefficient of 0.60 across the testing sets (**Supplementary Table 1**). The model performed best on genes *KRT14*, *CXCL14*, and *COL1A2* (involved in epithelial cell integrity[17], keratinocyte function[18], and collagen synthesis[19], respectively) and worst on genes *CKM*, *MYLPF*, and *ODF21* (the former two are known to be involved in energy homeostasis[20] and muscle development[21]). Continuous model predictions for genes *KRT14*, *CXCL14*, *PI16* were visualized across samples 107 and 167 (**Supplementary Figure 2**), demonstrating spatial concordance between true and predicted expression. Note that models in both the dichotomized and continuous prediction tasks were trained to predict the same set of 1000 spatially variable genes.

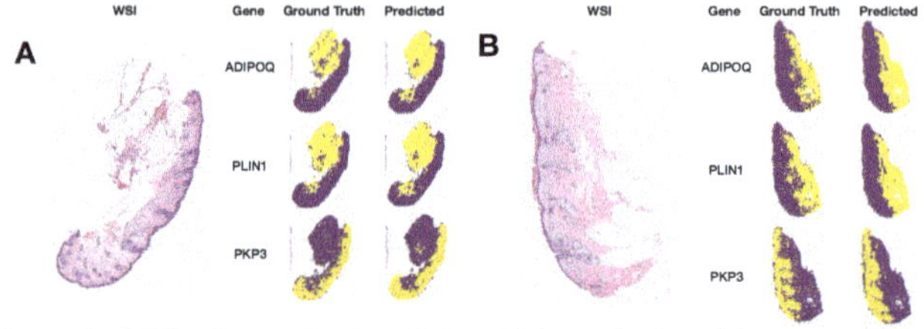

Figure 2: Dichotomized RNA Expression Prediction. Dichotomized spatial gene expression was inferred for **(A)** samples #14 and **(B)** #167 and compared with the respective ground truths. A spot is colored yellow if gene expression in this spot exceeds global mean gene expression. Performance is displayed for the top performing genes, ADIPOQ, PLIN1, and PKP3, which achieved macro-averaged AUC values of 0.942, 0.938, and 0.918, respectively.

[†] Supplementary materials can be found at the following DOI: https://doi.org/10.5281/zenodo.8197850

3.2. *Pathway Analysis*

For each performance decile, the top 3 most salient biological pathways by combined score were determined (**Supplementary Table 2**). Across both the dichotomized and continuous prediction tasks, biological pathways associated with the top performance decile (i.e., 90th to 100th percentile genes ranked by performance) pertained to skin and epidermis development and maintenance, skin cell proliferation, and the regulation of extracellular matrix and cell-cell adhesion (**Table 1**). By contrast, biological pathways associated with genes in the worst performance decile (i.e., 0th to 10th percentile genes ranked by performance) across both the dichotomized and continuous prediction were far less associated with relevant biological phenomena, pertaining to immune signaling, cell-turnover regulation, gas transport, and muscle cell development (**Table 1**). More generally, biological pathways associated with higher-performing genes tended to be more closely related to skin development, differentiation, pigmentation, and fat metabolism, while distinct trends were less clear for those biological pathways associated with lower-performing genes (**Supplementary Table 2**).

Table 1: Performance Pathway Analysis. Combined performance statistics for both the dichotomized and continuous models were used to perform a performance-stratified pathway analysis. AUC and Spearman coefficient were used to stratify genes in the dichotomized and continuous tasks, respectively. The top 3 pathways, measured via EnrichR using the Go Biological Process 2023 database, are reported for the highest and lowest performance deciles. Refer to **Supplementary Table 2** for an extended version of this table detailing all performance deciles.

Gene Performance	Task	Pathway	Score	Overlap	P-value
Top performing genes: 90-100th Percentile Genes	**Dichotomized**	Establishment of Skin Barrier	2127	6/19	6.9E-10
		Skin Epidermis Development	1785	6/21	4.0E-04
		Keratinocyte Proliferation	803	2/6	1.4E-11
	Continuous	Positive Regulation of Epidermis Development	2725	4/9	7.3E-08
		Desmosome Organization	2654	3/6	2.4E-06
		Intermediate Filament Bundle Assembly	1905	4/7	4.2E-06
Bottom performing genes: 0-10th Percentile Genes	**Dichotomized**	Interleukin-2-Mediated Signaling Pathway	615	2/7	2.8E-04
		Cellular Response to Interleukin-2	615	2/7	5.1E-04
		Negative Regulation of T Cell Apoptotic Process	615	2/7	5.1E-04
	Continuous	Gas Transport	504	3/15	5.3E-05
		Positive Regulation of Respiratory Burst	493	2/8	6.7E-04
		Regulation Of Skeletal Muscle Cell Differentiation	493	2/8	6.7E-04

The top 100 differentially expressed genes for each histological sub-type were determined for both ground truth and predicted data samples, and these genes were leveraged for further pathway analyses. The pathway analysis results for the top 100 differentially expressed genes for dichotomized and continuous gene expression data are reported in **Supplementary Table 3** and **Supplementary Table 4**. Across both dichotomized and continuous gene expression data, pathways associated with the formation of the sebaceous gland and epidermis were found to be in high agreement between ground truth and predicted expression, while the agreement was more modest other in histological features (**Table 1; Supplementary Table 3**).

3.3. *Topological Consistency*

3.3.1. *Leiden Clustering*

A visual inspection of the aligned-UMAP diagrams demonstrates similar clustering patterns and topological consistency between the predicted and the ground truth expression data across both models trained for dichotomized and continuous regression tasks (**Figure 3; Supplementary Figure 2**). We noted that the Leiden clusters assigned to the ground truth expression were similar to those assigned to predicted expression embeddings. Nonetheless, differences remained. We did not observe complete separation in the predicted expression embeddings, representing a fuzzier or more connected/intermediate topological structure (**Figure 3; Supplementary Figure 2**). These spots in the predicted data were, accordingly, located between Leiden clusters more often than spots in the ground truth genetic data, where Leiden clusters tended to be far more spatially distinct. This feature was noted for both dichotomized and continuous expression models, although this pattern was more prevalent for dichotomized expression (**Figure 3; Supplementary Figure 2**).

Model predictions in both the dichotomized and continuous expression tasks also preserved the general shape of ground truth genetic data while plotted across each whole slide image (**Supplementary Figures 3 and 4**). We further observed, however, that predicted data tended to contain genetically intermediate states, as evidenced by the greater number of Leiden clusters produced in the predicted data compared to the ground truth data while using the same Leiden clustering resolution (**Supplementary Figures 3 and 4**). Though models in both tasks produced data that captured larger macro-architectural differences in gene expression found across skin tissue, the dichotomized model tended to produce data that more closely preserved the relationships determined using Leiden clustering plotted across the slide found across the ground truth data (**Supplementary Figure 3**). Models in the continuous expression task, though high performing, tended to produce data that recapitulated the spatial genetic variation of macro-architectural features in skin tissue less well, evidenced, again, by disparities in the number and placement of Leiden clusters when comparing the predicted and ground truth data (**Supplementary Figure 4**).

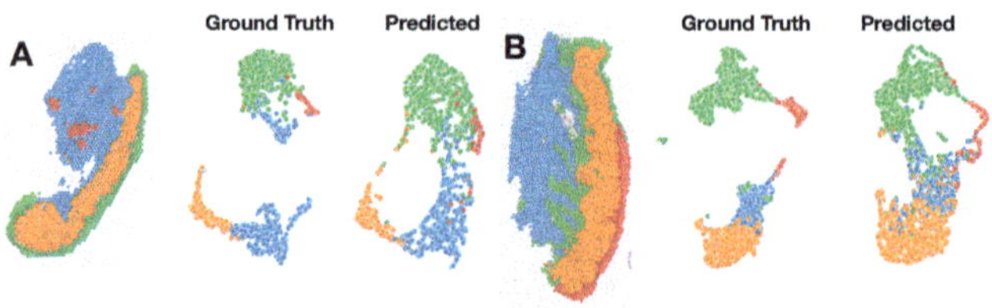

Figure 3: Dichotomized Expression Topological Analysis. (A) From left to right, ground truth spatial Leiden clustering, ground truth aligned UMAP, and predicted aligned UMAP for sample #14. **(B)** From left to right, ground truth spatial Leiden clustering, ground truth aligned UMAP, and predicted aligned UMAP for sample #178. In both rows, the same spots are colored identically according to their ground truth gene expression profiles after Leiden clustering analysis. Dichotomized gene expression data was used here. The Leiden clustering resolution was set to 0.2.

3.3.2. *Histological Annotations*

Performing Aligned-UMAP on the ground truth and predicted expression data for Visium spots tagged by histological structures demonstrated that embeddings in both groups clustered by distinct histological regions of skin tissue (**Figure 4; Supplementary Figure 5**). That is, Visium spots corresponding to similar histological structures clustered in similar locations across both UMAP plots, preserving the genetic relationships between these histological architectures. The distinctness of these clusters was preserved for both dichotomized and continuous gene expression predictions, though predicted continuous expression data appeared to preserve the topology better than dichotomized gene expression data (**Figure 4; Supplementary Figure 5**).

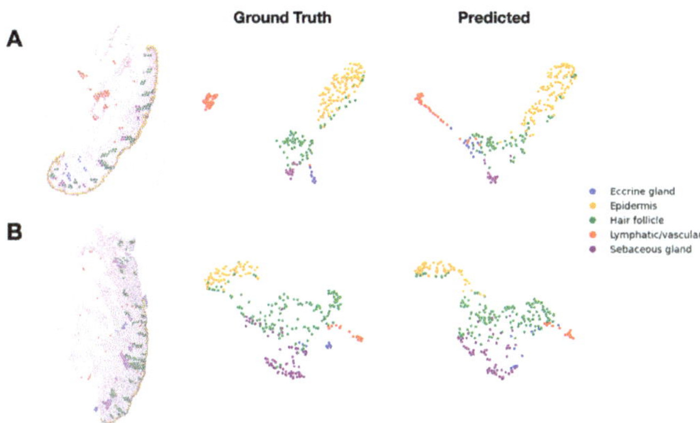

Figure 4: Dichotomized Expression Histological Analysis. Aligned-UMAP procedure was used to reduce the dimensionality of both the ground truth and predicted dichotomized gene expression vectors for **(A)** sample #14 and **(B)** sample #178. Spots are colored according to their histological annotations.

4. Discussion

In this work, we developed a set of spatial gene expression inference models for histopathologically normal skin tissue in the context of molecular changes associated with photoaging. We make use of the novel CytAssist co-registration/imaging technology, allowing for 40X resolution imaging of tissue slides. Beyond quantitative validation of performance (e.g., AUC, F1, Spearman coefficient), we also reaffirmed the biological relevance of the predicted expression pattern. In particular, extracted histological features from our models remained faithful to underlying biological pathways, buttressing their potential use across a range of biological inference tasks, and lending credibility to their role in democratizing the spatial transcriptomics paradigm to the broader research community.

With the Visium CytAssist technology, our models were trained with histological information at >4 times the spatial resolution of previous studies. We achieved comparable performance to a prior study that utilized an Inception convolutional neural network for both dichotomized and continuous prediction of gene expression.[5] While acknowledging the need for caution when comparing these models, as they represent different biological domains (Skin vs Colon; different genes used in the models), observed performance disparities may arise from variations in methodology. These differences could include the utilization of distinct imaging resolutions (40X vs 20X) or the selection of different modeling approaches (ZINB vs raw log gene expression). Since our predictions were made within a more focused visual receptive field, disregarding the surrounding wider tissue architecture, future work can explore the examination of larger-scale histological context.

The pathway and topological analysis provided useful insights on the nature of spatial RNA inference from histology. It is important to highlight that our findings suggest that genes with a clear histological basis are more likely to be accurately predicted compared to genes lacking such a theoretical histological foundation. Our results also demonstrate the importance of developing models germane to the biological question at hand: a model trained on colon tissue is not expected to perform well on skin tissue. Hence, investigating modeling approaches that prioritize specific biological phenomena emerges as a promising direction for future research. Genes that demonstrate good performance, or effectively recapitulate histological patterns, could potentially be utilized for further research applications across larger cohorts.

The topological analysis demonstrated that predicted expression profiles did not cluster as distinctly as the original expression patterns. When coloring by Leiden cluster affiliation and histological association, the predicted spot level gene expression fell between ground truth clusters, representing intermediate histological states learned by the neural networks. Future work may seek to understand which histomorphological features relate to different molecular pathways through newly established interpretation approaches[22]. The integration of coregistered slide imaging with spatial molecular information can facilitate such analyses. Moreover, by subsetting predicted and true gene expression based on shared molecular pathways (such as genes involved in epithelium development, cell-cell junctions, immune function, etc.) and conducting comparable topological analyses, it is possible to identify the molecular pathways that exhibit the highest degree of topological distinctiveness. Nevertheless, topological analyses have emerged as a timely and relevant topic in the realm of single-cell and spatial analyses, offering the potential to uncover additional dimensions of cellular and histological heterogeneity[23–25].

This study reinforces the potential of spatial transcriptomics approaches for research and clinical applications. For example, photoaging, which is linked to skin cancer risk, lacks reliable measurement tools due to variations in histological assessments and self-reported UV exposure. Existing analyses typically focus on specific cellular components (e.g., dermal fibroblasts, elastosis, keratoses), often disregarding or unaware of photoaging-related factors. Expanding spatial molecular findings through RNA inference to a larger cohort can help identify cell-type specific sources of photoaging in specific tissue architectures while controlling for numerous potential confounders and presents an intriguing area of follow up given the models established in this study. Spatial RNA inference can uncover novel cellular components related to precancerous alterations resulting from chronic sun exposure. By targeting profiling of these tissue regions, researchers can explore residual heterogeneity, while examining cell-type specific alterations and additional factors related to accelerated aging, with the caveat that tissue from this cohort is histologically adjacent to surgical site of repair, potentially harboring a cancerization field effect.

In clinical practice, the use of virtual RNA models has the potential to inform treatment planning and assess treatment response. If these models can identify proxy measures of photoaging, spatial molecular inference can be employed to evaluate the effectiveness of skin therapeutics through quantitative assessment of biomolecular changes at screening, baseline, and endpoint. This approach offers a more objective and quantitative measurement of the impact of treatments on skin health and can enhance the validity of therapeutic interventions. Similarly, applications are envisioned for treatment of non-healing skin ulcers and separately hair loss driven by an autoimmune response (e.g., alopecia areata), revealing potential components of relevant immune polarity (e.g., M1/M2 macrophage balance, etc.).[26,27] Additionally, virtual RNA inference models can inform disease management options for various solid tumors, functioning similar to immunohistochemical assays (e.g., immunoscore) that shed light on the infiltration of cytotoxic immune cell lineages, identifying independent risk factors of tumor recurrence and survival. Spatial molecular assessments can identify targetable therapeutic pathways for personalized treatment options.

This study is not without limitations that can direct for future research. Our sample size was small, limiting our ability to account for potential variability in histology and surgical sites. Additionally, the non-biopsied nature of the samples and their proximity to potentially precancerous tumor tissue may introduce differences in gene expression related to factors other than UV exposure. Introducing matching normal control tissue, considering factors like limited sun exposure and low field effect potential, along with expanding the cohort to control for additional age ranges, sex, and tissue site, could help reveal photoaging differences specific to these groups. Skin tone is another confounding factor that should be addressed, and it can be controlled using measures such as the Fitzpatrick skin phototype scale or derived continuous measures. To improve rigorous quantitative photoaging assessments, various measures of photoaging can be combined using factor analyses, leading to meaningful composite measures, such as DNA methylation, age-related measures, elastosis, and UV questionnaires. Additionally, one general limitation of our topological analyses included their more qualitative, rather than quantitative, nature. Shifts in distribution between ground truth and predicted Visium spot topology could also be captured using more nuanced mathematical notions such as the KL-divergence, Wasserstein distance, maximum mean discrepancy, and silhouette score. Addressing these limitations and incorporating a more diverse and extensive sample size can enhance the reliability and applicability of future studies in this field.

5. Conclusion

Machine learning technologies that can infer spatial molecular information from routine tissue stains have the potential to facilitate low-cost accessible spatial transcriptomic assessments for large scale molecular epidemiological studies. Such studies can uncover novel risk factors of early photocarcinogenesis or inform relevant treatment/therapeutic options by expanding the set of targetable molecular pathways within specific tissue architectures. Our skin study sets the stage for larger-scale studies to identify spatial molecular correlates of skin sun damage and evaluate novel therapeutics that may reverse this damage. While our models exhibited impressive performance in predicting dichotomized and continuous gene expression within tissue slides, it is crucial to acknowledge the need for further development and validation of this approach. When utilizing these algorithms, it is important to consider the genes that are known to be influenced by histological characteristics. Additionally, any novel findings obtained through these tools should be corroborated and validated using well-established immunostaining techniques, ensuring the reliability and robustness of the results.

6. Acknowledgements and Location of Supplementary Materials

The authors acknowledge the support of the Center for Clinical Genomics and Advanced Technology in the Department of Pathology and Laboratory Medicine of the Dartmouth Hitchcock Health System which includes the Pathology Shared Resource, at the Dartmouth Cancer Center with

NCI Cancer Center Support Grant 5P30 CA023108-37. Spatial transcriptomics assays were carried out in the Genomics and Molecular Biology Shared Resource (GMBSR) at Dartmouth which is supported by NCI Cancer Center Support Grant 5P30CA023108 and NIH S10 (1S10OD030242) awards. Spatial studies were conducted through the Dartmouth Center for Quantitative Biology in collaboration with the GMBSR with support from NIGMS (P20GM130454) and NIH S10 (S10OD025235) awards. Supplementary materials can be found at the following DOI: https://doi.org/10.5281/zenodo.8197850 .

References

1. Zou, Z. *et al.* A Single-Cell Transcriptomic Atlas of Human Skin Aging. *Developmental Cell* **56**, 383-397.e8 (2021).
2. Fuchs, E. Epithelial Skin Biology: Three Decades of Developmental Biology, a Hundred Questions Answered and a Thousand New Ones to Address. *Curr Top Dev Biol* **116**, 357–374 (2016).
3. Keyes, B. E. & Fuchs, E. Stem cells: Aging and transcriptional fingerprints. *J Cell Biol* **217**, 79–92 (2018).
4. Monjo, T., Koido, M., Nagasawa, S., Suzuki, Y. & Kamatani, Y. Efficient prediction of a spatial transcriptomics profile better characterizes breast cancer tissue sections without costly experimentation. *Sci Rep* **12**, 4133 (2022).
5. Fatemi, M. *et al.* Inferring spatial transcriptomics markers from whole slide images to characterize metastasis-related spatial heterogeneity of colorectal tumors: A pilot study. *Journal of Pathology Informatics* **14**, 100308 (2023).
6. Zeng, Y. *et al.* Spatial transcriptomics prediction from histology jointly through Transformer and graph neural networks. *Briefings in Bioinformatics* **23**, bbac297 (2022).
7. Liu, Z. *et al.* Swin Transformer V2: Scaling Up Capacity and Resolution. Preprint at http://arxiv.org/abs/2111.09883 (2022).
8. Buslaev, A. *et al.* Albumentations: Fast and Flexible Image Augmentations. *Information* **11**, 125 (2020).
9. Chen, X. *et al.* Symbolic Discovery of Optimization Algorithms. Preprint at http://arxiv.org/abs/2302.06675 (2023).
10. Kuleshov, M. V. *et al.* Enrichr: a comprehensive gene set enrichment analysis web server 2016 update. *Nucleic Acids Res* **44**, W90–W97 (2016).
11. The Gene Ontology Consortium *et al.* The Gene Ontology knowledgebase in 2023. *Genetics* **224**, iyad031 (2023).
12. Díez, J. J. & Iglesias, P. The role of the novel adipocyte-derived hormone adiponectin in human disease. *European Journal of Endocrinology* **148**, 293–300 (2003).
13. Brasaemle, D. L., Subramanian, V., Garcia, A., Marcinkiewicz, A. & Rothenberg, A. Perilipin A and the control of triacylglycerol metabolism. *Mol Cell Biochem* **326**, 15–21 (2009).
14. Gurjar, M. *et al.* Plakophilin3 increases desmosome assembly, size and stability by increasing expression of desmocollin2. *Biochem Biophys Res Commun* **495**, 768–774 (2018).
15. Kubota, Y. *et al.* Novel Mechanisms for Heme-dependent Degradation of ALAS1 Protein as a Component of Negative Feedback Regulation of Heme Biosynthesis. *J Biol Chem* **291**, 20516–20529 (2016).
16. Poser, I., Tatzel, J., Kuphal, S. & Bosserhoff, A. K. Functional role of MIA in melanocytes and early development of melanoma. *Oncogene* **23**, 6115–6124 (2004).
17. Bardhan, A. *et al.* Epidermolysis bullosa. *Nat Rev Dis Primers* **6**, 1–27 (2020).
18. Westrich, J. A., Vermeer, D. W., Colbert, P. L., Spanos, W. C. & Pyeon, D. The Multifarious Roles of the Chemokine CXCL14 in Cancer Progression and Immune Responses. *Mol Carcinog* **59**, 794–806 (2020).
19. Rivadeneira, F. & Uitterlinden, A. G. Chapter 22 - Osteoporosis Genes Identified by Genome-Wide Association Studies. in *Genetics of Bone Biology and Skeletal Disease (Second Edition)* (eds. Thakker, R. V., Whyte, M. P., Eisman, J. A. & Igarashi, T.) 377–395 (Academic Press, 2018). doi:10.1016/B978-0-12-804182-6.00022-8.
20. Zhao, T.-J., Yan, Y.-B., Liu, Y. & Zhou, H.-M. The Generation of the Oxidized Form of Creatine Kinase Is a Negative Regulation on Muscle Creatine Kinase *. *Journal of Biological Chemistry* **282**, 12022–12029 (2007).
21. Chong, J. X. *et al.* Mutations in MYLPF Cause a Novel Segmental Amyoplasia that Manifests as Distal Arthrogryposis. *The American Journal of Human Genetics* **107**, 293–310 (2020).

22. Chen, R. J. *et al.* Multimodal Co-Attention Transformer for Survival Prediction in Gigapixel Whole Slide Images. in *2021 IEEE/CVF International Conference on Computer Vision (ICCV)* 3995–4005 (2021). doi:10.1109/ICCV48922.2021.00398.
23. Levy, J., Haudenschild, C., Barwick, C., Christensen, B. & Vaickus, L. Topological Feature Extraction and Visualization of Whole Slide Images using Graph Neural Networks. *Pac Symp Biocomput* **26**, 285–296 (2021).
24. Rizvi, A. H. *et al.* Single-cell topological RNA-seq analysis reveals insights into cellular differentiation and development. *Nat Biotechnol* **35**, 551–560 (2017).
25. Wang, T., Johnson, T., Zhang, J. & Huang, K. Topological Methods for Visualization and Analysis of High Dimensional Single-Cell RNA Sequencing Data. *Pac Symp Biocomput* **24**, 350–361 (2019).
26. Theocharidis, G. *et al.* Single cell transcriptomic landscape of diabetic foot ulcers. *Nat Commun* **13**, 181 (2022).
27. Mariottoni, P. *et al.* Single-Cell RNA Sequencing Reveals Cellular and Transcriptional Changes Associated With M1 Macrophage Polarization in Hidradenitis Suppurativa. *Frontiers in Medicine* **8**, (2021).

PEPSI: Polarity measurements from spatial proteomics imaging suggest immune cell engagement

Eric Wu[1,2*], Zhenqin Wu[2], Aaron T. Mayer[2], Alexandro E. Trevino[2], James Zou[1,2,3]

[1]*Department of Electrical Engineering, Stanford University, Stanford, CA, USA*
[2]*Enable Medicine, Inc., Menlo Park, CA, USA*
[3]*Department of Biomedical Data Science, Stanford University, Stanford, CA, USA*
**Email: wue@stanford.edu*

Subcellular protein localization is important for understanding functional states of cells, but measuring and quantifying this information can be difficult and typically requires high-resolution microscopy. In this work, we develop a metric to define surface protein polarity from immunofluorescence (IF) imaging data and use it to identify distinct immune cell states within tumor microenvironments. We apply this metric to characterize over two million cells across 600 patient samples and find that cells identified as having polar expression exhibit characteristics relating to tumor-immune cell engagement. Additionally, we show that incorporating these polarity-defined cell subtypes improves the performance of deep learning models trained to predict patient survival outcomes. This method provides a first look at using subcellular protein expression patterns to phenotype immune cell functional states with applications to precision medicine.

Keywords: subcellular localization, proteomics, multi-plex immunofluorescence

1. Introduction

Spatial proteomics methods such as immunofluorescence (IF) and immunohistochemistry (IHC) enable an unprecedented view of tumor microenvironments by preserving the spatial structure of tissues at subcellular resolution[1]. However, standard analyses aggregate and average protein expression within segmented single cells, discarding sub-cellular and morphological signals[2]. This approach introduces a number of analytical limitations. First, segmentation can be imprecise. Second, subcellular protein expression patterns could allow the inference of cellular functional states (i.e. polarized vs. uniform). Thus, while cells can be phenotyped in the context of their spatial neighbors, cells that exhibit differential protein localization are not differentiated.

The relationship between protein localization and function is well-established in many contexts. For instance, during T cell engagement with presented antigens (e.g., on tumor cells), the CD4 and CD8 coreceptors are recruited to the immune synapse, while they present uniformly on the surface of a cell in a naive or exhausted state [3–5]. The immune synapse, also known as the supramolecular activation cluster, is a specialized junction formed by many proteins during an immune response. In the context of a tumor immune response, active and engaged T cells could correlate with better patient survival, whereas the presence of exhausted or inactive T cells may be indicative of worse outcomes [6–8]. However, it is unknown to what extent this or other such dynamic subcellular localization events are discernable from whole slide scale histology images.

Common analyses of cell morphology [9–13] utilize computer vision models for the automatic extraction of image features from tissue patches. However, interrogating such models for specific cell-cell interactions is difficult. Previous work toward characterizing surface protein localization includes statistical methods for identifying ligand-receptor pairs in transcriptomics [14,15], polarity localization measurements in mRNA [16,17], and co-localization with protein expression [18].

In this paper, we present a novel approach, PEPSI (**P**rotein **E**xpression **P**olarity **S**ubtyping in **I**mmunostains), for measuring subcellular protein localization toward characterizing the tumor microenvironment. We describe a simple, explainable method for computing the polarity of cell surface biomarkers. We apply this metric on multiple large-scale CODEX (co-detection by indexing) datasets spanning over two million cells, three clinical sites, and 600 patient samples. We focus on several key immune cells that are well-characterized and known to express polarized surface protein markers during activation/engagement. We define additional cell subtypes relating to morphology (polarized, uniform) for representative biomarkers (CD8, CD4, CD20) of immune cells (T cells and B cells). We find that surface protein marker polarity is significantly correlated with positive patient outcomes, even after controlling for various technical artifacts, suggesting that this may be important for characterizing the functional state of immune cells. We believe that inferring functional subtypes of cells can offer a better understanding of patient response to drug treatments and disease prognostic indicators.

2. Results

2.1. *Polarity measurement*

We describe a straightforward method for extracting polarity measurements for a given cell based on a polar transformation of the IF signal with respect to the cell centroid (**Figure 1A**, **Methods**). Plotting the distribution of scores for four markers in their cognate expression cell types - CD8 in CD8 T cells, CD4 in CD4 T cells, CD20 in B cells, and PanCK in tumor cells - shows that the scores exhibit continuous distributions (**Figure 1B**). The first three biomarkers, which are known to polarize in cells undergoing immune activation, show significantly higher average polarity scores versus PanCK, which is not known to polarize as such. To obtain a discrete polarity classification, we threshold the raw scores based on an empirical heuristic (**Methods**), obtaining proportions for polar, uniform, and 'other' cells for each of the three immune cell types (**Figure 1C**). For instance, polar cells account for 3.7%, 3.0%, and 2.6% of CD8 T cells, CD4 T cells, and

B cells, respectively. Example cells were randomly inspected in their cell contexts to visually validate the classifications (**Figure 1D**).

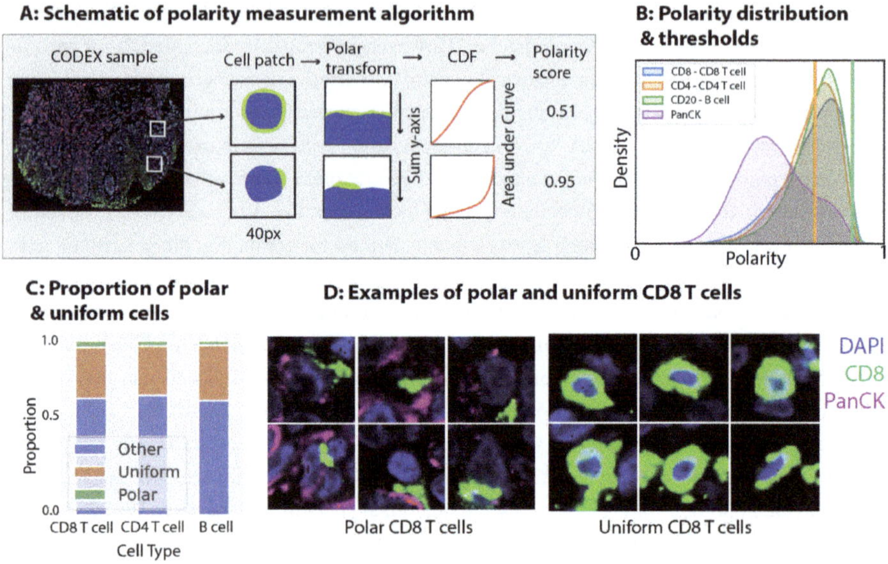

Fig. 1: Overview of the PEPSI polarity measurement framework. **Panel A:** Schematic of polarity measurement algorithm. For a given mIF sample, patches (40px by 40px) are extracted around each cell. For each cell, a polar transform is computed on the patch, followed by summing along the y-axis and then computing the area under the CDF curve, yielding a polarity score (from 0 to 1). **Panel B:** The polarity score histograms are shown for CD8, CD4, CD20, and PanCK biomarkers in CD8 T cells, CD4 T cells, B cells, and tumors, respectively. The orange (left) line indicates the upper threshold chosen for identifying uniform cells, whereas the green (right) line indicates the lower threshold chosen for identifying polar cells. Cells in between two thresholds are indicated as 'Other'. **Panel C:** The percent proportions of polar, uniform, and other cells across the three cell types and their key biomarkers. **Panel D:** For CD8 T cells, representative examples of polar and uniform CD8 T cells are shown, and color-coded by relevant biomarkers.

2.2. *Polarized cell neighborhoods are more enriched with tumors*

Next, we explore whether polar immune cells might exhibit differences in their cellular neighborhoods with respect to uniform cells of the same type. Examples of a polarized CD8 T cell adjacent to a tumor cell (left) or not (right) are shown in **Figure 2A**. We found that, for CD8 T cells, CD4 T cells, and B cells, tumor cells were consistently enriched in the immediate cell neighborhoods of polar cells versus uniform cells (**Figure 2B**). Conversely, we also find that cells with tumor cell neighbors are more likely to be polar (**Supp. Table 1**). Given that polar expression can indicate antigen engagement during contact with tumor cells [19], this provides evidence that polarity is a biologically significant biomarker.

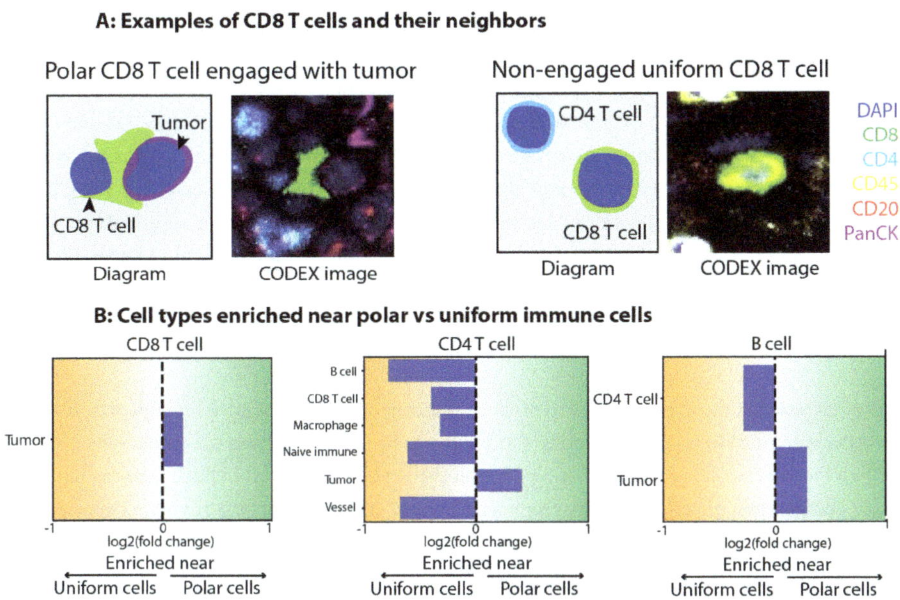

Fig 2: Tumor cells are more present next to polar cells versus uniform cells. **Panel A:** Diagrams and mIF images illustrating two possible states of CD8 T cells. Left: A CD8 T cell engaged with a tumor cell, with polar expression of CD8 at the immunological synapse. Right: A uniformly expressed CD8 T cell, with no tumor engagement. **Panel B:** Since polarity may be indicative of tumor engagement, we measure the cell type composition of neighborhoods around polar versus uniform cell types. We find that tumor cells are consistently and significantly more enriched in polar cell neighborhoods versus uniform cell neighborhoods for all three immune cell types. We compute bootstrapped 95% confidence intervals for each neighboring cell type and only show cell types with significant log fold changes.

2.3. *Metric control experiments*

In addition to visual validation, we perform tests to validate that the metric distribution is not explained by simple technical covariates or noise. We find that polarity cannot be simply explained by significantly more crowded cell neighborhoods (**Supp Figure 1A**) or differences in cell size (R^2 of 0.08, **Supp Figure 1B**). During antigen engagement, multiple biomarkers are known to jointly express at the site of the immunological synapse [20]. **Supp. Figure 1C** measures the correlation of polarity scores between all pairs of biomarkers as expressed in all T cells, B cells, and tumor cells. CD3e, a biomarker known to express during engagement, is jointly polarized with CD4 and CD8, while, PanCK, a biomarker not known to be active during antigen engagement, does not correlate with CD20, CD3e, CD4, or CD8. We note that the observed co-polarity between CD4 and CD8 is likely due to expression from neighboring cells that are being captured by our algorithm as originating from the same cell, an artifact that occurs in a small fraction of T cells (**Supp. Figure 1D**).

2.4. *Polarity-defined cell types improve model prediction of survival outcomes*

To demonstrate that the newly classified polar or uniform cell subtypes have biological or clinical relevance, we utilize deep learning models to predict patient survival from cell phenotypes. We train two models: a 3-layer multi-layer perceptron (MLP) neural network, which takes as input the percent composition of cell types per sample and predicts a binary outcome (five-year survival); and a graph-based neural network (GNN) that takes as input a 3-hop neighborhood of cells centered around a single cell, and predicts the survival status of the sample from which the neighborhood of cells originated. Both models show modest but consistent improvement in performance across three distinct studies and two disease types after including the 6 new cell types (**Table 1**). **Supp. Table 2** shows ablations where the MLP model is trained on each polar/uniform cell type individually. Of note, a model is trained with Ki67 polarity in tumor cells as a negative control (since Ki67 is not known to express polarly) and demonstrates no improvement over the baseline. Finally, we use the percent of polar cells per sample and compute the AUROC in its usefulness in predicting survival outcomes in **Supp. Table 3** and find that even this simple metric alone has predictive accuracy above chance.

Survival status	UPMC-HNC	Stanford-CRC	DFCI-HNC (using UPMC-HNC model) generalization
MLP			
- Baseline	0.759 (0.053)	0.538 (0.132)	0.655 (0.074)
- Adding polarity-defined cell subtypes	**0.803 (0.049)**	**0.559 (0.135)**	**0.672 (0.072)**
GNN			
- Baseline	0.839 (0.048)	0.684 (0.092)	0.853 (0.046)
- Adding polarity-defined cell subtypes	**0.856 (0.045)**	**0.743 (0.090)**	**0.880 (0.051)**

Table 1: Adding polarity-specific cell types improves patient survival prediction in machine learning models. To validate the usefulness of the polarity-specific cell types derived from our polarity measurement method, we train two models to predict patient survival status, with and without the additional cell types (polar and uniform CD8 T cells, CD4 T cells, B cells). Each cell represents the AUC of the model's prediction of patient survival. We observe that adding the additional cell types improves model performance across three datasets, three clinical sites, and two disease types. Standard deviations of bootstrapped samples are reported in parentheses. Predictions are generated at the sample level.

2.5. *Presence of polar cells improves patient survival with* in silico *models under label and spatial permutations*

In the permutation experiments shown in **Figure 3** and **Supp. Table 4,** the GNN model predicts significantly worse survival in tumor microenvironments where the subtype of the immune cells are flipped from polar to uniform **(Figure 3B)**. The inverse is also true; the predicted survival improves when cells are flipped from uniform to polar. Even when fixing the cell type composition, dispersing the location of the immune cells away from the tumor cells results in a decrease in predicted survival (and vice versa) **(Figure 3C)**. These results suggest that polar immune cells are important not simply for their presence in a sample, but for their proximity to tumor cells.

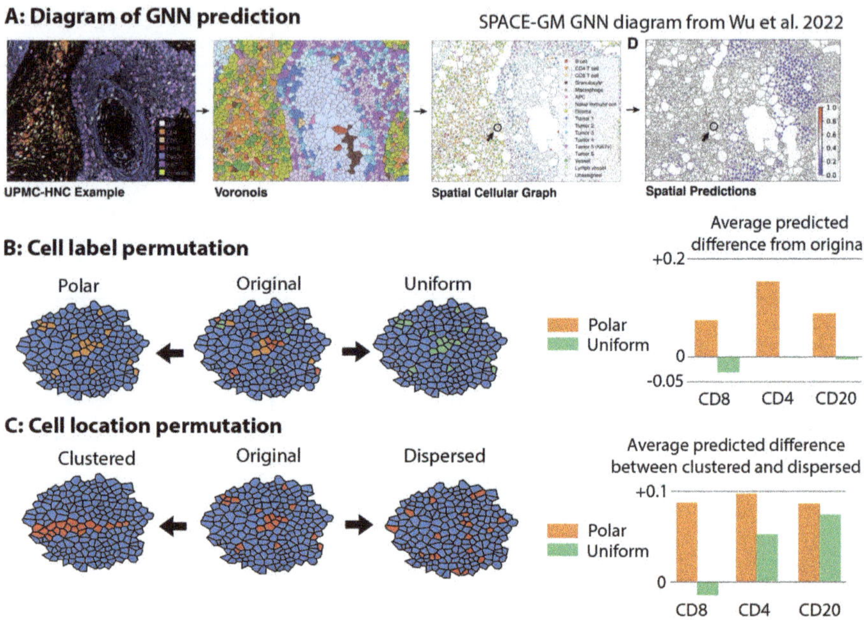

Fig 3: *In silico* experimentation reveals that polar cells are correlated with positive patient outcomes. **Panel A:** A schematic of a graph neural network described in *Wu et al. 2022.* A mIF sample is represented by a Voronoi diagram, which is projected into a spatial graph. A graph neural network is trained to predict survival outcomes based on 3-hop cellular neighborhoods. **Panel B:** Using a trained GNN, we perform label permutation on each sample graph, where the subtype of each immune cell is flipped to either polar or uniform, and the averaged model prediction is measured. Even when fixing spatial neighborhoods, we observe an increased predicted survival probability when cells are polarized, and a slight decrease when cells are turned into uniform states. **Panel C**: Now, fixing the cell types, polar and uniform cell neighborhoods are sampled and spatially permuted. We observe, on average, a larger increase in predicted survival when polar cells are dispersed from the clustered state than with uniform cells.

3. Methods

3.1. *Datasets*

Our primary dataset consists of 308 samples from 81 patients with head and neck squamous cell carcinomas at the University of Pittsburgh Medical Center (UPMC-HNC). Two external validation datasets are used: a colorectal cancer dataset with 292 samples from 161 patients from Stanford

University (Stanford-CRC) to demonstrate generalization to another disease; and a head and neck squamous cell carcinomas dataset with 112 samples from 29 patients from Dana Farber Cancer Institute (DFCI-HNC) to demonstrate generalization to an additional clinical site. The number of samples, patients, coverslips, and total cells in each dataset is described in **Supp. Table 5A**. Phenotype annotations for UPMC-HNC are described in **Supp. Table 5B**. Full CODEX data acquisition and preparation details are described in **Supp. Methods**. UPMC-HNC is chosen as the primary training and evaluation dataset as it contains the largest number of samples, coverslips, and total cells. We evaluate our models on held-out coverslips not seen during training to assess model robustness to technical artifacts across coverslips.

The UPMC-HNC and Stanford-CRC datasets have one held-out coverslip for model validation and one held-out coverslip for model evaluation. The Stanford-CRC dataset has half of one coverslip randomly split and held out for model validation and one held out for model evaluation. The DFCI-HNC dataset has one coverslip randomly split by patients for model evaluation.

3.2. Biomarker expression preprocessing

Single-cell expression was computed for each biomarker by 1. applying a deep learning cell segmentation algorithm (DeepCell) [21] on the DAPI biomarker channel (nuclear stain) to obtain nuclear segmentation masks; 2. successively dilating segmentation masks by flipping pixels each time with a probability equal to the fraction of positive neighboring pixels (repeated 9 times); 3. computing the mean expression value across pixels within the single cell; and 4. normalizing the expression values across all cells in a sample using quantile normalization and arcsinh transformation followed by a z-score normalization:

$$zscore(arcsinh(\frac{x}{5q_{0.2}(x)}))$$

Where $zscore$ is defined given μ and σ, the mean and standard deviation across all cell expression values in the sample:

$$zscore(x) \;=\; \frac{x-\mu}{\sigma}$$

x is the vector of a biomarker's values in a sample, $arcsinh$ is the inverse hyperbolic sine function; and $q_{0.2}(x)$ is the 20th percentile of x.

3.3. Image patch generation

After preprocessing (tile & cycle alignment, stitching, deconvolution, and background correction) CODEX data is available as multichannel OME-TIFF files, with each image channel corresponding to the fluorescence signal (expression) of a distinct biomarker probe. To prepare the input image patches for the deep learning model, we perform the following: All pixel values for a biomarker in a sample are normalized using ImageJ's AutoAdjust function.

3.4. Cell type ground truth and predictions

To produce cell type labels, we first obtained a cells-by-features biomarker expression matrix - for each marker, we took the average signal across all pixels in a segmented cell. This matrix was normalized and scaled as described above, and then principal component (PC) analysis was performed. We constructed a nearest-neighbor graph (k = 30) of cell expression in PC space with the top 20 PCs, then performed self-supervised graph clustering [22] on the result. Clusters were manually annotated according to their cell biomarker expression patterns. This procedure was performed on a subset of 10,000 cells and subsequently used to train a kNN algorithm. This algorithm was used to transfer labels to the entire dataset. The cell type labels that were used are: Tumor (CD15+, CD20+, CD21+, Ki67+, Podo+, Other), Naive immune cell, Granulocyte, Vessel, CD4 T cell, Macrophage, CD8 T cell, Stromal / Fibroblast, APC, Lymph vessel, and B cell.

3.5. Calculating polarity score

Our polarity measurement methodology is described in **Figure 1**. The segmentations are used to calculate cell center coordinates. For each cell, a 40px square patch is extracted around the center pixel. Several de-noising steps are first taken: 1. low/background values are zeroed out (values < 0.1), 2. biomarker expressions that spatially overlap with the DAPI channel are subtracted out in both the center and neighborhood cells.

We then transform the patch from cartesian coordinates to polar coordinates using the scikit package (*skimage.transform.warp_polar*). The polar image is then summed along the y-axis, producing a 1-dimensional vector. An additional refining step is taken where cells are assigned 'other' if the vector 1. sums to 0 either along the x- or y-axis, 2. does not contain multiple unique values, or 3. has a mean less than 0.02. Finally, the vector is normalized within a [0,1] range and sorted in ascending order, and a score is computed by subtracting the AUC of the sorted vector from 1.

On its own, the polarity score is difficult to interpret and incorporate into existing analysis pipelines that rely on discrete cell phenotypes. Thus, we define three cell subtypes based on the polarity score value: uniform (cells with a polarity score below a threshold), polar (cells with a polarity score above a threshold), and other (cells that fall in between both thresholds). To obtain optimal thresholds for defining polar and uniform cell types from the polarity scores, we perform a two-dimensional grid search on the MLP model and select the pair of values that yielded the highest validation AUC score in the survival prediction task. From this process, we obtain 0.94 and 0.8 as the polar and uniform threshold cutoffs, respectively (**Figure 1B**). **Figure 1C** shows the polarity distributions after thresholding.

These thresholds are used to define six new cell subtypes for polar and uniform CD8 T cells, CD4 T cells, and B cells. CD8, CD4, and CD20 were used as the representative surface biomarkers for each of the three cell types, respectively. These three cells and biomarkers were chosen as they are known in the literature to exhibit polar expression during engagement [19].

3.6. Machine learning models

We train two machine learning models to evaluate the benefit of including the six newly defined cell subtypes. First, we use a 3-layer multilayer perceptron (MLP) neural network that accepts the percent composition of cell types per sample and predicts binary 60-month survival. Each layer contains 256 nodes followed by a LeakyReLU [23] activation function. Each model is trained with binary cross-entropy loss across 200 epochs and a learning rate of 0.001.

Second, we train a graph-based neural network (GNN) [24] that takes as input 3-hop cell neighborhoods and predicts neighborhood-level survival status (**Figure 2A**). This model transforms the structure of each sample into a graph network, where cells are connected by edges to neighboring cells. It then pools information about the neighboring cells' cell types to output an outcome probability score for each cell. The sample predictions are generated by averaging the scores across all cells in that sample. Model training details follow the procedures described in Wu et al. [24].

Each of the models is trained first on the original 16 cell types (baseline) and then trained using the 6 additional cell types. In both of these settings, each model is trained and evaluated on the UPMC-HNC and Stanford-CRC datasets. An additional evaluation is performed on the DFCI-HNC dataset using the UPMC-HNC trained model.

3.7. Permutation experiments

To assess the effect of polar/uniform cell types on the GNN model's survival predictions, we perform several permutation experiments. In the first experiment (**Figure 3A**), we flip the cell type label of all immune cells to either polar or uniform and evaluate the predicted survival probability in each scenario. In the second experiment, we sample random subgraphs containing immune cells and tumor cells for CD4 T cells, CD8 T cells, and B cells. Then, we flip all immune cells to either polar or uniform and perform a spatial permutation, where we shuffle the immune cells into either clustered (where all immune cells are neighbors) or dispersed (where immune cells are randomly located in the subgraph) orientations and evaluate the predicted probabilities for each orientation.

4. Discussion

We describe a robust, interpretable subcellular morphology metric that reflects macro-biological states. Although our results do not conclude that these polarity events definitively quantify immune synapses, they do suggest that such measurements represent biologically relevant signals in tumor microenvironments. Though our described method is performed on CODEX data, it can similarly be applied to other lower-plexed imaging techniques like IHC that include a nuclear marker (i.e. DAPI) and one or more surface biomarkers.

To date, there has not been prior consensus demonstrating that biological events like engagement and activation or exhaustion can be reliably observed at the standard resolution of mIF imaging. Potential confounders include bleedover, sample slicing artifacts, measurement noise, cell size, and density of neighboring cells. We address this by conducting several negative control experiments and find that these factors alone do not adequately explain the signal present in our polarity measurements.

One counter-hypothesis is that polarity measurements serve as a proxy for neighborhood information -- i.e. the presence of certain cell types or spatial arrangements. Another possibility is that they are primarily an imaging artifact (for instance, irregular borders due to slicing). To test these, we trained a GNN that incorporates local neighborhood information into its predictions and then introduced the polar and uniform cell types. The results show that the new cell types improve performance even in models that have access to cell neighborhood information, suggesting that it introduces information beyond the neighborhood cell type composition or spatial arrangement of cells.

Further experimental evidence is required to define these observations as a specific biological phenomenon, i.e. T cell engagement. However, we believe that this work provides evidence of the importance of measuring and incorporating subcellular polarity information into tissue microenvironment analyses, and represents an important step toward a personalized understanding of disease states, drug response, and patient prognosis.

Supplementary Materials: All supplementary tables, figures, and data are available at: https://docs.google.com/document/d/1n97PEC2kq41fNOWMXrOASyd42DZ0_nUapnEs6HoeJ8c
Code Availability: Code for replicating the experiments in this paper is present in this code repository: https://gitlab.com/enable-medicine-public/polarity

References

1. Goltsev, Y. *et al.* Deep Profiling of Mouse Splenic Architecture with CODEX Multiplexed Imaging. *Cell* **174**, 968–981.e15 (2018).
2. Hickey, J. W., Tan, Y., Nolan, G. P. & Goltsev, Y. Strategies for Accurate Cell Type Identification in CODEX Multiplexed Imaging Data. *Front. Immunol.* **12**, 727626 (2021).
3. Garcia, E. & Ismail, S. Spatiotemporal Regulation of Signaling: Focus on T Cell Activation and the Immunological Synapse. *Int. J. Mol. Sci.* **21**, (2020).
4. Weber, M. S. *et al.* B-cell activation influences T-cell polarization and outcome of anti-CD20 B-cell depletion in central nervous system autoimmunity. *Ann. Neurol.* **68**, 369–383 (2010).
5. Negulescu, P. A., Krasieva, T. B., Khan, A., Kerschbaum, H. H. & Cahalan, M. D. Polarity of T cell shape, motility, and sensitivity to antigen. *Immunity* **4**, 421–430 (1996).
6. Talhouni, S. *et al.* Activated tissue resident memory T-cells (CD8+CD103+CD39+) uniquely predict survival in left sided 'immune-hot' colorectal cancers. *Front. Immunol.* **14**, 1057292 (2023).
7. Ma, J. *et al.* PD1Hi CD8+ T cells correlate with exhausted signature and poor clinical outcome in hepatocellular carcinoma. *J Immunother Cancer* **7**, 331 (2019).
8. Cheng, D. *et al.* Proliferative exhausted CD8+ T cells exacerbate long-lasting anti-tumor effects in human papillomavirus-positive head and neck squamous cell carcinoma. *Elife* **12**, (2023).
9. Zhu, X., Yao, J. & Huang, J. Deep convolutional neural network for survival analysis with pathological images. in *2016 IEEE International Conference on Bioinformatics and Biomedicine (BIBM)* 544–547 (2016).
10. Alom, M. Z. *et al.* Advanced Deep Convolutional Neural Network Approaches for Digital

Pathology Image Analysis: a comprehensive evaluation with different use cases. *arXiv [cs.CV]* (2019).

11. Lu, M. Y. *et al.* AI-based pathology predicts origins for cancers of unknown primary. *Nature* **594**, 106–110 (2021).
12. Wang, H. *et al.* Mitosis detection in breast cancer pathology images by combining handcrafted and convolutional neural network features. *J Med Imaging (Bellingham)* **1**, 034003 (2014).
13. Zhang, W. *et al.* Identification of cell types in multiplexed in situ images by combining protein expression and spatial information using CELESTA reveals novel spatial biology. *bioRxiv* 2022.02.02.478888 (2022) doi:10.1101/2022.02.02.478888.
14. Li, Z., Wang, T., Liu, P. & Huang, Y. SpatialDM for rapid identification of spatially co-expressed ligand-receptor and revealing cell-cell communication patterns. *Nat. Commun.* **14**, 3995 (2023).
15. Wilk, A. J., Shalek, A. K., Holmes, S. & Blish, C. A. Comparative analysis of cell-cell communication at single-cell resolution. *Nat. Biotechnol.* (2023) doi:10.1038/s41587-023-01782-z.
16. Park, H. Y., Trcek, T., Wells, A. L., Chao, J. A. & Singer, R. H. An unbiased analysis method to quantify mRNA localization reveals its correlation with cell motility. *Cell Rep.* **1**, 179–184 (2012).
17. Stueland, M., Wang, T., Park, H. Y. & Mili, S. RDI Calculator: An Analysis Tool to Assess RNA Distributions in Cells. *Sci. Rep.* **9**, 8267 (2019).
18. Savulescu, A. F. *et al.* Interrogating RNA and protein spatial subcellular distribution in smFISH data with DypFISH. *Cell Rep Methods* **1**, 100068 (2021).

19. Juzans, M., Cuche, C., Di Bartolo, V. & Alcover, A. Imaging polarized granule release at the cytotoxic T cell immunological synapse using TIRF microscopy: Control by polarity regulators. *Methods Cell Biol.* **173**, 1–13 (2023).

20. Hong, M. M. Y. & Maleki Vareki, S. Addressing the Elephant in the Immunotherapy Room: Effector T-Cell Priming versus Depletion of Regulatory T-Cells by Anti-CTLA-4 Therapy. *Cancers* **14**, (2022).

21. Greenwald, N. F. *et al.* Whole-cell segmentation of tissue images with human-level performance using large-scale data annotation and deep learning. *Nat. Biotechnol.* **40**, 555–565 (2022).

22. Traag, V. A., Waltman, L. & van Eck, N. J. From Louvain to Leiden: guaranteeing well-connected communities. *Sci. Rep.* **9**, 5233 (2019).

23. Xu, B., Wang, N., Chen, T. & Li, M. Empirical Evaluation of Rectified Activations in Convolutional Network. *arXiv [cs.LG]* (2015).

24. Wu, Z. *et al.* SPACE-GM: geometric deep learning of disease-associated microenvironments from multiplex spatial protein profiles. *bioRxiv* 2022.05.12.491707 (2022) doi:10.1101/2022.05.12.491707.

KombOver: Efficient k-core and K-truss based characterization of perturbations within the human gut microbiome

Nicolae Sapoval†, Marko Tanevski and Todd J. Treangen
Department of Computer Science, Rice University,
Houston, TX 77005, USA
†*E-mail: nsapoval@rice.edu*

The microbes present in the human gastrointestinal tract are regularly linked to human health and disease outcomes. Thanks to technological and methodological advances in recent years, metagenomic sequencing data, and computational methods designed to analyze metagenomic data, have contributed to improved understanding of the link between the human gut microbiome and disease. However, while numerous methods have been recently developed to extract quantitative and qualitative results from host-associated microbiome data, improved computational tools are still needed to track microbiome dynamics with short-read sequencing data. Previously we have proposed KOMB as a *de novo* tool for identifying copy number variations in metagenomes for characterizing microbial genome dynamics in response to perturbations. In this work, we present KombOver (KO), which includes four key contributions with respect to our previous work: (i) it scales to large microbiome study cohorts, (ii) it includes both k-core and K-truss based analysis, (iii) we provide the foundation of a theoretical understanding of the relation between various graph-based metagenome representations, and (iv) we provide an improved user experience with easier-to-run code and more descriptive outputs/results. To highlight the aforementioned benefits, we applied KO to nearly 1000 human microbiome samples, requiring less than 10 minutes and 10 GB RAM per sample to process these data. Furthermore, we highlight how graph-based approaches such as k-core and K-truss can be informative for pinpointing microbial community dynamics within a myalgic encephalomyelitis/chronic fatigue syndrome (ME/CFS) cohort. KO is open source and available for download/use at: https://github.com/treangenlab/komb

Keywords: metagenomics; graph based methods; anomaly detection.

1. Introduction

Metagenomics, the study of the genomes of microbes that inhabit a microbiome, offers an unprecedented and highly granular view into the interaction between host-associated microbiomes and host disease phenotypes. Numerous computational tools now exist to uncover the taxonomic composition and functional profiles of human host associated microbiomes [1–4]. Of particular relevance to this work, higher taxonomic and functional diversity of the microbiota is associated with healthy individuals, while lower diversity correlates with disease states [5–8]. Furthermore, with the growing number of metagenome assembled genomes (MAGs) [9, 10] the association between the genomic composition of microbial communities and the host health has become better quantified and understood [11, 12]. However, metagenomic assembly from short reads remains a challenge in highly repetitive regions of bacterial genomes [13–15], and among closely related strains of a given bacterial species [16, 17]. Genomic repeats arising from horizontal gene transfer or duplication events have been associated with bacterial adaptation and evolution [18–20], functional diversification [20], and pathogenesis [18]. Recent advances in long-read sequencing offer a path to resolution of complex inter- and intra-genomic repeats in microbial communities [21, 22]. However, limitations in high molecular weight DNA extraction [23] and financial cost of hybrid or high-quality long-read approaches poses a roadblock for large scale studies involving long-read sequencing. Additionally, a large existing corpus of metagenomic sequencing data consists predominantly of short paired-end reads, thus warranting the development of novel methods that can better capture and quantify inter- and intra-genomic repeat dynamics and flux.

To address this challenge, we have previously proposed the software KOMB [24] to extract high copy number sequences of potential biological significance in the microbial communities from the short paired-end read metagenomic sequencing data, expanding on prior approaches [25–27]. As the genomic diversity of a bacterial community has been correlated with host health, we hypothesize that the corresponding inter- and intra-genomic repeat structures can act as a "biomarker" for host health. Our prior work highlighted the ability of KOMB to detect shifts in the microbial community associated with antibiotic treatment and bowel cleanse, as well as identify associations between observed shifts and key bacterial members of pre- and post-FMT bacterial communities. Additionally, similarly to *de novo* assembly methods, KOMB is a database independent tool, and hence it avoids database biases [28]. However, unlike the common assembly approaches [29, 30] KOMB does not simplify the compacted de Bruijn graph, thus retaining the diversity originally present in the sequencing data. Furthermore, in contrast to k-mer profiling methods [1, 2], KOMB offers a set of genomic sequences that can be annotated for downstream analyses. Thus, KOMB bridges the gap between fast profiling methods that either require a database or do not yield sequence units that can be readily annotated, and computationally expensive assembly-based approaches.

For the purpose of identifying key sequences in the graph, KOMB employs the graph mining concept of k-core decomposition, which iteratively determines densely connected graph components. Previously, we had not investigated the set of sequences contained in the core of the graph as a whole, only focusing on sequences with high Core-A anomaly score [31] which captures deviations in coreness/degree ratios of a vertex. In KombOver (KO), we introduce and implement analysis of the maximal K-truss subgraph. Similarly to the vertices of the maximal k-core, the vertices of the maximal K-truss have been shown to have strong spreading (i.e. centrality) [32] which can be relevant in certain biological contexts as an alternative to betweenness centrality measure [25, 33].

One of the limitations of our prior work was its scalability to large metagenomic studies. In particular, the construction of the main data structure employed by KOMB, the hybrid unitig graph (HUG) incurred a high computational cost. It resulted in run times ranging from over an hour per single metagenomic sample, resulting in overwhelming computational costs for thousands to tens of thousands of samples. To address this limitation, in this work we propose a set of improvements to the HUG construction and analysis in KO aimed at enabling large-scale processing of genomic data and characterization of phenotype-associated dynamics. Furthermore, in addition to computational improvements, we provide a more extensive characterization of HUGs within the context of bacterial pangenomics, and draw parallels between pangenome graphs constructed from MAGs and HUGs. In order, to assess our tool, we analyzed short-read metagenomic sequencing data from a cohort of controls and patients with myalgic encephalomyelitis/chronic fatigue syndrome (ME/CFS), previously published by Xiong *et al.* [34]. Additionally, we have benchmarked KOMB on integrative human microbiome project [35] inflammatory bowel disease (IBD) cohort samples, as well as human genome sequencing data from Genome in a Bottle project [36], to demonstrate KO's scalability to both large data volumes, and complex repeat architectures.

2. Results

2.1. *Hybrid unitig graphs*

Hybrid unitig graphs (HUGs) are an extension of compacted de Bruijn graphs [37–39] used as a primary data structure in *de novo* de Bruijn graph assembly approaches [29, 30, 40]. The key addition in HUGs is presence of paired-end edges coming from the sequencing read data. Hence, while during assembly, the de Bruijn graphs are iteratively simplified to construct MAGs [29, 30] in KO denser and more complex HUGs are analyzed directly to facilitate the capture of repeat dynamics within microbial communities. Conversely, pangenome graphs are typically constructed from annotated genome assemblies, and capture high-level variation in synteny and copy numbers of gene clusters across related microbial genomes. In this context, HUGs bridge the gap between exact compaction achieved in the compacted de Bruijn graphs and high-level genomic variation representation of pangenome graphs [41–43].

Thus, compared to compacted de Bruijn graphs (Figure 1c) HUGs offer additional connectivity information based on local similarity and inferred adjacency between unitigs. Compared to the pangenome graphs, HUGs do not require neither complete genome assembly nor identification of putative gene clusters (Figure 1d) and hence can be constructed more efficiently from short paired-end read data.

2.2. *Analysis of an ME/CFS cohort*

First, we compared the overall distributions of the number of unitigs reconstructed from control samples with the ones from patients with short and long-duration ME/CFS. We observe that all samples in the control cohort contain more than 50,000 unitigs per sample, with 85 out of 92 samples containing between 50,000 and 400,000 unitigs (Figure 2). In contrast, 3 samples derived from patients with short-term ME/CFS contain less than 50,000 unitigs, and 62 out of 73 samples in this category contain up to 250,000 unitigs (Figure 2). Similarly, the data for long-term ME/CFS contains 6 samples with less than 50,000 unitigs, and 68 out of 73 the samples fall into the 0 to 300,000 unitigs range (Figure 2).

Next, we have designated unitigs with Core-A anomaly scores three standard deviations (3σ) above their corresponding sample's mean (μ) as the anomalous unitigs for the corresponding samples. We have explored the distribution of degrees (Figure 3A) in the anomalous unitigs

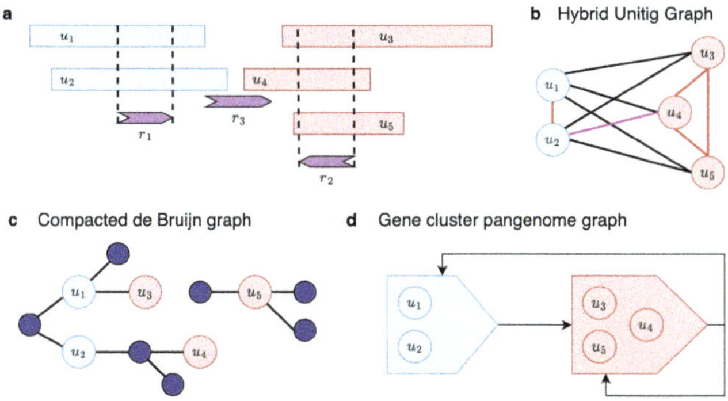

Fig. 1. **(a)** A set of 5 unitigs labeled u_1, u_2, u_3, u_4, and u_5 with the corresponding read mappings of r_1, r_2, and r_3. Note, that the read r_3 maps to both the end of unitig u_2 and the start of the unitig u_4. **(b)** A HUG corresponding to the unitigs and reads in (a). Edges marked in red are local similarity edges, and edges marked in black are adjacency edges arising from the paired reads (r_1, r_2). The magenta edge $\{u_2, u_4\}$ is an adjacency edge arising as a result of multi-mapping of a single read. **(c)** A schematic representation of a compacted de Bruijn graph. Dark blue nodes represent k-mers, while light blue and red nodes represent unitigs that have been compacted from unambiguous paths. **(d)** A schematic representation of a pangenome graph. Colored blocks represent gene clusters and arrows indicate possible paths through the gene sequences as indicated by corresponding genome assemblies.

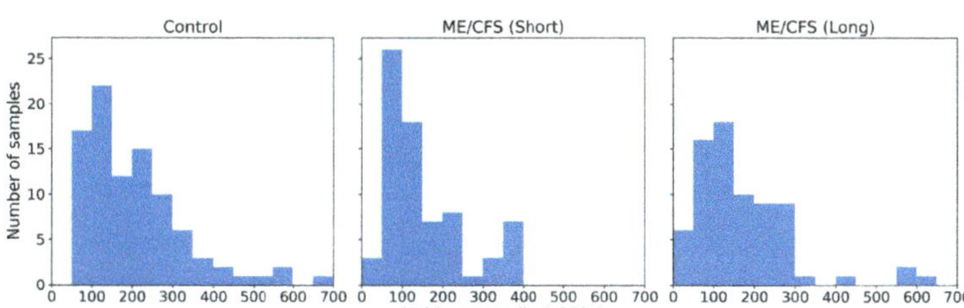

Fig. 2. Distribution of total unitig counts in HUGs constructed from control (left), ME/CFS short duration (center), and ME/CFS long duration (right) subjects's gut microbiome samples. Samples corresponding to short ME/CFS condition show lower absolute counts of unitigs, while those corresponding to the long ME/CFS are more similar to the controls. Both short and long ME/CFS associated samples have less high unitig count representatives.

based on the sample type and noted that the overall distributions are skewed to the left for all sample types. However, in the range of degrees from 250 to 1250, short-term ME/CFS samples exhibit sharper concentration towards the lower degree values than long-term ME/CFS and control samples. Additionally, in the 380-500 range of degrees long-term ME/CFS samples exhibit a more uniform distribution. The distributions in Figure 3A were tested for statistical difference using Kolmogorov-Smirnov (KS) test. All three pairwise distribution comparisons were significant with p-value $< 10^{-9}$. Since the degree of a unitig in an HUG depends on the number of locally similar unitigs and potential genomic adjacencies of it, this indicates that long-term ME/CFS communities have more anomalous highly connected unitigs.

We also investigated the distribution of coreness values in the anomalous unitigs grouped by condition (Figure 3B). Similarly to the degree distributions for the low coreness values (0-100) all three sample types agree. Analogously, short-term ME/CFS samples also have the distribution of the coreness skewed towards lower values. This agrees with the observations in Figure 2 and Figure 3A, as the lower overall unitig count (and hence a smaller graph), and lower degrees (which provide an upper bound on coreness) would result in lower coreness values. Long-term ME/CFS samples have several peaks in the distribution (coreness 180-200, 280-320) that are not observed in the controls. The distributions in Figure 3B were tested for statistical difference using KS test. All three pairwise distribution comparisons were significant with p-value $< 10^{-9}$. Since coreness is a proxy for the level of interconnectedness in a group of unitigs, this can indicate the presence of clusters of unitigs corresponding to either a complex repeat architecture or a high abundance of closely related organisms.

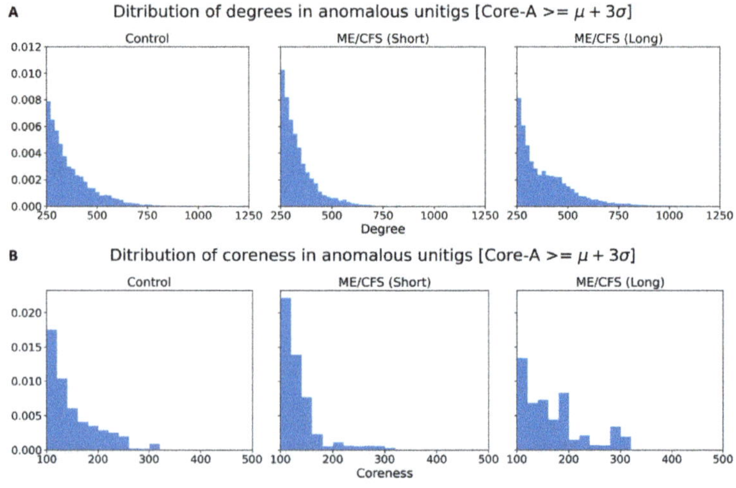

Fig. 3. **(A)** Distribution of the degrees of the unitigs that have Core-A anomaly score above $\mu + 3\sigma$ for their corresponding samples. Distributions for unitigs of degrees 0-250 are omitted for clarity. We note the samples corresponding to short-term ME/CFS have a distribution more skewed to the left. **(B)** Distribution of the coreness of the unitigs that have Core-A anomaly score above $\mu + 3\sigma$ for their corresponding samples. Distributions for unitigs of coreness 0-100 are omitted for clarity. Similarly to degree distribution, short-term ME/CFS samples show a skew towards lower coreness values. Additionally, long-term ME/CFS samples have more uniform distribution in the 100-200 range compared to controls and more unitigs in the 250-350 coreness range.

Next, we investigated the α-diversity at the species and genus level, as well as the overlaps between species and genus level classifications based on Kraken 2 [44] predictions for the anomalous unitigs in the three groups (Figure 4). We note that at both species and genus level the short-term ME/CFS samples exhibit a lower average α-diversity which can be indicative of dysbiosis. Additionally, at both species and genus levels, most taxonomic annotations are shared among the three cohorts. Still, the control group has consistently more unique taxa identified, further supporting the role of diverse microbial community composition in healthy individuals. Similarly, the long-term ME/CFS cohort has more unique taxa than the short-term ME/CFS cohort, indicating partial recovery from the dysbiosis.

We next compared the results for α-diversity and the overlaps obtained from anomalous unitigs, to the same information computed for the unitigs in the highest K-truss (Figure 5). We note that unlike in the case of general anomalous unitigs, those that belong to the highest K-truss show more similarity in the α-diversity between control and short-term ME/CFS samples, with long-term ME/CFS sample being the outlier (Figure 5A, B). Additionally, the total α-diversity in the trusses (Figure 5A, B) is noticeably lower than in general anomalous unitigs (Figure 4A, B). This is expected given trusses are densely connected subgraphs of a HUG, and hence have higher propensity to represent closely related genomic segments. Higher α-diversity in the long-term ME/CFS trusses can be a potential indicator for functional enrichment with multiple taxa coding for the same function in the long-term ME/CFS microbiota. Furthermore, we observed that, while the number of species and genera shared between all three cohorts makes up a smaller fraction of the total classifications. Namely, while species shared between all three categories make up 27.5% (1808 of 6567) of all species identified in the anomalous unitigs of the three cohorts (Figure 4C), they make up only 24.1% (177 of 735) of all species identified in the trusses of the samples (Figure 5C). Analogously, the shared genera make up 55.0% (1008 out of 1834) of all classifications for anomalous unitigs, and only 33.0% (156 out of 473) of all classifications for truss unitigs.

Additionally, when individual KO profiles are visualized for samples matched by age, gender, and race (Figure 6) we observe more compact profiles for the disease-associated samples. This matches the dysbiosis hypothesis, with the long-term ME/CFS sample showing a more complex profile than the short-term one. In all samples shown in Figure 6 unitigs with high anomaly scores are the ones for which the degree is larger than the expected coreness. This pattern occurs when a unitig is flanked by varying genomic contexts across the metagenome, and hence indicate unitigs with high inter- and intra-genomic copy numbers.

2.3. *Computational performance*

The k-core decomposition algorithm runs in $O(|V| + |E|)$ time [45] and hence scales linearly with the size of the graph. This scaling is particularly attractive in metagenomic communities, where the number of edges $|E|$ is proportional to the number of vertices $|V|$. In the case of more complex repeat architectures, such as Alu repeats in human genome, the number of edges is be proportional to the square of the number of vertices. Compared to Brandes's algorithm for betweenness centrality (a common algorithm for detecting influential nodes in a network) [46] k-core decomposition algorithm is significantly faster. The asymptotic time complexity of

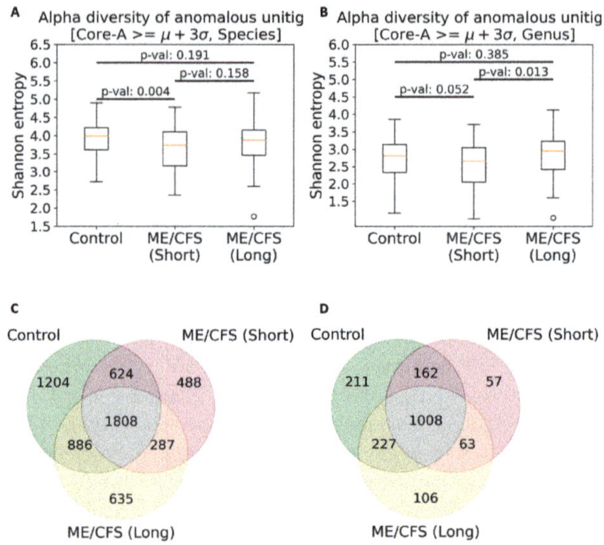

Fig. 4. **(A, B)** Distribution of alpha diversity (Shannon entropy) for anomalous unitigs that have anomaly score three standard deviations above the mean for the respective sample grouped by the condition and duration. p-values from Welch's t-test for equality of means are displayed above the boxplots. **(A)** Alpha diversity of samples associated with the ME/CFS condition is lower than that of control samples. Long ME/CFS samples on the other hand do not appear to be noticeably distinct from the control ones. Entropy was calculated based on the unitigs for which Kraken 2 provided species-level classification. **(B)** Alpha diversity of samples associated with the ME/CFS condition is lower than that of control samples and long ME/CFS samples. Long ME/CFS samples on the other hand do not appear to be noticeably distinct from the control ones. Entropy was calculated based on the unitigs for which Kraken 2 provided genus level classification (species annotations are rolled up into respective genus). **(C)** Venn diagram representing intersections between sets of species identified in the control, short ME/CFS, and long ME/CFS sample collections. **(D)** Venn diagram representing intersections between sets of genera identified in the control, short ME/CFS, and long ME/CFS sample collections.

Brandes's algorithm for unweighted graphs is $O(|E||V|)$, which even in the $|E| \sim \alpha|V|$ regime, leads to $O(|V|^2)$ complexity compared to $O(|V|)$ for the k-core decomposition.

The K-truss decomposition has an asymptotic time complexity of $O(|E|^{1.5})$ [47], making it slower than the k-core decomposition. Nevertheless, since we are only interested in the vertices contained in the maximal K-truss, we make the simplification of running the K-truss decomposition on only the maximal k-core of the graph, similar to the prior work [32].

Empirically, in addition to analyzing the ME/CFS data from Xiong *et al.* study [34], we have also benchmarked KO on the IBD data [35] from integrative HMP, as well as chromosome

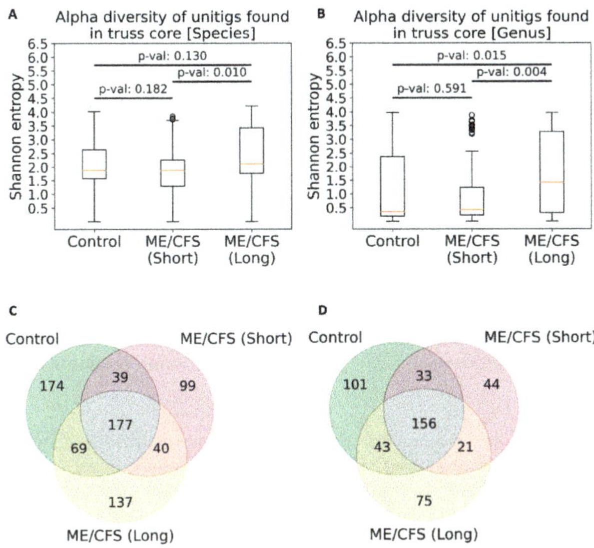

Fig. 5. **(A, B)** Distribution of alpha diversity (Shannon entropy) for anomalous unitigs that belong to the highest K-truss. p-values from Welch's t-test for equality of means are displayed above the boxplots. **(A)** Alpha diversity of samples associated with the ME/CFS condition is lower than that of control samples. Long ME/CFS samples on the other hand do not appear to be noticeably distinct from the control ones. Entropy was calculated based on the unitigs for which Kraken 2 provided species level classification. **(B)** Alpha diversity of samples associated with the ME/CFS condition is lower than that of control samples and long ME/CFS samples. Long ME/CFS samples on the other hand do not appear to be noticeably distinct from the control ones. Entropy was calculated based on the unitigs for which Kraken 2 provided genus level classification (species annotations are rolled up into respective genus). **(C)** Venn diagram representing intersections between sets of species identified in the control, short ME/CFS, and long ME/CFS sample collections. **(D)** Venn diagram representing intersections between sets of genera identified in the control, short ME/CFS, and long ME/CFS sample collections.

21 and chromosome 11 aligned reads from human genome HG002 from the Genome in a Bottle project [36]. The choice of human genome sequencing data is motivated by highly repetitive complex Alu regions present in the genome, hence the regime in which $|E| \sim \alpha|V|^2$ is in the HUG. All benchmarking was performed on a Ubuntu 18.04.6 LTS system with Intel(R) Xeon(R) Gold 5218 CPUs and 312GB of RAM and all runs used 60 threads. The results of benchmarking are summarized in Table 1.

The results in Table 1 showcase that resulting graph edge density is an important component of the overall computational performance, as indicated by a high run time value for the HG002 chromosome 11 experiment. Compared to the original KOMB implementation, we

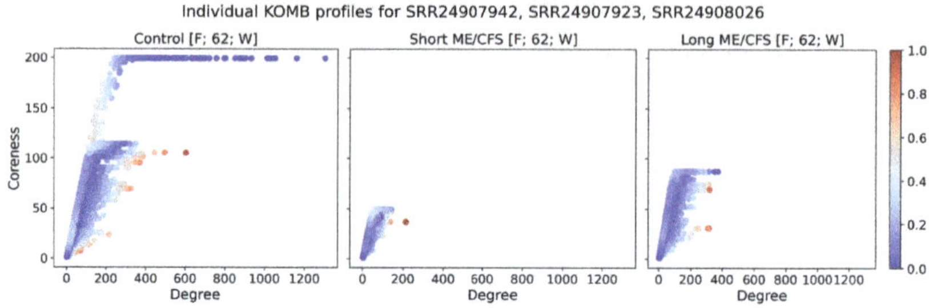

Fig. 6. Individual KO profiles (color: normalized Core-A anomaly score) for three samples from the ME/CFS cohort matched by age, gender and race. We observe a more complex profile in the control sample, while short- and long-term ME/CFS show compact profiles associated with lower bacterial genomic diversity. In all three samples, anomalous unitigs are predominantly high in degree compared to their coreness.

Table 1. Performance of KO on metagenomic and human genome datasets. Total dataset size refers to the cumulative size of all data processed, while average sample size describes the mean size of a single sample in a dataset. Analogously average runtime refers to mean time to process a single sample, while total runtime refers to cumulative time spent analyzing the dataset sequentially.

Dataset	# samples	Total dataset size (GB)	Average sample size (GB)	Average wall clock runtime (hrs)	Total wall clock runtime (hrs)
ME/CFS cohort	238	2,422	10.18	0.11	26.42
iHMP IBD	540	3,120	5.78	0.19	104.82
HG002 chr21 (300x)	1	26.17	-	-	0.71
HG002 chr21 (250bp)	1	5.90	-	-	0.14
HG002 chr11 (250bp)	1	20.43	-	-	1.70

achieve up to a 3-fold speed up for metagenomic samples containing an average of 16 million reads [24]. Additionally, we have performed a head-to-head comparison of KOMB and KO on a Zymo mock community sequenced by DOE Joint Genome Institute (BioProject Accession: PRJNA699918). We chose the Zymo mock community due to the large sample size (42 GB) and relatively simple genomic structure, allowing us to focus on the HUG construction performance, which we identified as a bottleneck, rather than the efficient k-core decomposition part of the analysis. On this data KOMB required a total of 7h16m of wall clock time (CPU time: 273h34m) and 48.28 GB of RAM to produce the final results, while KO required a total of 1h4m of wall clock time (CPU time: 14h34m) and 156.18 GB of RAM, note that both versions were ran with 40 threads for this experiment. The speedup is the result of three major changes in KO: (1) replacement of ABySS [48] with GGCAT [39] for the unitig construction, (2) change from BWA MEM to BWA MEM 2 for read mapping, and (3) improved parallelization in the KOMB codebase.

3. Discussion

In this work, we have provided a set of computational improvements to KOMB implemented in KO, and theoretical analysis of connections between HUGs and pangenome graphs. Additionally, we showcased the usage of KO on a ME/CFS patient cohort and identified disease-associated patterns. The dynamics inferred from KO profiles and taxa associated with important unitigs are concordant with the observations from a prior study [34]. Namely, we observe pronounced dysbiosis in short-term ME/CFS patients gut, and partial recovery from the dysbiosis in the long-term ME/CFS patients. We envision KO as an important tool to be integrated alongside existing approaches into clinically relevant microbiome studies. The key benefits of KO are: (a) selection of a small set of anomalous sequences without relying on taxonomy nor functional annotation, which can allow *de novo* analyses of these sequences and more sensitive detection of perturbations to host microbiome health, and (b) rapid profiling of a large number of samples, which can aid in the exploration of genotype-phenotype connections for large study cohorts.

An important next step is extending KO to an integrated approach that can annotate unitigs within the graph with associated transcriptomic or metabolomic information. Enriching the graph with multi-omic annotations can provide additional context for the nodes identified by KOMB as anomalous, enabling further functional associations to be extracted from the HUG structures. We believe that by adding -omics annotations KO can be further used to select genomic features relevant to the pathology, and hence enable better machine learning diagnostic tools. We also plan to add ability to distinguish between the edge types described in the Methods section and add the multi-omics annotations to the HUGs to directly extract hubs of functionally important genomic regions of a microbiome.

Additionally, it can be of interest to construct HUGs based on publicly available MAG catalogs as an annotation-rich reference for common community patterns identified in previous studies. We believe that this integrative large-scale approach can further illuminate mechanistic associations between microbiome and disease phenotypes.

4. Methods

4.1. *Hybrid unitig graph construction*

We begin construction of the HUG by constructing the underlying de Bruijn graph with a user-specified k-mer size parameter. The construction is done with the GGCAT [39], and the user can control the parameters exposed by the GGCAT command line interface. GGCAT produces a FASTA output file containing all maximal non-branching paths through the de Bruijn graph (unitigs). After unitigs are constructed the user has an option to specify an additional length based filtering step. Our recommended choice is setting this filter to be equal to the read length.

After construction and filtering, the final set of unitigs becomes the set of vertices in the HUG. Next, we perform read mapping of the input paired-end reads to the set of unitigs using BWA MEM 2 [49] v2.2.1 with the default parameters. We retain all read mappings for constructing the edges of the HUG. An edge is constructed between two unitigs u_1 and u_2 if either the same read maps to both of the unitigs, or, one read in the pair maps to u_1 and the other read in the pair maps to u_2. More precisely, let r_1 and r_2 be two paired end reads and let $M(r_1), M(r_2)$ be the sets of unitigs that r_1 and r_2 are mapped to. Then the initially empty set of edges (E) in the HUG is united with the set of newly created edges, i.e.

(a) $E \leftarrow E \cup \{\{u, v\} : u \in M(r_i) \text{ and } v \in M(r_i) \setminus u \text{ for } i = 1, 2\}$
(b) $E \leftarrow E \cup \{\{u, v\} : u \in M(r_1) \text{ and } v \in M(r_2) \setminus u\}$

Conceptually two kinds of edges arise from this construction: (a) local similarity edges, which capture subregions of unitigs that are similar as evidenced by the read mapping, and (b) adjacency edges which have potential proximity of two unitigs with a genome. While it is natural to expect that single read multi-mapping corresponds to similarity edges and paired-end information corresponds to the adjacency ones, it is worth noting that single read mapping also can contribute to the adjacency edge formation (see Figure 1a, b). We currently do not distinguish the two edge types (local similarity vs adjacency) in implementation.

4.2. *k-core decomposition and Core-A anomaly score*

The k-core of a graph is the maximal induced subgraph in which each node has a degree of at least k. If the vertices of the k-core of a graph are represented by V_k, then the coreness of vertex v is defined as $coreness(v) = \max\{k : v \in V_k\}$. Computing the coreness of each vertex is called k-core decomposition. Once the HUG is constructed, we perform a k-core decomposition of it using the igraph C library [50] implementation of the linear time Batagelj-Zaversnik [45] algorithm, which assigns a coreness to each vertex in the HUG. Subsequently, for each unitig a Core-A anomaly score is computed as specified in previous work on anomaly detection in networks [31]. In particular, for each vertex v we compute its rank based on the degree $\text{rank}_d(v)$, and its rank based on coreness $\text{rank}_c(v)$. The Core-A anomaly score is then defined as the absolute value of the difference of the log of the two ranks, i.e. $\text{Core-A}(v) = |\log \text{rank}_d(v) - \log \text{rank}_c(v)|$.

There are two key groups of unitigs with high anomaly scores: (a) individual anomalies and (b) anomalous clusters. In general, for any vertex v the shell number is upper bounded by the degree of that vertex. Thus, individual anomalies are nodes with a large discrepancy between their degree and coreness. In particular, this is can be described by the individual influence, ii value, defined as $ii = 1 - \text{coreness}(v)/\text{deg}(v)$ that is equal to 0 if the degree and shell number are equal, and approaches 1 for values of degree significantly larger than that of coreness. Individual anomalies are unitigs likely to have varying genomic contexts in the metagenome. Thus, individual anomalies are good candidates for mobile genetic elements or duplicated genes. Anomalous clusters on the other hand are more likely to arise due to shared local similarities between a large group of unitigs. Those can be nearly identical repeats, such as Alu elements in human genomes, or hypervariable regions of ribosomal proteins in bacterial genomes.

4.3. *K-truss computation*

A K-truss of a graph is an induced subgraph in which every edge is present in at least $K - 2$ triangles. A method proposed by Malliaros *et al.* [32] computes the maximal K-truss by computing it for the k-core of the graph, since the K-truss of a graph is always a subgraph of its $K - 1$-core. Thus, as the k-core decomposition of the HUG is computed, we select the k-core subgraph of the HUG and then compute its K-truss decomposition using the igraph

C library's implementation of Wang and Cheng's algorithm [47]. The algorithm assigns a trussness value to each edge in the subgraph, representing the maximum value of K for which the edge is present in the K-truss.

Now, let V_k be the set of nodes and let E_k be the set of edges of the maximal k-core subgraph of the HUG, and define $\tau : E_k \to \mathbf{N}$ to be the mapping realized by the igraph algorithm, whose time complexity is $O(|E_k|^{1.5})$. We then set $K = \max\{\tau(e) : e \in E_k\}$ and select the vertices (i.e. unitigs) in the maximal K-truss to be those in $\{v \in V_k : \tau(e) = K \text{ for some } e \text{ incident to } v\}$.

4.4. *Taxonomic classification and α-diversity calculations*

Taxonomic classification of the unitigs was performed with Kraken 2 [2] with the standard parameters ($k = 35$, $\ell = 31$) and the standard Kraken 2 database consisting of RefSeq viral, bacterial, and archeal genomes, as well as human genome and known vector sequences from UniVec_Core. For α-diversity computations, the unclassified portion of unitigs was discarded, and the remaining fractions were re-normalized to add up to 1. The α-diversity was defined as the Shannon entropy of the classified unitig fractions $H = -\sum_{i \in T} f_i \log f_i$, where f_i is the fraction of unitigs classified as taxa i.

5. Data availability

This work has not produced any new sequencing data, and relied on publicly available datasets. Details for accessing these datasets are specified below.

ME/CFS metagenomic sequencing data. Illumina short paired-end sequences (150bp) from stool samples of 92 controls, 73 short-term ME/CFS, and 73 long-term ME/CFS patients were analyzed [34]. Original sequencing data was deposited into SRA under BioProject accession PRJNA878603.

IBD data from integrative HMP. Illumina short paired-end sequences from stool samples of patients with IDB were analyzed [35]. We analyzed a subset of 540 out of 1,613 available samples. Data is available from the HMP portal (https://portal.hmpdacc.org/) via study IBDMDB.

Human genome dataset. We have used Illumina short paired-end reads (150bp and 250bp) from Genome in a Bottle project. We used aligned reads for HG002 genome that can be accessed via the index hosted on GitHub: https://github.com/genome-in-a-bottle/giab_data_indexes.

6. Code availability

KOMB source code is publicly available on GitHub: https://github.com/treangenlab/komb.

7. Acknowledgements

The authors acknowledge helpful discussions and advice on statistical testing provided by Dr. Michael Nute. N.S. is supported by the Ken Kennedy Institute Andrew Ladd Memorial Excellence in Computer Science Fellowship. N.S. and T.J.T. were supported in part by the P01-AI152999 NIH award. T.J.T. was also supported by National Science Foundation grant EF-2126387, and NSF CAREER award (IIS-2239114).

References

[1] D. E. Wood and S. L. Salzberg, Kraken: ultrafast metagenomic sequence classification using exact alignments, *Genome biology* **15**, 1 (2014).

[2] D. E. Wood, J. Lu and B. Langmead, Improved metagenomic analysis with kraken 2, *Genome Biology* **20**, p. 257 (2019).

[3] K. D. Curry, Q. Wang, M. G. Nute, A. Tyshaieva, E. Reeves, S. Soriano, Q. Wu, E. Graeber, P. Finzer, W. Mendling *et al.*, Emu: species-level microbial community profiling of full-length 16s rrna oxford nanopore sequencing data, *Nature methods* **19**, 845 (2022).

[4] A. Blanco-Míguez, F. Beghini, F. Cumbo, L. J. McIver, K. N. Thompson, M. Zolfo, P. Manghi, L. Dubois, K. D. Huang, A. M. Thomas *et al.*, Extending and improving metagenomic taxonomic profiling with uncharacterized species using metaphlan 4, *Nature Biotechnology* , 1 (2023).

[5] O. Manor, C. L. Dai, S. A. Kornilov, B. Smith, N. D. Price, J. C. Lovejoy, S. M. Gibbons and A. T. Magis, Health and disease markers correlate with gut microbiome composition across thousands of people, *Nature communications* **11**, p. 5206 (2020).

[6] P. Scepanovic, F. Hodel, S. Mondot, V. Partula, A. Byrd, C. Hammer, C. Alanio, J. Bergstedt, E. Patin, M. Touvier *et al.*, A comprehensive assessment of demographic, environmental, and host genetic associations with gut microbiome diversity in healthy individuals, *Microbiome* **7**, 1 (2019).

[7] E. Castro-Nallar, M. L. Bendall, M. Pérez-Losada, S. Sabuncyan, E. G. Severance, F. B. Dickerson, J. R. Schroeder, R. H. Yolken and K. A. Crandall, Composition, taxonomy and functional diversity of the oropharynx microbiome in individuals with schizophrenia and controls, *PeerJ* **3**, p. e1140 (2015).

[8] K. Lange, M. Buerger, A. Stallmach and T. Bruns, Effects of antibiotics on gut microbiota, *Digestive Diseases* **34**, 260 (2016).

[9] A. Almeida, A. L. Mitchell, M. Boland, S. C. Forster, G. B. Gloor, A. Tarkowska, T. D. Lawley and R. D. Finn, A new genomic blueprint of the human gut microbiota, *Nature* **568**, 499 (2019).

[10] A. Almeida, S. Nayfach, M. Boland, F. Strozzi, M. Beracochea, Z. J. Shi, K. S. Pollard, E. Sakharova, D. H. Parks, P. Hugenholtz *et al.*, A unified catalog of 204,938 reference genomes from the human gut microbiome, *Nature biotechnology* **39**, 105 (2021).

[11] P. G. Wolf, E. S. Cowley, A. Breister, S. Matatov, L. Lucio, P. Polak, J. M. Ridlon, H. R. Gaskins and K. Anantharaman, Diversity and distribution of sulfur metabolic genes in the human gut microbiome and their association with colorectal cancer, *Microbiome* **10**, 1 (2022).

[12] K. Lee, S. Raguideau, K. Sirén, F. Asnicar, F. Cumbo, F. Hildebrand, N. Segata, C.-J. Cha and C. Quince, Population-level impacts of antibiotic usage on the human gut microbiome, *Nature Communications* **14**, p. 1191 (2023).

[13] A. M. Phillippy, M. C. Schatz and M. Pop, Genome assembly forensics: finding the elusive mis-assembly, *Genome biology* **9**, 1 (2008).

[14] M. Pop, Genome assembly reborn: recent computational challenges, *Briefings in bioinformatics* **10**, 354 (2009).

[15] N. D. Olson, T. J. Treangen, C. M. Hill, V. Cepeda-Espinoza, J. Ghurye, S. Koren and M. Pop, Metagenomic assembly through the lens of validation: recent advances in assessing and improving the quality of genomes assembled from metagenomes, *Briefings in bioinformatics* **20**, 1140 (2019).

[16] C. Quince, S. Nurk, S. Raguideau, R. James, O. S. Soyer, J. K. Summers, A. Limasset, A. M. Eren, R. Chikhi and A. E. Darling, Strong: metagenomics strain resolution on assembly graphs, *Genome biology* **22**, 1 (2021).

[17] F. Meyer, A. Fritz, Z.-L. Deng, D. Koslicki, T. R. Lesker, A. Gurevich, G. Robertson, M. Alser, D. Antipov, F. Beghini *et al.*, Critical assessment of metagenome interpretation: the second

round of challenges, *Nature methods* **19**, 429 (2022).
[18] H. Ochman, J. G. Lawrence and E. A. Groisman, Lateral gene transfer and the nature of bacterial innovation, *nature* **405**, 299 (2000).
[19] F. A. Kondrashov, Gene duplication as a mechanism of genomic adaptation to a changing environment, *Proceedings of the Royal Society B: Biological Sciences* **279**, 5048 (2012).
[20] T. J. Treangen and E. P. Rocha, Horizontal transfer, not duplication, drives the expansion of protein families in prokaryotes, *PLoS genetics* **7**, p. e1001284 (2011).
[21] R. R. Wick and K. E. Holt, Polypolish: short-read polishing of long-read bacterial genome assemblies, *PLoS computational biology* **18**, p. e1009802 (2022).
[22] C. Y. Kim, J. Ma and I. Lee, Hifi metagenomic sequencing enables assembly of accurate and complete genomes from human gut microbiota, *Nature communications* **13**, p. 6367 (2022).
[23] F. Trigodet, K. Lolans, E. Fogarty, A. Shaiber, H. G. Morrison, L. Barreiro, B. Jabri and A. M. Eren, High molecular weight dna extraction strategies for long-read sequencing of complex metagenomes, *Molecular Ecology Resources* **22**, 1786 (2022).
[24] A. Balaji, N. Sapoval, C. Seto, R. L. Elworth, Y. Fu, M. G. Nute, T. Savidge, S. Segarra and T. J. Treangen, Komb: K-core based de novo characterization of copy number variation in microbiomes, *Computational and Structural Biotechnology Journal* **20**, 3208 (2022).
[25] S. Koren, T. J. Treangen and M. Pop, Bambus 2: scaffolding metagenomes, *Bioinformatics* **27**, 2964 (2011).
[26] G. Gautreau, A. Bazin, M. Gachet, R. Planel, L. Burlot, M. Dubois, A. Perrin, C. Médigue, A. Calteau, S. Cruveiller, C. Matias, C. Ambroise, E. P. C. Rocha and D. Vallenet, PPanGGOLiN: Depicting microbial diversity via a partitioned pangenome graph, *PLoS Comput. Biol.* **16**, p. e1007732 (2020).
[27] J. Ghurye, T. Treangen, M. Fedarko, W. J. Hervey, 4th and M. Pop, MetaCarvel: linking assembly graph motifs to biological variants, *Genome Biol.* **20**, p. 174 (2019).
[28] D. J. Nasko, S. Koren, A. M. Phillippy and T. J. Treangen, Refseq database growth influences the accuracy of k-mer-based lowest common ancestor species identification, *Genome biology* **19**, 1 (2018).
[29] D. Li, C.-M. Liu, R. Luo, K. Sadakane and T.-W. Lam, Megahit: an ultra-fast single-node solution for large and complex metagenomics assembly via succinct de bruijn graph, *Bioinformatics* **31**, 1674 (2015).
[30] S. Nurk, D. Meleshko, A. Korobeynikov and P. A. Pevzner, metaspades: a new versatile metagenomic assembler, *Genome research* **27**, 824 (2017).
[31] K. Shin, T. Eliassi-Rad and C. Faloutsos, Corescope: Graph mining using k-core analysis—patterns, anomalies and algorithms, *2016 IEEE 16th international conference on data mining (ICDM)* , 469 (2016).
[32] F. D. Malliaros, M.-E. G. Rossi and M. Vazirgiannis, Locating influential nodes in complex networks, *Sci. Rep.* **6**, p. 19307 (January 2016).
[33] M. Kitsak, L. K. Gallos, S. Havlin, F. Liljeros, L. Muchnik, H. E. Stanley and H. A. Makse, Identification of influential spreaders in complex networks, *Nature Physics* **6**, 888 (Nov 2010).
[34] R. Xiong, C. Gunter, E. Fleming, S. D. Vernon, L. Bateman, D. Unutmaz and J. Oh, Multi-'omics of gut microbiome-host interactions in short-and long-term myalgic encephalomyelitis/chronic fatigue syndrome patients, *Cell Host & Microbe* **31**, 273 (2023).
[35] J. Lloyd-Price, C. Arze, A. N. Ananthakrishnan, M. Schirmer, J. Avila-Pacheco, T. W. Poon, E. Andrews, N. J. Ajami, K. S. Bonham, C. J. Brislawn *et al.*, Multi-omics of the gut microbial ecosystem in inflammatory bowel diseases, *Nature* **569**, 655 (2019).
[36] E. D. Jarvis, G. Formenti, A. Rhie, A. Guarracino, C. Yang, J. Wood, A. Tracey, F. Thibaud-Nissen, M. R. Vollger, D. Porubsky *et al.*, Semi-automated assembly of high-quality diploid human reference genomes, *Nature* **611**, 519 (2022).

[37] J. Khan and R. Patro, Cuttlefish: fast, parallel and low-memory compaction of de bruijn graphs from large-scale genome collections, *Bioinformatics* **37**, i177 (2021).
[38] J. Khan, M. Kokot, S. Deorowicz and R. Patro, Scalable, ultra-fast, and low-memory construction of compacted de bruijn graphs with cuttlefish 2, *Genome biology* **23**, p. 190 (2022).
[39] A. Cracco and A. I. Tomescu, Extremely-fast construction and querying of compacted and colored de bruijn graphs with ggcat, *bioRxiv* (2022).
[40] P. E. Compeau, P. A. Pevzner and G. Tesler, How to apply de bruijn graphs to genome assembly, *Nature biotechnology* **29**, 987 (2011).
[41] Z. Iqbal, I. Turner and G. McVean, High-throughput microbial population genomics using the cortex variation assembler, *Bioinformatics* **29**, 275 (2013).
[42] R. M. Colquhoun, M. B. Hall, L. Lima, L. W. Roberts, K. M. Malone, M. Hunt, B. Letcher, J. Hawkey, S. George, L. Pankhurst *et al.*, Pandora: nucleotide-resolution bacterial pan-genomics with reference graphs, *Genome biology* **22**, 1 (2021).
[43] S. Martin, M. Ayling, L. Patrono, M. Caccamo, P. Murcia and R. M. Leggett, Capturing variation in metagenomic assembly graphs with metacortex, *Bioinformatics* **39**, p. btad020 (2023).
[44] D. E. Wood, J. Lu and B. Langmead, Improved metagenomic analysis with kraken 2, *Genome biology* **20**, 1 (2019).
[45] V. Batagelj and M. Zaversnik, An $O(m)$ algorithm for cores decomposition of networks, *arXiv preprint cs/0310049* (2003).
[46] U. Brandes, A faster algorithm for betweenness centrality, *Journal of mathematical sociology* **25**, 163 (2001).
[47] J. Wang and J. Cheng, Truss decomposition in massive networks, **5** (2012).
[48] S. D. Jackman, B. P. Vandervalk, H. Mohamadi, J. Chu, S. Yeo, S. A. Hammond, G. Jahesh, H. Khan, L. Coombe, R. L. Warren *et al.*, Abyss 2.0: resource-efficient assembly of large genomes using a bloom filter, *Genome research* **27**, 768 (2017).
[49] M. Vasimuddin, S. Misra, H. Li and S. Aluru, Efficient architecture-aware acceleration of bwa-mem for multicore systems, *2019 IEEE international parallel and distributed processing symposium (IPDPS)* , 314 (2019).
[50] G. Csardi, T. Nepusz *et al.*, The igraph software package for complex network research, *InterJournal, complex systems* **1695**, 1 (2006).

nSEA: n-Node Subnetwork Enumeration Algorithm Identifies Lower Grade Glioma Subtypes with Altered Subnetworks and Distinct Prognostics

Zhihan Zhang[1], Christiana Wang[1], Ziyin Zhao[1], Ziyue Yi[1], Arda Durmaz[1,2], Jennifer S. Yu[2], and Gurkan Bebek[1,3†]

[1] Systems Biology and Bioinformatics Graduate Program,
Case Western Reserve University, Cleveland OH 44106, USA
[2] Center for Cancer Stem Cell Research,
Cleveland Clinic Lerner Research Institute, Cleveland OH 44195, USA
[3] Center for Proteomics and Bioinformatics, Department of Nutrition,
Department of Computer and Data Sciences,
Case Western Reserve University, Cleveland OH 44106, USA
zhihan.zhang@case.edu, christiana.wang@case.edu, ziyin.zhao2@case.edu,
ziyue.yi@case.edu, arda.durmaz2@case.edu, yuj2@ccf.org,
† Corresponding Author: gurkan.bebek@case.edu

Abstract. Advances in molecular characterization have reshaped our understanding of low-grade glioma (LGG) subtypes, emphasizing the need for comprehensive classification beyond histology. Leveraging this, we present a novel approach, network-based Subnetwork Enumeration, and Analysis (*nSEA*), to identify distinct LGG patient groups based on dysregulated molecular pathways. Using gene expression profiles from 516 patients and a protein-protein interaction network we generated 25 million subnetworks. Through our unsupervised bottom-up approach, we selected 92 subnetworks that categorized LGG patients into five groups. Notably, a new LGG patient group with a lack of mutations in *EGFR*, *NF1*, and *PTEN* emerged as a previously unidentified patient subgroup with unique clinical features and subnetwork states. Validation of the patient groups on an independent dataset demonstrated the robustness of our approach and revealed consistent survival traits across different patient populations. This study offers a comprehensive molecular classification of LGG, providing insights beyond traditional genetic markers. By integrating network analysis with patient clustering, we unveil a previously overlooked patient subgroup with potential implications for prognosis and treatment strategies. Our approach sheds light on the synergistic nature of driver genes and highlights the biological relevance of the identified subnetworks. With broad implications for glioma research, our findings pave the way for further investigations into the mechanistic underpinnings of LGG subtypes and their clinical relevance.
Availability: Source code and supplementary data are available at https://github.com/bebeklab/nSEA

Keywords: Cancer Systems Biology · Network Analysis · Protein-protein Interaction Networks.

1 Introduction

Lower-grade gliomas (LGG) are brain neoplasms classified into 3 grades by the World Health Organization (WHO), where grades 2 and 3 present an infiltrative phenotype. While some LGGs remain stable, others progress to grade 4 gliomas (grade 4 astrocytoma [*IDH*-mutant tumors] and glioblastoma [*IDH*-wildtype tumors]), resulting in survival ranges between 1 and 15 years. Common treatment options include resection, chemotherapy, and radiation therapy. Based on the origin of glial cells, LGG can be classified into two subtypes: astrocytomas and oligodendrogliomas. Molecular features are also associated with clinical outcomes; for example, LGG with both an *IDH* mutation (*IDH1* or *IDH2*) and deletion of chromosome arms 1p and 19q (1p/19q codeletion) show a better response to radiochemotherapy and are associated with longer survival. However, neither grade-based stratification nor molecular features can fully capture the complex architecture of LGG.

Gliomas are histopathologically classified into four grades associated with a worse prognosis. While this classification has prognostic value, investigating the complex molecular alterations within gliomas can lead to a better understanding of the biology behind the tumor types. For instance, some low-grade gliomas behave like malignant glioblastoma, while others have a favorable outcome similar to low-grade gliomas. Identifying genetic and epigenetic alterations in these tumors can reveal biomarkers with both prognostic value and the potential to guide therapeutic decisions [1].

Recently, studies by The Cancer Genome Atlas (TCGA) on lower-grade diffuse gliomas defined disease classification based on genetic and epigenetic alterations, providing biological justification for the utility of these features over histologic ones. Integrated genome-wide data analysis from multiple platforms delineated three molecular classes of lower-grade gliomas that were more concordant with *IDH*, *1p/19q*, and *TP53* status than with histologic classes [2].

In recent years, various approaches have been proposed for finding disease-related sub-networks [3–7] or predicting disease-causing genes [8–11] from large knowledge bases, such as protein-protein interaction (PPI) networks or signaling pathway databases. Most of these methods integrate systems-level measurements of gene and/or protein expression to prioritize networks [12–17]. A scoring function is combined with a search strategy to evaluate identified sub-networks. However, since finding sub-networks is an NP-hard problem [12], long run times and sub-optimal solutions are major drawbacks of these applications. Among all applicable methods, Kernel clustering, modularity optimization, random-walk-based, and local network search methods outperform others [6]. While some of these approaches can identify prognostic modules or disease-relevant pathways [12, 18, 6], they lack the ability to prioritize modules for disease subtype identification and subsequent survival analyses.

Enrichment-based pathway analyses are also commonly used to identify biological functions related to biomarkers and study disease subtypes in cancer [19–21]. However, since such approaches depend on previously selected genes, these analyses may lead to biased results. For instance, Sanchez-Vega et al. [22] analyzed the mechanisms and patterns of somatic alterations in ten canonical pathways and mapped them to multiple tumor types to discover pan-cancer subtypes and link them to possible drug targets. This supervised approach easily captured *known* subtypes with *known* disease pathways. In contrast, Durmaz et al. [23] reported an unsupervised approach that repeated this identification process using frequent subgraph mining with sampling and identified 106 clusters from 43K sub-networks mined from patient-specific networks. However, the former approach lacks the freedom to discover new subtypes, while the latter randomized approach requires careful filtering and repeated trials to arrive at robust discoveries.

In this paper, we introduce a novel network analysis algorithm known as the *n-Node Subnetwork Enumeration Algorithm* (*nSEA*). Our aim is to address challenges encountered by disease classification methods, which often rely on disease-associated genes or subnetworks for patient characterization and prognostics. Here, we discern robust patient subtypes based on functional variations in gene/protein expression within each sample and their interactions. This approach enables us to establish a patient classification framework that not only enhances prognostic accuracy but also elucidates the distinct pathway-level differences among patient subgroups. Such an approach holds the promise of improved prognostication for future patients, along with opportunities for enhanced treatment strategies and personalized interventions.

The (*nSEA*) algorithm takes a protein-protein interaction (PPI) network and system-level measurements of gene expression profiles as inputs. The goal of *nSEA* is to identify differentiating patterns among disease samples in an unsupervised manner. The algorithm is based on a bottom-up methodology in which a large sparse biological network (a PPI network filtered by patient gene expression profiles) is exhaustively enumerated and decomposed into n-node sub-networks (Figure 1A and 1B). These sub-networks are then evaluated, ranked, and filtered based on their inner-pattern consistency and network topology (Figure 1C). In simple terms, the presented method aims to exhaustively identify n-node sub-networks that exhibit consistent expression patterns of network *edges*, quantified by the delta of gene expressions. The selected n-node sub-networks are expanded to include their neighboring nodes, forming more stable network structures (Figure 1D). By applying principal component analysis to network states, we identified sub-networks capable of discriminating disease states (Figure 2A-E) [24, 25]. The final set of sub-networks represents the major dynamics in the PPI network and provides a global picture of pathway dysfunction across cancer subtypes.

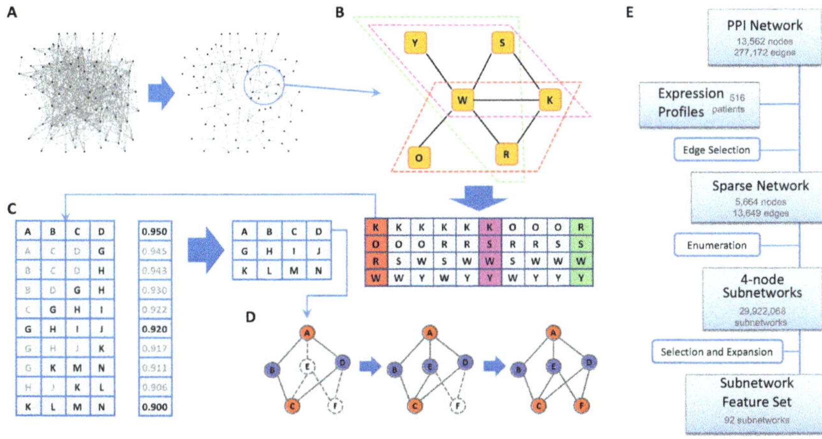

Fig. 1: **Diagram of the *nSEA* algorithm.** The algorithm takes a protein-protein interaction (PPI) network and gene expression profiles of samples as inputs. **(A)** The PPI network is converted into a sparse network. Edges are filtered based on the expression difference of their corresponding node pairs. **(B)** Network enumeration concept: All possible 4-node sub-networks are extracted from the original network, forming a list. Letters represent proteins. Three 4-node sub-networks and their positions in the list are annotated in colors as examples. **(C)** Feature selection based on the sub-network list. Sub-networks are ranked according to their inner-pattern consistency in a decreasing manner. They are then scanned and tested for topology (not shown in the diagram) from top to bottom. If a sub-network is selected into the feature set, it will exclude other sub-networks that share any node with it. **(D)** Selected sub-networks are expanded to neighboring nodes that share similar patterns, forming larger sub-networks. Solid lines represent edges at the current step, while dashed lines represent potential edges that can be added during expansion. Non-significant edges are omitted in this figure. **(E)** Specific application of *nSEA* to Lower grade gliomas (this study). Data is represented by a square and the process is represented by a "squircle." The basic properties of the data between each step were also annotated.

We applied *nSEA* to LGG samples and identified 5 latent groups/subtypes. We compared our subtypes with the current classification and identified significant sub-networks related to our clustering. We also explored the mutation, copy-number variation, and methylation features driving the force behind this classification and discussed several hypotheses based on these results. Furthermore, we compared our method with existing disease classification methods and validated our classification using an independent LGG cohort.

2 Methods

2.1 nSEA algorithm

The nSEA algorithm is based on a bottom-up methodology with which a large sparse biological network, $G(V, E)$, is enumerated and decomposed into n-node subnetworks exhaustively. The goal of the algorithm is to identify subnetworks that can classify patients into subgroups and also provide distinctive biological states for each patient group based on these subnetworks. The first step is to create a network that is sparse enough for further processing. The PPI networks available today are too large for any enumeration algorithm to complete in a reasonable time. We create a sparse network to speed up the process while preserving relevance to disease classification by utilizing gene expression profiles. This is accomplished by using a protein-protein interaction (PPI) network and system-level measurements of gene expression profiles as inputs. Since the subnetwork vector we will calculate in the next steps represents the first principal component or the largest variance of the expression values within the subnetwork, edge filtration should also facilitate achieving this (See a toy example of how this vector is generated in Section S1.1). Let $e \in E$ and $v \in V$ of the PPI network $G = (V, E)$. We define an edge score S_{e_k} between nodes (genes/proteins) v_i and v_j as:

$$S_{e_k} = \sigma(g_{v_i} - g_{v_j}), \quad e_k = (v_i, v_j), \quad i > j \tag{1}$$

where σ is the standard deviation and g is the expression vector of the gene (Figure 2). Edge filtration was done by selecting the top 5% edges ranked by the edge score S_{e_k}.

Enumeration was done by generating up to 4-node connected subnetworks from the filtered dataset. While larger n is possible to use, due to exponential increase in size, we only generated up to 4-node subnetworks only (See Section S1.2). Enumeration of all possible subnetworks was done to exhaustively identify and rank all possible subnetworks. To filter out insignificant subnetworks, the subnetwork score (inner-pattern consistency) of each n-node subnetwork was calculated:

$$\Delta g_{e_k} = g_{v_i} - g_{v_j}, \quad e_k = (v_i, v_j), \quad i > j \tag{2}$$

$$S_{Sbn} = \frac{\sum |\ cor(\Delta g_{e_x}, \Delta g_{e_y})\ |}{|e|}, \quad x > y \tag{3}$$

where g_{v_i} denotes expression vector of node (gene) v_i and $\Delta\ g_{e_k}$ denotes edge vector of edge e_k. cor denotes Pearson correlation. $|e|$ denotes the total edge count in the subnetwork. S_{Sbn} denotes score for subnetwork. To avoid extreme cases when only one node has a degree larger than 1, 4-node subnetworks with an average degree less or equal to 0.75 were discarded. A threshold of the subnetwork score was set and all subnetworks with a score below the threshold were discarded.

Feature selection for the subnetwork list $\mathbb{L}$ was done using Algorithm 1. First, all subnetworks are ranked in descending order and placed in an array. While there are subnetworks in this array, the top network is saved as a feature and removed from the array. The feature network is then compared against the other subnetworks in the array. If any subnetwork has shared genes with the selected feature, it is removed from the array. The final set of subnetwork features is returned.

Algorithm 1: Feature selection for n-node subnetworks

```
Data: Set of subnetworks 𝕃, scoring function S
Result: Feature Set 𝔽, a set of subnetworks with unique nodes
𝕊 ← rank(𝕃 , S) // rank subnetworks with score function S from Eq. 3
𝔽 ← ∅ // Feature set is empty;
while 𝕊 ≠ ∅ do
    t    max(𝕊) // first subnetwork in the ranked list is t ;
    𝕊 ← 𝕊 - t ;
    foreach u ∈ 𝕊 do
        // check if any nodes (genes) are shared
        if V(u) ∩ V(t) ≠ ∅ then
            𝕊 ← 𝕊 − u;
        end
    end
    𝔽 ← 𝔽 ∪ t // add t to Feature set ;
end
```

For subnetwork expansion, nodes (genes) neighboring the subnetwork (u) were added to the subnetwork one by one (Algorithm 2). At each iteration for each neighboring node, we test:

$$S(u) \geqslant S_T, \qquad S(u) - S(j) \geqslant T, \qquad |E(j)| - |E(u)| \geqslant a|E(u)| \tag{4}$$

where $S(u)$ denotes the subnetwork score at the expansion step. S_T denotes the minimum threshold for subnetwork score expansion, which is set to be 0.87. T is a threshold for the tolerance of score decrease. $|E(u)|$ denotes the total number of edges in the subnetwork. a is a constant coefficient, where the set of nodes in the network will not grow in size more than this ratio. j is the network state assuming the node being considered is added to the subnetwork. The purpose of these two rules is to prevent the subnetwork from infinite expansion. If the rules are not satisfied, the expansion will stop. In this study, we set T to 0.05 and a to 0.25. We then select the neighboring node (gene) which has the largest score and add that node to the subnetwork. This process is repeated until no node can be added due to constraints.

Algorithm 2: The subnetwork expansion algorithm

```
Data: Set of feature subnetworks F, where u ∈ F, and networks are scored by function S
S_T denotes the minimum threshold constant for subnetwork score expansion (see Section 2.2)
G is the protein-protein interaction network.
Result: Expanded subnetwork u
foreach u ∈ F do
    repeat
        foreach v' ∈ V(G) , v ∈ V(u) , (v, v') ∈ E(G) do
            j ← u ∪ {v'} ;
            if S(u) ⩾ S_T , S(u) − S(j) ⩾ T, |E(u_j)| − |E(u)| ⩾ a|E(u)| then
                if maxj < S(j) then
                    maxj ←S(j)
                    v'' ← v'
                end
            else
                break;
            end
        end
        u    u ∪ {v''};
    until S(u) ⩾ S_T , S(u) − S(j) ⩾ T, |E(u_j)| − |E(u)| ⩾ a|E(u)|;
end
```

2.2 Parameter Tuning

The aforementioned values of parameters were determined by parameter tuning. These include the edge selection proportion (a), the low threshold of subnetwork score (S_T), and the number of clusters for patient clustering (N_C). First, S_T and N_C were tuned while a was fixed to 5%. Two indicators were used to optimize S_T and N_C. One was the clustering stability (C_S), and the other one was the distance from the background (D_B). C_S is the mean of cluster-consensus values calculated by the `ConsensusClusterPlus` package. D_B is defined as the distance from background clustering, the clustering result generated by setting S_T to 0. Specifically, the distance is defined as:

$$D_B = 1 - FM_{index}(C_{ST}, C_0) \tag{5}$$

where C_{ST} is the clustering labels from threshold S_T and C_0 is the clustering labels when $S_T = 0$. Fowlkes-Mallows index (FM_{index}) is a measurement of similarity between two clustering results [26]. By gradually increasing S_T, for each number of clusters (k), the relationship between S_T and two indicators, C_S and D_B, was explored (Figure S1A and S1B). Noticeably, D_B increases with S_T, which indicates that the feature selection step is necessary in order to generate different clustering results from the background. For C_S, it is interesting that C_S reaches its maximum value when N_C is 5. We then further explored the relationship between C_S and D_B (Figure S1C). By considering both indicators, three S_T values from $N_C = 5$ were very prominent. Among 0.83, 0.85, and 0.87, we chose 0.87 as the final S_T value since when both D_B and C_S are similar, C_S is a more important parameter than D_B.

Second, the proportion of edge selection (a) was evaluated. Due to the limitation of computation power, 5% is almost the maximum percentage of edges we can keep. We then gradually decreased a to inspect its influence on patient clustering. By fixing D_B and C_S as mentioned above, FM indices between each clustering result caused by different a values were calculated. In addition, we fixed a to 5% but sampled its

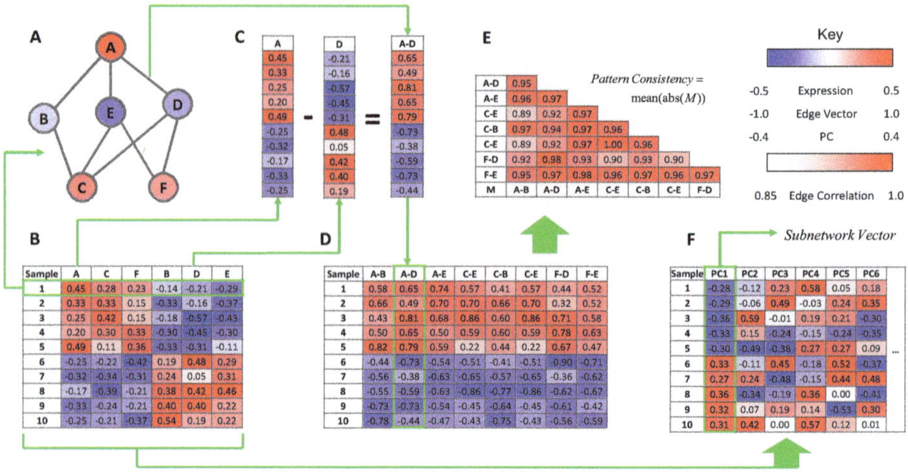

Fig. 2: **Subnetwork variables and their relationships** A subnetwork consisting of 6 nodes and 8 edges. The subnetwork state, which represents the expression pattern of this subnetwork in sample 1, is colored according to gene expression levels. Expression matrix of the subnetwork in (A) with 10 samples. Expression values are centered and scaled. Edge vector is defined as the difference between expression vectors of the corresponding node pair. Edge A-D is used here as an example. The edge matrix combines all edge vectors from the subnetwork. The edge correlation matrix is calculated from the edge matrix. The lower triangle (diagonal excluded) of the matrix is used to calculate the Pattern Consistency score which is defined as the mean of the absolute values of the correlations. The subnetwork vector is defined as the first principal component of the expression matrix. It is used as the summary of the patterns of this subnetwork across all samples. It is also used to cluster samples in the following steps.

subnetwork features (using 80% of all the features each time) to evaluate the error of clustering caused by random sampling (Figure S1D). It was interesting that the clustering difference caused by PE was even less than the clustering difference caused by 80% random sampling. Based on these results, a did not have a significant impact on patient clustering. Therefore, in this project, a was set to 5% since including more edges would produce more subnetwork features and therefore provide a better view of the underlying biological background.

2.3 Clustering of LGG patients and subnetworks

Subnetwork vector was calculated by the `prcomp` function from R package `stats`. Consensus clustering of patients and subnetworks were done with R package `consensusplus`. Clustering stability was defined as the mean of cluster-consensus values. *Fowlkes-Mallows* index was used to measure the distance of current clustering from the background. Consensus clustering of patients and subnetworks was done for 10,000 iterations with sampling proportion set to 0.75 and hierarchical clustering (Ward's method). The self-organizing map was done using R `som`.

2.4 Clinical analysis and tree models

Survival difference (including p-value) was calculated by `survdiff` function from R package `survival`. Distances between patient groups and previous subtypes were defined as the mean Euclidean distance of all possible patient pairs from the two clusters. Correlation between subnetwork cluster vectors and telomere length or Karnofsky score was calculated with `cor.test` function with Spearman's method and exact set to false. GO term (biological process) of subnetwork groups were annotated with `enrichgo` function from R package `clusterProfiler`. Mutation fold change was defined as the actual mutation count divided by the expected count.

Tree models were trained with `rpart` function from R package `rpart`. For binary classification of LG3, the parameter *minbucket* was set to 10, and parameter *maxdepth* was set to 2. For multi-label classification, *minbucket* was set to 22 to simplify the model and *maxdepth* was left as default (30).

Random forest model is trained with TCGA data using the subset function in R. The training process used 1000 trees and tried 8 variables at each split, while the importance of the predictor is set to be true.

Oncogenes and driver genes within each group were identified according to CCGD [27] and Uniprot [28] (Supplementary Table S4). Each subnetwork group was annotated by its corresponding activated oncogenes as well as the signs of the subnetwork vectors.

2.5 Comparison with existing methods

Clustering without gene selection and also nearest shrunken centroid-based gene selection [29] followed by network integration was used to compare with the nSEA approach. First, utilizing Consensus clustering, hierarchical clustering, principle component analysis, and k-means clustering we grouped patients and investigated the patient groups by running survival analysis and investigating clinical variables. Secondly, we trained a nearest shrunken centroid classifier. This widely used approach [30–33] is used to identify genes that stratify LGG samples. Subsequently, a protein-protein interaction (PPI) subnetwork was generated by overlaying the gene expression profiles with a network downloaded from STRING (Section 2.6), followed by node pruning and edge filtration. Networks were scored similar to nSEA approach as described in Section 2.1. PCA scores were subjected to various clustering techniques, including consensus clustering, K-means clustering, hierarchical clustering, and PCA, to classify individuals into multiple distinct classes. The Kaplan–Meier plots are generated based on the clustering results.

2.6 Data preparation

Gene expression data were downloaded from previously published studies by TCGA [34] and CGGA [35–37]. The TCGA datasets were generated by Illumina HiSeq 2000 platform. The level-3 expression data was obtained from UCSC Xena Portal [38]. Non-tumor samples were removed from the data resulting in data for 516 patients. Gene expression matrix was already log_2 transformed. Genes were normalized using z-score normalization across all patients. Outliers were identified by `adjboxStats` from `robustbase` R package. The CGGA datasets were genereted by Illumina HiSeq platform. The raw gene counts were downloaded from CGGA portal from the 'mRNAseq_693' dataset. CGGA data is log-transformed and normalized similar to the TCGA dataset. PPI data were downloaded from String PPI Database [39]. PPI network was filtered by eliminating edges with a combined evidence score of less than 0.7. The PPI network we downloaded had 13,562 nodes and 277,172 edges.

3 Results

3.1 Subnetworks Classify LGG Samples into 5 Groups

We employed the *n-Node Subnetwork Enumeration Algorithm* (*nSEA*) to analyze LGG gene expression profiles [40], comprising 516 patients categorized as astrocytoma (33%), oligodendroglioma (34%), and oligoastrocytoma (22%). A protein-protein interaction (PPI) network was derived from the STRING database using a threshold of combined evidence score set to 0.7 [39], resulting in an undirected PPI network with 13,562 nodes and 277,172 edges (Figure 1E). A sparse network was constructed by retaining the top 5% edges based on edge vector deviation (Figure 1A; Figure 2C), yielding 5,681 nodes and 13,643 edges. The subnetwork size (n) was set to 4 for balance between robustness and computational efficiency, generating a total of 25,413,392 4-node subnetworks through subnetwork enumeration.

We investigated diverse properties of subnetwork feature sets to determine the optimal threshold for inner-pattern consistency in subnetwork selection. Decreasing the threshold led to an incremental rise in

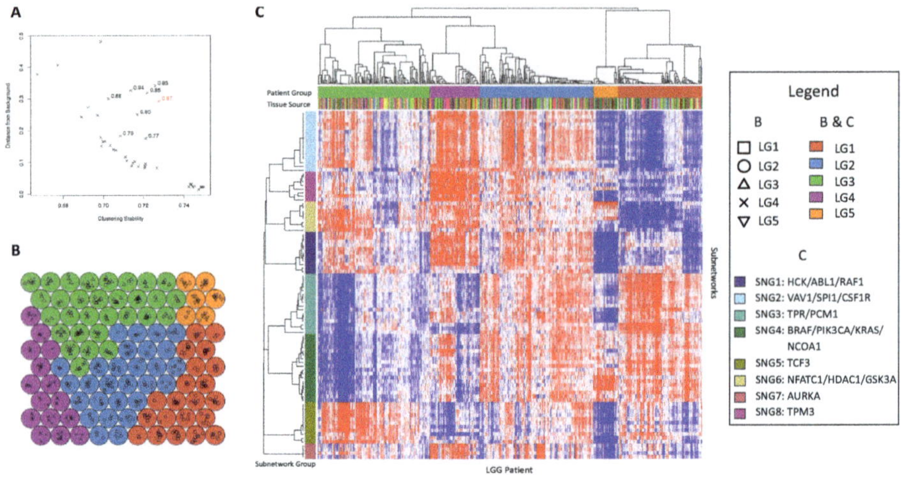

Fig. 3: **Patient Groups and Subnetwork Clusters (A)** Distance from background versus clustering stability from different inner-pattern consistency thresholds. 0.87 is highlighted in red. **(B)** Self-organizing map with 100 units. Patients were mapped to the units, with different shapes representing different patient groups. Units were also annotated with groups by majority voting. **(C)** Heatmap of subnetwork versus patients. LGG patients were clustered into 5 groups (LG1∼5) by consensus clustering using Euclidean distance. Subnetworks were clustered into 8 clusters by consensus clustering using absolute Pearson correlation distance. The sign of each subnetwork vector was adjusted to positively correlate with selected oncogenes or driver genes.

subnetwork inclusion in each feature set until saturation (Figure 3A). Clustering, based on subnetwork state matrices formed from the first principal component of subnetwork expression (Figure 2F), was then assessed for stability across thresholds. Interestingly, clustering stability peaked at both ends of the threshold curve for cluster numbers between 4 and 7 (Figure S1B), indicating distinct clustering patterns between high and low-threshold feature sets. Employing stability curves, we selected 5 clusters based on the relative change of cumulative distribution function (CDF) area (Figure S2E) [41].

Upon fixing the cluster number at 5, we applied the selection algorithm without a threshold to create a background for comparison against feature-based clustering (Figure S2C). The transition from background to high-threshold clustering was evident by a sharp increase around threshold 0.8. Examining the relationship between clustering stability and distance from the background revealed optimal thresholds (0.80 to 0.87) with high stability and separation (Figure 2A). Opting for 0.87 over 0.83 and 0.85, we selected a threshold conducive to subsequent steps.

Patient samples were clustered based on subnetwork state matrices derived from a final feature set of 92 subnetworks. Subnetwork sizes ranged from 6 to 11 nodes, predominantly comprising 6-node subnetworks (57%). Consensus clustering with Ward's method (10,000

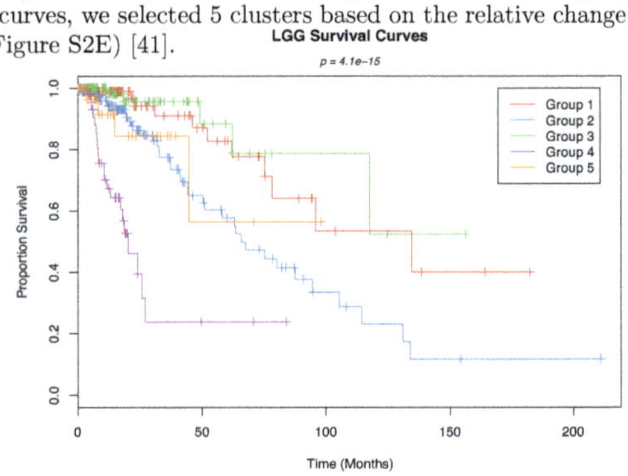

Fig. 4: The Kaplan Meier Plot shows the survival analysis for the TGGA patient groups based on TCGA prognostic networks. The $p - value < 4.1 - e15$ show that groups have distinct survival patterns.

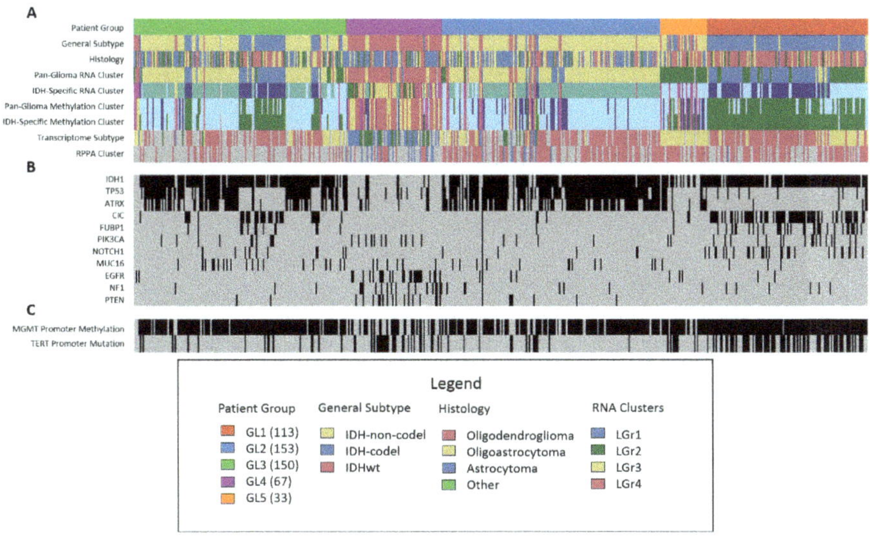

Fig. 5: **Characterization of Patient Groups (A)** Comparison of patient groups with current subtypes and clusters. **(B)** Relationship between patient groups and significant gene mutations. **(C)** Methylation of MGMT promoter and mutation of TERT promoter ordered by patient groups.

iterations) generated a heatmap ordered by clustering dendrogram, revealing 5 patient groups exhibiting distinct subnetwork state patterns (Figure 3). Validation of the consensus clustering approach using unsupervised self-organizing map affirmed unbiased clustering (Figure 3B).

To annotate subnetworks, we performed consensus clustering on subnetwork vectors, identifying 8 subnetwork groups (SNG1∼8). Genes within each group were divided into 2 clusters by correlations. Notably, SNG3 and SNG4 were enriched in cancer driver genes, with SNG4 housing 4 oncogenes associated with the p53 pathway. Protein classes and biological processes analysis further revealed significant associations with specific subnetwork groups, illuminating potential biological implications (Supplementary Table S2-S3).

Additionally, we explored the correlation between subnetwork vectors and clinical attributes like Karnofsky performance score and telomere length (Supplementary Table S6). Remarkably, SNG5 and SNG8 were significantly correlated with Karnofsky scores ($p-value < 8.5e-06$ and $p-value < 5.0e-03$, respectively). Further, gene cluster 2 of SNG5 contained driver genes linked to mental illnesses (Supplementary Table S7). Telomere length showed significant association with SNG3, SNG6, and SNG8 ($p-value < 0.021$), reinforcing links between chromatin remodeling and telomere regulation. Notably, *NIPBL* and *KALRN* emerged as promising gene candidates correlated with distinct patient subgroups, emphasizing their potential roles in promoter regulation and neuropathological disorders.

3.2 LG3: A Previously Unidentified Patient Group with Distinct Features

A comparison of our patient groups with TCGA subtypes and clusters demonstrated LG1-3's alignment with known LGG subtypes. However, LG3 defied such classification, signifying a novel patient group unnoticed in prior TCGA studies (Table S5). Intriguingly, LG3 exhibited a unique clinical profile and subnetwork state pattern.

LG4 exhibited the highest proportion of grade-3 tumors and the oldest mean age (Figure S3A-B), accompanied by the worst Karnofsky performance score (Table S6). LG2 included relatively younger patients compared to LG1, LG3, and LG5. Telomere length analysis showcased pronounced shortening in LG4, consistent with previous research (Figure S3C) [42]. Notably, LG3 displayed a distinct advantage with the highest proportion of patients exhibiting high Karnofsky scores (≥ 90).

Survival analysis further underscored the significance of LG3, presenting improved survival compared to other groups, including LG1, LG2, and LG4, which mirrored *IDHmut-codel*, *IDHmut-non-codel*, and *IDHwt* subtypes (Figure 4). Decision tree modeling unveiled key subnetworks (SNG4 and SNG5) driving LG3's unique clinical outcome (Figure S4).

Methylation analysis elucidated distinct genomic characteristics of LG3, marked by a scarcity of *EGFR*, *NF1*, and *PTEN* mutations, which could potentially contribute to its favorable prognosis. Additionally, supervised learning revealed methylation of *NIPBL* and *KALRN* as distinguishing features of LG3, offering novel insights into regulatory mechanisms and neuropathological associations.

3.3 Comparison with existing methods

First, we employed K-means clustering, hierarchical clustering, Principle Component Analysis and Consensus Clustering to determine subtypes of diseases based on mRNA gene expression profiles alone. While the groups had significant survival differences, the clusters did not follow any particular pattern and the number of genes was extremely high to discover any particular pattern from these analysis (Figure S5).

We also compared our method to sample classification from gene expression data by the method of nearest shrunken centroids [29]. We were able to stratify the samples into four distinct classes by utilizing sample differences based on correlation analysis. This classification informed the selection of an optimal gene inclusion threshold through a rigorous cross-validation procedure (PAMR package in R). Subsequently, we refined our original genomic matrix to incorporate only these curated genes. A tailored Protein-Protein Interaction (PPI) subnetwork was generated. This started with integrating the genomic expression matrix with the PPI network, followed by node pruning and edge filtration. High-correlation edges were selected using a stringent threshold to create subnetworks, revealing gene pairs with potential interconnected functionalities. While consensus-based clustering for both the PAMR-refined matrix and the PPI subnetwork yielded Kaplan Meier Plots with statistically significant survival differences, (Figure S6), the clusters had no discernible feature to study (Figure S7).

3.4 Validation of LGG Patient Groups

To ascertain the robustness of our patient groups, we validated our findings using an independent dataset, $CGGA_{693}$. Through this validation, we verified the consistent clustering of LGG patients into LG1-5, confirming the existence and preservation of distinct subnetwork-based patient groups across different datasets and platforms. Further survival analysis validated the prognostic significance of these patient groups (Figure 6).

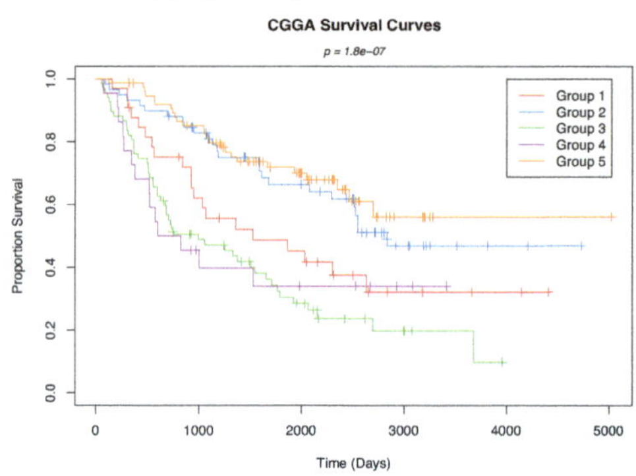

Fig. 6: The Kaplan Meier Plot shows the survival analysis for the CGGA patient groups based on TCGA prognostic network. The $p - value < 1.8 - e07$ show that groups have distinct survival patterns in this secondary data as well.

The subnetwork feature vectors from the TCGA dataset retained their ability to characterize the $CGGA_{693}$ dataset (Figure 7), solidifying the robustness and generalizability of our approach. The relationship between TCGA groups (LG1-5) and CGGA groups further confirmed the concordance between these datasets. Importantly, the conserved survival traits of LG1-5 across datasets validated the clinical relevance of our patient groups, offering a promising avenue for refined LGG prognosis and treatment strategies.

4 Discussion

Many researchers have proposed subtypes of LGG over the last decade. Classification based on genetic features rather than histological features has been demonstrated to be more biologically relevant. The most widely accepted classification is based on molecular subtypes, which classify LGG patients into three clusters

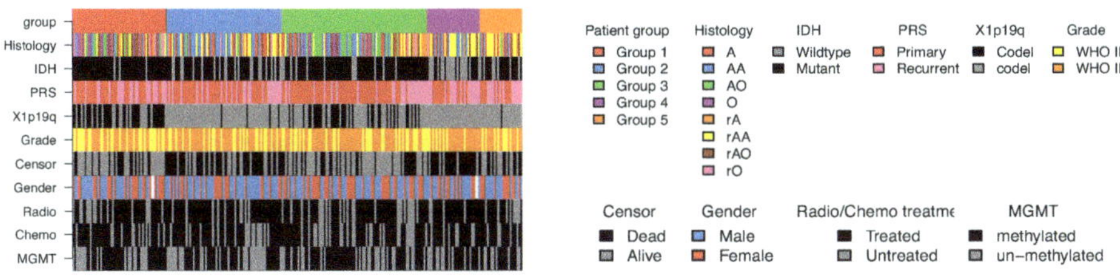

Fig. 7: The CGGA patient groups are based on the random forest model trained by the 92 prognostic networks of TCGA data. The 257 lower-grade glioma patient samples (filtered by WHO grade) were clustered into 5 groups (group 1~5) by consensus clustering using Euclidean distance and the same 92 network measures calculated from the expression data provided. The barplot shows clinical features reported by CGGA [35–37]. Note that IDH1 wildtype group is identified as LG4 in this unsupervised approach once more.

based on *IDH* mutation and chromosome *1p/19q* co-deletion. However, recent studies have challenged this classification by suggesting that *TERT* may play an important role in glioma development. Despite the increasing specificity of LGG classification, the underlying mechanisms of these biomarkers remain unclear. For instance, patients with *IDH* wildtype genotype experience the worst survival outcomes. However, if they have both *TERT* and *IDH* mutations, their survival length is significantly extended, forming the best survival group. This suggests the existence of synergistic relationships among driver genes in LGG.

In this context, our developed algorithm, *nSEA*, offers insight into characterizing these tumors by capturing dysregulation within pathways. Unlike common bioinformatics approaches that focus on mutations, methylation, and copy-number variation, our approach employs a different methodology. By scanning over nearly thirty million 4-node subnetworks, we provide a comprehensive view of subnetwork states within LGG. Through feature selection based on clustering statistics, we identify 92 subnetworks that categorize LGG patients into 5 groups. Three of these groups can be mapped to the general subtypes, demonstrating the ability of our algorithm to capture biologically significant signals. Additionally, we uncover one patient group, LG3, which not only exhibits distinct subnetwork states but also holds clinical significance. We further validate these patient subtype groups using a second cohort, showing that survival traits are conserved even across different patient populations.

Further analysis reveals that compared to other groups, LG3 demonstrates the best survival and Karnofsky performance score. The decision tree model trained on LG3 suggests that SNG4 and SNG5, enriched with oncogenes and associated with mental disorders respectively, can effectively distinguish LG3 from other patients with high accuracy. Mutation analysis indicates that LG3's improved clinical performance may be attributed to the absence of mutations in *EGFR*, *NF1*, and *PTEN*. Moreover, a tree model based on methylation data highlights *NIPBL* and *KALRN* as two genes responsible for the primary and secondary splits of the tree respectively. Apart from their roles in transcription regulation through promoters, *NIPBL* has been linked to various types of cancers [43], suggesting its potential importance in gliomagenesis. The protein encoded by *KALRN*, Kalirin, belongs to the RhoGEF protein family, several members of which have been identified as cancer driver genes [44]. The Dbl-homologous domain of this protein could potentially become a target for future drug development [45].

The unsupervised *nSEA* approach also identified high percentages of cancer driver genes in each subnetwork group. These networks underscore the biological significance of the subnetworks captured by *nSEA*. The synergistic nature of driver genes has been extensively studied in the past, and *nSEA* networks provide insights into how driver genes synergistically contribute to tumor progression. Our findings offer valuable insights based on correlation analysis. However, it is imperative to establish causative relationships in order to gain a deeper understanding of each subtype. Driver mutations and epigenetic events warrant further investigation to delineate these causative relationships. While our approach involved feature selection to categorize patients into groups, numerous driver genes that could differentiate patient groups were identified. Any drivers not included could be further explored using *nSEA* networks to better understand their roles in gliomagenesis.

References

1. Susan M. Chang, Daniel P. Cahill, Kenneth D. Aldape, and Minesh P. Mehta. Treatment of adult lower-grade glioma in the era of genomic medicine. *Am Soc Clin Oncol Educ Book*, (36):75–81, 2016.
2. Cancer Genome Atlas Research Network, Daniel J Brat, Roel G W Verhaak, et al. Comprehensive, integrative genomic analysis of diffuse lower-grade gliomas. *N Engl J Med*, 372(26):2481–98, Jun 2015.
3. M.A. Pujana, J.D. Han, L.M. Starita, K.N. Stevens, M. Tewari, J.S. Ahn, G. Rennert, V. Moreno, T. Kirchhoff, and B. Gold. Network modeling links breast cancer susceptibility and centrosome dysfunction. *Nat. Genet*, 39:1338–1349, 2007.
4. R.K. Nibbe, M. Koyuturk, and M.R. Chance. An integrative -omics approach to identify functional sub-networks in human colorectal cancer. *PLoS Comput. Biol*, 6:1000639, 2010.
5. David: database for annotation, visualization, and integrated discovery. *Genome Biol*, 4:3, 2003.
6. Sarvenaz Choobdar, Mehmet E. Ahsen, Jake Crawford, et al. Assessment of network module identification across complex diseases. *Nature Methods*, 16(9):843–852, 2019.
7. Luz García-Alonso, Roberto Alonso, Enrique Vidal, Alicia Amadoz, Alejandro de María, Pablo Minguez, Ignacio Medina, and Joaquín Dopazo. Discovering the hidden sub-network component in a ranked list of genes or proteins derived from genomic experiments. *Nucleic Acids Research*, 40(20):e158–e158, 07 2012.
8. M. Oti, B. Snel, M.A. Huynen, and H.G. Brunner. Predicting disease genes using protein-protein interactions. *J. Med. Genet*, 43:691–698, 2006.
9. L. Franke, H. Bakel, L. Fokkens, E.D. Jong, M. EgmontPetersen, and C. Wijmenga. Reconstruction of a functional human gene network, with an application for prioritizing positional candidate genes. *Am. J. Hum. Genet*, 78:1011–1025, 2006.
10. K. Lage, E.O. Karlberg, Z.M. Storling, P.I. Olason, A.G. Pedersen, O. Rigina, A.M. Hinsby, Z. Tumer, F. Pociot, and N. Tommerup. A human phenome-interactome network of protein complexes implicated in genetic disorders. *Nat. Biotechnol*, 25:309–316, 2007.
11. M. Vidal, M.E. Cusick, and A.L. Barabasi. Interactome networks and human disease. *Cell*, 144:986–998, 2011.
12. T. Ideker, O. Ozier, B. Schwikowski, and A.F. Siegel. Discovering regulatory and signalling circuits in molecular interaction networks. *Bioinformatics*, 18(Suppl. 1):233– 240, 2002.
13. I. Ulitsky, A. Krishnamurthy, R.M. Karp, and R. Shamir. Degas: de novo discovery of dysregulated pathways in human diseases. *PLoS One*, 5:13367, 2010.
14. M.T. Dittrich, G.W. Klau, A. Rosenwald, T. Dandekar, and T. Muller. Identifying functional modules in protein-protein interaction networks: an integrated exact approach. *Bioinformatics*, 24:223– 231, 2008.
15. S. Nacu, R. Critchley-Thorne, P. Lee, and S. Holmes. Gene expression network analysis and applications to immunology. *Bioinformatics*, 23:850–858, 2007.
16. Z. Guo, Y. Li, X. Gong, C. Yao, W. Ma, D. Wang, Y. Li, J. Zhu, M. Zhang, and D. Yang. Edge-based scoring and searching method for identifying condition-responsive protein protein interaction sub-network. *Bioinformatics*, 23:2121–2128, 2007.
17. H. Ma, E.E. Schadt, L.M. Kaplan, and H. Zhao. Cosine: Condition-specific sub-network identification using a global optimization method. *Bioinformatics*, 27:1290–1298, 2011.
18. Guanming Wu and Lincoln Stein. A network module-based method for identifying cancer prognostic signatures. *Genome Biol*, 13(12):R112, Dec 2012.
19. Michele Ceccarelli, Floris P Barthel, Tathiane M Malta, et al. Molecular profiling reveals biologically discrete subsets and pathways of progression in diffuse glioma. *Cell*, 164(3):550–63, Jan 2016.
20. Christina Curtis, Sohrab P Shah, Suet-Feung Chin, et al. The genomic and transcriptomic architecture of 2,000 breast tumours reveals novel subgroups. *Nature*, 486(7403):346–52, Apr 2012.
21. Cancer Genome Atlas Research Network, W Marston Linehan, Paul T Spellman, et al. Comprehensive molecular characterization of papillary renal-cell carcinoma. *N Engl J Med*, 374(2):135–45, Jan 2016.
22. Francisco Sanchez-Vega, Marco Mina, Joshua Armenia, et al. Oncogenic signaling pathways in the cancer genome atlas. *Cell*, 173(2):321–337.e10, 04 2018.
23. Arda Durmaz, Tim A D Henderson, and Gurkan Bebek. Frequent subgraph mining of functional interaction patterns across multiple cancers. *Pac Symp Biocomput*, 26:261–272, 2021.
24. Vishal N. Patel, Giridharan Gokulrangan, Salim A. Chowdhury, Yanwen Chen, Andrew E. Sloan, Mehmet Koyutürk, Jill Barnholtz-Sloan, and Mark R. Chance. Network signatures of survival in glioblastoma multiforme. *PLoS Computational Biology*, 9(9):e1003237, sep 2013.
25. Salim A. Chowdhury, Rod K. Nibbe, Mark R. Chance, and Mehmet Koyutürk. Subnetwork state functions define dysregulated subnetworks in cancer. In *Lecture Notes in Computer Science*, pages 80–95. Springer Berlin Heidelberg, 2010.
26. Marina Meila. Comparing clusterings: an axiomatic view. In Luc De Raedt and Stefan Wrobel, editors, *Machine Learning, Proceedings of the Twenty-Second International Conference (ICML 2005), Bonn, Germany, August 7-11, 2005*, volume 119 of *ACM International Conference Proceeding Series*, pages 577–584. ACM, 2005.

27. Kenneth L. Abbott, Erik T. Nyre, Juan Abrahante, Yen-Yi Ho, Rachel Isaksson Vogel, and Timothy K. Starr. The candidate cancer gene database: a database of cancer driver genes from forward genetic screens in mice. *Nucleic Acids Research*, 43(D1):D844–D848, sep 2014.
28. UniProt: a hub for protein information. *Nucleic Acids Research*, 43(D1):D204–D212, oct 2014.
29. Robert Tibshirani, Trevor Hastie, Balasubramanian Narasimhan, and Gilbert Chu. Diagnosis of multiple cancer types by shrunken centroids of gene expression. *Proceedings of the National Academy of Sciences*, 99(10):6567–6572, 2002.
30. L. J. van't Veer, H. Dai, M. J. van de Vijver, Y. D. He, A. A. Hart, M. Mao, H. Peterse, K. van der Kooy, M. J. Marton, A. T. Witteveen, G. J. Schreiber, R. M. Kerkhoven, C. Roberts, P. S. Linsley, R. Bernards, S. H. Friend, R. Verhoeven, C. J. F. M. van de Velde, H. Bartelink, M. van der Est, J. L. Peterse, L. F. A. Wessels, P. J. Lamy, L. J. van't Veer, H. Dai, M. J. van de Vijver, Y. D. He, A. A. Hart, M. Mao, H. Peterse, K. van der Kooy, M. J. Marton, A. T. Witteveen, G. J. Schreiber, R. M. Kerkhoven, C. Roberts, P. S. Linsley, R. Bernards, S. H. Friend, R. Verhoeven, C. J. F. M. van de Velde, H. Bartelink, M. van der Est, J. L. Peterse, L. F. A. Wessels, and P. J. Lamy. Gene expression profiling predicts clinical outcome of breast cancer. *Nature*, 415(6871):530–536, 2002.
31. Sandrine Dudoit, Jane Fridlyand, and Terence P Speed. Comparison of discrimination methods for the classification of tumors using gene expression data. *Journal of the American Statistical Association*, 97(457):77–87, 2002.
32. Balasubramanian Narasimhan, Robert Tibshirani, Trevor Hastie, Gavin Sherlock, Michael B Eisen, Patrick O Brown, and David Botstein. Gene expression profiling and statistical pattern recognition for cancer classification. *Genome biology*, 3(12):research0065.1, 2002.
33. Matthew A. Scott, Amelia R. Woolums, Cyprianna E. Swiderski, Andy D. Perkins, and Bindu Nanduri. Genes and regulatory mechanisms associated with experimentally-induced bovine respiratory disease identified using supervised machine learning methodology. *Scientific Reports*, 11(1):22916, 2021.
34. J. Zachary Sanborn, Stephen C. Benz, Brian Craft, Christopher Szeto, Kord M. Kober, Laurence Meyer, Charles J. Vaske, Mary Goldman, Kayla E. Smith, Robert M. Kuhn, Donna Karolchik, W. James Kent, Joshua M. Stuart, David Haussler, and Jingchun Zhu. The UCSC cancer genomics browser: update 2011. *Nucleic Acids Research*, 39(suppl_1):D951–D959, 11 2010.
35. Xing Liu, Yiming Li, Zenghui Qian, Zhiyan Sun, Kaibin Xu, Kai Wang, Shuai Liu, Xing Fan, Shaowu Li, Zhong Zhang, Tao Jiang, and Yinyan Wang. A radiomic signature as a non-invasive predictor of progression-free survival in patients with lower-grade gliomas. *Neuroimage Clin*, 20:1070–1077, 2018.
36. Zheng Zhao, Ke-Nan Zhang, Qiangwei Wang, Guanzhang Li, Fan Zeng, Ying Zhang, Fan Wu, Ruichao Chai, Zheng Wang, Chuanbao Zhang, Wei Zhang, Zhaoshi Bao, and Tao Jiang. Chinese glioma genome atlas (cgga): A comprehensive resource with functional genomic data from chinese glioma patients. *Genomics Proteomics Bioinformatics*, 19(1):1–12, 02 2021.
37. Yinyan Wang, Tianyi Qian, Gan You, Xiaoxia Peng, Clark Chen, Yongping You, Kun Yao, Chenxing Wu, Jun Ma, Zhiyi Sha, Sonya Wang, and Tao Jiang. Localizing seizure-susceptible brain regions associated with low-grade gliomas using voxel-based lesion-symptom mapping. *Neuro Oncol*, 17(2):282–8, Feb 2015.
38. Mary J Goldman, Brian Craft, Mim Hastie, Kristupas Repečka, Fran McDade, Akhil Kamath, Ayan Banerjee, Yunhai Luo, Dave Rogers, Angela N Brooks, Jingchun Zhu, and David Haussler. Visualizing and interpreting cancer genomics data via the xena platform. *Nat Biotechnol*, 38(6):675–678, 06 2020.
39. Damian Szklarczyk, Andrea Franceschini, Stefan Wyder, Kristoffer Forslund, Davide Heller, Jaime Huerta-Cepas, Milan Simonovic, Alexander Roth, Alberto Santos, Kalliopi P. Tsafou, Michael Kuhn, Peer Bork, Lars J. Jensen, and Christian von Mering. STRING v10: protein–protein interaction networks, integrated over the tree of life. *Nucleic Acids Research*, 43(D1):D447–D452, oct 2014.
40. Mary Goldman, Brian Craft, Teresa Swatloski, Kyle Ellrott, Melissa Cline, Mark Diekhans, Singer Ma, Chris Wilks, Josh Stuart, David Haussler, and Jingchun Zhu. The UCSC cancer genomics browser: update 2013. *Nucleic Acids Research*, 41(D1):D949–D954, oct 2012.
41. Matthew D. Wilkerson and D. Neil Hayes. ConsensusClusterPlus: a class discovery tool with confidence assessments and item tracking. *Bioinformatics*, 26(12):1572–1573, apr 2010.
42. Michele Ceccarelli, Floris P. Barthel, Tathiane M. Malta, Thais S. Sabedot, Sofie R. Salama, et al. Molecular profiling reveals biologically discrete subsets and pathways of progression in diffuse glioma. *Cell*, 164(3):550–563, jan 2016.
43. David A. Solomon, Jung-Sik Kim, and Todd Waldman. Cohesin gene mutations in tumorigenesis: from discovery to clinical significance. *BMB Reports*, 47(6):299–310, jun 2014.
44. D R Cook, K L Rossman, and C J Der. Rho guanine nucleotide exchange factors: regulators of rho GTPase activity in development and disease. *Oncogene*, 33(31):4021–4035, sep 2013.
45. X. Shang, F. Marchioni, C. R. Evelyn, N. Sipes, X. Zhou, W. Seibel, M. Wortman, and Y. Zheng. Small-molecule inhibitors targeting g-protein-coupled rho guanine nucleotide exchange factors. *Proceedings of the National Academy of Sciences*, 110(8):3155–3160, feb 2013.

APPLICATION OF QUANTILE DISCRETIZATION AND BAYESIAN NETWORK ANALYSIS TO PUBLICLY AVAILABLE CYSTIC FIBROSIS DATA SETS[a]

Kiyoshi Ferreira Fukutani and Thomas H. Hampton
Geisel School of Medicine, Dartmouth College, 1 Rope Ferry Road
Hanover, 03755, NH, USA
Email: kiyoshi.ferreira.fukutani@dartmouth.edu and Thomas.H.Hampton@dartmouth.edu

Carly A. Bobak
Research Computing and Data Services, Dartmouth College, 1 Rope Ferry Road
Hanover, 03755, NH, USA
Email: carlybobak@dartmouth.edu

Todd A. MacKenzie
The Dartmouth Institute for Health Policy and Clinical Practice, Dartmouth College, 1 Rope Ferry Road
Hanover, 03755, NH, USA
Email: Todd.A.MacKenzie@dartmouth.edu

Bruce A. Stanton
Geisel School of Medicine, Dartmouth College, 1 Rope Ferry Road
Hanover, 03755, NH, USA
Email: Bruce.A.Stanton@dartmouth.edu

The availability of multiple publicly-available datasets studying the same phenomenon has the promise of accelerating scientific discovery. Meta-analysis can address issues of reproducibility and often increase power. The promise of meta-analysis is especially germane to rarer diseases like cystic fibrosis (CF), which affects roughly 100,000 people worldwide. A recent search of the National Institute of Health's Gene Expression Omnibus revealed 1.3 million data sets related to cancer compared to about 2,000 related to CF. These studies are highly diverse, involving different tissues, animal models, treatments, and clinical covariates. In our search for gene expression studies of primary human airway epithelial cells, we identified three studies with compatible methodologies and sufficient metadata: GSE139078, Sala Study, and PRJEB9292. Even so, experimental designs were not identical, and we identified significant batch effects that would have complicated functional analysis. Here we present quantile discretization and Bayesian network construction using the Hill climb method as a powerful tool to overcome experimental differences and reveal biologically relevant responses to the CF genotype itself, exposure to virus, bacteria, and drugs used to treat CF. Functional patterns revealed by cluster Profiler included interferon signaling, interferon gamma signaling, interleukins 4 and 13 signaling, interleukin 6 signaling, interleukin 21 signaling, and inactivation of CSF3/G-CSF signaling pathways showing significant alterations. These pathways were consistently associated with higher gene expression in CF epithelial cells compared to non-CF cells, suggesting that targeting these pathways could improve clinical outcomes. The success of quantile discretization and Bayesian network analysis in the context of CF suggests that these approaches might be applicable to other contexts where exactly comparable data sets are hard to find.

Keywords: Cystic Fibrosis, Bayesian Network, Data.

[a] This work was supported by funding from the Cystic Fibrosis Foundation to B.A.S. (STANTO19G0, STANTO20P0, STANTO23R0 and STANTO19R0), the National Institutes of Health to B.A.S (P30-DK117469 and R01HL151385) and the Flatley Foundation.

1. Introduction

Worldwide initiatives are currently discussing the principles of acquiring, standardizing, storing, and making scientifically produced data accessible for reuse. However, one of the key difficulties is addressing the heterogeneity of the data, which is called batch effects. These batch effects occur when we compare multiple datasets obtained from different laboratories, platforms, or processed at different time points. These internal differences can lead to misinterpretations of the results and it is not only a common issue in omics data analysis but in many cross-study comparisons.[1,2] In recent years, there has been increasing consideration of batch effects in data analysis and several approaches have been proposed to address them.[3] The simplest way to handle batch effects is to include them in the statistical model during analysis. Other approaches involve estimating and creating a new dataset adjusted by batch effects, to perform the statistical analyses.[4] However, it is important to note that this technique can reduce statistical power, particularly when the batch-group is unbalanced, meaning that batch differences may be influenced by group differences. This correction can either diminish group differences or introduce new batch effects due to errors in batch effect estimation that may be inflated by false positives.[5]

Cystic fibrosis (CF) is a recessive genetic disorder characterized by alterations in electrolyte transport across polarized epithelia resulting from mutations in the CF transmembrane conductance regulator gene (*CFTR*).[6] Numerous studies on CF have identified similarities or specific gene signatures that are closely related.[7,8] However, the amount of available transcriptomic datasets for reanalysis and comparison is continually growing.[9] Integrating data from diverse sources can provide a more comprehensive understanding of underlying biological processes that may not be evident from individual studies alone, especially when dealing with multiple conditions and distinct variables.[10] The Meta-analysis instrument of individual microarray studies on CF can help assess the connections between respiratory disorders at the transcriptomic level and provide insights for pathway analysis, but deal with several conditions like: usage of antibiotics, type of mutations, infections by virus or bacteria.[10]

Meta-analysis is a statistical tool that allows the analysis of results from different scientific studies conducted in different locations or by using different methods.[11] In the late 1990s, network meta-analysis (NMA), also known as multiple-treatments or mixed-treatment comparison meta-analysis was introduced as an extension to standard meta-analysis[12]. NMA can compare multiple treatments simultaneously, even when direct comparisons are lacking in existing studies.[12] One systematic review of NMA methods found that around two-thirds of NMA studies utilized a Bayesian approach.[13] The Bayesian network (BN) models are promising in the medical field because they represent the relationships between variables based on real-world, making them more contextually meaningful than purely numeric associations[14] It has been used in various areas of medical science and can include different types of variables, such as clinical, diagnosis, prognosis, and symptoms.[15] This versatility allows researchers to integrate prior beliefs with sample data and BN analysis has recently been utilized in epidemiology, public health, and medicine.[13,16] On the other hand, there is limited knowledge about BN meta-analysis, which may be attributed to researchers' lack of understanding or familiarity with Bayesian methods. Nevertheless, there is significant potential for the application of BN meta-analysis in medicine.[12]

Standard meta-analysis only allows for comparing two interventions at a time, whereas BN Meta-analysis enables the inclusion of evidence from both direct and indirect comparisons in a single analysis.[12] However, BN analysis interpretations still require specific assumptions for accuracy of the algorithm learning and interpretation of network structure, making it a challenging task.[17] To address these issues inherent in Meta analysis, our study proposes a novel approach to pairing multiple transcriptomic datasets by quantile discretization and integrating metadata variables in a new BN Meta Transcriptomic analysis. This approach aims to provide new and valuable insights into understanding the complexities of a multifactorial disease like CF.

2. Methods

2.1. *Data Selection*

We accessed datasets available in the Gene Expression Omnibus (GEO) database (https://www.ncbi.nlm.nih.gov/geo/) by searching the keyword "cystic fibrosis". A total of 17 datasets were returned by this query, which was performed in November 2022. Nine datasets were excluded from further analysis due to methodological incompatibility or insufficient metadata, which involved the use of different cell tissues or experimental designs and did not measure the same patients variables. We retrieved metadata for these three studies. Three of these studies measured gene expression in airway epithelial cells. The first dataset (PRJEB9292), published by Balloy et al.,[18] included both non CF and CF epithelial cells infected with Pseudomonas aeruginosa for different time points. The second dataset (GSE139078) [19] involved epithelial cells from CF patients infected by Rhinovirus or control and treated with Ivacaftor or Lumacaftor/ivacaftor, modulator drugs used to enhance the functional of CFTR. The third study[20] included two datasets: a pilot dataset with 13 samples and a validation dataset contained 35 samples. All datasets provided patient genotype, modulator information, and infection status with either *Pseudomonas aeruginosa* or Rhinovirus.

2.2. *Data Harmonization and Analysis*

The metadata description included means and standard deviation for numeric variables and frequencies and percentages for categorical data. RNAseq datasets were individually normalized by library size and log CPM (count per million) transformation and differential expression analyses were performed individually for each dataset using deseq2.[21] In the Balloy dataset, we compared CF vs. non-CF infected or not infected with *Pseudomonas aeruginosa*; in the De Jong dataset, CF epithelial cells infected with virus or not infected with virus; and in the Salas dataset, epithelial cells of CF patients compared to non-CF subjects. In this exploratory design, the DEGs were used to filter the large number of targets, and they were determined by applying specific criteria: genes with a P-value less than 0.05 and a log2 expression fold change greater than 1 or less than -1 were considered as differentially expressed. These criteria were chosen to serve as a filter and help reduce processing time. Each study was normalized individually, and each gene was discretized according to sample distributions. The count table with filtered genes were discretized into quartiles (1st - Minimal to 25%, 2nd - 25% to 50%, 3rd - 50% to 75%, and 4th - 75% to maximum values by sample distribution) using Hartemink's algorithm, which is available in the bnlearn package.[22,23] Afterward, all the transformed transcriptomic datasets were merged into a single discretized dataset, to which metadata was added. The learning algorithm used to establish the Bayesian network structure was based on the heuristic Hill climb method.[24,25] Bayesian network learning was used to visualize conditional dependencies between multiple clinical and transcriptome variables.[26] The dependencies are represented qualitatively by a directed acyclic graph where each node corresponds to a variable and a direct arc between nodes represents a direct influence. Robustness of the arcs was scored by a non-parametric bootstrap test (100×replicates).[27] For functional analysis of genes related to CF, virus infection, bacterial infection, and use of modulators, enrichment pathway analysis was performed using the clusterProfiler package and REACTOME geneset.[28,29] For the Pathway meta-analysis we use the qusage package.[30] All analyses were performed in R version 4.0.2[24] and the Bayesian network and discretization scripts are available in github (https://github.com/FfKB/BNCF). Figure 1. presents a summary of the study selection process and experimental design.

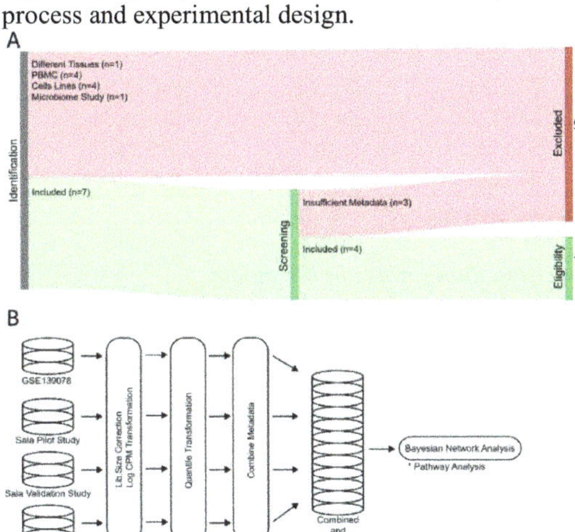

Figure 1. Experimental Design. A) Diagram illustrating the study selection process using a Sankey diagram. The excluded datasets are highlighted in red, while the eligible datasets are highlighted in green. B) Flowchart depicting the data processing steps in the study.

3. Results

3.1. *Study descriptions*

A total of three studies comprising four datasets were considered for analysis: GSE139078, Sala Study, and PRJEB9292. The GSE139078 dataset consists of CF patients who were infected with rhinovirus (RHV). The PRJEB9292 dataset includes four patients divided into four time points, enabling a comparison between gene expression in non CF subjects and CF patients infected with *Pseudomonas aeruginosa*. The Sala study included two datasets: the pilot study and the validation study, which involved a comparison of gene expression profiles between CF patients and non CF subjects. The analysis also includes the assessment of modulator use (Lumacaftor and Ivacftor; and Ivacftor alone) in three datasets (GSE139078, Sala Pilot, and Sala Validation). All CF patients included in these studies have the F508del/F508del genotype, a common genetic mutation (~50%) associated with CF. However, sex and age data were not available for all the datasets, thus, that metadata was not included in the Bayesian Network Analysis. These carefully selected datasets provide comprehensive insights into gene expression patterns related to CF, considering factors such as viral and bacterial infections and the influence of modulators (Table 1).

Table 1. The characteristics of subjects from the selected datasets.

	GSE139078	Sala Pilot	Sala Validation	PRJEB9292
Male sex, n(%)	48 (84.2)	-	-	-
Age, mean (SD)	3.4 (1.4)	35.3 (5.3)	34.1 (8.2)	-
Infection by virus, n(%)	38 (66.7)	-	-	-
Infection by *P. aeruginosa*, n(%)	-	-	-	32 (100)
Cystic Fibrosis, n(%)	57 (100)	7 (53.8)	24 (68.6)	4 (50)*
Modulators (Luma/Iva), n(%)	10 (17.5)	2 (15.4)	10 (28.6)	-
Modulators (Ivacaftor), n(%)	9 (15.8)	0 (0)	2 (5.7)	-
Genotypes F508del, n(%)	57 (100)	7 (53.8)	24 (68.6)	16 (50)

* = 4 Patients in 4 different timepoints (0, 2, 4 and 6).

3.2. *Filtering gene expression data for use in the model*

We began by selecting significant genes through a conventional RNAseq comparison within each dataset. In the Sala Pilot and Validation studies, we compared patients with CF against non-CF individuals to identify genes associated with CF in these datasets. The De Jong datasets exclusively included CF samples, so we compared the presence or absence of virus infection. Lastly, the Balloy dataset consisted of different time points of infection by *Pseudomonas aeruginosa,* with an uninfected control established as point zero for comparison. In all of the studies, we observed changes in gene expression across various comparisons, such as CF versus non-CF, presence or absence of virus, and infection by *Pseudomonas aeruginosa*. It gave us an idea about which genes should be integrated in our Bayesian Network Model. In the De Jong study, we identified 280 genes (220 up-regulated and 60 down-regulated) (Figure 2A). In the Balloy study, we identified 350 genes (221 up-regulated and 129 down-regulated) (Figure 2B). In the Sala pilot study, we identified 789 genes (639 up-regulated and 150 down-regulated) (Figure 2C), and in the Sala validation study, we identified 2716 genes (2114 up-regulated and 602 down-regulated) (Figure 2D). The differences between all the comparisons can be accessed for both up-regulated genes (Figure 2E) and down-regulated genes (Figure 2F).

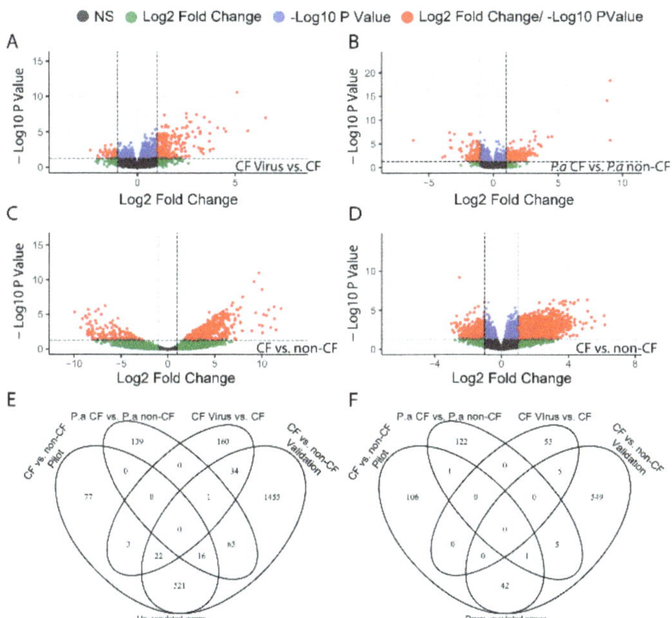

Figure 2. Differential gene expression analysis of epithelial cell datasets. A) GSE139078 shows gene expression changes in cystic fibrosis (CF) patients infected with a virus compared to non-infected CF patients. B) PRJEB9292 compares gene expression in CF patients and controls infected with *Pseudomonas aeruginosa* (P.a). C and D) The Sala Cohort dataset compares gene expression between CF patients and non CF subjects in a pilot study (C) and validation study (D). Red dots represent significant genes with fold changes above or below ±0.5, blue dots represent significant genes without fold change variation, and green dots represent non-significant genes with fold change variation. E and F) Venn diagrams represent the overlap and exclusivity of differentially expressed genes (DEGs) in each comparison, using the upregulated (>1 fold change and p-value <0.05) and downregulated (<-1 fold change and p-value <0.05) DEGs.

3.3. *The Bayesian network is capable of identifying genes associated with all conditions and covariates.*

To circumvent experimental design limitations and to measure the relationship between all conditions and covariates present, we discretized the log CPM table and retrieved all the significant genes obtained from all comparisons of each dataset combined with its respective metadata (infection type (viral or bacterial), CF, modulators (Luma/Iva or Ivacaftor) and genotype (F508del or non CF controls) to create a new dataset. In total we included 1976 genesin the Bayesian network model. As a result, the Bayesian network reveals which genes have a direct relationship with the presence of bacteria, virus, usage of modulators, CF, and the genotype (F508del). Each condition has its own network community despite the genotype, and it is associated with the presence of CF (Figure 3A). Genes present in each network community were used for functional analysis. The functional analysis revealed an Interferon signaling (alpha/beta and gamma) associated with CF, virus, and bacterial network communities. However, IL-9, IL-21, and IL-6 signaling were exclusively related to CF. Virus exposure was exclusively associated with the TGF-beta pathway, and the bacterial exposure did not have any exclusive pathway. Modulator treatment was associated with the response of EIF2AK1 to heme deficiency, late endosomal microautophagy, and IL-1 signaling (Figure 3B).

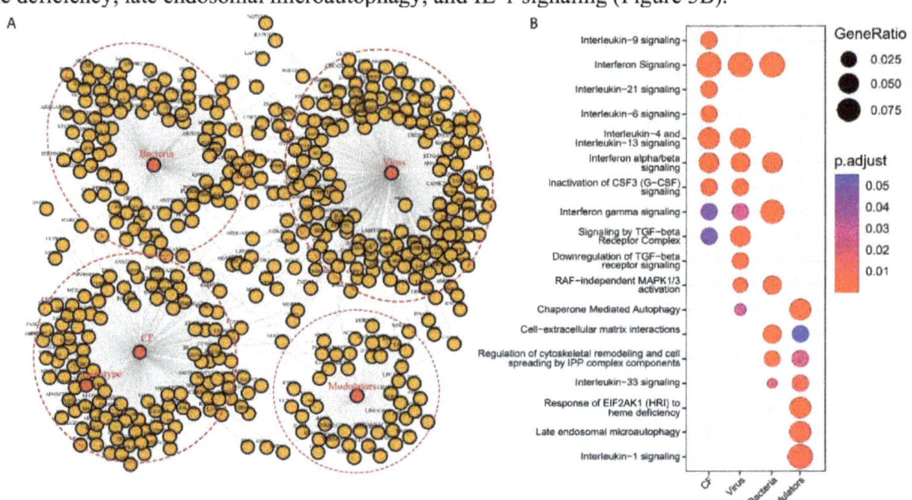

Figure 3. Bayesian Network signatures associated with cystic fibrosis (CF), infection, and mediators. Associations were extracted using Bayesian Network analysis and reconstructed using the "igraph" package in R. A) The main variables (CF, mutations, mediators, and infection) are represented by red nodes and clusters are depicted with red dotted lines. B) Genes presented in each cluster were used for over-represented pathway analysis.

3.4. *The CF Bayesian signature pathway is consistent across all datasets and shows higher expression levels when compared to non-CF epithelial cells.*

The pathways that were discovered in the Bayesian Network Analysis, related to CF were subjected to qusage pathway meta-analysis to measure their activation levels in each study individually, as well as their combination across all studies. As a result, the Interferon signaling, interferon gamma signaling, interleukin 4 and 13 signaling, interleukin 6 signaling, interleukin 21 signaling, and Inactivation of CSF3 G-CSF signaling pathways exhibited an overall alteration across all studies with significant p-values, while the pathways Interleukin 9 signaling and Signaling of TBF-g receptor complex were not significant (Figure 4). We investigated the gene composition of these significant pathways in CF and non-CF to understand their expression. Across all significant pathways investigated (Figure 5). A) Interferon signaling, B) Interferon gamma signaling, C) Interleukin 4 and interleukin 13 signaling, D) Interleukin 6 signaling, E) Inactivation of CSF3 G CSF signaling, and F) Interleukin 21 signaling. The analysis revealed a considerable proportion of epithelial cells derived from CF patients displayed a heightened expression of these genes present in the upper quartile (+75%), in comparison with non-CF. These genes were poorly expressed in all samples in the quantile transformed integrated dataset (Figure 5).

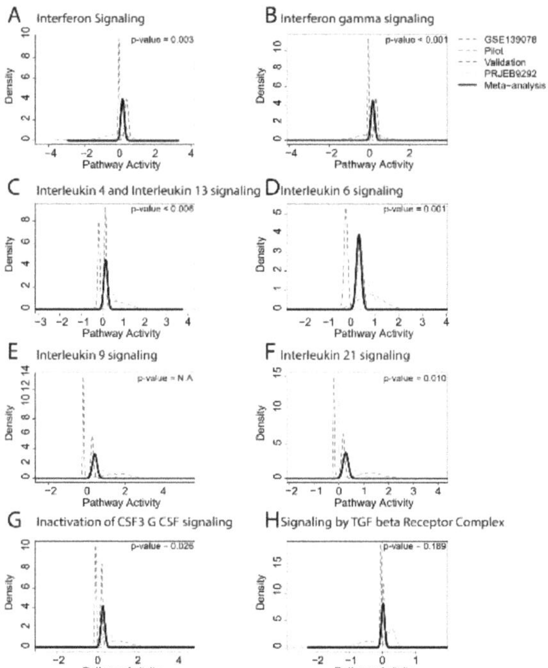

Figure 4. Meta-analysis of pathway enrichment across datasets. The accumulated pathway analysis between all studies was conducted using the pipeline available in the qusage package.

Dotted lines separate studies by color: red for GSE139078, blue for Sala pilot study, purple for Sala validation study, and green for PJREB292. Significant pathways increased related to cystic fibrosis (CF) were identified, including A) Interferon signaling, B) Interferon gamma signaling, C) Interleukin 4 signaling, D) Interleukin 6 signaling, E) Interleukin 9 signaling, F) Interleukin 21 signaling. Pathways decreased in CF include: G) Inactivation of CSF3 and G-CSF signaling, and H) Signaling by TGF-beta receptor complex.

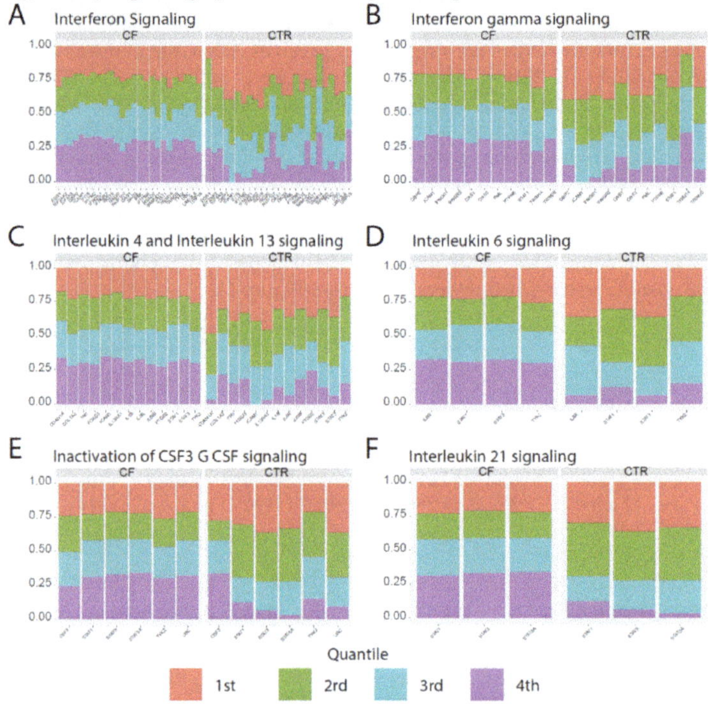

Figure 5. Quantile distribution of expressed genes in each significant pathway related to cystic fibrosis (CF). A) Interferon signaling, B) Interferon gamma signaling, C) Interleukin 4 and Interleukin 13 signaling, D) Interleukin 6 signaling, E) Inactivation of CSF3 and G-CSF signaling, and F) Interleukin 21 signaling.

4. Discussion

Integrating data from transcriptomics or other high-throughput systems, such as proteomics, metabolomics, and lipidomics, is expected to yield new insights. Unfortunately, it also introduces significant heterogeneity arising from various designs or methodologies, commonly known as batch effects. Batch effects are pervasive across all types of high-throughput biological platforms, including single measurement methods like PCR or ELISA.[31] When performing a meta-analysis, batch effects may create bias and reduce statistical power, making it challenging to detect all relevant features, especially those with small effect sizes or in unbalanced samples.[4] On the other hand, integrating several smaller datasets theoretically improves statistical power, provided that technical heterogeneity, including batch effects, is effectively resolved.

Efforts to mitigate batch effects have been proposed, as they are known to interfere with downstream statistical analysis, potentially introducing false significance between groups that only exist between batches without biological meaning.[32,33] Batch effects can also lead to the loss of biological signals contained in the data.[34,35] The proposed quantile transform approach tends to be respectful of each dataset's characteristics, and by mapping each variable's probability in a probabilistic graphical model, it can handle variables present in the metadata, such as group allocation, clinical data, and dichotomous variables, which can be added and probabilistically related to each other.[36] To achieve this, we evaluated four distinct Cystic Fibrosis Datasets with CF genotype, modulator therapy, and different types of infection, incorporating gene expression with these variables, without applying any batch correction while respecting each dataset's individuality. This approach has demonstrated a high level of accuracy in classifying cancer types when applied to expression datasets. [37]

To reduce processing time, we filtered the genes by selecting those that were differentially expressed in all datasets. For the Baloy dataset, we identified 350 differentially expressed genes (221 upregulated and 129 downregulated genes). In their original publication,[18] the authors found a significantly higher number of upregulated genes than down regulated genes compared to noninfected control cells, although their comparisons were done at each time point. In our study, we bulked the controls and the *Pseudomonas aeruginosa* infection time point 0 as a control and compared to *Pseudomonas aeruginosa* infection. In De Jong's study,[19] the author separated the cells by classes and made two different comparisons: virus infection versus controls and virus infections plus modulator with either Ivacaftor or Ivacaftor/Lumacaftor. We compared all cells together against the controls and identified 195 upregulated genes and 60 downregulated genes. In the study by Sala et al.,[20] our comparisons were similar, with 639 and 2114 upregulated genes in the pilot and validation datasets, respectively, and 568 and 1834 downregulated genes, and 150 and 112 upregulated genes, and 320 and 403 downregulated genes in our analysis, respectively. Differences can be noticed between the studies not only in how the comparisons were done, but also in the methods used for comparisons. In our study, all the analyses were performed with the DESEQ2 package,[21] whereas Sala and De Jong's studies used edgeR.[38]

The pathway analysis performed by Balloy[18] and Sala[20] did not use the same geneset. In our study, we used the Reactome geneset [39], and only De Jong[19] used Reactome geneset as well. However, the inflammatory responses were similar in all studies. In Sala's study, they associated the chaperone pathway in CF, while in our study, it was associated with the modulators. Other pathways, such as Interleukin 6, 9, and 21, were exclusively associated with CF in our analysis. The role of IL-6 is controversial; however, it participates in proinflammatory responses with TNF-a and interleukin-1b. IL-6 is a regulator of the host inflammatory response and is negatively associated with pulmonary function in chronic infection in CF and during acute exacerbation of respiratory symptoms or during a period of apparent clinical stability. In bronchoalveolar lavage fluid, IL-6 was significantly elevated in infants with CF.[40] Increased expression of IL-9 and IL-9R is responsible for the mucus-overproducing in the lung epithelium of patients with cystic fibrosis[41] and IL-21 is a multifunctional cytokine that acts on various immune cells.[42] Interestingly, in mice fibroblasts, IL-21R is expressed and upregulates matrix metalloproteinases in response to IL-21 by CD8+ T cells.[43]

When it comes to viral infection, we found that viruses have only one exclusive pathway associated with our analysis, which is related to TGF-b signaling. This pathway is involved in pulmonary fibrosis and other organ-related processes. Viruses utilize various mechanisms to modulate this pathway, including altering TGF-b protein expression and its receptors, as well as modulating the SMAD cascades, TGF-b lead to enhanced cell growth and induction of fibrosis.[44] On the other hand, bacterial infection does not influence any pathways in our analysis. As for the use of modulators, we identified three exclusive pathways: "Response of EIF2AK1 to heme deficiency," "late endosomal microautophagy," and "IL-1 signaling". The HRI kinase (or EIF2AK1) plays two main roles during development: it ensures a balanced synthesis of globin and heme and promotes the survival of erythroid precursors during iron deficiency.[45] Inhibitors of P-gp (P-Glycoprotein) such as fostamatinib[46] and Ivacaftor can be associated with various stress conditions, including oxidative stress, heme deficiency, osmotic shock, and heat shock.[47] In the context of CF, the usage of modulators is associated with an autophagy pathway, which compromises CFTR recycling to lysosomal degradation.[48] Moreover, in our study, the genes associated with modulators were linked to this pathway. In CF patients, CFTR modulators have been shown to increase airway nitric oxide (NO) by increasing the concentrations of IL-1α, IL-1β, and other Th17-associated cytokines in sputum, which is related to NO metabolism.[49]

The overall pathway activation in all studies discovered by the Bayesian network approach in CF confirms previous studies describing a hyperinflammatory state in CF, as well as the participation of other pathways such as interleukin 4, 6, 13, and 21. Notably, interleukin 4 and 13 were not exclusively associated with CF status. The roles of IL-4 and IL-13 in the epithelium of CF patients share several biological properties, including chloride secretion.[50] On the other hand, IL-4 inhibits antiviral immunity,[51] and neutralization of IL-13 reduces death and disease severity in COVID-19 without affecting viral load, indicating an immunopathogenic role for this cytokine.[52] Additionally, G-CSF and GM-CSF can induce elastase and MMP-9 release by neutrophils [53]. Interestingly, all the genes presented in the pathway analysis were in the last quantile of expression in our dataset. The main limitation of this study is that it serves as the initial proof of concept for quantile discretization in the integration of raw datasets. A comparison with different methods should be conducted. Additionally, clinical non-numeric data were included in a single analysis. Therefore, this analysis must be interpreted carefully and should serve as a guide for future models aiming to integrate all datasets and variables in a similar manner. Unfortunately, this study was limited to using only four CF datasets due to the considerable challenge of aligning complete metadata, which encompasses treatment, genotype mutation profiling, and infection status. It is uncommon to find metadata with all these features available, and new studies using this approach must be conducted to assess its efficacy. Despite these limitations, this study sheds light on various biological processes related to CF, particularly concerning viral and bacterial infections, as well as the impact of modulators on epithelial cells within a single assessment, providing valuable insights into these complex.

5. Conclusion

The analysis of integrated data remains a powerful hypothesis generation tool among data scientists. However, dealing with the heterogeneity of multiple datasets poses real challenges. In this study, we proposed a novel approach to integrate several datasets while respecting the unique characteristics of each individual dataset. By applying quantile transformation to multiple datasets and integrating them, we obtained biologically meaningful results that align with existing literature and established associations with other variables such as modulators, virus, and bacterial infections, and included access to good quality metadata. Our analysis revealed an inflammatory signature in CF patients, with exclusive associations observed in interleukin 4, 6, 13, and 21 pathways. Furthermore, we identified potential links between virus infections and the TGF-b pathway, as well as associations between modulators and pathways such as "Response of EIF2AK1 to heme deficiency," "late endosomal microautophagy," and "IL-1 signaling." These findings contribute to a better understanding of the complex interactions in CF and highlight potential targets for further research and development of new integration protocols. Nonetheless, additional studies employing this methodology are imperative to determine the extent to which this innovative approach can uncover novel associations compared to traditional methods.

References

1. Lin, S. et al. Comparison of the transcriptional landscapes between human and mouse tissues. Proc. Natl. Acad. Sci. U. S. A. 111, 17224–17229 (2014).

2. Fei, T., Zhang, T., Shi, W. & Yu, T. Mitigating the adverse impact of batch effects in sample pattern detection. Bioinformatics 34, 2634–2641 (2018).

3. Hicks, S. C., Townes, F. W., Teng, M. & Irizarry, R. A. Missing data and technical variability in single-cell RNA-sequencing experiments. Biostatistics 19, 562–578 (2018).

4. Nygaard, V., Rødland, E. A. & Hovig, E. Methods that remove batch effects while retaining group differences may lead to exaggerated confidence in downstream analyses. Biostatistics 17, 29–39 (2016).

5. Buhule, O. D. et al. Stratified randomization controls better for batch effects in 450K methylation analysis: a cautionary tale. Front. Genet. 5, 354 (2014).

6. Carraro, G. et al. Transcriptional analysis of cystic fibrosis airways at single-cell resolution reveals altered epithelial cell states and composition. Nat. Med. 27, 806–814 (2021).

7. Clarke, L. A., Sousa, L., Barreto, C. & Amaral, M. D. Changes in transcriptome of native nasal epithelium expressing F508del-CFTR and intersecting data from comparable studies. Respir. Res. 14, 38 (2013).

8. Hampton, T. H. & Stanton, B. A. A novel approach to analyze gene expression data demonstrates that the DeltaF508 mutation in CFTR downregulates the antigen presentation pathway. Am. J. Physiol. Lung Cell. Mol. Physiol. 298, L473–82 (2010).

9. Brazma, A. Minimum Information About a Microarray Experiment (MIAME)--successes, failures, challenges. ScientificWorldJournal 9, 420–423 (2009).

10. Clarke, L. A., Botelho, H. M., Sousa, L., Falcao, A. O. & Amaral, M. D. Transcriptome meta-analysis reveals common differential and global gene expression profiles in cystic fibrosis and other respiratory disorders and identifies CFTR regulators. Genomics 106, 268–277 (2015).

11. Hackenberger, B. K. Bayesian meta-analysis now - let's do it. Croat. Med. J. 61, 564–568 (2020).

12. Liu, Y. et al. A Gentle Introduction to Bayesian Network Meta-Analysis Using an Automated R Package. Multivariate Behav. Res. 1–17 (2022).

13. Chambers, J. D. et al. An assessment of the methodological quality of published network meta-analyses: a systematic review. PLoS One 10, e0121715 (2015).

14. Huang, S. et al. Applications of Support Vector Machine (SVM) Learning in Cancer Genomics. Cancer Genomics Proteomics 15, 41–51 (2018).

15. McLachlan, S., Dube, K., Hitman, G. A., Fenton, N. E. & Kyrimi, E. Bayesian networks in healthcare: Distribution by medical condition. Artif. Intell. Med. 107, 101912 (2020).

16. Lee, A. W. Review of mixed treatment comparisons in published systematic reviews shows marked increase since 2009. J. Clin. Epidemiol. 67, 138–143 (2014).

17. Briganti, G., Scutari, M. & Linkowski, P. Network Structures of Symptoms From the Zung Depression Scale. Psychol. Rep. 124, 1897–1911 (2021).

18. Balloy, V. et al. Normal and Cystic Fibrosis Human Bronchial Epithelial Cells Infected with Pseudomonas aeruginosa Exhibit Distinct Gene Activation Patterns. PLoS One 10, e0140979 (2015).

19. De Jong, E. et al. Ivacaftor or lumacaftor/ivacaftor treatment does not alter the core CF airway epithelial gene response to rhinovirus. J. Cyst. Fibros. 20, 97–105 (2021).

20. Sala, M. A. et al. The proteostatic network chaperome is downregulated in F508del homozygote cystic fibrosis. J. Cyst. Fibros. 20, 356–363 (2021).

21. Love, M. I., Huber, W. & Anders, S. Moderated estimation of fold change and dispersion for RNA-seq data with DESeq2. Genome Biol. 15, 550 (2014).

22. Hartemink, A. J., Gifford, D. K., Jaakkola, T. S. & Young, R. A. Using graphical models and genomic expression data to statistically validate models of genetic regulatory networks. Pac. Symp. Biocomput. 422–433 (2001).

23. Scutari, M. Bayesian network constraint-based structure learning algorithms: Parallel and optimized implementations in the bnlearn R package. J. Stat. Softw. 77, (2017).

24. R. Core Team. An Introduction to R. (Samurai Media Limited, 2015).

25. Liu, Y., Wang, L. & Sun, M. Efficient Heuristics for Structure Learning of -Dependence Bayesian Classifier. Entropy 20, (2018).

26. Prada-Medina, C. A. et al. Systems Immunology of Diabetes-Tuberculosis Comorbidity Reveals Signatures of Disease Complications. Sci. Rep. 7, 1999 (2017).

27. Friedman, N., Goldszmidt, M. & Wyner, A. Data analysis with Bayesian networks: A bootstrap approach. (2013) doi:10.48550/ARXIV.1301.6695.

28. Wu, T. et al. clusterProfiler 4.0: A universal enrichment tool for interpreting omics data. Innovation (Camb) 2, 100141 (2021).

29. Yaari, G., Bolen, C. R., Thakar, J. & Kleinstein, S. H. Quantitative set analysis for gene expression: a method to quantify gene set differential expression including gene-gene correlations. Nucleic Acids Res. 41, e170 (2013).

30. Meng, H., Yaari, G., Bolen, C. R., Avey, S. & Kleinstein, S. H. Gene set meta-analysis with Quantitative Set Analysis for Gene Expression (QuSAGE). PLoS Comput. Biol. 15, e1006899 (2019).

31. Goh, W. W. B., Wang, W. & Wong, L. Why Batch Effects Matter in Omics Data, and How to Avoid Them. Trends Biotechnol. 35, 498–507 (2017).

32. Leek, J. T. et al. Tackling the widespread and critical impact of batch effects in high-throughput data. Nat. Rev. Genet. 11, 733–739 (2010).

33. Li, T., Zhang, Y., Patil, P. & Johnson, W. E. Overcoming the impacts of two-step batch effect correction on gene expression estimation and inference. Biostatistics 24, 635–652 (2023).

34. Nyamundanda, G., Poudel, P., Patil, Y. & Sadanandam, A. A Novel Statistical Method to Diagnose, Quantify and Correct Batch Effects in Genomic Studies. Sci. Rep. 7, 10849 (2017).

35. Cai, H. et al. Identifying differentially expressed genes from cross-site integrated data based on relative expression orderings. Int. J. Biol. Sci. 14, 892–900 (2018).

36. Guha, N., Baladandayuthapani, V. & Mallick, B. K. Quantile Graphical Models: Bayesian Approaches. J. Mach. Learn. Res. 21, 1–47 (2020).

37. Jung, S., Bi, Y. & Davuluri, R. V. Evaluation of data discretization methods to derive platform independent isoform expression signatures for multi-class tumor subtyping. *BMC Genomics* 16 Suppl 11, S3 (2015).

38. Robinson, M. D., McCarthy, D. J. & Smyth, G. K. edgeR: a Bioconductor package for differential expression analysis of digital gene expression data. Bioinformatics 26, 139–140 (2010).

39. Cold Spring Harbor Laboratory. Reactome, a Knowledgebase of Biological Processes. (2003).

40. Nixon, L. S., Yung, B., Bell, S. C., Elborn, J. S. & Shale, D. J. Circulating immunoreactive interleukin-6 in cystic fibrosis. Am. J. Respir. Crit. Care Med. 157, 1764–1769 (1998).

41. Hauber, H.-P. et al. Increased expression of interleukin-9, interleukin-9 receptor, and the calcium-activated chloride channel hCLCA1 in the upper airways of patients with cystic fibrosis. Laryngoscope 113, 1037–1042 (2003).

42. Asao, H. Interleukin-21 in Viral Infections. Int. J. Mol. Sci. 22, (2021).

43. Brodeur, T. Y. et al. IL-21 Promotes Pulmonary Fibrosis through the Induction of Profibrotic CD8+ T Cells. J. Immunol. 195, 5251–5260 (2015).

44. Mirzaei, H. & Faghihloo, E. Viruses as key modulators of the TGF-β pathway; a double-edged sword involved in cancer. Rev. Med. Virol. 28, (2018).

45. Rios-Fuller, T. J. et al. Translation Regulation by eIF2α Phosphorylation and mTORC1 Signaling Pathways in Non-Communicable Diseases (NCDs). Int. J. Mol. Sci. 21, (2020).

46. Duran, G. E. & Sikic, B. I. The Syk inhibitor R406 is a modulator of P-glycoprotein (ABCB1)-mediated multidrug resistance. PLoS One 14, e0210879 (2019).

47. Rolf, M. G. et al. In vitro pharmacological profiling of R406 identifies molecular targets underlying the clinical effects of fostamatinib. Pharmacol Res Perspect 3, e00175 (2015).

48. Maiuri, L., Raia, V. & Kroemer, G. Strategies for the etiological therapy of cystic fibrosis. Cell Death Differ. 24, 1825–1844 (2017).

49. Nissen, G. et al. Interleukin-1 beta is a potential mediator of airway nitric oxide deficiency in cystic fibrosis. J. Cyst. Fibros. 21, 623–625 (2022).

50. Zünd, G., Madara, J. L., Dzus, A. L., Awtrey, C. S. & Colgan, S. P. Interleukin-4 and interleukin-13 differentially regulate epithelial chloride secretion. J. Biol. Chem. 271, 7460–7464 (1996).

51. Moran, T. M., Isobe, H., Fernandez-Sesma, A. & Schulman, J. L. Interleukin-4 causes delayed virus clearance in influenza virus-infected mice. J. Virol. 70, 5230–5235 (1996).

52. Donlan, A. N. et al. IL-13 is a driver of COVID-19 severity. JCI Insight 6, (2021).

53. Castellani, S. et al. G-CSF and GM-CSF Modify Neutrophil Functions at Concentrations found in Cystic Fibrosis. Sci. Rep. 9, 12937 (2019).

Low- and high-level information analyses of transcriptome connecting endometrial-decidua-placental origin of preeclampsia subtypes: A preliminary study

Herdiantri Sufriyana[a], Yu-Wei Wu[b], and Emily Chia-Yu Su[cd]

Graduate Institute of Biomedical Informatics, College of Medical Science and Technology, Taipei Medical University, 250 Wu-Xing Street, Taipei 11031, Taiwan

Email: emilysu@tmu.edu.tw

Background. Existing proposed pathogenesis for preeclampsia (PE) was only applied for early onset subtype and did not consider pre-pregnancy and competing risks. We aimed to decipher PE subtypes by identifying related transcriptome that represents endometrial maturation and histologic chorioamnionitis. **Methods.** We utilized eight arrays of mRNA expression for discovery (n=289), and other eight arrays for validation (n=352). Differentially expressed genes (DEGs) were overlapped between those of: (1) healthy samples from endometrium, decidua, and placenta, and placenta samples under histologic chorioamnionitis; and (2) placenta samples for each of the subtypes. They were all possible combinations based on four axes: (1) pregnancy-induced hypertension; (2) placental dysfunction-related diseases (e.g., fetal growth restriction [FGR]); (3) onset; and (4) severity. **Results.** The DEGs of endometrium at late-secretory phase, but none of decidua, significantly overlapped with those of any subtypes with: (1) early onset (p-values $\leq$0.008); (2) severe hypertension and proteinuria (p-values $\leq$0.042); or (3) chronic hypertension and/or severe PE with FGR (p-values $\leq$0.042). Although sharing the same subtypes whose DEGs with which significantly overlap, the gene regulation was mostly counter-expressed in placenta under chorioamnionitis (n=13/18, 72.22%; odds ratio [OR] upper bounds $\leq$0.21) but co-expressed in late-secretory endometrium (n=3/9, 66.67%; OR lower bounds $\geq$1.17). Neither the placental DEGs at first- nor second-trimester under normotensive pregnancy significantly overlapped with those under late-onset, severe PE without FGR. **Conclusions.** We identified the transcriptome of endometrial maturation in placental dysfunction that distinguished early- and late-onset PE, and indicated chorioamnionitis as a PE competing risk. This study implied a feasibility to develop and validate the pathogenesis models that include pre-pregnancy and competing risks to decide if it is needed to collect prospective data for PE starting from pre-pregnancy including chorioamnionitis information.

Keywords: Preeclampsia; Endometrial maturation; Chorioamnionitis; Microarray analysis; Gene regulation.

[a] Department of Medical Physiology, Faculty of Medicine, Universitas Nahdlatul Ulama Surabaya, 57 Raya Jemursari Road, Surabaya 60237, Indonesia.

[b] Clinical Big Data Research Center, Taipei Medical University Hospital, 250 Wu-Xing Street, Taipei 11031, Taiwan.

[c] Clinical Big Data Research Center, Taipei Medical University Hospital, 250 Wu-Xing Street, Taipei 11031, Taiwan; Research Center for Artificial Intelligence in Medicine, Taipei Medical University, 250 Wu-Xing Street, Taipei 11031, Taiwan.

[d] Corresponding author.

1. Introduction

Preeclampsia (PE) is one of pregnancy-induced hypertension (PIH) subtypes related to placenta and endothelial dysfunction [1, 2]. This disease makes the survival more susceptible to cardiovascular diseases later in life [3]. Many studies have proposed pathogeneses for PE [4]. Most of these were typical for the early-onset subtype and shared with those of fetal growth restriction (FGR) [5], whereas both PE and FGR were placenta dysfunction-related diseases (PDDs). Meanwhile, the early-onset subtype only contributed to <30% cases of PE [6]. Therefore, regardless of numerous proposed pathogeneses, the common etiology for most of the PE subtypes is still unclear.

Worldwide, PE affected 3–8% pregnant women [7] and contributed to 11–18% maternal deaths. The risk of hypertension later in life increased 3.7 times for women with a history of PE and the onset was 7.7 years earlier with that of PIH, compared to women without PE or PIH, respectively [8]. Hypertension contributed to all-cause morbidities and mortalities in one fourth adults worldwide, although it is a modifiable risk factor [9]. This disease was more common in postmenopausal women compared to either men or the premenopausal counterparts and only 50% have controlled their blood pressure despite the well-awareness of necessary medications [10]. In addition, since the only cure is early delivery, PE was also the major contributor of prematurity and low-birth-weight infants [11], which led to neonatal deaths [12]. The preterm infants increased neonatal intensive care unit utilization and it was not reduced by the infants born from preeclamptic mother given preventive intervention using low-dose aspirin at 11–13 weeks' gestation. Infants born from the preeclamptic mother also demonstrated signs of cardiac injury [13], which may increase risk of cardiovascular diseases later in life. Therefore, PE prevention has several impacts to mother and child healthcare, including the mortalities, morbidities, resources utilization, and cardiovascular diseases later in life.

Improvements of prevention strategy for any subtypes of PE need understanding of the pathogeneses. These were commonly believed to occur in the first trimester of pregnancy based on timing for the most successful prediction [14]. Yet, most of the comparisons were made against those conducted at the next trimesters without considering pre-pregnancy period [15]. Enormous theories have been proposed for the first-trimester pathogenesis, culminated into 2-stage theory [16, 17]. This consisted of two sequential dysfunctions in placenta and endothelium. However, the cause of pathophysiological derangement in placenta was still unclear [5]. Antecedents of this event were revealed by association between PE with either endometrial maturation [18] or metagenomics profiling of placenta [19]. There was a significant number of differentially expressed genes (DEGs) overlapped between those from preeclamptic chorionic villi sampling (CVS) and those from pathological endometrium [20]. But, the overlapped DEGs were regulated in the same direction instead of opposite ones, indicating the likelihood of co-occurrence instead of potential causal-effect relationship. Meanwhile, many publications describing association between microbiome and PE were proof-of-concept reviews instead of research articles. Eventually, PE remains a vascular disease with unknown etiology. This study aimed to identify transcriptome representing endometrial maturation and histologic chorioamnionitis, enriched by DEGs of the PE subtypes, using microarray meta-analysis workflow at low- and high-level information.

2. Materials and Methods

2.1. *Dataset integration*

A previous workflow on microarray dataset integration was applied [21]. We utilized 15 publicly-accessed microarray experiments of mRNA expression (*n*=653). The datasets were queried in Gene Expression Omnibus (GEO) and Array Express databases. To understand how these datasets helped in achieving the objective of our study, it is important to describe the spatial and temporal contexts of datasets in this study and the conditions they represented (Figure 1). Our datasets covered from pre- to post-pregnancy (post-partum) period. Pre-pregnancy period was represented by endometrial samples, while the pregnancy period was represented by either decidual (maternal side) or placental (fetal side) samples. The placental samples also represented the post-partum period in term of chronic/gestational hypertension phenotype, as defined by the original study. Gestational hypertension starts from 20 weeks' gestation to 6 weeks after delivery, but this condition is considered as chronic hypertension if the elevated blood pressure persisted more than 6 weeks after delivery. Furthermore, our datasets also covered histologic chorioamnionitis, the PE subtypes, and hemolysis, elevated liver enzymes, and low platelets (HELLP) syndrome (mostly preceded PE, but may occur without PE).

We utilized the datasets 1 to 8 for DEGs discovery sets (Figure 1; Tables S1 and S2), which were GSE4888 (*n*=27; dataset 1) and GSE6364 (*n*=37; dataset 2) for endometrium, E-MTAB-680 (*n*=24; dataset 3) for decidua, GSE12767 (*n*=12; dataset 4) and GSE9984 (*n*=12; dataset 5) for CVS (i.e., the first-trimester placenta), and GSE75010 (*n*=157; dataset 6), GSE98224 (*n*=48; dataset 7), and GSE100415 (*n*=20; dataset 8) for third-trimester placenta of normotensive pregnant women and those with several subtypes of PE, other PIH, and other PDDs. The placental samples also consisted of those with and without either histologic chorioamnionitis or HELLP syndrome, but we included second-trimester placenta in addition to third-trimester placenta. All the discovery sets applied total RNA extraction. Endometrium datasets covered proliferative, and early-, mid-, and late-secretory phases. These four phases respectively represented endometrial maturation. Meanwhile, decidua datasets consisted of ectopic pregnancy (implantation site outside endometrium) without or with intermediate decidualization, and intrauterine pregnancy (implantation site inside endometrium) with intermediate and confluent decidualization. These four conditions represented decidualization from the lowest to the highest degree. For endometrium datasets, we excluded subjects with endometriosis and ambiguous histology reading of endometrial phases. For placenta datasets, we only included subjects with phenotypes that fitted the group definitions (see 2.2 Group definition). There were overlapped samples between GSE75010 and GSE98224 (*n*=48), but the duplicates were removed. There were no additional eligibility criteria applied for decidua and CVS datasets beyond those from the original datasets.

For validation sets (Figure 1; Table S1 and S2), we utilized the datasets 9 to 15, which were GSE30186 (*n*=12; dataset 9) with GPL10558 Illumina HumanHT-12 V4.0 expression beadchip, GSE10588 (*n*=43; dataset 10) with GPL2986 ABI Human Genome Survey Microarray Version 2, GSE24129 (*n*=24; dataset 11) with GPL6244 Affymetrix Human Gene 1.0 ST Array (transcript [gene] version), GSE25906 (*n*=60; dataset 12) with GPL6102 Illumina human-6 v2.0 expression beadchip, GSE4707 (*n*=14; dataset 13) with GPL1708 Agilent-012391 Whole Human Genome

Oligo Microarray G4112A (Feature Number version), GSE44711 (n=16; dataset 14) with GPL10558 Illumina HumanHT-12 V4.0 expression beadchip, and GSE128381 (n=183; dataset 15) with GPL17077 Agilent-039494 SurePrint G3 Human GE v2 8x60K Microarray 039381 (Probe Name version). All the validation sets also applied total RNA extraction. Since most of the platforms were different, several DEGs might not be included in both discovery and validation sets corresponding the same subtype. Thus, we only used the intersected genes among them.

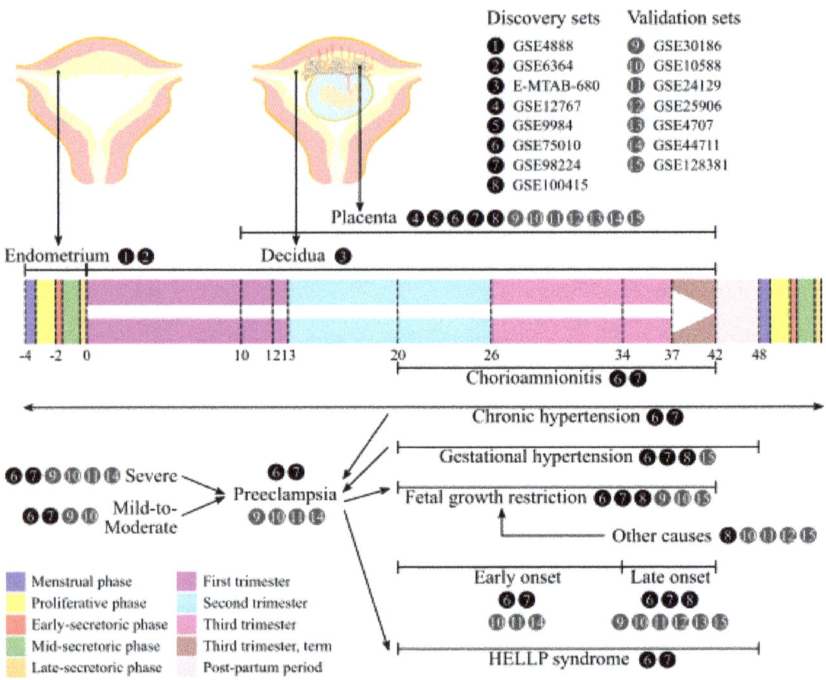

Figure 1. The spatial and temporal contexts of datasets in this study and the conditions they represented. HELLP, hemolysis, elevated liver enzymes, and low platelets.

We did not apply data integration for discovery sets. The experiments using third-trimester placenta were conducted by the same microarray platform of GPL6244 Affymetrix Human Gene 1.0 ST Array (transcript [gene] version). Meanwhile, the remaining experiments using endometrium, decidua, and the first-trimester placenta were conducted by the other platform, which was GPL570 Affymetrix Human Genome U133 Plus 2.0 Array. Therefore, we identified DEGs separately for each tissue that used the same platform (Figure 2A).

Several pathogenesis models for PE subtypes will be developed in the extension work of this preliminary study according to the discovery sets without merging experiments by different platforms. To get comparable gene expression between discovery and validation sets after determining the DEGs and before developing the models, we normalized the validation sets according to the quantile distribution of the discovery controls, as previously described [22]. Therefore, the expression values are centered to those of the discovery controls. To ensure the comparable expression is achieved, we conducted principal component analysis. The samples were not separated among the experiment groups (Figure S1); thus, we could use the validation sets.

For the downstream analysis, the transcripts were summarized into genes. The raw expression data were combined according to the group definition. Outliers were estimated according to relative log expression before normalization and hierarchical clustering of sample-to-sample distances after normalization [23]. After removing outliers, the raw expression data were background-corrected and normalized using robust multi-array average algorithm. Quality control was conducted by data visualization using boxplot, quantile-to-quantile plot, and the MA plot, and confounder identification by surrogate variable analysis.

2.2. *Group definition*

PE is a syndrome characterized by both chronic/gestational hypertension and gestational proteinuria. PE subtypes in these datasets were all possible combinations based on four axes (Figures 1 and 2B): (1) PIH; (2) PDDs; (3) onset; and (4) severity. By PIH, there were two PE subtypes: (1) PE (i.e., gestational hypertension with gestational proteinuria); and (2) superimposed PE (i.e., chronic hypertension with gestational proteinuria). By PDDs, there were PE subtypes without and with FGR. By onset, there were two PE subtypes: (1) early onset (<34 weeks' gestation); and (2) late onset (≥34 weeks' gestation). By severity, there were two PE subtypes: (1) mild-to-moderate (i.e., systolic and diastolic blood pressures [SBP/DBP] of 140/90 to <160/<110 mm Hg with proteinuria of 300 to 2000 mg/24 h, and HELLP negative); and (2) severe (i.e., SBP/DBP of ≥160/110 mm Hg with proteinuria of >2000 mg/24 h, or HELLP positive). As comparators, we also included either chronic/gestational hypertension without gestational proteinuria, of which subtypes were also defined by PDDs and onset but not severity axes.

Using the distribution variances of gene expressions from controls in each tissue dataset, power analysis was conducted to estimate sample size for differential expression analysis with multiple testing by Benjamin-Hochberg false discovery rate (FDR) [24]. We also conducted sample size estimation using the intersected genes among the discovery and validation sets for all possible group combinations. After considering the sample size estimation, the group combinations and the intersected genes were selected for the downstream analysis (Figure 2B).

2.3. *Transcriptome analysis*

Transcriptome analysis was conducted in two stages. In stage 1 (Figure 2B), we used low-level information in each dataset with same experiment and platform to identify a gene set by differential expression analysis. In stage 2 (Figure 2C), high-level information was used across datasets with different experiments and platforms by gene set overlap analysis.

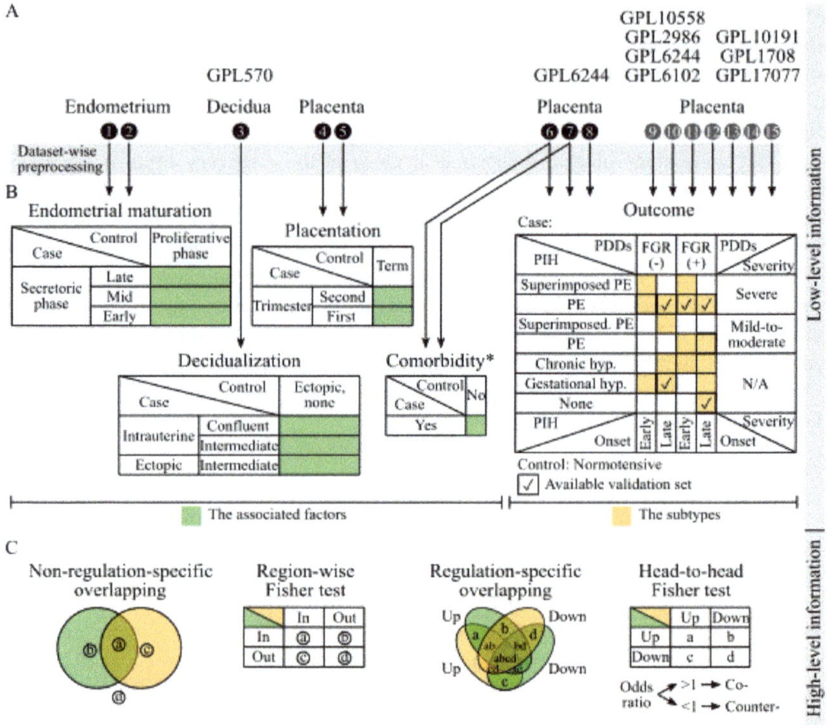

Figure 2. The analytical pipeline: (A) Data preprocessing, including quality control, for each dataset; (B) Differential expression analysis for discovery sets and data integration for validation sets; (C) Gene-set overlap analysis. *, conducted for chorioamnionitis/HELLP in either second or third trimester; FGR, fetal growth restriction; HELLP, hemolysis, elevated liver enzymes, and low platelets; hyp.; hypertension; N/A, not applicable; PDDs, placenta dysfunction-related diseases; PE, preeclampsia; PIH, pregnancy-induced hypertension.

2.3.1. *Low-level information analysis to identify differential expression*

In stage 1, we only conducted a differential expression analysis for each grouping by utilizing transcriptomic data from the same tissue type, i.e., placenta, and the same platform (Figures 2A and 2B). Before differential expression analysis, we filtered out transcripts that were expressed less than 20th percentile. Transcript expression modelling was conducted. We applied moderated *t*-statistics and multiple testing by Benjamini-Hochberg method. The groups were pairs of the subtypes versus control of microarray data from the same platform. Differentially-expressed transcripts were selected if the FDR was less than 5%. Up- and downregulated transcripts were determined based on positive and negative $\log_2$ fold change, respectively.

2.3.2. *High-level information analysis to identify gene set overlap*

In stage 2, since the experiments were conducted with different platforms between the associated factors and the subtypes, to identify association between them we applied association test at high-level information by set operation., i.e., gene set overlap analysis. There were two approaches to overlap a pair of gene sets before applying the Fisher test (Figure 2C): (1) non-regulation-specific overlapping for the region-wise Fisher test; and (2) regulation-specific overlapping for the head-to-head Fisher test. Region-wise Fisher test identified whether an overlap between a pair of gene sets was statistically significant or simply by chance. This test was computed between DEGs of the associated factor with those of the subtype, taking total number of genes of interest into account. Meanwhile, head-to-head Fisher test identified whether co- or counter-expression, indicated by overlaps between a pair of gene sets, were statistically significant or simply by chance (undifferentiated). This test was computed between up- and down regulated DEGs of the associated factor with those of the subtype. If the p-value <0.05, the odds ratio (OR) >1 or and <1 concluded more overlapped DEGs respectively with co-expression and counter expression compared to the opposite regulation. This test determined regulation-specific overlap of interests between the associated factor and the subtype. This approach, however, could not identify the causes of the subtypes; yet, the goal of this study was to gain insights of the possible causes, particularly those related to the pre-pregnancy period and microbial community. For the downstream analysis, we only selected DEGs from the significant non-regulation-specific overlap of interest for each of the subtypes, regardless the significance of the regulation-specific overlap.

2.4. *Code availability*

We used R 4.2.2. To synchronize all the package versions and their dependencies, we used Bioconductor 3.16. All analytical codes are available in https://github.com/herdiantrisufriyana/pec.

3. Results

3.1. *Sample characteristics*

From the publicly-accessed microarray datasets collected for the associated factors, the phenotype characteristics were described (Table S3). Leiomyomata and other non-endometrial conditions were found in the endometrium datasets, but this lesion was beyond the tissue of interest. For placenta datasets, the first- and second-trimester samples were taken from pregnant women with gestational ages of respectively 8.43 ± 0 and 16.43 ± 0 weeks on average, which would be compared with those at 41 ± 0 weeks' gestation on average. Almost all the placenta samples with chorioamnionitis were taken from non-preeclamptic pregnant women without HELLP. Meanwhile, all the placenta samples with HELLP were taken from preeclamptic pregnant women without chorioamnionitis.

The phenotype characteristics were also described for the subtypes (Table S4). No chorioamnionitis was found for placenta samples from pregnant women with PE without or with FGR, regardless of the onset, severity and previous hypertension (i.e., superimposed PE). This situation was the same with the controls of any subtypes, i.e., normotensive pregnancy, and other

PIH subtypes, except early-onset gestational hypertension. Pregnant women with HELLP were only found in the severe subtypes of PE, except superimposed, early-onset, severe PE with FGR.

We conducted differential expression analysis within each experiment of the same tissue (Table S5) with 5118 background genes. They were intersected among the microarray datasets with sufficient, gene-wise sample size. Since we did not find a cohort including all the associated factors and the subtypes, we could only identify high-level associations by overlapping the gene sets determined by different cohorts (Figure 2). The role in the placental dysfunction of the PE subtypes was indicated for endometrial maturation but not decidualization (Figure 3), in addition to placentation. Opposite gene regulation was also indicated between the PE subtypes and chorioamnionitis but not between the PE subtypes and HELLP syndrome (Figure 4; Table S6). We also gained insights related to late-onset PE.

3.2. *Role in placental dysfunction of PE by endometrial maturation but not decidualization*

Endometrial maturation, particularly late-secretory phase, showed a potential role in several subtypes. Significant overlaps were found between DEGs of late-secretory endometrium and placenta under any subtypes with: (1) early onset (p-values $\leq$0.008); (2) severe hypertension and proteinuria (p-values $\leq$0.042); or (3) chronic hypertension and/or severe PE with FGR (p-values $\leq$0.042). Among these overlaps, placenta under early-onset, gestational hypertension also indicated significant counter-expression (OR 0.15, 95% confidence interval [CI] 0.04 to 0.44; p-value <0.001). Meanwhile, a significant co-expression was identified if the subtypes fulfilled criteria of: (1) early-onset, severe superimposed PE (OR 2.43, 95% CI 1.51 to 3.95; p-value <0.001); or (2) early-onset FGR with either severe PE (OR 1.72, 95% CI 1.17 to 2.54; p=0.005) or chronic hypertension only (OR 2.85, 95% CI 1.38 to 6.03; p-value 0.003) but not both (i.e., superimposed PE; OR 1.15, 95% CI 0.69 to 1.93; p-value >0.05). These findings implied endometrial maturation play the putative role by sharing the same up- and down-regulated genes with placenta under those subtypes. Meanwhile, the opposite regulation would lead to gestational hypertension alone without affecting the early onset. In addition, unlike late-secretory endometrium, the overlap patterns were inconclusive between early- and mid-secretory endometrium with the subtypes.

While decidualization is considered a subsequent process of endometrial maturation, our finding did not show its potential role in almost all the subtypes, except late-onset FGR with chronic hypertension. A significant overlap was only found between the DEGs of intrauterine, confluent decidua and placenta under late-onset FGR with chronic hypertension, in which all the decidual DEGs were included in the placental ones (OR ∞, 95% CI 2.09 to ∞; p-value 0.025). Neither significant co- nor counter-expression was identified between both sets of DEGs. Nevertheless, since decidua is a pregnancy version of endometrium, we believe this tissue may connect endometrial maturation and placentation by a process that cannot be identified by gene expression.

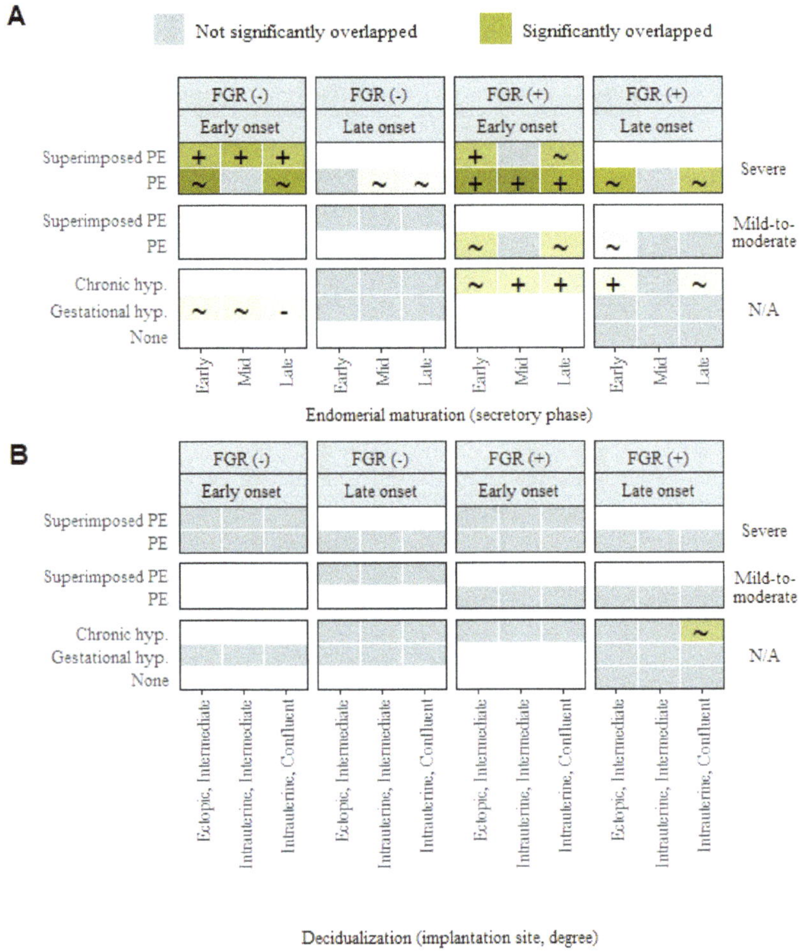

Figure 3. Transcriptome analysis by gene-set overlapping between the subtypes and the associated factors of: (A) Endometrial maturation; and (B) Decidualization. Significances of undifferentiated, co-, and counter-expression are respectively indicated by ~, +, and -. The color gradation represents the number of overlapped DEGs for undifferentiated expression and the number of either co- or counter-expressed DEGs. All non-grey tiles are significant for the non-regulation-specific overlapping. DEGs, differentially-expressed genes; FGR, fetal growth restriction; hyp.; hypertension; PE, preeclampsia; N/A, not applicable.

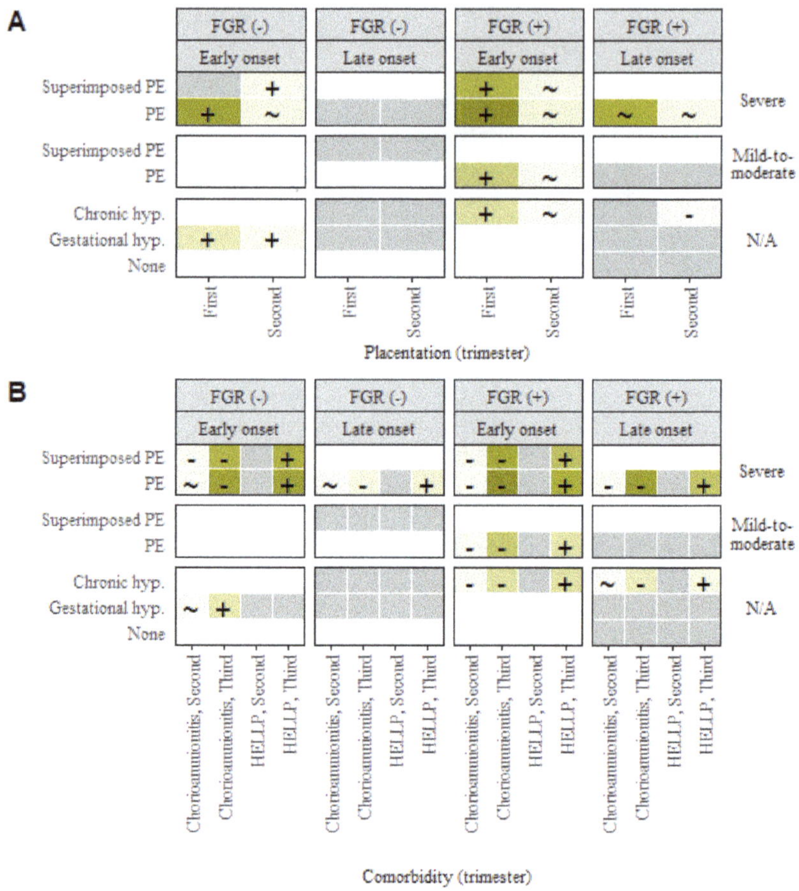

Figure 4. Transcriptome analysis by gene-set overlapping between the subtypes and the associated factors of: (A) Placentation; and (B) Comorbidity. Significances of undifferentiated, co-, and counter-expression are respectively indicated by ~, +, and -. The color gradation represents the number of overlapped DEGs for undifferentiated expression and the number of either co- or counter-expressed DEGs. All non-grey tiles are significant for the non-regulation-specific overlapping. DEGs, differentially-expressed genes; FGR, fetal growth restriction; HELLP, hemolysis, elevated liver enzymes, and low platelets; hyp.; hypertension; PE, preeclampsia; N/A, not applicable.

Several findings were indeed implying the role of endometrial maturation in placentation. Significant overlaps were also found between DEGs of first- and third-trimester placentas respectively under normotensive pregnancy and any subtypes with chronic hypertension and/or PE with FGR (p-values ≤ 0.035). However, the third-trimester placenta was under neither always early onset nor severe hypertension and proteinuria. This exception differed gene expression of first-trimester placenta from that of late-secretory endometrium in terms of overlaps with placenta of the subtypes. The same overlapping patterns were also applied to second-trimester placenta but there was no regulation-specific overlapping. Meanwhile, significant co-expressions (OR lower bounds ≥ 1.56) were identified in the aforementioned overlaps of first-trimester placenta. The role of endometrial maturation in placentation might be related to the impact of chronic hypertension and/or PE on fetal growth more than onset and severity.

3.3. *Competing risk of chorioamnionitis and PE with opposite gene regulation*

Furthermore, endometrial maturation implied a putative role in differing PE from chorioamnionitis. Late-secretory endometrium and placenta under chorioamnionitis shared the same subtypes whose placental DEGs overlapped with their respective ones. However, the gene regulations were significantly counter-expressed in majority for chorioamnionitis (n=13/18, 72.22%; OR upper bounds ≤ 0.21). The remaining overlaps were neither co- nor counter-expressed, except third-trimester placentas under early-onset, gestational hypertension. Its DEGs indicated significant co-expression with those under chorioamnionitis (OR ∞, 95% CI 157.41 to ∞; p-value <0.001) but counter-expression with those of late-secretory endometrium (OR 0.15, 95% CI 0.04 to 0.44; p-value <0.001). These findings implied that PE and chorioamnionitis might have different endometrial maturation.

3.4. *Role in HELLP syndrome by endometrial maturation*

Similar to placenta under chorioamnionitis, endometrial maturation also implied a putative role in differing PE from HELLP. The similar subtypes were also shared, whose placental DEGs overlapped with those of late-secretory endometrium and placenta under HELLP. However, there was an exception, i.e., early-onset, gestational hypertension. The significant overlaps of HELLP were also only in third- (p-values ≤ 0.001) but not second-trimester placenta. This finding implied that a HELLP syndrome might be a competing risk of PE if it occurs alone in earlier trimester. The role of endometrial maturation in HELLP syndrome might be mediated by PE only. In addition, unlike chorioamnionitis, all the gene regulations were significantly co-expressed in the overlaps between the third-trimester, placental DEGs of the shared subtypes and HELLP syndrome (OR lower bounds ≥ 12.79). These findings gain insights in differentiating between HELLP that occurs alone and with PE.

3.5. *Insights related to late-onset, severe PE*

Eventually, it is important to point out the difference between any PE subtypes with early and late onset. Most findings up to this point were related to any PE subtypes with early onset. While we identify a significant overlap between the placental DEGs under late-onset, severe PE without FGR and the endometrial ones at mid- (OR 1.7, 95% CI 1.21 to ∞; p-value 0.004) and late-secretory phases (OR 1.47, 95% CI 1.06 to ∞; p-value 0.026), we did not identify any significant overlaps between these DEGs with the placental DEGs under normotensive pregnancy. Meanwhile, differential expression analysis identified the placental DEGs for: (1) first- and second-trimester versus term placentas; and (2) late-onset, severe PE without FGR versus normotensive pregnancy. These findings implied that placentation did not play any role in late-onset PE without FGR but endometrial maturation did. However, the significant overlaps were identified if this condition was accompanied by FGR (p-values $\leq$0.041); which indicated that placentation might still play a role in FGR under late-onset PE. These findings implied that impaired placentation after impaired endometrial maturation might lead to either an earlier PE or a PE impact on fetal growth.

4. Discussion

4.1. *Interpretation and comparison to previous works*

Transcriptome analysis indicated that the role in placental dysfunction of chronic hypertension and/or severe PE with FGR was potentially played by endometrial maturation but not decidualization, particularly gene expression during late-secretory endometrium. The role of endometrial maturation via placentation for these subtypes was considered smaller in affecting their onset and severity. Meanwhile, only the role of endometrial maturation but none of placentation was indicated in late-onset, severe PE without FGR, distinguishing this subtype among others. In addition, endometrial maturation was also indicated to play a role in HELLP syndrome as a subsequent of PE (e.g., during third trimester) but not as its antecedent (e.g., during second trimester). Furthermore, the decision to terminate the pregnancy was likely taken if the preeclamptic pregnant women was accompanied by HELLP, since most HELLP placenta were found in preeclamptic pregnant women without chorioamnionitis. Therefore, endometrial maturation alone in pre-pregnancy period might play a putative role in PE pathogenesis.

For histologic chorioamnionitis, its gene set also overlapped with those of the chronic hypertension and/or severe PE with FGR. It was similar to late-secretory endometrium but the gene regulation was the opposite. While endometrial maturation might also play a role in histologic chorioamnionitis, its gene expression was regulated in the opposite to those subtypes. This finding might be related to the phenotype data, which implied that chorioamnionitis was likely a competing risk of PE. Either PE or chorioamnionitis was probably diagnosed earlier, and in turn, this resulted in early termination before onset of the other condition. Similarly, the competing risk approach have been used for predicting PE with satisfying accuracy [25]. Furthermore, histological changes of placenta in chorioamnionitis were only increased fetal capillaries without villous remodeling as observed in those of PE, but the changes were more acute [26], probably preceding PE. Hence, the gene regulation of endometrial maturation probably determined if a pregnancy would end up with either PE or chorioamnionitis by affecting microbial community in endometrium.

Furthermore, the role of endometrial maturation in pre-pregnancy period was specific to PE among other PIH/PDDs, e.g., early-onset, gestational hypertension. It shared similar characteristics to chorioamnionitis in term of gene regulations which were the opposites to that of late-secretory endometrium. A previous study showed gestational hypertension had the highest risk of acute chorioamnionitis (n=29/91, 31.9%; p-value <0.001), compared to the other PIH [27]. Nevertheless, what differ early-onset, gestational hypertension from chorioamnionitis is still unclear.

4.2. *Strength and limitation*

By our workflow, an expensive, time-consuming wet lab experiment could be well-prepared not only by literature review but also by a data-driven approach utilizing either at low- or high-level data. This preliminary study also demonstrated how off-the-shelf tools might be variably applied to answer diverse questions in a similar topic. This kind of secondary data analysis across different tissues and timing is inevitable, particularly in pregnancy-related research, because of ethical reasons. Similar situations may also be applied other conditions with long interval time in which a primary data collection is difficult and expensive.

Several limitations are considered in this study. Microarray dataset of gene expression may not help to reveal all parts of the pathogenesis. Nevertheless, we differentiated the several subtypes of PE, and identified the novel, data-driven pathways, using the microarray dataset only. Additional information from the next-generation sequencing data may reveal new perspectives to the proposed pathogenesis of PE in this study, by including non-coding genes. We also could not apply the results directly to develop screening and preventive strategies in clinical settings. This is because we used microarray from tissues by invasive sampling, unlike blood sampling or other methods which are routinely used in clinical settings. Yet, these give a specific direction for the variables and the study design for the next investigation to support the clinical implementation.

5. Conclusions

The role in placental dysfunction was potentially played by endometrial maturation, but not decidualization, for any subtypes with early onset, severe hypertension and proteinuria, or chronic hypertension and/or severe PE with FGR. However, no role of placentation was indicated in late-onset, severe PE without FGR. Both phenotype and genotype also implied that histologic chorioamnionitis was likely a competing risk of PE, in which the gene regulation of endometrial maturation might affect surrounding microbial community to determine if a pregnancy ends up with either PE or chorioamnionitis. In addition, our preliminary results showed the feasibility of developing and validating pathogenesis models of PE subtypes which will be the focus of the extension work of this preliminary study. Eventually, this study will help to decide if future studies need prospective, pre-pregnancy and chorioamnionitis data for preeclampsia.

Acknowledgments

Preprint of an article published in Pacific Symposium on Biocomputing © 2023 World Scientific Publishing Co., Singapore, http://psb.stanford.edu/. This study was funded by: (1) the Postdoctoral Accompanies Research Project from the National Science and Technology Council (NSTC) in Taiwan (grant no.: NSTC111-2811-E-038-003-MY2) to Herdiantri Sufriyana; and (2) the Ministry of Science and Technology (MOST) in Taiwan (grant nos.: MOST110-2628-E-038-001 and MOST111-2628-E-038-001-MY2), and the Higher Education Sprout Project from the Ministry of Education (MOE) in Taiwan (grant no.: DP2-111-21121-01-A-05) to Emily Chia-Yu Su.

Supplementary material

https://github.com/herdiantrisufriyana/pec/blob/master/supplementary_material.pdf.

References

1. Schneider H. Placental Dysfunction as a Key Element in the Pathogenesis of Preeclampsia. Dev Period Med. 2017;21(4):309-16.
2. Nissaisorakarn P, Sharif S, Jim B. Hypertension in Pregnancy: Defining Blood Pressure Goals and the Value of Biomarkers for Preeclampsia. Curr Cardiol Rep. 2016;18(12):131.
3. Vahedi FA, Gholizadeh L, Heydari M. Hypertensive Disorders of Pregnancy and Risk of Future Cardiovascular Disease in Women. Nurs Womens Health. 2020.
4. Rahnemaei FA, Fashami MA, Abdi F, Abbasi M. Factors effective in the prevention of Preeclampsia:A systematic review. Taiwan J Obstet Gynecol. 2020;59(2):173-82.
5. Jim B, Karumanchi SA. Preeclampsia: Pathogenesis, Prevention, and Long-Term Complications. Semin Nephrol. 2017;37(4):386-97.
6. Huluta I, Panaitescu AM. Prediction of preeclampsia developing at term. Ginekol Pol. 2018;89(4):217-20.
7. Abalos E, Cuesta C, Grosso AL, Chou D, Say L. Global and regional estimates of preeclampsia and eclampsia: a systematic review. Eur J Obstet Gynecol Reprod Biol. 2013;170(1):1-7.
8. Bokslag A, van Weissenbruch M, Mol BW, de Groot CJ. Preeclampsia; short and long-term consequences for mother and neonate. Early Hum Dev. 2016;102:47-50.
9. Oparil S, Acelajado MC, Bakris GL, Berlowitz DR, Cífková R, Dominiczak AF, et al. Hypertension. Nat Rev Dis Primers. 2018;4:18014.
10. Dorobantu M, Onciul S, Tautu OF, Cenko E. Hypertension and Ischemic Heart Disease in Women. Curr Pharm Des. 2016;22(25):3885-92.
11. Nijkamp JW, Sebire NJ, Bouman K, Korteweg FJ, Erwich J, Gordijn SJ. Perinatal death investigations: What is current practice? Semin Fetal Neonatal Med. 2017;22(3):167-75.
12. Yu J, Flatley C, Greer RM, Kumar S. Birth-weight centiles and the risk of serious adverse neonatal outcomes at term. J Perinat Med. 2018;46(9):1048-56.
13. Mutlu K, Karadas U, Yozgat Y, Meşe T, Demirol M, Coban S, et al. Echocardiographic evaluation of cardiac functions in newborns of mildly preeclamptic pregnant women within postnatal 24-48 hours. J Obstet Gynaecol. 2018;38(1):16-21.
14. Poon LC, McIntyre HD, Hyett JA, da Fonseca EB, Hod M. The first-trimester of pregnancy - A window of opportunity for prediction and prevention of pregnancy complications and future life. Diabetes Res Clin Pract. 2018;145:20-30.
15. Townsend R, Khalil A, Premakumar Y, Allotey J, Snell KIE, Chan C, et al. Prediction of pre-eclampsia: review of reviews. Ultrasound Obstet Gynecol. 2019;54(1):16-27.
16. Fisher SJ. Why is placentation abnormal in preeclampsia? Am J Obstet Gynecol. 2015;213(4 Suppl):S115-22.
17. El-Sayed AAF. Preeclampsia: A review of the pathogenesis and possible management strategies based on its pathophysiological derangements. Taiwan J Obstet Gynecol. 2017;56(5):593-8.
18. Rabaglino MB, Post Uiterweer ED, Jeyabalan A, Hogge WA, Conrad KP. Bioinformatics approach reveals evidence for impaired endometrial maturation before and during early pregnancy in women who developed preeclampsia. Hypertension. 2015;65(2):421-9.
19. Amarasekara R, Jayasekara RW, Senanayake H, Dissanayake VH. Microbiome of the placenta in pre-eclampsia supports the role of bacteria in the multifactorial cause of pre-eclampsia. J Obstet Gynaecol Res. 2015;41(5):662-9.

20. Rabaglino MB, Conrad KP. Evidence for shared molecular pathways of dysregulated decidualization in preeclampsia and endometrial disorders revealed by microarray data integration. Faseb j. 2019;33(11):11682-95.
21. Planey CR, Butte AJ. Database integration of 4923 publicly-available samples of breast cancer molecular and clinical data. AMIA Jt Summits Transl Sci Proc. 2013;2013:138-42.
22. Heider A, Alt R. virtualArray: a R/bioconductor package to merge raw data from different microarray platforms. BMC Bioinformatics. 2013;14:75.
23. Klaus B, Reisenauer S. An end to end workflow for differential gene expression using Affymetrix microarrays. 2018. https://www.bioconductor.org/packages/devel/workflows/vignettes/maEndToEnd/inst/doc/MA-Workflow.html. Accessed January, 3rd 2020.
24. Warnes GR, Liu P, Li F. ssize: Estimate Microarray Sample Size. R package version 1.60.0. 2019. https://www.bioconductor.org/packages/release/bioc/html/ssize.html. Accessed January, 3rd 2019.
25. Wright D, Wright A, Nicolaides KH. The competing risk approach for prediction of preeclampsia. Am J Obstet Gynecol. 2020;223(1):12-23.e7.
26. Vangrieken P, Vanterpool SF, van Schooten FJ, Al-Nasiry S, Andriessen P, Degreef E, et al. Histological villous maturation in placentas of complicated pregnancies. Histol Histopathol. 2020;35(8):849-62.
27. Stanek J. Placental pathology varies in hypertensive conditions of pregnancy. Virchows Arch. 2018;472(3):415-23.

Deconvolution of Nascent Sequencing Data Using Transcriptional Regulatory Elements

Zachary Maas
Department of Computer Science, BioFrontiers Institute, University of Colorado Boulder
596 UCB
Boulder, CO 80309, USA
E-mail: zachary.maas@colorado.edu

Rutendo Sigauke
BioFrontiers Institute, University of Colorado Boulder
596 UCB
Boulder, CO 80309, USA
E-mail: rutendo.sigauke@colorado.edu

Robin Dowell
Department of Molecular, Cellular, and Developmental Biology,
BioFrontiers Institute, Department of Computer Science, University of Colorado Boulder
596 UCB
Boulder, CO 80309, USA
E-mail: robin.dowell@colorado.edu

The problem of microdissection of heterogeneous tissue samples is of great interest for both fundamental biology and biomedical research. Until now, microdissection in the form of supervised deconvolution of mixed sequencing samples has been limited to assays measuring gene expression (RNA-seq) or chromatin accessibility (ATAC-seq). We present here the first attempt at solving the supervised deconvolution problem for run-on nascent sequencing data (GRO-seq and PRO-seq), a readout of active transcription. Then, we develop a novel filtering method suited to the mixed set of promoter and enhancer regions provided by nascent sequencing, and apply best-practice standards from the RNA-seq literature, using *in-silico* mixtures of cells. Using these methods, we find that enhancer RNAs are highly informative features for supervised deconvolution. In most cases, simple deconvolution methods perform better than more complex ones for solving the nascent deconvolution problem. Furthermore, undifferentiated cell types confound deconvolution of nascent sequencing data, likely as a consequence of transcriptional activity over the highly open chromatin regions of undifferentiated cell types. Our results suggest that while the problem of nascent deconvolution is generally tractable, stronger approaches integrating other sequencing protocols may be required to solve mixtures containing undifferentiated celltypes.

Keywords: Nascent Sequencing, Deconvolution

1. Introduction

One key problem of interest when studying transcription is the ability to capture the heterogeneity that exists in true biological samples.[1] Bulk sequencing samples from cells are an aggregate across a cellular population, and thus average out differences between individual cells to capture only an ensemble profile of a given sample. Notably in the case of samples taken from tissues composed of heterogeneous constituent cells, any celltype specific differences are not necessarily discernible in the heterogeneous mixture of expression data.

To some extent, this problem has been at least partially solved in the context of RNA-seq with the emergence of single cell RNA-seq protocols which allow for RNA content at the level of the individual cell to be measured.[2] However, the relatively high cost of sampling deeply limits the use of scRNA-seq in many contexts. Consequently, a great deal of work has been done to separate samples into constituent cell types *in silico*. This task is interchangeably referred to as deconvolution or microdissection. Deconvolution has been studied extensively in the context of both microarray data and in RNA-seq,[1,3–6] but has seen only limited application to other high throughput genomic data.

Nascent transcription protocols[7,8] are of particular interest for studies into transcriptional regulation.[9,10] Nascent sequencing protocols profile active RNA Polymerase II activity, which captures enhancer associated RNAs (eRNAs), short unstable transcripts that are often associated with transcription factor binding sites.[11] These eRNA transcript have proven to be highly informative markers of transcription factor activity.[9,10,12–16] Unfortunately RNA-seq, whether bulk or single cell, does not capture enhancer associated transcripts due to the fact they are unstable and not polyadenylated.[11] For this reason, the theoretical possibility of single cell measures of nascent transcription has tremendous potential for understanding regulation and transcription factor activity in key biological processes including development and disease progression.

Today, nascent sequencing protocols still operate only on the bulk level, largely because nascent protocols are relatively onerous, taking up to a week to process a set of samples.[7,8,17] Because nascent protocols capture RNA production, many of the signals arise from lowly abundant, highly unstable RNAs.[11] Furthermore, with current biochemical efficiencies, a single cell nascent sequencing protocol is likely infeasible, and thus deconvolution is needed to dissect nascent transcription profiles within tissues.

Nascent transcription data has relatively unique properties compared to RNA-seq. First, RNA-seq measures steady state mature, stable RNA levels which tend to be of relatively high abundance. In contrast, nascent sequencing protocols cover a much larger proportion of the genome ($\sim 40\%$ as opposed to $\sim 8\%$).[17] The consequence is that the average sequencing depth per transcript is typically lower in nascent data, in spite of often sequencing samples to a higher depth. Second, many transcripts measured in nascent protocols are unannotated, lowly transcribed, unstable eRNAs (Figure 1A).[11,17] In development, enhancer activities are the first changes detectable when a cell undergoes state change, suggesting their associated eRNAs have high potential as cell type markers.[18] Furthermore, enhancer associated RNAs tend to be more cell type specific than protein coding genes.[19] However, their low transcription levels lead to issues of reliable detection.[17] Thus methods developed for RNA-seq must be appropriately adapted to use with nascent sequencing data.

Here, we use standardized methods for supervised deconvolution to nascent sequencing data, applying a newly developed filtering technique to solve problems presented by nascent data in the deconvolution context. We show that deconvolution of nascent sequencing data works reliably, albeit with different model performance than in RNA-seq. We find that eRNAs present an informative set of information for deconvolution that can be inferred without a reference annotation. Furthermore, we find that undifferentiated celltypes confound deconvolution of nascent sequencing data, likely because their transcriptional expression resembles that of an aggregate of different differentiated celltypes.

2. Results

The problem of supervised deconvolution with sequencing data is formulated as follows: *Given sequencing samples from homogenous cell types and a heterogenous sample made up of those cell types, can we estimate the mixing proportions of those constituent cell types?* The problem of supervised (or partial) deconvolution is typically formulated as a linear system (Equation 1).[5,20]

$$X = AS \tag{1}$$

Here, X is a single-row matrix with one column per region of interest (ROI) $(1 \times g)$, A is a single row matrix with one column per reference homogenous cell type $(1 \times s)$, and S is a matrix with one row per sample and one column per ROI $(s \times g)$. In most contexts, regions of interest (ROIs) correspond to annotated genes.

This is a overdetermined linear system, since the number of ROIs far exceeds the number of constituent cell types. Additionally, because these are biological values sampled from a noisy process, the key challenge is minimizing errors when solving the system. Most work in the literature has sought to solve the issues of this system in the context of RNA-seq or microarray[1,3–6,20–22] data, with limited applications of this approach to other kinds of sequencing data.

For RNA-seq, a large variety of tools and approaches have been developed,[1,3–5,5,6,20–22] which approach the problem using different models, constraints, and regularization approaches, as well as different ways to shrink the linear system. Many of these approaches claim to be the state-of-the-art, with most tools providing good performance. Consequently, we first examine the deconvolution problem on nascent sequencing using annotated genes and methods developed for RNA-seq.

2.1. *Deconvolution on annotated genes*

To evaluate existing deconvolution methods on nascent sequencing data, we first identified a number of high quality nascent sequencing data sets from a variety of cell types (see Table 2.1). Samples were processed using a standardized analysis pipeline[23] which includes quality control, mapping and bidirectional transcript identification. These bidirectional transcripts originate from both gene start sites and regulatory elements such as enhancers (Figure 1A). The non-gene associated bidirectionals are often referred to as enhancer associated RNAs, or eRNAs.

As a first test, we examined only annotated protein coding genes to mimic deconvolution analysis typically done in RNA-seq. Notably, nascent data differs from RNA-seq in that splicing information is not present in nascent sequencing experiments, as RNA is collected pre-splicing. Furthermore, consistent with standards in nascent transcription analysis,[17] we exclude the +300 initiation region of each gene when using featureCounts[24] to count reads (see Figure 1A), as this avoids the 5′ bidirectional peak.

To simulate a mixed sample, we generated 128 randomly mixed samples by subsampling reads from each reference sample. Samples used for all *in-silico* experiments in this paper were mixed proportionally from raw reads using samtools,[25] and are listed in Table 2.1. With these randomly mixed samples, we then performed supervised deconvolution using 4 different methods which are commonly discussed in the literature — Nonnegative-Least Squares Regression (NNLS), Ridge Regression, LASSO Regression, and ϵ-Support Vector Regression (SVR). For all methods tested, we apply a nonnegativity constraint (all mixing proportions must be at least zero) and a sum-to-one constraint (all mixing proportions must sum to one), as suggested in prior work.[1] These constraints serve to make results from various deconvolution procedures interpretable as mixing weights for the linear deconvolution system. Code and supplemental materials for this project are available at `https://github.com/Dowell-Lab/DeconvolutionNascent`. We find that these methods provide generally good accuracy on

Study	GEO Accession	SRR	Cell Type
Samples used in Figure 1–3			
Jiang 2018[26]	GSM3025555	SRR6789175	HCT116
Fei 2018[27]	GSM3100195	SRR7010982	HeLA
Andrysik 2017[28]	GSM2296635	SRR4090102	MCF7
Dukler 2017[29]	GSM2545324	SRR5364303	K562
Zhao 2016[30]	GSM2212033	SRR3713700	Kasumi-1
Danko 2018[31]	GSM3021718	SRR6780907	CD4+-T-cell
Chu 2018[32]	GSM3309955	SRR7616132	Jurkat-T-cell
Samples added for Figure 4			
Core 2014[33]	GSM1480326	SRR1552485	GM12878
Smith 2021[34]	GSM4214080	SRR10669536	ESC
Ikegami 2020[35]	GSM4207079	SRR10601203	BJ5ta

Table 2.1: Samples used in this study.

deconvolution on our 128 randomly generated mixtures, although it appears that regularized methods perform more poorly than naive NNLS (Figure 1B,C) in certain celltypes across these mixtures. In this context, it appears that regularization does not improve accuracy at the cost of significant computational slowdowns relative to NNLS. Given these promising initial results, we next sought to shift the focus away from annotated genes to the unannotated bidirectional transcripts present at both promoters and enhancers.

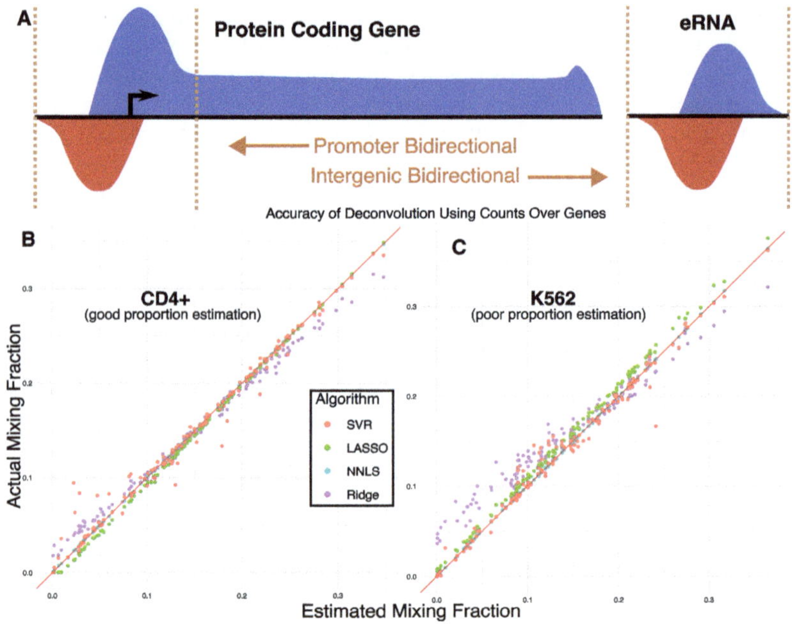

Figure 1. *A:* Nascent transcripts accumulate in a known bidirectional pattern around promoter sites as well as at enhancers.[7,36] These bidirectional regions are counted by convention around ±300bp from the site of RNA Polymerase initiation (roughly the center of the bidirectional).[9,10,36] For annotated genes, we exclude the initiation peak by counting +300 to the annotated transcription end site. *B:* Deconvolution was performed on random mixtures of cells from Table subsection 2.1. Some celltypes show highly accurate estimation of mixing proportion when doing deconvolution over all annotated genes, with most methods showing good linearity in their estimation. *C:* Other celltypes confound the regularized models used here, suggesting a systematic failure of regularization for proper estimation mixing proportion in this naive analysis. This failure appears to be more pronounced with L2 regularized methods and appears in all analyses conducted in this work, to some extent.

2.2. *Identifying bidirectionals as regions of interest*

In addition to transcription at annotated genes, nascent transcription data contains bidirectional transcription at both promoters and regulatory elements. While annotated genes are widely studied and the typical target for this class of deconvolution algorithms, the study of enhancer associated RNAs is important for understanding the regulatory landscape of the cell. Various methods exist to identify sites of bidirectional transcription[36–39] and to combine them across different samples.[10] As such, bidirectionals are an additional region of interest that we now consider in our deconvolution framework.

To this end, we use a combined set of 485,688 bidirectionals, identified by Tfit and dReg within the Nascent-flow framework, capturing both enhancer RNAs and promoter regions, for all samples in Table 2.1.[36,38] Notably, this system is significantly larger than the set of protein coding genes (approximately 490,000 vs 20,000). In this work, we use the following terminology in reference to subsets of this system — Bidirectionals refers to any site of RNA polymerase II initiation and generally includes both promoters and enhancers; any bidirectionals whose $5'$ end (+/-300bp annotated TSS) overlaps an annotated $5'$ gene in the RefSeq hg38 annotation is called a promoter; all other bidirectionals are called enhancers. Given the large size of this system, we next turn our attention to filtering the set of bidirectionals, to shrink the size of the overdetermined system to make deconvolution more computationally feasible.

2.3. *Filtering methods are useful for shrinking the system*

In traditional deconvolution contexts like microarray and RNA-seq, patterns of differential expression are often leveraged to shrink the system. For example, CIBERSORT[40] uses an adaptive filtering method based on DESeq2 to find genes most indicative of specific celltypes. In the context of nascent sequencing data, however, tools like DEseq2 are problematic. The relatively low read coverage and cell type specificity of bidirectionals (e.g. inherent variability) leads DESeq2 to distrust these regions. To counter this, we developed a naive filtering scheme, selecting a fixed number of ROIs defined by the user for each homogenous reference sample where the reads for that sample were most different compared to all other samples. More formally, we define an algorithm for pruning the system of ROIs to a tractable level:

- Filter all ROIs to restrict them to regions where all celltypes have counts lower than the 99th percentile of reads in the sample. We do this to remove outliers whose extreme values could break the assumptions of a linear system.
- Generate transformed ratio T such that for each ROI (row), for each celltype (column), that entry is the log2 ratio of the count at that ROI over the maximum count for that ROI not in that celltype. This step generates a log2 transformed list of the ROIs that are the most specific to a single celltype.
- Order this list by the largest log2 ratio in any celltype in any ROI. Then, walk down this list keeping ROIs such that the number of ROIs for each celltype is approximately equal, up to some limit of elements. This generates a subset of the full system with the most celltype specific elements for each cell. The number of ROIs is approximate because the number of celltype specific elements varies per-celltype and can be exhausted at larger system sizes.

2.4. *Most linear methods perform with high accuracy on synthetic nascent data*

Given that bidirectional regions have distinct transcription characteristics compared to more robustly transcribed annotated genes, we first sought to assess deconvolution methods on the filtered bidirectional set. Using this set, we find that deconvolution achieves a high degree of accuracy (Figure 2A). Unexpectedly, we observe that across all sizes of system tested (including systems far in excess of the total number of genes in the human genome), non-negative least squares (NNLS) regression performs with the highest degree of accuracy. LASSO

(L1 regularized linear regression) has a close second in performance. This is likely because LASSO regularization will only drop out cell types that are unlikely to be present in the mixture. In contrast, Ridge Regression (L2 regularized linear regression) performs worse than all other tested methods for most system sizes. Similarly, ϵ-Support Vector Regression (ϵ-SVR) with L2 regularization also performs relatively poorly compared to NNLS, but relatively well compared to Ridge regression. Despite these differences in accuracy, all models perform reasonably well on our synthetic mixtures, achieving accuracy to within a few percent on randomized mixtures. This is notable because these deconvolution methods perform well both on systems much smaller and much larger than those typically used for deconvolution of RNA-seq data.

Interestingly, we find our subsetting method consistently selects a mixture of enhancers and promoters that does not significantly differ from the distribution expected by random chance (Figure 2B). Consequently, this procedure captures mostly eRNAs and not promoters, since the number of eRNAs far outnumbers the number of promoters. This suggests that certain enhancer-driven regulatory elements are highly informative in identifying celltype.

We next sought to determine which ROIs were most informative to the deconvolution problem. To answer this question, we utilized NNLS, the best performing method in our prior tests. Using NNLS, we compared the performance on bidirectionals (as in Figure 2A), to annotated genes (as in Figure 1B,C) and a combination of these features – selected using our region filtering approach (Figure 3). We find that these methods achieve high accuracy for both genes and bidirectionals across a number of system sizes, with somewhat reduced accuracy when combining these two sets of ROIs. This reduction in accuracy could be a result of colinearity in the combined set of ROIs, as some bidirectionals may be intronic and thus they are not a strictly non-overlapping set relative to annotated genes.

For the data tested and the size of system used, we found that certain methods in the literature were prohibitively slow for the large linear systems we tested. For example, a ν support vector regression (ν-SVR) approach as suggested by CIBERSORT[40] was too computationally expensive to test or benchmark reliably, taking more than 24 hours to do deconvolution on a single mixture of cells at large system sizes (approximately 100k ROIs or more). Due to these poor scaling characteristics, we instead chose to use an optimized implementation of the primal version of ϵ-SVR. This was chosen instead of a dual formulation to maintain computational tractability for the large number of samples relative to the number of features. In the context of nascent sequencing data, NNLS is likely the best model to use based on our benchmarking.

2.5. *Undifferentiated celltypes confound deconvolution of mixtures*

In the course of testing our model, we observed that certain celltypes strongly confounded all deconvolution models tested when using bidirectionals. To understand this puzzling behavior, we examined deconvolution in the presence and absence of these cell types. To do deconvolution of this system, we generated a titration curve, mixing celltypes from distinct separate mixing proportions into equivalent proportions for all celltypes.

We observed that both ESC cell lines and BJ5TA cell lines caused deconvolution to fail (Figure 4A,B). Specifically, inclusion of either cell line results in an overestimation of the mixing proportion for those cell types. We carefully examined these two cell lines to identify

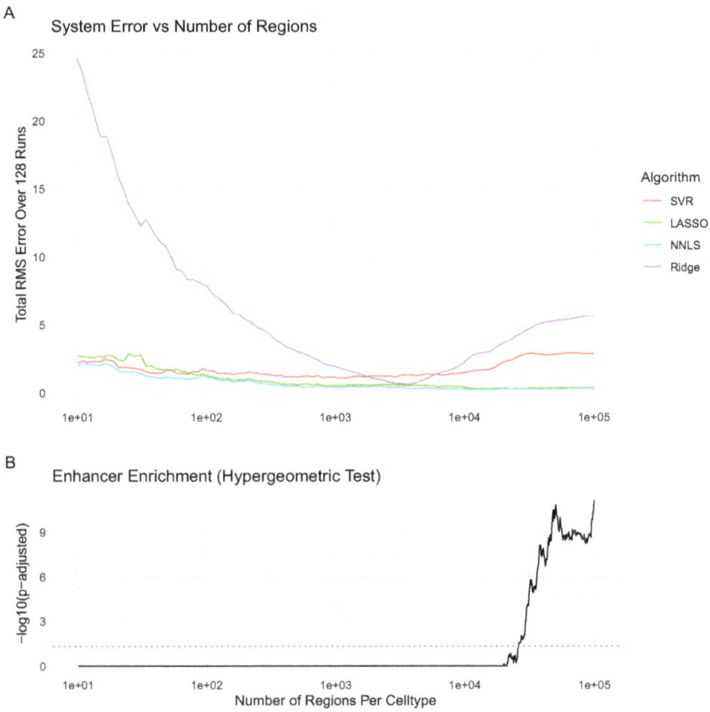

Figure 2. *A:* Models were tested using standard library implementations on a set of 128 randomly generated sets of mixing parameters. Each model was tested on 100 different subsets of ROIs selecting 10^n-many points for $n \in [1, 5]$ using linear spacing between subsequent n. Most models perform well in the intermediate region of 10^3–10^4 points selected per-sample, but diverge outside of that regime. For each set of ROIs selected, the same 128 randomly generated sets of mixing parameters were used as in Figure 2. We observe that for essentially all points, NNLS outperforms more complex models. *B:* To understand the selection process of our subsetting algorithm, we tested whether enhancers were selected from the full ROI set at a greater rate than would be expected by random. To do so, we performed a hypergeometric test with Bonferroni correction over all trials of our ROI subsets. We observe that for smaller system sizes the enhancer/promoter sampling ratio does not differ dramatically from that expected by random sampling. When the system size increases, enhancers become preferentially selected over promoters ($p < 0.05$), but this increase in the rate of enhancer selection does not correlate with the accuracy of any model.

distinguishing features relative to the other cell lines.

To determine whether the number of cell lines or cell line immortalization differences could be the source of the problem, we added lymphoblastoid cell lines immortalized (LCL) by EBV. Notably, LCLs do not confound the model and show excellent performance (Figure 4C). Both

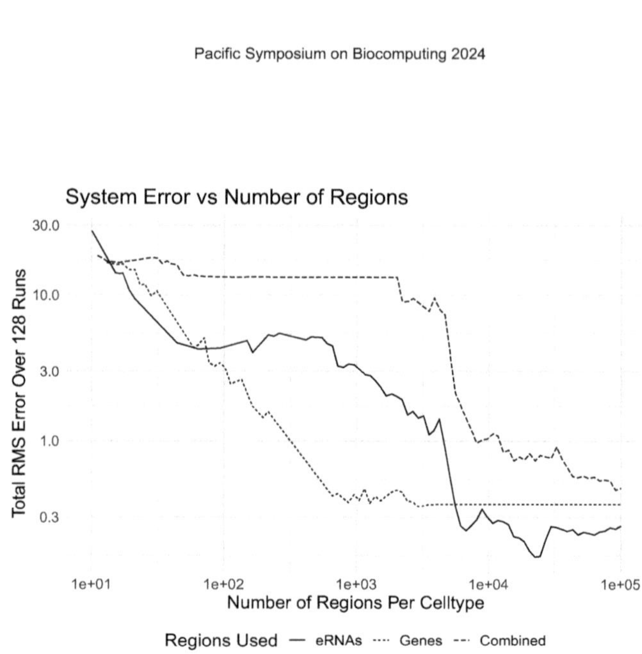

Figure 3. To compare the maximum theoretical accuracy of our system, we conducted the same analysis as in Figure 2 using either the region sets of bidirectionals, annotated genes, or a combination of the two, performing the same subsetting procedure as before. We observe that at smaller region sizes using genes alone provides a higher degree of accuracy than just bidirectionals, but that at larger sets of ROIs bidirectionals alone can achieve a higher absolute degree of accuracy. Somewhat unexpectedly, the combination of both sets of regions performs more poorly than each separate subset. Note that as system size increases, the accuracy of the set using annotated genes reaches a constant level purely because the total size of that system is exhausted by virtue of being an order of magnitude smaller than that of the bidirectionals or combined set.

ESC (embryonic stem cells) and BJ5TA (fibroblast derived) are non-terminally differentiated and non-oncogenic (Figure 4D). Furthermore, we see that even without regularization, NNLS successfully removes non-present celltypes (Figure 4A-C), meaning that undifferentiated celltypes will not be inferred in the mixing proportion if they are not present at all in the mixture. Furthermore, regularization techniques are not required to accomplish this removal of celltypes that are absent.

One alternative hypothesis to the source of this problem is that heterogeneity in the population of undifferentiated celltypes is the source. However, this would suggest that more heterogeneous cell populations should perform worse in deconvolution, as should cells from similar tissue types. Yet based on this data, this seems unlikely, given that both CD4+ and Jurkat cells, both peripheral blood mononuclear cells (PBMC) derived, are present in the mixture and are successfully estimated by our models. Since the addition of a lymphoblast cell line immortalized using EBV (GM12878) does not result in system failure in the same way that is observed with the non-differentiated cell-lines, we suspect that differentiation is the key issue here as opposed to heterogeneity. Our work suggests that undifferentiated or partially differentiated cell types pose a key challenge to the deconvolution of nascent sequencing data

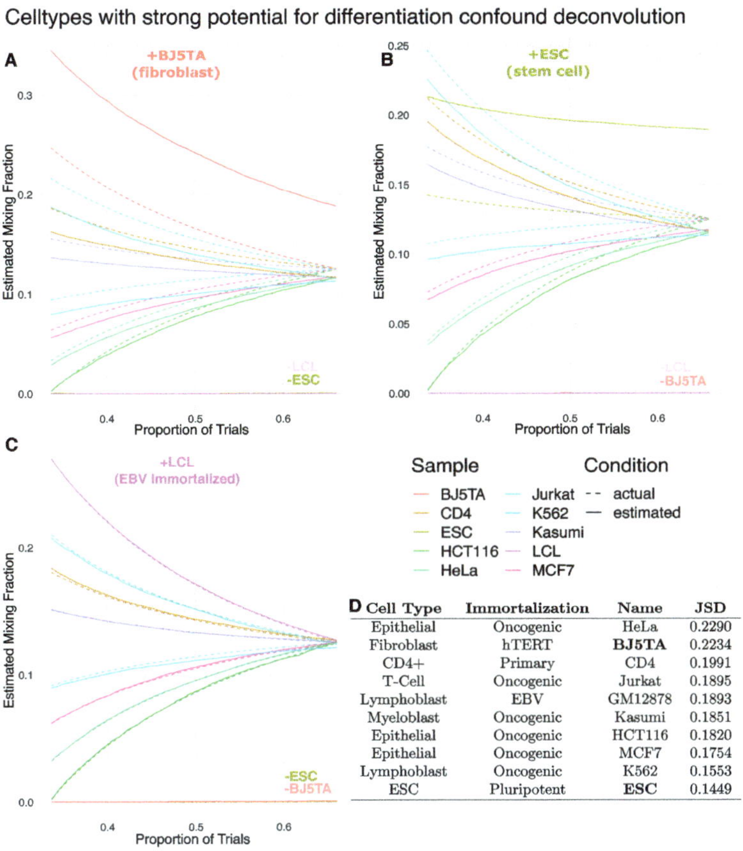

Cell Type	Immortalization	Name	JSD
Epithelial	Oncogenic	HeLa	0.2290
Fibroblast	hTERT	**BJ5TA**	0.2234
CD4+	Primary	CD4	0.1991
T-Cell	Oncogenic	Jurkat	0.1895
Lymphoblast	EBV	GM12878	0.1893
Myeloblast	Oncogenic	Kasumi	0.1851
Epithelial	Oncogenic	HCT116	0.1820
Epithelial	Oncogenic	MCF7	0.1754
Lymphoblast	Oncogenic	K562	0.1553
ESC	Pluripotent	**ESC**	0.1449

Figure 4. To interrogate the effect of undifferentiated and partially differentiated celltypes on the performance of deconvolution, we performed a titration experiment, estimating mixing parameters for 100 different mixtures of celltypes as mixing proportions were taken from maximally separated to equivalent. For each trial n, the mixing proportions are equally spaced points in $[n\frac{1}{n_{\text{tot}}}, 1]$ that are then rescaled to sum to one. Each subset (A,B,C) was generated by holding out one celltype from the full mixture and renormalizing the adjusted mixing proportions to sum to one. *A,B:* Adding either BJ5TA or ESC cells into the mixture causes a higher-than-true proportion of those cells to be estimated. Neither of these cell lines are terminally differentiated. *C:* Addition of EBV immortalized LCL cells into the mixture does not result in failure of the deconvolution model, suggesting that the observed failures are not a function of how cells were immortalized. *D:* To understand if this failure could be attributed to celltype specificity, we calculated the mean Jensen-Shannon Divergence for each sample compared to all others. The pluripotent ESC cells show the lowest celltype specificity while the partially differentiated BJ5TA cells show the highest celltype specificity, with the exception of HeLa cells.

when using enhancers because their regulatory profile, particularly that of their enhancer regions, resemble an ensemble profile of multiple differentiated celltypes. In support of this, the problem does not seem to occur when using genes alone, suggesting that undifferentiated cells may lack the same level of specificity at bidirectionals as terminally differentiated cell types.

Our results suggest either very low or very high celltype specificity when looking at these samples' bidirectional ROIs (Figure 4D). When looking at the mean Jensen Shannon Divergence for each sample compared to all others, we observe that our undifferentiated cell lines are either the least specific (ESC) or the most specific (BJ5TA). Although HeLa cells show the highest degree of celltype specificity by this measure, HeLa cells are not representative of human cells, exhibiting notably different expression patterns[41] which would lead to a high degree of cell type specificity. Past work has shown that ESC cell lines have genome-wide transcriptional hyperactivity[42] that narrows as differentiation progresses. Additionally, work in hematopoetic cells has suggested that these undifferentiated cell lines are characterized by a high degree of fluidity in chromatin modification.[43] More work is required to definitively establish that differentiation is the source of the breakdown of deconvolution in this system, and will likely require significant work outside the scope of this preliminary study.

3. Conclusion

This work is the first to examine supervised deconvolution of heterogenous mixtures of nascent sequencing data. Deconvolution is an essential tool for the study of heterogenous samples, whether cell lines or tissues. While most work on deconvolution of heterogenous samples has moved on to focusing on single cell protocols, a single cell nascent sequencing protocol currently seems infeasible. Thus, nascent sequencing is limited to bulk experiments, which appear to be reliably separable by supervised deconvolution. We present here the use of nascent sequencing data as a testbed for this supervised deconvolution problem. We integrate best practices from the literature and develop new techniques to handle characteristics in nascent sequencing data where assumptions from the RNA-seq deconvolution literature do not hold.

To benchmark various deconvolution algorithms, we first developed a new algorithm to filter ROIs to only use regions with the most celltype specific expression. We find that this selection process does not preferentially select enhancer or promoter ROIs. That said, the number of enhancer associated bidirectionals far exceeds annotated genes, providing ample features from which to select regions of interest. Our proposed algorithm is simple, fast, and reliable, and establishes a strong first basis for the development of more specific ROI filtering tools for nascent deconvolution.

Using this algorithm, we compared standard methods used for solving the deconvolution problem. Specifically, we tested NNLS, Ridge, LASSO, as well as ϵ-SVR. We found that all methods reliably separate the nascent deconvolution system, with L2-regularized methods achieving comparatively poor performance to NNLS. Furthermore, we found that even a simple method like NNLS could reliably eliminate celltypes that were not present in the sample, suggesting regularization is not necessary for solving the deconvolution problem here. While we find that both annotated genes and bidirectionals can achieve high accuracy in supervised deconvolution (with bidirectionals having an edge in absolute accuracy), it is worth emphasizing that bidirectionals are distinctly advantageous in that they are annotation-independent and discovered *de-novo* for each sample.

We show that the addition of undifferentiated samples to a nascent deconvolution system results in highly skewed mixing estimates, with undifferentiated celltypes predicted as far more likely than their actual frequency in the mixture. One possible reason for this is that undifferentated celltypes tend to show regulatory patterns akin to a combination of the regulatory patterns of each constituent celltype. It appears to be a necessary condition for some amount of the undifferentiated celltype to be present in the mixture in order for the system to fail.

One key issue in this work is the lack of availability of diverse high quality nascent sequencing data to perform simulations against. Although a large amount of nascent sequencing data is available and published, the number of cell types available is somewhat limited. Protocols aimed at extending run-on sequencing to a broader base of samples, such as ChRO-seq[44] show promise in alleviating this bottleneck. Importantly, many of the earliest nascent data sets lacked replicates – which excluded their usage here. Data quality and availability is often a limiting factor in computational studies, and this work is not an exception to that rule.

In this work, and generally for the supervised deconvolution problem, we assume that all cells in a sample are taken from an approximately homogeneous population. This is sometimes a reasonable assumption but is often not. One future frontier that could be highly beneficial to this project is the incorporation of single cell ATAC-seq (scATAC) as a secondary source of information to augment bulk nascent sequencing data. scATAC combines the chromatin accessibility readout provided by ATAC-seq (indicative of regions open to transcription) with the cell-specific information provided by modern single cell sequencing protocols. Tools are already well defined for clustering single cell sequencing data into constituent cell types, as individual cells can typically be separated using dimensionality reduction methods like PCA, tSNE, or UMAP.[45,46] Because transcription occurs in regions of open chromatin, which is what ATAC-seq measures, mixing fractions and celltype specific transcripts could be estimated more reliably using combined data from both protocols. Future work combining pairing single cell ATAC-seq data and nascent sequencing data could leverage techniques used by existing tools[21] to do deconvolution on a more granular level for individual samples, providing a strong complementary tool to the bulk deconvolution discussed here. While single cell approaches remain comparatively expensive, this combination would be a powerful tool for looking at transcriptional regulatory networks at the level of sub populations of samples.

Nascent sequencing is a powerful tool for the assessment of transcriptional regulatory networks, and when paired with deconvolution tools will also facilitate deeper understanding of those regulatory networks in heterogeneous cell populations. Leveraging a transcription oriented sequencing approach instead of an expression oriented (e.g. steady state) one provides myriad benefits — more thorough coverage of the genome, understanding of regulatory elements, and a deep view of underlying transcriptional dynamics — all of which can be integrated with different sequencing protocols to great effect. Supervised deconvolution represents an important preliminary foothold into this space, and this work shows that nascent sequencing data is well suited for that class of problems.

Bibliography

1. S. Mohammadi, N. Zuckerman, A. Goldsmith and A. Grama, A Critical Survey of Deconvolution Methods for Separating Cell Types in Complex Tissues, *Proceedings of the IEEE* **105**, 340 (February 2017).
2. B. Hwang, J. H. Lee and D. Bang, Single-cell RNA sequencing technologies and bioinformatics pipelines, *Experimental & Molecular Medicine* **50**, 1 (August 2018).
3. S. S. Shen-Orr, R. Tibshirani, P. Khatri, D. L. Bodian, F. Staedtler, N. M. Perry, T. Hastie, M. M. Sarwal, M. M. Davis and A. J. Butte, Cell type–specific gene expression differences in complex tissues, *Nature Methods* **7**, 287 (April 2010).
4. Y. Zhong, Y.-W. Wan, K. Pang, L. M. Chow and Z. Liu, Digital sorting of complex tissues for cell type-specific gene expression profiles, *BMC Bioinformatics* **14**, p. 89 (March 2013).
5. F. Avila Cobos, J. Vandesompele, P. Mestdagh and K. De Preter, Computational deconvolution of transcriptomics data from mixed cell populations, *Bioinformatics* **34**, 1969 (June 2018).
6. A. R. Abbas, K. Wolslegel, D. Seshasayee, Z. Modrusan and H. F. Clark, Deconvolution of Blood Microarray Data Identifies Cellular Activation Patterns in Systemic Lupus Erythematosus, *PLOS ONE* **4**, p. e6098 (July 2009).
7. L. J. Core, J. J. Waterfall and J. T. Lis, Nascent RNA sequencing reveals widespread pausing and divergent initiation at human promoters, *Science (New York, N.Y.)* **322**, 1845 (December 2008).
8. D. B. Mahat, H. Kwak, G. T. Booth, I. H. Jonkers, C. G. Danko, R. K. Patel, C. T. Waters, K. Munson, L. J. Core and J. T. Lis, Base-pair-resolution genome-wide mapping of active RNA polymerases using precision nuclear run-on (PRO-seq), *Nature Protocols* **11**, p. 1455 (August 2016).
9. J. G. Azofeifa, M. A. Allen, J. R. Hendrix, T. Read, J. D. Rubin and R. D. Dowell, Enhancer RNA profiling predicts transcription factor activity, *Genome Research* **28**, 334 (March 2018).
10. J. D. Rubin, J. T. Stanley, R. F. Sigauke, C. B. Levandowski, Z. L. Maas, J. Westfall, D. J. Taatjes and R. D. Dowell, Transcription factor enrichment analysis (TFEA) quantifies the activity of multiple transcription factors from a single experiment, *Communications Biology* **4**, 1 (June 2021).
11. J. F. Cardiello, G. J. Sanchez, M. A. Allen and R. D. Dowell, Lessons from eRNAs: Understanding transcriptional regulation through the lens of nascent RNAs, *Transcription* **11**, 3 (January 2020).
12. T.-K. Kim, M. Hemberg, J. M. Gray, A. M. Costa, D. M. Bear, J. Wu, D. A. Harmin, M. Laptewicz, K. Barbara-Haley, S. Kuersten, E. Markenscoff-Papadimitriou, D. Kuhl, H. Bito, P. F. Worley, G. Kreiman and M. E. Greenberg, Widespread transcription at neuronal activity-regulated enhancers, *Nature* **465**, 182 (2010).
13. Z. Wang, T. Chu, L. A. Choate and C. G. Danko, Identification of regulatory elements from nascent transcription using dREG, *Genome Research* **29**, 293 (February 2019).
14. M. U. Kaikkonen, N. J. Spann, S. Heinz, C. E. Romanoski, K. A. Allison, J. D. Stender, H. B. Chun, D. F. Tough, R. K. Prinjha, C. Benner and C. K. Glass, Remodeling of the enhancer landscape during macrophage activation is coupled to enhancer transcription, *Molecular Cell* **51**, 310 (August 2013).
15. K. Kristjánsdóttir, A. Dziubek, H. M. Kang and H. Kwak, Population-scale study of eRNA transcription reveals bipartite functional enhancer architecture, *Nature Communications* **11**, p. 5963 (November 2020).
16. S. Bae, K. Kim, K. Kang, H. Kim, M. Lee, B. Oh, K. Kaneko, S. Ma, J. H. Choi, H. Kwak, E. Y. Lee, S. H. Park and K.-H. Park-Min, Rankl-responsive epigenetic mechanism reprograms macrophages into bone-resorbing osteoclasts, *Cellular & Molecular Immunology* **20**, 94 (2023).
17. S. Hunter, R. F. Sigauke, J. T. Stanley, M. A. Allen and R. D. Dowell, Protocol variations in

run-on transcription dataset preparation produce detectable signatures in sequencing libraries, *BMC Genomics* **23**, p. 187 (March 2022).
18. F. Spitz and E. E. M. Furlong, Transcription factors: From enhancer binding to developmental control, *Nature Reviews Genetics* **13**, 613 (September 2012).
19. K. Lidschreiber, L. A. Jung, H. von der Emde, K. Dave, J. Taipale, P. Cramer and M. Lidschreiber, Transcriptionally active enhancers in human cancer cells, *Molecular Systems Biology* **17**, p. e9873 (January 2021).
20. T. Gong, N. Hartmann, I. S. Kohane, V. Brinkmann, F. Staedtler, M. Letzkus, S. Bongiovanni and J. D. Szustakowski, Optimal Deconvolution of Transcriptional Profiling Data Using Quadratic Programming with Application to Complex Clinical Blood Samples, *PLOS ONE* **6**, p. e27156 (November 2011).
21. H. A. Pliner, J. S. Packer, J. L. McFaline-Figueroa, D. A. Cusanovich, R. M. Daza, D. Aghamirzaie, S. Srivatsan, X. Qiu, D. Jackson, A. Minkina, A. C. Adey, F. J. Steemers, J. Shendure and C. Trapnell, Cicero Predicts cis-Regulatory DNA Interactions from Single-Cell Chromatin Accessibility Data, *Molecular Cell* **71**, 858 (September 2018).
22. D. D. Erdmann-Pham, J. Fischer, J. Hong and Y. S. Song, Likelihood-based deconvolution of bulk gene expression data using single-cell references, *Genome Research* **31**, 1794 (October 2021).
23. I. J. Tripodi and M. A. Gruca, Nascent-Flow (December 2018).
24. Y. Liao, G. K. Smyth and W. Shi, featureCounts: An efficient general purpose program for assigning sequence reads to genomic features, *Bioinformatics (Oxford, England)* **30**, 923 (April 2014).
25. H. Li, B. Handsaker, A. Wysoker, T. Fennell, J. Ruan, N. Homer, G. Marth, G. Abecasis and R. Durbin, The Sequence Alignment/Map format and SAMtools, *Bioinformatics* **25**, 2078 (August 2009).
26. W. Jiang, Z. Guo, N. Lages, W. J. Zheng, D. Feliers, F. Zhang and D. Wang, A Multi-Parameter Analysis of Cellular Coordination of Major Transcriptome Regulation Mechanisms, *Scientific Reports* **8**, p. 5742 (April 2018).
27. J. Fei, H. Ishii, M. A. Hoeksema, F. Meitinger, G. A. Kassavetis, C. K. Glass, B. Ren and J. T. Kadonaga, NDF, a nucleosome-destabilizing factor that facilitates transcription through nucleosomes, *Genes & Development* **32**, 682 (May 2018).
28. Z. Andrysik, M. D. Galbraith, A. L. Guarnieri, S. Zaccara, K. D. Sullivan, A. Pandey, M. MacBeth, A. Inga and J. M. Espinosa, Identification of a core TP53 transcriptional program with highly distributed tumor suppressive activity, *Genome Research* **27**, 1645 (October 2017).
29. N. Dukler, G. T. Booth, Y.-F. Huang, N. Tippens, C. T. Waters, C. G. Danko, J. T. Lis and A. Siepel, Nascent RNA sequencing reveals a dynamic global transcriptional response at genes and enhancers to the natural medicinal compound celastrol, *Genome Research* **27** (October 2017).
30. Y. Zhao, Q. Liu, P. Acharya, K. Stengel, Q. Sheng, X. Zhou, H. Kwak, M. Fischer, J. Bradner, S. Strickland, S. Mohan, M. Savona, B. Venters, M.-M. Zhou, J. Lis and S. Hiebert, High-resolution mapping of RNA polymerases identifies mechanisms of sensitivity and resistance to BET inhibitors in t(8;21) AML, *Cell Reports* **16**, 2003 (August 2016).
31. C. G. Danko, L. A. Choate, B. A. Marks, E. J. Rice, Z. Wang, T. Chu, A. L. Martins, N. Dukler, S. A. Coonrod, E. D. Tait Wojno, J. T. Lis, W. L. Kraus and A. Siepel, Dynamic evolution of regulatory element ensembles in primate CD4+ T cells, *Nature Ecology & Evolution* **2**, 537 (2018).
32. T. Chu, E. J. Rice, G. T. Booth, H. H. Salamanca, Z. Wang, L. J. Core, S. L. Longo, R. J. Corona, L. S. Chin, J. T. Lis *et al.*, Chromatin run-on and sequencing maps the transcriptional regulatory landscape of glioblastoma multiforme, *Nature genetics* **50**, 1553 (2018).
33. L. J. Core, A. L. Martins, C. G. Danko, C. T. Waters, A. Siepel and J. T. Lis, Analysis of nascent RNA identifies a unified architecture of initiation regions at mammalian promoters and

enhancers, *Nature Genetics* **46**, 1311 (December 2014).
34. J. P. Smith, A. B. Dutta, K. M. Sathyan, M. J. Guertin and N. C. Sheffield, Quality control and processing of nascent RNA profiling data, *bioRxiv* **22**, p. 2020.02.27.956110 (February 2020).
35. K. Ikegami, S. Secchia, O. Almakki, J. D. Lieb and I. P. Moskowitz, Phosphorylated Lamin A/C in the Nuclear Interior Binds Active Enhancers Associated with Abnormal Transcription in Progeria, *Developmental Cell* **52**, 699 (March 2020).
36. J. G. Azofeifa and R. D. Dowell, A generative model for the behavior of RNA polymerase, *Bioinformatics* **33**, 227 (January 2017).
37. J. Azofeifa, M. A. Allen, M. Lladser and R. Dowell, FStitch: A Fast and Simple Algorithm for Detecting Nascent RNA TranscriptsBCB '14 (ACM, New York, NY, USA, Sept 2014).
38. C. G. Danko, S. L. Hyland, L. J. Core, A. L. Martins, C. T. Waters, H. W. Lee, V. G. Cheung, W. L. Kraus, J. T. Lis and A. Siepel, Identification of active transcriptional regulatory elements from GRO-seq data, *Nature Methods* **12**, 433 (May 2015).
39. Y. Zhao, N. Dukler, G. Barshad, S. Toneyan, C. G. Danko and A. Siepel, Deconvolution of expression for nascent RNA-sequencing data (DENR) highlights pre-RNA isoform diversity in human cells, *Bioinformatics* **37** (August 2021).
40. A. M. Newman, C. L. Liu, M. R. Green, A. J. Gentles, W. Feng, Y. Xu, C. D. Hoang, M. Diehn and A. A. Alizadeh, Robust enumeration of cell subsets from tissue expression profiles, *Nature Methods* **12**, 453 (May 2015).
41. J. J. M. Landry, P. T. Pyl, T. Rausch, T. Zichner, M. M. Tekkedil, A. M. Stütz, A. Jauch, R. S. Aiyar, G. Pau, N. Delhomme, J. Gagneur, J. O. Korbel, W. Huber and L. M. Steinmetz, The Genomic and Transcriptomic Landscape of a HeLa Cell Line, *G3: Genes—Genomes—Genetics* **3**, 1213 (March 2013).
42. S. Efroni, R. Duttagupta, J. Cheng, H. Dehghani, D. J. Hoeppner, C. Dash, D. P. Bazett-Jones, S. Le Grice, R. D. G. McKay, K. H. Buetow, T. R. Gingeras, T. Misteli and E. Meshorer, Global transcription in pluripotent embryonic stem cells, *Cell stem cell* **2**, 437 (May 2008).
43. Y. S. Chung, H. J. Kim, T.-M. Kim, S.-H. Hong, K.-R. Kwon, S. An, J.-H. Park, S. Lee and I.-H. Oh, Undifferentiated hematopoietic cells are characterized by a genome-wide undermethylation dip around the transcription start site and a hierarchical epigenetic plasticity, *Blood* **114**, 4968 (December 2009).
44. T. Chu, E. J. Rice, G. T. Booth, H. H. Salamanca, Z. Wang, L. J. Core, S. L. Longo, R. J. Corona, L. S. Chin, J. T. Lis, H. Kwak and C. G. Danko, Chromatin run-on and sequencing maps the transcriptional regulatory landscape of glioblastoma multiforme, *Nature Genetics* **50**, 1553 (2018).
45. L. van der Maaten and G. Hinton, Visualizing High-Dimensional Data Using t-SNE, *Journal of Machine Learning Research* **9**, 2579 (2008).
46. L. McInnes, J. Healy and J. Melville, UMAP: Uniform Manifold Approximation and Projection for Dimension Reduction (September 2020).

Splitpea: quantifying protein interaction network rewiring changes due to alternative splicing in cancer

Ruth Dannenfelser and Vicky Yao†
Department of Computer Science, Rice University,
Houston, TX 77005, USA
†*E-mail: vy@rice.edu*

Protein-protein interactions play an essential role in nearly all biological processes, and it has become increasingly clear that in order to better understand the fundamental processes that underlie disease, we must develop a strong understanding of both their context specificity (e.g., tissue-specificity) as well as their dynamic nature (e.g., how they respond to environmental changes). While network-based approaches have found much initial success in the application of protein-protein interactions (PPIs) towards systems-level explorations of biology, they often overlook the fact that large numbers of proteins undergo alternative splicing. Alternative splicing has not only been shown to diversify protein function through the generation of multiple protein isoforms, but also remodel PPIs and affect a wide range diseases, including cancer. Isoform-specific interactions are not well characterized, so we develop a computational approach that uses domain-domain interactions in concert with differential exon usage data from The Cancer Genome Atlas (TCGA) and the Genotype-Tissue Expression project (GTEx). Using this approach, we can characterize PPIs likely disrupted or possibly even increased due to splicing events for individual TCGA cancer patient samples relative to a matched GTEx normal tissue background.

Keywords: alternative splicing; protein-protein interaction networks; protein network rewiring

1. Introduction

Alternative splicing is a crucial mechanism that underlies the increased complexity of higher eukaryotes. It is now estimated that ~95% of human genes[1,2] undergo splicing changes, and the increase in protein diversity that results from splicing has been put forth as one of the primary explanations for the apparent mismatch between species complexity and their genome size.[3,4] Importantly, alternative isoforms of the same gene can exhibit highly different interaction profiles and thus affect the dynamics of protein interaction networks.[5] Splicing has been shown to be a key regulator of tissue specificity (especially in the brain),[2,6] and dysregulation has been increasingly implicated in a wide array of diseases,[7] from cancer[8,9] to neurodegenerative diseases.[10] Thus, it is critical to understand the changes in protein interactions due to splicing that underlie cellular function and dysfunction.

However, a systematic study of splicing-related protein network dynamics is hampered by multiple challenges. Although emergent experimental approaches to directly screen for isoform-level protein-protein interactions are promising,[5] they are very early in development and highly restricted in resolution. Furthermore, all such screens are naturally bounded by not only a combination of technical and cost constraints, but also the inherent complexity of the underlying networks and the vast number of potential cell types and conditions of interest. Fortunately, the now standard use of RNA-sequencing provides a window into the exploration of splicing patterns across varied conditions. While RNA-seq data alone is still insufficient to chart out the entirety of any particular splicing interaction network, it can be used to understand condition-specific splicing dynamics.

Here, we present Splitpea (SPLicing InTeractions PErsonAlized), a method for detecting sample-specific PPI network rewiring events. Splitpea takes advantage of the key insight that splicing can disrupt critical *protein domains* that mediate PPIs through domain-domain interactions (DDIs), which have been derived based on a mix of structural, evolutionary, and computational approaches.[11–14] Splitpea integrates PPI and DDI information with sample-specific differential splicing events, and can be used easily in concert with existing, established computational approaches for the identification and quantification of differential splicing.[15] In the scenario where only an individual sample is available or a different background context is preferable (versus existing control samples), Splitpea provides functionality to use a separate reference database of background splice events; for example, one can choose to use normal GTEx data as background for individual TCGA cancer samples (matched by tissue type). Furthermore, as part of Splitpea's characterization of the potential downstream interaction network changes, Splitpea indicates likely direction: gain, loss, or chaos (mixed / unclear).

Thus, to our knowledge, Splitpea is the first general tool to characterize potential direction of protein interaction rewiring due to splicing for individual samples. We demonstrate the utility of Splitpea on breast and pancreatic cancer samples from TCGA, using matched normal tissue samples (breast and pancreas) from GTEx. All source code for Splitpea and the corresponding analyses are available via Github (`https://github.com/ylaboratory/splitpea`), with additional links to download all data and associated networks.

1.1. *Prior work*

Prior work considering domain-domain interactions in the context of splicing have mostly focused on query-based or visualization interfaces. Many consider interactions at the isoform level, aiming to provide a context-specific isoform interaction graph.[16–18] There has been relatively less work focusing on characterizing network rewiring events. Recently, the first tool to characterize the mechanistic effects of splicing on downstream PPIs was proposed,[19] but this tool is unable to differentiate between the potential directionality of interaction rewiring (likely gain or loss events). Specifically for the study of cancer, there has also been large-scale analysis efforts to characterize the impact of splicing on PPIs across patients.[8] Though this work was not patient-specific, it provided strong evidence to demonstrate that there exists a large catalog of isoform changes (with potential downstream impacts on PPIs and regulatory networks) that exist independently of expression changes in cancer. Beyond using PPI networks, there have also been exciting efforts integrating cancer RNA-seq together with somatic mutation data and using functional networks to interpret the downstream impact of splicing.[20]

2. Methods

2.1. *Protein interaction and domain interaction data*

Human protein-protein interactions were downloaded from BioGRID (v4.4.207),[21] DIP (2017-02-05),[22] HIPPIE (v2.2),[23] HPRD (Release 9),[24] Human Interactome (HI-II),[25] IntAct (2022-04-18),[26] iRefIndex (v18.0),[27] and MIPS (Nov 2014).[28] All proteins were mapped to Entrez Gene IDs.[29]

Known and predicted domain-domain interactions were downloaded from 3did (v2017_06),[11] DOMINE (v2.0),[12] IDDI (2011.05.16),[13] and iPFAM (v1.0).[14] For predicted DDIs, only interactions with confidence > 0.5 were used in downstream analyses.

Protein domain locations were translated to genomic locations using the Ensembl BioMart API and the biomaRt R package[30] and indexed using tabix[31] to facilitate fast retrieval given a set of genomic coordinates.

2.2. *Tissue and tumor splicing data processing*

Spliced exon values in the form of percent spliced in (PSI or ψ) were obtained for both normal pancreas and breast tissue samples from the Genotype-Tissue Expression (GTEx) project and pancreatic cancer and breast cancer samples from The Cancer Genome Atlas (TCGA) using the IRIS database.[32] IRIS uses rMATS[33] to tabulate ψ values for skipped exon events (the most abundant splicing event). Though we use rMATS ψ values in this study, Splitpea is agnostic to the choice of upstream differential splicing analysis tool and can easily be applied in concert with other tools that use a form of ψ as their quantification metric.[34–37]

Specifically, we delineate ψ_i as the ψ value for exon $i = 1, ..., n_E$, where there are n_E total exons that had a reported exon skipping event. Note that the precise exons captured in the sample of interest and the background samples are typically non-identical. We are only able to estimate ψ for exons that are captured in both, and thus, n_E represents the number of exons that lie at the intersection of the two larger sets of exons. In the scenario where a background reference distribution of ψ values are provided, we calculate $\Delta\psi_i$ as the following:

$$\Delta\psi_i^{(s)} = \psi_i^{(s)} - \frac{1}{n_B}\sum_{b=1}^{n_B} \psi_i^{(b)} \tag{1}$$

where $\psi_i^{(s)}$ is the ψ for exon i in our sample of interest s (e.g., a cancerous pancreatic sample from TCGA), while $\psi_i^{(b)}$ is the ψ for the same exon i in an individual background sample b (e.g., a normal pancreatic sample from GTEx), and n_B is total number of background samples. Intuitively, larger n_B will provide better estimates of the background distribution, especially if there is large variability in splicing patterns. We recommend assembling backgrounds with at least $n_B \geq 30$ for the empirical cumulative density function estimate below.

ψ values lie in the range $[0, 1]$; thus $\Delta\psi \in [-1, 1]$, and we are naturally primarily interested in significant events for large $|\Delta\psi|$ values (cases where exons are significantly skipped or significantly retained relative to reference). To calculate an estimated significance level for $|\Delta\psi_i^{(s)}|$, we rely on similar intuition as used in previous studies,[8,38] that the normal reference samples can be used to construct an empirical cumulative density function for each exon:

$$\hat{F}_{n_E}(t_i) = \frac{1}{n_B}\sum_{b=1}^{n_B} \mathbf{1}_{|\psi_i^{(b)}|\leq t_i} \tag{2}$$

where $\mathbf{1_A}$ is the indicator function for event A. Given this exon-specific $\hat{F}_{n_E}(t_i)$, we can estimate an empirical p-value for each exon i in sample s

$$\hat{p}_i^{(s)} = \frac{1}{2}(1 - \hat{F}_{n_E}(|\Delta\psi_i^{(s)}|)) \tag{3}$$

Finally, as input to Splitpea, we filtered exons to only those that are significantly different from background ($\hat{p}_i^{(s)} < 0.05$) and those with a $\Delta\psi$ change bigger than 0.05 ($|\Delta\psi| > 0.05$), defined as ψ below. We chose to use a p-value cutoff here as opposed to a multiple hypothesis corrected value to reduce false negatives, because we are interested in any possible rewiring events. We hope that this will better enable Splitpea's use for hypothesis generation tasks. In general, these thresholds can be easily varied depending on the downstream purpose.

2.3. *Clustering $\Delta\psi$ values*

For each cancer type, we remove any exons that had missing values in any of the samples, then filtered the exons by variance, keeping only those with variance greater than 0.01. The final set of $\Delta\psi$ for each cancer type were clustered using the complete hiearchical clustering algorithm and plotted with the heatmap.2 function in the gplots R package.[39] Clinical annotations for TCGA samples were obtained from the Genomic Data Commons portal with Pam50 calls from Netanely et al.[40]

2.4. *Network rewiring algorithm*

There is inherent complexity in considering the impact of exon changes on protein domains, and finally, proteins, as there are several many-to-many relationships. A single exon can include multiple protein domains, but a single protein domain can also span multiple exons; proteins can thus consist of multiple exons as well as multiple protein domains. Splitpea hones in on potentially domain-mediated protein interactions by first overlaying DDIs on the aggregated PPI network based on the presence of each of the domains that constitute the pair of interactors in the protein. In other words, for a pair of proteins g_1 and g_2, we consider protein domain d_1 in g_1 and domain d_2 in g_2 as potentially mediating a known PPI between g_1 and g_2 if a DDI has been reported between d_1 and d_2. Fig. 1A depicts an example interaction where several DDIs potentially mediate the same PPI.

In the event that there are multiple exons within the same protein domain, we attribute the *minimum* $\Delta\psi$ value to the entire protein domain. The underlying assumption here is that loss of any portion of a particular protein domain may potentially negatively impact the protein domain's downstream capacity to interact with other domains. Splitpea then determines the directionality of change based on whether or not there is consistency across the changing domains. In the event that there are mixed exon changes, the directionality is labeled as "chaos," or undetermined (Fig. 1B). The weight of the edge is calculated as

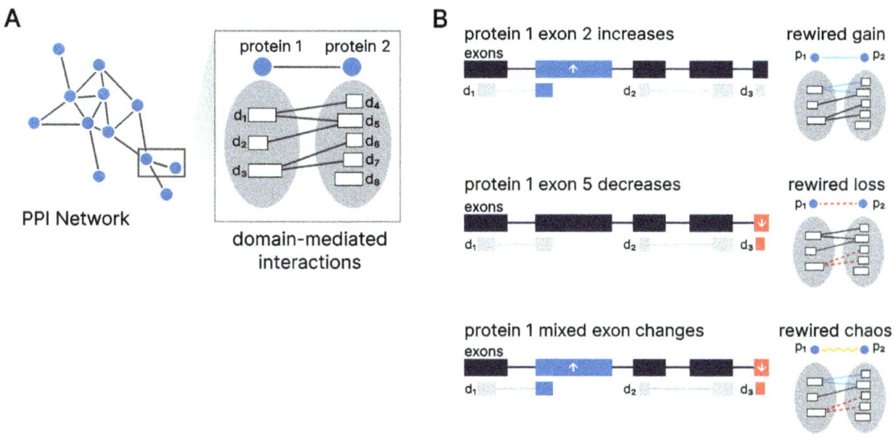

Fig. 1. *Overview of Splitpea.* Splitpea combines prior knowledge in the form of protein-protein and domain-domain interactions with splicing changes to provide a view of a rewired network for a given experimental context. Splitpea defines a rewiring event when exon changes affect an underlying domain-domain interaction. Toy scenarios that would result in the three possible rewiring events predicted by Splitpea are illustrated in B.

the mean domain-level $\Delta\psi$ values. Essentially, the following pseudocode describes the crux of Splitpea's algorithm for a given sample with a set of exons with associated $\Delta\psi$ values:

for each PPI between g_u, g_v **do**
  $\Psi^{(u)}$:= significant exons for gene u
  $\Psi^{(v)}$:= significant exons for gene v
  $D^{(u)} := \{d_u | d_u \in g_u, \exists \text{ exon}_i \text{ s.t. exon}_i \in \Psi^{(u)} \;\&\; \text{exon}_i \in d_u\}$
  $D^{(v)} := \{d_v | d_v \in g_v, \exists \text{ exon}_i \text{ s.t. exon}_i \in \Psi^{(v)} \;\&\; \text{exon}_i \in d_v\}$
  w_{uv} := network rewiring edge weight between g_u, g_v
  δ_{uv} := direction classification of network rewiring between g_u, g_v
  for each DDI between $d_u \in D^{(u)}, d_v \in D^{(v)}$ **do**
    $\Delta\psi_{d_u} := \min(\{\Delta\psi_i | \text{ exon}_i \text{ s.t. exon}_i \in \Psi^{(u)} \;\&\; \text{exon}_i \in d_u\})$
    $\Delta\psi_{d_v} := \min(\{\Delta\psi_i | \text{ exon}_i \text{ s.t. exon}_i \in \Psi^{(v)} \;\&\; \text{exon}_i \in d_v\})$

  if $\forall d_u, d_v \in \text{DDI}(d_u, d_v), \Delta\psi_{d_u} > 0, \Delta\psi_{d_v} > 0$ **then**
    δ_{uv} = positive
  else if $\forall d_u, d_v \in \text{DDI}(d_u, d_v), \Delta\psi_{d_u} < 0, \Delta\psi_{d_v} < 0$ **then**
    δ_{uv} = negative
  else
    δ_{uv} = chaos
  $w_{uv} = \frac{1}{|D^{(u)}|+|D^{(v)}|}(\sum_{d \in D^{(u)}} \Psi^{(u)} + \sum_{d \in D^{(v)}} \Psi^{(v)})$

  return w_{uv}, δ_{uv}

w_{uv} and δ_{uv} are reported as long as $\Psi^{(u)}$ *or* $\Psi^{(v)}$ is non-empty. Please note that the w_{uv} calculation only includes domains that have a DDI that is considered to be mediating the PPI between g_u, g_v. For readability, the equation above omits the removal of non-DDI pairs.

2.5. *Consensus network*

The main factor to consider when aggregating several sample-specific Splitpea networks into a consensus network is whether the directionality of edges agree. Thus, a "positive" consensus network and "negative" consensus network are built separately. "Chaos" edges are ignored since they are of ambiguous state. For each consensus network, two factors are considered for the edge weight: the sum of the original edge weights w_{uv} and how many networks support the same directionality δ_{uv}. The downstream analysis with each consensus network focuses on the largest connected component. As is common in biological networks, we found that the largest connected component covers the majority of the edges of the complete consensus network (breast cancer: 96.4% edges retained in negative consensus, 89.5% edges retained in positive consensus; pancreatic cancer: 96.1% edges retained in negative consensus, 88.8% edges retained in positive consensus).

2.6. *Network embedding and clustering*

To enable network clustering and other downstream uses of the Splitpea patient-specific networks, we created whole graph level embeddings. Here, we chose to focus only on potential gain-of-interaction edges and first filtered each patient-specific network accordingly. Taking the largest connected component, we applied the FEATHER[41] algorithm from the KarateClub NetworkX extension library[42] to generate an embedding for each network.

We clustered the resulting embeddings for each cancer type using hierarchical density-based clustering (HDBSCAN)[43] with minimum cluster sizes of 10. Clustering results were generally robust to the choice of the minimum cluster size parameter; 10 was chosen for downstream interpretability (and we would consider samples with fewer neighbors as outliers). Final plots were produced using principal component analysis (PCA), plotting all embeddings by their first two components.

3. Results

3.1. *Quantifying splicing changes in pancreatic and breast tumors*

In total, we collected data from TCGA covering 177 pancreatic primary tumors and 1,088 breast primary tumors, together with 192 normal pancreatic tissue and 218 normal breast tissue samples from GTEx that were used as a reference distribution of normal splicing variation for each respective cancer type. With these data, we calculated a $\Delta\psi$ value corresponding to the change in exon splicing in each tumor sample relative to its normal tissue background, resulting in $\Delta\psi$ estimates for a total of 139,661 unique exons across all breast cancer samples and 98,761 unique exons across the pancreatic cancer samples. Furthermore, we calculated an accompanying p-value that compares how extreme the observed ψ value for each exon in each cancer sample is relative to the corresponding background distribution of ψ values for normal tissue samples (see Methods).

3.2. $\Delta\psi$ *values primarily reflect primary diagnoses*

We then clustered the $\Delta\psi$ matrices for each tumor type and checked whether they corresponded to relevant clinical and pathological tumor features for both breast cancer (Fig. 2A) (pam50 subtypes, diagnosed type, pathologic stage, and age) and pancreatic cancer (Fig. 2B) (site of origin, diagnosed type, pathologic stage, age, and sex). While the majority of clinical features are not meaningfully clustered with $\Delta\psi$ values, we do observe that the most unique patient cluster for pancreatic cancer (far right columns in Fig. 2B) are all pancreatic neuroendocrine tumors. Neuroendocrine tumors are a rare subset of pancreatic cancers that originate not in the cells of the pancreas but in neuroendocrine cells. Interestingly, this cell type has commonality with neurons which are known to undergo more splicing changes.[44] For breast cancer, we see some clustering of lobular carcinomas (red cluster in "type" bar Fig. 2A), but otherwise do not see obvious patterns of clinical or pathological separation with $\Delta\psi$ values alone.

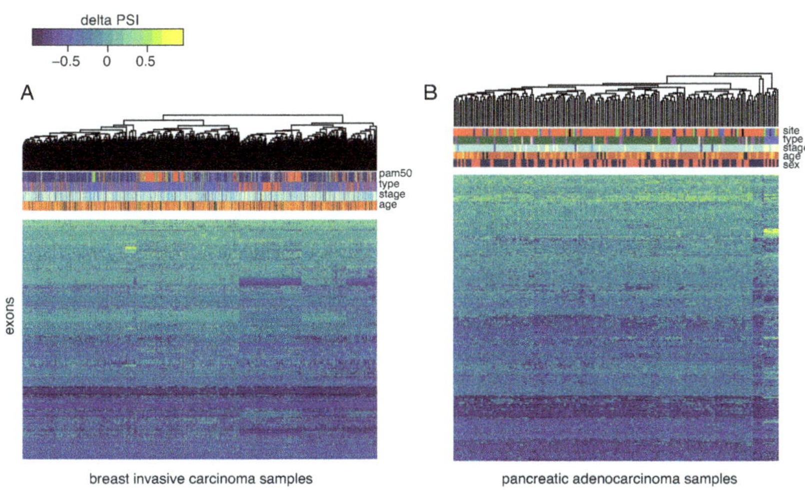

Fig. 2. *Clustering on* $\Delta\psi$ *values.* We cluster the $\Delta\psi$ values showing different sample groups for different spliced exons. Heatmaps depict splicing changes relative to average normal tissue background. Bar columns show known clinical information about each sample. In general, there are more subgroup level exon changes for breast cancer, (A) but these are not strongly correlated with any clinical variable. In pancreatic cancer, a small subset of neuroendocrine samples (B, dark blue) share similar splicing patterns. All other samples do not have obvious meaningful structure.

3.3. *Quantifying rewired protein-protein interactions for pancreatic and breast tumors*

We applied Splitpea to build patient-specific rewired PPI networks for 177 pancreatic and 1,088 breast primary tumor samples. Each PPI network contains three types of edges (gain, loss, or chaotic (mixed)) based on how underlying splicing changes may affect the individual protein-protein interaction (Fig. 3). In general, most splicing changes cause potential loss of protein interactions, though breast cancer had relatively fewer loss of edges proportionally on average (76% edges) than pancreatic cancer (84% of edges). Chaos (mixed) edges, where domain interactions have inconsistent directions per protein are relatively uncommon and comprise on average less than 2% of total edges for pancreatic and breast cancer. Between the two cancer types, breast cancer has more potential gain-of-interaction edges and a lower proportion of potential lost edges relative to pancreatic cancer. Interestingly, there is also more variability across edge types per sample in breast cancer samples.

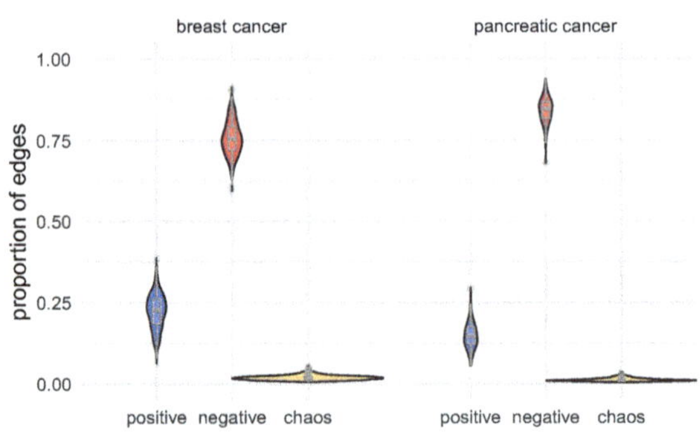

Fig. 3. *Proportion of relative gain and loss in edges across breast cancer and pancreatic cancer samples.* Breast cancer samples have proportionally more "gain of interactions" than pancreatic cancer samples, but in both cancer types, interaction loss is much more prevalent. For each TCGA cancer sample, the proportion of edges gained versus lost is calculated using the total number of edges in the largest connected component of the entire Splitpea rewired network (both directions) as the denominator. To be conservative, the number of edges retained in the largest connected components for the gain-only subnetwork and loss-only subnetworks are used as numerators.

Looking at individual patient networks (Fig. 4), we can see potential hubs and protein clusters that undergo extensive remodeling. In Fig. 4A, we show an example of one pancreatic tumor network with the most remodeling changes in the oncogene, RAB35, proto-oncogenes, HRAS and FYN, the signaling protein, MAPK3, the cell cycle and growth genes, NEDD8 and PRKAA1, among others. Breast cancer patient-specific networks have a different topology (Fig. 4C), though there is also overlap of proto-oncogenes HRAS and FYN.

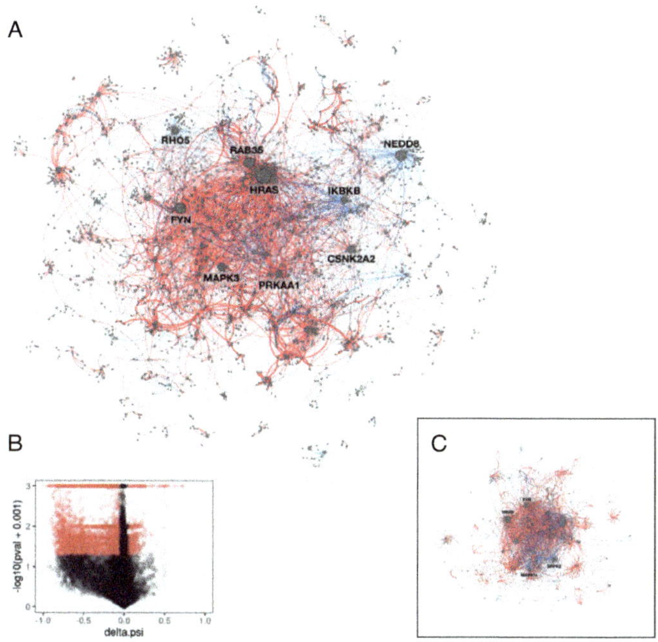

Fig. 4. *Patient specific rewired networks.* Here, we show two sample network outputs from Splitpea and the accompanying exon value cutoff. The large network (A) depicts pancreatic patient sample (TCGA-HZ-7918-01A-11R-2156-07), with edge losses in red and gains in blue. The corresponding volcano plot is shown in (B), where exons with significant $\Delta\psi$ ($\hat{p} < 0.05$) as well as absolute change ($|\Delta\psi| > 0.05$) are shown in red. Box (C) shows a patient-specific network for an example breast cancer sample, TCGA-BH-A0BG-01A-11R-A115-07, which exhibits a very different topology from the pancreatic sample in A.

3.4. *A consensus network of changes across breast cancer patients*

While patient-specific networks highlight network rewiring at the level of individual tumor samples, we also sought to look for more general cancer level patterns of PPI rewiring. Towards this end, we assembled a consensus rewiring network for breast cancer by taking splicing rewiring events conserved across 80% of patient samples and assembling a meta-network of these events. Edges were only preserved when their type (gain, loss) was consistent. Chaos edges were not included in the consensus network. Naturally, as the threshold increases, the number of genes preserved in the network decreases (Fig. 5A). Interestingly, up through the 80% threshold, gained edges are relatively more consistently preserved (Fig. 5B). Visualizing the breast cancer consensus network (Fig. 5C) revealed that the most gained interaction

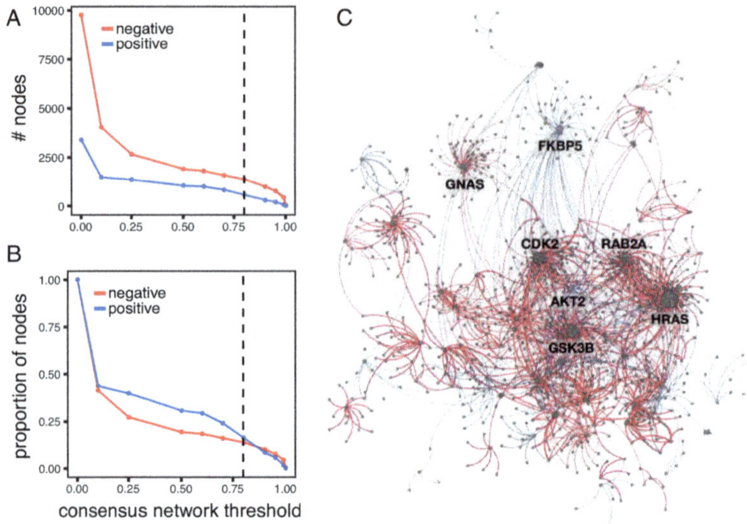

Fig. 5. *Meta-network of breast cancer patients.* The line graphs show the number of nodes preserved for different consensus thresholds (A) or the proportion of nodes relative to the non-thresholded consensus network (B) for edge loss (negative, red) and edge gain (positive, blue) events. The dashed line in both graphs denotes a threshold of 80%, corresponding to the visualization of the consensus network of splicing rewiring events conserved across 80% of breast cancer patient samples (C, red: edge loss; blue: edge gain).

involved the gene, FKBP5, which is an immune regulator responsible for protein trafficking and folding. This protein has been studied in breast cancer for its various hormone receptor signaling functions.[45]

3.5. *Network clusters reveal novel patient subgroups*

The patient-specific networks generated by Splitpea have many downstream applications, especially when the networks are used as features for other machine learning tasks. Here, we demonstrate their utility by finding patient subgroups across both breast and pancreatic cancer when the networks are clustered (Fig. 6). Specifically, we use a state-of-the-art graph embedding method, FEATHER,[41] which calculates characteristic functions using different random walk weights for node features, but any graph embedding method could be used for this type of analysis. For each cancer type, we clustered the network embeddings using HDBSCAN (see Methods). Interestingly, three distinct groups emerged across the cancer types (Fig. 6A). The dominant source of variation across the networks is the gain or loss of PPIs involving KRAS (Fig. 6B). Mutations in KRAS are known to affect subgroups of both pancreatic and breast cancer[46] with ties to prognosis. It is possible that splicing changes in interacting partner genes also induce changes to KRAS that may have yet unknown interaction effects with these somatic mutations, highlighting the potential of Splitpea to find additional disease subtypes. Furthermore, other interesting cancer drivers have distinct patterns of gains and losses, including RAB5A, which appears to have PPI gains in the BRCA outliers, and IKBKB, which is enriched for gains in the predominantly pancreatic cancer cluster 3.

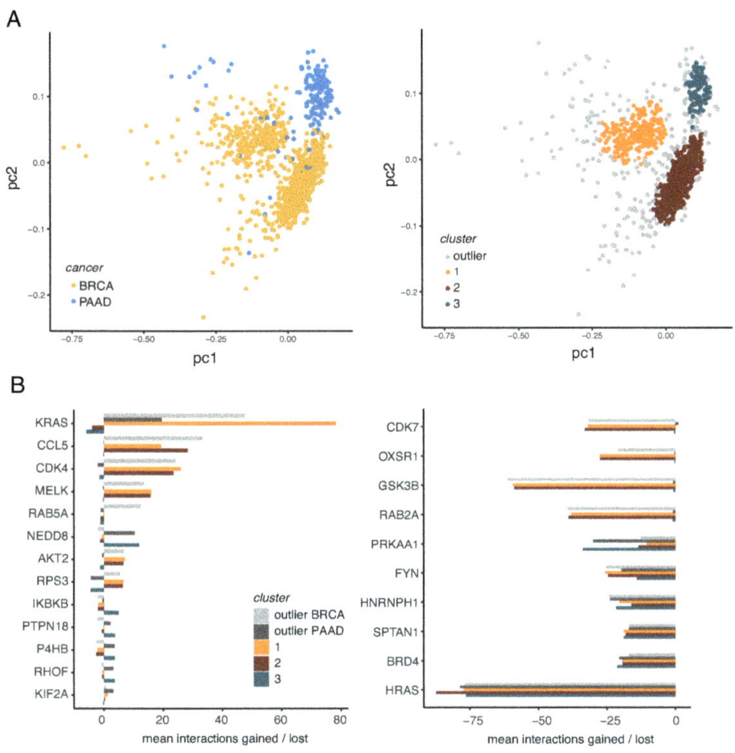

Fig. 6. *Splitpea networks cluster into distinct subgroups.* (A) PCA plots of graph embeddings of each patient-specific Splitpea network, with samples colored by either cancer type (left) or cluster (right). Clusters were assigned using HDBSCAN, with outliers colored in grey. (B) For each cluster, the top nodes undergoing the most changes (mean interactions gained or lost) were also identified. The bar graphs are roughly separated by genes that have the most gain of interactions (left) versus those that have primarily losses (right). Interestingly, the main variation captured in PC1 seems to be defined by networks that change in KRAS. Other cancer driver genes also undergo distinct patterns of gains and losses that drive clustering patterns.

4. Discussion and conclusion

We present a new method, Splitpea, for characterizing protein-protein network rewiring events. Splitpea is flexible and can be applied with different background contexts to highlight splicing changes between a disease and relevant background context of interest. We applied Splitpea to breast and pancreatic cancer samples to highlight the potential of Splitpea to find new and relevant cancer biology, both on an individual patient sample level and more broadly across samples of a single tumor type. To our knowledge, Splitpea is the first systematic method for identifying both potential gains in addition to PPIs lost for individual experimental samples.

Splitpea makes heavy use of existing knowledge of protein-protein interactions. Because of this, our method is inherently limited by the availability of known PPIs (which are largely incomplete), as well as DDIs, which are even less complete. As more of these are experimentally characterized, Splitpea will continue to improve, capturing more accurate and comprehensive sets of network rewiring events. Since we wrote Splitpea to be modular, updates to PPIs and DDIs can be easily integrated once they become available. Specifically, study bias is a well-reported issue in PPIs, and thus there is a large amount of overlap between well-studied nodes (including many cancer driver genes) with nodes of high degree in PPI networks, and given the dependency of Splitpea on reported PPIs, this also affects our results. As more systematic experimental PPI screens and more reliable PPI predictions become available, we can also readily adapt Splitpea.

We have only scratched the surface of cancer biology here. In our initial exploration of breast and pancreatic cancer, we have discovered subgroups and outliers within each cancer type that can be characterized by different network hubs. We believe this merits more thorough exploration, as it may carry important implications for precision medicine efforts. Beyond this, it will also be interesting to apply Splitpea to more cancer types and look for pan-cancer conservation patterns.

Acknowledgments

This work was supported by the Cancer Prevention & Research Institute of Texas (RR190065). VY is a CPRIT Scholar in Cancer Research.

References

1. Q. Pan, O. Shai, L. J. Lee, B. J. Frey and B. J. Blencowe, Deep surveying of alternative splicing complexity in the human transcriptome by high-throughput sequencing, *Nature Genetics* **40**, 1413 (December 2008).
2. E. T. Wang, R. Sandberg, S. Luo, I. Khrebtukova, L. Zhang, C. Mayr, S. F. Kingsmore, G. P. Schroth and C. B. Burge, Alternative isoform regulation in human tissue transcriptomes, *Nature* **456**, 470 (November 2008).
3. D. L. Black, Protein diversity from alternative splicing: a challenge for bioinformatics and post-genome biology, *Cell* **103**, 367 (October 2000).
4. T. W. Nilsen and B. R. Graveley, Expansion of the eukaryotic proteome by alternative splicing, *Nature* **463**, 457 (January 2010).
5. X. Yang, J. Coulombe-Huntington, S. Kang, G. M. Sheynkman, T. Hao, A. Richardson, S. Sun, F. Yang, Y. A. Shen, R. R. Murray, K. Spirohn, B. E. Begg, M. Duran-Frigola, A. MacWilliams,

S. J. Pevzner, Q. Zhong, S. A. Trigg, S. Tam, L. Ghamsari, N. Sahni, S. Yi, M. D. Rodriguez, D. Balcha, G. Tan, M. Costanzo, B. Andrews, C. Boone, X. J. Zhou, K. Salehi-Ashtiani, B. Charloteaux, A. A. Chen, M. A. Calderwood, P. Aloy, F. P. Roth, D. E. Hill, L. M. Iakoucheva, Y. Xia and M. Vidal, Widespread Expansion of Protein Interaction Capabilities by Alternative Splicing, *Cell* **164**, 805 (February 2016).

6. F. E. Baralle and J. Giudice, Alternative splicing as a regulator of development and tissue identity, *Nature Reviews. Molecular Cell Biology* **18**, 437 (2017).
7. M. M. Scotti and M. S. Swanson, RNA mis-splicing in disease, *Nature Reviews. Genetics* **17**, 19 (January 2016).
8. H. Climente-González, E. Porta-Pardo, A. Godzik and E. Eyras, The Functional Impact of Alternative Splicing in Cancer, *Cell Reports* **20**, 2215 (August 2017).
9. A. Kahles, K.-V. Lehmann, N. C. Toussaint, M. Hüser, S. G. Stark, T. Sachsenberg, O. Stegle, O. Kohlbacher, C. Sander, S. J. Caesar-Johnson *et al.*, Comprehensive analysis of alternative splicing across tumors from 8,705 patients, *Cancer cell* **34**, 211 (2018).
10. J. E. Love, E. J. Hayden and T. T. Rohn, Alternative Splicing in Alzheimer's Disease, *Journal of Parkinson's Disease and Alzheimer's Disease* **2** (August 2015).
11. R. Mosca, A. Céol, A. Stein, R. Olivella and P. Aloy, 3did: a catalog of domain-based interactions of known three-dimensional structure, *Nucleic Acids Research* **42**, D374 (January 2014).
12. S. Yellaboina, A. Tasneem, D. V. Zaykin, B. Raghavachari and R. Jothi, DOMINE: a comprehensive collection of known and predicted domain-domain interactions, *Nucleic Acids Research* **39**, D730 (January 2011).
13. Y. Kim, B. Min and G.-S. Yi, IDDI: integrated domain-domain interaction and protein interaction analysis system, *Proteome Science* **10 Suppl 1**, p. S9 (June 2012).
14. R. D. Finn, B. L. Miller, J. Clements and A. Bateman, iPfam: a database of protein family and domain interactions found in the Protein Data Bank, *Nucleic Acids Research* **42**, D364 (January 2014).
15. A. Mehmood, A. Laiho, M. S. Venäläinen, A. J. McGlinchey, N. Wang and L. L. Elo, Systematic evaluation of differential splicing tools for rna-seq studies, *Briefings in bioinformatics* **21**, 2052 (2020).
16. M. A. Ghadie, L. Lambourne, M. Vidal and Y. Xia, Domain-based prediction of the human isoform interactome provides insights into the functional impact of alternative splicing, *PLoS computational biology* **13**, p. e1005717 (2017).
17. T. Will and V. Helms, Ppixpress: construction of condition-specific protein interaction networks based on transcript expression, *Bioinformatics* **32**, 571 (2016).
18. Z. Louadi, K. Yuan, A. Gress, O. Tsoy, O. V. Kalinina, J. Baumbach, T. Kacprowski and M. List, Digger: exploring the functional role of alternative splicing in protein interactions, *Nucleic acids research* **49**, D309 (2021).
19. E. Gjerga, I. S. Naarmann-de Vries and C. Dieterich, Characterizing alternative splicing effects on protein interaction networks with linda, *Bioinformatics* **39**, i458 (2023).
20. Y. Li, N. Sahni, R. Pancsa, D. J. McGrail, J. Xu, X. Hua, J. Coulombe-Huntington, M. Ryan, B. Tychhon, D. Sudhakar *et al.*, Revealing the determinants of widespread alternative splicing perturbation in cancer, *Cell reports* **21**, 798 (2017).
21. A. Chatr-aryamontri, R. Oughtred, L. Boucher, J. Rust, C. Chang, N. K. Kolas, L. O'Donnell, S. Oster, C. Theesfeld, A. Sellam, C. Stark, B.-J. Breitkreutz, K. Dolinski and M. Tyers, The BioGRID interaction database: 2017 update, *Nucleic Acids Research* **45**, D369 (January 2017).
22. L. Salwinski, C. S. Miller, A. J. Smith, F. K. Pettit, J. U. Bowie and D. Eisenberg, The Database of Interacting Proteins: 2004 update, *Nucleic Acids Research* **32**, D449 (January 2004).
23. G. Alanis-Lobato, M. A. Andrade-Navarro and M. H. Schaefer, HIPPIE v2.0: enhancing meaningfulness and reliability of protein–protein interaction networks, *Nucleic Acids Research* **45**,

D408 (January 2017).
24. T. S. Keshava Prasad, R. Goel, K. Kandasamy, S. Keerthikumar, S. Kumar, S. Mathivanan, D. Telikicherla, R. Raju, B. Shafreen, A. Venugopal, L. Balakrishnan, A. Marimuthu, S. Banerjee, D. S. Somanathan, A. Sebastian, S. Rani, S. Ray, C. J. Harrys Kishore, S. Kanth, M. Ahmed, M. K. Kashyap, R. Mohmood, Y. L. Ramachandra, V. Krishna, B. A. Rahiman, S. Mohan, P. Ranganathan, S. Ramabadran, R. Chaerkady and A. Pandey, Human Protein Reference Database–2009 update, *Nucleic Acids Research* **37**, D767 (January 2009).
25. T. Rolland, M. Taşan, B. Charloteaux, S. J. Pevzner, Q. Zhong, N. Sahni, S. Yi, I. Lemmens, C. Fontanillo, R. Mosca, A. Kamburov, S. D. Ghiassian, X. Yang, L. Ghamsari, D. Balcha, B. E. Begg, P. Braun, M. Brehme, M. P. Broly, A.-R. Carvunis, D. Convery-Zupan, R. Corominas, J. Coulombe-Huntington, E. Dann, M. Dreze, A. Dricot, C. Fan, E. Franzosa, F. Gebreab, B. J. Gutierrez, M. F. Hardy, M. Jin, S. Kang, R. Kiros, G. N. Lin, K. Luck, A. MacWilliams, J. Menche, R. R. Murray, A. Palagi, M. M. Poulin, X. Rambout, J. Rasla, P. Reichert, V. Romero, E. Ruyssinck, J. M. Sahalie, A. Scholz, A. A. Shah, A. Sharma, Y. Shen, K. Spirohn, S. Tam, A. O. Tejeda, S. A. Trigg, J.-C. Twizere, K. Vega, J. Walsh, M. E. Cusick, Y. Xia, A.-L. Barabási, L. M. Iakoucheva, P. Aloy, J. De Las Rivas, J. Tavernier, M. A. Calderwood, D. E. Hill, T. Hao, F. P. Roth and M. Vidal, A proteome-scale map of the human interactome network, *Cell* **159**, 1212 (November 2014).
26. S. Orchard, M. Ammari, B. Aranda, L. Breuza, L. Briganti, F. Broackes-Carter, N. H. Campbell, G. Chavali, C. Chen, N. del Toro, M. Duesbury, M. Dumousseau, E. Galeota, U. Hinz, M. Iannuccelli, S. Jagannathan, R. Jimenez, J. Khadake, A. Lagreid, L. Licata, R. C. Lovering, B. Meldal, A. N. Melidoni, M. Milagros, D. Peluso, L. Perfetto, P. Porras, A. Raghunath, S. Ricard-Blum, B. Roechert, A. Stutz, M. Tognolli, K. van Roey, G. Cesareni and H. Hermjakob, The MIntAct project–IntAct as a common curation platform for 11 molecular interaction databases, *Nucleic Acids Research* **42**, D358 (January 2014).
27. S. Razick, G. Magklaras and I. M. Donaldson, iRefIndex: a consolidated protein interaction database with provenance, *BMC bioinformatics* **9**, p. 405 (September 2008).
28. P. Pagel, S. Kovac, M. Oesterheld, B. Brauner, I. Dunger-Kaltenbach, G. Frishman, C. Montrone, P. Mark, V. Stümpflen, H.-W. Mewes, A. Ruepp and D. Frishman, The MIPS mammalian protein-protein interaction database, *Bioinformatics (Oxford, England)* **21**, 832 (March 2005).
29. G. R. Brown, V. Hem, K. S. Katz, M. Ovetsky, C. Wallin, O. Ermolaeva, I. Tolstoy, T. Tatusova, K. D. Pruitt, D. R. Maglott and T. D. Murphy, Gene: a gene-centered information resource at NCBI, *Nucleic Acids Research* **43**, D36 (January 2015).
30. S. Durinck, P. Spellman, E. Birney and W. Huber, Mapping identifiers for the integration of genomic datasets with the r/bioconductor package biomart., *Nature Protocols* **4**, 1184 (2009).
31. H. Li, Tabix: fast retrieval of sequence features from generic TAB-delimited files, *Bioinformatics (Oxford, England)* **27**, 718 (March 2011).
32. Y. Pan, J. W. Phillips, B. D. Zhang, M. Noguchi, E. Kutschera, J. McLaughlin, P. A. Nesterenko, Z. Mao, N. J. Bangayan, R. Wang, W. Tran, H. T. Yang, Y. Wang, Y. Xu, M. B. Obusan, D. Cheng, A. H. Lee, K. E. Kadash-Edmondson, A. Champhekar, C. Puig-Saus, A. Ribas, R. M. Prins, C. S. Seet, G. M. Crooks, O. N. Witte and Y. Xing, IRIS: Discovery of cancer immunotherapy targets arising from pre-mRNA alternative splicing, *Proceedings of the National Academy of Sciences* **120**, p. e2221116120 (May 2023), Publisher: Proceedings of the National Academy of Sciences.
33. S. Shen, J. W. Park, Z.-x. Lu, L. Lin, M. D. Henry, Y. N. Wu, Q. Zhou and Y. Xing, rmats: robust and flexible detection of differential alternative splicing from replicate rna-seq data, *Proceedings of the National Academy of Sciences* **111**, E5593 (2014).
34. A. Kahles, C. S. Ong, Y. Zhong and G. Rätsch, Spladder: identification, quantification and testing of alternative splicing events from rna-seq data, *Bioinformatics* **32**, 1840 (2016).

35. J. Vaquero-Garcia, A. Barrera, M. R. Gazzara, J. Gonzalez-Vallinas, N. F. Lahens, J. B. Hogenesch, K. W. Lynch and Y. Barash, A new view of transcriptome complexity and regulation through the lens of local splicing variations, *elife* **5**, p. e11752 (2016).
36. Y. I. Li, D. A. Knowles, J. Humphrey, A. N. Barbeira, S. P. Dickinson, H. K. Im and J. K. Pritchard, Annotation-free quantification of rna splicing using leafcutter, *Nature genetics* **50**, 151 (2018).
37. J. L. Trincado, J. C. Entizne, G. Hysenaj, B. Singh, M. Skalic, D. J. Elliott and E. Eyras, SUPPA2: fast, accurate, and uncertainty-aware differential splicing analysis across multiple conditions, *Genome Biology* **19**, p. 40 (March 2018).
38. J. L. Trincado, J. C. Entizne, G. Hysenaj, B. Singh, M. Skalic, D. J. Elliott and E. Eyras, Suppa2: fast, accurate, and uncertainty-aware differential splicing analysis across multiple conditions, *Genome biology* **19**, 1 (2018).
39. G. R. Warnes, B. Bolker, L. Bonebakker, R. Gentleman, W. Huber, A. Liaw, T. Lumley, M. Maechler, A. Magnusson, S. Moeller, M. Schwartz and B. Venables, gplots: Various r programming tools for plotting data (2015).
40. D. Netanely, A. Avraham, A. Ben-Baruch, E. Evron and R. Shamir, Expression and methylation patterns partition luminal-A breast tumors into distinct prognostic subgroups, *Breast Cancer Research* **18**, p. 74 (July 2016).
41. B. Rozemberczki and R. Sarkar, Characteristic functions on graphs: Birds of a feather, from statistical descriptors to parametric models, in *Proceedings of the 29th ACM international conference on information & knowledge management*, 2020.
42. B. Rozemberczki, O. Kiss and R. Sarkar, Karate club: An api oriented open-source python framework for unsupervised learning on graphs, in *Proceedings of the 29th ACM International Conference on Information and Knowledge Management (CIKM '20)*, 2020.
43. R. J. Campello, D. Moulavi and J. Sander, Density-based clustering based on hierarchical density estimates, in *Pacific-Asia conference on knowledge discovery and data mining*, 2013.
44. G. Yeo, D. Holste, G. Kreiman and C. B. Burge, Variation in alternative splicing across human tissues, *Genome Biology* **5**, p. R74 (September 2004).
45. L. Li, Z. Lou and L. Wang, The role of FKBP5 in cancer aetiology and chemoresistance, *British Journal of Cancer* **104**, 19 (January 2011).
46. I. A. Prior, F. E. Hood and J. L. Hartley, The frequency of Ras mutations in cancer, *Cancer research* **80**, 2969 (July 2020).

Lymphocyte Count Derived Polygenic Score and Interindividual Variability in CD4 T-cell Recovery in Response to Antiretroviral Therapy

Kathleen M. Cardone[1], Scott Dudek[1], Karl Keat[2], Yuki Bradford[1], Zinhle Cindi[1,7], Eric S. Daar[3], Roy Gulick[4], Sharon A. Riddler[5], Jeffrey L. Lennox[6], Phumla Sinxadi[7], David W. Haas[8,9], Marylyn D. Ritchie[1,10]
[1]Department of Genetics, [2]Genomics and Computational Biology Graduate Program
University of Pennsylvania, Philadelphia, PA, USA
[3]Lundquist Institute at Harbor-UCLA Medical Center, Torrance, CA, USA
[4]Weill Cornell Medicine New York, New York, NY, USA
[5]University of Pittsburgh, Pittsburgh, PA, USA
[6]Emory University School of Medicine, Atlanta, GA, USA
[7]Division of Clinical Pharmacology, Department of Medicine
University of Cape Town, Cape Town, South Africa
[8]Vanderbilt University Medical Center, Nashville, TN, USA
[9]Meharry Medical College, Nashville, TN, USA
[10]Institute for Biomedical Informatics
University of Pennsylvania, Philadelphia, PA, USA
**Email: marylyn@pennmedicine.upenn.edu*

Access to safe and effective antiretroviral therapy (ART) is a cornerstone in the global response to the HIV pandemic. Among people living with HIV, there is considerable interindividual variability in absolute CD4 T-cell recovery following initiation of virally suppressive ART. The contribution of host genetics to this variability is not well understood. We explored the contribution of a polygenic score which was derived from large, publicly available summary statistics for absolute lymphocyte count from individuals in the general population (PGS_{lymph}) due to a lack of publicly available summary statistics for CD4 T-cell count. We explored associations with baseline CD4 T-cell count prior to ART initiation (n=4959) and change from baseline to week 48 on ART (n=3274) among treatment-naïve participants in prospective, randomized ART studies of the AIDS Clinical Trials Group. We separately examined an African-ancestry-derived and a European-ancestry-derived PGS_{lymph}, and evaluated their performance across all participants, and also in the African and European ancestral groups separately. Multivariate models that included PGS_{lymph}, baseline plasma HIV-1 RNA, age, sex, and 15 principal components (PCs) of genetic similarity explained ~26-27% of variability in baseline CD4 T-cell count, but PGS_{lymph} accounted for <1% of this variability. Models that also included baseline CD4 T-cell count explained ~7-9% of variability in CD4 T-cell count increase on ART, but PGS_{lymph} accounted for <1% of this variability. In univariate analyses, PGS_{lymph} was not significantly associated with baseline or change in CD4 T-cell count. Among individuals of African ancestry, the African PGS_{lymph} term in the multivariate model was significantly associated with change in CD4 T-cell count while not significant in the univariate model. When applied to lymphocyte count in a general medical biobank population (Penn Medicine BioBank), PGS_{lymph} explained ~6-10% of variability in multivariate models (including age, sex, and PCs) but only ~1% in univariate models. In summary, a lymphocyte count PGS derived from the general population was not consistently associated with CD4 T-cell recovery on ART. Nonetheless, adjusting for clinical covariates is quite important when estimating such polygenic effects.

Keywords: HIV; Polygenic Scores; Lymphocyte Count; CD4 T-Cell Count; Pharmacogenomics

1. Introduction

1.1. *Incomplete CD4 T-Cell Recovery in Response to Antiretroviral Therapy*

Human immunodeficiency virus type 1 (HIV-1) is a global health challenge, with 38.4 million individuals worldwide living with HIV[1], including nearly 1.2 million in the United States[2]. This virus depletes CD4 T lymphocytes (hereafter referred to as CD4 cells), a critical component of the immune system[3]. Effective antiretroviral therapy (ART) controls viral replication, improves health and prevents transmission[4]. With viral load reduction, CD4 cell counts may return to normal levels, but in many individuals this is not achieved[5–7]. Understanding the etiology of CD4 cell recovery is important because individuals with lower CD4 cell counts may be at increased risk for non-AIDS conditions such as hepatic cirrhosis, cardiovascular disease, kidney disease, and cancer[8].

The etiology of incomplete CD4 cell recovery has not been fully elucidated, but many biological, demographic, treatment, and genetic factors have been associated[9]. Individuals who begin ART with CD4 cell counts <200 cells/mm^3 are less likely to achieve normal CD4 cell counts >500 cells/mm[5–7]. Other biological factors associated with this treatment response include higher body mass index (BMI), lower naïve/memory CD4$^+$ cell ratio, lower CD4/CD8 cell ratios, and other immunological factors[9]. Demographic factors have also been associated with poor CD4 cell recovery including older age, male sex, and Eastern African ancestry, as well as specific ART regimens[9,10]. Additionally, variants that influence the absorption, distribution, metabolism, and elimination of ART may also play a role[11]. Genes with single nucleotide polymorphisms (SNPs) reported to be associated with CD4 cell recovery on ART have included *IL-2*, *IL-2Rβ*, *IL-2Rγ*, *IL-15*, *IL-15Rα*, *TRAIL*, *Bim*, *TNF-α*, and *IFN-γ*[12]. One particular SNP (rs6897932) in *IL7RA* was associated with a faster CD4 cell count increase in individuals of both European and African ancestry, but another SNP in this gene (rs3194051) was only associated with this response in individuals of African ancestry[13,14]. Another study suggested that differences in *CCR5* genotype and *CCL3L1* dosage were associated with the extent and rate of CD4 cell recovery[15]. Additionally, *HLA-Bw4* homozygosity was associated with impaired CD4 cell recovery[16]. Particular mitochondrial DNA haplogroups were associated with CD4 cell recovery in individuals of European and African ancestry[17,18]. More recently, whole exome sequencing associated 41 genes with CD4 cell response in females[19].

Although multiple genes and SNPs have been associated with poor CD4 cell count recovery on ART, these explain a small fraction of the variance. Previous studies considered effects of SNPs individually, which fails to consider whether combinations of many SNPs may explain a larger portion of the variance. Many conditions are polygenic (e.g., coronary artery disease), meaning that many genes and variants have impact[20]. It is conceivable that CD4 cell recovery on ART is also polygenic, so it is worth exploring whether polygenic scores may explain a larger portion of the genetic variance, which has never been investigated for this treatment response. Furthermore, understanding the pharmacogenomic underpinnings of treatment response has the potential to better individualize therapy[21].

1.2 *Polygenic Scores May Predict Complex Treatment Responses*

One way to assess the contribution of many variants in combination is by applying Polygenic Scores (PGS), which are the mathematical, cumulative aggregation of risk derived from the total contribution of numerous variants in the genome[22]. PGS effectively predict phenotypes such as schizophrenia[23–27], bipolar disorder[23,28,29], breast cancer[30–33], type 2 diabetes[30,34,35], coronary artery disease [30,34,36], and atrial fibrillation[30,34,37]. Given their success in other disease areas, it is plausible that PGS could predict poor CD4 cell recovery in response to ART.

When using PGS, it is important to consider the potential for ancestral health disparity. Across many phenotypes, PGS is more predictive for individuals of European ancestry because this population has more readily available summary statistics from large genome-wide association studies (GWAS)[38]. An ultimate goal of PGS is clinical implementation so that patients can be informed of their genetic risk for disease[38]. However, clinical implementation could create a larger health disparity whereby individuals of European ancestry may more readily benefit from these risk prediction models[38]. Thus, it is important to improve risk prediction for global populations. This is particularly important for HIV given its global distribution of prevalence, particularly in Africa. We hope to better predict genetic risk in individuals of African ancestry by generating a PGS based on summary statistics generated in a dataset of individuals largely of African ancestry, in addition to a PGS generated in a dataset of individuals largely of European ancestry. Additionally, we plan to use PRScsx, a method that more effectively predicts polygenic risk in global populations[39].

In this study, we assess whether the PGS generated from a general population is predictive of CD4 cell recovery in persons living with HIV (PWH). A similar approach used a body mass index PGS generated from a general population to study ART-associated weight gain[40]. As there are no large GWAS studies of CD4 cell count, either in the general population or in PWH, we generate statistical power by using summary statistics on total lymphocyte count from a general population, for which large sample sizes are publicly available. Finally, the principle of predicting phenotypic effects in a population affected by a health condition by using genetics from the general population was effective in one study that found that variants associated with cardiac QRS duration in individuals without cardiac diseases were also associated with arrhythmia and atrial fibrillation[41]. We assess whether this same principle applies to treatment response by testing whether the genetic underpinnings of lymphocyte count in a general population predicts CD4 cell recovery in PWH. We hypothesize that cumulative genetic variants that affect total lymphocyte count also affect recovery of the CD4 T cell subset in response to ART (i.e., that a lymphocyte count PGS [PGS_{lymph}] generated from the general population will be associated with CD4 cell recovery on ART). We also hypothesize that PGS_{lymph} will be associated with CD4 cell counts prior to initiating ART.

2. Methods

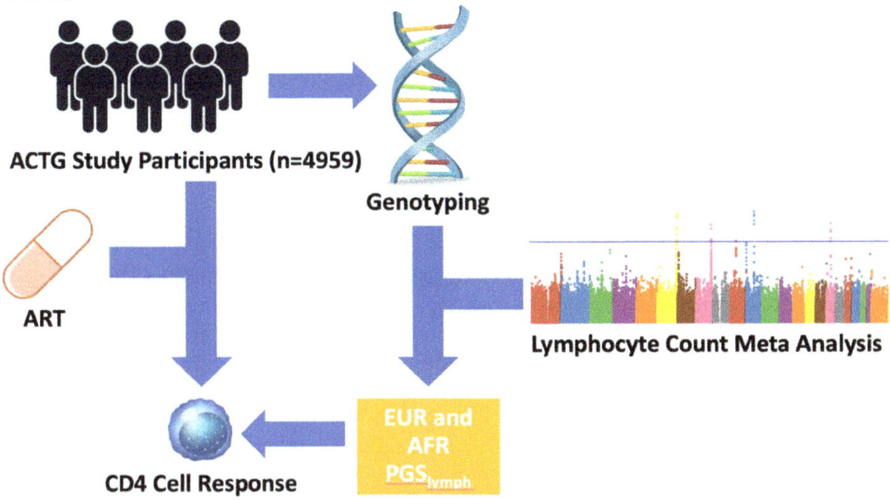

Figure 1: Study Overview: EUR and AFR PGS_{lymph} were trained using lymphocyte count GWAS summary statistics. Both PGS_{lymph} were applied to individuals in the AIDS Clinical Trials Group (ACTG) to assess its predictability of CD4 cell response to ART.

2.1 *Data and Study Participants*

2.1.1 *Lymphocyte Count Meta Analysis*

We used publicly available summary statistics from a published meta-analysis of existing GWAS for lymphocyte count in populations of European and African ancestry in the general population[42]. The meta-analysis included 524,923 individuals of European ancestry with 47,264,266 SNPs, and 13,477 individuals of African ancestry with 34,121,887 SNPs[42]. The European ancestry summary statistics were subset to 1,120,498 SNPs that were present on the European linkage disequilibrium (LD) panels and the African ancestry summary statistics were subset to 1,225,091 SNPs that were present on the African LD reference panels.

2.1.2 *AIDS Clinical Trials Group*

Participants were ART-naïve individuals who had initiated ART in prospective, randomized clinical trials of the AIDS Clinical Trials Group (ACTG), and had consented to genetic research and provided DNA under ACTG protocol A5128[43]. Data were generated by conducting a retrospective analysis of this cohort. Individuals had initiated ART in the United States in studies ACTG384, A5095 (NCT00013520), A5142 (NCT00050895), A5202 (NCT00118898), and A5257 (NCT25285539)[44–47]. All participants provided written, informed consent for genetic testing. Drug class components of regimens were randomly assigned except for nucleoside reverse transcriptase inhibitor (NRTI) choice in A5142. Included individuals had the following data: imputed genotype, sex, genetically inferred ancestry (GIA), lymphocyte count or CD4 cell count data. Additional eligibility criteria included HIV-1 RNA <400 copies/mL at week 48 on ART.

2.1.3 *Penn Medicine BioBank*

The Penn Medicine BioBank (PMBB) is an electronic health record (EHR)-linked biobank research program at the University of Pennsylvania[48]. PMBB participants included in this study provided consent for research including access to their medical records, blood sample collection, and generation of genetic data[48]. Individuals with both imputed genotype data from PMBB v2.0 and with lymphocyte count data were included in PGS analysis as a positive control. Included individuals had the following data: imputed genotype, lymphocyte count, sex, and GIA.

2.2 *Genotyping and Quality Control*

2.2.1 *AIDS Clinical Trials Group*

DNA extracted from whole blood was labeled with coded identifiers and genotyped in seven phases. Phases 1-3 were genotyped at the Broad Institute (Phases 1 and 2 with HumanHap650Yv3_A, and Phase 3 with Human1M-Duov3_B). Phases 4-7 were genotyped at the Vanderbilt Technologies for Advanced Genomics (VANTAGE) facility (Phase 4 using the Human Core Exome chip, phase 5 with the HumanOmni2.5Exome-8-v1.1_A1 chip, Phase 6 with the HumanOmni25-8v1-2_A1 chip, and phase 7 with the Illumina Infinium Multi-Ethnic Global BeadChip ($MEGA^{EX}$).

Post-genotype quality control procedures utilizing PLINK v1.9[49] were conducted by Vanderbilt Technologies for Advanced Genomics Analysis and Research Design (VANGARD). Prior to imputation, samples with genotyping efficiency < 99% or with discordance between genotype sex and reported sex were removed. After completing these quality control procedures, each genotyping phase was imputed separately utilizing the TOPMed reference panel, which was parallelized by chromosome to increase computational efficiency[50]. During the imputation process, liftOver was used to transform genotype data to genome build 38[50]. After imputation, PLINK was used to merge the seven imputed datasets, and variants with imputation R^2 scores < 0.3, genotyping call rates < 95%, or minor allele frequency (MAF) < 0.05 were dropped[49]. GIA was determined using principal component analysis (PCA) with 1000 Genomes as the reference, subsequently assigning each participant to one of six superpopulations: African (AFR), Admixed American (AMR), East Asian (EAS), European (EUR), South Asian (SAS), and Other.

2.2.2 *Penn Medicine BioBank*

DNA was extracted from blood samples. Approximately 80% of samples were genotyped by the Regeneron Genomics Center (RGC) using an Illumina Global Screening Array v.2.0 (GSAv2)[48], while the remaining 20% were genotyped by the Center for Applied Genomics (CAG) at the Children's Hospital of Philadelphia using the GSAv1 and GSAv2 genotyping array[48].

Prior to imputation, sample level quality control was conducted[48]. Using PLINK v1.9, variants with genotyping call rates < 95%, individuals with sample call rates < 90%, and individuals with discordance between reported sex and genotype sex were dropped[48]. Autosomes were imputed utilizing a TOPMed version R2 genome build 38 reference panel[48,50]. After imputation, variants with imputation R^2 scores < 0.3, genotype call rate < 99%, MAF < 1%, and/or were multi-allelic were dropped using PLINK v1.9[48]. Individuals with sample call rate < 99% or discordant sex information were also dropped[48]. PCA was done to identify GIA using 1000 Genomes as the reference and subsequently separated individuals into six superpopulations: African (AFR), Admixed American (AMR), East Asian (EAS), European (EUR), South Asian (SAS), Other[48].

2.3 *Polygenic Score Calculation*

The PGS_{lymph} was constructed using PRScsx (version released on July 29 2021), which integrates summary statistics and LD panels across genetically diverse populations to better predict polygenic risk in global populations[39]. 1000 Genomes phase 3 LD reference panels were used in the calculation[51]. Summary statistics from the lymphocyte count meta-analysis were used to train the PGS_{lymph}[42]. The PGS_{lymph} was applied to ACTG study participants with CD4 cell count data using PLINK2 "--score" function[49]. As positive controls, the PGS_{lymph} was also applied to individuals with lymphocyte count data in ACTG as well as individuals with lymphocyte count data in PMBB.

2.4 *Statistical Analysis*

The results were analyzed to assess model predictability across all ancestries combined, and in European and African ancestries separately. Linear regressions were calculated, and performance was assessed with an R^2 value generated from a multivariate linear regression between the phenotype of interest and the PGS_{lymph}. Additionally, performance of individual covariates was assessed with effect sizes generated from these regressions. We used a p-value threshold of 0.05 to assess significance. Regressions were calculated in individuals of European and African ancestry only, as well as individuals of all superpopulations combined. PGS_{lymph} was applied to two different cohorts, ACTG and PMBB. In ACTG, the predictability of the PGS_{lymph} for three different phenotypes was assessed: the square root (SQRT) of CD4 cell count at study entry prior to ART (baseline), change in CD4 cell count from study entry to 48 weeks of ART (a measure of treatment response), and inverse normal lymphocyte count prior to ART (a control variable). We performed two regressions for each phenotype, one without correcting for any covariates, and one correcting for age, sex, principal components (PC) of genetic similarity 1-15, as well as log_{10}-HIV-1 RNA (a measure of viral load). Additionally, we adjusted for SQRT of baseline CD4 cell count in regression models between PGS_{lymph} and change in CD4 cell count on ART. In addition to these regressions, we also evaluated interactions between the PGS_{lymph} and age, sex, viral load, and baseline CD4 cell count to identify whether PGS_{lymph} interacts with any covariate. In PMBB, the predictability of PGS_{lymph} for inverse normal lymphocyte count was assessed as a positive control and to understand predictability in a general medical biobank population. Similarly, two regressions were performed, one without correcting for covariates, and one correcting for age, sex, and PC1-15. These results were visualized using SynthesisView[52].

3. Results

Table 1: ACTG Participant Demographics at Baseline

	Lymphocyte Count Data	Baseline CD4 Cell Count Data	On-Treatment CD4 Cel Count Data
Total, N	4680	4959	3274
European ancestry, n (%)	1835 (39.2%)	1958 (39.4%)	1319 (40.3%)
African ancestry, n (%)	1721 (36.8%)	1826 (36.8%)	1154 (35.2%)
Male/Female, n (%)	3824/856 (81.7%/18.3%)	4051/908 (81.7%/18.3%)	2715/559 (82.9%/17.1%
Age, mean (range)	37.9 (17.0-77.0)	38.0 (17.0-77.0)	38.2 (17.0-76.0)

Table 2: PMBB Demographics

	Lymphocyte Count Data
Total, N	37211
European ancestry, n (%)	25330 (68.1%)
African ancestry, n (%)	10217 (27.5%)
Male/Female, n (%)	18215/18996 (49.0%/51.0%)
Mean Age (Range)	55.6 (13.9-101.7)

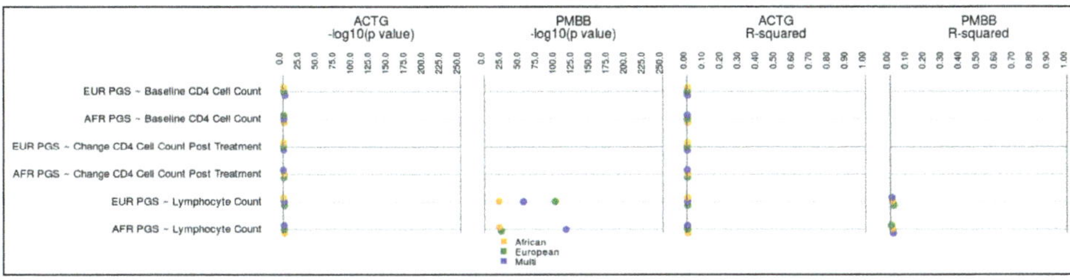

Figure 2: Summary of Regression Results Between PGS_{lymph} and Phenotype without Controlling for Covariates (Age, Sex, PC1-15, log_{10}HIV-1 RNA (viral load), and SQRT of baseline CD4 cell count)

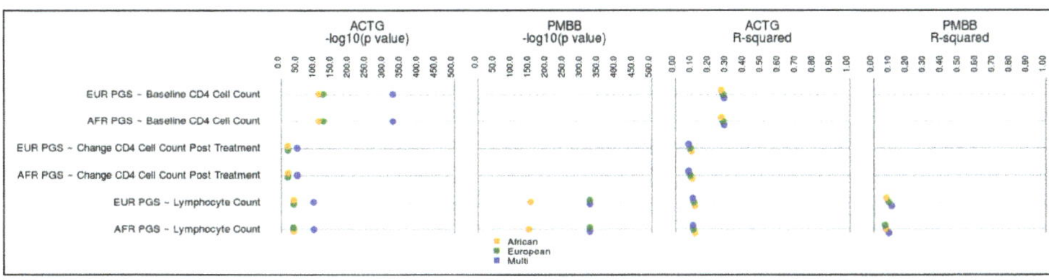

Figure 3: Summary of Regression Results Between PGS_{lymph} and Phenotype While Controlling for Covariates (Age, Sex, PC1-15, log_{10}HIV-1 RNA (viral load), and SQRT of baseline CD4 cell count)

4. Discussion

A lymphocyte count PGS trained in the general population did not effectively predict baseline CD4 cell count or change in CD4 cell count in response to ART, leading to rejection of our hypothesis that poor CD4 cell recovery in response to ART is dependent on each individual's overall genetic predisposition to this outcome. When running regressions without correcting for covariates, R^2 values were low across all ancestry groups and most regressions were not statistically significant (Figure 2, Supplementary Table 1. In contrast, clinical covariates were predictive of these phenotypes. When correcting for covariates, performance of the model improved markedly. Baseline regressions performed modestly ($R^2 = 0.278$) while on-treatment regressions were not very predictive ($R^2 = 0.073$), although all values were statistically significant (Figure 3, Supplementary Table 2). However, because the PGS_{lymph} itself was not highly predictive, the success of this model was mostly due to the contribution of covariates. Additionally, when including covariates in the model, the model including the African PGS_{lymph} better predicted change in CD4 cell count on-treatment in individuals of African ancestry than the model including the European PGS_{lymph} (R^2 was greater by 0.003) (Figure 3, Supplementary Table 2). This is the only case where we see improved performance by an AFR PGS_{lymph} compared to a EUR PGS_{lymph}. Interestingly, when considering effects of individual covariates in this model, the influence of the AFR

PGS_{lymph} is significant ($p = 0.044$) in individuals of African ancestry with an effect size of -2.062 (Supplementary Table 3). In comparison to other covariates, this effect size is minimal, but suggests that the AFR PGS_{lymph} is playing a role. Furthermore, this shows that our methods improved PGS_{lymph} performance in individuals of African ancestry, which was likely because of a combination of a PGS_{lymph} based on African ancestry summary statistics and utilizing PRScsx for calculation.

In univariate analyses, lymphocyte count PGS did not effectively predict baseline lymphocyte count in ACTG participants. R^2 values were also low and insignificant (Figure 2, Supplementary Table 5). Performance improved when including covariates in this model, as R^2 values rose to ~0.10 and regressions became statistically significant (Figure 3, Supplementary Table 6). Within the covariate models, the influence of the EUR PGS_{lymph} is significant in individuals of European ancestry ($p = 0.018$) with a minimal effect size of 0.025 (Supplementary Table 7). However, as the effect size is small, though significant, the EUR PGS_{lymph} is not adding much to this model. Still, this significant effect is exhibited as the R^2 value of the EUR PGS_{lymph} covariate model in individuals of European ancestry (0.103) is slightly higher than the R^2 value of the AFR PGS_{lymph} covariate model in individuals of European ancestry (0.101) (Figure 3, Supplementary Table 6). Additionally, in the multivariate model, the influence of the AFR PGS_{lymph} is significant in the multi-ancestry group (p=8.7e-3) with an effect size of 8.3e-3 (Supplementary Table 9). Although this evidently did not have a large impact on the model, the effects of this are still present as the R^2 value of the AFR PGS_{lymph} covariate model in the multi-ancestry group (0.098) is slightly higher than the R^2 value of the EUR PGS_{lymph} covariate model in the multi-ancestry group (0.097) (Figure 3, Supplementary Table 6). Also, it is interesting that the R^2 value did not increase as high as in CD4 cell count regressions, perhaps because viral load was the greatest contributing covariate (viral load had the lowest p-value of all variables in all CD4 cell count regressions), and total lymphocyte counts are not greatly affected by viral load, in contrast to CD4 cell counts[53] (Supplementary Table 3).

Although this model did not perform well in PWH, it performed slightly better when applied to a general medical biobank population. The PGS_{lymph} best predicted lymphocyte count in a general medical biobank population. Regressions were highly statistically significant, likely due to a large sample size (~37,000 individuals). In the univariate model, the African PGS_{lymph} applied to the multi-ancestry group and the European PGS_{lymph} applied to the European population had the highest R^2 values (~0.01) (Figure 2, Supplementary Table 11). It is interesting that these regressions had the highest R^2 values, as these are the only ACTG lymphocyte count regressions that had a significant contribution from PGS_{lymph} in the multivariable model. Seeing these patterns across the general population and PWH shows that the AFR PGS_{lymph} performs best in a multi-ancestry group and the EUR PGS_{lymph} performs best in individuals of European ancestry. When controlling for covariates, performance of the model increased. R^2 values rose to ~0.06-0.10 and p-values dropped even lower (Figure 3, Supplementary Table 12). This mirrors the impact of covariates seen in PWH. The effect size of the EUR PGS_{lymph} was ~0.01 in all ancestry groups (Supplementary Table 13). It is interesting that without covariates, the EUR PGS_{lymph} in individuals of European ancestry was the only regression mirroring this effect size (Figure

2, Supplementary Table 11). The effect size of the AFR PGS_{lymph} was much lower, ~-5e-3 (Supplementary Table 14). This effect size was mirrored in the AFR PGS_{lymph} regressions without covariates in European and African ancestry, as the R^2 values were also low (~$3e^{-3}$ or $8e^{-3}$), but interestingly the R^2 value was higher when the AFR PGS_{lymph} was applied to the multi-ancestry group (~0.01) (Figure 2, Supplementary Table 11).

Although these results showed that PGS_{lymph} itself is not predictive of this treatment response, some results show that in combination with covariates, the impact of PGS_{lymph} can become significant, suggesting a possible synergistic effect between PGS_{lymph} and clinical covariates in the model. In the regressions between AFR PGS_{lymph} and change in CD4 cell count in individuals of African ancestry, the impact of the PGS_{lymph} was insignificant, but when including clinical covariates in the regression, the impact of the PGS_{lymph} became significant (Supplementary Table 3). However, the AFR PGS_{lymph} did not significantly interact with any covariates, eliminating the possibility of a synergistic effect (Supplementary Table 4). Additionally, in the regressions between the AFR PGS_{lymph} and baseline lymphocyte count in PWH of all ancestry groups, as well as in the regressions between the EUR PGS_{lymph} and baseline lymphocyte count in individuals of European ancestry, the same patterns were observed (Supplementary Table 7, Supplementary Table 9). Similarly, the AFR PGS_{lymph} did not significantly interact with any covariates, but the EUR PGS_{lymph} significantly interacted with age (Supplementary Table 8, Supplementary Table 10). Thus, it is possible that in PWH, there are synergistic effects between the EUR PGS_{lymph} and covariates, thus leading the PGS_{lymph} to become significant. These findings highlight the importance of including clinical covariates in PGS analyses, not only because the covariates themselves very predictive of treatment response, but also because they seem to interact with the PGS_{lymph} in some way. Another explanation for this observation is that covariates with strong effects overshadow the effects of PGS_{lymph} when not controlled for. Covariates such as viral load have such high significance and large effect sizes, that the effects of smaller impact variables such as PGS_{lymph} are not seen unless these covariates were controlled for. Thus, it is important to consider clinical covariates when implementing PGS in a clinical setting.

This study had several limitations. First, the sample size of the African ancestry summary statistics that were used to generate the African PGS_{lymph} were small (~13,000 individuals), which is due to the lack of availability of lymphocyte count summary statistics for individuals of African ancestry. To improve these results, more lymphocyte count GWAS data are needed in future studies, as it is possible that the AFR PGS_{lymph} could have performed better with a larger base sample size. Additionally, the ACTG sample size was modest (~4600 individuals) which was subset to even smaller groups when stratified by ancestry. It is possible that associations with PGS_{lymph} may have become statistically significant with a larger sample size. Subsequent work in this area could investigate whether this model is predictive of other drug response traits, specifically other ART treatment responses.

Polygenic scores have the potential to leverage large, publicly available datasets to find novel genetic discoveries in pharmacogenomic cohorts. This study utilized a novel method to predict CD4 cell recovery in response to ART and illustrated the importance of including clinical covariates in a

PGS model. As more associations or lack thereof are found, we continue to narrow down the biological underpinnings of responses to ART including suboptimal CD4 cell recovery.

5. Acknowledgements

The authors are grateful to the many persons living with HIV who volunteered for ACTG protocols ACTG384, A5095, A5142, A5202 and A5257. In addition, they acknowledge the contributions of study teams and site staff for these protocols. We thank Paul J. McLaren, PhD (Public Health Agency of Canada, Winnipeg, Canada) for prior involvement and collaborations that used these genome-wide genotype data. Study drugs were provided by Bristol-Myers Squibb, Inc., Gilead Sciences, Inc., GlaxoSmithKline, Inc.. The clinical trials were A5095 (NCT00013520), A5142 (NCT00050895), and A5202 (NCT00118898).

Research reported in this publication was supported by the National Institute of Allergy and Infectious Diseases of the National Institutes of Health under Award Number UM1 AI068634, UM1 AI068636 and UM1 AI106701. The content is solely the responsibility of the authors and does not necessarily represent the official views of the National Institutes of Health. Research reported in this publication was also supported in part by grants funded by the National Center for Research Resources and the National Center for Advancing Translational Sciences.

Grant support included TR000124 (to E.S.D.); AI110527, AI077505, TR000445, and AI069439 (to D.W.H.); AI077505 (to K.M.C, S.D., and M.D.R). This work was supported by the Tennessee Center for AIDS Research (P30) AI110527.

Clinical research sites that participated in ACTG protocols A5095, A5142 and/or A5202, and collected DNA under protocol A5128 were supported by the following grants from the National Institutes of Health (NIH): A1069412, A1069423, A1069424, A1069503, AI025859, AI025868, AI027658, AI027661, AI027666, AI027675, AI032782, AI034853, AI038858, AI045008, AI046370, AI046376, AI050409, AI050410, AI050410, AI058740, AI060354, AI068636, AI069412, AI069415, AI069418, AI069419, AI069423, AI069424, AI069428, AI069432, AI069432, AI069434, AI069439, AI069447, AI069450, AI069452, AI069465, AI069467, AI069470, AI069471, AI069472,AI069474, AI069477, AI069481, AI069484, AI069494, AI069495, AI069496, AI069501, AI069501, AI069502, AI069503, AI069511, AI069513, AI069532, AI069534, AI069556, AI072626, AI073961, RR000046, RR000425, RR023561, RR024156, RR024160, RR024996, RR025008, RR025747, RR025777, RR025780, TR000004, TR000058, TR000124, TR000170, TR000439, TR000445, TR000457, TR001079, TR001082, TR001111, and TR024160.

We acknowledge the Penn Medicine BioBank (PMBB) for providing data and thank the patient-participants of Penn Medicine who consented to participate in this research program. We would also like to thank the Penn Medicine BioBank team and Regeneron Genetics Center for providing genetic variant data for analysis. The PMBB is approved under IRB protocol# 813913 and supported by Perelman School of Medicine at University of Pennsylvania, a gift from the Smilow family, and the

National Center for Advancing Translational Sciences of the National Institutes of Health under CTSA award number UL1TR001878.

6. Supplementary Tables
All supplemental data can be found at:
https://ritchielab.org/files/PSB_supplemental_data/PSB_LymphocytePGSHIV_Cardone_2023_Supplementary.pdf

References

1. HIV. https://www.who.int/data/gho/data/themes/hiv-aids#cms.
2. Basic Statistics | HIV Basics | HIV/AIDS | CDC. https://www.cdc.gov/hiv/basics/statistics.html (2022).
3. Vidya Vijayan, K. K., Karthigeyan, K. P., Tripathi, S. P. & Hanna, L. E. Pathophysiology of CD4+ T-Cell Depletion in HIV-1 and HIV-2 Infections. *Front. Immunol.* **8**, 580 (2017).
4. Menéndez-Arias, L. & Delgado, R. Update and latest advances in antiretroviral therapy. *Trends Pharmacol. Sci.* **43**, 16–29 (2022).
5. Kelley, C. F. *et al.* Incomplete Peripheral CD4 $^+$ Cell Count Restoration in HIV-Infected Patients Receiving Long-Term Antiretroviral Treatment. *Clin. Infect. Dis.* **48**, 787–794 (2009).
6. Lok, J. J. *et al.* Long-term increase in CD4+ T-cell counts during combination antiretroviral therapy for HIV-1 infection. *AIDS* **24**, 1867–1876 (2010).
7. Moore, R. D. & Keruly, J. C. CD4+ Cell Count 6 Years after Commencement of Highly Active Antiretroviral Therapy in Persons with Sustained Virologic Suppression. *Clin. Infect. Dis.* **44**, 441–446 (2007).
8. Baker, J. V. *et al.* CD4+ count and risk of non-AIDS diseases following initial treatment for HIV infection. *AIDS* **22**, 841–848 (2008).

9. Yang, X. *et al.* Incomplete immune reconstitution in HIV/AIDS patients on antiretroviral therapy: Challenges of immunological non-responders. *J. Leukoc. Biol.* **107**, 597–612 (2020).
10. Geng, E. H. *et al.* CD4+ T cell recovery during suppression of HIV replication: an international comparison of the immunological efficacy of antiretroviral therapy in North America, Asia and Africa. *Int. J. Epidemiol.* **44**, 251–263 (2015).
11. Haas, D. W. & Tarr, P. E. Perspectives on pharmacogenomics of antiretroviral medications and HIV-associated comorbidities: *Curr. Opin. HIV AIDS* **10**, 116–122 (2015).
12. Haas, D. W. *et al.* Immunogenetics of CD4 Lymphocyte Count Recovery during Antiretroviral Therapy: An AIDS Clinical Trials Group Study. *J. Infect. Dis.* **194**, 1098–1107 (2006).
13. Hartling, H. J. *et al.* Polymorphism in interleukin-7 receptor α gene is associated with faster CD4+ T-cell recovery after initiation of combination antiretroviral therapy. *AIDS* **28**, 1739–1748 (2014).
14. Rajasuriar, R. *et al.* The role of SNPs in the α-chain of the IL-7R gene in CD4+ T-cell recovery in HIV-infected African patients receiving suppressive cART. *Genes Immun.* **13**, 83–93 (2012).
15. Ahuja, S. K. *et al.* CCL3L1-CCR5 genotype influences durability of immune recovery during antiretroviral therapy of HIV-1–infected individuals. *Nat. Med.* **14**, 413–420 (2008).
16. Rauch, A. *et al.* *HLA-Bw4* Homozygosity Is Associated with an Impaired CD4 T Cell Recovery after Initiation of Antiretroviral Therapy. *Clin. Infect. Dis.* **46**, 1921–1925 (2008).
17. Grady, B. J. *et al.* Mitochondrial Genomics and CD4 T-Cell Count Recovery After Antiretroviral Therapy Initiation in AIDS Clinical Trials Group Study 384. *JAIDS J. Acquir. Immune Defic. Syndr.* **58**, 363–370 (2011).

18. Guzmán-Fulgencio, M. *et al.* European mitochondrial haplogroups are associated with CD4+ T cell recovery in HIV-infected patients on combination antiretroviral therapy. *J. Antimicrob. Chemother.* **68**, 2349–2357 (2013).
19. Greenblatt, R. *et al.* Genetic and clinical predictors of CD4 lymphocyte recovery during suppressive antiretroviral therapy: Whole exome sequencing and antiretroviral therapy response phenotypes. *PLOS ONE* **14**, e0219201 (2019).
20. Polygenic Risk Scores. *Genome.gov* https://www.genome.gov/Health/Genomics-and-Medicine/Polygenic-risk-scores (2022).
21. National Institute of General Medical Sciences. *National Institute of General Medical Sciences (NIGMS)* https://nigms.nih.gov/.
22. Choi, S. W., Mak, T. S.-H. & O'Reilly, P. F. Tutorial: a guide to performing polygenic risk score analyses. *Nat. Protoc.* **15**, 2759–2772 (2020).
23. Calafato, M. S. *et al.* Use of schizophrenia and bipolar disorder polygenic scores to identify psychotic disorders. *Br. J. Psychiatry J. Ment. Sci.* **213**, 535–541 (2018).
24. Jonas, K. G. *et al.* Schizophrenia polygenic risk score and 20-year course of illness in psychotic disorders. *Transl. Psychiatry* **9**, 300 (2019).
25. Zheutlin, A. B. *et al.* Penetrance and Pleiotropy of Polygenic Risk Scores for Schizophrenia in 106,160 Patients Across Four Health Care Systems. *Am. J. Psychiatry* **176**, 846–855 (2019).
26. Alnæs, D. *et al.* Brain Heterogeneity in Schizophrenia and Its Association With Polygenic Risk. *JAMA Psychiatry* **76**, 739–748 (2019).

27. Mas-Bermejo, P. *et al.* Schizophrenia polygenic risk score in psychosis proneness. *Eur. Arch. Psychiatry Clin. Neurosci.* (2023) doi:10.1007/s00406-023-01633-7.
28. Mistry, S., Harrison, J. R., Smith, D. J., Escott-Price, V. & Zammit, S. The use of polygenic risk scores to identify phenotypes associated with genetic risk of bipolar disorder and depression: A systematic review. *J. Affect. Disord.* **234**, 148–155 (2018).
29. Hasseris, S. *et al.* Polygenic Risk and Episode Polarity Among Individuals With Bipolar Disorder. *Am. J. Psychiatry* **180**, 200–208 (2023).
30. Khera, A. V. *et al.* Genome-wide polygenic scores for common diseases identify individuals with risk equivalent to monogenic mutations. *Nat. Genet.* **50**, 1219–1224 (2018).
31. Li, H. *et al.* Breast cancer risk prediction using a polygenic risk score in the familial setting: a prospective study from the Breast Cancer Family Registry and kConFab. *Genet. Med.* **19**, 30–35 (2017).
32. Mavaddat, N. *et al.* Polygenic Risk Scores for Prediction of Breast Cancer and Breast Cancer Subtypes. *Am. J. Hum. Genet.* **104**, 21–34 (2019).
33. Roberts, E., Howell, S. & Evans, D. G. Polygenic risk scores and breast cancer risk prediction. *Breast Edinb. Scotl.* **67**, 71–77 (2023).
34. O'Sullivan, J. W., Ashley, E. A. & Elliott, P. M. Polygenic risk scores for the prediction of cardiometabolic disease. *Eur. Heart J.* **44**, 89–99 (2023).
35. McCarthy, M. I. & Mahajan, A. The value of genetic risk scores in precision medicine for diabetes. *Expert Rev. Precis. Med. Drug Dev.* **3**, 279–281 (2018).

36. Rao, A. S. & Knowles, J. W. Polygenic risk scores in coronary artery disease. *Curr. Opin. Cardiol.* **34**, 435–440 (2019).

37. Pulit, S. L. *et al.* Atrial fibrillation genetic risk differentiates cardioembolic stroke from other stroke subtypes. *Neurol. Genet.* **4**, e293 (2018).

38. Martin, A. R. *et al.* Clinical use of current polygenic risk scores may exacerbate health disparities. *Nat. Genet.* **51**, 584–591 (2019).

39. Ruan, Y. *et al.* Improving polygenic prediction in ancestrally diverse populations. *Nat. Genet.* **54**, 573–580 (2022).

40. Keat, K. *et al.* Leveraging Multi-Ancestry Polygenic Risk Scores for Body Mass Index to Predict Antiretroviral Therapy-Induced Weight Gain. *Pac. Symp. Biocomput. Pac. Symp. Biocomput.* **28**, 233–244 (2023).

41. Ritchie, M. D. *et al.* Genome- and Phenome-Wide Analyses of Cardiac Conduction Identifies Markers of Arrhythmia Risk. *Circulation* **127**, 1377–1385 (2013).

42. Chen, M.-H. *et al.* Trans-ethnic and Ancestry-Specific Blood-Cell Genetics in 746,667 Individuals from 5 Global Populations. *Cell* **182**, 1198-1213.e14 (2020).

43. Haas, D. W. *et al.* A Multi-Investigator/Institutional DNA Bank for AIDS-Related Human Genetic Studies: AACTG Protocol A5128. *HIV Clin. Trials* **4**, 287–300 (2003).

44. Daar, E. S. Atazanavir Plus Ritonavir or Efavirenz as Part of a 3-Drug Regimen for Initial Treatment of HIV-1: A Randomized Trial. *Ann. Intern. Med.* **154**, 445 (2011).

45. Gulick, R. M. Three- vs Four-Drug Antiretroviral Regimens for the Initial Treatment of HIV-1 InfectionA Randomized Controlled Trial. *JAMA* **296**, 769 (2006).

46. Gulick, R. M. *et al.* Triple-Nucleoside Regimens versus Efavirenz-Containing Regimens for the Initial Treatment of HIV-1 Infection. *N. Engl. J. Med.* **350**, 1850–1861 (2004).
47. Riddler, S. A. *et al.* Class-Sparing Regimens for Initial Treatment of HIV-1 Infection. *N. Engl. J. Med.* **358**, 2095–2106 (2008).
48. Verma, A. *et al.* The Penn Medicine BioBank: Towards a Genomics-Enabled Learning Healthcare System to Accelerate Precision Medicine in a Diverse Population. *J. Pers. Med.* **12**, 1974 (2022).
49. Chang, C. C. *et al.* Second-generation PLINK: rising to the challenge of larger and richer datasets. *GigaScience* **4**, 7 (2015).
50. Hinrichs, A. S. The UCSC Genome Browser Database: update 2006. *Nucleic Acids Res.* **34**, D590–D598 (2006).
51. The 1000 Genomes Project Consortium *et al.* A global reference for human genetic variation. *Nature* **526**, 68–74 (2015).
52. Pendergrass, S. A., Dudek, S. M., Crawford, D. C. & Ritchie, M. D. Synthesis-View: visualization and interpretation of SNP association results for multi-cohort, multi-phenotype data and meta-analysis. *BioData Min.* **3**, 10 (2010).
53. 2.14 How CD4 and viral load are related | Training manual | HIV i-Base. https://i-base.info/ttfa/section-2/14-how-cd4-and-viral-load-are-related/.

Polygenic risk scores for cardiometabolic traits demonstrate importance of ancestry for predictive precision medicine

Rachel L. Kember*
Department of Psychiatry, University of Pennsylvania, 3535 Market Street
Philadelphia, PA 19104, USA
Email: rkember@pennmedicine.upenn.edu

Shefali S. Verma*
Department of Pathology and Laboratory Medicine, University of Pennsylvania, 3700 Hamilton Walk
Philadelphia, PA 19104, USA
Email: shefali.setiaverma@pennmedicine.upenn.edu

Anurag Verma*
Department of Medicine, University of Pennsylvania, 3700 Hamilton Walk
Philadelphia, PA 19104, USA
Email: anurag.verma@pennmedicine.upenn.edu

Brenda Xiao
Graduate Program in Genomics and Computational Biology, University of Pennsylvania,
Philadelphia, PA 19104, USA
Email: brendax@pennmedicine.upenn.edu

Anastasia Lucas
Graduate Program in Genomics and Computational Biology, University of Pennsylvania,
Philadelphia, PA 19104, USA
Email: anastasia.lucas@pennmedicine.upenn.edu

Colleen M. Kripke
Institute for Translational Medicine and Therapeutics, University of Pennsylvania,
Philadelphia, PA 19104, USA.
Email: colleen.morse@pennmedicine.upenn.edu

Renae Judy
Department of Surgery, Division of Vascular Surgery and Endovascular Therapy, University of Pennsylvania, Philadelphia, PA 19104, USA.
Email: rjudy@pennmedicine.upenn.edu

Jinbo Chen
Department of Biostatistics and Epidemiology, University of Pennsylvania,
203 Blockley Hall, Philadelphia, PA 19104, USA
Email: jinboche@pennmedicine.upenn.edu

Scott M. Damrauer
Department of Surgery, Division of Vascular Surgery and Endovascular Therapy, University of Pennsylvania, Philadelphia, PA 19104, USA.
Email: scott.damrauer@pennmedicine.upenn.edu

Daniel J. Rader
Department of Medicine and Genetics, Institute for Translational Medicine and Therapeutics, University of Pennsylvania, 3801 Filbert St
Philadelphia, PA 19104, USA
Email: rader@pennmedicine.upenn.edu

Marylyn D. Ritchie
Department of Genetics, Institute for Biomedical Informatics, University of Pennsylvania, Perelman School of Medicine, 3700 Hamilton Walk
Philadelphia, PA 19104, USA
Email: marylyn@pennmedicine.upenn.edu

Polygenic risk scores (PRS) have predominantly been derived from genome-wide association studies (GWAS) conducted in European ancestry (EUR) individuals. In this study, we present an in-depth evaluation of PRS based on multi-ancestry GWAS for five cardiometabolic phenotypes in the Penn Medicine BioBank (PMBB) followed by a phenome-wide association study (PheWAS). We examine the PRS performance across all individuals and separately in African ancestry (AFR) and EUR ancestry groups. For AFR individuals, PRS derived using the multi-ancestry LD panel showed a higher effect size for four out of five PRSs (DBP, SBP, T2D, and BMI) than those derived from the AFR LD panel. In contrast, for EUR individuals, the multi-ancestry LD panel PRS demonstrated a higher effect size for two out of five PRSs (SBP and T2D) compared to the EUR LD panel. These findings underscore the potential benefits of utilizing a multi-ancestry LD panel for PRS derivation in diverse genetic backgrounds and demonstrate overall robustness in all individuals. Our results also revealed significant associations between PRS and various phenotypic categories. For instance, CAD PRS was linked with 18 phenotypes in AFR and 82 in EUR, while T2D PRS correlated with 84 phenotypes in AFR and 78 in EUR. Notably, associations like hyperlipidemia, renal failure, atrial fibrillation, coronary atherosclerosis, obesity, and hypertension were observed across different PRSs in both AFR and EUR groups, with varying effect sizes and significance levels. However, in AFR individuals, the strength and number of PRS associations with other phenotypes were generally reduced compared to EUR individuals. Our study underscores the need for future research to prioritize 1) conducting GWAS in diverse ancestry groups and 2) creating a cosmopolitan PRS methodology that is universally applicable across all genetic backgrounds. Such advances will foster a more equitable and personalized approach to precision medicine.

Keywords: Polygenic risk scores, multi-ancestry GWAS, cardiometabolic phenotypes, precision medicine

1. Introduction

The era of precision medicine has been marked by significant efforts to identify the genetic and environmental factors that influence the risk of disease as well as the disease prognosis and treatment. Advance knowledge of these factors can provide a major health benefit to individuals, as preventative strategies and tailored therapies can be targeted toward individuals at higher risk. Results from genome-wide association studies (GWAS) have highlighted the polygenic nature of most common, complex diseases in that they have identified a large number of loci with small genetic effects[1,2]. The polygenic risk score (PRS) has thus emerged as a promising factor for predicting disease risk. PRS is the cumulative, mathematical aggregation of risk derived from the contributions of many DNA variants across the genome[3].

Recent studies have shown the high prevalence of cardiometabolic conditions among adults in the United States[4], and together they are the leading cause of mortality around the world[5,6]. GWAS have identified hundreds of loci associated with common diseases such as coronary artery disease (CAD)[7], obesity[8], hypertension[9] (measured using systolic blood pressure [SBP] and diastolic blood pressure [DBP]), and type 2 diabetes (T2D)[10]. Among the individuals that are diagnosed with one disease (for example, T2D), the prevalence of comorbidities such as hypertension, CAD, heart failure, and chronic kidney disease is also increased. To fully evaluate disease risk in an individual, it is therefore essential to also consider comorbid or secondary conditions related to the primary disease. There are several GWAS that have identified shared genetic associations between cardiometabolic conditions, demonstrating similarity in the underlying genetic architecture[11,12]. Pathophysiology of these conditions also shows the *cross-talk* between organ systems and its effect on disease progression, such as hemodynamic interaction between heart and kidney in heart failure[13]. With PRS, it is possible to derive an individuals' disease risk for each cardiometabolic condition using GWAS summary statistics. PRS represents an aggregate measure of the cumulative effect of numerous genetic variants on a particular disease, capturing an individual's genetic predisposition. As such, PRS can be instrumental in assessing the genetic interplay among coexisting or comorbid conditions.

Numerous methodologies exist for constructing PRS targeted at specific diseases. Conventionally, genetic risk scores (GRS) were derived using the genome-wide significant SNPs from a GWAS; however, recent studies show that using association results with much lower p-value significance ($p<0.05$) segregate individuals risk with better accuracy[1]. The development and clinical utility of PRS is under active investigation, especially in globally diverse populations[14–16]. Most large-scale GWAS have been conducted in individuals from European ancestry populations and most PRS are derived from these studies. Subsequently, the majority of PRS investigations published to date have been conducted in populations of European ancestry[17]. There can be several differences such as linkage disequilibrium (LD) structure and allele frequency of the variants, which can lead to inaccurate PRS for non-European populations[17]. This is not unique to PRS studies, and the majority of human genetic research suffers from this same phenomenon[18]. To ensure the successful clinical implementation of PRS, it is imperative to evaluate its performance in diverse global populations that closely reflect the healthcare population being treated. Moreover, for PRS to become a truly inclusive and effective tool for precision medicine, they must be applicable to individuals of all genetic backgrounds, including those with mixed ancestral backgrounds. Achieving this level of equity and broad usability will contribute significantly to the advancement of personalized healthcare practices.

In this study, we investigated the implementation of PRS for cardiometabolic conditions in individuals in the Penn Medicine BioBank (PMBB). PMBB is a cohort of >250,000 individuals established for genomic and precision medicine research. Approximately 45,000 of the individuals have genetic data imputed using the Trans-Omics for Precision Medicine (TOPMed) v2 dataset[19]. 20% of the PMBB study population is classified as African (AFR) ancestry based on genetic similarity to the 1000 genome (1KGP)[20] AFR superpopulation group. We calculated PRS in the PMBB based on GWAS summary statistics generated in multi-ancestry data to evaluate 1) risk prediction accuracy among all individuals, and among AFR and European (EUR) subpopulations; and 2) the utility of PRS in determining genetic overlap among cardiometabolic conditions.

2. Methods

2.1. Penn Medicine BioBank

The Penn Medicine BioBank (PMBB) recruits participants through the University of Pennsylvania Health System by enrolling at the time of appointment[21]. Patients participate by donating either blood or a tissue sample and allowing researchers access to their electronic health record (EHR) information. This academic biobank provides researchers with centralized access to a large number of blood and tissue samples with extensive health information from the EHR. The facility banks both blood specimens (i.e., whole blood, plasma, serum, buffy coat, and DNA isolated from leukocytes) and tissues (i.e., formalin-fixed paraffin-embedded, fresh, and flash frozen).

2.2. Genotyping and Quality Control and Imputation

The DNA extracted from blood samples was genotyped using the Illumina Global Screening Array. To ensure data integrity, we conducted quality control measures, excluding SNPs with a marker call rate of less than 95% and samples with a call rate of less than 90%. Additionally, individuals with sex discrepancies were removed from the analysis. Imputation was carried out using the Michigan Imputation server, leveraging the TOPMed Reference panel[19]. To determine genetic ancestry, we employed principal component analysis (PCA) using the smartpca tool[22] and the 1KGP dataset[20]. Genetic ancestry was inferred through a k-means clustering approach, utilizing the 1KGP super populations as genetic ancestry labels.

2.3. Polygenic Risk Scores

To derive PRS, we used the multi-ancestry summary statistics from the largest and/or most recent GWAS studies for each trait (See Table 1).

Table 1. Multi-ancestry GWAS

Phenotype	Sample size (N cases)	PMID
BMI	241,258	28443625[8]
CAD	547,261 (122,733)	29212778[7]
Hypertension (DBP, SBP)	318,891	30578418[9]
T2D	1,407,282 (228,499)	32541925[10]

Weights for each SNP were calculated using PRS-CS[23] (version from April 24, 2020), a method that performs Polygenic Prediction via Bayesian regression and continuous shrinkage priors. PRS-CS requires a reference panel that matches the ancestry distribution of the target data set. We generated multiple reference panels for analyses: a multi-ancestry LD reference panel using the HapMap SNPs from the entire 1KGP populations (2504 individuals), an African-only reference panel from the 1KGP African ancestry population, and a European-only reference panel from 1KGP European ancestry population. We identified LD patterns within the 1KGP population by using PLINK (version 1.90) to determine LD blocks and calculate the LD between the SNPs in each block. For PRS-CS, the global shrinkage parameter φ was fixed to 0.01, and default values were selected for all other parameters. PRSs were then calculated using the weights with PLINK. Only the SNPs in the target data set, summary statistics, and LD reference panel were included in the PRSs.

2.4. Phenotypes

We focused on four primary phenotypes to derive and evaluate the PRS association: CAD, hypertension (for DBP and SBP PRS), T2D, and BMI. Cases and controls for each binary phenotype were defined using International Classification of Diseases (ICD-9 and ICD-10) diagnosis codes (CAD: 414.0*, I25.1*; T2D: 250*, E11*; hypertension: 401*, I10*). Participants were coded as cases of a given phenotype if their records contained at least 1 of the corresponding ICD-9 or ICD-10 codes. The median value for BMI was extracted from the EHR.

For Phenome-wide Association Study (PheWAS) analysis, we derived phenotypes using ICD-9 and ICD-10 data from individuals from the Penn Medicine EHR. ICD-9 codes were aggregated to phecodes using the phecode ICD-9 map 1.2[24,25]; ICD-10 codes were aggregated to phecodes using the phecode ICD-10 map 1.2 (beta)[26]. Individuals are considered cases for the phenotype if they have at least 2 instances of the phecode on unique dates, controls if they have no instance of the phecode, and 'other/missing' if they have one instance of the phecode or a related phecode.

2.5. Statistical Analysis

PRS were normalized (mean of 0 and standard deviation of 1) for each analysis separately (stratified by ancestry and overall). Logistic or linear regression models accounting for age, sex, and the first 5 within-ancestry principal components (PCs) were used to test for association of PRS with each of the primary phenotypes (T2D, BMI, hypertension, and CAD). Area under the receiver operator curve (AUC) and DeLong test was determined using the R package pROC, using the full logistic regression model as above. AUC was also calculated for a reduced logistic regression model including covariates alone (age, sex, and the first 5 PCs). The DeLong test[27] is a non-parametric approach used to compare the AUCs of two correlated ROC curves, especially when the models are applied to the same set of samples. This test was used to compare null model and full model that includes PRS and obtain a p-value indicating the statistical significance of the difference between the two AUCs. For BMI, we treated it as a continuous trait and provided the R^2 value for all analyses.

A PheWAS was performed using logistic regression models with each PRS as the independent variable, phecodes as the dependent variables, and age, sex, and the first 10 PCs as covariates. A phenome-wide Bonferroni significance threshold of 4.2×10^{-5} (0.05/1190) in AFR and 3.6×10^{-5} (0.05/1377) in EUR was applied to account for multiple testing.

3. Results

3.1. Penn Medicine BioBank (PMBB) Demographics

PMBB currently consists of >250,000 consented individuals. Approximately 45,000 of these participants have been genotyped to date. Demographics of the sample included in this study are shown in Table 2.

Table 2. Demographics of PMBB sample

	All	AFR	EUR
Total patients	43,530	11,189	30,094
% Female	50.1%	62.8%	44.9%
Mean age	55.2	51.7	57.3
% CAD	23.8%	18.8%	26.4%
% Hypertension	54.4%	65.2%	51.7%
% T2D	23.5%	35.1%	19.3%
Patients with BMI data	40,043	10,619	27,489
% Female	50.4%	63.4%	44.9%
Mean age	55.6	51.9	57.7

3.2. Determining the effect of linkage disequilibrium panel on PRS in the overall sample

Using publicly available multi-ancestry GWAS data (Table 1), we generated a PRS for each primary phenotype of interest: type 2 diabetes, body mass index, hypertension (SBP and DBP), and coronary artery disease. We assessed the impact of using a multi-ancestry LD panel, akin to the GWAS data, and compared it with an AFR LD panel (in all PMBB individuals and in AFR PMBB individuals) and an EUR LD panel (in all PMBB individuals and in EUR PMBB individuals). AUC values were computed for each binary phenotype PRS in all individuals (Table 3) and contrasted between the full model (AUC, covariates + PRS) and the model containing covariates alone (AUC Null). The addition of PRS consistently improved the covariate model for all phenotypes, showing an average AUC improvement of 0.014. Across the entire dataset, the PRS created with the multi-ancestry LD panel (DBP, BMI) or the EUR LD panel (CAD, SBP, T2D) demonstrated the strongest association with their respective primary phenotypes (Table 3).

Table 3. Comparison of LD panel for PRS in all

PRS	LD Panel	AUC[1] Null	AUC[1]	DeLong P	Model OR	Model P-value
	Multi-ancestry		0.808	1.22E-53	1.495	5.82E-186
CAD	AFR	0.795	0.807	1.22E-52	1.472	7.11E-182
	EUR		0.807	2.33E-52	1.515	1.00E-184
	Multi-ancestry		0.773	8.90E-06	1.236	1.65E-49
DBP	AFR	0.770	0.772	1.32E-15	1.219	1.59E-49
	EUR		0.772	6.15E-14	1.226	6.32E-43
	Multi-ancestry		0.775	4.47E-23	1.365	2.48E-83
SBP	AFR	0.770	0.775	3.74E-22	1.338	2.78E-80
	EUR		0.775	7.40E-23	1.376	2.31E-83
	Multi-ancestry		0.730	5.41E-88	2.223	1.24E-286
T2D	AFR	0.695	0.727	2.68E-79	2.095	3.18E-266
	EUR		0.731	2.44E-91	2.263	1.46E-297
PRS	**LD Panel**	**R^2 Null**	**R^2**	**R^2 difference**	**Model Beta**	**Model P-value**
	Multi-ancestry		0.110	0.043	2.205	0
BMI	AFR	0.067	0.110	0.043	2.125	0
	EUR		0.108	0.042	2.198	0

3.3. Determining the effect of linkage disequilibrium panel on PRS within ancestry

In both AFR (Table 4) and EUR (Table 5) individuals, the addition of PRS to the covariate model enhances model performance. However, it is noteworthy that PRS performance was relatively stronger in EUR individuals compared to AFR individuals. In AFR, the full model shows a somewhat smaller improvement over the covariate-based model (average improvement in AUC=0.011) compared to the improvement observed in EUR (average improvement in AUC=0.021).

Notably, in AFR individuals, the PRS calculated using the multi-ancestry LD panel exhibited a higher effect size in four out of the five PRSs (DBP, SBP, T2D, and BMI) compared to the AFR LD panel (Table 4). This indicates the potential benefits of using a multi-ancestry LD panel to derive PRS in populations with diverse genetic backgrounds.

[1] AUC rounded to three decimal points

Table 4. Comparison of LD panel for PRS in AFR individuals

PRS	LD Panel	AUC Null	AUC	DeLong P	Model OR	Model P-value
CAD	AFR	0.764	0.770	1.33E-06	1.261	2.75E-18
	Multi-ancestry		0.770	4.52E-06	1.253	2.45E-17
DBP	AFR	0.793	0.797	1.72E-05	1.208	4.56E-15
	Multi-ancestry		0.797	1.25E-05	1.214	2.56E-15
SBP	AFR	0.793	0.797	3.82E-06	1.252	3.00E-18
	Multi-ancestry		0.797	1.11E-06	1.277	9.65E-20
T2D	AFR	0.681	0.710	3.03E-25	1.630	5.73E-77
	Multi-ancestry		0.711	4.21E-26	1.689	1.73E-79
PRS	**LD Panel**	**R^2 Null**	**R^2**	**R^2 difference**	**Model Beta**	**Model P-value**
BMI	AFR	0.041	0.065	0.024	1.449	1.02E-59
	Multi-ancestry		0.063	0.022	1.462	6.84E-56

In EUR individuals, the PRS calculated using the multi-ancestry LD panel demonstrated a higher effect size in two out of the five PRSs (SBP and T2D) when compared to the EUR LD panel (Table 5). This observation highlights the potential advantages of leveraging a multi-ancestry LD panel in deriving PRS for certain phenotypes in populations with European ancestry.

Table 5. Comparison of LD panel for PRS in EUR individuals

PRS	LD Panel	AUC Null	AUC	DeLong P	Model OR	Model P-value
CAD	EUR	0.796	0.812	9.49E-48	1.533	5.65E-166
	Multi-ancestry		0.812	2.38E-48	1.531	5.73E-165
DBP	EUR	0.747	0.750	6.17E-11	1.173	9.17E-34
	Multi-ancestry		0.750	1.51E-12	1.158	9.43E-29
SBP	EUR	0.747	0.753	6.64E-21	1.251	1.49E-64
	Multi-ancestry		0.753	1.61E-20	1.255	2.40E-66
T2D	EUR	0.651	0.708	8.26E-87	1.721	5.68E-243
	Multi-ancestry		0.710	1.12E-82	1.757	8.59E-258
PRS	**LD Panel**	**R^2 Null**	**R^2**	**R^2 difference**	**Model Beta**	**Model P-value**
BMI	EUR	0.006	0.076	0.070	1.637	0
	Multi-ancestry		0.075	0.069	1.626	0

3.4 PheWAS of polygenic risk scores

We conducted a PheWAS of each multi-ancestry LD panel PRS in AFR and EUR individuals, identifying additional phenotypes associated with the PRS for our primary phenotypes (Figure 1, full results in Supplemental Tables Online: https://shorturl.at/uBDSX). The results reveal significant associations between the PRS and various phenotypic categories, shedding light on the potential implications of PRS in predicting disease susceptibility. All PRS exhibited associations with other phenotypes. However, in AFR individuals, the strength and number of PRS associations with other phenotypes were generally reduced compared to EUR individuals.

In our analysis, the CAD PRS in AFR individuals was associated with 18 distinct phenotypes, including notable associations with hyperlipidemia (OR=1.12, $p=1.1x10^{-6}$) and renal failure (OR=1.12, $p=1.0x10^{-5}$). In contrast, EUR individuals exhibited associations with a broader range of 82 phenotypes, with hyperlipidemia (OR=1.23, p=7.3x10-45) and renal failure (OR=1.10, $p=2.1x10^{-8}$) being among them.

For the DBP and SBP PRS, AFR individuals showed associations with 9 and 20 phenotypes respectively. Specific associations of interest included atrial fibrillation for DBP (OR=1.20, $p=1.4x10^{-5}$) and both coronary atherosclerosis (OR=1.20, $p=3.7x10^{-7}$) and T2D (OR=1.12, $p=3.2x10^{-5}$) for SBP. EUR individuals, on the other hand, had DBP and SBP PRS associated with 12 and 27 phenotypes, respectively. This encompassed associations like coronary atherosclerosis for both DBP (OR=1.09, $p=4.9x10^{-7}$) and SBP (OR=1.13, $p=1.6x10^{-13}$), and T2D specifically for SBP (OR=1.17, $p=1.0x10^{-17}$).

The T2D PRS in AFR individuals was linked with a vast array of 84 phenotypes. Key associations here were hyperlipidemia (OR=1.30, $p=6.0x10^{-16}$), obesity (OR=1.20, $p=6.6x10^{-10}$), and hypertension (OR=1.22, $p=4.5x10^{-9}$). EUR individuals had a slightly lesser range with 78 phenotypes, but with significant associations like hyperlipidemia (OR=1.31, $p=9.2x10^{-17}$), obesity (OR=1.29, $p=9.9x10^{-57}$), and hypertension (OR=1.22, $p=3.2x10^{-38}$). Lastly, the BMI PRS in AFR was associated with 19 phenotypes, including T2D (OR=1.17, $p=1.6x10^{-8}$) and hypertension (OR=1.18, $p=8.6x10^{-8}$). In EUR individuals, this PRS was linked with a more extensive 72 phenotypes, with notable associations being T2D (OR=1.26, $p=4.6x10^{-39}$) and hypertension (OR=1.19, p=2.2x10-32).

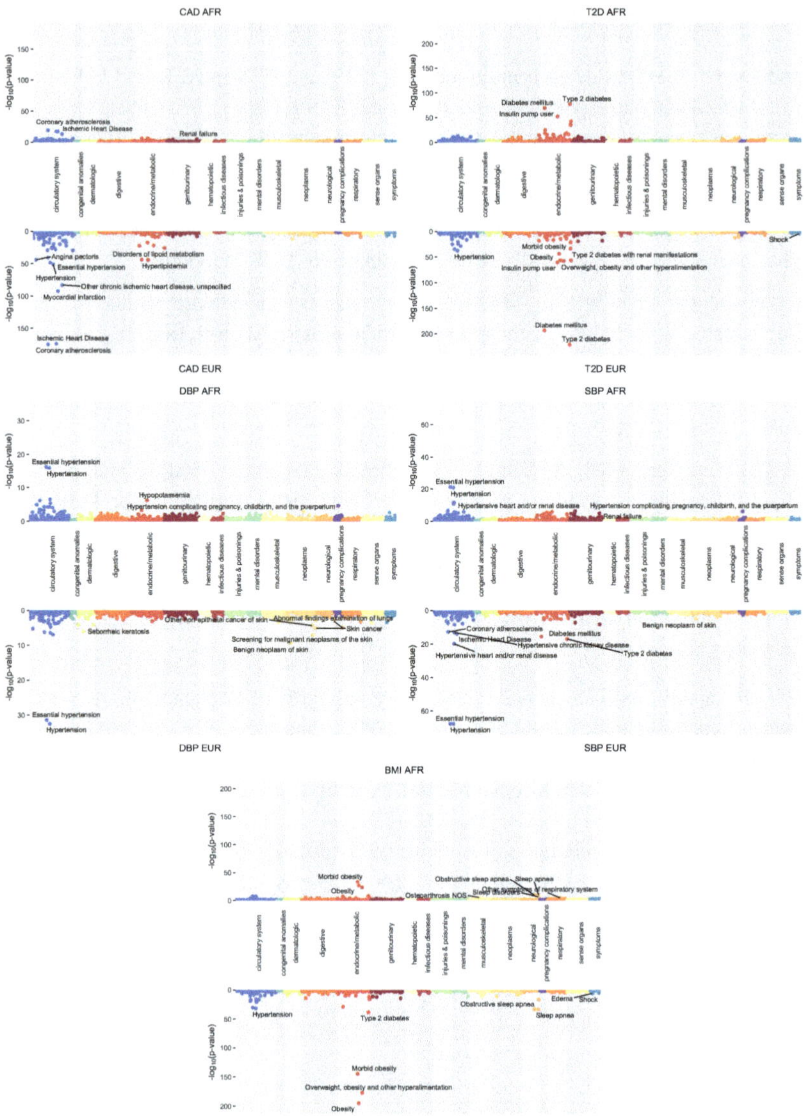

Figure 1. Phenome-wide Association Study (PheWAS) Results for Polygenic Risk Scores (PRS) for coronary artery disease (CAD), Diastolic Blood Pressure (DBP), Systolic Blood Pressure (SBP), Type 2 Diabetes (T2D), and Body Mass Index (BMI). The x-axis represents the phecode categories, and the y-axis shows the -log10 p-values, color-coded by category.

4. Discussion

We generated five polygenic risk scores representing genetic liability for cardiometabolic diseases and assessed their performance across different ancestry groups in the Penn Medicine BioBank (PMBB), a biobank including DNA linked with electronic health records. For all PRS tested, we identified a statistically significant association with the primary phenotype in both ancestry groups, as validated by the DeLong test comparing the null and the full model.

Type 2 diabetes consistently exhibited the highest effect size, reflecting the large number of cases in the GWAS used to generate this PRS and the PMBB dataset. Contrarily, the hypertension PRSs (DBP and SBP) showed a weaker effect size, even with a larger GWAS and over 50% of PMBB patient participants with hypertension. These observations suggest that factors beyond sample size, such as disease heterogeneity, prevalence, and non-additive effects, influence PRS associations. Consequently, understanding the interplay of these factors will be pivotal in refining and optimizing the application of PRS in disease prediction and risk assessment.

Our PheWAS analyses were conducted to explore the broader phenotypic landscape associated with each PRS with an EHR-linked biobank. Many of the identified phenotypes could be linked to broader effects of known disease risk factors and established comorbidities. For instance, risk for Type 2 diabetes was associated with hypertension, a known commonly co-occurring trait[28]. Similarly, the BMI PRS was associated with sleep apnea, diabetes, and hypertension, all of which are known to be more prevalent in individuals with higher BMI[29–32]. However, these associations don't necessarily imply causality. The high prevalence of comorbidities among these phenotypes complicates the task of discerning whether the genetic risk for one condition directly influences the onset of another.

Our findings underscore a significant challenge in the future implementation of PRS into routine clinical care. While PRS derived from multi-ancestry GWAS can be associated with phenotypes in individuals of African ancestry (AFR), their impact is not as pronounced as those generated in European ancestry (EUR). This observation, although expected, has been a topic of extensive discussion in recent years, emphasizing a notable disparity in genetic research[15,17]. Our results here affirm that these expectations persist even in large-scale, diverse ancestry datasets. Furthermore, our study suggests that PRS for cardiometabolic diseases based on multi-ancestry GWAS data might not perform as robustly for the primary disease and its associated secondary cardiometabolic traits.

Our utilization of a multi-ancestry LD panel to compute PRS for all individuals from multi-ancestry GWAS demonstrated robust performance across all populations. This was especially true for African ancestry individuals, emphasizing the potential advantages of leveraging a multi-ancestry reference panel in PRS generation. As the field of precision medicine continues to evolve, advocating for the adoption of such panels becomes increasingly important. By addressing these challenges, we can pave the way for more inclusive and accurate personalized healthcare strategies.

One notable limitation of our study is the modest gain in predictive performance over the null model across all categories, as reflected in the AUC values. While we observed differences in AUC between the ancestry groups, the absolute increase in AUC over the null model was relatively small. This underscores the need for further refinement in PRS methodologies to achieve more substantial improvements in predictive performance. Additionally, in our PheWAS approach, there are inherent challenges when comparing results between AFR and EUR groups. The difference in sample sizes between these groups can lead to variations in statistical power, potentially influencing the observed associations. Moreover, the generally lower PRS performance in the AFR group, as highlighted in our results, can further compound these challenges. It's essential to interpret the PheWAS results with these considerations in mind.

In conclusion, while there's considerable enthusiasm surrounding PRS in clinical care, there remains a significant amount of research to be conducted to determine its optimal implementation. It is essential to explore how PRS can be incorporated alongside other commonly used predictors[33], such as family history, clinical comorbidities, and environmental/lifestyle factors. By combining PRS with established clinical guidelines, we can aim for a more comprehensive risk assessment, leading to personalized interventions. Another important issue to address is whether we will ultimately need ancestry-specific PRS models or if we can develop the statistical framework to integrate global and local LD patterns into the PRS model to produce a cosmopolitan PRS approach. For clinical implementation, a cosmopolitan PRS approach will be easier for clinicians to adopt; however, it is unclear how this can be done effectively, given the heterogeneity in LD patterns, effect sizes, and causal variants in different ancestry groups. Our work here suggests that the use of multi-ancestry GWAS and LD panels may be a step towards this goal. The ultimate success of PRS in precision medicine lies in integrating it seamlessly with published clinical guidelines and incorporating an individual's ancestry within the PRS framework. This integration will empower clinicians to make informed decisions based on a comprehensive and personalized risk profile for each patient. By addressing these key aspects and enhancing our understanding of PRSs role in precision medicine, we can unlock its full potential as a transformative tool in healthcare, facilitating early interventions and preventive measures that cater to each individual's unique genetic makeup and health needs.

5. Acknowledgements

We acknowledge the Penn Medicine BioBank (PMBB) for providing data and thank the patient-participants of Penn Medicine who consented to participate in this research program. We would also like to thank the Penn Medicine BioBank team and Regeneron Genetics Center for providing genetic variant data for analysis. The PMBB is approved under IRB protocol# 813913 and supported by Perelman School of Medicine at University of Pennsylvania, a gift from the Smilow family, and the National Center for Advancing Translational Sciences of the National Institutes of Health under CTSA award number UL1TR001878.

References

1. Abraham, G. *et al.* Genomic prediction of coronary heart disease. *Eur. Heart J.* **37**, 3267–3278 (2016).
2. Tada, H. *et al.* Risk prediction by genetic risk scores for coronary heart disease is independent of self-reported family history. *Eur. Heart J.* **37**, 561–567 (2016).
3. International Schizophrenia Consortium *et al.* Common polygenic variation contributes to risk of schizophrenia and bipolar disorder. *Nature* **460**, 748–752 (2009).
4. Arnold, S. V. *et al.* Burden of cardio-renal-metabolic conditions in adults with type 2 diabetes within the Diabetes Collaborative Registry. *Diabetes Obes. Metab.* **20**, 2000–2003 (2018).
5. Wang, H. *et al.* Global, regional, and national life expectancy, all-cause mortality, and cause-specific mortality for 249 causes of death, 1980–2015: a systematic analysis for the Global Burden of Disease Study 2015. *The Lancet* **388**, 1459–1544 (2016).
6. Ogurtsova, K. *et al.* IDF Diabetes Atlas: Global estimates for the prevalence of diabetes for 2015 and 2040. *Diabetes Res. Clin. Pract.* **128**, 40–50 (2017).
7. van der Harst, P. & Verweij, N. Identification of 64 Novel Genetic Loci Provides an Expanded View on the Genetic Architecture of Coronary Artery Disease. *Circ. Res.* **122**, 433–443 (2018).
8. Justice, A. E. *et al.* Genome-wide meta-analysis of 241,258 adults accounting for smoking behaviour identifies novel loci for obesity traits. *Nat. Commun.* **8**, 14977 (2017).
9. Giri, A. *et al.* Trans-ethnic association study of blood pressure determinants in over 750,000 individuals. *Nat. Genet.* **51**, 51–62 (2019).

10. Vujkovic, M. *et al.* Discovery of 318 new risk loci for type 2 diabetes and related vascular outcomes among 1.4 million participants in a multi-ancestry meta-analysis. *Nat. Genet.* **52**, 680–691 (2020).

11. Ma, R. C. Genetics of cardiovascular and renal complications in diabetes. *J. Diabetes Investig.* **7**, 139–154 (2016).

12. Regele, F. *et al.* Genome-wide studies to identify risk factors for kidney disease with a focus on patients with diabetes. *Nephrol. Dial. Transplant.* **30**, iv26–iv34 (2015).

13. Rangaswami, J. *et al.* Cardiorenal Syndrome: Classification, Pathophysiology, Diagnosis, and Treatment Strategies: A Scientific Statement From the American Heart Association. *Circulation* **139**, (2019).

14. Kim, M. S., Patel, K. P., Teng, A. K., Berens, A. J. & Lachance, J. Genetic disease risks can be misestimated across global populations. *Genome Biol.* **19**, (2018).

15. De La Vega, F. M. & Bustamante, C. D. Polygenic risk scores: a biased prediction? *Genome Med.* **10**, (2018).

16. Popejoy, A. B. & Fullerton, S. M. Genomics is failing on diversity. *Nature* **538**, 161–164 (2016).

17. Martin, A. R. *et al.* Clinical use of current polygenic risk scores may exacerbate health disparities. *Nat. Genet.* **51**, 584–591 (2019).

18. Gurdasani, D., Barroso, I., Zeggini, E. & Sandhu, M. S. Genomics of disease risk in globally diverse populations. *Nat. Rev. Genet.* (2019) doi:10.1038/s41576-019-0144-0.

19. Taliun, D. *et al.* Sequencing of 53,831 diverse genomes from the NHLBI TOPMed Program. *Nature* **590**, 290–299 (2021).

20. Auton, A. *et al.* A global reference for human genetic variation. *Nature* **526**, 68–74 (2015).

21. Verma, A. *et al.* The Penn Medicine BioBank: Towards a Genomics-Enabled Learning Healthcare System to Accelerate Precision Medicine in a Diverse Population. *J. Pers. Med.* **12**, 1974 (2022).

22. Price, A. L. *et al.* Principal components analysis corrects for stratification in genome-wide association studies. *Nat. Genet.* **38**, 904–909 (2006).

23. Ge, T., Chen, C.-Y., Ni, Y., Feng, Y.-C. A. & Smoller, J. W. Polygenic prediction via Bayesian regression and continuous shrinkage priors. *Nat. Commun.* **10**, 1776 (2019).

24. Denny, J. C. *et al.* PheWAS: demonstrating the feasibility of a phenome-wide scan to discover gene-disease associations. *Bioinformatics* **26**, 1205–1210 (2010).

25. Denny, J. C. *et al.* Systematic comparison of phenome-wide association study of electronic medical record data and genome-wide association study data. *Nat. Biotechnol.* **31**, 1102–1110 (2013).

26. Wu, P. *et al.* Developing and Evaluating Mappings of ICD-10 and ICD-10-CM Codes to PheCodes. *bioRxiv* (2019) doi:10.1101/462077.

27. DeLong, E. R., DeLong, D. M. & Clarke-Pearson, D. L. Comparing the areas under two or more correlated receiver operating characteristic curves: a nonparametric approach. *Biometrics* **44**, 837–845 (1988).

28. Sun, D. *et al.* Type 2 Diabetes and Hypertension: A Study on Bidirectional Causality. *Circ. Res.* **124**, 930–937 (2019).

29. Romero-Corral, A., Caples, S. M., Lopez-Jimenez, F. & Somers, V. K. Interactions Between Obesity and Obstructive Sleep Apnea. *Chest* **137**, 711–719 (2010).

30. Dua, S., Bhuker, M., Sharma, P., Dhall, M. & Kapoor, S. Body mass index relates to blood pressure among adults. *North Am. J. Med. Sci.* **6**, 89 (2014).

31. Xiang, B.-Y. *et al.* Body mass index and the risk of low bone mass–related fractures in women compared with men: A PRISMA-compliant meta-analysis of prospective cohort studies. *Medicine (Baltimore)* **96**, e5290 (2017).

32. Gray, N., Picone, G., Sloan, F. & Yashkin, A. Relation between BMI and Diabetes Mellitus and Its Complications among US Older Adults. *South. Med. J.* **108**, 29–36 (2015).

33. Arnett, D. K. *et al.* 2019 ACC/AHA Guideline on the Primary Prevention of Cardiovascular Disease: A Report of the American College of Cardiology/American Heart Association Task Force on Clinical Practice Guidelines. *Circulation* (2019) doi:10.1161/CIR.0000000000000678.

ıtCC: An efficient weighted integrative consensus clustering of multimodal data

Can Huang and Pei Fen Kuan*
Department of Applied Mathematics and Statistics,
Stony Brook University,
Stony Brook, NY 11794, USA
**E-mail: peifen.kuan@stonybrook.edu*
www.ams.sunysb.edu

High throughput profiling of multiomics data provides a valuable resource to better understand the complex human disease such as cancer and to potentially uncover new subtypes. Integrative clustering has emerged as a powerful unsupervised learning framework for subtype discovery. In this paper, we propose an efficient weighted integrative clustering called intCC by combining ensemble method, consensus clustering and kernel learning integrative clustering. We illustrate that intCC can accurately uncover the latent cluster structures via extensive simulation studies and a case study on the TCGA pan cancer datasets. An R package intCC implementing our proposed method is available at `https://github.com/candsj/intCC`.

Keywords: Integrative clustering; Consensus clustering; Multiomics data; Ensemble learning.

. Introduction

ecent advancements in high throughput technologies have enabled rapid profiling of different mics data, including genomics, epigenomics, transcriptomics, proteomics and metabolomics hich allow for in-depth study of the complex regulatory patterns from a systems biology perpective. For example, the Cancer Genome Atlas (TCGA) has generated over 2.5 petabytes of ıultiomics data. Such datasets offer the opportunity to explore the heterogeneity underpinning diseases such as cancer via unsupervised learning based on clustering framework, which ould help define cancer subtypes, bringing us a step closer towards personalized medicine.

In multimodal data structure, e.g., the different omics data, a key challenge in data analysis in identifying the most appropriate approach for data integration. For unsupervised clusterıg over multimodal data, these include the choice of a single step *versus* two-step approach. single step approach is also known as joint modeling which combines all datasets together. wo-step approach works by clustering each dataset separately, followed by integration of ıese clusters.

A number of integrative clustering methods and tools have been proposed to date. This ıcludes Bayesian Consensus Clustering (BCC[1]), iCluster,[2] iClusterPlus,[3] Cluster Of Clus-ers Analysis (COCA[4]), Clusternomics[5] and kernel learning integrative clustering (KLIC[6]).

BCC, Clusternomics and iClusterPlus are based on Bayesian modeling framework and rely on Markov Chain Monte Carlo (MCMC) algorithm for fitting the model. These methods also assume that the probability model for each dataset is specified. However, softwares for BCC and Clusternomics currently only implement the algorithms for Gaussian distributed dataset, thus limiting the applicability of these methods to non-Gaussian datasets such as SNPs, mutation or copy number datasets.

On the other hand, iCluster works by assuming a Gaussian latent variable model for inferring the cluster structures, whereas iClusterPlus increases the versatility of iCluster by incorporating statistical models for continuous, binary, multinomial count datasets via a Bayesian latent variable model and employs MCMC algorithm for sampling from its posterior distribution for statistical inference. However, software implementation of iClusterPlus currently is limited to integrative clustering of at most four datasets. Since the model involves tuning a number of parameters, the bottleneck is the computational time when the number of datasets or features increases.

Another popular integrative clustering approach is COCA[4] which was first introduced to define cancer subtypes by clustering six different datasets, namely DNA copy number, DNA methylation, mRNA expression, microRNA expression, protein expression, and somatic point mutation. COCA works by first clustering each dataset using consensus clustering,[7] followed by clustering the binary matrix generated by aggregating the clusters obtained from each dataset. While this approach is robust and easily scalable to a large number of datasets, a limitation of COCA is that all datasets contribute equally to the final clustering which affects the accuracy of the clusters obtained, especially in scenario in which certain dataset is less reliable.

Taking inspiration from COCA and multiple kernel learning,[8,9] KLIC[6] was developed to address the pitfall of COCA. Similar to COCA, KLIC works by first applying consensus clustering to each dataset. The authors proved that these consensus matrices are positive semi-definite kernels, which can then be used as input in multiple kernel k-means clustering and allows for weights to be estimated for each kernel via a two-step optimization strategy and convex quadratic programming. This approach allows for more informative dataset to contribute more to the overall clustering. Currently, KLIC runs one clustering algorithm on each dataset to generate the consensus matrix.

In this paper, we seek to extend the KLIC framework to a more robust integrative clustering by proposing a two layer weighted integrative clustering which allows for more than one clustering algorithm to be run on each dataset, i.e, ensemble clustering and aggregated together via an efficient weight estimation.

2. Methods

Our proposed method can be viewed as a combination of (a) ensemble clustering, i.e, aggregating multiple clustering algorithms, (b) consensus clustering, i.e., resampling, and (c) kernel learning integrative clustering. While some papers use ensemble and consensus clustering interchangeably, in this paper, we refer to ensemble clustering as a collection of multiple clustering algorithms, e.g., k-means, hierarchical clustering or partitioning around medoid (PAM), whereas consensus clustering as a framework which draws a random sample from either the sample or feature space. We now briefly describe the consensus clustering and kernel learning integrative clustering framework.

Consensus clustering was originally proposed by Monti et al (2003).[7] The main idea behind consensus clustering is to apply a resampling scheme on the sample or feature dimension under the assumption that different subsamples drawn from the dataset should not differ much in the clustering results. The resampling scheme allows one to assess the stability of the cluster assignments and the robustness of the dataset to perturbations, thus could aid in deriving a more stable and reliable result that reveals the real structure underlying the dataset.

A key element derived from the consensus clustering is the consensus matrix which measures the agreement among samples. For a dataset with N samples, the consensus matrix $\mathcal{M}$ is a $N \times N$ matrix whose element $M(i, j)$ denotes the proportion of sample i and sample j in the same cluster during the resampling iterations. Values which are close to 1 (and vice versa 0) indicate that the two samples are always assigned to the same cluster (and vice versa different clusters). $1 - \mathcal{M}$ is a distance measure which can be used to derive a final clustering result.

Cabassi and Kirk (2020)[6] proved that the consensus matrix is positive semi-definite and thus can be used as input in kernel learning integrative clustering via the application of multiple kernel k-means algorithm. The kernel k-means algorithm utilizes the kernel trick by projecting the data into a non-linear feature space via a kernel. This overcomes the drawback of regular k-means clustering which cannot identify clusters that are not linearly separable in the original input space. The integration of the multimodal data within the kernel learning integrative clustering involves a convex sum of the kernels, i.e., consensus matrix from each dataset, and the estimation of the weights in the convex sum. In the KLIC integrative clustering algorithm of Cabassi and Kirk (2020),[6] the authors adopted the optimization strategy proposed by Gonen and Margolin (2014)[10] which involves a convex quadratic programming.

In this paper, we reason that the weights in the kernel learning integrative clustering can be estimated by utilizing the fuzziness in the consensus matrix. Furthermore, we extend the framework of KLIC by allowing multiple base clustering algorithms, e.g., k-means, hierarchical clustering, PAM, to be applied within each dataset and aggregated, i.e., ensemble clustering[11] which has been shown to enhance the robustness of clustering results compared to individual clustering algorithm. To this end, we propose an efficient weight estimation method and a two layer weighted integrative consensus clustering.

2.1. *Weight estimation*

The consensus matrix can be used to assess cluster stability and composition. As a motivating example, we generate two datasets, each with 10 features and 100 samples. For both datasets, we assume that there are 3 clusters with cluster sizes 20, 30 and 50. All the features are generated from the Gaussian distribution. For dataset 1, 9 out of the 10 features are informative, where the means of cluster 1, 2 and 3 are 1, -1 and 0, respectively with unit variance. For dataset 2, 3 out of the 10 features are informative, where the means of cluster 1, 2 and 3 are 0.2, -0.2 and 0, respectively with unit variance. Non-informative features are generated from standard Gaussian distribution. We designate datasets 1 and 2 as having high and and low signal-to-noise-ratio (SNR), respectively and run consensus clustering on both datasets using 100 iterations of k-means and resampling 80% of samples and features in each iteration. Figure 1 shows the heatmaps of the consensus matrices. The diagonal blocks plot the in-cluster values, whereas the off diagonals blocks plot the out-of-cluster values. For the low SNR dataset, the off diagonal blocks are much noisier compared to the high SNR dataset. We argue that this can be used to derive the weights in the multiple kernel integrative clustering. Specifically, we define the weights based on the ratio of in-cluster proportion to out-of-cluster proportion using the cluster estimated by the algorithm itself. Clustering result closer to the real structure tends to have higher in-cluster proportion and lower out-of-cluster proportion. In other words, datasets with a higher ratio of in-cluster proportion to out-of-cluster proportion will be assigned larger weights.

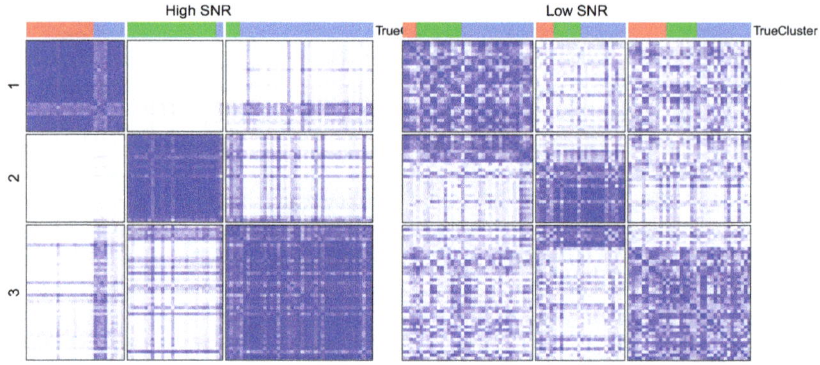

Fig. 1. Heatmaps of consensus matrices for high and low signal-to-noise ratio (SNR) datasets. True cluster membership is given in the annotation above each heatmap. Predicted cluster membership corresponds to the three gap-separated blocks in each heatmap.

Without loss of generality, we consider $\mathcal{P}$ consensus matrices $\mathcal{M}_1, ..., \mathcal{M}_P$ for number of clusters K. Here, the consensus matrices could arise by applying different clustering algorithms to the same dataset or could denote consensus matrices derived from different datasets. We further define:

$W^p_{in}(k)$: in-cluster proportion for cluster k of consensus matrix $\mathcal{M}_p$.
$W^p_{out}(k)$: out-of-cluster proportion for cluster k of consensus matrix $\mathcal{M}_p$.
W^p_{in}: average in-cluster proportion across all clusters of consensus matrix $\mathcal{M}_p$.
W^p_{out}: average out-of-cluster proportion across all clusters of consensus matrix $\mathcal{M}_p$.
R_p: ratio of in-cluster proportion to out-of-cluster proportion for consensus matrix $\mathcal{M}_p$.
W_p: weight for consensus matrix $\mathcal{M}_p$.

We propose calculating the weights as follows:

$$W_{in}^{p}(k) = \frac{\sum_{i \in k, j \in k} M_p(i,j)}{\sum I\{i \in k, j \in k\}}, \quad W_{in}^{p} = \frac{\sum_{k=1}^{K} W_{in}^{p}(k)}{K}$$

$$W_{out}^{p}(k) = \frac{\sum_{i \in k, j \notin k} M_p(i,j)}{\sum I\{i \in k, j \notin k\}}, \quad W_{out}^{p} = \frac{\sum_{k=1}^{K} W_{out}^{p}(k)}{K}$$

$$R_p = \frac{W_{in}^{p}}{W_{out}^{p}}$$

$$W_p = \frac{R^p}{\sum_{i=1}^{P} R^i}$$

In practice, true cluster membership is unknown, thus the weights will be computed based on predicted cluster membership. Using this formula, $W_1 = 0.726$ and $W_2 = 0.274$ for the consensus matrices derived based on predicted cluster membership of the two datasets above.

2.2. *Two Layer Weighted Integrative Consensus Clustering*

We now describe our proposed two layer weighted integrative consensus clustering. We assume that there are D datasets, X_1,...,X_D, and number of clusters K.

Layer 1: For each dataset X_d where $d = 1, 2, ..., D$:

(1) Perform ensemble clustering using P different clustering methods, where $p = 1, 2, ..., P$. This will generate consensus matrices $\mathcal{M}_p^d$, where $p = 1, 2, ..., P$.
(2) Compute the weights $w_1^d, w_2^d, ..., w_P^d$ for each consensus matrix $\mathcal{M}_1^d, \mathcal{M}_2^d, ..., \mathcal{M}_P^d$.
(3) Define the weighted consensus matrix $\mathcal{M}_{weight}^d$ as $\mathcal{M}_{weight}^d = \sum_{p=1}^{P} w_p^d \times \mathcal{M}_p^d$.
(4) Apply a clustering algorithm, e.g., PAM or hierarchical clustering, to each weighted consensus matrix $\mathcal{M}_{weight}^d$.

Layer 2:

1) For the weighted consensus matrix $\mathcal{M}_{weight}^1, \mathcal{M}_{weight}^2, ..., \mathcal{M}_{weight}^D$, compute the weights $W_1, W_2, ..., W_D$.
2) Define the weighted of weighted consensus matrix $\mathcal{M}_{weight}$ as $\mathcal{M}_{weight} = \sum_{d=1}^{D} W_d \times \mathcal{M}_{weight}^d$.
3) Apply a clustering algorithm, e.g., PAM or hierarchical clustering, to $\mathcal{M}_{weight}$ to derive a final clustering result.

We provide a flowchart in Figure 2 summarizing our proposed two layer weighted integrative consensus clustering. Our method is implemented as a GitHub R package intCC available at `https://github.com/candsj/intCC`.

3. Simulation studies

We conduct simulation studies to compare the performance of our proposed two layer weighted integrative consensus clustering intCC against other integrative clustering methods which are implemented for both Gaussian and non-Gaussian distributed datasets, namely KLIC[6] and ClusterPlus.[3]

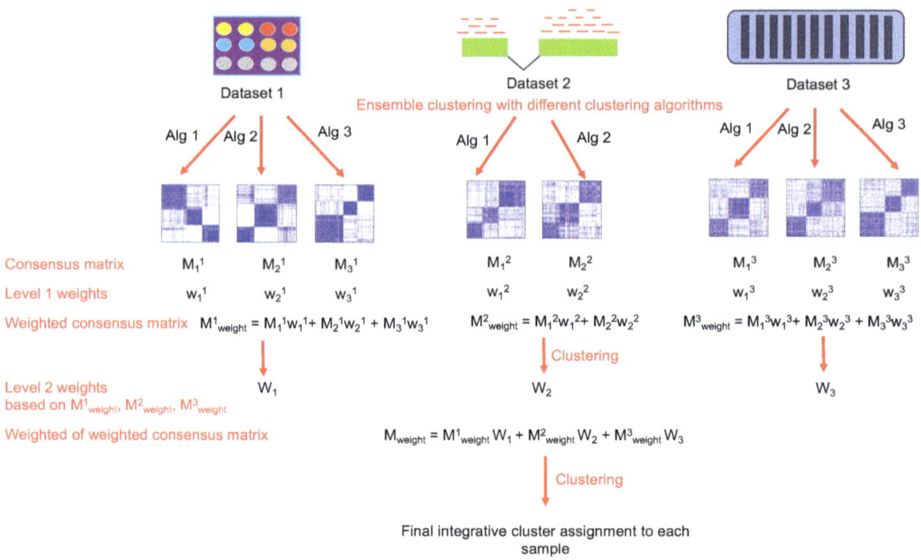

Fig. 2. Flowchat describing our proposed algorithm.

3.1. *Datasets*

Unlike Cabassi and Kirk (2020)[6] which only considered data simulated from Gaussian distributions, we follow the strategy of Mo et al. (2013)[3] where we generate datasets from different distributions, including Gaussian (e.g., M-values from DNA methylation, microarray data such as gene expression), binomial (e.g., somatic mutations), Poisson (e.g., count data from sequencing technologies such as RNA-Seq data or copy number data represented as number of copies gained or lost) and multinomial (e.g., copy number data states represented as gain, normal or loss, or SNP data) distributions. This is to ensure that our proposed method is applicable to integration of continuous, binary, count and categorical types of datasets. For Settings 1-6, we set the sample size and the true number of clusters to be 60 and 3, respectively in which each cluster consists of 20 samples. We vary the number of informative and non-informative, i.e., noise features. The parameters used in our simulations for Settings 1-6 are provided in Supplementary Table 1. Settings 7-9 follow from the simulation setup of of Cabassi and Kirk (2020).[6] We consider several simulation settings, namely:

(1) Setting 1: 4 datasets following Gaussian, binomial, Poisson and multinomial distribution, respectively. Each dataset has 30 features, in which 15 features are informative and the rest are noise features.
(2) Setting 2: 4 datasets includes normal, binomial, Poisson and multinomial distribution, respectively. Each dataset has 30 features, in which 15 features are informative and the

rest are noise features. Informative features have slighly lower signal compared to the Setting 1.

(3) Setting 3: 3 datasets following Gaussian, binomial and Poisson distribution, respectively. Each dataset has 30 features, in which 15 features are informative and the rest are noise features.
(4) Setting 4: 5 datasets following Gaussian, binomial, Poisson, multinomial and Gaussian distribution, respectively. Each dataset has 30 features. For the first 4 datasets, 15 features are informative and the rest are noise features. The 5th dataset follows a Gaussian distribution in which all features are noise features.
(5) Setting 5: 4 datasets following Gaussian, binomial, Poisson and multinomial distribution, respectively. Each dataset has 500 features, in which 100 features are informative and the rest are noise features.
(6) Setting 6: 4 datasets following Gaussian, binomial, Poisson and multinomial distribution, respectively. Each dataset has 500 features, in which 250 features are informative and the rest are noise features.
(7) Setting 7: 4 datasets following Gaussian distribution with similar parameter setting. Each dataset consists of 300 samples with 6 clusters of size 50 samples each. There are 2 features with no noise feature. For cluster k, $\mu = k \times (\text{separation level} - 1)/2, \sigma = 1$, $k = 1, 2, 3, 4, 5, 6$. Separation level $= 4$ is used in this setting.
(8) Setting 8: 4 datasets following Gaussian distribution with different parameter setting. Each dataset consists of 300 samples with 6 clusters of size 50 samples each. There are 2 features with no noise feature. For cluster k, $\mu = k \times (\text{separation level} - 1)/2, \sigma = 1$, $k = 1, 2, 3, 4, 5, 6$. Varying separation levels $= 1, 2, 3, 4$ are used in this setting. Only 3 datasets are used as input. We consider 4 dataset combinations, namely 123, 124, 134, 234. Here 123 implies that the clustering algorithms are applied to only datasets 1, 2 and 3.
(9) Setting 9 (nested cluster structure): 2 datasets following Gaussian distribution, in which each dataset consists of 300 samples. There are 2 features with no noise feature. Dataset 1 has 6 clusters of size 50 samples each. Dataset 2 has 3 clusters of size 100 samples each. For cluster k, $\mu = k \times (\text{separation level} - 1)/2, \sigma = 1$, $k = 1, 2, 3, 4, 5, 6$ for dataset 1 and $k = 1, 2, 3$ for dataset 2. Separation level $= 4$ is used in this setting.

Each setting is repeated 100 times. Additional simulation settings including multivariate Gaussian distribution are provided in Supplementary Material.

3.2. *Clustering algorithms*

We apply several clustering strategies based on our proposed method intCC, KLIC[6] and iClusterPlus.[3] To evaluate the advantage of ensemble clustering, i.e., applying multiple clustering algorithms to each dataset, we also include our proposed method which only runs a single clustering algorithm to each dataset. We denote this as one layer weighted integrative consensus clustering. We also compare application of PAM and hierarchical clustering to the weighted consensus matrix in deriving a final clustering result. These methods are denoted as:

(1) iClusterPlus: applying iClusterPlus with the data type specified.

(2) KLIC-k-means: KLIC by applying k-means to each dataset for generating the consensus matrix.
(3) KLIC-Hclust: KLIC by applying hierarchical clustering to each dataset for generating the consensus matrix.
(4) 1 layer intCC-k-means (PAM): One layer weighted integrative consensus clustering by applying k-means to each dataset for generating the consensus matrix, followed by PAM to derive a final clustering result.
(5) 1 layer intCC-Hclust (PAM): One layer weighted integrative consensus clustering by applying hierarchical clustering to each dataset for generating the consensus matrix, followed by PAM to derive a final clustering result.
(6) 1 layer intCC-k-means (Hclust): One layer weighted integrative consensus clustering by applying k-means to each dataset for generating the consensus matrix, followed by hierarchical clustering to derive a final clustering result.
(7) 1 layer intCC-Hclust (Hclust): One layer weighted integrative consensus clustering by applying hierarchical clustering to each dataset for generating the consensus matrix, followed by hierarchical clustering to derive a final clustering result.

To obtain an unbiased comparison to our two layer approach, we also apply KLIC with multiple clustering algorithms. In other words, suppose there are 4 datasets and two clustering algorithms are applied to each dataset, there will be a total of 8 consensus matrices, i.e., akin to applying KLIC to 8 datasets. KLIC is applied using these 8 consensus matrices as input in the multiple kernel integrative clustering. Additionally, to illustrate the advantage of two layer approach, we also include another one layer approach in which we apply a single layer weight estimation to the 8 consensus matrices. These methods are denoted as:

(8) 2 layer intCC-2 methods (PAM): Two layer weighted integrative consensus clustering by applying both k-means and hierarchical clustering to each dataset for generating the consensus matrices, followed by PAM to derive a final clustering result.
(9) 2 layer intCC-2 methods (Hclust): Two layer weighted integrative consensus clustering by applying both k-means and hierarchical clustering to each dataset for generating the consensus matrices, followed by hierarchical clustering to derive a final clustering result.
(10) KLIC-2-methods: KLIC by applying both k-means and hierarchical clustering to each dataset for generating the consensus matrices.
(11) 1 layer intCC-2 methods (PAM): One layer weighted integrative consensus clustering by applying both k-means and hierarchical clustering to each dataset for generating the consensus matrices, followed by PAM to derive a final clustering result.
(12) 1 layer intCC-2 methods (Hclust): One layer weighted integrative consensus clustering by applying both k-means and hierarchical clustering to each dataset for generating the consensus matrices, followed by hierarchical clustering to derive a final clustering result.

For Settings 1-8, we apply each method by setting the number of clusters to be the true number of clusters. In practice, one can tune the optimal number of clusters using criteria such as the silhouette method,[12] gap statistics,[13] Dunn index[14] or the delta K method.[7] For Setting 9, we consider (a) global clustering, where we set the number of clusters to be the same throughout for both individual dataset and final integrative clustering, i.e., either 3 or 6 throughout (we denote these strategies as "Global K=3" and "Global K=6"), and (b) separate clustering, where we use the true number of clusters for individual dataset, i.e., 6 for dataset 1 and 3 for dataset 2, and consider both $K = 3$ and $K = 6$ in the final integrative clustering (we denote these strategies as "Separate K=3" and "Separate K=6"). Additionally, due to the poor performance of iClusterPlus and the long computational time, we omit iClusterPlus for Settings 4-6. We compare the performance of the clustering methods via the average adjusted rand index (ARI). We also report the weight estimation time of intCC and KLIC.

3.3. *Results*

We summarize the ARI for each simulation setting in Figure 3. Overall, results show that our proposed methods, namely 2 layer intCC-2 methods (PAM) and 1 layer intCC-k-means (PAM) perform well across all simulation settings. To explain this observation, without loss of generality, we summarize the ARI within each simulated dataset of Setting 4 in Figures 4A and 4B. The ARI by applying k-means as the base algorithm in the consensus clustering within each dataset is significantly better than hierarchical clustering in the simulated datasets considered in this paper. Thus, it is not surprising that methods which use k-means as the base clustering algorithm in the consensus clustering yield better performance. However, in practice the best base clustering algorithm is sometimes unknown. Thus, the 2 layer intCC which aggregates multiple base clustering algorithms can automatically assign higher weights to the better algorithm as shown in our simulation studies, as evident from the estimated weights in Figures 4C and 4D. It is also worth noting that our method assigns significantly smaller weights to the 5th dataset in which all the features are noise features. Additionally, using PAM to derive a final clustering result in general yields better performance compared to hierarchical clustering. We also note that the performance of iClusterPlus is significantly poorer compared to other methods, consistent with the findings of Cabassi and Kirk (2020).[6] Moreover, extending KLIC to run multiple base clustering algorithms, i.e., KLIC-2-methods has lower ARI compared to our proposed method, implying that the current KLIC framework does not yield a straightforward extension to incorporate ensemble clustering.

Without loss of generality, we also report the weight calculation time for KLIC and our proposed method intCC for Setting 1 (60 samples) and Setting 7 (300 samples) in Table 1, which shows that our proposed weight calculation is computationally efficient and yields good operating characteristics.

4. Case study

We illustrate our proposed method intCC on the TCGA pan cancer datasets.[15] There are 5 datasets across 12 cancer types which represent different tissues of origin, including DNA copy number, DNA methylation, mRNA expression, microRNA expression and protein expression data. To minimize bias in the comparison, we use the same preprocessing pipeline as previously described.[6,15]

Fig. 3. Distribution of ARI across all methods and simulation settings. Blue (red) boxplots are methods which apply one (two) clustering algorithm(s) per dataset. A-G. Settings 1-7. H-K. Setting 8 with different dataset combinations as input. L-O. Setting 9 with different strategies for setting number of clusters for individual dataset and final integrative clustering.

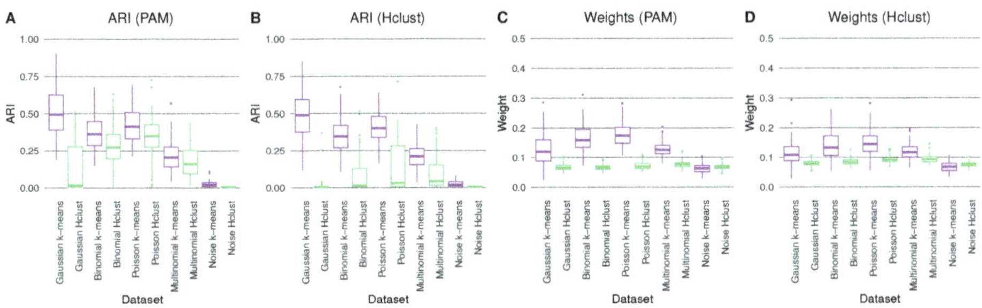

Fig. 4. A-B. Distribution of ARI within each simulated dataset of Setting 4. C-D. Distribution of estimated weights from intCC within each simulated dataset of Setting 4. Purple (green) boxplots are results by applying k-means (hierarchical clustering) algorithm in the consensus clustering. A, C. Using PAM to derive a final clustering result. B, D. Using hierarchical clustering to derive a final clustering result.

Table 1. Weight calculation time comparison.

Method	Setting 1 (seconds)	Setting 7 (seconds)
KLIC-k-means	0.541	7.209
KLIC-Hclust	0.791	8.241
1 layer intCC-k-means (PAM)	0.000879	0.00330
1 layer intCC-Hclust (PAM)	0.000882	0.00335
1 layer intCC-k-means (Hclust)	0.000876	0.00333
1 layer intCC-Hclust (Hclust)	0.000909	0.00332
2 layer intCC-2 methods (PAM)	0.00273	0.0103
2 layer intCC-2 methods (Hclust)	0.00265	0.0102
KLIC-2-methods	2.428	27.592
1 layer intCC-2 methods (PAM)	0.00155	0.00673
1 layer intCC-2 methods (Hclust)	0.00152	0.00686

Cabassi and Kirk (2020)[6] followed the same procedures described in Hoadley et al. (2014)[15] in setting the number of clusters for each dataset, except for microRNA expression in which the authors identified 8 as the number of clusters. We also set the number of clusters for each dataset following Cabassi and Kirk (2020).[6] Subsequently, we apply our proposed method intCC to obtain an integrative clustering across these datasets using the PAM algorithm to derive a final clustering result. Our method also selects 10 as the optimal number of clusters based on the average silhouette criterion, similar to KLIC.[6] Figure 5A compares the cluster membership of our method intCC against the results of KLIC, with ARI 0.693, whereas Figures 5B and 5C compare the cluster membership of intCC and KLIC against the 12 cancer type annotation, respectively. The ARI between intCC and cancer type annotation associated with tissues of origin is 0.754, whereas the ARI between KLIC and cancer type annotation is

0.585, indicating that the cluster membership of intCC yields a higher consistency with tissues of origin in the TCGA pan cancer datasets. Further investigation into the clusters obtained by intCC versus KLIC among subset of breast invasive carcinoma (BRCA) indicates that the results from intCC yield a higher consistency with the TCGA-BRCA molecular subtypes compared to the results from KLIC (Supplementary Material).

The estimated weights of each dataset for intCC and KLIC are (DNA copy number, DNA methylation, mRNA expression, miRNA expression, protein expression) =(0.073, 0.401,0.045, 0.272, 0.209) and (0.309, 0.192, 0.168, 0.183, 0.148), respectively. intCC assigns a higher weight to DNA methylation data, whereas KLIC assigns a higher weight to the copy number data, which could explain the differences observed in cluster memberships obtained by these two methods. Finally, the weight calculation time for intCC is 0.43 second, whereas the weight calculation time for KLIC via quadratic programming is $>$ 10 hours on an Intel(R) Xeon(R) CPU E5-1650 v3 @ 3.50GHz.

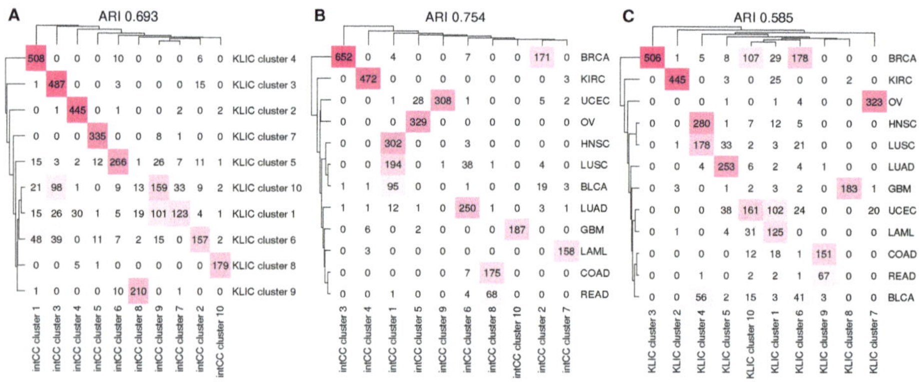

Fig. 5. Heatmaps of coincidence matrices comparing A. intCC clusters to KLIC clusters, B. intCC clusters to cancer type annotation, C. KLIC clusters to cancer type annotation. The ARI is reported in the header of each plot.

5. Discussion

The rapid development of high throughput technologies has provided an avenue to scientists to decipher the complex human diseases from a systems biology perspective via multiomics profiling. Integrative clustering has become a powerful approach to dissect the heterogeneity underpinning these diseases, e.g., to define new cancer subtypes which may help inform treatment efforts. In this paper, we extend the framework of KLIC[6] which recasts the integrative clustering model into multiple kernel learning framework by utilizing the consensus matrices estimated from consensus clustering as input. Specifically, our model further incorporates the ensemble learning via an aggregation of multiple base clustering algorithms to enhance the robustness of multiple kernel integrative clustering model. This is to safeguard against applying a single base clustering algorithm that performs poorly on the dataset. Additionally, we also propose an efficient weight estimation to combine the consensus matrices. Our simulation studies show that the proposed two layer weighted integrative clustering yields better performance overall.

Conceptually, the weight estimation is analogous to the heuristics of multiple kernel support vector machine (MKL-SVM) based on kernel-target alignment.[16–18] Specifically, MKL-SVM is developed for supervised learning and the kernel-target alignment depends on the true binary class labels. For a fixed cluster membership, this is equivalent to multi-class classification. One can extend the kernel-target alignment for multi-class classification by dividing the problem into several binary classification subproblems (e.g., one-versus-all or all-pairs). However, how to optimally combine the results across these binary subproblems is not trivial and may require longer computational time compared to our proposed method.

Besides identifying appropriate and robust clustering algorithms, another important research question in unsupervised learning is in tuning the optimal number of clusters. Several metrics have been proposed for this task, including the silhouette method,[12] gap statistics,[13] Dunn index[14] and the delta K method.[7] An immediate extension to our intCC framework is to aggregate the different metrics/criteria for selecting the optimal number of clusters.

Supplementary Material and Code

Supplementary Material is available online at
`http://www.ams.sunysb.edu/~pfkuan/PDF/SM_PSB2024.pdf`.
The R code implementing intCC is available online at `https://github.com/candsj/intCC`.

Acknowledgments

This work is supported in part by CDC/NIOSH award U01OH012257. The findings and conclusions presented in this article are those of the authors and do not represent the official position of NIOSH, the CDC or the U.S. Public Health Service.
Conflict of Interest: None declared.

References

1. E. F. Lock and D. B. Dunson, Bayesian consensus clustering, *Bioinformatics* **29**, 2610 (2013).
2. R. Shen, A. B. Olshen and M. Ladanyi, Integrative clustering of multiple genomic data types using a joint latent variable model with application to breast and lung cancer subtype analysis, *Bioinformatics* **25**, 2906 (2009).
3. Q. Mo, S. Wang, V. E. Seshan, A. B. Olshen, N. Schultz, C. Sander, R. S. Powers, M. Ladanyi and R. Shen, Pattern discovery and cancer gene identification in integrated cancer genomic data, *Proceedings of the National Academy of Sciences* **110**, 4245 (2013).
4. B. . W. H. . H. M. S. C. L. . . P. P. J. . K. R. 13, G. data analysis: Baylor College of Medicine Creighton Chad J. 22 23 Donehower Lawrence A. 22 23 24 25, I. for Systems Biology Reynolds Sheila 31 Kreisberg Richard B. 31 Bernard Brady 31 Bressler Ryan 31 Erkkila Timo 32 Lin Jake 31 Thorsson Vesteinn 31 Zhang Wei 33 Shmulevich Ilya 31 *et al.*, Comprehensive molecular portraits of human breast tumours, *Nature* **490**, 61 (2012).

5. E. Gabasova, J. Reid and L. Wernisch, Clusternomics: Integrative context-dependent clustering for heterogeneous datasets, *PLoS Computational Biology* **13**, p. e1005781 (2017).
6. A. Cabassi and P. D. Kirk, Multiple kernel learning for integrative consensus clustering of omic datasets, *Bioinformatics* **36**, 4789 (2020).
7. S. Monti, P. Tamayo, J. Mesirov and T. Golub, Consensus clustering: a resampling-based method for class discovery and visualization of gene expression microarray data, *Machine Learning* **52**, 91 (2003).
8. F. R. Bach, G. R. Lanckriet and M. I. Jordan, Multiple kernel learning, conic duality, and the smo algorithm, in *Proceedings of the Twenty-First International Conference on Machine Learning*, (Banff, Canada, 2004).
9. G. R. Lanckriet, N. Cristianini, P. Bartlett, L. E. Ghaoui and M. I. Jordan, Learning the kernel matrix with semidefinite programming, *Journal of Machine Learning Research* **5**, 27 (2004).
10. M. Gönen and A. A. Margolin, Localized data fusion for kernel k-means clustering with application to cancer biology, *Advances in Neural Information Processing Systems* **27** (2014).
11. O. Sagi and L. Rokach, Ensemble learning: A survey, *Wiley Interdisciplinary Reviews: Data Mining and Knowledge Discovery* **8**, p. e1249 (2018).
12. P. J. Rousseeuw, Silhouettes: a graphical aid to the interpretation and validation of cluster analysis, *Journal of Computational and Applied Mathematics* **20**, 53 (1987).
13. R. Tibshirani, G. Walther and T. Hastie, Estimating the number of clusters in a data set via the gap statistic, *Journal of the Royal Statistical Society: Series B (Statistical Methodology)* **63**, 411 (2001).
14. M. Halkidi, Y. Batistakis and M. Vazirgiannis, On clustering validation techniques, *Journal of Intelligent Information Systems* **17**, 107 (2001).
15. K. A. Hoadley, C. Yau, D. M. Wolf, A. D. Cherniack, D. Tamborero, S. Ng, M. D. Leiserson, B. Niu, M. D. McLellan, V. Uzunangelov *et al.*, Multiplatform analysis of 12 cancer types reveals molecular classification within and across tissues of origin, *Cell* **158**, 929 (2014).
16. N. Cristianini, J. Shawe-Taylor, A. Elisseeff and J. Kandola, On kernel-target alignment, *Advances in neural information processing systems* **14** (2001).
17. C. Cortes, M. Mohri and A. Rostamizadeh, Two-stage learning kernel algorithms, *Proceedings of the 27 th International Conference on Machine Learning* , 239 (2010).
18. S. Qiu and T. Lane, A framework for multiple kernel support vector regression and its applications to sirna efficacy prediction, *IEEE/ACM Transactions on Computational Biology and Bioinformatics* **6**, 190 (2008).

LARGE LANGUAGE MODELS (LLMS) AND CHATGPT FOR BIOMEDICINE

Cecilia Arighi[†]
Department of Computer and Information Sciences, University of Delaware, Ammon-Pinizzotto Biopharmaceutical Innovation Building, 590 Avenue 1743
Newark, DE19713, US
Email: arighi@udel.edu

Steven Brenner
Department of Plant & Microbial Biology, UC Berkeley, 461 Koshland Hall
Berkeley, CA94720
Email: brenner@compbio.berkeley.edu

Zhiyong Lu[*]
NCBI, NLM, NIH, Bethesda, MD 20894
Bethesda, MD20894, US
Email: zhiyong.lu@nih.gov

Large Language Models (LLMs) are a type of artificial intelligence that has been revolutionizing various fields, including biomedicine. They have the capability to process and analyze large amounts of data, understand natural language, and generate new content, making them highly desirable in many biomedical applications and beyond. In this workshop, we aim to introduce the attendees to an in-depth understanding of the rise of LLMs in biomedicine, and how they are being used to drive innovation and improve outcomes in the field, along with associated challenges and pitfalls.

Keywords: ChatGPT; large language model; LLM; generative AI; biomedicine and health; education; ethics.

1. Background

A language model (LM) is a machine learning technique for natural language processing tasks. LMs typically predict the probability of a word appearing in a text sequence based on the previous word, modeling linguistic intuition (like completing a missing word in a sentence). One of the key advances in LM was the introduction of the transformer architecture [1], which became the cornerstone for many of the large language models (LLMs) that followed. In brief, the transformer architecture includes two modules, namely, an encoder of bidirectional attention blocks and a decoder of unidirectional attention blocks. Based on which modules are used, the LLMs are classified as encoder-only (e.g., BERT, Bidirectional Encoder Representations from Transformers [2]), encoder-decoder (e.g., T5 , Text-to-Text Transfer Transformer [3]), or decoder-only (e.g.,

[†] Work partially supported by NIH award U24 HG007822.
[*] This research is supported by the NIH Intramural Research Program, National Library of Medicine

GPT, Generative Pre-trained Transformer, series [4]–[6]). The latter class are able to use billions (or even trillions) of parameters and trained on massive amounts of unlabeled text, providing the ability to generate human-like text [7]. In addition to capturing the language, these models can "memorize" facts during training. Thus, LLMs have the capacity to efficiently handle and analyze extensive text data and generate fresh content, demonstrating significant promise in diverse applications.

The launch of ChatGPT, the LLM-based chatbot developed by OpenAI [8], to the public in late 2022 has sparked a number of exciting opportunities, but also some challenges and ethical concerns. It was recently reported that a keyword search for "large language models" OR "ChatGPT" in PubMed returned 582 articles by the end of May 2023 [9]. The same search conducted at the end of September 2023 returned 1,495 articles, more than doubled in a short period. Publications include research and review articles as well as relevant commentaries on how LLMs are reshaping biomedicine, healthcare and education [9]–[16]. The extent of LLM applications goes beyond language, with active research in the field of protein annotation [17], [18], function [19], and structure prediction [20]. While LLMs offer substantial benefits, it is important to acknowledge its limitations such as hallucinations and key ethical challenges including: perpetuating biases present in the training data, thus efforts are needed to ensure fairness and equity in their applications; privacy issues when handling sensitive data; transparency and plagiarism, among others.

2. *LLM and ChatGPT in Biomedicine Workshop*

Given the rapid evolution and dissemination of the LLMs, and more specifically ChatGPT, the proposed workshop aims at introducing and discussing latest developments in the first year surrounding this new technology in biomedicine. The workshop will consist of talks spanning the following topics:

- Introduction to LLM Technology: This talk will provide an overview of LLMs, including their architecture, training process, and how they work. It will help attendees understand the basics of this technology and why it is relevant to biomedicine.

- Use of Standard LLMs in Scientific Research: This talk will focus on the use of standard LLMs in research, and how they can support researchers in various ways, including helping design and analyze experiments, writing code, brainstorming, and writing papers.

- Use of LLMs in the Education and Academic Writing: This talk will discuss the use of LLMs in the classroom, both for teachers and students. Highlighting the benefits of using LLMs in teaching and learning and provide examples (e.g., as a writing assistant) of how they are being used to enhance the educational experience.

- Applications of LLMs in Healthcare: This talk will showcase the use of LLM technologies in custom methods development for existing/new problems in clinical informatics research and healthcare.

- Ethics of using LLMs: This talk will feature topics surrounding the use of such a chatbot technology in medical care and scientific research, including but not limited to privacy and ethical concerns, AI bias, and legal liabilities.

3. Conclusion

We envision that similarly to previous new technology disruptions in society (e.g., calculator, computer or the internet), LLMs will become integral part of our lives, and the discussions in this workshop will help to shape the landscape ahead.

4. Acknowledgments

We would like to thank the Organizing committee of the Pacific Symposium for Bioinformatics 2024 for giving us the opportunity of organizing the proposed workshop.

References

[1] A. Vaswani *et al.*, "Attention is All you Need," in *Advances in Neural Information Processing Systems*, I. Guyon, U. V. Luxburg, S. Bengio, H. Wallach, R. Fergus, S. Vishwanathan, and R. Garnett, Eds., Curran Associates, Inc., 2017. [Online]. Available: https://proceedings.neurips.cc/paper_files/paper/2017/file/3f5ee243547dee91fbd053c1c4a845aa-Paper.pdf

[2] J. Devlin, M.-W. Chang, K. Lee, and K. Toutanova, "BERT: Pre-training of Deep Bidirectional Transformers for Language Understanding," in *Proceedings of the 2019 Conference of the North American Chapter of the Association for Computational Linguistics: Human Language Technologies, NAACL-HLT 2019, Minneapolis, MN, USA, June 2-7, 2019, Volume 1 (Long and Short Papers)*, J. Burstein, C. Doran, and T. Solorio, Eds., Association for Computational Linguistics, 2019, pp. 4171–4186. doi: 10.18653/v1/n19-1423.

[3] C. Raffel *et al.*, "Exploring the Limits of Transfer Learning with a Unified Text-to-Text Transformer," *J Mach Learn Res*, vol. 21, no. 1, Jan. 2020.

[4] A. Radford and K. Narasimhan, "Improving Language Understanding by Generative Pre-Training," 2018. [Online]. Available: https://api.semanticscholar.org/CorpusID:49313245

[5] A. Radford, J. Wu, R. Child, D. Luan, D. Amodei, and I. Sutskever, "Language Models are Unsupervised Multitask Learners," 2019. [Online]. Available: https://api.semanticscholar.org/CorpusID:160025533

[6] T. Brown *et al.*, "Language Models are Few-Shot Learners," in *Advances in Neural Information Processing Systems*, H. Larochelle, M. Ranzato, R. Hadsell, M. F. Balcan, and H. Lin, Eds., Curran Associates, Inc., 2020, pp. 1877–1901. [Online]. Available: https://proceedings.neurips.cc/paper_files/paper/2020/file/1457c0d6bfcb4967418bfb8ac142f64a-Paper.pdf

[7] W. X. Zhao *et al.*, "A Survey of Large Language Models." 2023.

[8] OpenAI, "Introducing ChatGPT." [Online]. Available: https://openai.com/blog/chatgpt

[9] S. Tian *et al.*, "Opportunities and Challenges for ChatGPT and Large Language Models in Biomedicine and Health." 2023.

[10] A. S. P. M. Arachchige, "Early applications of ChatGPT in medical practice, education and research.," *Clin. Med. Lond. Engl.*, vol. 23, no. 4, pp. 429–430, Jul. 2023, doi: 10.7861/clinmed.Let.23.4.2.
[11] S. Koga, "The Integration of Large Language Models Such as ChatGPT in Scientific Writing: Harnessing Potential and Addressing Pitfalls.," *Korean J. Radiol.*, vol. 24, no. 9, pp. 924–925, Sep. 2023, doi: 10.3348/kjr.2023.0738.
[12] P. P. Ray and P. Majumder, "The Potential of ChatGPT to Transform Healthcare and Address Ethical Challenges in Artificial Intelligence-Driven Medicine.," *J. Clin. Neurol. Seoul Korea*, vol. 19, no. 5, pp. 509–511, Sep. 2023, doi: 10.3988/jcn.2023.0158.
[13] H. Zhang, Y. Guan, J. Chen, and W. Tong, "Commentary: AI-based online chat and the future of oncology care: a promising technology or a solution in search of a problem?," *Front. Oncol.*, vol. 13, p. 1239932, 2023, doi: 10.3389/fonc.2023.1239932.
[14] Q. Jin, R. Leaman, and Z. Lu, "Retrieve, Summarize, and Verify: How Will ChatGPT Affect Information Seeking from the Medical Literature?," *J. Am. Soc. Nephrol. JASN*, vol. 34, no. 8, pp. 1302–1304, Aug. 2023, doi: 10.1681/ASN.0000000000000166.
[15] Q. Jin, Z. Wang, C. S. Floudas, J. Sun, and Z. Lu, "Matching Patients to Clinical Trials with Large Language Models." United States, Jul. 28, 2023.
[16] Q. Jin, Y. Yang, Q. Chen, and Z. Lu, "GeneGPT: Augmenting Large Language Models with Domain Tools for Improved Access to Biomedical Information." United States, May 16, 2023.
[17] A. Gane, "How artificial intelligence can help us annotate protein names." [Online]. Available: https://insideuniprot.blogspot.com/2022/12/how-artificial-intelligence-can-help-us.html
[18] L. Kelly, Z. Flamholz, and S. Biller, "Large language models improve annotation of viral proteins.," *Research square*. United States, p. rs.3.rs-2852098, May 02, 2023. doi: 10.21203/rs.3.rs-2852098/v1.
[19] A. Elnaggar *et al.*, "ProtTrans: Toward Understanding the Language of Life Through Self-Supervised Learning.," *IEEE Trans. Pattern Anal. Mach. Intell.*, vol. 44, no. 10, pp. 7112–7127, Oct. 2022, doi: 10.1109/TPAMI.2021.3095381.
[20] Z. Lin *et al.*, "Evolutionary-scale prediction of atomic-level protein structure with a language model," *Science*, vol. 379, no. 6637, pp. 1123–1130, 2023, doi: 10.1126/science.ade2574.

Practical Approaches to Enhancing Fairness, Social Responsibility and the Inclusion of Diverse Viewpoints in Biomedicine

Daphne O. Martschenko, Nicole Martinez-Martin, Meghan Halley
Stanford Center for Biomedical Ethics
Stanford, CA 94305 USA
Email: nicolemz@stanford.edu

Workshop Description

In biomedical research and clinical medicine, many of the ethical frameworks and processes focus on benefits and harms at the individual level. However, in biomedicine, there is increasing recognition of a need to implement frameworks and processes that address the social impacts of technologies, such as genomics and AI technologies, and their social benefit for underrepresented populations and communities. For example, studies demonstrating the potential for bias in AI shed light on the need to develop processes to more effectively identify and address downstream impacts of medical AI, as well as engage communities who are stakeholders in the research. Privacy is often envisioned as an individual right, but the collection and use of data also have repercussions at the level of groups and communities. For that reason, there have been recent efforts to arrive at models for data stewardship and data sovereignty. This workshop will provide a forum for discussion of practical approaches to enhancing fairness, social responsibility and inclusion of diverse viewpoints in biomedicine. Interdisciplinary research on ethics and how fairness, social responsibility, and community engagement can be operationalized in biomedical research will provide a foundation for robust discussion on these issues.

The 3-hour workshop will consist of two parts:

- The first part will include a series of 15-minute talks that address fairness, social responsibility & inclusion/community engagement for different areas of biomedicine, followed by an audience Q& A and discussion of the topics such as diversity in precision medicine, ethical and sustainable data stewardship, and public engagement with social and behavioral genomics.
- For the second half of the workshop, weare conducting an interactive exercise with the audience. Focusing on case studies, based on topics from the first half of the workshop, such as community engagement and data stewardship, we will use smartphone-based polling to facilitate feedback from the audience on approaches, challenges and solutions for addressing the ethical issues from the case study.

Learning Objectives

By the end of this workshop attendees will be able to:

1. Understand the social and political context that underlays the need for frameworks and processes that more effectively address the impacts of these technologies on individuals and communities.
2. Explore and analyze efforts to identify and address the downstream harms and benefits of biomedical technologies
3. Locate actors that have the ability to mitigate the downstream harms of biomedical technologies and/or the ability to promote its downstream benefits.

Presenter Information

This workshop brings together rich and interdisciplinary perspectives from medical anthropology, biomedical engineering, education, and bioethics, as well as, legal perspectives. Importantly, our multidisciplinary and multi-institution workshop aims to do more than provide the PSB community with the opportunity to come together to analyze and evaluate efforts to enhance social responsibility and the inclusion of diverse viewpoints in biomedicine. We offer workshop attendees strategies for intervening to assist with promoting fairness, social responsibility, inclusion, and justice in biomedical research and practice.

About the Workshop Organizers

Daphne Martschenko, Ph.D., is an Assistant Professor at the Stanford University Center for Biomedical Ethics and a co-organizer of the international Race, Empire, and Education Research Collective. Dr. Martschenko holds an MPhil from the University of Cambridge in Politics, Development, and Democratic Education and in 2019 received a Ph.D. in Education, also from the University of Cambridge. Dr. Martschenko's work advocates for and facilitates the ethical and responsible conduct of and public engagement with genetic/genomic research.

Nicole Martinez-Martin, JD, Ph.D., is an Assistant Professor at the Stanford Center for Biomedical Ethics. She received her JD from Harvard Law School and her doctorate in social sciences (comparative development/medical anthropology) from the University of Chicago. Her broader research interests concern the impact of new technologies on the treatment of vulnerable populations. Her recent work in bioethics and neuroethics has focused on the ethics of AI and digital health technology, such as digital phenotyping or computer vision, for medical and behavioral applications.

Meghan Halley, PhD, MPH, is a Senior Research Scholar in the Stanford Center for Biomedical Ethics (SCBE) at Stanford University. She completed her doctorate in medical anthropology from Case Western Reserve University in 2012, and additional training in health services research at the Palo Alto Medical Foundation Research Institute from 2012 through 2016. Her current research focuses at the intersection of the ethics and economics of new genomic technologies. Her current projects include examining ethical issues related to sustainability and governance of patient data and relationships when large clinical genomic studies transition to new models of funding; ethnographic work exploring how diverse stakeholders perceive value in the use of genome sequencing for diagnosis of rare diseases; and the development of new measures for assessing patient-centered outcomes in pediatric rare diseases.

Presentations

- Daphne Martschenko, PhD, Assistant Professor in Biomedical Ethics, Stanford University: "Wrestling with Public Input on Social and Behavioral Genomics" reporting on scholarship gathering the perspectives of members of the public on the risks and potential benefits of social and behavioral genomics.
- Mildred Cho, PhD, Professor in Biomedical Ethics, Stanford University, reporting on the use of hypothetical design exercises in order to examine values in biomedical AI/ML development
- Meghan Halley, PhD, MPH, Senior Research Scholar in Biomedical Ethics, Stanford University: "Toward more ethical and sustainable data stewardship in rare disease research" reporting on the parameters of ethical data sharing and sustainability in rare disease research, involving perspectives on cloud-based genomic databases.
- Krystal Tsosie, PhD, MPH, Assistant Professor, School of Life Sciences, Center for Biology and Society, Arizona State University: "Platforms Not Platitudes: Operationalizing Ethics and Advancing Indigenous Data and Digital Sovereignty" on community data governance and stewardship with digital data tools rooted in machine learning and dynamic consent e-platforms
- Carole Federico, PhD, GSK.ai-Stanford Ethics Fellow, Stanford University: "Synthetic Data for Biomedicine: Epistemic and Ethical Challenges".

  Interactive Hypothetical Design Case Study Presentation:
- Nicole Martinez-Martin, JD, PhD, Assistant Professor in Biomedical Ethics, Stanford University
- Mildred Cho, PhD, Professor in Biomedical Ethics, Stanford University
- Tiffany Bright, Co-Director Center for Artificial Intelligence Research
  Cedars-Sinai, Computational Biomedicine

Speaker Presentations

The speaker presentations will provide examples of how issues of diversity and inclusion, as well as social responsibility, are being engaged in the fields of genomics and machine learning in medicine.

Genes, and the social narratives we tell about them, continue to grip the popular imagination. In particular, claims regarding genetic differences in human behavior and social outcomes have been a pervasive and often ugly feature of American society since the eugenics movement of the twentieth century. Today, researchers in the rapidly growing field of social/behavioral genomics investigate whether and how genetic differences between individuals relate to differences in behaviors (e.g., aggressive behavior) and social outcomes (e.g., educational attainment), as well as how genetic information can inform the design of social/behavioral studies. There is staunch and polarizing academic debate about the risks and benefits of this science. Many researchers are optimistic that this work will increase understanding of human behavior, improve health and well-being, and reduce societal inequality. Others worry about its potential to be misused in service of racist, classist, and ableist claims.

Defining the harms and benefits of research has traditionally been left to researchers, professional societies, and regulatory bodies. In the US, researchers are regulated by policies such as the Common Rule (45 CFR 46), research ethics committees, and Institutional Review Boards (IRBs). These systems of regulation guide the ethical conduct of research by ensuring studies have an acceptable risk-benefit profile such that potential harms (i.e., risks) are minimized and potential benefits enhanced.

Confining debate about the threats and promises of social and behavioral genomics to the research community is limiting. Academic considerations of the harms and benefits of research, generally neglect to consider the broader social impacts. IRBs are expressly prohibited by the Common Rule from considering any broad social or policy risks. IRBs generally don't regulate risks other than those directly encountered by research participants. However, per the Common Rule, IRBs are allowed to judge the broader social benefits of research; that is, whether research has the potential to enhance health or knowledge. As a result, existing mechanisms for regulating the ethical conduct of research are limited in their ability to appraise the downstream implications of research, especially the potential social harms.

Daphne Martschenko, PhD (Stanford University) will present the results of an 18-month effort to gather input from an 11-member Community Sounding Board comprised of individuals from across the United States on the risks, benefits, and ethical responsibilities of social and behavioral genomics. Attendees will leave this presentation with tools that can help them better elicit and engage public perspectives to produce socially and ethically informed decisions about whether and how to conduct biomedical research, as well as socially and ethically responsible policy decisions and research communication.

The presentation by **Krystal Tsosie, PhD, MPH** (Arizona State University) will provide an overview of how community data governance and stewardship with digital data tools rooted in machine learning and dynamic consent e-platforms have been applied to advance Indigenous Data and Digital Sovereignty.

Mildred K. Cho, PhD (Stanford University) has conducted research regarding the integration of ethical values into medical AI/ML. Her most recent work examines the use of hypothetical design exercises in order to support ethics in the development of AI/ML applications in medicine. **Carole Federico, PhD** (GSK.ai-Stanford Ethics Fellow) will discuss ethical issues relevant to synthetic data, with a focus on representativeness and fairness in synthetic data and practical challenges in applying existing ethical frameworks to synthetic data.

Machine learning predictive analytics (MLPA) are increasingly utilized in health care to reduce costs and improve efficacy. The growth of MLPA could be fueled by payment reforms that hold health care organizations responsible for providing high-quality, cost-effective care. At the same time, policy analysts, ethicists, and computer scientists have identified unique ethical and regulatory challenges from the use of

MLPA in health care, and they have also proposed a variety of principles and guidelines focused on confronting these challenges.

However, critical gaps in knowledge have challenged our ability to assess these potential solutions. Understanding the perspectives of MLPA developers is essential for overcoming the “principles-to-practice” gap. **Meghan Halley, MPH, PhD** (Stanford University) will present a study that sought to better characterize available MLPA health care products, identifying and characterizing claims about products recently or currently in use in US health care settings that are marketed as tools to improve health care efficiency by improving quality of care while reducing costs. The research team conducted systematic database searches of relevant business news and academic research to identify MLPA products for health care efficiency meeting our inclusion and exclusion criteria. Their findings provide a foundational reference to inform the analysis of specific ethical and regulatory challenges arising from the use of MLPA to improve healthcare efficiency.

Mildred Cho, **PhD** (Stanford University) has conducted research examining how developers of machine learning applications in healthcare envision and put values into practice in their work. Using a case study approach that draws from issues from the workshop presentations, Dr. Cho, **Nicole Martinez-Martin, JD, PhD** (Stanford University) and **Tiffany Bright**, **PhD** (Center for Artificial Intelligence Research Cedars-Sinai) will lead the audience in an interactive discussion regarding how values of diversity, representation and social responsibility are put into practice in the work of researchers in genomics and computational biomedicine.

Risk prediction: Methods, Challenges, and Opportunities

Ruowang Li
Department of Computational Biomedicine, Cedars-Sinai Medical Center, West Hollywood, California, USA
Email: ruowang.li@cshs.org

Rui Duan
Department of Biostatistics, Harvard T.H. Chan School of Public Health, Boston, Massachusetts, USA
Email: rduan@hsph.harvard.edu

Lifang He
Department of Computer Science and Engineering, Lehigh University, Bethlehem, Pennsylvania, USA
Email: lih319@lehigh.edu

Jason H. Moore
Department of Computational Biomedicine, Cedars-Sinai Medical Center, West Hollywood, California, USA
Email: jason.moore@csmc.edu

1. Introduction to the workshop

The objective of this workshop is to delve into the current and future landscape of risk prediction within the realm of disease and epidemiological research. Discussion topics encompass everything from data sources to model implementation. The workshop will feature speakers addressing commonly used data sources—genetics, imaging, clinical, and epidemiological data—in developing prediction models. Moreover, the workshop will cover model-based and post-hoc analyses, delving into biases, uncertainty quantification, model interpretation, fairness, diversity of prediction results, and the transferability and generalizability of models across different populations and datasets. The moderated discussion session will offer a future perspective on the validation and implementation of risk prediction models. The workshop will maintain a balanced focus across all stages of risk prediction model development and validation. By emphasizing a well-rounded workshop theme instead of exclusively delving into methodologies, we aim to create an environment that fosters the exchange of ideas and viewpoints among speakers and audiences.

2. Workshop Presenters

The three-hour workshop will have a total of six presentations followed by a moderated panel discussion session. The workshop speakers are:

Randi Foraker, PhD, is the Director of the Center for Population Health Informatics (CPHI) at the Institute for Informatics (I2) and a Professor of Medicine within the Division of General Medical Sciences at Washington University in St. Louis. As director of the CPHI, she aims to improve the health of the community through data and support data access, analytics, and dissemination efforts. Her own work specializes in the design of population-based studies and the integration of electronic health record data with socioeconomic indicators, and her research portfolio has been supported by a combination of governmental and industry grants and contracts. Her most recent research has focused on the application of clinical decision support to complement risk scoring in primary care, cardiology, and oncology. Dr. Foraker also serves as Director of the Public Health Data and Training Center for the Institute for Public Health. As director of the Data and Training Center, she aims to amplify public health knowledge through data sharing, strategic partnerships with the community, and the training of future public health leaders. During the COVID pandemic, she has served as PI of the COVID umbrella IRB leveraging electronic health record data at Washington University in St. Louis and works closely with investigators who conduct research using data from our COVID Data Commons, which is maintained by I2. Dr. Foraker chairs the Epidemiology Strike Force and convenes members of the St. Louis City, St. Louis County, Jefferson County, Franklin County, and St. Charles County Departments of Public Health on a weekly basis along with academic, health system, and business partners to assist with their data architecture, management, and analytic needs during the pandemic and beyond.

Yong Chen, PhD, is Professor of Biostatistics at University of Pennsylvania. He directs a Computing, Inference and Learning Lab at University of Pennsylvania (https://penncil.med.upenn.edu/about-pi/), which focuses on integrating fundamental principles and wisdoms of statistics into quantitative methods for tackling key challenges in modern biomedical data. Dr. Chen is an expert in synthesis of evidence from multiple data sources, including systematic review and meta-analysis, distributed algorithms, and data integration, with applications to comparative effectiveness studies, health policy, and precision medicine. He is also working on developing methods to deal with suboptimal data quality issues in health system data, dynamic risk prediction, pharmacovigilance, and personalized health management. He has over 100 publications in a wide spectrum of methodological and clinical areas. Dr. Chen has been principal investigator on a number of grants, including R01s from the National Library of Medicine and National Institute of Allergy and Infectious Diseases, and Improving Methods for Conducting Patient-Centered Outcomes Research grant from Patient-Centered Outcomes Research Institute. Dr. Chen received his bachelor's degree in Mathematics at the University of Science and Technology of China, Master degree in Pure Mathematics and Ph.D. in Biostatistics at the Johns Hopkins University. He is an elected fellow of the Society for Research Synthesis Methodology, and the International Statistical Institute. He is a recipient of Best Paper Award by the International Medical Informatics Association (IMIA) Yearbook Section on Clinical Research Informatics, Institute of Mathematical Statistics Travel Award, Margaret Merrell Award for excellence in research at the Johns Hopkins University, and Distinguished Faculty Award at the University of Pennsylvania.

Graciela Gonzalez-Hernandez, PhD, is Vice Chair for Research and Education in the new Department of Computational Biomedicine at Cedars-Sinai Medical Center. Prior to joining Cedars-Sinai in May 2022, Dr Gonzalez-Hernandez was an Associate Professor of Informatics in the Department of Biostatistics, Epidemiology and Informatics (DBEI) of the Perelman School of Medicine, University of Pennsylvania. She transferred her Health Language Processing (HLP) Lab to Cedars-Sinai, which focuses on natural language processing (NLP) and machine learning for knowledge discovery, extracting unstructured information from clinical records, journal articles, and social media postings to elucidate data patterns, trends, and relationships that can aid the discovery process in areas such as pharmacoepidemiology, clinical research, or public health monitoring and surveillance. Dr Gonzalez-Hernandez and her team have made available to the health research community novel approaches to complete pipelines for information extraction from different sources using NLP, such as the DeepADRMiner pipeline for extracting and normalizing adverse effects from social media – a unique end-to-end system that makes it possible to tap into the value of direct reports by patients. She has published over 220 peer-reviewed articles in prestigious journals and conferences, routinely making code and datasets available to other researchers, and ensuring reproducibility. These publications span multiple areas of Biomedical Informatics, including natural language processing, bioinformatics, biomedical ontologies, information retrieval, MS and machine learning, as well as domain-specific publications in collaboration with clinicians and epidemiologists. Her work has appeared in the top peer-reviewed journals, including Nature Digital Medicine, JAMA Network Open, Bioinformatics, BMC Bioinformatics, the Journal of the American Medical Informatics Association, and the Journal of Biomedical Informatics, among others, as well as in numerous informatics conference proceedings.

Bogdan Pasaniuc, PhD, is a professor of Computational Medicine, Human Genetics and Pathology&Laboratory Medicine at UCLA. Dr Pasaniuc develops statistical and computational methods to understand the genetic basis of disease, focusing on under-represented populations, integrative genomics, and biobank studies. Dr Pasaniuc group developed machine learning methods to integrate epigenetic profiles within trans-ancestry studies to localize disease variants and genes; his group introduced transcriptome-wide association studies (TWAS) using predicted gene expression as a principled approach to identify disease genes for many traits such as Schizophrenia, Ovarian Cancer and Prostate Cancer. Dr Pasaniuc serves as Associate Director of Population Genetics of the Institute for Precision Health at UCLA that links the genetics of more than 150k patients with their electronic health record to predict health outcomes, to stratify patients based on their genetic risk to disease and to translate genomics to the clinic. Dr Pasaniuc also serves as PI for the Center for Admixed populations and Health Equity and for the Biomedical Data Science Training Program for Precision Health Equity at UCLA.

John Witte, PhD, is serving as Vice Chair and professor in the Department of Epidemiology & Population Health, and as a professor of Biomedical Data Science and, by courtesy, of Genetics, he will also serve as a member of the Stanford Cancer Institute. Dr. Witte is an internationally recognized expert in genetic epidemiology. His scholarly contributions include deciphering the genetic and environmental basis of prostate cancer and developing widely used methods for the genetic epidemiologic study of disease. His prostate cancer work has used comprehensive genome-wide studies of germline genetics, transcriptomics, and somatic genomics to successfully detect novel variants underlying the risk and aggressiveness of this common disease. A key aspect of this work has been distinguishing genetic factors that may drive increased prostate cancer risk and mortality among African American men. Providing an avenue to determine which men are more likely to be diagnosed with clinically relevant prostate cancer and require additional screening or specific treatment can help reduce disparities in disease prevalence and outcomes across populations. Dr. Witte has also developed novel hierarchical and polygenic risk score modeling for undertaking genetic epidemiology studies. These advances significantly improve our ability to detect disease-causing genes and to translate genetic epidemiologic findings into medical practice. Dr. Witte has received the Leadership Award from the International Genetic Epidemiology Society (highest award), and the Stephen B. Hulley Award for Excellence in Teaching. His extensive teaching portfolio includes a series of courses in genetic and molecular epidemiology. He has mentored over 50 graduate students and postdoctoral fellows, serves on the executive committees of multiple graduate programs, and has directed a National Institutes of Health funded post-doctoral training program in genetic epidemiology for over 20 years. Recently appointed to the National Cancer Institute Board of Scientific Counselors, Dr. Witte has been continuously supported by the National Institutes of Health.

Marinka Zitnik, PhD, is an Assistant Professor at Harvard University in the Department of Biomedical Informatics. Dr. Zitnik is Associate Faculty at the Kempner Institute for the Study of Natural and Artificial Intelligence, Broad Institute of MIT and Harvard, and Harvard Data Science. Dr. Zitnik investigates foundations of AI to enhance scientific discovery and facilitate individualized diagnosis and treatment in medicine. Her algorithms and methods have had a tangible impact, which has garnered interests of government, academic, and industry researchers and has put new tools in the hands of practitioners. Some of her methods are used by major biomedical institutions, including Baylor College of Medicine, Karolinska Institute, Stanford Medical School, and Massachusetts General Hospital.

Statistical analysis of single-cell protein data

Brooke L. Fridley, PhD
Department of Biostatistics and Bioinformatics, Moffitt Cancer Center
Tampa, FL 33612, USA
Biostatistics and Epidemiology Core, Children's Mercy Hospital
Kansas City, MO 64108, USA
Email: Brooke.Fridley@Moffitt.org; Fridley.Brooke@gmail.com

Simon Vandekar, PhD
Department of Biostatistics, Vanderbilt University Medical Center
Nashville, TN 37203, USA
Email: Simon.Vandekar@VUMC.org

Inna Chervoneva, PhD
Division of Biostatistics, Thomas Jefferson University
Philadelphia, PA 19107, USA
Email: Inna.Chervoneva@Jefferson.edu

Julia Wrobel, PhD
Department of Biostatistics and Bioinformatics, Emory University
Atlanta, GA 30322, USA
Email: Julia.Wrobel@Emory.edu

Siyuan Ma, PhD
Department of Biostatistics, Vanderbilt University Medical Center
Nashville, TN 37203, USA
Email: Siyuan.Ma@VUMC.org

Immune modulation is considered a hallmark of cancer initiation and progression, with immune cell density being consistently associated with clinical outcomes of individuals with cancer. Multiplex immunofluorescence (mIF) microscopy combined with automated image analysis is a novel and increasingly used technique that allows for the assessment and visualization of the tumor microenvironment (TME). Recently, application of this new technology to tissue microarrays (TMAs) or whole tissue sections from large cancer studies has been used to characterize different cell populations in the TME with enhanced reproducibility and accuracy. Generally, mIF data has been used to examine the presence and abundance of immune cells in the tumor and stroma compartments; however, this aggregate measure assumes uniform patterns of immune cells throughout the TME and overlooks spatial heterogeneity. Recently, the spatial contexture of the TME has been explored with a variety of statistical methods. In this PSB workshop, speakers will present some of the state-of-the-art statistical methods for assessing the TIME from mIF data.

Keywords: spatial biology, multiplex immunofluorescence, single-cell protein, tumor microenvironment, biostatistical analysis, spatial analysis

1. Introduction, Background and Motivation

The treatment of cancers has been revolutionized in recent years with the advent of immunotherapies[1-6]. However, not all patients respond to immunotherapies and a subset of patients that initially respond to immunotherapy go on to develop resistance. To understand why some patients do not respond to immunotherapies, much research has been devoted to understanding the role of the immune contexture of the tumor immune microenvironment (TIME) and its association with clinical outcomes[4,7-9]. Thus, immune profiling using a variety of approaches has become an important part of immuno-oncology.

Some commonly used approaches for studying the tumor immune microenvironment include (but are not limited to): flow cytometry[10], imaging mass cytometry[11], immunohistochemistry (IHC)[12], immune cell devolution of bulk RNA-seq data[13], single-cell RNA-seq[14], spatial transcriptomics[15] and multiplex immunofluorescence (mIF)[16]. Multiplex immunofluorescence microscopy combined with automated image analysis is a novel and increasingly used technique that allows for the assessment and visualization of the TME. This technology has been applied to a variety of sample types, from whole slide images to regions of interest (ROIs)[17] and tissue microarrays (TMAs)[18,19].

As with any new technology, there are inevitability challenges with the statistical analysis of the single-cell imaging data[20,21]. Some of the challenges come from cell phenotyping, which is labeling cells as positive or negative for each antibody of interest. This is a necessary preprocessing step that occurs before spatial data analysis that is critical for accurately estimating immune cell abundance in the TIME. After phenotyping, it is typical to measure immune cell abundance, typically calculated as percent or proportions of specific cell types in the tumor compartment of the tissue. A challenge of this task is that many cell types are often observed at low-abundance (i.e., zero-inflated), particularly in low immune infiltrated tumors (e.g., immune "cold" tumors).

Besides the protein markers used for phenotyping cells, it is often of important to quantify the actual levels of proteins of interest in all or some cell types. Such quantitative functional markers may include proliferation markers (e.g., Ki-67, PCNA), checkpoint proteins (e.g., PD-1, PD-L1, CTLA-4) and growth factors and receptors (e.g., EGFR, HER2). Traditionally, a single mean expression level across the cells of interest is computed and considered as a biomarker. This approach ignores important tumor heterogeneity and has low sensitivity for detecting high expression in some portion but not all cells of interest. Alternative approaches have been recently developed[22,23] using the entire distributions of single-cell protein expression levels in a tumor tissue to derive quantitative functional markers.

Finally, there is growing evidence that the spatial architecture of the TIME has high impact on disease progression and response to immunotherapy. Generally, mIF data has been used to examine the presence and abundance of immune cells in the TIME; however, this aggregate measure assumes uniform patterns of immune cells throughout the tumor and overlooks spatial heterogeneity. Recently, the spatial contexture of the TIME has been explored with a variety of spatial statistical methods, including those for assessing co-localization. In this session, speakers will present some of the state-of-the-art statistical methods for assessing the TIME from mIF data. All slides and R code presented during the workshop can be found at http://juliawrobel.com/PSB_scProteomics .

2. Speaker Abstracts

Overview of abundance-based and spatial-based analysis approaches for multiplex imaging data

Brooke L Fridley

With the advent of immunotherapies for the treatment of cancer, much research is being conducted to understand the tumor immune microenvironment (TIME). To date, much of the research completed has focused on understanding the abundance of different immune cell subsets in the TIME using either single-cell RNA-seq or multiplex immunofluorescence (mIF). One benefit in using mIF based technologies is that, in addition to abundance of immune cells, one is also able to get the spatial location of these cells within the TIME. Thus, researchers can answer question that relate to the spatial architecture or contexture of the TIME and how this might impact clinical outcomes. In this presentation, we provide an overview of how mIF data is generated and analysis methods used for assessing the non-spatial aspects of the TIME (i.e., abundance level analyses). After providing an overview of mIF data and abundance-based analysis approaches, we will review a variety of spatial statistical approaches for analyzing the spatial contexture. To facilitate spatial analyses, we will also present on an R package, *spatialTIME,* developed to generate these spatial statistics on large sets of samples[17,24].

Normalization and Cell Phenotyping for mIF data

Simon Vandekar

Normalization and cell phenotyping are critical steps in the multiplexed image analysis pipeline prior to performing downstream statistical analysis because they remove batch effects and identify consistent cell types across slides. These analysis steps are particularly challenging for mIF data due to the unique heterogeneity of the image intensities across slides and overlapping cell distributions. We review some recently proposed normalization methods [25,26] and discuss the three main procedures for cell phenotyping (marker gating, unsupervised clustering, and supervised algorithms), in the context of mIF imaging[27], including our recently developed semi-supervised algorithm, *GammaGateR*. The R package *GammaGateR* focuses on efficiently estimating the marginal distributions of single-cell marker intensities using a novel closed-form Gamma mixture model to identify marker positive cells. It incorporates biological constraints to improve consistency across a large number of slides and allows users to interactively curate the model fit. We compare several cell phenotyping algorithms developed for multiplexed imaging and demonstrate how to use the results to perform spatial analyses of mIF imaging data.

Quantile biomarkers based on single-cell multiplex immunofluorescence imaging data

Inna Chervoneva

Modern pathology platforms for multiplex fluorescence-based immunohistochemistry provide distributions of cellular signal intensity (CSI) levels of proteins across the entire cell populations within the sampled tumor tissue. However, heterogeneity of CSI levels is usually ignored, and the simple mean signal intensity (MSI) value is considered as a cancer biomarker. To account for tumor heterogeneity, we consider the entire CSI distribution as a predictor of clinical outcome. This allows retaining all quantitative information at the single-cell level by considering the values of the quantile function (inverse of the cumulative distribution function) estimated from a sample of CSI levels in a tumor tissue.

A simple and intuitive approach is to select an optimal quantile of the CSI distribution as the best predictor of clinical outcome of interest. In Yi et al (2023)[23], we developed an algorithm, implemented in the R package *Qindex,* for selecting optimal CSI distribution quantiles as best predictors of outcome. The proposed algorithm was used to select optimal quantile biomarkers of progression-free survival in a large cohort of breast cancer patients and validated in an independent external validation cohort. The optimal quantile protein biomarkers yielded generally improved prognostic value as compared to the standard MSI biomarkers.

A more comprehensive approach is to derive new biomarkers as single-index predictors based on the entire CSI distribution summarized as a quantile function. [22] The proposed Quantile Index (QI) biomarker is defined as a linear or nonlinear functional regression predictor of outcome. The linear functional regression quantile Index (FR-QI) is the integral of subject-specific CSI quantile function multiplied by the common weight function[22]. The nonlinear functional regression quantile index (nFR-QI) is computed as the integral of unspecified bivariate twice differentiable function with probability p and subject-specific quantile function as arguments. The weight and nonlinear bivariate function are represented by penalized splines and estimated by fitting suitable functional regression models to a clinical outcome. The proposed QI biomarkers were derived for proteins expressed in cancer cells of malignant breast tumors and compared to the standard MSI predictors and optimal quantile protein biomarkers[23]. The R package *Qindex* implements the optimization of QI biomarkers and their evaluation in an independent test set.

Tools and software for functional data analysis of multiplexed imaging data
Julia Wrobel

The TME, which characterizes the tumor and its surroundings, plays a critical role in understanding cancer development and progression. Recent advances in imaging techniques enable researchers to study spatial structure of the TME at a single-cell level. Many popular approaches for analyzing spatial relationships between cell types or quantifying spatial co-expression of biological markers in multiplex imaging data are based on point process theory. The location of cells in mIF data are treated as following a point process, realizations of a point process are called "point patterns", and point process models seek to understand correlations in the spatial distributions of cells. Under the assumption that the rate of a cell is constant over an entire region of interest a point pattern will exhibit complete spatial randomness (CSR), and it is often of interest to model whether cells deviate from CSR either through clustering or repulsion.

Spatial summary functions characterize the degree of spatial interaction among cells across different radii, however, these are often evaluated at a single arbitrarily chosen cellular distance. Using techniques from functional data analysis, we introduce an approach to model the association between these summary spatial functions and patient-level survival outcomes across all radii simultaneously, while controlling for other clinical scalar predictors such as age and disease stage. In addition, we introduce a novel hypothesis test to what level of model flexibility is most appropriate for a given multiplex imaging dataset. Finally, our methods are implemented in *mxfda,* a general-purpose R package for functional data analysis of multiplex imaging data.

A Flexible Generalized Linear Mixed Effects Model for Testing Cell-Cell Colocalization in Spatial Immunofluorescent Data
Siyuan Ma

mIF data analysis is interested in characterizing the nuanced spatial context of tissue microenvironments, such as the infiltration or exclusion of certain immune cell populations in tumor tissues. To test for cell colocalization or exclusion events, existing methods often rely on image-wide statistics to create null distributions for cell colocalization events and evaluate their statistical significance[28]. Given that tissue characteristics can be image-specific (i.e., size of images, the local topology of tissue organization), this type of approach does not generalize well for comparisons between images/conditions. We show that, by examining cell colocalization events on a per-cell basis, they can be modeled with common count-based distributions such as the binomial. As such, cell colocalization or exclusion can be practically analyzed with generalized linear mixed effects models with spatially correlated error terms. This allows flexible inclusion and testing of image/condition effects and subject-specific correlations, because they can be easily modeled as fixed or random regression effects. We demonstrate that this model relies on essentially the same assumptions as existing image-wide modeling approaches. In practice, it can be implemented with the readily available R package *spaMM.* We exemplify the utility of such a model with an application in protein immunofluorescent imaging of inflammatory bowel disease tissues[29].

3. Acknowledgments

This research was supported in part by the National Institutes of Health (R01 CA279065, R01 CA222847).

References

1. Ribas A, Wolchok JD. Cancer immunotherapy using checkpoint blockade. *Science.* Mar 23 2018;359(6382):1350-1355. doi:10.1126/science.aar4060
2. Couzin-Frankel J. Breakthrough of the year 2013. Cancer immunotherapy. *Science.* Dec 20 2013;342(6165):1432-3. doi:10.1126/science.342.6165.1432
3. Drake CG, Lipson EJ, Brahmer JR. Breathing new life into immunotherapy: review of melanoma, lung and kidney cancer. *Nat Rev Clin Oncol.* Jan 2014;11(1):24-37. doi:10.1038/nrclinonc.2013.208
4. Havel JJ, Chowell D, Chan TA. The evolving landscape of biomarkers for checkpoint inhibitor immunotherapy. *Nat Rev Cancer.* Mar 2019;19(3):133-150. doi:10.1038/s41568-019-0116-x
5. Thorsson V, Gibbs DL, Brown SD, et al. The Immune Landscape of Cancer. *Immunity.* Apr 17 2018;48(4):812-830 e14. doi:10.1016/j.immuni.2018.03.023
6. Pardoll DM. The blockade of immune checkpoints in cancer immunotherapy. *Nat Rev Cancer.* Mar 22 2012;12(4):252-64. doi:10.1038/nrc3239
7. Martinez-Morilla S, Villarroel-Espindola F, Wong PF, et al. Biomarker Discovery in Patients with Immunotherapy-Treated Melanoma with Imaging Mass Cytometry. *Clin Cancer Res.* Apr 1 2021;27(7):1987-1996. doi:10.1158/1078-0432.CCR-20-3340
8. Thurin M, Cesano A, Marincola, eds. *Biomarkers for immunotherapy of cancer* Springer; 2020. *Methods in Molecular Biology*

9. Fridman WH, Zitvogel L, Sautes-Fridman C, Kroemer G. The immune contexture in cancer prognosis and treatment. *Nat Rev Clin Oncol*. Jul 25 2017;doi:10.1038/nrclinonc.2017.101
10. Cossarizza A, Chang HD, Radbruch A, et al. Guidelines for the use of flow cytometry and cell sorting in immunological studies (second edition). *Eur J Immunol*. Oct 2019;49(10):1457-1973. doi:10.1002/eji.201970107
11. Baharlou H, Canete NP, Cunningham AL, Harman AN, Patrick E. Mass Cytometry Imaging for the Study of Human Diseases-Applications and Data Analysis Strategies. *Front Immunol*. 2019;10:2657. doi:10.3389/fimmu.2019.02657
12. Magaki S, Hojat SA, Wei B, So A, Yong WH. An Introduction to the Performance of Immunohistochemistry. *Methods in molecular biology*. 2019;1897:289-298. doi:10.1007/978-1-4939-8935-5_25
13. Sturm G, Finotello F, Petitprez F, et al. Comprehensive evaluation of transcriptome-based cell-type quantification methods for immuno-oncology. *Bioinformatics*. Jul 15 2019;35(14):i436-i445. doi:10.1093/bioinformatics/btz363
14. Haque A, Engel J, Teichmann SA, Lonnberg T. A practical guide to single-cell RNA-sequencing for biomedical research and clinical applications. *Genome medicine*. Aug 18 2017;9(1):75. doi:10.1186/s13073-017-0467-4
15. Burgess DJ. Spatial transcriptomics coming of age. *Nature reviews*. Jun 2019;20(6):317. doi:10.1038/s41576-019-0129-z
16. Tan WCC, Nerurkar SN, Cai HY, et al. Overview of multiplex immunohistochemistry/immunofluorescence techniques in the era of cancer immunotherapy. *Cancer Commun (Lond)*. Apr 2020;40(4):135-153. doi:10.1002/cac2.12023
17. Wilson C, Soupir AC, Thapa R, et al. Tumor immune cell clustering and its association with survival in African American women with ovarian cancer. *PLoS Comput Biol*. Mar 2022;18(3):e1009900. doi:10.1371/journal.pcbi.1009900
18. Hathaway CA, Wang T, Townsend MK, et al. Lifetime Exposure to Cigarette Smoke and Risk of Ovarian Cancer by T-cell Tumor Immune Infiltration. *Cancer Epidemiol Biomarkers Prev*. Jan 9 2023;32(1):66-73. doi:10.1158/1055-9965.EPI-22-0877
19. Hathaway CA, Conejo-Garcia JR, Fridley BL, et al. Measurement of ovarian tumor immune profiles by multiplex immunohistochemistry: implications for epidemiologic studies. *Cancer Epidemiol Biomarkers Prev*. Mar 20 2023;doi:10.1158/1055-9965.EPI-22-1285
20. Wilson CM, Ospina OE, Townsend MK, et al. Challenges and Opportunities in the Statistical Analysis of Multiplex Immunofluorescence Data. *Cancers (Basel)*. Jun 17 2021;13(12)doi:10.3390/cancers13123031
21. Wrobel J, Harris C, Vandekar S. Statistical Analysis of Multiplex Immunofluorescence and Immunohistochemistry Imaging Data. *Methods in molecular biology*. 2023;2629:141-168. doi:10.1007/978-1-0716-2986-4_8
22. Yi M, Zhan T, Peck AR, et al. Quantile Index Biomarkers Based on Single-Cell Expression Data. *Laboratory investigation; a journal of technical methods and pathology*. Aug 2023;103(8):100158. doi:10.1016/j.labinv.2023.100158
23. Yi M, Zhan T, Peck AR, et al. Selection of optimal quantile protein biomarkers based on cell-level immunohistochemistry data. *BMC Bioinformatics*. Jul 22 2023;24(1):298. doi:10.1186/s12859-023-05408-8
24. Creed JH, Wilson CM, Soupir AC, et al. spatialTIME and iTIME: R package and Shiny application for visualization and analysis of immunofluorescence data. *Bioinformatics*. Nov 4 2021;doi:10.1093/bioinformatics/btab757

25. Harris CR, McKinley ET, Roland JT, et al. Quantifying and correcting slide-to-slide variation in multiplexed immunofluorescence images. *Bioinformatics*. Mar 4 2022;38(6):1700-1707. doi:10.1093/bioinformatics/btab877
26. Graf J, Cho S, McDonough E, et al. FLINO: a new method for immunofluorescence bioimage normalization. *Bioinformatics*. Jan 3 2022;38(2):520-526. doi:10.1093/bioinformatics/btab686
27. Geuenich MJ, Hou J, Lee S, Ayub S, Jackson HW, Campbell KR. Automated assignment of cell identity from single-cell multiplexed imaging and proteomic data. *Cell Syst*. Dec 15 2021;12(12):1173-1186 e5. doi:10.1016/j.cels.2021.08.012
28. Schapiro D, Jackson HW, Raghuraman S, et al. histoCAT: analysis of cell phenotypes and interactions in multiplex image cytometry data. *Nat Methods*. Sep 2017;14(9):873-876. doi:10.1038/nmeth.4391
29. Kondo A, Ma S, Lee MYY, et al. Highly Multiplexed Image Analysis of Intestinal Tissue Sections in Patients With Inflammatory Bowel Disease. *Gastroenterology*. Dec 2021;161(6):1940-1952. doi:10.1053/j.gastro.2021.08.055

Tools for assembling the cell: Towards the era of cell structural bioinformatics

Mengzhou Hu[1†], Xikun Zhang[2†], Andrew Latham[3†],
Andrej Šali[3*], Trey Ideker[1*], and Emma Lundberg[2*]
[1]*Department of Medicine, University of California San Diego, San Diego, CA, USA*
[2]*Department of Bioengineering, Stanford University, Stanford, CA, USA*
[3]*Department of Bioengineering and Therapeutic Sciences, University of California, San Francisco, San Francisco, CA, USA.*

† These authors contributed equally

* To whom correspondence should be addressed:
sali@salilab.org, tideker@health.ucsd.edu, emmalu@stanford.edu

Cells consist of large components, such as organelles, that recursively factor into smaller systems, such as condensates and protein complexes, forming a dynamic multi-scale structure of the cell. Recent technological innovations have paved the way for systematic interrogation of subcellular structures, yielding unprecedented insights into their roles and interactions. In this workshop, we discuss progress, challenges, and collaboration to marshal various computational approaches toward assembling an integrated structural map of the human cell.

Keywords: cell mapping, subcellular structures, computational modeling of cell

1. Overview

A fundamental objective of cell biology is to decode the intricate multi-scale structures within cells, ranging from macroscopic organelles to microscopic condensates and protein complexes. This goal necessitates a comprehensive understanding of the spatial and functional organizations of subcellular components, particularly within the context of cell function and diseases.

In recent years, a plethora of advanced technologies have emerged, enabling systematic interrogation of subcellular structures and providing unprecedented insights into their functional significance. For example, immunofluorescence imaging[1–3] facilitates the real-time visualization of static and dynamic subcellular interactions at high resolution. Similarly, cryo-electron tomography[4,5] and microscopy[6–9] capture intricate structural details of subcellular components in their native, hydrated state, thus preserving their functional context. On the biochemical front, affinity purification,[10,11] co-elution,[12] and crosslinking mass spectrometry[13,14] techniques have provided avenues for elucidating the complex networks of protein interactions within cells. The emerging machine learning pipelines[15–18] associated with these technologies have further augmented the systematic interpretation of cell architecture and association with diseases.

The integration of these complementary technologies represents a promising avenue for mapping the architecture of cells across a broad range of scales. Using two of these techniques, protein imaging and affinity purification, the session organizers have recently published a novel framework, called MuSIC (Multi-Scale Integrated Cell),[19] for assembling hierarchical maps of human subcellular components spanning the multiple scales of cell biology. The timely and distinct opportunity that emerges from this work is to assemble a key group of thought leaders in a suitable location to discuss progress, open challenges, and, most importantly, how collaborative teams can be established to marshal the various technologies toward an integrated structural map of the human cell.

Hence, this workshop, "Tools for assembling the cell: Towards the era of cell structural bioinformatics," aims to be a catalyst for scientific discourse and collaboration, providing a platform for eminent professionals from varying domains to explore and strategize the future of subcellular structure mapping. We invited seven distinguished speakers (Drs. David Baker, Markus Covert, Jan Ellenberg, Rachel Karchin, Tychele Turner, Aubrey Weigel, and Marinka Zitnik) to share insights on data acquisition and computational approaches for cellular modeling. This workshop is designed to provide attendees with a deep dive into the present technological innovations and highlight avenues for potential collaboration and exploration.

2. Navigating the Workshop

Developing a spatiotemporal map of the cell necessitates integrating various sources of data into a single model. To enable communication and synergy between experimental scientists and computational modelers, this workshop features incisive talks from seven experts in procuring spatiotemporal biological data and advancing computational modeling of cellular architecture across multiple scales.

Dr. David Baker is a Henrietta and Aubrey Davis Endowed Professor of Biochemistry at University of Washington, Director of Institute for Protein Design and an Investigator at Howard Hughes Medical Institute. His research focuses on developing protein design software and using it to create molecules that solve challenges in medicine, technology and sustainability. His group developed the Rosetta algorithm for ab initio protein structure prediction.[16,20] Most recently, his group has developed RoseTTAFold, a three-track network to process sequence, distance, and coordinate information simultaneously, and achieved more accurate protein structure prediction.[21]

Dr. Markus Covert, a Professor of Bioengineering and, by courtesy, of Chemical and Systems Biology at Stanford University, focuses on building computational models of complex biological processes and using these models to guide an experimental program. His lab pioneered the "whole-cell" model encoding all known information about each gene and molecule to predict cell behaviors.[22] His lab has also made significant contributions to live-cell imaging of immune signaling, including a game-changing method to analyze microscopy images using deep learning[15] and a technique that traces cellular behavior from the initial stimulus, through the signaling pathways, down to genome-wide changes in gene expression, within the single cell.[23]

Dr. Jan Ellenberg is Head of Cell Biology and Biophysics, and Head of the European

Molecular Biology Laboratory (EMBL) Imaging Center, at EMBL Heidelberg. He has developed state-of-the-art quantitative fluorescence-based imaging techniques,[2] and combined these technologies with subsequent automation and analysis platforms.[3] His lab leveraged these four-dimensional imaging approaches to enable characterization of processes within human cells, such as protein localization during cell division[24] and nuclear pore complex assembly.[25]

Dr. Rachel Karchin is a professor at Johns Hopkins University and the Institute for Computational Medicine. She has made significant contributions to the field of cancer genomics by leveraging 3D protein structure for variant interpretation developing tools to detect somatic mutation hotspot regions in 3D protein structures.[17,26] Similarly, **Dr. Tychele Turner**, an Assistant Professor at Washington University in St. Louis, worked on precision genomics in neurodevelopmental disorders, determining all possible relevant variations within an individual to the precise nucleotide.[27] Together, both of them focused on mapping mutations in 3D and aimed to compare the 3D mutation clusters between neurodevelopmental diseases and cancers, bringing new insight into genomics research.

Dr. Aubrey Weigel is a Project Scientist of the Cellular Organelle Segmentation in Electron Microscopy (COSEM) Project Team at Howard Hughes Medical Institute (HHMI) - Janelia Research Campus. She has pioneered a pipeline that combines focused ion beam scanning electron microscopy (FIB-SEM) with deep learning annotation methods to reconstruct maps of entire cells at 4-8 nm resolution.[8,9] Such data and models are available to the scientific community through an open-sourced platform, called OpenOrganelle. These data acquisition and analysis techniques can provide insight into complicated cellular processes, and similar analyses revealed the dynamics of endoplasmic reticulum (ER)-to-Golgi protein delivery.[28]

Dr. Marinka Zitnik is an Assistant Professor at Harvard Medical School, and affiliated with several Harvard-based institutes. She investigates the foundations of AI to enhance scientific discovery and to realize individualized diagnosis and treatment. She proposed Decagon, a graph-convolution-network-based model to model polypharmacy side effects.[18] She also founded Therapeutics Data Commons (TDC), an initiative to access and evaluate AI capability across therapeutic modalities and stages of discovery. Their aim is to establish which AI methods are most suitable for advancing therapeutic science and why these techniques are advantageous.[29,30]

3. Discussion and Implications

In this workshop, we delve into cutting-edge technologies designed to illuminate the spatial and functional organizations of subcellular components. Drs. David Baker, Markus Covert, Jan Ellenberg, Rachel Karchin, Tychele Turner, Aubrey Weigel, and Marinka Zitnik are the distinguished speakers contributing their extensive knowledge to this workshop. They elucidate the advancements in data acquisition, sophisticated analysis techniques, and computational tools essential for the assembly of human subcellular components at various scales. This workshop provides a platform not only as a repository of knowledge but also as a forum for academic exchange. Scientists are welcome to discuss the promises, pitfalls, and challenges of modeling the subcellular structures. In addition, the insights of the distinguished speakers can foster the promise of interdisciplinary projects using cell mapping techniques, encouraging potential collaborations to drive cell structural biology further.

References

1. P. J. Thul, *et al.*, A subcellular map of the human proteome, *Science* **356** (2017).
2. J. K. Hériché, *et al.*, Integrating Imaging and Omics: Computational Methods and Challenges, *Annual Review of Biomedical Data Science* **2**, 175 (2019).
3. D. D. Nogare, *et al.*, Using AI in bioimage analysis to elevate the rate of scientific discovery as a community, *Nature methods* **20**, 973 (2023).
4. M. Turk, *et al.*, The promise and the challenges of cryo-electron tomography, *FEBS Letters* **594**, 3243 (2020).
5. S. Zheng, *et al.*, AreTomo: An integrated software package for automated marker-free, motion-corrected cryo-electron tomographic alignment and reconstruction, *Journal of Structural Biology: X* **6** (2022).
6. X. Li, *et al.*, Electron counting and beam-induced motion correction enable near-atomic-resolution single-particle cryo-EM, *Nature Methods* **10**, 584 (2013).
7. S. Q. Zheng, *et al.*, MotionCor2: Anisotropic correction of beam-induced motion for improved cryo-electron microscopy, *Nature Methods* **14**, 331 (2017).
8. L. Heinrich, *et al.*, Whole-cell organelle segmentation in volume electron microscopy, *Nature* **599**, 141 (2021).
9. C. S. Xu, *et al.*, An open-access volume electron microscopy atlas of whole cells and tissues, *Nature* **599**, 147 (2021).
10. J. H. Morris, *et al.*, Affinity purification-mass spectrometry and network analysis to understand protein-protein interactions, *Nat. Protoc.* **9**, 2539 (2014).
11. E. L. Huttlin, *et al.*, Architecture of the human interactome defines protein communities and disease networks, *Nature* **545**, 505 (2017).
12. D. Salas, *et al.*, Next-generation interactomics: Considerations for the use of co-elution to measure protein interaction networks, *Mol. Cell. Proteomics* **19**, 1 (2020).
13. F. Liu, *et al.*, Proteome-wide profiling of protein assemblies by cross-linking mass spectrometry, *Nat. Methods* **12**, 1179 (2015).
14. L. Piersimoni, *et al.*, Cross-linking mass spectrometry for investigating protein conformations and protein–protein interactions-a method for all seasons, *Chemical Reviews* **122**, 7500 (2022), PMID: 34797068.
15. D. A. Van Valen, *et al.*, Deep learning automates the quantitative analysis of individual cells in live-cell imaging experiments, *PLoS computational biology* **12**, p. e1005177 (2016).
16. C. A. Rohl, *et al.*, Protein structure prediction using rosetta, in *Methods in enzymology*, eds. L. Brand, *et al.* (Academic Press, 2004) pp. 66–93.
17. C. Tokheim, *et al.*, Exome-Scale Discovery of Hotspot Mutation Regions in Human Cancer Using 3D Protein Structure, *Cancer Research* **76**, 3719 (2016).
18. M. Zitnik, *et al.*, Modeling polypharmacy side effects with graph convolutional networks, *Bioinformatics* **34**, i457 (2018).
19. Y. Qin, *et al.*, A multi-scale map of cell structure fusing protein images and interactions, *Nature* **600**, 536 (2021).
20. D. E. Kim, *et al.*, Protein structure prediction and analysis using the robetta server, *Nucleic acids research* **32**, W526 (2004).
21. M. Baek, *et al.*, Accurate prediction of protein structures and interactions using a three-track neural network, *Science* **373**, 871 (2021).
22. J. R. Karr, *et al.*, A whole-cell computational model predicts phenotype from genotype, *Cell* **150**, 389 (2012).

23. K. Lane, *et al.*, Measuring signaling and rna-seq in the same cell links gene expression to dynamic patterns of nf-κb activation, *Cell systems* **4**, 458 (2017).
24. Y. Cai, *et al.*, Experimental and computational framework for a dynamic protein atlas of human cell division, *Nature* **561**, 411 (2018).
25. S. Otsuka, *et al.*, A quantitative map of nuclear pore assembly reveals two distinct mechanisms, *Nature* **613**, 575 (2023).
26. N. Niknafs, *et al.*, Mupit interactive: webserver for mapping variant positions to annotated, interactive 3d structures, *Human genetics* **132**, 1235 (2013).
27. T. N. Turner, *et al.*, Genomic patterns of de novo mutation in simplex autism, *Cell* **171**, 710 (2017).
28. A. V. Weigel, *et al.*, ER-to-Golgi protein delivery through an interwoven, tubular network extending from ER, *Cell* **184**, 2412 (2021).
29. K. Huang, *et al.*, Therapeutics data commons: Machine learning datasets and tasks for drug discovery and development, *Proceedings of Neural Information Processing Systems, NeurIPS Datasets and Benchmarks* (2021).
30. K. Huang, *et al.*, Artificial intelligence foundation for therapeutic science, *Nature Chemical Biology* **18**, 1033 (2022).

ERRATUM

How Fitbit data are being made available to registered researchers in All of Us Research Program

Hiral Master, Aymone Kouame, Kayla Marginean, Melissa Basford, Paul Harris
Vanderbilt University Medical Center Nashville TN, USA
Email: hiral.master@vumc.org,aymone.kouame@vumc.org,kayla.marginean@vumc.org, melissa.basford@vumc.org,paul.a.harris@vumc.org

Michelle Holko
Google Public Sector Washington DC, USA
Email: michelle.holko@gmail.com

In the above PSB article published in Biocomputing 2023: Proceedings of the Pacific Symposium, pp. 19-30; PMCID: PMC9811842; PMID: 36540961
The following correction has been made.

Author Correction

In the version of this PSB 2023 conference full-length paper, the authors had mistakenly shown counts in the figures such that counts <20 for some categories could be derived using mathematical formula. The authors apologize for this unintentional error. Therefore, authors have updated figures 4a, 4b and 4d to ensure exact participants counts for categories <20 cannot be derived using mathematical formula. This update was done to ensure participants' privacy and follow *All of Us* Data and Statistics Dissemination Policy.

4a

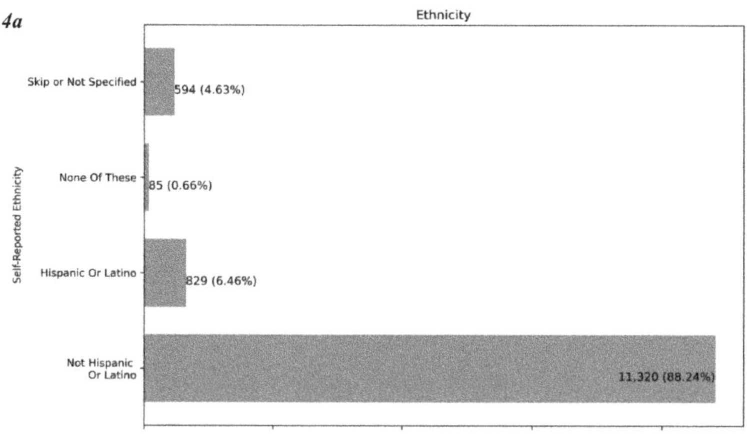

4b

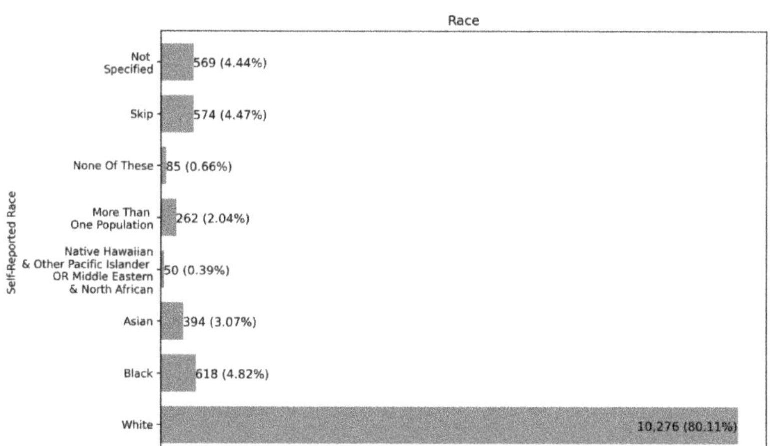

4d

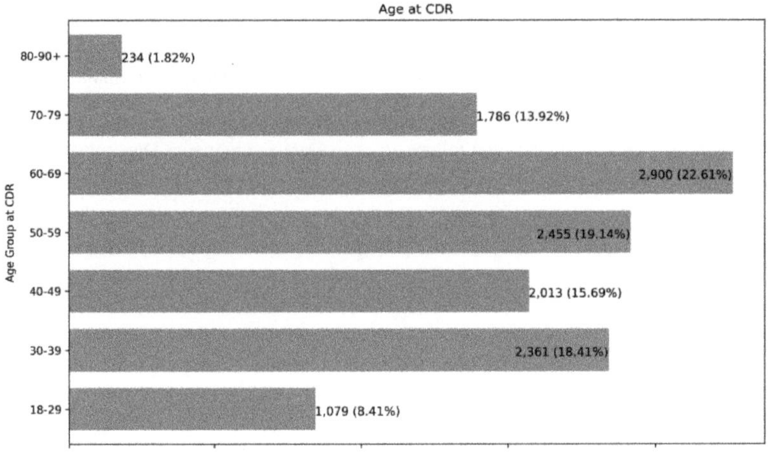

Fig. 4. Self-reported a) ethnicity, b) race, d) age of participants with Fitbit data in June 2022 curated data repository, which can be accessed by registered users on Researcher Workbench.
Figures have been updated to ensure exact participants counts for categories <20 cannot be derived using mathematical formula as per the *All of Us* Data and Statistics Dissemination Policy.

www.ingramcontent.com/pod-product-compliance
Ingram Content Group UK Ltd.
Pitfield, Milton Keynes, MK11 3LW, UK
UKHW021834100926
8556IPUK00004B/343

9 789811 28641